APPLIED DIMENSIONAL ANALYSIS AND MODELING

SECOND EDITION

APPLIED DIMENSIONAL ANALYSIS AND MODELING

SECOND EDITION

Thomas Szirtes, Ph.D., P.Eng.
Thomas Szirtes and Associates Inc., Toronto, Ontario, Canada
Senior Staff Engineer (ret.) SPAR Aerospace Ltd.
Toronto, Ontario, Canada
(Predecessor of McDonald Dettwiler and Associates
Brampton, Ontario, Canada)

with a chapter on mathematical preliminaries

by

Pál Rózsa, D.Sc.
Professor and former Chairman
Department of Mathematics and Computer Science
Technical University of Budapest, Hungary

AMSTERDAM • BOSTON • HEIDELBERG • LONDON
NEW YORK • OXFORD • PARIS • SAN DIEGO
SAN FRANCISCO • SINGAPORE • SYDNEY • TOKYO

Butterworth-Heinemann is an imprint of Elsevier

Butterworth–Heinemann is an imprint of Elsevier
30 Corporate Drive, Suite 400, Burlington, MA 01803, USA
Linacre House, Jordan Hill, Oxford OX2 8DP, UK

Copyright © 2007, Elsevier Inc. All rights reserved.

No part of this publication may be reproduced, stored in a retrieval system, or transmitted in any form or by any means, electronic, mechanical, photocopying, recording, or otherwise, without the prior written permission of the publisher.

Permissions may be sought directly from Elsevier's Science & Technology Rights Department in Oxford, UK: phone: (+44) 1865 843830, fax: (+44) 1865 853333, E-mail: permissions@elsevier.com. You may also complete your request on-line via the Elsevier homepage (http://elsevier.com), by selecting "Support & Contact" then "Copyright and Permission" and then "Obtaining Permissions."

 Recognizing the importance of preserving what has been written, Elsevier prints its books on acid-free paper whenever possible.

Library of Congress Cataloging-in-Publication Data

Szirtes, Thomas.
　Applied dimensional analysis and modeling / Thomas Szirtes.
　　p. cm.
　Includes bibliographical references and index.
　ISBN 0-07-062811-4
　1. Dimensional analysis.　2. Engineering models.　I. Title
TA347.D5S95　1997
530.8—dc21　　　　　　　　　　　　　　　　97-26056
　　　　　　　　　　　　　　　　　　　　　　　CIP

British Library Cataloguing-in-Publication Data
A catalogue record for this book is available from the British Library.

ISBN 13: 978-0-12-370620-1
ISBN 10: 0-12-370620-3

For information on all Butterworth–Heinemann publications visit our Web site at www.books.elsevier.com

Printed in the United States of America
06　07　08　09　10　11　12　13　　10 9 8 7 6 5 4 3 2 1

Working together to grow libraries in developing countries

www.elsevier.com | www.bookaid.org | www.sabre.org

ELSEVIER　　BOOK AID International　　Sabre Foundation

I dedicate this book to the memory of my late teacher, Professor Ádám Muttnyánszky of the Technical University of Budapest, who, by his style, supreme educating skill, and human greatness, taught me to love my profession.

Thomas Szirtes

The author with a small-scale model

Thomas Szirtes is a professional engineer who devotes his career to consulting, writing, and teaching. He is a former Senior Staff Engineer at SPAR Aerospace Ltd. in Toronto, and was one of the Project Engineers of the Shuttle Robotic Manipulator Arm (Canadarm), for which he received NASA's Achievement Award.

Dr. Szirtes has published over 60 scientific and engineering papers, as well as a college text on mathematical logic. He has taught at the Technical University of Budapest, McGill University, and Loyola College (Montreal) and was the founding editor of the *SPAR Journal of Engineering and Technology*.

CONTENTS

List of Titled Examples and Problems   xiii
Foreword to the First Edition *by Michael Isaacson*   xix
Foreword to the Second Edition *by Michael Isaacson*   xxi
Acknowledgments   xxiii
Preface to the First Edition   xxv
Preface to the Second Edition   xxvii
Organization, Notation, and Conventions   xxix

Chapter 1   Mathematical Preliminaries *by Pál Rózsa*   **1**
     1.1   Matrices and Determinants   1
     1.2   Operations with Matrices   6
     1.3   The Rank of a Matrix   14
     1.4   Systems of Linear Equations   19
          1.4.1   Homogeneous Case   19
          1.4.2   Nonhomogeneous Case   23
     1.5   List of Selected Publications Dealing with Linear Algebra and Matrices   26

Chapter 2   Formats and Classification   **27**
     2.1   Formats for Physical Relations   27
          2.1.1   Numeric Format   27
          2.1.2   Symbolic Format   29
          2.1.3   Mixed Format   30
     2.2   Classification of Physical Quantities   32
          2.2.1   Variability   32
          2.2.2   Dimensionality   33

Chapter 3   Dimensional Systems   **37**
     3.1   General Statements   37
          3.1.1   Monodimensional System   38
          3.1.2   Omnidimensional System   41
          3.1.3   Multidimensional System   42
     3.2   Classification   42
     3.3   The SI   43
          3.3.1   Preliminary Remarks   43

	3.3.2	Structure	44
		(a) Fundamental Dimensions	44
		(b) Derived Dimensionless Units	50
		(c) Derived Dimensional Units with Specific Names	53
		(d) Derived Dimensional Units without Specific Names	53
		(e) Non-SI Units Permanently Permitted to be Used with SI	56
		(f) Non-SI Units Temporarily Permitted to be Used with SI	59
		(g) Prohibited Units	59
	3.3.3	Prefixes	59
	3.3.4	Rules of Etiquette in Writing Dimensions	63
		3.3.4.1 Problems	65
3.4	Other Than SI Dimensional Systems		66
	3.4.1	Metric, Mass-based Systems	66
		(a) CGS System	66
		(b) SI (for reference only)	67
	3.4.2	Metric, Force-based System	67
	3.4.3	American/British Force (Engineering) System	67
	3.4.4	American/British Mass (Scientific) System	67
3.5	A Note on the Classification of Dimensional Systems		68

Chapter 4 Transformation of Dimensions 69
4.1 Numerical Equivalences 69
4.2 Technique 73
4.3 Examples 74
4.4 Problems 91

Chapter 5 Arithmetic of Dimensions 95

Chapter 6 Dimensional Homogeneity 99
6.1 Equations 99
6.2 Graphs 110
6.3 Problems 126

Chapter 7 Structure of Physical Relations 133
7.1 Monomial Power Form 133
7.2 The Dimensional Matrix 134
7.3 Generating Products of Variables of Desired Dimension 135
7.4 Number of Independent Sets of Products of Given Dimension (I) 139
7.5 Completeness of the Set of Products of Variables 143
7.6 Special Case: Matrix **A** is Singular 144
7.7 Number of Independent Sets of Products of Given Dimension (II); Buckingham's Theorem 148

7.8	Selectable and Nonselectable Dimensions in a Product of Variables	151
7.9	Minimum Number of Independent Products of Variables of Given Dimension	152
7.10	Constancy of the Sole Dimensionless Product	153
7.11	Number of Dimensions Equals or Exceeds the Number of Variables	157
	7.11.1 Number of Dimensions Equals the Number of Variables	157
	7.11.2 Number of Dimensions Exceeds the Number of Variables	159
7.12	Problems	160

Chapter 8 Systematic Determination of Complete Set of Products of Variables — 163

8.1	Dimensional Set; Derivation of Products of Variables of a Given Dimension	163
8.2	Checking the Results	171
8.3	The Fundamental Formula	176

Chapter 9 Transformations — 181

9.1	Theorems Related to Some Specific Transformations	181
9.2	Transformation Between Systems of Different **D** Matrices	202
9.3	Transformation Between Dimensional Sets	211
9.4	Independence of Dimensionless Products of the Dimensional System Used	225

Chapter 10 Number of Sets of Dimensionless Products of Variables — 229

10.1	Distinct and Equivalent Sets	229
10.2	Changes in a Dimensional Set Not Affecting the Dimensionless Variables	231
10.3	Prohibited Changes in a Dimensional Set	238
	10.3.1 Duplications	242
10.4	Number of Distinct Sets	244
10.5	Exceptions	250
	10.5.1 Dimensionally Irrelevant Variable	250
	10.5.2 In Matrix **C**, One Row is a Multiple of Another Row	255
10.6	Problems	259

Chapter 11 Relevancy of Variables — 263

11.1	Dimensional Irrelevancy	263
	11.1.1 Condition	263
	11.1.2 Adding a Dimensionally Irrelevant Variable to a Set of Relevant Variables	267
	11.1.3 The Cascading Effect	268

	11.2 Physical Irrelevancy	274
	11.2.1 Condition	274
	11.2.2 Techniques to Identify a Physically Irrelevant Variable	276
	Common Sense	277
	Existence of Dimensional Irrelevancy	277
	Heuristic Reasoning	280
	Tests Combined with Deft Interpretation of Results	296
	11.3 Problems	312

Chapter 12 Economy of Graphical Presentation — 317

	12.1 Number of Curves and Charts	317
	12.2 Problems	329

Chapter 13 Forms of Dimensionless Relations — 333

	13.1 General Classification	333
	13.2 Monomial is Mandatory	335
	13.3 Monomial is Impossible—Proven	338
	13.4 Monomial is Impossible—Not Proven	348
	13.5 Reconstructions	353
	13.5.1 Determination of Exponents of Monomials	353
	The Measurement Method	353
	The Analytic Method	354
	The Heuristic Reasoning Method	354
	13.5.2 Determination of Some Nonmonomials	366
	13.6 Problems	373

Chapter 14 Sequence of Variables in the Dimensional Set — 381

	14.1 Dimensionless Physical Variable is Present	381
	14.2 Physical Variables of Identical Dimensions are Present	385
	14.3 Independent and Dependent Variables	389
	14.4 Problems	399

Chapter 15 Alternate Dimensions — 401

Chapter 16 Methods of Reducing the Number of Dimensionless Variables — 413

	16.1 Reduction of the Number of Physical Variables	414
	16.2 Fusion of Dimensionless Variables	427
	16.3 Increasing the Number of Dimensions	433
	16.3.1 Dimension Splitting	433
	16.3.2 Importation of New Dimensions	449
	16.3.3 Using Both Mass and Force Dimensions	454
	16.4 Problems	460

Chapter 17	**Dimensional Modeling**	**463**
	17.1 Introductory Remarks	463
	17.2 Homology	467
	17.3 Specific Similarities	468
	17.3.1 Geometric Similarity	468
	17.3.2 Kinematic Similarity	471
	17.3.3 Dynamic Similarity	471
	17.3.4 Thermal (or Thermic) Similarity	472
	17.4 Dimensional Similarity	472
	17.4.1 Scale Factors	479
	17.4.2 Model Law	479
	17.4.3 Categories and Relations	489
	Categories	489
	Relations	492
	17.4.4 Modeling Data Table	495
	17.5 Scale Effects	511
	17.6 Problems	523
Chapter 18	**Fifty-two Additional Applications**	**527**

References:

	Numerical Order	659
	Alphabetical Order of Authors' Surnames	667

Appendices: **675**

	1	Recommended Names and Symbols for Some Physical Quantities	677
	2	Some More-Important Physical Constants	681
	3	Some More-Important *Named* Dimensionless Variables	683
	4	Notes Attached to Figures	693
	5	Acronyms	721
	6	Solutions of Problems	723
	7	Proofs of Selected Theorems and Equations	797
	8	Blank Modeling Data Table	803

Indices

	Subject Index	805
	Surname Index	817

LIST OF TITLED EXAMPLES AND PROBLEMS

Solutions to problems are in Appendix 6. Some titles are listed in abbreviated form. "Ex." = example; "Pr." = problem.

Ex. 2-1	Surface area of a right circular cylinder	30
Ex. 4-8	Strain energy in a cantilever loaded laterally by a concentrated force	78
Ex. 4-10	Fuel consumption of cars	80
Ex. 4-13	Lengths vs. weights of dolphins	84
Ex. 4-14	End slope of a simply supported beam subjected to distributed load	85
Ex. 4-15	An unusual dimensional system (I)	85
Ex. 4-16	An unusual dimensional system (II)	87
Ex. 4-17	An impossible dimensional system	89
Pr. 4/3	Writers' efficiency	91
Pr. 4/7	Cost to build a medicinal bathhouse	91
Pr. 4/12	Numerical measures of importance of public demonstrations	92
Pr. 4/13	Ferguson's lubrication problem	93
Pr. 4/14	An unusual dimensional system (III)	94
Pr. 4/15	The Kapitza dimensional system	94
Pr. 4/16	Heart rate of mammals	94
Ex. 5-1	Dimension of "jerk"	97
Ex. 6-7	Heat loss through a pipe wall	105
Ex. 6-8	Lengths of dog tails in urban areas	106
Ex. 6-9	Variation of atmospheric pressure with altitude	107
Ex. 6-10	Quantity of paint on a sphere	109
Ex. 6-15	Deflection of a cantilever upon a concentrated lateral load (I)	116
Ex. 6-19	Jumping heights of animals (I)	124
Ex. 6-20	Urine secretion rate of mammals	125
Pr. 6/7	Volume of a barrel	127
Pr. 6/9	Circulation velocity of money	128
Pr. 6/11	Bending of light in gravitational field	129
Pr. 6/14	Bending a uniform cross-section cantilever	131
Pr. 6/16	Body surface area of slim individuals	131

LIST OF TITLED EXAMPLES AND PROBLEMS

Pr. 6/17	Frequency of feeding versus mass of warm-blooded animals	132
Ex. 7-5	Deflection of a cantilever upon a concentrated lateral load (II)	141
Ex. 7-6	Deflection of a cantilever upon a concentrated lateral load (III)	143
Ex. 7-8	Deflection of a cantilever upon a concentrated lateral load (IV)	146
Ex. 7-9	Deflection of a cantilever upon a concentrated lateral load (V)	149
Ex. 7-10	Deflection of a cantilever upon a concentrated lateral load (VI)	150
Ex. 7-11	Propagation of the wavefront in an atomic explosion	154
Ex. 7-12	Relativistic energy-mass equivalence	157
Ex. 8-5	Power requirement of a rotating-blade mixer	172
Ex. 8-6	Resistance of a flat surface moving tangentially on water	175
Ex. 8-7	Radiation pressure on satellites	177
Ex. 8-8	Mass of a drop of liquid slowly emerging from a pipe	178
Ex. 9-24	Beach profile characteristics	208
Ex. 9-32	Kinetic energy of a moving ball	227
Ex. 10-16	Oscillating period of a simple pendulum	250
Ex. 11-1	Free vibration of a massive body on a weightless spring (I)	264
Ex. 11-5	Critical sliding friction of a sphere rolling down an incline	270
Ex. 11-6	Elongation by its own weight of a suspended bar	271
Ex. 11-8	Gravitational pull by a solid sphere on an external material point (I)	275
Ex. 11-9	Hydroplaning of tires on flooded surfaces	277
Ex. 11-10	Diameter of a crater caused by an underground explosion	279
Ex. 11-11	Gravitational pull by a flat infinite plate	280
Ex. 11-12	Gravitational pull by a solid sphere on an external material point (II)	282
Ex. 11-13	Deformation by wind of an air-supported radome	284
Ex. 11-14	Axial reaction force generated by thermal load on a straight bar	287
Ex. 11-15	Maximum running speed of animals	289
Ex. 11-16	Jumping heights of animals (II)	290
Ex. 11-17	Velocity of surface waves	291
Ex. 11-18	Frequency of vibration of a sphere of liquid	294
Ex. 11-19	Minimum pressure in an inflated radome (I)	297
Ex. 11-20	Linear speed of a sphere rolling down an incline	299
Ex. 11-21	Velocity of fluid flowing through an orifice	302
Ex. 11-22	Minimum pressure in an inflatable radome (II)	304
Ex. 11-23	Period of a conical pendulum	305
Ex. 11-24	Kepler's third law	308
Ex. 11-25	Frequency of transversal vibration of a stretched wire	310
Pr. 11/1	Period of oscillation of a fluid in a U tube	312
Pr. 11/2	Velocity of disturbance along a stretched wire	312
Pr. 11/3	Geometry of a catenary	312
Pr. 11/4	Stress in glass windows by wind	313
Pr. 11/5	Energy of a laterally vibrating stretched wire	313
Pr. 11/6	Sloshing frequency of fuel during take-off in missile tanks	313
Pr. 11/7	Time scale of the universe	313
Pr. 11/8	Linear momentum of a quantum	313

LIST OF TITLED EXAMPLES AND PROBLEMS

Pr. 11/9	A general physical system	314
Pr. 11/10	Stresses generated by the collision of two steel balls	314
Ex. 12-2	Volume of a right circular cone	318
Ex. 12-3	Gravitational acceleration on a celestial body as a function of altitude (positive or negative)	321
Ex. 12-4	Radial deflection of a semicircular ring by a concentrated radial force	322
Ex. 12-5	Blackbody radiation law	324
Ex. 12-6	Deformation of an elastic foundation under a dropped mass	326
Pr. 12/3	Power of a dynamo	330
Pr. 12/4	Critical axial load on columns	330
Pr. 12/5	Relativistic mass	331
Ex. 13-4	Flow of fluid over a spillway	335
Ex. 13-5	Drag on a moving body immersed in a fluid	336
Ex. 13-6	Impact velocity of a meteorite	339
Ex. 13-7	Area of a triangle whose side-lengths form a geometric progression	341
Ex. 13-8	Boy meets girl	343
Ex. 13-9	Speed of a vertically ejected projectile (I)	346
Ex. 13-10	Capstan drive	348
Ex. 13-11	Time to clear out from the path of a falling brick upon hearing a warning	350
Ex. 13-13	Deflection of a simply supported uniform beam loaded by its own weight	354
Ex. 13-14	The coffee warmer (I)	356
Ex. 13-15	Relativistic red shift	358
Ex. 13-16	Kepler's second law	360
Ex. 13-17	Relative load-carrying capabilities of animals	363
Ex. 13-18	Injuries to animals falling to the ground	364
Ex. 13-19	Power required to lift a given mass to a given height within a given time	367
Ex. 13-20	Deflection of a simply supported beam by two symmetrically placed forces	370
Pr. 13/1	Deflection of a simply supported beam loaded by a lateral force	373
Pr. 13/2	Fundamental lateral frequencies of geometrically similar cantilevers	374
Pr. 13/3	Area of an elliptic segment	374
Pr. 13/4	Torus volume—normal, degenerating, and degenerated	375
Pr. 13/5	Bolometric luminosity of a star	376
Pr. 13/6	Volume of a conical wedge	376
Pr. 13/7	Volume of the frustum of a right circular cone	377
Pr. 13/8	Discharge of a capacitor	378
Pr. 13/9	Area of a triangle whose side-lengths form an arithmetic progression	379
Ex. 14-1	Mass of electrochemically deposited material	381
Ex. 14-2	Maximum thickness of wet paint on an inclined surface	383

Ex. 14-4	The coffee warmer (II)	386
Ex. 14-5	Bursting speed of a flywheel	389
Ex. 14-6	Longitudinal deformation of an axially loaded bar	391
Ex. 14-7	Volume and surface of the frustum of a cone	393
Ex. 14-8	Free fall	395
Ex. 14-9	Stresses in a structure loaded by its own weight, external forces, and moments	397
Pr. 14/1	Terminal speed of a mass sliding down on a nonfrictionless incline	399
Pr. 14/3	Surface area, volume, and weight of a right circular cylinder	400
Ex. 15-1	Charge of free electrons in a wire	402
Ex. 15-2	Compound interest	403
Ex. 15-3	Optimal density and location pattern for retail shops	406
Ex. 15-4	Land price	410
Ex. 16-1	Terminal velocity of a raindrop	414
Ex. 16-2	Mass of fluid flowing through a tube (I)	415
Ex. 16-3	Moment of inertia of a thin lamina	417
Ex. 16-4	Path of an electron in a magnetic field	418
Ex. 16-5	Density and size of the universe	420
Ex. 16-6	Deflection of a cantilever upon a concentrated lateral load (VII)	422
Ex. 16-7	Terminal velocity of a sinking ball in a viscous liquid (I)	425
Ex. 16-8	Jumping heights of animals (III) (air resistance is not ignored)	428
Ex. 16-9	Transient velocity of a sinking ball in a viscous liquid	430
Ex. 16-10	Rise of fluid in a capillary tube	434
Ex. 16-11	Range of a horizontally ejected bullet	436
Ex. 16-12	Tangential speed of a conical pendulum	438
Ex. 16-13	An elegant method to determine the kinetic friction coefficient	440
Ex. 16-14	Mass of fluid flowing through a tube (II)	443
Ex. 16-15	Torsion of a prismatic bar	444
Ex. 16-16	Number of free electrons in a metal conductor	449
Ex. 16-17	Heat energy input by pouring hot liquid into a tank	450
Ex. 16-18	Free vibration of a mass suspended on a weightless spring (II)	452
Ex. 16-22	Terminal velocity of a sphere slowly descending in a viscous liquid (II)	458
Pr. 16/1	Terminal velocity of a sphere slowly descending in a viscous liquid (III)	460
Pr. 16/6	Curvature of a bimetallic thermometer	461
Ex. 17-2	Deflection of a cantilever upon a concentrated lateral load (VIII)	474
Ex. 17-3	Thermally induced reaction force in a curved beam	476
Ex. 17-4	Power required to tow a barge	480
Ex. 17-5	Deflection of a curved bar upon a concentrated force	483
Ex. 17-6	Frequency of respiration of warm-blooded animals	485
Ex. 17-7	Torsional pendulum of double suspension	487
Ex. 17-8	Deflection of a cantilever upon a concentrated lateral load (IX)	490
Ex. 17-9	Deflection of a cantilever upon a concentrated lateral load (X)	495
Ex. 17-10	Size and impact velocity of a meteorite	505

Ex. 17-11	Roasting time for turkey	508
Ex. 17-12	Relative mass of mammalian skeletons	513
Ex. 17-13	Ballet on the Moon	515
Ex. 17-14	Most comfortable walking speed	518
Pr. 17/1	To crack a window	523
Pr. 17/4	Contact time of impacting balls	524
Pr. 17/5	Gravitational collapse of a star	525
Ex. 18-1	Proof of Pythagorean Theorem	528
Ex. 18-2	Time to fall a given distance	530
Ex. 18-3	Tension in a wire ring rotated about its central axis perpendicular to its plane	530
Ex. 18-4	Velocity of disturbance in a liquid	531
Ex. 18-5	Reverberation period in a room	532
Ex. 18-6	Flying time of a bullet landing on an inclined plane	533
Ex. 18-7	Heat transfer in a calorifer	535
Ex. 18-8	Electric field by a dipole at a point that is on line joining them	537
Ex. 18-9	Maximum velocity of an electron in a vacuum tube	538
Ex. 18-10	Period of oscillation of a magnetic torsional dipole	539
Ex. 18-11	Force on a straight wire carrying current in a magnetic field	541
Ex. 18-12	Force between two parallel wires carrying a current	542
Ex. 18-13	Buckling load of a hinged column	543
Ex. 18-14	Centrifugal force acting on a point mass	544
Ex. 18-15	Work done by a piston compressing ideal gas	546
Ex. 18-16	Energy levels of electrons in a Bohr atom	547
Ex. 18-17	Existence criteria for black holes	548
Ex. 18-18	Axial thrust of a screw propeller	550
Ex. 18-19	Boussinesq's problem	551
Ex. 18-20	Rolling resistance of automobile tires	555
Ex. 18-21	General fluid flow characteristics	557
Ex. 18-22	Heat transfer to a fluid flowing in a pipe	558
Ex. 18-23	Volume of a torus of arbitrary cross-section	561
Ex. 18-24	Pitch of a kettledrum	563
Ex. 18-25	Volume of spherical segment	564
Ex. 18-26	Force-related characteristics and geometry of a catenary	566
Ex. 18-27	Speed of a vertically ejected projectile (II)	580
Ex. 18-28	Drag on a flat plate in a parallel fluid flow	582
Ex. 18-29	Lateral natural frequency of a simply supported beam	584
Ex. 18-30	Buckling of a vertical rod under its own weight	587
Ex. 18-31	Lateral natural frequency of a cantilever (I)	590
Ex. 18-32	Lateral natural frequency of a cantilever (II)	593
Ex. 18-33	Velocity of sound in a liquid	595
Ex. 18-34	Diameter of a soap bubble	596
Ex. 18-35	Velocity of collapse of a row of dominoes	598
Ex. 18-36	Generated pressure by an underwater explosion	599
Ex. 18-37	Operational characteristics of an aircraft	601
Ex. 18-38	Volume of fluid flowing in a horizontal pipe	606

xviii　　　　　　LIST OF TITLED EXAMPLES AND PROBLEMS

Ex. 18-39	Vertical penetration of a vehicle's wheels into soft soil	607
Ex. 18-40	Deflection of a simply supported beam upon a centrally dropped mass	609
Ex. 18-41	Minimum deflection cantilevers	615
Ex. 18-42	Choosing the right shock absorber	621
Ex. 18-43	Jamming a circular plug into a nonsmooth circular hole	627
Ex. 18-44	Size of the human foot and hand	633
Ex. 18-45	Depression of an inflated balloon floating on oil	637
Ex. 18-46	Resonant frequency of an L,C circuit	640
Ex. 18-47	Energy of water waves	641
Ex. 18-48	Vertical oscillation of an immersed Nicholson hydrometer	644
Ex. 18-49	Relativistic kinetic energy	646
Ex. 18-50	Penetration of a bullet	650
Ex. 18-51	Sag of a wire-rope under general load	652
Ex. 18-52	Proof that the time needed for a mass-point to descend on a straight line from the top of a vertical circle to any point on its perimeter is independent of this line's slope	655

FOREWORD TO THE FIRST EDITION

The student being introduced to dimensional analysis for the first time is always amazed by the demonstration, without recourse to full physical analysis, that the period of oscillation of a simple pendulum must be proportional to the square root of the pendulum length and independent of its mass. The rationale for this relationship is, of course, based on the simple argument that each term of a "properly" constructed physical equation needs to be dimensionally homogeneous with the others. Likewise, the student is also impressed by the application of such results to predicting full-scale behavior from measurements using a scale model. From this simple example, dimensional arguments can be taken to increasing levels of complexity, and can be applied to a wide range of situations in science and engineering.

This book develops the ideas of dimensions, dimensional homogeneity, and model laws to considerable depth and with admirable rigor, enabling Dr. Szirtes to provide us with intriguing insights into an impressive range of physical examples—spanning such topics as the impact velocity of a meteorite, the lengths of dogs' tails in urban areas, the price of land, the cracking of window glass, and so on—and all without the tedium so often involved in a conventional step-by-step physical analysis.

Dr. Szirtes makes it clear that he regards dimensional analysis and modeling as an art. His intention here is to remind us that the practitioner of dimensional analysis is offered a refreshing scope for personal initiative in the approach to any given problem. It is, then, inevitable that the reader will be inspired by the creative spirit that illuminates the following pages, and will be struck by the elegance of the techniques that are developed.

This book gives an in-depth treatment of all aspects of dimensional analysis and modeling, and includes an extensive selection of examples from a variety of fields in science and engineering. I am sure that this text will be well received and will prove to be an invaluable reference to researchers and students with an interest in dimensional analysis and modeling and those who are engaged in design, testing, and performance evaluation of engineering and physical systems.

MICHAEL ISAACSON, Ph.D., P.Eng.
Dean, Faculty of Applied Science, and
Professor of Civil Engineering
University of British Columbia
Vancouver, B.C., Canada

FOREWORD TO THE SECOND EDITION

I was honoured to have written the foreword to the first edition of the book *Applied Dimensional Analysis and Modeling*, and I am equally honoured to be invited to write the foreword to the second edition. The first edition was very well received by the engineering and scientific communities world-wide. As one indication of this, a Hungarian translation of the first edition is under development and will serve as a university text in Hungary.

Dimensional analysis is often indispensable in treating problems that would otherwise prove intractable. One of its most attractive features is the manner in which it may be applied across a broad range of disciplines—including engineering, physics, biometry, physiology, economics, astronomy, and even social sciences. As a welcome feature of the book, the first edition included more than 250 examples drawn from this broad range of fields.

It would be difficult to improve on the comprehensive, in-depth treatment of the subject, including the extensive selection of examples that appeared in the first edition. But remarkably Dr. Szirtes has managed to accomplish this. The second edition contains a number of changes and improvements to the original version, including a range of additional applications of dimensional analysis. These include examples from biometry, mechanics, and fluid dynamics. For instance, Dr. Szirtes elegantly demonstrates that the size of the human foot is not proportional to body height, but instead is proportional to the body height to the power of 1.5!

Overall, the second edition of the book once more provides an in-depth treatment of all aspects of dimensional analysis, and includes an extensive selection of examples drawn from a variety of fields. I am sure that the second edition will be well received, and once more will prove to be an invaluable reference to students, researchers, and professionals across a range of disciplines.

> MICHAEL ISAACSON, Ph.D., P.Eng.
> *Dean, Faculty of Applied Science, and*
> *Professor of Civil Engineering*
> *University of British Columbia*
> *Vancouver, B.C., Canada*

ACKNOWLEDGMENTS

The author wishes to thank his former colleague Bruce Sayer, of SPAR Aerospace Ltd., Toronto, Ontario, for the expert preparation of line illustrations, and John H. Connor, of the Centre for Research in Earth and Space Technology, Toronto, Ontario, for his reading of the first draft of this book and for his subsequent valuable comments and suggestions.

Bertrand Russell reputedly once said: *"An interesting problem has more value than an interesting solution; for while the former generates the thinking of many, the latter registers the thinking of only one."* With this in mind, the author is appreciative of another former colleague, Robert Ferguson, P.Eng., now of Taiga Engineering Group, Inc., Bolton, Ontario, for proposing two interesting problems that appear as Problem 4/13 and Example 18-43. The latter was found to be especially stimulating, mainly for its seemingly intricate, but in fact relatively simple, outcome.

Sincere appreciation is expressed to Robert Houserman, former Senior Editor of McGraw-Hill, for his initiative in proposing this book, for his subsequent and sustained general helpfulness, and, in particular, for his two argument-free acquiescences to extensions of the manuscript's deadline.

For his help in securing reference material and subsequent detailed observations and suggestions, the author conveys his indebtedness to Eugene Brach, P.Eng., Senior Technical Adviser for the Government of Canada. His constructive scrutiny greatly reduced the numerous minor, and not so minor, numerical errors and inconsistencies in the draft text.

For his assistance in reading the manuscript from a professional mathematician's perspective and for his numerous and detailed suggestions and comments, much indebtedness is expressed to Prof. Pál Rózsa of the Technical University of Budapest, Hungary, whose teaching during the author's undergraduate years greatly enhanced his affinity for mathematical techniques.

The author acknowledges the enthusiastic, conscientious, and timely efforts of Dr. Simon Yiu of Seneca College, Toronto, in checking the text for errors in arithmetic, references, and typography.

The author offers his special thanks and great appreciation to his wife, Penny, who meticulously corrected the large number of misplaced or misused definite articles, inappropriate pronouns, faulty pluralities, flawed use of tenses, wrongly placed adjectives, erroneous adverbs, dangling modifiers, and other serious linguistic offenses.

Ideas for a number of examples and problems appearing in this book were obtained from a multitude of sources, and where appropriate, credit is given. However, the author may have missed mentioning some sources, and for this he sincerely apologizes to those who may feel slighted.

PREFACE TO THE FIRST EDITION

> I have often been impressed by the scanty attention paid even by original workers in physics to the great principle of similitude. It happens not infrequently that results in the form of "laws" are put forward as novelties on the basis of elaborate experiments, which might have been predicted *a priori* after a few minutes of consideration.
>
> Lord Rayleigh (Ref. 71)

In the late sixties at RCA Victor Ltd. in Montreal, where the author was working as an engineer, Peter Foldes—the "antenna man" in the organization—gave him a problem: he was to determine the deformation and force-related characteristics of a very large, "log-periodic" and as yet nonexistent antenna for a set of widely differing load conditions (a log-periodic antenna is essentially a three-dimensional cable structure of a complex V-shaped configuration).

An analytical approach was out of the question, due to the geometric intricacies and uneven concentrated and distributed mechanical load patterns. So, what to do? In quasidesperation, the author went to the McGill University Engineering Library to find some material—*any* material—that would help to extricate him from his predicament. What greatly hampered his effort in the library, of course, was the fact that he did not have any notion of what he wanted to find! It is always difficult to find something if the seeker has no idea what he is looking for. But, in this instance, by luck and good fortune, he found H. Langhaar's great book *Dimensional Analysis and Theory of Models* (Ref. 19) from which he could, albeit with some effort, learn how to size and construct a *small scale model* of the antenna, determine the loads to be put on this model, and on it *measure* the corresponding mechanical parameters. Then, with these results and by formulating the appropriate scale factors, he was able to determine all the deformations and reactive forces for the full-scale product.

When the results were presented to the in-house "customer," he asked how the "impressive" set of data was obtained. He was told that they were derived by modeling experimentation. His response was: "Very good, and since it seems that you have such specialized knowledge, you should be a specialist here!" And indeed, within a couple of weeks your chronicler was promoted to "Engineering Specialist," at that time a rather exalted title at RCA. This episode proves that the dimensional technique can be beneficial to its user in more than one way!

Since then, the author has been hooked on the fine art of dimensional techniques and modeling. He has used the method a great many times and in widely varying circumstances. It has unfailingly provided him with reliable results, often achieved with astonishing speed and little effort. He has also found the method to be potent, since it could usually produce results even when all other means failed. In short, the "dimensional method" has proven to be an extremely *elegant* technique, and elegance is perhaps the most important characteristic in a technical or scientific discipline.

This book is the result of the author's study of existing dimensional technology, his practical experience in the field, and his own independent work in further developing and refining the technique. These sources have enabled him to construct and include in the text over two hundred illustrative examples and problems, which he strongly advises the motivated reader to study carefully. For these examples and problems not only demonstrate in detail the great diversity of the practical applications of the technique, but they also serve as powerful learning routines that will develop and further illuminate the essential ideas.

Since the chapter titles in this book are rather descriptive and self-explanatory, the author feels he does not need to further explain the contents of each chapter here, as is frequently done in prefaces to technical books. It is pointed out, nevertheless, that the author's main contribution to the dimensional method is in Chapter 8, where the compact and efficient technique, culminating in the *Fundamental Formula,* is presented. By this formula, the *Dimensional Set* is constructed, from which the generated *dimensionless variables* can be simply read off. These variables are the indispensable building blocks of all dimensional methods, including analysis and modeling. The introduced technique is then consistently employed throughout the balance of the book.

Apart from an inquisitive mind and a knowledge of basic mechanics and electricity, the only requisite to understanding and fully benefiting from the material presented is some familiarity with determinants and elementary matrix arithmetic, especially inversion and multiplication. Although these operations can be easily performed on most hand-held scientific calculators, for those readers who wish to go a little deeper into the subject and who feel that their relevant proficiency might require some brushing-up, an outline of the essentials of matrices and determinants is presented in Chapter 1. This chapter was written by Prof. Pál Rózsa, whose contributions to the field of linear algebra are known internationally, and whose roster of students includes the author.

Considerable effort was made to find and eliminate numerical errors, omissions, etc. from the text. However, in a work of this size it is virtually unavoidable that some mistakes will remain undetected. Therefore, the author would be grateful to anyone who, finding an error, communicates this to him through the publisher.

This book contains a great deal of information, and at times demands concentration. But this should not discourage the reader, for the effort spent will be amply rewarded by the acquired skills, proficiency, and even pleasure derived from possessing an extremely powerful tool in engineering and scientific disciplines.

The author wishes the reader an enriching, stimulating and rewarding journey through the ideas and thoughts presented in these pages.

Thomas Szirtes
Toronto, Ontario, Canada
1996 May

PREFACE TO THE SECOND EDITION

Apart from the correction of minor errors and misprints in the first edition, this second edition contains (in Chapter 18) nine more applications of dimensional analysis. These new examples are from biometry, applied and theoretical physics, electric circuitry, mechanics, hydrodynamics, and geometry.

In facilitating the publication of this book, the author greatly appreciates the expert and generous assistance of Joel Stein, Senior Editor of Elsevier Science, Highland Park, NJ, USA.

The author sincerely thanks Lev Gorelic, D.Sc., Chicago, IL, USA, and Eugene Brach, P.Eng., Ottawa, Canada, for their proficient scrutiny of the first edition, and as a result the identification of the numerous typographical and other similar errors.

Also, for expediting and preparing the graphics in the added text, the author acknowledges the valued contributions of Lynne Vanin and Richard McDonald at MacDonald Dettwiler and Associates, Brampton, Ontario, Canada.

Finally, the author sincerely thanks his wife, Penny, for careful scrutiny of the appended text, resulting in the elimination of grammatical and punctuation errors and other grave linguistic misdeeds.

Thomas Szirtes
Toronto, Ontario, Canada
2006 April

ORGANIZATION, NOTATION, AND CONVENTIONS

This book is composed of *chapters* and *appendices*. The chapters are consecutively numbered from 1 to 18 and are frequently composed of *articles*. Articles are denoted by single or, if necessary, multiple digits separated by periods. For example "3.4.2" designates Article 4.2 in Chapter 3. To identify or refer to this article, we write "Art. 3.4.2." Appendices are consecutively numbered from 1 to 8. For example, the fifth appendix is referred to as "Appendix 5."

Examples are consecutively numbered starting with 1 in each chapter. The designation consists of *two* (single or double-digit) numbers separated by a hyphen. These numbers identify, in order, the chapter and the example's sequential position in the chapter. Thus, Example 18-12 is the twelfth example in Chapter 18. Note that the numbers are *not* in parentheses. The *last line* of every example is double underlined ═══ followed by the symbol ⇑.

Problems are consecutively numbered starting from 1 in each chapter. The designation of a typical problem consists *two* (single or double-digit) numbers separated by a solidus (/). The first and second numbers identify the chapter and the problem's sequential position in the chapter, respectively. Thus, "Problem 4/15" designates the fifteenth problem in Chapter 4.

Figures in *chapters* are consecutively numbered starting with 1 in each chapter. The designation consists of *two* (single or double-digit) numbers separated by a hyphen. The first number indicates the chapter; the second number indicates the figure's sequential position in that chapter. Thus, Fig. 4-2 designates the second figure in Chapter 4. Note that the numbers are *not* enclosed in parentheses and the word "Figure" is abbreviated to "Fig."

Figures in *appendices* are consecutively numbered starting with 1 in each appendix. The designation is of the format Ax-n, where x is the appendix number and n is a number indicating the figure's sequential position in that appendix. Thus, Fig. A3-1 designates the first figure in Appendix 3. Note that the word "Figure" is abbreviated to "Fig."

Equations in the *body* of the text—*not* in the examples, problems, or appendices—are designated by *two* numbers separated by a hyphen. The first (single or double-digit) number indicates the chapter; the second (single or double-digit) number indicates the equation's sequential position starting with 1 in each chapter. Equation numbers are *always* in parentheses. Thus, "(7-26)" designates the twenty-sixth equation in Chapter 7. On rare occasions, the second number is augmented by a lower-case letter, for example (7-19/a).

Equations in *examples* are designated by either a single lower-case letter or, rarely, a single character immediately followed by a one-digit number. These symbols are in alphabetic and, if applicable, numeric order, and are *always* in parentheses. Thus, in Example 17-9, the equations are designated by (a), (b), ..., (x), whereas in Example 18-26, by (a1), (b1), ..., (z1), (a2), (b2), ..., (i2).

Equations in *problems* are designated by the symbol #, whose number starts with 1 in each problem. These symbols are always enclosed in parentheses.

Equations in the *solutions* of problems (in Appendix 6) are designated by asterisks whose number starts with 1 in each solution. These asterisks are *always* enclosed in parentheses. Thus, in the solution of Problem 4/10 (Appendix 6) there are two equations, and hence they are consecutively denoted by (*) and (**).

Equations in the *proofs* for selected theorems and formulas (in Appendix 7) are designated by a single one- or two-digit number in parentheses. Numeration starts with 1 for each proof. Thus, one will find the designations (1), (2), etc.

When referring to an equation, the word "equation" is usually omitted. Thus the sentence "The value of $x = 8$ is now substituted into (16-10)" means "The value of $x = 8$ is now substituted into *equation* (16-10)." Similarly, the sentence "... rearranging (6-12), we obtain ..." is equivalent to "... rearranging *equation* (6-12), we obtain ...."

Theorems are consecutively numbered starting with 1 in each chapter. The designation consists of *two* (single or double-digit) numbers separated by a hyphen. The first and second numbers identify the chapter and the theorem's sequential position in the chapter, respectively. Thus, Theorem 7-5 is the fifth theorem in Chapter 7. Note that the numbers are *not* in parentheses.

Definitions are consecutively numbered starting with 1 in each chapter. The designation consists of *two* (single or double-digit) numbers separated by a hyphen. The first and second numbers identify the chapter and the definition's sequential position in the chapter, respectively. Thus, Definition 10-1 is the first definition in Chapter 10. Note that the numbers are *not* in parentheses.

Matrices and *vectors* are always symbolized by **bold** characters. Thus **B** is the symbol for a matrix, but B is the symbol for a scalar quantity. Similarly, **u** is a matrix (or a vector), but u is a scalar. Symbols for matrices are capital or (rarely) lower-case characters, for vectors they are always lower-case letters.

A *determinant* is denoted by $|\mathbf{U}|$, where **U** is any *square* matrix. For example, if $\mathbf{U} = \begin{bmatrix} 9 & 4 \\ 3 & 2 \end{bmatrix}$, then $|\mathbf{U}| = \begin{vmatrix} 9 & 4 \\ 3 & 2 \end{vmatrix} = 6$. If **U** is not a square matrix, then it does not have a determinant.

The *dimension* of any quantity z (variable or constant) is denoted by $[z]$. Thus, $[\text{speed}] = \text{m/s}$, $[\text{mass}] = \text{kg}$, $[\text{force}] = (\text{m·kg})/\text{s}^2$, etc. If z is dimensionless, then $[z] = 1$, thus $[7] = 1$.

Dimensionless variables are denoted by the Greek letter π_x, where x stands for a *mandatory* subscript. Thus, $\pi_1, \pi_5, \pi_{21}, \pi_c$ are dimensionless *variables*, but $\pi = 3.1415$ (no subscript) is a dimensionless *constant*; the former have no connection to the latter, although of course $[\pi_1] = 1$ and $[\pi] = 1$.

In tables where space is limited, the *E-notation* is used. Thus, the number 3.98×10^{-23} is written 3.98E-23, and similarly 5.4×10^3 is written 5.4E3.

Finally, in writing this book the author has exerted some effort to avoid using clichés, modish expressions, and engineering jargon. Thus, in the following pages the word "impact" *always* means the collision of two bodies, and "ongoing" is completely absent. About the latter abomination, the reader may find the following anecdote to be both appropriate and amusing. A noted editor of a major New York magazine sitting in his penthouse office remarked upon reading a submitted manuscript: "If I find the word *ongoing* one more time, I will be *downgoing* and somebody will be *outgoing*."

CHAPTER 1
MATHEMATICAL PRELIMINARIES

By Pál Rózsa

This chapter deals with the basic concepts of matrices, determinants and their applications in linear systems of equations. It is not the aim of this concise survey to teach mathematics, but to provide a very brief recapitulation so that the main text will be more easily understood.

More comprehensive treatment on linear systems can be found in a great many books; a selected list of such publications recommended by the writer is presented in Art. 1.5.

1.1. MATRICES AND DETERMINANTS

The array of $m \times n$ real numbers a_{ij} consisting of m *rows* and n *columns*

$$\begin{bmatrix} a_{11} & a_{12} & \cdots & a_{1n} \\ a_{21} & a_{22} & \cdots & a_{2n} \\ \vdots & \vdots & \vdots & \vdots \\ a_{m1} & a_{m2} & \cdots & a_{mn} \end{bmatrix}$$

is called an $m \times n$ *matrix* with the numbers a_{ij} as its *elements*. The following notation will be used for a matrix:

$$\mathbf{A} = [a_{ij}] \quad i = 1, 2, \ldots, m; \quad j = 1, 2, \ldots, n$$

where subscripts i and j denote the rows and columns, respectively. By interchanging the rows and columns of a matrix we get the *transpose* of the matrix denoted by $\mathbf{A}^T$. Thus we write $\mathbf{A}^T = [a_{ji}]$. As an example, let us consider the 3×4 matrix $\mathbf{A}$

$$\mathbf{A} = \begin{bmatrix} 1 & 2 & 3 & 5 \\ 4 & 0 & -1 & 2 \\ -1 & 1 & 7 & 6 \end{bmatrix}$$

the transpose of which is a 4 × 3 matrix

$$\mathbf{A}^\mathrm{T} = \begin{bmatrix} 1 & 4 & -1 \\ 2 & 0 & 1 \\ 3 & -1 & 7 \\ 5 & 2 & 6 \end{bmatrix}$$

If $m = n$, the matrix is called a *square* matrix of *order n*. Elements with subscripts *ii* of a *square* matrix constitute the *main diagonal* of the matrix. If all the elements not in the main diagonal of a matrix are zero, then the matrix is a *diagonal* one. For example $\begin{bmatrix} 1 & 0 & 0 \\ 0 & 2 & 0 \\ 0 & 0 & 5 \end{bmatrix}$ is a diagonal matrix of order 3.

The *unit* matrix, or *identity* matrix **I** is defined as a diagonal matrix whose nonzero elements are all "1." For example

$$\mathbf{I} = \begin{bmatrix} 1 & 0 & 0 & 0 \\ 0 & 1 & 0 & 0 \\ 0 & 0 & 1 & 0 \\ 0 & 0 & 0 & 1 \end{bmatrix}$$

is an identity matrix of order 4.

If all the elements of a matrix are zero, we get the *zero* or *null* matrix, which is denoted by **0**.

The $m \times 1$ and $1 \times n$ matrices are called *column vectors* and *row vectors*, respectively. For example,

$$\mathbf{a} = \begin{bmatrix} a_1 \\ a_2 \\ \vdots \\ a_n \end{bmatrix}$$

is a column vector, and $\mathbf{a}^\mathrm{T} = [a_1, a_2, \ldots, a_n]$ is a row vector.

A matrix can be split into several parts by means of horizontal and vertical lines. A matrix obtained in such a way is called a *partitioned* matrix. For example, let us consider matrix **A** partitioned as follows:

$$\mathbf{A} = \left[\begin{array}{cc|ccc} 1 & 2 & 3 & 5 & 6 \\ 4 & -1 & 7 & 0 & 1 \\ \hline 6 & 2 & 1 & 5 & -3 \\ 1 & 4 & 2 & 4 & -1 \\ 6 & 5 & 9 & -4 & 1 \end{array}\right] = \begin{bmatrix} \mathbf{A}_{11} & \mathbf{A}_{12} \\ \mathbf{A}_{21} & \mathbf{A}_{22} \end{bmatrix} \qquad (1\text{-}1)$$

where $\mathbf{A}_{11}, \mathbf{A}_{12}, \mathbf{A}_{21}, \mathbf{A}_{22}$ are *submatrices* of **A**. Thus, by (1-1), the submatrices of **A** are

$$\mathbf{A}_{11} = \begin{bmatrix} 1 & 2 \\ 4 & -1 \end{bmatrix}; \quad \mathbf{A}_{12} = \begin{bmatrix} 3 & 5 & 6 \\ 7 & 0 & 1 \end{bmatrix}; \quad \mathbf{A}_{21} = \begin{bmatrix} 6 & 2 \\ 1 & 4 \\ 6 & 5 \end{bmatrix}; \quad \mathbf{A}_{22} = \begin{bmatrix} 1 & 5 & -3 \\ 2 & 4 & -1 \\ 9 & -4 & 1 \end{bmatrix}$$

MATHEMATICAL PRELIMINARIES 3

If a square matrix is partitioned in such a way that the submatrices along the main diagonal—i.e., submatrices with subscripts *ii*—are square matrices, then the given matrix is *symmetrically* partitioned.

A *number* can be assigned to any *square* matrix of order n. This number is called the *determinant* of the matrix and is denoted by $|\mathbf{A}|$. We now give the *formal* definition of this number. The reader is advised *not* to be frightened off; a *more palatable* explanation is to follow!

Definition 1-1. *The determinant of a matrix* $\mathbf{A}$ *is*

$$|\mathbf{A}| = \sum_{n!} (-1)^q \cdot a_{1j_1} \cdot a_{2j_2} \ldots a_{nj_n}$$

where q *is the number of inversions in the permutation set* $j_1\, j_2 \ldots j_n$ *for numbers 1, 2, . . . , n, which are summed over all* n! *permutations of the first* n *natural numbers.* (For example, the number of inversions in the permutation 35214 is 6, since 3 is in inversion with 2 and 1; 5 with 2, 1, 4; and 2 with 1).

According to this definition, any determinant of order 3 can be calculated in the following way:

$$\begin{vmatrix} a_{11} & a_{12} & a_{13} \\ a_{21} & a_{22} & a_{23} \\ a_{31} & a_{32} & a_{33} \end{vmatrix} = a_{11} \cdot a_{22} \cdot a_{33} - a_{11} \cdot a_{23} \cdot a_{32} - a_{12} \cdot a_{21} \cdot a_{33} + a_{12} \cdot a_{23} \cdot a_{31} + a_{13} \cdot a_{21} \cdot a_{32} - a_{13} \cdot a_{22} \cdot a_{31}$$

Factoring out a_{11}, a_{12} and a_{13}, we can write for the above determinant

$$|\mathbf{A}| = a_{11} \cdot (a_{22} \cdot a_{33} - a_{23} \cdot a_{32}) - a_{12} \cdot (a_{21} \cdot a_{33} - a_{23} \cdot a_{31}) + a_{13} \cdot (a_{21} \cdot a_{32} - a_{22} \cdot a_{31})$$

The expressions in parentheses are the *determinants* of the second order, namely

$$a_{22} \cdot a_{33} - a_{23} \cdot a_{32} = \begin{vmatrix} a_{22} & a_{23} \\ a_{32} & a_{33} \end{vmatrix};\ a_{21} \cdot a_{33} - a_{23} \cdot a_{31} = \begin{vmatrix} a_{21} & a_{23} \\ a_{31} & a_{33} \end{vmatrix};$$

$$a_{21} \cdot a_{32} - a_{22} \cdot a_{31} = \begin{vmatrix} a_{21} & a_{22} \\ a_{31} & a_{32} \end{vmatrix}$$

Therefore determinant $|\mathbf{A}|$ can be written as

$$|\mathbf{A}| = a_{11} \cdot \begin{vmatrix} a_{22} & a_{23} \\ a_{32} & a_{33} \end{vmatrix} - a_{12} \cdot \begin{vmatrix} a_{21} & a_{23} \\ a_{31} & a_{33} \end{vmatrix} + a_{13} \cdot \begin{vmatrix} a_{21} & a_{22} \\ a_{31} & a_{32} \end{vmatrix} \qquad (1\text{-}2)$$

Note that the elements of the first row of $|\mathbf{A}|$ are multiplied by the second-order determinants, which are obtained by *omitting* the first row and the corresponding column of $|\mathbf{A}|$, and then affixing a negative sign to the second determinant. Expression (1-2) is called the *expansion* of the determinant by its first row. This technique can be generalized for any determinant of arbitrary order n, and, in fact, it is more useful, understandable, and much more practical to use than Definition 1-1. Thus, for the general case we proceed as follows:

First, we define the concept of *cofactors*. To any element a_{ij} of a determinant of order n can be assigned a *subdeterminant* of order $n-1$, by *omitting* the *i*th row and the jth column of the determinant. Then we assign the sign $(-1)^{i+j}$ to it (the "chessboard" rule: white squares are positive, black squares are negative) and the result is the *cofactor* denoted by A_{ij}. Note that the cofactor A_{ij} already includes the appropriate sign!

Example 1-1

Given determinant $|\mathbf{A}| = \begin{vmatrix} 1 & 2 & 3 \\ 4 & 5 & 6 \\ 7 & 8 & 9 \end{vmatrix}$, what are the cofactors of its elements?

$$A_{11} = \begin{vmatrix} 5 & 6 \\ 8 & 9 \end{vmatrix}; A_{12} = -\begin{vmatrix} 4 & 6 \\ 7 & 9 \end{vmatrix}; A_{13} = \begin{vmatrix} 4 & 5 \\ 7 & 8 \end{vmatrix}; A_{21} = -\begin{vmatrix} 2 & 3 \\ 8 & 9 \end{vmatrix}; A_{22} = \begin{vmatrix} 1 & 3 \\ 7 & 9 \end{vmatrix}$$

$$A_{23} = -\begin{vmatrix} 1 & 2 \\ 7 & 8 \end{vmatrix}; A_{31} = \begin{vmatrix} 2 & 3 \\ 5 & 6 \end{vmatrix}; A_{32} = -\begin{vmatrix} 1 & 3 \\ 4 & 6 \end{vmatrix}; A_{33} = \begin{vmatrix} 1 & 2 \\ 4 & 5 \end{vmatrix}$$

⇑

On the basis of Definition 1-1 and with the help of cofactors, the expression of a determinant can be formulated by the following theorem:

Theorem 1-1. *Any determinant $|\mathbf{A}|$ of order n can be expanded by its ith row (or jth column) by multiplying each element of the row (or column) by its cofactor and then summing the results. Thus,*

if $|\mathbf{A}|$ is expanded by its ith row, then $\quad |\mathbf{A}| = \sum_{j=1}^{n} a_{ij} \cdot A_{ij};$ \hfill (1-3)

if $|\mathbf{A}|$ is expanded by its jth column, then $\quad |\mathbf{A}| = \sum_{i=1}^{n} a_{ij} \cdot A_{ij}.$ \hfill (1-4)

An interesting characteristic of a determinant is that the sum of the products of the elements of any row and the cofactors of another row is zero (the same is true for columns). That is

$$\sum_{j=1}^{n} a_{ij} \cdot A_{kj} = 0 \text{ if } i \neq k; \quad \sum_{i=1}^{n} a_{ij} \cdot A_{ik} = 0 \quad \text{if } j \neq k \quad (1-5)$$

The following two examples demonstrate the use of these methods.

Example 1-2

We evaluate the determinant $|\mathbf{A}| = \begin{vmatrix} 1 & 2 & 3 \\ 4 & 5 & 6 \\ 1 & 3 & 4 \end{vmatrix}$. Expanding it by the first row, we obtain

$$|\mathbf{A}| = 1 \cdot \begin{vmatrix} 5 & 6 \\ 3 & 4 \end{vmatrix} - 2 \cdot \begin{vmatrix} 4 & 6 \\ 1 & 4 \end{vmatrix} + 3 \cdot \begin{vmatrix} 4 & 5 \\ 1 & 3 \end{vmatrix} = 1 \cdot (20 - 18) - 2 \cdot (16 - 6) + 3 \cdot (12 - 5) = 3$$

and by the second column

$$|\mathbf{A}| = -2 \cdot \begin{vmatrix} 4 & 6 \\ 1 & 4 \end{vmatrix} + 5 \cdot \begin{vmatrix} 1 & 3 \\ 1 & 4 \end{vmatrix} - 3 \cdot \begin{vmatrix} 1 & 3 \\ 4 & 6 \end{vmatrix} = -2 \cdot (16 - 6) + 5 \cdot (4 - 3) - 3 \cdot (6 - 12) = 3$$

These two values are identical, of course. The reader may calculate the value by expanding the determinant by its third row, or by any column; the value in all cases will be 3.

Now we multiply the elements of the *second* row by the *cofactors* of the elements of the *third* row:

$$-4 \cdot \begin{vmatrix} 2 & 3 \\ 5 & 6 \end{vmatrix} + 5 \cdot \begin{vmatrix} 1 & 3 \\ 4 & 6 \end{vmatrix} - 6 \cdot \begin{vmatrix} 1 & 2 \\ 4 & 5 \end{vmatrix} = -4 \cdot (12 - 15) + 5 \cdot (6 - 12) - 6 \cdot (5 - 8) = 0$$

We see that this expansion can be written as a third-order determinant $\begin{vmatrix} 1 & 2 & 3 \\ 4 & 5 & 6 \\ 4 & 5 & 6 \end{vmatrix}$ in which the second and third rows are equal.

Example 1-3

We evaluate the determinant $\mathbf{A} = \begin{vmatrix} 3 & 2 & 1 & -6 \\ -2 & 3 & 1 & -1 \\ 2 & 3 & 4 & 0 \\ 5 & 3 & 1 & -3 \end{vmatrix}$. Expanding it by the fourth column, we obtain

$$|\mathbf{A}| = -(-6) \cdot \begin{vmatrix} -2 & 3 & 1 \\ 2 & 3 & 4 \\ 5 & 3 & 1 \end{vmatrix} + (-1) \cdot \begin{vmatrix} 3 & 2 & 1 \\ 2 & 3 & 4 \\ 5 & 3 & 1 \end{vmatrix} - (0) \cdot \begin{vmatrix} 3 & 2 & 1 \\ -2 & 3 & 1 \\ 5 & 3 & 1 \end{vmatrix} + (-3) \cdot \begin{vmatrix} 3 & 2 & 1 \\ -2 & 3 & 1 \\ 2 & 3 & 4 \end{vmatrix} \quad (a)$$

Next, we evaluate the above four third-order determinants by the method shown in Example 1-2. Hence

$$\begin{vmatrix} -2 & 3 & 1 \\ 2 & 3 & 4 \\ 5 & 3 & 1 \end{vmatrix} = 63; \quad \begin{vmatrix} 3 & 2 & 1 \\ 2 & 3 & 4 \\ 5 & 3 & 1 \end{vmatrix} = 0; \quad \begin{vmatrix} 3 & 2 & 1 \\ -2 & 3 & 1 \\ 5 & 3 & 1 \end{vmatrix} = -7; \quad \begin{vmatrix} 3 & 2 & 1 \\ -2 & 3 & 1 \\ 2 & 3 & 4 \end{vmatrix} = 35 \quad (b)$$

and therefore, by (a) and (b), $|\mathbf{A}| = -(-6) \cdot (63) + (-1) \cdot (0) - (0) \cdot (-7) + (-3) \cdot (35) = 273$.

From Definition 1-1, the following important properties of a determinant can be derived:

- If all elements of a row are multiplied by a constant c, the determinant's value is multiplied by c.
- If two rows are interchanged, the sign of the determinant's value is changed.
- If a row of a determinant is a multiple of any other row, the determinant's value is zero.
- If the multiple of a row is added to any other row, the determinant's value does not change.
- If all the elements below (or above) the main diagonal are zero, the determinant's value is the product of the elements of the main diagonal.
- The value of a determinant is a linear function of any of its elements (provided, of course, that all the other elements remain unchanged). That is, if z is any element, then the value of the determinant is $D = k_1 \cdot z + k_2$, where k_1 and k_2 are constants. In particular, if D_0 and D_1 are the values of the determinant if $z = 0$ and

$z = 1$, respectively, then the value of the element z at which the determinant vanishes is

$$z_0 = \frac{D_0}{D_0 - D_1} \qquad (1\text{-}6)$$

Remark: All of the above statements are true if the word "row" is changed to "column."

1.2. OPERATIONS WITH MATRICES

Matrices $\mathbf{A} = [a_{ij}]$ and $\mathbf{B} = [b_{ij}]$ are *equal* if they have the same number of rows and columns, and if $a_{ij} = b_{ij}$ for all i and j. If $\mathbf{A}$ is a square matrix and $\mathbf{A} = \mathbf{A}^T$, then $\mathbf{A}$ is called a *symmetric* matrix. If $\mathbf{A} = -\mathbf{A}^T$, then $\mathbf{A}$ is called a *skew-symmetric* matrix. For example

$\mathbf{A} = \begin{bmatrix} 1 & 5 & -4 \\ 5 & 2 & 7 \\ -4 & 7 & 3 \end{bmatrix}$ is a symmetric matrix, whereas $\mathbf{B} = \begin{bmatrix} 0 & 3 & 4 \\ -3 & 0 & -1 \\ -4 & 1 & 0 \end{bmatrix}$ is a skew-symmetric matrix.

Addition of two matrices and the multiplication of a matrix by a scalar are defined, respectively, as follows:

$$\mathbf{A} + \mathbf{B} = [a_{ij} + b_{ij}] \text{ and } c \cdot \mathbf{A} = [c \cdot a_{ij}]$$

From these definitions, the *distributive* property follows:

$$c \cdot (\mathbf{A} + \mathbf{B}) = c \cdot \mathbf{A} + c \cdot \mathbf{B} \text{ and } (p + q) \cdot \mathbf{C} = p \cdot \mathbf{C} + q \cdot \mathbf{C}$$

We also have the following properties

$$(\mathbf{A} + \mathbf{B})^T = \mathbf{A}^T + \mathbf{B}^T; \ (c \cdot \mathbf{A})^T = c \cdot \mathbf{A}^T; \ |c \cdot \mathbf{A}| = c^n \cdot |\mathbf{A}| \quad (n = \text{order of the determinant})$$

Example 1-4

If $\mathbf{A} = \begin{bmatrix} 5 & 8 \\ 3 & 2 \end{bmatrix}$, $\mathbf{B} = \begin{bmatrix} -10 & 6 \\ 2 & 19 \end{bmatrix}$ and $c = 8$, then $\mathbf{A} + \mathbf{B} = \begin{bmatrix} -5 & 14 \\ 5 & 21 \end{bmatrix}$. Moreover,

$$c \cdot (\mathbf{A} + \mathbf{B}) = 8 \begin{bmatrix} -5 & 14 \\ 5 & 21 \end{bmatrix} = \begin{bmatrix} -40 & 112 \\ 40 & 168 \end{bmatrix} = 8 \begin{bmatrix} 5 & 8 \\ 3 & 2 \end{bmatrix} + 8 \begin{bmatrix} -10 & 6 \\ 2 & 19 \end{bmatrix}$$

$$= \begin{bmatrix} 40 & 64 \\ 24 & 16 \end{bmatrix} + \begin{bmatrix} -80 & 48 \\ 16 & 152 \end{bmatrix} = \begin{bmatrix} -40 & 112 \\ 40 & 168 \end{bmatrix}$$

⇑

The product of two matrices $\mathbf{A} \cdot \mathbf{B}$ are defined only if they are *compatible*. Matrices $\mathbf{A}$ and $\mathbf{B}$ are compatible if the number of *columns* of $\mathbf{A}$ equals the number of *rows* of $\mathbf{B}$.

MATHEMATICAL PRELIMINARIES

Definition 1-2. *If* **A** *is an* m × p *matrix and* **B** *is a* p × n *matrix, then the product* **A·B** *is an* m × n *matrix, the elements of which are the scalar products of the rows of* **A** *and the columns of* **B**. *Thus*

$$\mathbf{A \cdot B} = \left[\sum_{k=1}^{p} a_{ik} \cdot b_{kj} \right] \quad \text{where } i = 1, 2, \ldots, m; \quad j = 1, 2, \ldots, n. \tag{1-7}$$

Example 1-5

Let us calculate product **A·B** if $\mathbf{A} = \begin{bmatrix} 1 & 2 & 3 & 4 \\ 2 & 5 & 0 & 3 \\ 4 & -2 & 1 & 7 \end{bmatrix}$ and $\mathbf{B} = \begin{bmatrix} 3 & -2 \\ 1 & 0 \\ 4 & 5 \\ 6 & 8 \end{bmatrix}$.

Since **A** is a 3 × 4 matrix and **B** is a 4 × 2 matrix, product **A·B** will be a 3 × 2 matrix. It is now advantageous to arrange **A** and **B** in the configuration shown in Fig. 1-1. For example, the top left element of matrix **A·B** is (1)·(3) + (2)·(1) + (3)·(4) + (4)·(6) = 41.

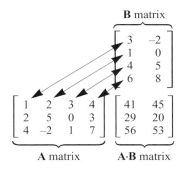

Figure 1-1
A didactically useful configuration to determine the product of matrices A and B
The elements of product **A·B** appear at the intersections
of the extensions of **A**'s rows and **B**'s columns

From Definition 1-2 it follows that the multiplication of matrices is *not commutative*, i.e., in general, **A·B** ≠ **B·A**. The lack of commutativity is the cause of most of the problems in matrix theory. On the other hand, this property gives the theory some of its beauty. Nevertheless, there are a few special matrices for which multiplication is commutative. Matrices with such a property are called *commutative matrices*. It is easy to verify that commutativity holds in the following cases:

- Any matrix is commutative with respect to the *identity* matrix.
- Any matrix is commutative with respect to the *null* matrix.
- If both **A** and **B** are *diagonal* matrices, then they are commutative.

By the repeated application of (1-7), it can be shown that the multiplication of three or more matrices is *associative*. That is

$$(\mathbf{A \cdot B}) \cdot \mathbf{C} = \mathbf{A} \cdot (\mathbf{B \cdot C}) \tag{1-8}$$

Moreover, the rule of *distributivity* holds, as well. That is

$$(A + B) \cdot C = A \cdot C + B \cdot C; \quad A \cdot (B + C) = A \cdot B + A \cdot C. \tag{1-9}$$

Because of the associativity of multiplication, the integer powers of a square matrix can be uniquely defined if we consider that any square matrix is commutative with itself: $A^p = \underbrace{A \cdot A \cdots A}_{p \text{ times}}$. Hence $A^p \cdot A^q = A^{p+q}$. The zero power of any square matrix of order n is defined to be the *identity* matrix of the same order. That is $A^0 = I$.

The *transpose* of the product of matrices A and B is determined by the formula

$$(A \cdot B)^T = B^T \cdot A^T \tag{1-10}$$

Example 1-6

We shall consider the matrices in Example 1-5 and then verify relation (1-10).

$$(A \cdot B)^T = \left(\underbrace{\begin{bmatrix} 1 & 2 & 3 & 4 \\ 2 & 5 & 0 & 3 \\ 4 & -2 & 1 & 7 \end{bmatrix}}_{A \text{ matrix}} \cdot \underbrace{\begin{bmatrix} 3 & -2 \\ 1 & 0 \\ 4 & 5 \\ 6 & 8 \end{bmatrix}}_{B \text{ matrix}} \right)^T = \underbrace{\begin{bmatrix} 41 & 45 \\ 29 & 20 \\ 56 & 53 \end{bmatrix}^T}_{A \cdot B \text{ matrix}} = \underbrace{\begin{bmatrix} 41 & 29 & 56 \\ 45 & 20 & 53 \end{bmatrix}}_{(A \cdot B)^T \text{ matrix}} \quad \text{(a)}$$

By (1-10),

$$B^T \cdot A^T = \begin{bmatrix} 3 & -2 \\ 1 & 0 \\ 4 & 5 \\ 6 & 8 \end{bmatrix}^T \cdot \begin{bmatrix} 1 & 2 & 3 & 4 \\ 2 & 5 & 0 & 3 \\ 4 & -2 & 1 & 7 \end{bmatrix}^T = \begin{bmatrix} 3 & 1 & 4 & 6 \\ -2 & 0 & 5 & 8 \end{bmatrix} \cdot \begin{bmatrix} 1 & 2 & 4 \\ 2 & 5 & -2 \\ 3 & 0 & 1 \\ 4 & 3 & 7 \end{bmatrix} = \begin{bmatrix} 41 & 29 & 56 \\ 45 & 20 & 53 \end{bmatrix} \quad \text{(b)}$$

and we see that (a) and (b) are indeed identical.

⇑

From the definition of multiplication, the rule for the multiplication of *partitioned* matrices can be easily obtained. If matrices A and B are compatible, and if A is partitioned with regard to its *columns* in the same way as B is with regard to its *rows*, then product $A \cdot B$ can be calculated in terms of the generated submatrices.

Example 1-7

Given the following matrices and their partitioning:

$$A = \left[\begin{array}{cc|cc} 1 & 0 & 2 & 2 \\ 0 & 1 & -1 & -1 \\ \hline 0 & 0 & 2 & 0 \\ 0 & 0 & 0 & 2 \end{array} \right] = \left[\begin{array}{c|c} A_1 & A_2 \\ \hline A_3 & A_4 \end{array} \right]; \quad B = \left[\begin{array}{cc|cc} 1 & 2 & 0 & 0 \\ 3 & 4 & 0 & 0 \\ \hline 1 & 0 & 2 & 2 \\ 0 & 1 & -1 & -1 \end{array} \right] = \left[\begin{array}{c|c} B_1 & B_2 \\ \hline B_3 & B_4 \end{array} \right] \quad \text{(a)}$$

Therefore

$$A_1 = \begin{bmatrix} 1 & 0 \\ 0 & 1 \end{bmatrix}; \quad A_2 = \begin{bmatrix} 2 & 2 \\ -1 & -1 \end{bmatrix}; \quad A_3 = \begin{bmatrix} 0 & 0 \\ 0 & 0 \end{bmatrix}; \quad A_4 = \begin{bmatrix} 2 & 0 \\ 0 & 2 \end{bmatrix} \quad \text{(b)}$$

and

$$\mathbf{B}_1 = \begin{bmatrix} 1 & 2 \\ 3 & 4 \end{bmatrix}; \mathbf{B}_2 = \begin{bmatrix} 0 & 0 \\ 0 & 0 \end{bmatrix}; \mathbf{B}_3 = \begin{bmatrix} 1 & 0 \\ 0 & 1 \end{bmatrix}; \mathbf{B}_4 = \begin{bmatrix} 2 & 2 \\ -1 & -1 \end{bmatrix} \quad (c)$$

Thus, we write

$$\mathbf{A} \cdot \mathbf{B} = \begin{bmatrix} \mathbf{A}_1 & \mathbf{A}_2 \\ \mathbf{A}_3 & \mathbf{A}_4 \end{bmatrix} \begin{bmatrix} \mathbf{B}_1 & \mathbf{B}_2 \\ \mathbf{B}_3 & \mathbf{B}_4 \end{bmatrix} = \begin{bmatrix} \mathbf{A}_1 \cdot \mathbf{B}_1 + \mathbf{A}_2 \cdot \mathbf{B}_3 & \mathbf{A}_1 \cdot \mathbf{B}_2 + \mathbf{A}_2 \cdot \mathbf{B}_4 \\ \mathbf{A}_3 \cdot \mathbf{B}_1 + \mathbf{A}_4 \cdot \mathbf{B}_3 & \mathbf{A}_3 \cdot \mathbf{B}_2 + \mathbf{A}_4 \cdot \mathbf{B}_4 \end{bmatrix} = \begin{bmatrix} 3 & 4 & | & 2 & 2 \\ 2 & 3 & | & -1 & -1 \\ \hline 2 & 0 & | & 4 & 4 \\ 0 & 2 & | & -2 & -2 \end{bmatrix} \quad (d)$$

and by direct multiplication

$$\mathbf{A} \cdot \mathbf{B} = \begin{bmatrix} 1 & 0 & 2 & 2 \\ 0 & 1 & -1 & -1 \\ 0 & 0 & 2 & 0 \\ 0 & 0 & 0 & 2 \end{bmatrix} \begin{bmatrix} 1 & 2 & 0 & 0 \\ 3 & 4 & 0 & 0 \\ 1 & 0 & 2 & 2 \\ 0 & 1 & -1 & -1 \end{bmatrix} = \begin{bmatrix} 3 & 4 & 2 & 2 \\ 2 & 3 & -1 & -1 \\ 2 & 0 & 4 & 4 \\ 0 & 2 & -2 & -2 \end{bmatrix} \quad (e)$$

We see that the results of (e) and (d) are indeed identical.

We obtain special matrix products if the factors of a product are vectors. Thus, we define the *inner* product of two vectors

$$\mathbf{a}^T \cdot \mathbf{b} = \sum_{k=1}^{n} a_k \cdot b_k \quad (1\text{-}11)$$

Obviously, the inner product produces a matrix of order 1, i.e., a *scalar* (this is a scalar product). Similarly, we define the *outer* product of two vectors, also called a *dyad* or dyadic product. Hence

$$\mathbf{a} \cdot \mathbf{b}^T = \begin{bmatrix} a_1 \cdot b_1 & a_1 \cdot b_2 & \cdots & a_1 \cdot b_n \\ a_2 \cdot b_1 & a_2 \cdot b_2 & \cdots & a_2 \cdot b_n \\ \vdots & \vdots & \vdots & \vdots \\ a_m \cdot b_1 & a_m \cdot b_2 & \cdots & a_m \cdot b_n \end{bmatrix} \quad (1\text{-}12)$$

If vectors **a** and **b** contain m and n elements, respectively, then the *outer* product $\mathbf{a} \cdot \mathbf{b}^T$ is an $m \times n$ matrix.

Example 1-8

Given column vectors $\mathbf{a} = \begin{bmatrix} 2 \\ -1 \\ 3 \end{bmatrix}$ and $\mathbf{b} = \begin{bmatrix} 1 \\ 4 \\ -3 \end{bmatrix}$, we wish to determine $\mathbf{a}^T \cdot \mathbf{b}$ and $\mathbf{a} \cdot \mathbf{b}^T$. Thus,

$$\mathbf{a}^T \cdot \mathbf{b} = \begin{bmatrix} 2 \\ -1 \\ 3 \end{bmatrix}^T \cdot \begin{bmatrix} 1 \\ 4 \\ -3 \end{bmatrix} = [2 \ -1 \ 3] \cdot \begin{bmatrix} 1 \\ 4 \\ -3 \end{bmatrix} = (2)(1) + (-1)(4) + (3)(-3) = -11$$

and

$$\mathbf{a} \cdot \mathbf{b}^T = \begin{bmatrix} 2 \\ -1 \\ 3 \end{bmatrix} \cdot [1 \ 4 \ -3] = \begin{bmatrix} 2 & 8 & -6 \\ -1 & -4 & 3 \\ 3 & 12 & -9 \end{bmatrix}$$

Utilizing now the outer products of vectors [see (1-12)], if in **A·B**, matrix **A** is partitioned into its columns and **B** into its rows, then the product obtained is the sum of the outer products, or dyads, formed by the columns of **A** and the rows of **B**. That is, the sum of dyads can always be written as the product of two matrices, where the first factor consists of the columns, the second factor the rows of the dyads. It follows that the problem of factoring a given matrix is equivalent to that of the *decomposition* of this matrix into dyads (i.e., outer products).

Example 1-9

Consider the matrix product $\mathbf{A}\cdot\mathbf{B} = \begin{bmatrix} 1 & 2 & 3 & 4 \\ 2 & 5 & 0 & 3 \\ 4 & -2 & 1 & 7 \end{bmatrix} \cdot \begin{bmatrix} 3 & -2 \\ 1 & 0 \\ 4 & 5 \\ 6 & 8 \end{bmatrix}$. The product written as the sum of outer products, or dyads, is

$$\mathbf{A}\cdot\mathbf{B} = \begin{bmatrix} 1 \\ 2 \\ 4 \end{bmatrix}\cdot[3\ -2] + \begin{bmatrix} 2 \\ 5 \\ -2 \end{bmatrix}\cdot[1\ 0] + \begin{bmatrix} 3 \\ 0 \\ 1 \end{bmatrix}\cdot[4\ 5] + \begin{bmatrix} 4 \\ 3 \\ 7 \end{bmatrix}\cdot[6\ 8]$$

or

$$\mathbf{A}\cdot\mathbf{B} = \begin{bmatrix} 3 & -2 \\ 6 & -4 \\ 12 & -8 \end{bmatrix} + \begin{bmatrix} 2 & 0 \\ 5 & 0 \\ -2 & 0 \end{bmatrix} + \begin{bmatrix} 12 & 15 \\ 0 & 0 \\ 4 & 5 \end{bmatrix} + \begin{bmatrix} 24 & 32 \\ 18 & 24 \\ 42 & 56 \end{bmatrix} = \begin{bmatrix} 41 & 45 \\ 29 & 20 \\ 56 & 53 \end{bmatrix}$$

This result is identical to that shown in Fig. 1-1.

⇑

Let us now introduce the concept of the *inverse* of a matrix.

Definition 1-3. *Given the square matrix* **A**. *If* **X** *satisfies the equation* **A·X** = **I**, *then* **X** *is called the right inverse of* **A**. *If* **Y** *satisfies the equation* **Y·A** = **I**, *then* **Y** *is called the left inverse of* **A**.

It can be shown that the right and left inverses exist if, and only if, the determinant of **A** is not zero. In this case **X** = **Y**.

Definition 1-4. *Matrix* **A** *is called "nonsingular" if* $|\mathbf{A}| \neq 0$; *if* $|\mathbf{A}| = 0$, *then* **A** *is "singular."*

Definition 1-5. *The transpose of the matrix formed by the* cofactors A_{ij} *of a matrix* $\mathbf{A} = [a_{ij}]$ *is called the* adjoint *of* **A** *and is denoted by* adj **A**; *i.e.,*

$$\text{adj } \mathbf{A} = [A_{ji}] \qquad (1\text{-}13)$$

The formulas (1-3), (1-4), and (1-5) imply that

$$\mathbf{A}\cdot\text{adj } \mathbf{A} = |\mathbf{A}|\cdot\mathbf{I} \qquad (1\text{-}14)$$

$$\text{adj } \mathbf{A}\cdot\mathbf{A} = |\mathbf{A}|\cdot\mathbf{I} \qquad (1\text{-}15)$$

Assuming now that **A** is nonsingular, we obtain from (1-14) and (1-15)

and
$$\mathbf{A} \cdot \frac{\text{adj } \mathbf{A}}{|\mathbf{A}|} = \mathbf{I} \tag{1-16}$$

$$\frac{\text{adj } \mathbf{A}}{|\mathbf{A}|} \cdot \mathbf{A} = \mathbf{I} \tag{1-17}$$

Based on the existence of the inverse of a matrix, we now state the following theorem:

Theorem 1-2. *If* **A** *is nonsingular, then its inverse is*

$$\mathbf{A}^{-1} = \frac{\text{adj } \mathbf{A}}{|\mathbf{A}|} \tag{1-18}$$

Proof. Postmultiplication both sides of (1-17) by $\mathbf{A}^{-1}$ yields

$$\frac{\text{adj } \mathbf{A}}{|\mathbf{A}|} \cdot \mathbf{A} \cdot \mathbf{A}^{-1} = \frac{\text{adj } \mathbf{A}}{|\mathbf{A}|} \cdot \mathbf{I} = \frac{\text{adj } \mathbf{A}}{|\mathbf{A}|} = \mathbf{I} \cdot \mathbf{A}^{-1} = \mathbf{A}^{-1}$$

This proves the theorem.

The inverse of a matrix product is equal to the product of the inverses taking the factors in reverse order. That is

$$(\mathbf{A} \cdot \mathbf{B})^{-1} = \mathbf{B}^{-1} \cdot \mathbf{A}^{-1} \tag{1-19}$$

To show this, by (1-8) we write

$$(\mathbf{B}^{-1} \cdot \mathbf{A}^{-1}) \cdot (\mathbf{A} \cdot \mathbf{B}) = \mathbf{B}^{-1} \cdot (\mathbf{A}^{-1} \cdot \mathbf{A}) \cdot \mathbf{B} = \mathbf{B}^{-1} \cdot \mathbf{I} \cdot \mathbf{B} = \mathbf{B}^{-1} \cdot \mathbf{B} = \mathbf{I}$$

therefore $(\mathbf{B}^{-1} \cdot \mathbf{A}^{-1}) = \mathbf{I} \cdot (\mathbf{A} \cdot \mathbf{B})^{-1} = (\mathbf{A} \cdot \mathbf{B})^{-1}$, which was to be demonstrated.

Example 1-10

Find the inverse of the matrix

$$\mathbf{A} = \begin{bmatrix} 1 & 2 & 4 \\ 3 & 5 & 6 \\ 1 & 1 & 1 \end{bmatrix}$$

By Definition 1-5, the *adjoint* matrix of **A** is

$$\text{adj } \mathbf{A} = \begin{bmatrix} -1 & 3 & -2 \\ 2 & -3 & 1 \\ -8 & 6 & -1 \end{bmatrix}^T = \begin{bmatrix} -1 & 2 & -8 \\ 3 & -3 & 6 \\ -2 & 1 & -1 \end{bmatrix}$$

Since the determinant of **A** is $|\mathbf{A}| = -3$, therefore, by (1-18), the *inverse* of **A** is

$$\mathbf{A}^{-1} = \frac{\text{adj } \mathbf{A}}{|\mathbf{A}|} = \frac{1}{-3} \begin{bmatrix} -1 & 2 & -8 \\ 3 & -3 & 6 \\ -2 & 1 & -1 \end{bmatrix} = \begin{bmatrix} \frac{1}{3} & -\frac{2}{3} & \frac{8}{3} \\ -1 & 1 & -2 \\ \frac{2}{3} & -\frac{1}{3} & \frac{1}{3} \end{bmatrix}$$

Example 1-11

Find the determinant of the matrix

$$\mathbf{A} = \begin{bmatrix} 1 & 2 & 3 \\ 4 & 5 & 6 \\ 7 & 8 & 9 \end{bmatrix}$$

Expanding the determinant by the first row, we get

$$|\mathbf{A}| = (1) \cdot \begin{vmatrix} 5 & 6 \\ 8 & 9 \end{vmatrix} - (2) \cdot \begin{vmatrix} 4 & 6 \\ 7 & 9 \end{vmatrix} + (3) \cdot \begin{vmatrix} 4 & 5 \\ 7 & 8 \end{vmatrix} = 0$$

Thus, matrix **A** is singular and hence its inverse does not exist. The adjoint of **A** is

$$\text{adj } \mathbf{A} = \begin{bmatrix} -3 & 6 & -3 \\ 6 & 12 & 6 \\ -3 & 6 & -3 \end{bmatrix}$$

and formula (1-14) gives

$$\mathbf{A} \cdot \text{adj } \mathbf{A} = \begin{bmatrix} 1 & 2 & 3 \\ 4 & 5 & 6 \\ 7 & 8 & 9 \end{bmatrix} \cdot \begin{bmatrix} -3 & 6 & -3 \\ 6 & 12 & 6 \\ -3 & 6 & -3 \end{bmatrix} = 0$$

⇑

Now we introduce the concept of a *permutation* matrix.

Definition 1-6. *If the columns of an identity matrix are permuted, a permutation matrix is obtained.* Accordingly,

$$\mathbf{P} = [\mathbf{e}_{j_1}, \mathbf{e}_{j_2}, \ldots, \mathbf{e}_{j_n}] \qquad (1\text{-}20)$$

where $\mathbf{e}_j$ is the *j*th column vector of the identity matrix and the indices $j_1, j_2, \ldots, j_n$ form a permutation of the natural numbers $1, 2, \ldots, n$. It follows that *postmultiplication* of an arbitrary matrix **A** by **P** results in a permutation of the *columns* of **A**.

Example 1-12

Consider

$$\underbrace{\begin{bmatrix} 4 & 6 & 8 & 1 \\ 5 & 3 & 1 & 0 \\ 7 & 21 & 0 & 9 \\ 3 & -4 & 1 & 3 \end{bmatrix}}_{\mathbf{A} \text{ matrix}} \cdot \underbrace{\begin{bmatrix} 0 & 1 & 0 & 0 \\ 0 & 0 & 1 & 0 \\ 0 & 0 & 0 & 1 \\ 1 & 0 & 0 & 0 \end{bmatrix}}_{\mathbf{P} \text{ matrix}} = \underbrace{\begin{bmatrix} 1 & 4 & 6 & 8 \\ 0 & 5 & 3 & 1 \\ 9 & 7 & 21 & 0 \\ 3 & 3 & -4 & 1 \end{bmatrix}}_{\mathbf{A} \cdot \mathbf{P} \text{ matrix}}$$

In this case permutation matrix **P** was generated from the fourth-order identity matrix **I** since

the first column of **I** became the second column of **P**,
the second column of **I** became the third column of **P**,
the third column of **I** became the fourth column of **P**,
the fourth column of **I** became the first column of **P**.

Therefore

> the first column of **A** became the second column of **A·P**,
> the second column of **A** became the third column of **A·P**,
> the third column of **A** became the fourth column of **A·P**,
> the fourth column of **A** became the first column of **A·P**.

In a similar way, the *premultiplication* of matrix **A** by **P**^{T} will produce the same permutations of the *rows* of **A**.

Example 1-13

Consider

$$\begin{bmatrix} 0 & 0 & 0 & 1 \\ 1 & 0 & 0 & 0 \\ 0 & 1 & 0 & 0 \\ 0 & 0 & 1 & 0 \end{bmatrix} \cdot \begin{bmatrix} 4 & 6 & 8 & 1 \\ 5 & 3 & 1 & 0 \\ 7 & 21 & 0 & 9 \\ 3 & -4 & 1 & 3 \end{bmatrix} = \begin{bmatrix} 3 & -4 & 1 & 3 \\ 4 & 6 & 8 & 1 \\ 5 & 3 & 1 & 0 \\ 7 & 21 & 0 & 9 \end{bmatrix}$$

$\quad\quad\quad\quad$ **P**^{T} matrix $\quad\quad$ **A** matrix $\quad\quad$ **P**^{T}·**A** matrix

Here permutation matrix **P**^{T} was generated from the fourth-order identity matrix **I** since

> the first row of **I** became the second row of **P**^{T},
> the second row of **I** became the third row of **P**^{T}
> the third row of **I** became the fourth row of **P**^{T},
> the fourth row of **I** became the first row of **P**^{T}.

Therefore

> the first row of **A** became the second row of **P**^{T}·**A**,
> the second row of **A** became the third row of **P**^{T}·**A**,
> the third row of **A** became the fourth row of **P**^{T}·**A**,
> the fourth row of **A** became the first row of **P**^{T}·**A**.

We now define the *orthogonality* of a matrix.

Definition 1-7. *If the inverse of matrix* **Q** *is equal to its transpose, i.e.,*

$$\mathbf{Q}^{-1} = \mathbf{Q}^T \quad\quad\quad (1\text{-}21)$$

then **Q** *is called an orthogonal matrix.*

It can be shown that every permutation matrix is orthogonal, i.e., **P**^{T} = **P**$^{-1}$.

Example 1-14

Consider *permutation* matrix $\mathbf{P} = \begin{bmatrix} 0 & 1 & 0 & 0 \\ 0 & 0 & 1 & 0 \\ 0 & 0 & 0 & 1 \\ 1 & 0 & 0 & 0 \end{bmatrix}$. Its *inverse* is $\mathbf{P}^{-1} = \begin{bmatrix} 0 & 0 & 0 & 1 \\ 1 & 0 & 0 & 0 \\ 0 & 1 & 0 & 0 \\ 0 & 0 & 1 & 0 \end{bmatrix}$ and its *transpose* $\mathbf{P}^T = \begin{bmatrix} 0 & 0 & 0 & 1 \\ 1 & 0 & 0 & 0 \\ 0 & 1 & 0 & 0 \\ 0 & 0 & 1 & 0 \end{bmatrix}$. Therefore its transpose and inverse are identical and hence, by Definition 1-7, the matrix is orthogonal.

⇑

Example 1-15

Prove that matrix $\mathbf{A} = \begin{bmatrix} 0.19863 & -0.95106 & 0.23672 \\ 0.61133 & 0.30902 & 0.72855 \\ 0.76604 & 0 & -0.64279 \end{bmatrix}$ is orthogonal.

We could first determine $\mathbf{A}^{-1}$, and then $\mathbf{A}^T$, to see whether they are identical. If they were, then by (1-21) $\mathbf{A}$ would be proven to be orthogonal. But to obtain the inverse of $\mathbf{A}$ is "labor intensive." So to alleviate our burden, we consider that by (1-21) we can write

$$\mathbf{A} \cdot \mathbf{A}^{-1} = \mathbf{A} \cdot \mathbf{A}^T = \mathbf{I}$$

Thus we only have to come up with $\mathbf{A}^T$, which is easy. Accordingly,

$$\mathbf{A} \cdot \mathbf{A}^T = \underbrace{\begin{bmatrix} 0.19863 & -0.95106 & 0.23672 \\ 0.61133 & 0.30902 & 0.72855 \\ 0.76604 & 0 & -0.64279 \end{bmatrix}}_{\mathbf{A} \text{ matrix}} \cdot \underbrace{\begin{bmatrix} 0.19863 & 0.61133 & 0.76604 \\ -0.95106 & 0.30902 & 0 \\ 0.23672 & 0.72855 & -0.64279 \end{bmatrix}}_{\mathbf{A}^T \text{ matrix}} = \begin{bmatrix} 1 & 0 & 0 \\ 0 & 1 & 0 \\ 0 & 0 & 1 \end{bmatrix} = \mathbf{I}$$

Thus $\mathbf{A}$ is indeed orthogonal.

⇑

1.3. THE RANK OF A MATRIX

In a sense rank serves as a measure of the singularity of a matrix. To define and determine a matrix's rank we shall use the notion of a *minor*. The determinant of a *square* submatrix of a given matrix is called a *minor* of that matrix. For example if

$\mathbf{G} = \begin{bmatrix} 4 & 6 & 1 \\ 8 & 7 & 6 \\ 3 & 6 & 2 \end{bmatrix}$, then $\begin{bmatrix} 7 & 6 \\ 6 & 2 \end{bmatrix}$ is a minor of matrix $\mathbf{G}$, and so is $\begin{bmatrix} 4 & 1 \\ 3 & 2 \end{bmatrix}$. The former

is associated with the element "4," the latter with the element "7." We have the following definition:

Definition 1-8. *The rank of matrix* $\mathbf{A}$ *is the order of its nonvanishing minor of highest order.*

The rank of matrix $\mathbf{A}$ shall be denoted by $R(\mathbf{A})$. Thus, $R(\mathbf{A}) = r$ means that all minors of order greater than r are equal to zero, but there is at least one minor of order r that is not zero.

It is now obvious that the outer product of two vectors (i.e., a dyad) gives a matrix of rank 1. Thus we write

$$R(\mathbf{u} \cdot \mathbf{v}^T) = 1 \qquad (1\text{-}22)$$

Example 1-16

Recall the dyad in Example 1-8 where $\mathbf{a}\cdot\mathbf{b}^T = \begin{bmatrix} 2 & 8 & -6 \\ -1 & -4 & 3 \\ 3 & 12 & -9 \end{bmatrix}$. Since here the rows (columns) are multiples of any other row (column), therefore any second-order minor (i.e., subdeterminant) must be zero, and hence the rank must be less than 2. On the other hand, since the elements (i.e., minors of order 1) are different from zero, the rank is 1 (it is sufficient if there is only one element different from zero). Therefore

$$R(\mathbf{a}\cdot\mathbf{b}^T) = R\left(\begin{bmatrix} 2 & 8 & -6 \\ -1 & -4 & 3 \\ 3 & 12 & -9 \end{bmatrix}\right) = R\left(\begin{bmatrix} 2 \\ -1 \\ 3 \end{bmatrix}\cdot[1\ 4\ -3]\right) = 1$$

In connection with this property of a dyad, another definition of the rank can be given; it is related to the *sum* of the dyads by which a matrix can be expressed.

Definition 1-9. If

$$\mathbf{A} = \sum_{k=1}^{r} \mathbf{u}_k \cdot \mathbf{v}_k^T \tag{1-23}$$

where r is the minimum number of dyads by which $\mathbf{A}$ can be expressed, then r is the rank of $\mathbf{A}$.

We now present an algorithm to find the rank of any matrix by the *minimal dyadic decomposition* of that matrix. In order to make this process easy to understand, we use a numerical example.

Example 1-17

Given matrix $\mathbf{A}_1 = \begin{bmatrix} ③ & 3 & 6 & 5 & 5 \\ 7 & 4 & 7 & 2 & 0 \\ -1 & -2 & -3 & -4 & -5 \\ -1 & -3 & -8 & -9 & -10 \end{bmatrix}$. What is its rank?

Step 1. Select any *nonzero* element of $\mathbf{A}_1$. Say we select the top left element "3," which is then *marked*, as above. We call this element the *pivot*.

Step 2. Generate matrix $\mathbf{A}_2$ such that

$$\mathbf{A}_2 = \mathbf{A}_1 - \frac{1}{③}\cdot\begin{bmatrix} 3 \\ 7 \\ -1 \\ -1 \end{bmatrix}\cdot\underbrace{[3\ 3\ 6\ 5\ 5]}_{\text{row of pivot}} = \mathbf{A}_1 - \frac{1}{3}\cdot\begin{bmatrix} 9 & 9 & 18 & 15 & 15 \\ 21 & 21 & 42 & 35 & 35 \\ -3 & -3 & -6 & -5 & -5 \\ -3 & -3 & -6 & -5 & -5 \end{bmatrix} = \begin{bmatrix} 0 & 0 & 0 & 0 & 0 \\ 0 & ③ & -7 & -\frac{29}{3} & -\frac{35}{3} \\ 0 & -1 & -1 & -\frac{7}{3} & -\frac{10}{3} \\ 0 & -2 & -6 & -\frac{22}{3} & -\frac{25}{3} \end{bmatrix}$$

Note that the *denominator* of the fraction (just before the pivot's column vector) is the pivot itself (in this case "3"). If $\mathbf{A}_2$ happens to be a null matrix, then the process terminates and the rank of $\mathbf{A}_1$ is 1, which is then the *largest subscript* of a *nonzero* matrix. However, in our case here, $\mathbf{A}_2$ is not zero, and so we continue with Step 3.

Step 3. Select any nonzero element of $\mathbf{A}_2$. Say we select the top left nonzero element "–3," which is then marked as above. Again, we call this element the *pivot*.

Step 4. Generate matrix $\mathbf{A}_3$

$$\mathbf{A}_3 = \mathbf{A}_2 - \frac{1}{\text{\textcircled{-3}}} \cdot \begin{bmatrix} 0 \\ -3 \\ -1 \\ -2 \end{bmatrix} \cdot \underbrace{\begin{bmatrix} 0 & -3 & -7 & -\frac{29}{3} & -\frac{35}{3} \end{bmatrix}}_{\text{pivot's row}} = \mathbf{A}_2 - \frac{1}{-3} \cdot \begin{bmatrix} 0 & 0 & 0 & 0 & 0 \\ 0 & 9 & 21 & 29 & 35 \\ 0 & 3 & 7 & \frac{29}{3} & \frac{35}{3} \\ 0 & 6 & 14 & \frac{58}{3} & \frac{70}{3} \end{bmatrix} = \begin{bmatrix} 0 & 0 & 0 & 0 & 0 \\ 0 & 0 & 0 & 0 & 0 \\ 0 & 0 & \frac{4}{3} & \frac{8}{9} & \text{\textcircled{$\frac{5}{9}$}} \\ 0 & 0 & -\frac{4}{3} & -\frac{8}{9} & -\frac{5}{9} \end{bmatrix}$$

Since $\mathbf{A}_3$ is still not zero, the process continues.

Step 5. Select another *pivot*, say $\frac{5}{9}$, which is marked. With this we now have

$$\mathbf{A}_4 = \mathbf{A}_3 - \frac{1}{\text{\textcircled{$\frac{5}{9}$}}} \cdot \begin{bmatrix} 0 \\ 0 \\ \frac{5}{9} \\ -\frac{5}{9} \end{bmatrix} \cdot \underbrace{\begin{bmatrix} 0 & 0 & \frac{4}{3} & \frac{8}{9} & \frac{5}{9} \end{bmatrix}}_{\text{pivot's row}} = \mathbf{A}_3 - \frac{9}{5} \cdot \begin{bmatrix} 0 & 0 & 0 & 0 & 0 \\ 0 & 0 & 0 & 0 & 0 \\ 0 & 0 & \frac{20}{27} & \frac{40}{81} & \frac{25}{81} \\ 0 & 0 & -\frac{20}{27} & -\frac{40}{81} & -\frac{25}{81} \end{bmatrix} = 0$$

Thus we got a null matrix, and hence in this example the *largest subscript* designating a *nonzero* matrix is 3; it follows that the rank of $\mathbf{A}_1$ is 3.

⇑

The above process can be condensed into the following compact protocol. We wish to find the rank of an arbitrary matrix $\mathbf{A}_1$.

Step 1. Set $n = 1$.

Step 2. Select a pivot "p" in matrix $\mathbf{A}_n$.

Step 3. Generate matrix $\mathbf{A}_{n+1} = \mathbf{A}_n - \dfrac{1}{p} \underbrace{\begin{bmatrix} \\ \\ \end{bmatrix}}_{\text{column of } p} \cdot \underbrace{[]}_{\text{row of } p}$

Step 4. If $\mathbf{A}_{n+1} = 0$, then the rank of $\mathbf{A}_1$ is n, otherwise set $n = n + 1$, then go to Step 2.

In order to minimize round-off errors, it is usually advisable to select a pivot which has the largest absolute value among the nonzero elements. An alternate approach is to select, if possible, "1" as a pivot. This will obviate the need to deal with fractions.

We now introduce the notion of *linear dependence*.

Definition 1-10. *Vectors* $\mathbf{a}_1, \mathbf{a}_2, \ldots, \mathbf{a}_r$ *are linearly dependent if there exist real numbers* $c_1, c_2, \ldots, c_r$ *not all zero such that*

$$c_1 \cdot \mathbf{a}_1 + c_2 \cdot \mathbf{a}_2 + \ldots + c_r \cdot \mathbf{a}_r = 0 \qquad (1\text{-}24)$$

Otherwise the set of vectors $\mathbf{a}_1, \mathbf{a}_2, \ldots, \mathbf{a}_r$ *are linearly independent.*

Expression (1-24) is called the *linear combination* of vectors $\mathbf{a}_1, \mathbf{a}_2, \ldots, \mathbf{a}_r$. Thus the linear independence of vectors $\mathbf{a}_1, \mathbf{a}_2, \ldots, \mathbf{a}_r$ can also be defined as follows:

Definition 1-11. *Vectors $\mathbf{a}_1, \mathbf{a}_2, \ldots, \mathbf{a}_r$ are linearly independent if their linear combination vanishes only in case of $c_1 = c_2 = \ldots = c_r = 0$.*

That is, vectors $\mathbf{a}_1, \mathbf{a}_2, \ldots, \mathbf{a}_r$ are linearly independent if (1-24) implies $c_1 = c_2 = \ldots = c_r = 0$.

Example 1-18

Vectors

$$\mathbf{a}_1 = \begin{bmatrix} 1 \\ 0 \\ 0 \end{bmatrix}; \quad \mathbf{a}_2 = \begin{bmatrix} 0 \\ 1 \\ 0 \end{bmatrix}; \quad \mathbf{a}_3 = \begin{bmatrix} 0 \\ 0 \\ 1 \end{bmatrix}$$

are *linearly independent* because their linear combination

$$c_1 \cdot \mathbf{a}_1 + c_2 \cdot \mathbf{a}_2 + \ldots + c_r \cdot \mathbf{a}_r = \begin{bmatrix} c_1 \\ 0 \\ 0 \end{bmatrix} + \begin{bmatrix} 0 \\ c_2 \\ 0 \end{bmatrix} + \begin{bmatrix} 0 \\ 0 \\ c_3 \end{bmatrix} = \begin{bmatrix} c_1 \\ c_2 \\ c_3 \end{bmatrix}$$

is zero *only* if $c_1 = c_2 = \ldots = c_r = 0$.

Example 1-19

We prove that vectors $\mathbf{a}_1 = \begin{bmatrix} 1 \\ 2 \\ 3 \end{bmatrix}$, $\mathbf{a}_2 = \begin{bmatrix} 4 \\ 5 \\ 6 \end{bmatrix}$ and $\mathbf{a}_3 = \begin{bmatrix} 7 \\ 8 \\ 9 \end{bmatrix}$ are linearly dependent. By inspection, we write $\mathbf{a}_1 - 2 \cdot \mathbf{a}_2 + \mathbf{a}_3 = 0$, hence, by Definition 1-10, the coefficients are $c_1 = 1$, $c_2 = -2$, $c_3 = 1$, which are not all zeros. Thus, vectors $\mathbf{a}_1, \mathbf{a}_2, \mathbf{a}_3$ are *linearly dependent*.

The definition of linear dependence and the independence of vectors can also be formulated in a concise form by using matrix $\mathbf{A}$ expressed by its column vectors $\mathbf{A} = [\mathbf{a}_1, \mathbf{a}_2, \ldots, \mathbf{a}_n]$ and by the column vector $\mathbf{x} = \begin{bmatrix} x_1 \\ x_2 \\ \vdots \\ x_n \end{bmatrix}$, where quantities $x_1, x_2 \ldots$ are *scalars*.

Definition 1-12. *If the equation*

$$\mathbf{A} \cdot \mathbf{x} = 0 \qquad (1\text{-}25)$$

has only the trivial solution

$$\mathbf{x} = 0 \qquad (1\text{-}26)$$

then the column vectors of matrix $\mathbf{A}$ are linearly independent.

Consequently, equation (1-25) has a *nontrivial* (i.e., **x** ≠ 0) solution if, and only if, the column vectors of **A** are *linearly dependent*.

We now state the following theorem without proof:

Theorem 1-3. *It is always possible to find* r *linearly independent vectors of a matrix of rank* r, *but any of its* r + 1 *columns — if such a number of columns exist — are necessarily linearly dependent.*

On the basis of this theorem, an equivalent definition can also be given for the rank of a matrix:

Definition 1-13. *The rank of a matrix is the maximum number of its linearly independent column vectors (or row vectors).*

From this definition it is obvious that the rank of a matrix cannot exceed the number of its rows (or columns). It also can be shown that the columns (rows) of a square matrix are linearly independent only if the matrix is nonsingular. In other words, the rank of any nonsingular matrix of order n is n.

Example 1-20

The column vectors of matrix $\begin{bmatrix} 1 & 4 & 7 \\ 2 & 5 & 8 \\ 3 & 6 & 9 \end{bmatrix}$ are *linearly dependent* since its determinant is zero, i.e., $\begin{vmatrix} 1 & 4 & 7 \\ 2 & 5 & 8 \\ 3 & 6 & 9 \end{vmatrix} = 0$; in other words, the matrix is *singular*.

⇑

Example 1-21

Consider the matrix

$$\mathbf{A} = \begin{bmatrix} 3 & 3 & 6 & 5 & 5 \\ 7 & 4 & 7 & 2 & 0 \\ -1 & -2 & -3 & -4 & -5 \\ -1 & -3 & -8 & -9 & -10 \end{bmatrix}$$

The rank of this matrix is 3, as was shown in Example 1-17. This means that any four of its columns are necessarily linearly dependent. There exist, however, three linearly independent columns among the five in **A**. For example, the first three columns are linearly independent.

⇑

Finally we present, without proofs, some basic theorems concerning the ranks of matrices.

Theorem 1-4. *The rank of the sum of two matrices cannot exceed the sum of their ranks; i.e.,*

$$R(\mathbf{A} + \mathbf{B}) \leq R(\mathbf{A}) + R(\mathbf{B}) \qquad (1\text{-}27)$$

Theorem 1-5. *The rank of a matrix cannot increase when the matrix is multiplied by another one, i.e.,*

$$R(\mathbf{A}\cdot\mathbf{B}) \leq R(\mathbf{A}) \text{ and } R(\mathbf{A}\cdot\mathbf{B}) \leq R(\mathbf{B}) \tag{1-28}$$

Theorem 1-6. *The rank of a matrix is invariant upon multiplication by a nonsingular matrix, i.e.,*

$$R(\mathbf{A}\cdot\mathbf{M}) = R(\mathbf{A}) \quad \text{if } |\mathbf{M}| \neq 0 \tag{1-29}$$

Theorem 1-7. *If the product of two square matrices of order* n *is zero, then the sum of their ranks cannot exceed their order, i.e.,*

$$R(\mathbf{A}) + R(\mathbf{B}) \leq n \quad \text{if } \mathbf{A}\cdot\mathbf{B} = 0 \tag{1-30}$$

1.4. SYSTEMS OF LINEAR EQUATIONS

Consider the system of linear equations

$$\left.\begin{array}{l} a_{11}\cdot x_1 + a_{12}\cdot x_2 + \ldots + a_{1n}\cdot x_n = b_1 \\ a_{21}\cdot x_1 + a_{22}\cdot x_2 + \ldots + a_{2n}\cdot x_n = b_2 \\ \ldots\ldots\ldots\ldots\ldots\ldots\ldots\ldots\ldots\ldots\ldots \\ a_{m1}\cdot x_1 + a_{m2}\cdot x_2 + \ldots + a_{mn}\cdot x_n = b_m \end{array}\right\} \tag{1-31}$$

where x_j ($j = 1, 2, \ldots, n$) are the unknowns, a_{ij} ($i = 1, 2, \ldots, m; j = 1, 2, \ldots, n$) are the coefficients of the unknowns and b_i ($i = 1, 2, \ldots, m$) are given numbers. If **A** is the *coefficient matrix*, and **x** and **b** are column vectors, then (1-31) can be written in the matrix form

$$\mathbf{A}\cdot\mathbf{x} = \mathbf{b} \tag{1-32}$$

If **b** = **0**, system (1-32) is called a *homogeneous* linear system of equations, otherwise it is a *nonhomogeneous* linear system of equations.

The main problem is to determine whether the given system has a solution at all, and if it does, whether the solution is unique. Moreover, an algorithm should be created by which solutions may be found.

1.4.1. Homogeneous Case

By (1-32), a *homogeneous* linear system of m equations and n unknowns can be written

$$\mathbf{A}\cdot\mathbf{x} = \mathbf{0} \tag{1-33}$$

where **A** is the $m \times n$ *coefficient* matrix, and **x** is the *unknown* $m \times 1$ *column vector*. That is, n is the number of unknowns and m is the number of equations in the linear system. Let us assume that $m \leq n$. If the rank of **A** is $r \leq m$, then of the n unknowns, $n - r$ are *selectable* and r are *dependent*, i.e., they are determined by the selectable ones. If $n = m = r$, then there exists only the *trivial* solution $\mathbf{x} = \mathbf{0}$. If $r < m$, then there exists always a *nontrivial* solution, i.e., $\mathbf{x} \neq \mathbf{0}$.

Example 1-22

Given the linear system

$$\left. \begin{array}{l} 5 \cdot x_1 + 6 \cdot x_2 + 7 \cdot x_3 = 0 \\ 2 \cdot x_1 + 2 \cdot x_2 + 4 \cdot x_3 = 0 \\ 7 \cdot x_1 + 8 \cdot x_2 + 11 \cdot x_3 = 0 \end{array} \right\} \quad (a)$$

By (1-33) and (a)

$$\mathbf{A} = \begin{bmatrix} 5 & 6 & 7 \\ 2 & 2 & 4 \\ 7 & 8 & 11 \end{bmatrix}; \quad \mathbf{x} = \begin{bmatrix} x_1 \\ x_2 \\ x_3 \end{bmatrix} \quad (b)$$

thus, $m = 3$ and $n = 3$. The rank of **A** cannot be 3, since the third row is the sum of rows 1 and 2.

That the rank is $r = 2$ can be quickly verified by noting that the value of the 2×2 determinant at the bottom right is $-10 \neq 0$. Therefore there is $n - r = 3 - 2 = 1$ *selectable* unknown and there are $r = 2$ *dependent* unknowns. The selectable unknown can be any one of the 3, namely x_1, x_2, or x_3. Let us choose x_3 as the selectable unknown. Then, in view of (a) and (b), we can write

$$\begin{bmatrix} 5 & 6 \\ 2 & 2 \end{bmatrix} \cdot \begin{bmatrix} x_1 \\ x_2 \end{bmatrix} + \begin{bmatrix} 7 \\ 4 \end{bmatrix} \cdot x_3 = \begin{bmatrix} 0 \\ 0 \end{bmatrix}$$

or

$$\begin{bmatrix} 5 & 6 \\ 2 & 2 \end{bmatrix} \cdot \begin{bmatrix} x_1 \\ x_2 \end{bmatrix} = -\begin{bmatrix} 7 \\ 4 \end{bmatrix} \cdot x_3$$

from which

$$\begin{bmatrix} x_1 \\ x_2 \end{bmatrix} = -\begin{bmatrix} 5 & 6 \\ 2 & 2 \end{bmatrix}^{-1} \cdot \begin{bmatrix} 7 \\ 4 \end{bmatrix} \cdot x_3 = -\begin{bmatrix} -1 & 3 \\ 1 & -2.5 \end{bmatrix} \cdot \begin{bmatrix} 7 \\ 4 \end{bmatrix} \cdot x_3 = \begin{bmatrix} -5 \\ 3 \end{bmatrix} \cdot x_3 \quad (c)$$

Say we *select* $x_3 = 8$. Then by (c), $x_1 = -40$ and $x_2 = 24$.

⇑

The important question is: How many *linearly independent* solutions exist? It can be shown that there are exactly $n - r$ independent solutions; i.e., the number of independent solutions equals the number of selectable unknowns. Thus, although in general an infinite number of solutions exists (if $n > r$), only $n - r$ of them are independent.

Example 1-23

How many *solutions* and how many *linearly independent solutions* exist for the system discussed in Example 1-22?

Since x_3 can be any number, therefore by (c) of Example 1-22, there are an infinite number of solutions. The number of selectable unknowns is $n - r = 3 - 2 = 1$, therefore the number of linearly independent solutions is 1.

⇑

Example 1-24

Given the system

$$\left.\begin{array}{l} 5 \cdot x_1 + 6 \cdot x_2 + 7 \cdot x_3 + 8 \cdot x_4 = 0 \\ 2 \cdot x_1 + 3 \cdot x_2 + 7 \cdot x_3 + 4 \cdot x_4 = 0 \\ 3 \cdot x_1 + 6 \cdot x_2 + 8 \cdot x_3 - 4 \cdot x_4 = 0 \\ 3 \cdot x_1 + 3 \cdot x_2 - 7 \cdot x_3 - 5 \cdot x_4 = 0 \end{array}\right\} \quad (a)$$

How many selectable and dependent unknowns exist, and what is the solution of this homogeneous system?

By (1-33) and (a)

$$\mathbf{A} = \begin{bmatrix} 5 & 6 & 7 & 8 \\ 2 & 3 & 7 & 4 \\ 3 & 6 & 8 & -4 \\ 3 & 3 & -7 & -5 \end{bmatrix}; \quad \mathbf{x} = \begin{bmatrix} x_1 \\ x_2 \\ x_3 \\ x_4 \end{bmatrix} \quad (b)$$

Thus, there are $n = 4$ unknowns, and $m = 4$ equations. The value of the determinant of matrix $\mathbf{A}$ is 99, which is not zero; therefore the rank of $\mathbf{A}$ is $r = 4$. Hence the number of *selectable* unknowns is $n - r = 4 - 4 = 0$; i.e., none of them is selectable. Since $r = m$, therefore only the *trivial* solution exists and hence $\mathbf{x} = 0$, i.e., $x_1 = x_2 = x_3 = x_4 = 0$.

⇑

Example 1-25

Given the linear system

$$\left.\begin{array}{l} 7 \cdot x_1 + 3 \cdot x_2 - 7 \cdot x_3 - 5 \cdot x_4 = 0 \\ 14 \cdot x_1 + 6 \cdot x_2 - 14 \cdot x_3 - 10 \cdot x_4 = 0 \\ 21 \cdot x_1 + 9 \cdot x_2 - 21 \cdot x_3 - 15 \cdot x_4 = 0 \\ 49 \cdot x_1 + 21 \cdot x_2 - 49 \cdot x_3 - 35 \cdot x_4 = 0 \end{array}\right\} \quad (a)$$

How many selectable and dependent unknowns are there, and what is the solution of this homogeneous system?

We have $m = 4$ equations and $n = 4$ unknowns. By (1-33) we write (a) in matrix form as

$$\begin{bmatrix} 7 & 3 & -7 & -5 \\ 14 & 6 & -14 & -10 \\ 21 & 9 & -21 & -15 \\ \boxed{49} & 21 & -49 & -35 \end{bmatrix} \cdot \begin{bmatrix} x_1 \\ x_2 \\ x_3 \\ x_4 \end{bmatrix} = \begin{bmatrix} 0 \\ 0 \\ 0 \\ 0 \end{bmatrix} \quad (b)$$

Thus the coefficient matrix and the unknown column vector are

$$\mathbf{A} = \begin{bmatrix} 7 & 3 & -7 & -5 \\ 14 & 6 & -14 & -10 \\ 21 & 9 & -21 & -15 \\ 49 & 21 & -49 & -35 \end{bmatrix}; \quad \mathbf{x} = \begin{bmatrix} x_1 \\ x_2 \\ x_3 \\ x_4 \end{bmatrix} \quad (c)$$

As a first step we must determine the rank of $\mathbf{A}$. We will use the method expounded in Art. 1.3. Thus, we rename $\mathbf{A}$ to $\mathbf{A}_1$ and select "49" as the *pivot* (marked). Accordingly,

$$\mathbf{A}_2 = \mathbf{A}_1 - \frac{1}{49} \cdot \begin{bmatrix} 7 \\ 14 \\ 21 \\ 49 \end{bmatrix} \cdot \underbrace{[49 \quad 21 \quad -49 \quad -35]}_{\text{row of pivot}} = \mathbf{A}_1 - \frac{1}{49} \cdot \begin{bmatrix} 343 & 147 & -343 & -245 \\ 686 & 294 & -686 & -490 \\ 1029 & 441 & -1029 & -735 \\ 2401 & 1029 & -2401 & -1715 \end{bmatrix} = 0$$

pivot ↖
column of pivot

We note that the largest subscript of a nonzero **A** matrix is 1. Therefore the rank of **A** in (c) is $r = 1$. Thus, we have $n - r = 4 - 1 = 3$ *selectable* unknowns, and $r = 1$ *dependent* unknown. Let the selectable unknowns be x_2, x_3, and x_4. We can write

$$7 \cdot x_1 + [3 \quad -7 \quad -5] \cdot \begin{bmatrix} x_2 \\ x_3 \\ x_4 \end{bmatrix} = 0$$

from which

$$x_1 = -\frac{1}{7} \cdot [3 \quad -7 \quad -5] \cdot \begin{bmatrix} x_2 \\ x_3 \\ x_4 \end{bmatrix} = -\frac{3}{7} \cdot x_2 + \frac{7}{7} \cdot x_3 + \frac{5}{7} \cdot x_4 \quad\quad (d)$$

We can now select x_2, x_3, and x_4, and then calculate x_1 by (d). Hence the number of solutions is infinite. The number of *independent* solutions is $n - r = 3$. Thus, although the number of solutions is infinite, there are only three independent solutions.

Remark: This method, i.e., the *minimal dyadic decomposition* of coefficient matrix, has the advantage that the dependent and the selectable unknowns are automatically separated, inasmuch as the columns of the pivot elements determine the *dependent* unknowns and the others do the *selectable* ones. Also note that the unknowns can be separated in different ways, since the choice of the pivot element is not unique.

If the number of unknowns n is larger than the number of equations m, then it is advisable to rearrange the coefficient matrix such that the columns corresponding to the *dependent* unknowns *precede* the columns corresponding to the *selectable* unknowns. In this way the manipulation of matrices is greatly simplified. The following example illustrates.

⇑

Example 1-26

Given the system

$$\begin{bmatrix} 1 & 3 & 2 & -1 & 2 & 5 \\ 0 & 1 & 2 & 4 & 1 & -2 \\ 0 & 0 & 0 & 0 & 2 & -1 \end{bmatrix} \cdot \begin{bmatrix} x_1 \\ x_2 \\ x_3 \\ x_4 \\ x_5 \\ x_6 \end{bmatrix} = \begin{bmatrix} 0 \\ 0 \\ 0 \end{bmatrix} \quad\quad (a)$$

The coefficient matrix has the rank 3. Therefore of the 6 variables (unknowns), 3 are *selectable* and 3 are *dependent*. In this example the decreasing number of zeros in the columns indicate which unknowns can be considered as dependent ones. Thus x_1 and x_2 are obviously the dependent unknowns, the next column with less number of zeros (i.e., with no zeros) is the fifth one, therefore x_5 is the third dependent unknown. Consequently, x_1, x_2, and x_5 are the dependent unknowns, and x_3, x_4, and x_6 are the *independent*—hence *selectable*—unknowns. Accordingly, we reconfigure (a) such that

$$\begin{bmatrix} 1 & 3 & 2 & 2 & -1 & 5 \\ 0 & 1 & 1 & 2 & 4 & -2 \\ 0 & 0 & 2 & 0 & 0 & -1 \end{bmatrix} \cdot \begin{bmatrix} x_1 \\ x_2 \\ x_5 \\ x_3 \\ x_4 \\ x_6 \end{bmatrix} = \begin{bmatrix} 0 \\ 0 \\ 0 \end{bmatrix} \quad\quad (b)$$

Thus, we can write

$$\begin{bmatrix} 1 & 3 & 2 \\ 0 & 1 & 1 \\ 0 & 0 & 2 \end{bmatrix} \begin{bmatrix} x_1 \\ x_2 \\ x_5 \end{bmatrix} + \begin{bmatrix} 2 & -1 & 5 \\ 2 & 4 & -2 \\ 0 & 0 & -1 \end{bmatrix} \begin{bmatrix} x_3 \\ x_4 \\ x_6 \end{bmatrix} = \begin{bmatrix} 0 \\ 0 \\ 0 \end{bmatrix} \qquad (c)$$

which yields

$$\begin{bmatrix} x_1 \\ x_2 \\ x_5 \end{bmatrix} = -\begin{bmatrix} 1 & 3 & 2 \\ 0 & 1 & 1 \\ 0 & 0 & 2 \end{bmatrix}^{-1} \begin{bmatrix} 2 & -1 & 5 \\ 2 & 4 & -2 \\ 0 & 0 & -1 \end{bmatrix} \begin{bmatrix} x_3 \\ x_4 \\ x_6 \end{bmatrix} = \frac{1}{2} \begin{bmatrix} 8 & 26 & -21 \\ -4 & -8 & 3 \\ 0 & 0 & 1 \end{bmatrix} \begin{bmatrix} x_3 \\ x_4 \\ x_6 \end{bmatrix} \qquad (d)$$

For example, let us *select* $x_3 = 12$, $x_4 = -22$, $x_6 = 17$. For this set, (d) provides

$$\begin{bmatrix} x_1 \\ x_2 \\ x_5 \end{bmatrix} = \frac{1}{2} \begin{bmatrix} 8 & 26 & -21 \\ -4 & -8 & 3 \\ 0 & 0 & 1 \end{bmatrix} \begin{bmatrix} 12 \\ -22 \\ 17 \end{bmatrix} = \begin{bmatrix} -416.5 \\ 89.5 \\ 8.5 \end{bmatrix} \qquad (e)$$

If we substitute $x_3 = 12$, $x_4 = -22$, $x_6 = 17$ and (e) into (b), we obtain

$$\begin{bmatrix} 1 & 3 & 2 & 2 & -1 & 5 \\ 0 & 1 & 1 & 2 & 4 & -2 \\ 0 & 0 & 2 & 0 & 0 & -1 \end{bmatrix} \begin{bmatrix} x_1 \\ x_2 \\ x_5 \\ x_3 \\ x_4 \\ x_6 \end{bmatrix} = \begin{bmatrix} 1 & 3 & 2 & 2 & -1 & 5 \\ 0 & 1 & 1 & 2 & 4 & -2 \\ 0 & 0 & 2 & 0 & 0 & -1 \end{bmatrix} \begin{bmatrix} -416.5 \\ 89.5 \\ 8.5 \\ 12 \\ -22 \\ 17 \end{bmatrix} = \begin{bmatrix} 0 \\ 0 \\ 0 \end{bmatrix}$$

as required.

1.4.2. Nonhomogeneous Case

If in relation (1-32)

$$\mathbf{A} \cdot \mathbf{x} = \mathbf{b} \qquad \text{repeated (1-32)}$$

the column vector **b** is not zero, then we have a *nonhomogeneous* linear system. Such a system can be reduced to a *homogeneous* one simply by putting (1-32) in the form of

$$\mathbf{A} \cdot \mathbf{x} + \mathbf{b} \cdot (-1) = 0 \qquad (1\text{-}34)$$

We recapitulate: relation (1-32) represents a linear system comprising n unknowns and m equations. Now we introduce the additional "unknown" $x_{n+1} = -1$, by which (1-34) can be written in matrix form

$$[\mathbf{A} \quad \mathbf{b}] \begin{bmatrix} \mathbf{x} \\ x_{n+1} \end{bmatrix} = 0 \qquad (1\text{-}35)$$

Thus the problem is reduced to that of the *homogeneous* case, with the important restriction that x_{n+1} must be a *selectable* unknown, for otherwise it would be impossible to substitute -1 for it. If x_{n+1} turned out to be a *dependent unknown*, then the rank of the *augmented* matrix [A b] would be $r + 1$. It follows that the *condition for the solvability* of the nonhomogeneous linear system of (1-35) is

$$R(\mathbf{A}) = R([\mathbf{A} \quad \mathbf{b}]) \tag{1-36}$$

i.e., the ranks of the *coefficient* matrix and the *augmented* matrix must be identical. From the above, the general solution of a nonhomogeneous linear system can be obtained as

$$\mathbf{x} = \mathbf{x}_0 + \mathbf{X} \tag{1-37}$$

where $\mathbf{x}_0$ is a *particular* solution of the *nonhomogeneous* system and $\mathbf{X}$ is the *general* solution of the *homogeneous* system.

Example 1-27

Solve the nonhomogeneous linear system

$$\begin{bmatrix} 5 & 3 & 6 \\ 6 & 2 & 2 \\ 10 & 6 & 12 \end{bmatrix} \cdot \begin{bmatrix} x_1 \\ x_2 \\ x_3 \end{bmatrix} = \begin{bmatrix} 5 \\ 6 \\ 10 \end{bmatrix} \tag{a}$$

By the form (1-32) we have $\mathbf{A} = \begin{bmatrix} 5 & 3 & 6 \\ 6 & 2 & 2 \\ 10 & 6 & 12 \end{bmatrix}$ and $\mathbf{b} = \begin{bmatrix} 5 \\ 6 \\ 10 \end{bmatrix}$. The first step is to ascertain if there is a solution. Therefore we check if condition (1-36) is satisfied. The rank of $\mathbf{A}$ cannot be 3 since the third row in $\mathbf{A}$ is twice the first row. That the rank is 2 can be verified at once by observing that the value of the top right 2×2 determinant is -6. Hence $R(\mathbf{A}) = 2$.

The *augmented* matrix is

$$[\mathbf{A} \quad \mathbf{b}] = \begin{bmatrix} 5 & 3 & 6 & 5 \\ 6 & 2 & 2 & 6 \\ 10 & 6 & 12 & 10 \end{bmatrix}$$

and we see that its rank cannot be 3 either, since the third row is twice the first row. That the rank is at least 2 follows from the fact that it cannot be less than that of $\mathbf{A}$, which is 2. Hence the rank is 2. Therefore condition (1-36) is satisfied, and thus (a) *has* a solution.

A *particular* solution of (a) is obviously

$$\mathbf{x}_0 = \begin{bmatrix} 1 \\ 0 \\ 0 \end{bmatrix} \tag{b}$$

since the first column of $\mathbf{A}$ is identical to column vector $\mathbf{b}$.

Next, we determine $\mathbf{X}$, i.e., the *general* solution of the *homogeneous* part of (a). To this end, we write (since the top left 2×2 determinant of $\mathbf{A}$ is not zero)

$$\begin{bmatrix} 5 & 3 \\ 6 & 2 \end{bmatrix} \cdot \begin{bmatrix} x_1 \\ x_2 \end{bmatrix} + \begin{bmatrix} 6 \\ 2 \end{bmatrix} \cdot x_3 = 0$$

from which

$$\begin{bmatrix} x_1 \\ x_2 \end{bmatrix} = -\begin{bmatrix} 5 & 3 \\ 6 & 2 \end{bmatrix}^{-1} \cdot \begin{bmatrix} 6 \\ 2 \end{bmatrix} \cdot x_3 = -\begin{bmatrix} \frac{2}{8} & -\frac{3}{8} \\ -\frac{6}{8} & \frac{5}{8} \end{bmatrix} \cdot \begin{bmatrix} 6 \\ 2 \end{bmatrix} \cdot x_3 = -\begin{bmatrix} \frac{3}{4} \\ -\frac{13}{4} \end{bmatrix} \cdot x_3$$

Hence, the *general* solution of the *homogeneous* part is

$$\mathbf{X} = \frac{1}{4} \cdot \begin{bmatrix} 3 \\ -13 \\ 4 \end{bmatrix} \cdot t \quad \text{(c)}$$

where t is an arbitrary scalar parameter. Finally, the *general* solution of the *nonhomogeneous* system (a) is by (1-37), (b) and (c)

$$\begin{bmatrix} x_1 \\ x_2 \\ x_3 \end{bmatrix} = \begin{bmatrix} 1 \\ 0 \\ 0 \end{bmatrix} + \frac{1}{4} \cdot \begin{bmatrix} 3 \\ -13 \\ 4 \end{bmatrix} \cdot t \quad \text{(d)}$$

For example, let $t = 52$. Then by (d)

$$\begin{bmatrix} x_1 \\ x_2 \\ x_3 \end{bmatrix} = \begin{bmatrix} 1 \\ 0 \\ 0 \end{bmatrix} + \frac{1}{4} \cdot \begin{bmatrix} 3 \\ -13 \\ 4 \end{bmatrix} \cdot (52) = \begin{bmatrix} 40 \\ -169 \\ 52 \end{bmatrix} \quad \text{(e)}$$

which, by (a), yields

$$\begin{bmatrix} 5 & 3 & 6 \\ 6 & 2 & 2 \\ 10 & 6 & 12 \end{bmatrix} \cdot \begin{bmatrix} 40 \\ -169 \\ 52 \end{bmatrix} = \begin{bmatrix} 5 \\ 6 \\ 10 \end{bmatrix}$$

which confirms that (e) is, indeed, a solution.

Example 1-28

Given the system

$$\begin{bmatrix} 3 & 11 & 10 & 5 & 10 \\ 2 & 7 & 6 & 2 & 5 \\ 1 & 3 & 2 & -1 & 2 \\ 4 & 10 & 4 & -12 & 4 \end{bmatrix} \cdot \begin{bmatrix} x_1 \\ x_2 \\ x_3 \\ x_4 \\ x_5 \end{bmatrix} = \begin{bmatrix} 10 \\ 8 \\ 5 \\ \lambda \end{bmatrix} \quad \text{(a)}$$

at what value of λ does this nonhomogeneous system have a solution?

Coefficient matrix $\mathbf{A}$ has the rank of $R(\mathbf{A}) = 3$, as can be ascertained by the method described in Art. 1.3 (see in particular Example 1-17). Therefore to have a solution at all, condition (1-36) must be satisfied. This means that the *augmented* matrix $[\mathbf{A} \ \mathbf{b}]$ must also have the rank 3. The augmented matrix is

$$[\mathbf{A} \ \mathbf{b}] = \begin{bmatrix} 3 & 11 & 10 & 5 & 10 & 10 \\ 2 & 7 & 6 & 2 & 5 & 8 \\ 1 & 3 & 2 & -1 & 2 & 5 \\ 4 & 10 & 4 & -12 & 4 & \lambda \end{bmatrix} \quad \text{(b)}$$

This matrix will have the rank 3 only if *every* one of its fourth-order determinants is zero. Hence the rightmost fourth-order determinant must also be zero, i.e.,

$$D = \begin{vmatrix} 10 & 5 & 10 & 10 \\ 6 & 2 & 5 & 8 \\ 2 & -1 & 2 & 5 \\ 4 & -12 & 4 & \lambda \end{vmatrix} = 0 \quad \text{(c)}$$

If $\lambda = 0$, then $D = D_0 = 500$; on the other hand, if $\lambda = 1$, then $D = D_1 = 480$. Therefore, by (1-6), the value of λ at which $D = 0$ is $\lambda_0 = \dfrac{D_0}{D_0 - D_1} = \dfrac{500}{500 - 480} = 25$. We conclude therefore that (a) has a solution *only* if $\lambda = \lambda_0 = 25$.

⇑

1.5. LIST OF SELECTED PUBLICATIONS DEALING WITH LINEAR ALGEBRA AND MATRICES

Linear Algebra with Applications, by W. Keith Nicholson; PWS Co., Boston, 1990.

Matrix Algebra for Engineers, by James M. Gere and William Weaver, Jr.; Van Nostrand Reinhold, New York, 1965.

Vectors and Matrices, by C. C. MacDuffee; The Mathematical Association of America, Buffalo, 1943.

Applied Linear Algebra, by Ben Noble; Prentice Hall, Englewood Cliffs, 1969.

Elementary Linear Algebra, by H. Anton; Wiley, Chichester, U.K., 1994.

Linear Algebra, Computer Application, by S. Barnett; Prentice Hall, Englewood Cliffs, 1987.

CHAPTER 2
FORMATS AND CLASSIFICATION

2.1. FORMATS FOR PHYSICAL RELATIONS

The magnitude of every numerically expressible variable or constant (collectively called *entities*) in engineering, physics, geometry, etc. must be ultimately expressed in a *numeric, symbolic,* or *mixed* format. In the following sections we discuss these three formats in sequence.

2.1.1. Numeric Format

A numeric format comprises exactly four elements: the entity's *name,* the equality sign "=," the entity's *magnitude,* and its *dimension*—in this order. For example, if the length of an object is 18 meters, then the name is "length," the magnitude is "18," and the dimension is "meter." Usually, the name and the dimension are abbreviated to symbols, as in Fig. 2-1. Let us now examine in more detail the elements present in both of the relations in Fig. 2-1.

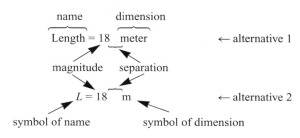

Figure 2-1
A physical entity expressed in numeric format
The two alternatives are functionally identical

Name (or symbol for the name). In principle this can be anything, although it is advisable to choose a descriptive word or symbol (i.e., a *mnemonic*) which the reader easily associates with the name in English (or sometimes in Latin or Greek). For example, we usually choose v as a symbol for velocity, L or λ (Greek lambda) for length, S for distance traveled (*spatium* = distance in Latin), although exceptions are also frequent. For example, the speed of light in vacuum is almost always symbolized by c, the electric charge by Q, etc. It is strongly advised to choose a single letter for a symbol, to avoid mistakenly interpreting a multiletter symbol for a product. For example, to use the symbol "*sp*" for speed is ambiguous, since it could be understood as the product of two quantities "*s*" and "*p*." If a differentiation between symbols designating similar quantities is necessary, the use of subscripts (numerical or alphabetical) is recommended. For example, use D_i and not *Di* for inside diameter. For technical literature dealing with engineering, physics, chemistry, etc., there exists a list of recommended names and their symbols for the frequently—and not so frequently—occurring entities. A reasonably comprehensive list of these items is presented in Appendix 1.

Note that the same symbol is sometimes used for markedly differing entities. The key phrase here is *markedly differing*, since by virtue of this discriminating feature, ambiguity is easily avoided. For example, the symbol E designates both the modulus of elasticity (Young's modulus), and electric field strength. There is little chance of mistaken identity here, however, because electric field strength has nothing to do with the elongation of a bar caused by an axial force. Therefore, it is unlikely that the two variables would appear together in a relation.

Equality Sign "=". This sign must always be between the name (or the symbol for the name) and the magnitude, and *never* between the magnitude and the dimension. Therefore, the expression "meter = 18" is utterly meaningless. If we want to read aloud a valid relationship, we should say, for example, "The length is 18 meters," or "The length equals 18 meters." In these two sentences the words "is" and "equals" stand for the equality sign in the numeric format.

In passing, we mention the regrettable practice by which a *dimension* with the suffix "ge" or "age" is substituted for a *name*. For example, one hears the linguistic abomination "mileage" for distance (or fuel consumption of a car), "acreage" for area, "frontage" for frontal length, "footage" for length in feet, etc. The absurdity of this custom is clearly evident when it is logically extended to include expressions like "dollarage" for salary or price, "yearage," "weekage," or "dayage" for time, etc. Using this parlance, "How old are you?" becomes "What is your yearage?"

Another important and easily overlooked characteristic of the numeric format is the necessary *space* between the magnitude and the dimension. This space is required to ensure that the expression is unambiguous. For example suppose someone sees the following

$$L = 18 \text{m}$$

This could mean that L (whatever it is) equals the *product* of 18 and the *quantity* "m" (whatever "m" may be). Therefore, it is essential to leave a space between the magnitude and the dimension.

If we want to show the *dimension* of an entity, then we put the name (or the symbol for the name) of that entity between square brackets [ ]. Thus we write

$$[\text{speed}] = \frac{m}{s}$$

which translates into "the dimension of speed is meter per second."

Magnitude. The magnitude of the length L in the example shown in Fig. 2-1 was 18, which is a *number*. A number can be alternatively designated a "pure number," an "absolute number," a "magnitude," etc.—all these mean the same thing. The number can be an integer, e.g., 18, a decimal fraction, e.g., 2.689, a proper fraction $3/4$, an improper fraction $4/3$, etc. It can also be a symbol representing a single number, e.g., π (=3.14159 . . .). Mathematical expressions resulting in the value of a number are theoretically permitted, but discouraged. For example, the relation $L = 6.375$ m is preferred over $L = \sqrt{(5^{2.3} + 6.0061)}/\ln \pi$ m, although they both mean the same thing.

Dimension. The "dimension" part of the formula shown in Fig. 2-1 identifies the "object" of the "magnitude." In other words, it tells us the "measuring unit" on which the magnitude acts. For example in the expression

$$\text{mass of George Schmideg} = 91 \text{ kg}$$

the mass of one "kg"—whatever it is—must be taken 91 times (the magnitude) to obtain the mass of the man named George Schmideg. The fact that we have not yet defined "kg" is immaterial. The important thing is that somewhere there is a physical object whose mass we call 1 (one) kg and we state that George Schmideg's mass is exactly 91 times as much.

In ordinary life, people usually interpret "dimension" as "length." For example the *dimension* of this sheet of paper on which I am now writing these words with my fountain pen is 8.5 by 11 inches. The *length* of my fountain pen is 6 inches, hence its dimension is 6 inches, etc. However, we choose not to be bound by this restriction. For our purpose the term "dimension" means this: *it is a collection of previously agreed upon base quantities, joined by (maybe) repeated multiplication and division, but not addition or subtraction, which permits a numerical expression of any physical or abstract quantity so expressible.* The restriction "numerical" is necessary, for otherwise we would be required to *quantify* fear, pain, love, jealousy, anxiety, laziness, apathy, etc.—an obviously impossible task.

The set from which we choose to select the *base* quantities is called the *Dimensional System*; we will discuss Dimensional Systems in detail in Chapter 3.

2.1.2. Symbolic Format

The symbolic format comprises exactly three elements which correspond to the first three elements of the numeric format discussed in Art. 2.1.1. The fourth element, "dimension," is missing from the symbolic format because it is neither necessary nor permitted here. The reason for the mandatory absence of dimension from the symbolic format is that the dimension in that format is not unique, since it was already determined by the dimensions in which the variables and constants are expressed.

As an illustration, consider the formula

$$v = \sqrt{2 \cdot g \cdot h}$$

in which v is the speed of a free-falling object without air resistance, g is the gravitational acceleration, and h is the height from which the object is falling. Now, if the dimension of h is "m" and that of g is m/s^2, then the dimension of v is m/s. If, however, the respective dimensions of h and g are "light-year" and "light-year/week^2," then the dimension of v will be "light-year/week." Therefore, we see that the dimension of v is set by the dimensions of g and h. Incidentally, the dimension of the number 2 in the formula is "1," i.e., it is a "pure number," and hence it does not play a role in dimensional considerations.

An important and interesting characteristic of a relation in symbolic format is that it is true irrespective of the dimensions used, provided these dimensions are used *consistently*. What this condition means is this: if, for example, the length occurs in more than one entity, then it must have the same dimension, say "meter," everywhere; if time occurs repeatedly, then it must have the same dimension, say "second," everywhere, etc. Otherwise, the symbolically given relation loses validity. To illustrate, suppose we have for the above relation for the speed of a free-falling object the consistent dimensions $h = 12$ m and $g = 9.81$ m/s^2. Using these data, the formula supplies $v = 15.344$ m/s speed. As an alternative, we can express everything consistently in inches and seconds. Hence, $h = 472.441$ in, $g = 386.22$ in/s^2, and by the formula $v = 604.097$ in/s—which is of course equivalent to the 15.344 m/s obtained previously. Here, coincidence exists because we used dimensions consistently, i.e., we used either inches and seconds everywhere, or meters and seconds everywhere. Everything fits nicely. But if we have (erroneously) $h = 472.441$ in $g = 9.81$ m/s^2 (i.e., for length we use *inch* in h, and *meter* in g), then the formula supplies $v = 96.277 \sqrt{\text{m·in}}$/s, which is a complete nonsense since the magnitude is wrong and the dimension is meaningless.

2.1.3. Mixed Format

In a mixed format the magnitude contains both numbers and symbols. Let us consider an example.

Example 2-1.   Surface area of a right circular cylinder

The cylinder shown in Fig. 2-2 has a surface area A. Using the notation of the figure, we write

$$A = \pi \cdot D \cdot \left(\frac{D}{2} + L\right) \tag{a}$$

Since this formula is in *symbolic* format, dimensions must not be indicated—and they were not. The quantities designated by the symbols D and L on the right side can be in any

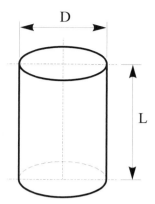

Figure 2-2
Surface area of a right circular cylinder

---

consistent dimensions, e.g., meters, inches, light-years, etc., and (a) will still supply the right answer given in the dimensions meters^2, inches^2, and light-years^2, respectively. Thus, if $D = 3$ m and $L = 9$ m, then $A = 98.9602$ m^2; and if $D = 5.3$ km and $L = 0.7$ km, then $A = 55.77898$ km^2.

But if we now select a particular *numeric* value for D, say $D = 3$ m, and leave L in a *symbolic* form, then our relation (a) will assume the form

$$A = 3\cdot\pi\cdot(1.5 + L) \tag{b}$$

In this case, it is important to realize that we are no longer free to choose a dimension for L, for if we do, we might select one that produces an erroneous result. For example, not knowing anything about the dimension of D, we measure our cylinder and find $L = 354.331$ in. Hence, we will calculate the area by (b) to be $A = 3353.628$ in$^2 = 2.1636$ m^2, which is wrong, since the area in fact is 98.9602 m^2, as given above. The fault, of course, is the omission of the dimension in relation (b). Because not *all* the variables in (b) are represented by symbols (i.e., at least one of them appears in numeric form), the dimensions of all the other variables in the relation are set, and hence may not be chosen at will. So, to avoid this kind of error, relation (b) should be written as

$$A = 3\cdot\pi\cdot(1.5 + L) \text{ m}^2 \tag{c}$$

which is then the proper format.

⇑

Relation (c) in the above example is in the *mixed format* since it includes at least one variable expressed by a *symbol*, and at least one variable expressed by its *numeric* value.

From the above example it is obvious that, to avoid error, the following simple *rule* must be followed:

Rule: *If a relation among physical variables or constants is in the mixed format, then the same rules apply as if it were in the numeric format.*

32   APPLIED DIMENSIONAL ANALYSIS AND MODELING

In other words, unless *all* the variables and dimensional constants appear as symbols—i.e., the relation is in the symbolic format—the *dimension element* of the numeric format must be present. Yet another way to say this is: *The rules of numeric and mixed formats are identical.* Therefore, relation (b) in Example 2-1 violates this rule, and hence the formula in question is meaningless and should not be used.

2.2. CLASSIFICATION OF PHYSICAL QUANTITIES

In this chapter we discuss the different types of physical quantities occurring in equations. Here the adjective "physical" must be interpreted very broadly, for it encompasses all of our physical world—physics, geometry, mathematics, biology, economics, etc.

Things in the universe can be classified—grouped—in an infinite number of ways. For example, *people* can be classified according to sex, height, weight, kidney size, occupation, marital status, income, their kitchen sink volume, hair count on chest, etc. It is a fine art to choose those—and only those—characteristics that are useful. The same thing can be said for *physical quantities*. They too can be divided into groups according to any one or more of their characteristics, but we have to choose those—and only those—that serve our purposes. Accordingly, we select the following two characteristics that are paramount in a physical quantity:

- variability;
- dimensionality.

We shall now elaborate on these two traits.

2.2.1. Variability

A physical quantity can be—in order of increasing variability—either a *constant*, a *parameter*, or a *variable*.

A *constant* is a physical quantity that *never* changes; it is unchanging in the domains of both space and time. For example, the speed of light in vacuum is such a constant; so is the universal gravitational constant, Planck's constant, etc. There are quite a few such constants in our world, and some of them are presented in Appendix 2 (but see Fig. 2-3, Note 3).

It is remarked that the accuracy by which the numerical values of these constants are known is essentially irrelevant. Some of these quantities are known exactly—or to any desired level of accuracy—since they are fixed by *definition* (for example, the permeability of vacuum, $4 \cdot \pi \times 10^{-7}$ m·kg/(A^2·s^2) in SI, discussed in Art. 3.3.); the majority of them, however, are known only to a fixed number of significant digits.

One particular kind of constant is derived from mathematical or geometric laws and definitions. For example, π (= 3.141 . . .) is the ratio of the length of the circumference of a circle to its diameter. Similarly, the base of natural logarithm,

$e = 2.718\ldots$, is such a number. An interesting characteristic of these constants is that they are independent from the physical properties of the universe. An individual sitting in a room completely devoid of contact with the outside world, and even without the opportunity to measure, could arrive at these figures by relying solely on his *speculative* mind.

A *parameter* is a physical quantity which is *constant* in the context in which it appears, but can be varied if so desired, or can assume a different value if circumstances change. In other words, parameters are not derived from the unvarying characteristics of the universe (as constants are), but are the result of some *reasonably* steady conditions or properties of the world. For example, gravitational acceleration, g, is such a number, for in most equations it appears as a constant, even though it is not exactly so, not even in a fixed location on Earth. Moreover, it also varies slightly (about 0.5%) with terrestrial latitude, and changes rather sensitively with altitude. Further, it varies even more significantly on different celestial bodies ($g = 9.81$ m/s^2 on Earth, 1.57 m/s^2 on the Moon, 25.89 m/s^2 on Jupiter, and 274.6 m/s^2 on the Sun). To put it paradoxically, *a parameter is a constant that can be varied*. Further examples are: modulus of elasticity (Young's modulus), thermal expansion coefficients, conductivities (both thermal and electric), etc.

A *variable* is a physical quantity that can be changed (varied) either directly or indirectly. In the former case it is called an *independent* variable; in the latter case (in which the value of the variable changes by way of its functional dependence on other—independent—variables) it is called a *dependent* variable. For example, in the formula $A = \pi \cdot R^2$ expressing the area A of a circle of radius R, the dependent variable is A, and the independent variable is R. Note that there should be only one dependent variable, but can be several independent variables in a relationship. Variables are the fundamental building elements of all relationships in the physical world.

2.2.2. Dimensionality

The dimensionality of a physical quantity can be one of two kinds: it can be *dimensional* or *dimensionless*. A *dimensional* quantity is a number (variable, parameter, or constant) connected to its dimension, which is different from 1. For example, in "speed = 30 m/s" the *speed* is a dimensional quantity since it does have a dimension different from 1, namely m/s.

The dimensions appearing in the expression of a dimensional quantity are those that are part of a (particular) *dimensional system*. There are an infinite number of such possible and different systems, some of which will be described in Chapter 3. For consistency, once such a system has been selected, all dimensions appearing in a relation that may consist, of variables, parameters, and constants, must come from that selected system. For example, in

$$v = \sqrt{2 \cdot a \cdot s}$$

"v" is the speed attained, "a" is the acceleration and "s" is the distance traveled. If now we use ft/s^2 for acceleration, we should not use "m" for distance.

On the other hand, there are quantities whose dimensions are 1 (one). These always occur when all the dimensions associated with the quantity cancel out. For example, the magnitude of a plane angle φ may be expressed in radians. But the radian (abbreviated "rad") itself is the ratio of the length of the subtended arc λ of a circle and that circle's radius R. Thus, we have $\varphi = \lambda/R$, and hence we can write for the dimension of the angle

$$[\varphi] = \text{rad} = \left[\frac{\lambda}{R}\right] = \frac{[\lambda]}{[R]} = \frac{\text{meter}}{\text{meter}} = 1$$

Therefore, the dimension of the radian is 1. In this respect, a dimension of 1 is analogous to a coefficient of 1, or an exponent of 1. They are not usually written out; they are "implied." Thus, in the relation $y = x^3 + 4 \cdot x^2 + 3 \cdot x$ the coefficients on the right side are 1, 4, 3, and the powers of x in the same sequential order are 3, 2, 1.

For some reason (completely unknown to this author), a physical quantity—be it a variable, a parameter, or a constant—whose dimension is 1 has acquired the adjective "dimensionless," or rarely "nondimensional." Thus, there are dimensionless constants, dimensionless parameters, and dimensionless variables. These designations are, of course, manifestly inappropriate, since the quantities involved *do* have a dimension, namely 1. Referring to the above-cited analogy about coefficients and exponents, one would *not* say that in the relation

$$y = 5 \cdot x^6 + x^4 + 8 \cdot x$$

the second member on the right is "coefficientless," and x in the third member is "exponentless." So why must our poor dimension suffer this indignity? Nevertheless, succumbing to accepted practice, in this book, a quantity whose dimension is 1 will be called—albeit reluctantly—a *dimensionless* quantity, variable, constant, etc.

In engineering, physics, geometry, etc. there are quite a few quantities whose dimension is 1. These quantities are frequently called a *ratio, factor, coefficient,* or just simply a *number.* Examples are: Mach number, slenderness ratio, Reynolds' number, and friction coefficient. The great advantage of using these dimensionless quantities is that their magnitudes are independent of the employed dimensions—provided these dimensions are consistent. For example, the Mach number, (ratio of the speed of a body to the speed of sound in the same medium) is the same, whether we express *both* these speeds in ft/s, or in km/h.

A few dimensionless variables, important in describing some particular characteristics of our physical world, are associated with names of scientists or engineers who either constructed these numbers, or are believed to have constructed them. These numbers are called *named dimensionless numbers*, and by international standards they are designated by the first *two* letters of the surname of the scientist involved. For example, the symbol for the *Reynolds' number* is Re, after Osborne Reynolds (1842–1912); Re is used, incidentally, to characterize the flow of a viscous liquid or gas. The Reynolds' number is defined as $\text{Re} = v \cdot L/v$, where v is the velocity of fluid, L is the representative linear dimension (e.g., the diameter of a sphere immersed in flowing liquid), and v is the kinematic viscosity of the liquid. If the respective *consistent* dimensions (expressed in any dimensional system) are substituted into this equation, all the dimensions will cancel out; hence we have $[\text{Re}] = 1$; i.e. Re is a dimensionless number, as expected. In Art. 3.3, dimensional systems are discussed and the dimensions of the more important physical quantities are given in the *Système International d'Unités* (designated *SI* in all languages).

A significant number of named dimensionless variables are used in the physical and engineering sciences, and the more important ones, with their short descriptions, are in Appendix 3. Also, the definitions of 210 of these named numbers can be found in Ref. 12.

We mentioned earlier that a dimensionless number can arise when all of its dimensions cancel out. We saw that the Reynolds' number is such a number. However, this is not the only mechanism by which dimensionless quantities are generated, for there are quantities which are, by their very nature, devoid of dimension, and hence are dimensionless to begin with. They are mathematical (or geometric) constants and cardinal numbers (or simply "cardinals"). The latter are also called "numerics." For example e (the base of natural logarithm) is such a dimensionless number, and so are 5, 38.9, –60.771, etc. That is $[e] = [5] = [38.9] = [-60.771] = 1$.

Some of the dimensionless numbers in mathematics (referred to above) have a dual ancestry. Take for example π (= 3.14159 . . .). On the one hand it arises from its original definition

$$[\pi] = \frac{\text{dimension of length of circumference a circle}}{\text{dimension of length of diameter of the same circle}} = \frac{m}{m} = 1$$

		Dimensionality	
		Dimensional	Dimensionless
Variability	Constant (value never changes)	speed of light, universal gravitational constant	$\pi, e, 5, \sqrt{-1}$
	Parameter (value rarely changes)	gravitational acceleration, Young's modulus, Earth's mean radius	Poisson's ratio, relative density
	Variable (value frequently changes)	speed, force, mass, magnetic flux	Mach's number, friction coefficient

Figure 2-3
Classification of physical and mathematical quantities according to their dimensionality and variability

Note 1: Demarcation between *parameter* and *variable* is rather fuzzy.
Note 2: Items in shaded areas are typical examples.
Note 3: Notwithstanding the above classification, a quantity that does not change during the application of a *particular* equation is usually designated a "constant," even though—strictly speaking—it is not. For example, a "constant of integration" will change from one integral to another, yet we call it a "constant." Fortunately, there is little likelihood that this relaxation of terminology will cause any confusion.

On the other hand, π can be generated without reference to geometry. For example, by Euler

$$\pi = \sqrt{6} \cdot \left(\sum_{j=1}^{\infty} \frac{1}{j^2} \right)^{1/2} = \sqrt{6} \cdot \left(1 + \frac{1}{2^2} + \frac{1}{3^2} + \ldots \right)^{1/2}$$

where of course both "6" and "j" are numerics. And hence again $[\pi] = 1$ (in his book, "Journey Through Genius," Penguin 1990, p. 215, William Dunham gives an absorbing, step-by-step account of how Euler derived this formula). The above is summarized in Fig. 2-3.

CHAPTER 3
DIMENSIONAL SYSTEMS

3.1. GENERAL STATEMENTS

A useful dimensional system must be comprised of a number of *fundamental* (base) entries (dimensions) that are *sufficient* to define the magnitude of any numerically expressible quantity. These fundamental dimensions *may* be chosen rather *arbitrarily*, but, for practical reasons, *should* be chosen *appropriately*. Thus, it is prudent to select this set of fundamental dimensions such that they will fall in the domain of everyday human experience, else the system will not acquire the needed popularity to become universally accepted. For example, the dimension of *light-year* as a fundamental dimension of *length* would neither be popular among textile merchants nor the public because it is much too large for practical use. One would *not* say to the salesman in a dry-goods store, "Please give me 2.1×10^{-16} light-years of fabric for a pair of trousers." Notwithstanding this impracticality, it is *possible* to create a dimensional system in which the fundamental dimension of length is the light-year; it is just that it is not practical to do so.

The selection of a dimensional system must be carried out in two steps. The first step is to select the *number* of fundamental dimensions, and second is to select the *standard magnitudes* for these dimensions. For instance, if *length, force,* and *time* are selected as fundamental dimensions, then the *number* is 3. The second step is to decide the *standard magnitude* of length, force, and time. For these we could select, for example, the *foot*, the *pound-force*, and the *second*; or, we could choose the *meter*, the *kilogram-force*, and the *year*, etc. These are all equally valid, although may not be equally practical.

It is important to realize that the number of fundamental dimensions may not be the *minimum* needed to express every possible physical quantity in the universe. For example, temperature, which is a fundamental dimension of all the dimensional systems described so far, is not a necessary dimension because it can be *uniquely* defined by the average kinetic energy of the moving molecules or atoms, and the energy in turn can be described by length and force alone. Therefore, temperature can also be described by length and force. Nevertheless, since temperature is so close to our everyday *experience*, it is *practical* to include it among the fundamentals.

The number of fundamental dimensions is not fixed by nature; for this number can be anything including 1. Dingle (Ref. 113, p. 331) quotes Milne, who proposed using only *one* fundamental dimension, viz., time, and to derive everything else from it. The other extreme is a dimensional system in which *every* dimension is a fundamental one. Fig. 3-1 shows the classification of dimensional systems according to the number of dimensions they use.

| monodimensional | multidimensional | omnidimensional |

Figure 3-1
Classification of dimensional systems by the number of dimensions they use

Let us now briefly demonstrate the *possibility* and the profound *impracticality* of such extreme systems. Imagine systems in which

1. All dimensions—except one—are derived; i.e., there is only one fundamental dimension. This is a *monodimensional* system.
2. Every dimension is fundamental; i.e., none is derived. This is an *omnidimensional* system.

3.1.1. Monodimensional System

We have a system in which there is only one fundamental dimension, say that of length (meter). To define the dimension of time now, we may use the duration it takes light to propagate 1 m in vacuum. Hence, if t is time, L is distance, and c is the speed of light, then $t = (1/c) \cdot L = A \cdot L$, where A, of course, is a dimensionless constant. So, $[L] = $ m and $[t] = $ m. Next, we consider speed defined as $v = dL/dt$. Therefore,

$$[v] = \frac{[L]}{[t]} = \frac{m}{m} = 1$$

i.e., *speed* is a dimensionless variable. Also if "a" is *acceleration*, then $a = dv/dt$ from which

$$a = \frac{[v]}{[t]} = \frac{1}{m} = m^{-1}$$

Now the dimension of *force* can be defined as the magnitude of the attraction generated by two unit masses 1 m apart. Hence

$$[F] = \frac{[M^2]}{[L^2]}$$

On the other hand, by Newton's second law $[F] = [M] \cdot [a]$. These last two formulas now yield

$$[M] \cdot [a] = \frac{[M]^2}{[L]^2}$$

from which $[M] = [a] \cdot [L]^2 = \mathrm{m}^{-1} \cdot \mathrm{m}^2 = \mathrm{m}$ and with this, the dimension of force will be

$$[F] = [M] \cdot [a] = \mathrm{m} \cdot \mathrm{m}^{-1} = 1$$

i.e., force—like speed—is a dimensionless variable. For the dimension of *energy*, E, we have

$$[E] = [F] \cdot [L] = 1 \cdot \mathrm{m} = \mathrm{m}.$$

Temperature, θ, is proportional to the average kinetic energy of the molecules per unit mass. Therefore, for its dimension we must have

$$[\theta] = \frac{[\text{energy}]}{[\text{mass}]} = \frac{\mathrm{m}}{\mathrm{m}} = 1$$

i.e., temperature is a dimensionless variable. Similarly to mass, *electric charge*, Q, has a dimension of $[Q] = \mathrm{m}$, and hence for electric current I, we have

$$[I] = \frac{[Q]}{[t]} = \frac{\mathrm{m}}{\mathrm{m}} = 1$$

For *pressure*, p, and mechanical normal and shear stresses, σ and τ, we can write

$$[\sigma] = [\tau] = [p] = \frac{[\text{force}]}{[\text{area}]} = \frac{1}{\mathrm{m}^2} = \mathrm{m}^{-2}$$

and for *power*, W

$$W = \frac{[\text{energy}]}{[\text{time}]} = \frac{\mathrm{m}}{\mathrm{m}} = 1$$

which is, therefore, dimensionless. For *electric potential*, V

$$[V] = \frac{[\text{energy}]}{[\text{electric charge}]} = \frac{\mathrm{m}}{\mathrm{m}} = 1$$

and for *electric resistance*, R

$$[R] = \frac{[V]}{[I]} = \frac{1}{1} = 1$$

Similarly, for *dynamic viscosity*, μ, *density*, ρ, and *universal gravitational constant*, k, we easily derive

$$[\mu] = \left[\frac{\tau \cdot dy}{dv}\right] = \frac{[\tau] \cdot [y]}{[v]} = \frac{\mathrm{m}^{-2} \cdot \mathrm{m}}{1} = \mathrm{m}^{-1}$$

$$[\rho] = \frac{[\text{mass}]}{[\text{volume}]} = \frac{\mathrm{m}}{\mathrm{m}^3} = \mathrm{m}^{-2}$$

$$[k] = \frac{[\text{force}] \cdot [\text{length}]^2}{[\text{mass}]^2} = \frac{1 \cdot \mathrm{m}^2}{\mathrm{m}^2} = 1$$

Continuing this process we can obtain without any problem the dimensions of every physical variable and constant.

A conspicuous question now arises: why is such a system utterly impractical? The most obvious reason, of course, is that this system forces us to use dimensions that are grotesquely inappropriate. We would have to express, for example, the flying *time* between New York and Chicago in terms of meters, and the speed of the airplane in absolute numbers. We also would go to the butcher and purchase pork chops by the meter, and so on.

But there is a much more consequential reason why a monodimensional system is inferior to a (moderately) multidimensional one. Consider a simple pendulum swinging with a small amplitude. Its period, T, i.e., the time between successive extreme positions on the same side of the swing, may be a function of its mass, M, length, L, and gravitational acceleration, g. Thus we write

$$T = k \cdot M^a \cdot L^b \cdot g^c \qquad (3\text{-}1)$$

where a, b, c, and k are as yet undetermined constants. Now we invoke the condition of dimensional homogeneity, discussed in more detail in Chapter 6. Here it is sufficient to say that this condition simply prescribes the obvious: in any true equation, the respective dimensions of both sides must be identical. That is, if meter appears to the power of 2 on the left, then it must appear to the power of 2 on the right, etc.

Suppose we now express (3-1) in a system using m, kg, and s as fundamental dimensions. Then

$$s = kg^a \cdot m^b \cdot m^c \cdot s^{-2 \cdot c}$$

from which, by inspection

$$a = 0; \quad b = \frac{1}{2}; \quad c = -\frac{1}{2}$$

Hence (3-1) becomes

$$T = k \cdot \sqrt{\frac{L}{g}} \qquad (3\text{-}2)$$

which says that for a small amplitude oscillation, the period of a simple pendulum is proportional to the square root of its length, inversely proportional to the square root of the gravitational acceleration, and—significantly—independent of its mass. We emphasize that these important results were obtained not by physical analyses or measurements, but by the most elementary—one might say primitive—dimensional considerations.

What would we have if we used the monodimensiona system having the single dimension "meter" for length? This system was previously described in some detail. Equation (3-1) would still be valid of course, and hence we would write—by substituting the appropriate dimension (singular!) for variables T, M, L, and g

$$m = m^a \cdot m^b \cdot (m^{-1})^c$$

which yields, by the condition of dimensional homogeneity, $1 = a + b - c$. From this formula the exponents *a, b,* and *c* cannot be uniquely determined since we have three unknowns, but only one equation. As a consequence, dimensional considerations do not supply enough information to arrive at a formula like (3-2).

The fact that dimensional conditions do not make the derivation of formulas possible has other—and unpleasant—consequences. As explained in detail in later chapters, our ability to verify relations (on account of the requirement of dimensional homogeneity) is also seriously curtailed, and the advantage of forming dimensionless variables (in order to reduce the total number of variables involved in a problem) is drastically decreased.

For the above reasons, a monodimensional system will remain forever no more than a curiosity, as it could never attain an accepted level of popularity, either among the general public or among scientists.

3.1.2. Omnidimensional System

Let us now investigate a dimensional system that comprises *only* fundamental dimensions; i.e., *every* dimension is fundamental. If the dimension of every variable is fundamental, then each of these variables must be measured entirely independently. This means that every physical relation would be a separate scientific discovery and in every one of these formulas there would be a mandatory dimensional constant (different in each case), making the formula dimensionally homogeneous. For example, speed v would be defined as

$$v = k_1 \cdot \frac{L}{T}$$

where L is distance, T is time and k_1 is a dimensional constant. From the formula, then

$$[k_1] = \frac{[v] \cdot [T]}{[L]}$$

Similarly, for force F we would have $F = k_2 \cdot M \cdot a$, in which M is mass, a is acceleration, and k_2 is a dimensional constant. Hence

$$[k_2] = \frac{[F]}{[M] \cdot [a]}$$

It is clear, then, that in every relation there would be at least one "constant of nature" that would have to be determined separately—a manifestly impractical and burdensome task. To illustrate, let us revisit the earlier pendulum problem. We now use the omnidimensional system, i.e., a system in which *every* dimension is fundamental. Equation (3-1) states that

$$T = k \cdot M^a \cdot L^b \cdot g^c \qquad \text{repeated (3-1)}$$

where k is a dimensionless constant and a, b, and c are as yet undetermined exponents. The dimension of k therefore is

$$[k] = \left[\frac{T}{M^a \cdot L^b \cdot g^c}\right] = \frac{[T]}{[M]^a \cdot [L]^b \cdot [g]^c}$$

That is, the dimension of k can be "adjusted" to "harmonize" with any preselected a, b, and c. This means, of course, that the exponents a, b, and c cannot be determined by dimensional considerations. Consequently, this omnidimensional system suffers from the same deficiency as the previously discussed monodimensional system does.

3.1.3 Multidimensional System

Thus, we must arrive at the obvious conclusion that the optimum number of fundamental dimensions must be somewhere between *one* and *all*. As it turns out, if we wish to maximize the usefulness of dimensional homogeneity and still avoid the many drawbacks resulting from the selection of too few or too many fundamental dimensions, the best choice is the traditional *length–mass–time–temperature* set for thermomechanical systems, augmented by three more dimensions (electric current, amount of substance, and luminous intensity) to deal with electrical, electrochemical, and optical subjects. Thus we have a *multidimensional* system.

A system like this is still not free of ambiguities, though. For example, the following pairs of widely *differing* physical quantities have *identical* dimensions:

- heat energy and mechanical torque;
- pressure and modulus of elasticity (Young's modulus);
- growth of mass per unit length (e.g., thickening of a tree branch) and dynamic viscosity;
- growth of mass per unit area (e.g., thickening of the skin during embryonic development) and mass of flow per unit cross section;
- deceleration of tree limb thickening with age and mechanical stress.

Once the fundamental dimensions are agreed upon (i.e., selected), all the other derived dimensions can be obtained by way of (maybe repeated) multiplication and division of these fundamentals. The idea is that, armed with the fundamental *and* derived dimensions, every numerically expressible quantity in the universe can be formulated. Carried further, this means that every possible relation, encompassing every facet of our knowledge about the physical world can be—at least in theory—uniquely defined.

3.2. CLASSIFICATION

Apart from some "fringe systems" (some of which are described in Ref. 10, p. 3 ff), essentially every dimensional system can be classified in two ways: whether it uses the *metric* or *American/British* (also called *Imperial*) system, *and* whether it uses *force* or *mass* for one of its fundamental dimensions. The table in Fig. 3-2 presents the names of these systems. Among these systems, the officially endorsed system (by more than 30 countries and certainly by all industrially developed countries) is the SI. It is described in detail in Art. 3.3; the other systems appearing in Fig. 3-2 are briefly described in Art. 3.4.

	metric	American/British
force based	MKS Force system	American/British Force (Engineering) system
mass based	SI	American/British Mass system

Figure 3-2
Classification of dimensional systems

3.3. THE SI

3.3.1. Preliminary Remarks

The name SI (used in all languages) is an abbreviated acronym derived from the French *Le Système International d'Unités*, the English translation of which is *International System of Units*. SI was adopted at the 11th *General Conference of Weights and Measures* in 1960, and by the 14th such conference for minor modifications in 1972. But even before these dates, metric dimensions were made legal in 1866 in the U.K. and in the U.S.A., and in 1873 in Canada. However, the origin of the system can be traced back to the late 18th century, when the French in 1791 adopted the *metre* (meter) as a dimension of length. They defined the metre as being the 1/40,000,000 part of the length the meridian passing through Paris. In turn, Lavoisier, the French chemist (1743–1794), established the dimension of mass based on the meter. He took 1 cm^3 (= 10^{-6} m^3) distilled water at its maximum density (at 4 °C) and called it a *gramme*, from the Greek *gramma*, meaning small weight. SI has a number of general and specific advantageous characteristics. Some of these are:

- It is inherently simple, and hence easy to learn.
- It is logically precise.
- It is a decimal system, and hence it simplifies technical calculations.
- It makes international communication easy.
- It has a small number of fundamental dimensions.
- It has a large number of derived dimensions.
- It is coherent.

The benefits of these characteristics will become obvious as we proceed with a detailed description of SI, but we pause here briefly to elaborate on the last-mentioned property: *coherence*.

If a dimensional system is coherent in its *fundamental* dimensions, then the *derived* dimensions are obtained by combinations of fundamental dimensions *without* having to use conversion factors different from 1. For example if *pascal* is the dimension of pressure, then it is the ratio of the dimension of force to the dimension of area. No numerical factor different from 1 is involved. In contrast, consider the notoriously noncoherent *in–lbf–s* system. Here, if the pressure is expressed in inches of mercury, h, the force in poundforce, and the area in square inches, then we have

$$p = 0.491154077497 \times h \text{ lbf/in}^2$$

For example, if $h = 20.3602096738$ inches, then $p = 10$ lbf/in^2. Since the coefficient in the above equation is different from 1, this system with the dimensions indicated is *noncoherent*. As we will shortly see, SI is coherent, i.e., all the coefficients in it are 1. To illustrate: if the dimension of *energy* is *joule,* of *force* is *newton,* of *pressure* is *pascal*, and of *power* is *watt*, then

- 1 watt is exactly 1 joule per second;
- 1 watt is exactly 1 newton times meter per second;
- 1 watt is exactly 1 pascal times cubic meter per second.

As we see, no numerical factor other than 1 is used in these relations.

3.3.2. Structure

SI is composed of seven *fundamental* dimensions and a very large number of *derived* units. Here the word "unit" has a special restricted meaning; it is (typically) a particular combination of fundamental dimensions that designates a particular physical or mathematical quantity (variable, parameter, or constant). A unit can have a specific name (usually the name of a physicist or mathematician working in the relevant discipline), or it can be *nameless*. For example, if *kilogram, meter,* and *second* are three of the seven fundamental dimensions in SI, then the unit of *force* is a quantity derived from Newton's second law: Force = (mass)·(acceleration). Hence, the *derived* dimension of force is kg·(m/s^2) or (m·kg)/s^2, and is called—appropriately—*newton* (lower case!). Therefore, newton is a *named* derived unit. In contrast, the dimension of speed is m/s, which is the unit of speed. However, this unit does not have a name, and so it is a *nameless* derived unit.

The table in Fig. 3-3 presents a possible classification of dimensions and units in SI.

(a) Fundamental Dimensions. Fig. 3-4, gives the seven *fundamental* dimensions of SI. What follows is a description of the entries in this table.

Quantity:	length
Dimension:	meter (or metre)
Symbol of dimension:	m

The meter was defined by the 17th *Conférence Générale des Poids et Mesures** (CGMP) in 1983 as follows: *The meter is the length of path traveled by light in vacuum during the time interval of 1/299,792,458 second.* Note that for this defini-

---

*Jerrard in Ref. 10, p. 197, gives a description of this organization.

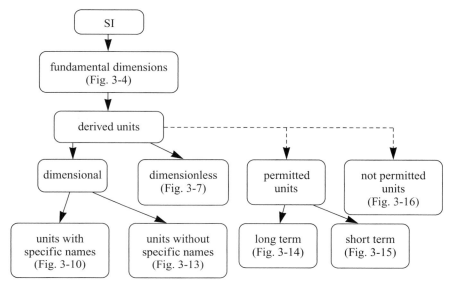

Figure 3-3
A possible classification of dimensions and units in SI

tion to be meaningful, the "second" must be defined such that it does not include—directly or indirectly—length, else the definition would be circular. The meter as defined here is the approximate length of $1/10^7$ part of the quadrant of the meridian in Paris; i.e., the length of the meridian in Paris is approximately 4×10^7 m.

Quantity: mass
Dimension: kilogram
Symbol of dimension: kg

Quantity	Dimension	Symbol
length	meter (or metre)	m
mass	kilogram	kg
time	second	s
electric current	ampere	A
thermodynamic temperature	kelvin	K
amount of substance	mole	mol
luminous intensity	candela	cd

Figure 3-4
Fundamental dimensions of SI

The kilogram, as a unit of mass, was originally defined in 1889 in terms of length (meter) and the density of distilled water at a temperature of maximum density (4 °C). Accordingly, the "kilogram" was defined as the mass of $1/1000$ m^3 water at 4 °C. A platinum–iridium cylinder was made as a *primary* standard; it is kept at the *International Bureau of Weights and Measures* in Sèvres, France. Several copies of this primary standard mass were made; these are the *secondary* standards. The secondary standards were verified, i.e., compared with the primary standard by means of a balance whose inaccuracy was less than $1/10^8$ part of the masses compared. The secondary standards were sequentially numbered, and then sent to different countries; copy No. 18 went to England, No. 20 to the U.S.A.

Note that kg is the dimension of *mass*, not of *weight*. The latter is the *force* generated by gravitation upon the mass. Thus, if g is the gravitational acceleration, then by Newton's second law

$$\text{weight} = g \cdot (\text{mass})$$

Hence, for its dimension we have

$$[\text{weight}] = [g] \cdot [\text{mass}] = \frac{m}{s^2} \cdot kg = \frac{m \cdot kg}{s^2}$$

which, of course, is identical to the dimension of force. If we now use the adopted value for g, viz. $g = 9.80665$ m/s^2, then the *unit of force* (corresponding to 1 kg mass) is obtained. This unit of force is called *newton* (symbol: N). Therefore

$$1 \text{ N} = 1 \, \frac{m \cdot kg}{s^2}$$

Quantity: time
Dimension: second
Symbol of dimension: s

By the decision of the 13th CGMP, held in 1967, the second is defined as the *duration of 9,192,631,770 periods of the radiation corresponding to the transition between two hyperfine levels of the ground state of cesium-133 atom.* The second is the most accurately known of the seven fundamental SI dimensions; its error of determination—with the best equipment—is less than 10^{-13} seconds.

Demand for precision, and the availability of technology to satisfy this demand, may necessitate relativistic corrections. Strictly speaking, to utilize the available precision, the user (observer) must be in proximity to the clock and at rest with respect to it. Usually, considering the size of the average laboratory, only the effect of *special* relativity may be significant. But when measurements of *distant* clocks are taken, the *general* theory of relativity must also be employed. In particular, at different heights clocks tick differently. Thus we have (Ref. 9, Vol. 2, p. 41-10)

$$\Delta t = t_0 \cdot \frac{g_0 \cdot h}{c^2} \tag{3-3}$$

where Δt = relativistic time correction due to height;
t_0 = time duration at sea level (the observer's level);
g_0 = gravitational acceleration at sea level;
h = height—relative to the observer;
c = speed of light.

For example, at 2000 m height, a clock runs about 2.2×10^{-11} percent *slower* than at sea level. Similarly, since the relativistic rate of clocks is influenced by the gravitational acceleration (which can vary as much as 0.5% at sea level), the rate of a clock will vary as the clock moves about on the surface of Earth. To find the magnitude of this effect, let us write (3-3) in the form

$$\Delta t = t_0 \cdot \frac{g}{c^2} \cdot \Delta R \qquad (3\text{-}4)$$

where ΔR is the change in distance between the clock (observer) and the center of Earth, as the clock moves. By Newton's law, the weight G of an object is

$$G = M \cdot g = k \cdot \frac{M_0 \cdot M}{R^2} \qquad (3\text{-}5)$$

where M = mass of the clock (object);
k = universal gravitational constant;
M_0 = mass of Earth;
R = radius of Earth.

So, by (3-5),

$$g = \frac{k \cdot M_0}{R^2} \qquad (3\text{-}6)$$

and hence, by differentiation (sign disregarded)

$$\Delta g = 2 \cdot \frac{k \cdot M_0}{R^3} \cdot \Delta R \qquad (3\text{-}7)$$

or

$$\Delta R = \frac{R^3}{2 \cdot k \cdot M_0} \cdot \Delta g \qquad (3\text{-}8)$$

Now we substitute g from (3-6) and ΔR from (3-8) into (3-4). This will yield—after some elementary simplification

$$\frac{\Delta t}{t_0} = \frac{R}{2 \cdot c^2} \cdot \Delta g \qquad (3\text{-}9)$$

But

$$\Delta g = 0.005 \cdot g_0 \qquad (3\text{-}10)$$

where $g_0 = 9.81$ m/s² is the gravitational acceleration at sea level, and the factor 0.005 is the aforementioned 0.5 % variation of g at sea level. Since $R = 6.367 \times 10^6$ m, relation (3-9)—with the appropriate numerical inputs of $c = 3 \times 10^8$ m/s and $k = 6.6726 \times 10^{-11}$ m³/(s²·kg)—will yield

$$\frac{\Delta t}{t_0} = 1.77 \times 10^{-13}$$

This means that there can be as much as a 1.77×10^{-11} percent relativistic difference in time lengths shown by two clocks that are at sea level, but at different locations.

Quantity: electric current
Dimension: ampere
Symbol of dimension: A

The 9th CGPM, in 1948, decided on the following definition of ampere: *The ampere is that constant current which, if maintained in two straight parallel conductors of infinite length and negligible cross section, and placed 1 meter apart in vacuum, would produce between these conductors a force equal to 2×10^{-7} newton per meter of length.*

The unit of force here is newton; it is the force required to impart 1 m/s² acceleration to a mass of 1 kg. Thus newton is a *named derived dimension* that is (m·kg)/s². The number 2×10^{-7} appearing above is exact, by definition.

The fundamental dimension *ampere* is named after André Marie Ampère (1775–1836) a French mathematician and physicist. He was one of the principal pioneers of electrodynamics, which, incidentally, he named as such. His father was guillotined in the French Revolution. Ampere was a tireless worker, but an unhappy man throughout his life. His tombstone bears the epitaph chosen by himself: *Tandem felix* (Happy at last). Incidentally, the ampere is one of the only two fundamental dimensions named after a scientist (the other is kelvin).

Quantity: thermodynamic temperature
Dimension: kelvin
Symbol of dimension: K

The definition of kelvin for thermodynamic temperature was adopted at the 16th CGMP (in 1967) as follows: *The kelvin is the fraction 1/273.16 of the thermodynamic temperature of the triple point of water.* The *triple point* of water is, by definition, that single *point* in the temperature-versus-pressure coordinate plane at which ice, water, and steam can coexist in equilibrium (see Fig. 3-5). This point is called a triple point because at this location (and only at *this* location) the three curves separating the three phases (ice, liquid, and vapor) intersect. The curves are the *fusion* curve A, the *vaporization* curve B, and the *sublimation* curve C (Fig. 3-5). The temperature at the intersect is *defined* as 271.16 K (exactly), and the pressure at this point is *measured* to be 610.483 pascal. This is a very low value, as it is only 1/166 part of the atmospheric pressure.

Here we should also mention the *International Practical Temperature Scale—*

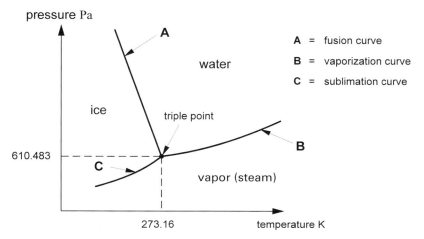

Figure 3-5
Triple point of water (diagram not to scale)

which is the *Celsius* scale. The relation between the Kelvin and the Celsius scales is shown in Fig. 3-6.

From the figure, if t and T are temperatures in the Celsius and Kelvin scales, respectively, then

$$t = T - 273.15 \text{ °C} \qquad (3\text{-}11)$$

Since by this equation $(dt/dT) = 1 = \text{const}$, therefore the *differences* expressed in both scales are numerically identical. For example, if the difference between two temperatures is $\Delta t = 33$ °C, then it is $\Delta T = 33$ kelvin. Note that the temperature expressed in the Kelvin scale (also called absolute scale) is called kelvin, abbreviated K. There is no such thing as "Kelvin degree," or "degree Kelvin," or °K. In contrast, we have "degree Celsius," abbreviated °C.

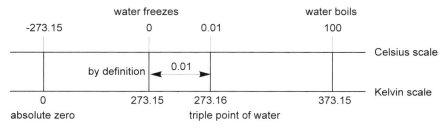

Figure 3-6
Relation between the absolute (Kelvin) and Celsius temperature scales
Normal atmospheric pressure (= 101,325 Pa) is assumed; diagram not to scale

The Kelvin temperature is named after Lord Kelvin (1824–1907), formerly William Thomson, a Scottish mathematician and physicist. The Celsius scale is named after Anders Celsius (1701–1744), a Swedish astronomer. In 1742, Celsius published his temperature-measuring technique by which—contrary to the popular belief—he defined the freezing point of water as 100 degrees, and the boiling point as 0 degree. This, however, was reversed the following year.

Quantity: amount of substance
Dimension: mole
Symbol of dimension: mol

For the mole, the 14th CGMP (1971) adopted the following definition: *The mole is the amount of substance (quantity of matter) that contains as many elementary entities as there are atoms in exactly 0.012 kg of carbon-12.*

Here the term "elementary entities" may mean molecules, atoms, ions, electrons, other particles, or specified groups of such particles. An example will illustrate: The atom of oxygen ^{16}O and the atom of carbon ^{12}C have a ratio of masses 15.9994/12.011. Therefore the molar mass of the molecular gas O_2 is

$$\text{mass}(O_2) = \frac{(2)\cdot(15.9994)}{12.011}\cdot(0.012) = 0.0319695 \text{ kg/mol}$$

So the amount of substance corresponding to a given mass of gas, say, $O_2 = 1.33$ kg is

$$Q = \frac{1.33}{0.0319695}\frac{\text{kg}}{(\text{kg/mol})} = 41.60215 \text{ mol}$$

Quantity: luminous intensity
Dimension: candela
Symbol of dimension: cd

The candela is the dimension of the intensity of a light source. It is defined by the 16th CGPM (1979) as follows: *The candela is the luminous intensity in a given direction, of a source that emits monochromatic radiation of frequency 5.4 × 10^{14} hertz and that has a radiant intensity in that direction of 1/683 watt per steradian.*

Here the following remarks should be made:

- The adjective "luminous" means "visible."
- The word "source" means "light source."
- The frequency given is exact; Hertz means oscillations per second.
- The phrase "radiant intensity" means "radiated power."
- The value 1/683 is exact.
- The steradian is the dimension of a solid angle (see below).

(b) Derived Dimensionless Units. *Derived* dimensions in SI are formed by (maybe repeated) multiplication and division (but not addition or subtraction) of *fundamental dimensions.* Any combination of fundamental dimensions generates

DIMENSIONAL SYSTEMS

either a dimensional or a dimensionless quantity (see Fig. 3-3). Dimensionless derived quantities in SI are also called *SI supplementary units* (see Fig. 3-7).

Quantity	Dimension	Symbol
plane angle	radian	rad
solid angle	steradian	sr

Figure 3-7
SI supplementary (dimensionless) units

Quantity: plane angle
Dimension: radian
Symbol of dimension: rad

In general the magnitude of a plane angle, φ, is defined as the ratio of the length of arc λ and radius R of a circle, as illustrated in Fig. 3-8. Thus, by the cited definition,

$$\varphi = \frac{\lambda}{R} \qquad (3\text{-}12)$$

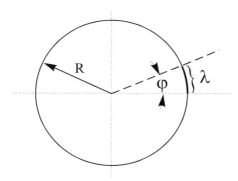

Figure 3-8
Definition of a plane angle

If now $\lambda = R$, then $\varphi = 1$ radian. Hence, a plane angle is 1 radian if the length of the arc subtended by the angle is identical to the radius generating that arc. Equation (3-12) provides the *dimension* of the radian, namely

$$[\text{radian}] = \frac{[\text{length of arc}]}{[\text{length of radius}]} = \frac{m}{m} = 1$$

52 APPLIED DIMENSIONAL ANALYSIS AND MODELING

i.e., it is, indeed, dimensionless. Since the perimeter of a full circle is $\lambda = 2 \cdot \pi \cdot R$, therefore the angle corresponding to a full circle is, by (3-12),

$$\lambda = \frac{2 \cdot \pi \cdot R}{R} = 2 \cdot \pi$$

Quantity: solid angle
Dimension: steradian
Symbol of dimension: sr

A solid angle σ can be visualized as an "angle of three dimensions" occurring at the vertex of a cone (Fig. 3-9). The vertex coincides with the center of a sphere of radius R. The surface cut out by the cone from the sphere is A. Then, by definition, the solid angle is

$$\sigma = \frac{A}{R^2} \tag{3-13}$$

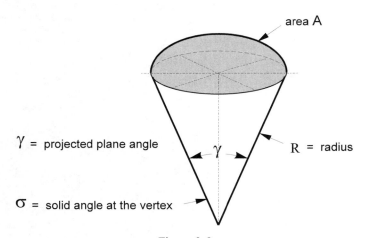

Figure 3-9
Definition of a solid angle. The *solid* angle is σ, the *projected* plane angle is γ

If $A = R^2$, then, by (3-13), $\sigma = 1$ steradian. Therefore, the solid angle is 1 steradian if the corresponding area cut from the sphere is R^2. It can be easily shown that the projected plane angle γ at the vertex of the cone is

$$\gamma = 2 \cdot \arccos\left(\frac{2 \cdot \pi - \sigma}{2 \cdot \pi}\right) \tag{3-14}$$

So that if $\sigma = 1$ sr, then $\gamma = 1.1439$ rad ($= 65.541$ deg).

(c) Derived Dimensional Units with Specific Names. As Fig. 3-3 shows, there are two types of dimensional derived units: those with specific names and those without. We wish to point out at this time that the use of these named derived units is merely for convenience, not necessity. These named units are nothing other than *abbreviations* of the combined fundamental dimensions of the physical quantities involved, which, otherwise would be too cumbersome to write out or pronounce. For example, the unit of force in SI is 1 m·kg/s^2, which is called *newton*. So, instead of saying "The tension force is one hundred meterkilogram per second squared," we say "The tension force is one hundred newtons"—which is definitely a labor-saving stratagem. Similarly, instead of saying "The electric resistance is five meters squared times kilogram per second cubed times ampere squared," we say "The electric resistance is five ohms"—which is an even more drastic improvement of efficiency in communication.

The table in Fig 3-10 lists the named SI derived units. The reader is advised to consult the Notes in Appendix 4.

(d) Derived Dimensional Units without Specific Names. It is evident that, in theory, there can be an infinite number of SI units derivable from the seven fundamental dimensions. For example, a dimension of $(m^{63} \cdot kg^4)/s^8$ for a unit is possible; so is $(A^6 \cdot mol^4)/(kg^{40} \cdot s^6)$. However, if we restrict the exponents of fundamental dimensions within reasonable limits, then the number of *possible* units becomes finite.

In general, if we have a set of elements $a_1, a_2, \ldots, a_n$, and element a_1 can assume m_1 distinct values, a_2 can assume m_2 distinct values, and so on, then the number of possible combinations of n distinct elements is

$$N = m_1 \cdot m_2 \cdot \ldots, m_n$$

If it happens that $m_1 = m_2 = \ldots = m_n$, then obviously $N = m^n$. For example, if two men, Harry and Adam, can wear green, red, or black hats, then $n = 2$, $m = 3$, and hence the number of different ways Harry and Adam can appear in public—regarding their hat colors—is $N = 3^2 = 9$. Note that if we have three men (instead of two), and two hat colors (instead of three), then the number of different ways our men can appear is only $N = 2^3 = 8$.

Now we employ this simple rule for combinations to find the number of derivable SI dimensional units. First we note that we have seven fundamental dimensions, hence $n = 7$. Next, we assume that exponents of fundamental dimensions are integers and that their maximum absolute value is p. For example, if $p = 3$, then the exponents can be $-3, -2, -1, 0, 1, 2, 3$. Therefore the number of different exponents is $m = 1 + 2 \cdot p$. Thus, if n is the number of fundamental dimensions, then the number of possible combinations is

$$N' = (1 + 2 \cdot p)^n \qquad (3\text{-}15)$$

However, this number must be diminished by 1, since the case when *all* exponents are zero cannot occur, because that would result in a *dimensionless* unit which, by de-

Name	Symbol	Typical form	In fundamental dimensions	Description	Note
becquerel	Bq	s^{-1}	s^{-1}	activity of a radionuclide	1
coulomb	C	s·A	s·A	quantity of electricity, electrical charge	2
degree Celsius	°C	K	K	Celsius temperature	3
farad	F	C/V	$m^{-2} \cdot kg^{-1} \cdot s^4 \cdot A^2$	capacitance	4
gray	Gy	J/kg	$m^2 \cdot s^{-2}$	absorbed dose of ionizing radiation	5
henry	H	Wb/A	$m^2 \cdot kg \cdot s^{-2} \cdot A^{-2}$	inductance	6
hertz	Hz	s^{-1}	1/s	frequency	7
joule	J	N·m	$m^2 \cdot kg \cdot s^{-2}$	energy, work, quantity of heat	8
lumen	lm	cd·sr	cd	luminous flux	9
lux	lx	lm·m^{-2}	$m^{-2} \cdot cd$	illuminance	10
newton	N	m·kg·s^{-2}	m·kg·s^{-2}	force	11
ohm	Ω	V/A	$m^2 \cdot kg \cdot s^{-3} \cdot A^{-2}$	electric resistance	12
pascal	Pa	N·m^{-2}	$m^{-1} \cdot kg \cdot s^{-2}$	pressure, stress	13
siemens	S	A/V	$m^{-2} \cdot kg^{-1} \cdot s^3 \cdot A^2$	electric conductance	14
sievert	Sv	J/kg	$m^2 \cdot s^{-2}$	dose equivalent of ionizing radiation	15
tesla	T	Wb·m^{-2}	$kg \cdot s^{-2} \cdot A^{-1}$	magnetic flux density	16
volt	V	W/A	$m^2 \cdot kg \cdot s^{-3} \cdot A^{-1}$	electric potential, potential difference	17
watt	W	J/s	$m^2 \cdot kg \cdot s^{-3}$	power, radiant flux	18
weber	Wb	V·s	$m^2 \cdot kg \cdot s^{-2} \cdot A^{-1}$	magnetic flux	19

Figure 3-10
Named derived SI units
(Notes are in Appendix 4)

finition, is excluded. Furthermore, cases in which (any) one dimension is raised to the power of 1 and all the others are raised to the power of zero must also be excluded, since they would result in units with dimensions identical to a *fundamental* dimension. It is easy to see that there are exactly n such combinations. Consequently, considering now the above 2 exclusions, we have $N = N' - n - 1$, which, by (3-15), can be written

DIMENSIONAL SYSTEMS

$$N = (1 + 2 \cdot p)^n - n - 1 \qquad (3\text{-}16)$$

where n = the number of fundamental dimensions in a dimensional system $[n > 0]$;
  p = the maximum absolute value of the integral exponent that any fundamental dimension can have $[p > 0]$;
  N = the number of possible derivable units in the dimensional system.

The restrictions indicated in brackets [ ] are necessary because

- the number of fundamental dimensions cannot be either zero or negative, since any dimensional system must comprise at least one fundamental dimension;
- the absolute value of the maximum exponent cannot be zero, since otherwise all exponents of fundamental dimensions would be zero, thereby causing the derived "dimensional" unit to be, in fact, dimensionless.

The table in Fig. 3-11 presents the number of derivable dimensional units in a dimensional system of n fundamental dimensions. The parameter p is the integral exponent's upper absolute value that any fundamental dimension can have. Entries in the table were calculated by formula (3-16). For example, if a dimensional system comprises $n = 4$ fundamental dimensions, and each dimension can have an integral exponent whose absolute value does not exceed $p = 3$, then this dimensional set can have exactly $N = 2396$ derived units.

		p					
		1	2	3	4	5	6
n	2	6	22	46	78	118	166
	3	23	121	339	725	1,327	2,193
	4	76	620	2,396	6,556	14,636	28,556
	5	237	3,119	16,801	59,043	161,045	371,287
	6	722	15,618	117,642	531,434	1,771,554	4,826,802
	7	2,179	78,117	823,535	4,782,961	19,487,163	62,748,509

Figure 3-11
The number of derivable dimensional units in a dimensional system of n fundamental dimensions
(p = maximum absolute value of integral exponent of any fundamental dimension)

To illustrate, Fig. 3-12 presents all the $N = 22$ derivable dimensional units in a dimensional system of $n = 2$ fundamental dimensions and the maximum exponent value of $p = 2$.

The SI dimensional system has $n = 7$ fundamental dimensions. Also, we should consider the maximum exponent to be not less than 4, since there are cases where a

56 APPLIED DIMENSIONAL ANALYSIS AND MODELING

$U_1 = \dfrac{1}{d_1^2 \cdot d_2^2}$	$U_6 = \dfrac{1}{d_1 \cdot d_2^2}$	$U_{11} = \dfrac{1}{d_2^2}$	$U_{16} = d_1 \cdot d_2$
$U_2 = \dfrac{1}{d_1^2 \cdot d_2}$	$U_7 = \dfrac{1}{d_1 \cdot d_2}$	$U_{12} = \dfrac{1}{d_2}$	$U_{17} = d_1 \cdot d_2^2$
$U_3 = \dfrac{1}{d_1^2}$	$U_8 = \dfrac{1}{d_1}$	$U_{13} = d_2^2$	$U_{18} = \dfrac{d_1^2}{d_2^2}$
$U_4 = \dfrac{d_2}{d_1^2}$	$U_9 = \dfrac{d_2}{d_1}$	$U_{14} = \dfrac{d_1}{d_2^2}$	$U_{19} = \dfrac{d_1^2}{d_2}$
$U_5 = \dfrac{d_2^2}{d_1^2}$	$U_{10} = \dfrac{d_2^2}{d_1}$	$U_{15} = \dfrac{d_1}{d_2}$	$U_{20} = d_1^2$
$U_{21} = d_1^2 \cdot d_2$		$U_{22} = d_1^2 \cdot d_2^2$	

Figure 3-12
Complete set of dimensional units in a dimensional system of two fundamental dimensions and maximum absolute integral exponent value of 2
(d_1, d_2 = fundamental dimensions, U_1, U_2, . . . , U_{22} = all the possible dimensional units)

fundamental dimension is raised to the fourth power, e.g., second moment of area of plane figures, radiative heat transfer, friction losses in pipes carrying viscous liquids, etc. Hence we should have $p = 4$. By these two values, relation (3-16) and Fig. 3-11 produce $N = 4,782,961$, which is then the number of possible derivable dimensional units in SI, considering of course *integral* exponents only. Obviously, only a small fraction of these will be ever utilized, but still, the size of this number clearly indicates the practically inexhaustible potential of the SI dimensional system. Tables in Figs. 3-13a, b, c, and d list the more important SI derived *nameless* dimensional units.

(e) Non-SI Units Permanently Permitted to be Used with SI. There are certain units outside of SI which must be retained simply because at present they enjoy wide usage, or acceptance, in some specialized fields. It is like smoking; we know it is bad for us, yet we cannot ban it outright because people would ignore the healthy prohibition. The same is true for non-SI units. Take for instance "degree" for plane angle, or "year" for time. They are so universally popular that banning them would have a negligible effect on their use. Therefore, the *International Committee of Weights and Measures* (CIPM) *wisely* decided to *permanently* allow their use but, to preserve *coherence* (see Art. 3.3.1), the combination of these units with SI units "should be used with discretion."

The table in Fig. 3-14 lists the units permitted for use with SI. The reader's attention is directed to Appendix 4, where the Notes are given in numerical order.

DIMENSIONAL SYSTEMS

Quantity	Dimension	Note	Quantity	Dimension	Note
area	m^2		thermal resistivity	$s^3 \cdot K/(m \cdot kg)$	32
volume	m^3		specific volume	m^3/kg	20
density	kg/m^3	20, 83	energy density	$kg/(m \cdot s^2)$	22
weight	$m \cdot kg/s^2$	21	surface tension	kg/s^2	33
heat transfer coefficient	$kg/(s^3 \cdot K)$	32	wave number	$1/m$	34
moment of force	$m^2 \cdot kg/s^2$		wave length	m	34
linear velocity	m/s		momentum	$m \cdot kg/s$	27
angular velocity	$1/s$	23	second moment of area	m^4	35
linear acceleration	m/s^2		stress (normal, shear)	$kg/(m \cdot s^2)$	36
angular acceleration	$1/s^2$	23	heat flux	$m^2 \cdot kg/s^3$	40
linear jerk	m/s^3	148	angular jerk	$1/s^3$	149
moment of inertia	$m^2 \cdot kg$	24	Young's modulus	$kg/(m \cdot s^2)$	36
gravitational acceleration	m/s^2	25	modulus of shear	$kg/(m \cdot s^2)$	36
dynamic viscosity	$kg/(m \cdot s)$	26	compressibility	$m \cdot s^2/kg$	37
kinematic viscosity	m^2/s	26	flow rate (mass)	kg/s	
impulse	$m \cdot kg/s$	27	flow rate (volume)	m^3/s	
moment of momentum	$m^2 \cdot kg/s$	28	specific energy	m^2/s^2	22
specific heat capacity	$m^2/(s^2 \cdot K)$	29	linear expansion coefficient	$1/K$	38
heat capacity	$m^2 \cdot kg/(s^2 \cdot K)$	29	enthalpy	$m^2 \cdot kg/s^2$	39
specific entropy	$m^2/(s^2 \cdot K)$	30	specific enthalpy	m^2/s^2	39
entropy	$m^2 \cdot kg/(s^2 \cdot K)$	30	volumetric exp. coefficient	$1/K$	38
universal gravitational constant	$m^3/(s^2 \cdot kg)$	31	linear density	kg/m	20
thermal conductivity	$m \cdot kg/(s^3 \cdot K)$	32	area density	kg/m^2	20
			material permeance	s/m	53

Figure 3-13a
Some nameless dimensional SI units related to mechanics and heat
(Notes are in Appendix 4)

Quantity	Dimension	Note	Quantity	Dimension	Note
magnetic field strength	A/m	46	magnetic vector potential	m·kg/(A·s^2)	
surface current density	A/m^2	42	electromagnetic moment	A·m^2	
electric flux density	s·A/m^2	56	magnetic polarization	kg/(A·s^2)	
permittivity	s^4·A^2/(m^3·kg)	44	magnetic dipole moment	m^3·kg/(A·s^2)	47
permeability	m·kg/(s^2·A^2)	59	reluctance	s^2·A^2/(m^2·kg)	51
electric field strength	m·kg/(s^3·A)	43	magnetic permeance	m^2·kg/(A^2·s^2)	52
magnetomotive force	A	50	resistivity	m^3·kg/(A^2·s^3)	45
surface charge density	A·s/m^2	41	reactance	m^2·kg/(A^2·s^3)	48
volume charge density	s·A/m^3	41	impedance	m^2·kg/(A^2·s^3)	49
electric flux	A·s	56	conductivity	s^3·A^2/(m^3·kg)	45
electric dipole moment	m·s·A	57	susceptance	s^3·A^2/(m^2·kg)	54
magnetic potential difference	A		admittance	s^3·A^2/(m^2·kg)	55
electric dipole potential	m^2·kg/(A·s^3)	58	electric power	m^2·kg/s^3	60

Figure 3-13b
Some nameless dimensional SI units related to electromagnetic characteristics
(Notes are in Appendix 4)

Quantity	Dimension	Note	Quantity	Dimension	Note
radiant energy	m^2·kg/s^2	62	sound pressure	kg/(m·s^2)	71
radiant power (flux)	m^2·kg/s^3	63	velocity of sound	m/s	
radiant intensity	m^2·kg/(s^3·sr)	64	sound energy flux	m^2·kg/s^3	73
radiance	kg/(s^3·sr)	65	sound power	m^2·kg/s^3	73
irradiance	kg/s^3	66	sound intensity	kg/s^3	74
quantity of light	s·cd	67	acoustic impedance	kg/(m^2·s)	75
luminance	cd/m^2	70	mechanical impedance	kg/s	76
light exposure	cd·s/m^2	68	wave number	1/m	61
luminous efficacy	s^3·cd/(m^2·kg)	69	sound pressure level	1	72

Figure 3-13c
Some nameless dimensional SI units related to acoustical characteristics and light
(Notes are in Appendix 4)

DIMENSIONAL SYSTEMS

Quantity	Dimension	Note	Quantity	Dimension	Note
molar mass	kg/mol	77, 83	mass attenuation coefficient	m²/kg	
molar volume	m³/mol	78, 83	ion number density	1/m³	90
molar energy	$\dfrac{m^2 \cdot kg}{s^2 \cdot mol}$	79	mean free path	m	91
molar heat capacity	$\dfrac{m^2 \cdot kg}{s^2 \cdot mol \cdot K}$	80	activity	1/s	92
molar entropy	$\dfrac{m^2 \cdot kg}{s^2 \cdot mol \cdot K}$	81	absorbed dose rate	m²/s³	93
concentration	mol/m³	82, 83	dose equivalent rate	m²/s³	94
molality	mol/kg	83	radiant exposure	kg/s²	95
diffusion coefficient	m²/s	84	particle fluence	1/m²	96
elementary charge	A·s	86	particle fluence rate	1/(m²·s)	97
electronvolt	m²·kg/s²	87	particle flux density	1/(m²·s)	98
energy flux density	kg/s³	88	transport diffusion coefficient	mol.s/(m·kg)	85
linear attenuation coefficient	1/m	89	energy fluence rate	kg/s³	88
atomic attenuation coefficient	m²				

Figure 3-13d
Some nameless dimensional SI units related to molecular physics and nuclear reactions
(Notes are in Appendix 4)

(f) Non-SI Units Temporarily Permitted to be Used with SI. Because of prevailing practices, certain units are accepted—but *only temporarily*—for use with SI. Fig. 3-15 lists these units.

(g) Prohibited units. Certain units presently in use are not part of SI, and hence they shall not be used. The tables in Fig. 3-16a and Fig. 3-16b give a representative collection of units whose use should be avoided.

3.3.3. Prefixes

In the Imperial (inch–pound–second) system, each *size* of a physical quantity has a *different* name. For example, 12 *inches* equal 1 *foot,* 3 feet equal 1 *yard,* 1760 yards

APPLIED DIMENSIONAL ANALYSIS AND MODELING

Quantity	Name	Recommended symbol	SI equivalent	Note
time	minute	min	60 s	99
	hour	h	3600 s	99
	day	d	86400 s	99
	year	a	31,536,000 s	99, 100
plane angle	degree	o	$\pi/180$ rad	101
	minute	'	$\pi/10800$ rad	101
	second	''	$\pi/648{,}000$ rad	101
	revolution	rev	$2 \cdot \pi$ rad	102
area	hectare	ha	10,000 m^2	103
	acre	ac	4,046.87261 m^2	103
volume	litre (or liter)	L or l	0.001 m^3	
mass	metric ton (or tonne)	t	1000 kg	104
linear density	tex	tex	10^{-6} kg/m	105
energy	electronvolt	eV	$1.60217733 \times 10^{-19}$ J	106
mass of an atom	unified atomic mass unit	u	$1.6605402 \times 10^{-27}$ kg	107
length	astronomical unit		1.495979×10^{11} m	108
	light-year	ly	$9.460528405 \times 10^{15}$ m	109
	parsec	pc	$3.085678186 \times 10^{16}$ m	110

Figure 3-14
Non-SI units that are permanently permitted to be used with SI
(Notes are in Appendix 4)

Quantity	Name	Recommended symbol	SI equivalent	Note
length	nautical mile		1852 m	111
velocity	knot		$463/900 = 0.51\dot{4}$ m/s	112
pressure	millibar	mbar	100 Pa	113
area	barn	b	10^{-28} m^2	114

Figure 3-15
Non-SI units that are temporarily permitted to be used with SI
(Notes are in Appendix 4)

Quantity	Name	Symbol	SI equivalent	Note
length	angström	Å	10^{-10} m	
	micron	μ	10^{-6} m	
	fermi	fm	10^{-15} m	
	x unit	xu	1.002×10^{-13} m	115
area	are	a	100 m^2	
volume	stere	st	1 m^3	116
acceleration	gal	Gal	0.01 m/s^2	
mass	carat		2×10^{-4} kg	117
	gamma	γ	10^{-9} kg	
force	kilogram-force	kgf	9.80665 N	118
	kilopond	kp	9.80665 N	119
	dyne	dyn	10^{-5} N	123
pressure	torr	Torr	133.322368 Pa	143
	millimeter of mercury	mmHg	133.322368 Pa	
	bar	bar	100,000 Pa	120
	standard atmosphere	atm	101,325 Pa	
energy	calorie (IT)	cal	4.18680 J	121
	erg	erg	10^{-7} J	122
dynamic viscosity	poise	p	0.1 kg/(m·s)	124
kinematic viscosity	stokes	st	10^{-4} m^2/s	124

Figure 3-16a
Mechanics and heat-related units that are not permitted to be used in SI
(Notes are in Appendix 4)

equal 1 *mile,* 16 *ounces* equal 1 *pound,* etc. Moreover, in these systems, the respective multipliers (factors) are all different: it is *12* inches to the foot, *3* feet to the yard, *4* quarts to the gallon, etc. These two characteristics make the Imperial system manifestly awkward; indeed, its still prevalent use in industry and commerce is not due to any logical, or even practical reason, but merely to human inertia, which opposes change however beneficial that change would be.

SI gracefully avoids these Imperial impediments, for it has the *same* name for each dimension, *regardless of size.* Moreover, the multipliers used to express different magnitudes are all *powers of ten.* These multipliers appear as *prefixes* attached to the respective dimensions. Thus, the multiplier and the fundamental dimension form a new integrated unit. The table in Fig. 3-17 lists the accepted prefixes in SI.

Quantity	Name	Usual Symbol	SI equivalent	Note
conductance	mho	mho	$1\ A^2 \cdot s^3/(m^2 \cdot kg)$	125
magnetic field strength	oersted	Oe	$(1000/(4\pi))\ A/m$	126
magnetic flux	maxwell	Mx	$10^{-8}\ m^2 \cdot kg/(s^2 \cdot A)$	127
magnetic flux density	gauss	Gs or G	$10^{-4}\ kg/(s^2 \cdot A)$	128
	gamma	γ	$10^{-9}\ kg/(s^2 \cdot A)$	
illuminance	phot	ph	$10^4\ cd/m^2$	129
luminance	stilb	sb	$10^4\ cd/m^2$	130
radioactive activity	curie	Ci	$3.7 \times 10^{10}\ 1/s$	131
radiation exposure	röntgen	R	$2.58/10{,}000\ A \cdot s/kg$	132
absorbed dose of ionizing radiation	rad	rad	$0.01\ m^2/s^2$	133
dose equivalent	rem	rem	$0.01\ m^2/s^2$	134

Figure 3-16b
Electromagnetism, light, and nuclear reactions-related units not permitted in SI
(Notes are in Appendix 4)

Multiplier power of 10	Prefix		Note
	Name	Symbol	
18	Exa	E	
15	Peta	P	
12	Tera	T	
9	Giga	G	
6	Mega	M	
3	kilo	k	
2	hecto	h	use is not recommended
1	deca	da	use is not recommended
0	—	—	principal value
–1	deci	d	use is not recommended
–2	centi	c	use is not recommended
–3	milli	m	
–6	micro	μ	symbol is the Greek mu
–9	nano	n	
–12	pico	p	
–15	femto	f	
–18	atto	a	

Figure 3-17
Prefixes used with dimensions in SI

For example 1000 meters = 1 kilometer and is written (abbreviated) as 1 km, with no gap between the prefix and the dimension. Similarly, 1 nm is a nanometer, which is $1/10^9$ of a meter. Some other useful rules to use prefixes to dimensions are given in Art. 3.3.4. Here we only remark that:

(a) In the case of a *positive* power of 10, the prefix is a Greek word; of a *negative* power, it is *Latin*. For example, *kilo* (power of ten is +3) is Greek, *pico* (power of ten is –12) is Latin.

(b) If the exponent (power of ten) is equal to or larger than 6, then the prefix's symbol is a *capital* letter, otherwise it is all *lowercase*. Thus, 10^9 m = 1 Gm, but 10^{-3} m = 1 mm.

(c) Prefixes are used *only* with *fundamental* dimensions (Art. 3.3.2a, Fig. 3-4) and *named dimensional derived* units (Art. 3.3.2c, Fig. 3-10). For example, we can write $F = 28$ MN (mega newton), but not $F = 28$ M(m·kg)/s^2, nor should we write $v = 33$ m(m/s) for milli m/s, because the above *derived* dimensional units are not the *named* type. In these cases we must attach the prefix to (any) one of the *fundamental* dimensions appearing in the *numerator*. Thus, in the latter cited instance we should write $v = 33$ mm/s.

3.3.4. Some Rules of Etiquette in Writing Dimensions

When writing dimensions it is extremely wise to follow some elementary rules. These rules exist not to make the life of the writer more difficult, but to make the life of the *reader* easier. In other words, it is the reader whose interest is paramount, not the writer's. Observance of a few, rather simple and altogether logical "prescriptions" will make the appearance of equations and dimensional quantities unambiguous, consistent, and even aesthetically pleasing. It is like a well-constructed English sentence; it is not only informative, but it is also a pleasure to read. This is why it is both prudent and efficient to abide by the following *rules of etiquette* in writing dimensions in equations and expressions.

(a) Do not abbreviate English text to express a dimension. For example, 28 cubic centimeters is 28 cm^3 not 28 cc; 44 square meters is 44 m^2, not 44 sq. m.

(b) Write (print) all SI symbols in upright (roman) type, irrespective of the typeface used in the surrounding text. Here "upright" excludes *italic* and other sloped typefaces. For example, it is *The table is 28.32* m *long*, not *The table is 28.32 m long*.

(c) Observe carefully the upper and lowercase letters for symbols. In general, symbols must be written in lowercase letters, except *named* SI units, which, when *abbreviated,* are written in capital letters. For example, *The current is 4 ampere,* or *The current is 4 A*, not *The current is 4 Ampere*. If the abbreviation of a named derived SI unit is composed of 2 characters, then the first character is uppercase and the second is lowercase. For example, *5 weber* = *5* Wb.

(d) Do not affix an "s" to any *symbol* of dimension to indicate plurality. For example, it is 19 kg, not 19 kgs, but *The length is 34 meters* is correct (meter is not a symbol).

(e) Do not put a period at the end of the abbreviation of a dimension, except if this abbreviation is at the end of a sentence. For example, it is *A 9.5 m long table*, not *A 9.5 m. long table.* But *The length of the table is 9.5 m.* is correct.

(f) Put a space (gap) between the last digit of a magnitude and its dimension, whether the latter is abbreviated or not. For example, it is *The mass is 33* kg, not *The mass is 33*kg. Exception: when writing degree Celsius, do not leave a space. For example, 28°C, not 28 °C.

(g) Do not begin a sentence with a symbol of dimension. Wrong: m/s *was the dimension in this case.* Recast the sentence to read: *In this case the dimension was* m/s.

(h) Use lowercase letters for unabbreviated named derived SI units. For example, it is *63 newton*, not *63 Newton* [see also (c), above]. Exception: the word Celsius is always capitalized. For example, it is *5 degree Celsius*, not *5 degree celsius*.

(i) Do not mix names and symbols in a dimension. For example, it is 5 N·m, not 5 N·meter or 5 newton·m.

(j) Do not attach a quantifier to a dimension. If necessary, attach the quantifier to the magnitude of the unit in question. For example, it is *The gauge pressure is 25000 Pa*, not *The pressure is 25000 Pag.*

(k) Do not put a space between the prefix and the symbol (or name) of a dimension. For example, it is *8 km*, not *8 k m*. The prefix and its SI symbol form a new symbol, which shall be treated as such. For example:

$$4 \text{ cm}^3 = 4(\text{cm})^3 = 4 \times (10^{-2} \text{ m})^3 = 4 \times (10^{-2})^3 \text{ m}^3 = 4 \times 10^{-6} \text{ m}^3$$

$$7.3 \text{ μs}^{-1} = 7.3 \times (\text{μs})^{-1} = 7.3 \times (10^{-6} \text{ s})^{-1} = 7.3 \times 10^6 \text{ s}^{-1} = 7.3 \times 10^6 \frac{1}{\text{s}}$$

(l) Do not compound prefixes, i.e., multiple prefixes are not allowed. For example, it is *3.3 pm*, not *3.3 μμm*.

(m) Do not use more than one unit (base, multiple, or submultiple) to describe a quantity. For example, it is *L = 3.896 m*, not *L = 3 m 89 cm 6 mm*. Exception: Plane angles and time units. For example, φ = 33° 48′ 29″ is permitted, so is *t = 3 h 42 m 23 s*.

(n) Do not use more than one prefix in a dimension, and—if possible—apply this single prefix in the numerator of the dimension.. For example, 0.003 m/s may be written as 3 mm/s, but not as 3 km/Ms, nor as 3 m/ks.

(o) To avoid ambiguity, place a dot between two units (dimensions) to indicate multiplication. For example, m·N means (meter) × (newton), but mN means millinewton. The presence of the dot is essential.

(p) Do not use the *solidus* (/) more than once in any dimensional expression, unless (to avoid ambiguity) parentheses are also used. For example, it is m/s², not m/s/s; it is m/(A/s), not m/A/s.

(q) Do not substitute the dimension of a quantity for its name. For example: it is *Frontal length is 120 feet*, not *Frontal footage is 120.*

(r) When the *name* of a dimension appears in a text and a division is indicated, use the *word* "per," not the *symbol* "/." For example, it is not *8 newton/square meter*, but *8 newton per square meter*. However, *8* N/m² is correct.

3.3.4.1. **Problems** (solutions are in Appendix 6).

The expressions below are written incorrectly. Identify the errors and correct them as appropriate.

3/1 The force experienced was 44.3 Newton.
3/2 John's height is 1.93m.
3/3 The temperature in the room is 22.3 °celsius.
3/4 The tire pressure is at least 206 kPag.
3/5 The speed of the car on impact was in excess of 29 m/second.
3/6 The average size of dust particles was 2.3 mμm.
3/7 Mary was walking at a steady speed of 5 kmh^{-1}.
3/8 Kg is the dimension of mass in SI.
3/9 Oliver's speed when he hit the pavement was at least 120 kilometer/hour.
3/10 My body's mass is 85 kg 300 grams.
3/11 The maximum climbing speed of a sloth is about 85 cm/ks.
3/12 In the formula $v = \sqrt{2 \cdot g \cdot h}$, g is the gravitational acceleration $g = 9.80665$ *m/s²*.
3/13 The daily volume of blood received was 260 cc, administered intravenously.
3/14 Which is a larger mass: 55 kg. or 122 lb?
3/15 The noise-temperature of the parametric amplifier was 4.5 °K.
3/16 The time duration was 20 M s.
3/17 The mass of the Sun is about 2×10^{12} E kg.
3/18 The mass of the Sun is about 2 TEkg.
3/19 George's height is 1 m 77 cm.
3/20 The dimension of force in the SI is mkg/s².
3/21 The car's deceleration on impact was about 55 m/s/s.
3/22 M, kg, s are 3 of the 7 fundamental dimensions of SI.
3/23 The dimension of capacitance in SI is A²·s⁴/m²/kg.
3/24 The definition of pressure is force/area.
3/25 The pressure on the floor was p = 0.03 N/mm².
3/26 The pressure on the floor was 0.03kPa.
3/27 Harry's weight is the equivalent of 93.5 kgs.
3/28 The megaparsec between the Earth and Sun is about 4.86×10^{-12}.
3/29 The magnetic flux is 49 wb.
3/30 The moment of inertia of the flywheel is 43.2 kg·meter².

3/31 The floor area of the house, excluding the basement and the attic, is 144.3 sq.m.

3/32 Tom asked Rosemary: "Is your height more than 1.55 *m*"?

3/33 "Please pass the salt," said Greg, leaning his dainty 140 kgs frame on the counter.

3/34 The acceleration is extremely small; it does not exceed 4.5 m/ks^2.

3/35 The length was 6 m.—not less!

3/36 The old steam engine performed well; it had a power output of 3600 Watt.

3.4. OTHER THAN SI DIMENSIONAL SYSTEMS

We mentioned in Art. 3.2. that, apart from SI, there are other dimensional systems, some of which are still quite popular, mainly in English-speaking countries, while others are invented mainly for specific and limited purposes and, as such, are understandably ignored by the general population.

Take, for example, the *Kiang* system. In this system the unit of time is 1.5×10^{10} years (Hubble time), and the unit of mass is that of a black hole, 1.99×10^{53} kg (Ref. 10, p. 6). One can safely predict the relatively limited acceptance of such a system by the public.

Nevertheless, as the table in Fig. 3-2 indicates, there are three dimensional systems—apart from SI—which, because of their historical significance and prevailing usage, merit at least cursory descriptions.

The table cited shows that essentially all systems appearing in it—including SI—can be classed according to whether they are *metric* or *nonmetric* (i.e., Imperial), and whether they are *force* or *mass* based. So, altogether we have four possible combinations. We now briefly describe the more important systems in each of these four categories.

3.4.1. Metric, Mass-based Systems

(a) CGS system. The CGS system is one of the earliest. In CGS the C, G, and S stand for, respectively, *centimeter, gram* (or gramme), and *second*. The unit of force in CGS is the *dyne*, which is defined as the force necessary to impart 1 cm/s^2 acceleration to a body of 1 gram mass. By Newton's second law, the weight W of a gram mass is W = m·g, where m = 1 gram is the mass and g = 9.80665 m/s^2 is the acceleration. Thus, the dimension of force, or weight, in the CGS is

$$[\text{weight}] = [\text{force}] = [\text{mass}] \cdot [\text{acceleration}] = \text{gr} \cdot \frac{\text{cm}}{\text{s}^2} = \frac{\text{gr} \cdot \text{cm}}{\text{s}^2}$$

(see the table in Fig. 3-16a and Note 123 in Appendix 4). The CGS system can be considered the noble precursor of SI.

(b) SI. The SI is currently the most accepted system. Its advantages are important and numerous; it is described in detail in Art. 3.3, and is only mentioned here for the sake of completeness.

3.4.2. Metric, Force-based System

This is commonly known as the MKS system. In this system the dimensions *meter* (m), *kilogram* (kg) and *second* (s) are *fundamental*. Here the kg is *force*. As a consequence, the mass has a *derived* dimension, which is, by Newton's second law, mass = force/acceleration. Therefore for the dimension of mass we have

$$[\text{mass}] = \frac{\text{kg}}{\left(\dfrac{\text{m}}{\text{s}^2}\right)} = \frac{\text{kg} \cdot \text{s}^2}{\text{m}}$$

The force dimension *kg* in MKS is sometimes written as *kgf* (kilogram-force) to distinguish it from the *mass* unit kg in SI. A peculiarity of the MKS system is that mass has no named unit; hence mass in MKS can only be described by its dimension.

The MKS system has been widely used in engineering in non–English-speaking countries, mainly in Europe, and is still favored by many.

3.4.3. American/British Force (Engineering) System

In this system the fundamental dimensions are the *foot* (ft) [or *inch* (in)], *pound* (lb), and *second* (s). Here lb means *force*. Accordingly, the dimension of mass—entirely analogously to the MKS system—is

$$[\text{mass}] = \frac{\text{lb}}{\left(\dfrac{\text{ft}}{\text{s}^2}\right)} = \frac{\text{lb} \cdot \text{s}^2}{\text{ft}}$$

This unit of mass is called the *slug*. By contrast, the mass expressed in units of (lb·s^2)/in has no name, although one can write

$$1 \frac{\text{lb} \cdot \text{s}^2}{\text{in}} = 12 \text{ slugs}$$

The *force* dimension lb is sometimes written lbf (pound-force) to distinguish it from the mass dimension lb in the *American/British Mass System* (see below). This system is also known as the *Imperial Engineering* (or *Technical*) *System*.

3.4.4. American/British Mass (Scientific) System

In this system the *pound* (lb) is the dimension of mass and is legally equivalent to 0.45359237 kg in SI. The dimension of force, therefore, is a derived one; its unit is the *poundal* and is equal to the force necessary to impart 1 ft/s² acceleration to a 1 pound mass. The force of 1 poundal is equivalent to 0.138254954376 N in SI, and 0.0310809501716 lbf in the American/British Engineering System. The American/British Mass System is sometimes referred to as the *Imperial Scientific System*.

3.5. A NOTE ON THE CLASSIFICATION OF DIMENSIONAL SYSTEMS

Before concluding this description of dimensional systems, it should be noted that all *mass-based* systems, such as SI, have the advantage that mass, which is a *fundamental* dimension in all these systems, is independent of the numerical value of the gravitational acceleration g. This is an important fact, because on Earth, even at sea level, g can vary with terrestrial latitude by as much as 0.53% (see Note 25 in Appendix 4). Therefore, the use of a mass-based dimensional system is uniform anywhere in the universe.

On the other hand—and this should also be pointed out—in everyday life, the use of mass as a fundamental dimension is manifestly impractical. This is because mass cannot be *directly* perceived, it cannot be "felt," and—most importantly—it cannot be measured. Nobody has ever measured mass, and nobody ever will! Mass must always be determined indirectly by subjecting it to a known inertial or gravitational* acceleration and then *measuring* the generated *force*. Therefore all past, present and future instruments can only measure force which, in a mass-based dimensional system, is a *derived* dimension. And even if all butcher shops were equipped with scales calibrated in newtons, these scales would still show only the *derived* unit, *force*.

---

*The *Equivalence Principle* is one of the bases of the *General Relativity Theory*. This principle posits that the inertial (i.e., accelerated) and gravitational (i.e., heavy) masses of any body are numerically *exactly* equal at all times. Accordingly, no experiment of any kind can be constructed to differentiate between the forces generated by an accelerating mass and a gravitationally attracted one. The Equivalence Principle was experimentally proved in Budapest between 1891 and 1908, by Baron Loránd Eötvös (1848–1919), a Hungarian physicist. The Eötvös experiment is famous for the significance of its *negative* results (in this respect it is similar to another famous experiment, by Michelson and Morley, which proved the *non-existence* of ether-wind). Eötvös demonstrated that, within a factor of 2×10^{-8}, inertial and gravitational masses are identical. This limit has been recently lowered to less than 10^{-11} by U.S. and Russian scientists.

CHAPTER 4
TRANSFORMATION OF DIMENSIONS

4.1. NUMERICAL EQUIVALENCES

Before starting to describe a simple technique to transform fundamental and derived dimensions among different dimensional systems, we list a subjective selection of basic equivalences (a more extensive list is easily available from many sources, for example from Ref. 10, pp. 200–220). Exact values are in **bold** typeface, otherwise data are presented to five decimal places. Where appropriate, values are also given in *exact* proper or improper fractions. The integer following the symbol "E" indicates the exponent of 10; for example, 4.33 E6 means 4.33×10^6.

	meter	inch	foot	yard	mile	nautical mile
meter	**1**	**5000/127** 39.37008	**1250/381** 3.28084	**1250/1143** 1.09361	**125/201,168** 6.21371 E-4	**1/1852** 5.39957 E-4
inch	**127/5000** **0.0254**	**1**	**1/12** 8.33333 E-2	**1/36** 2.77778 E-2	**1/63,360** 1.57828 E-5	**127/9,260,000** 1.37149 E-5
foot	**381/1250** **0.3048**	**12**	**1**	**1/3** 0.33333	**1/5280** 1.89394 E-4	**381/2,315,000** 1.64579 E-4
yard	**1143/1250** **0.9144**	**36**	**3**	**1**	**1/1760** 5.68182 E-4	**1143/2,315,000** 4.93737 E-4
mile	**201,168/125** **1609.344**	**63360**	**5280**	**1760**	**1**	**50,292/57,875** 0.86898
nautical mile [111]	**1852**	**9,260,000/127** 72,913.38583	**2,315,000/381** 6076.11549	**2,315,000/1143** 2025.37183	**57,875/50,292** 1.15078	**1**

Figure 4-1a
Length-related dimensional equivalents
Quantities in any *row* are equivalent; numbers in [ ] are for notes in Appendix 4; examples (highlighted): 1 yard = 0.9144 m exactly, 1 light-year = 0.30659 parsec.
(Figure is continued on next page)

	meter	astronomical unit	light-year	parsec
meter	1	6.68459 E-12	1.05702 E-16	3.24078 E-17
astronomical unit [108]	1.49598 E11	1	1.58128 E-5	4.84814 E-6
light-year [109]	9.46053 E15	6.32397 E4	1	0.30659
parsec [110]	3.08568 E16	2.06265 E5	3.26163	1

Figure 4-1a
(Continued from previous page)

	liter	cubic meter	UK gallon	US gallon	US pint	US quart
liter	1	0.001	0.21997	0.26417	2.11338	1.05669
cubic meter	1000	1	219.96915	264.17205	2113.37642	1056.68821
UK gallon [135]	**4.546092**	**4.546092 E-3**	1	1.20095	9.6076	4.8038
US gallon [136]	3.78541	3.78541 E-3	0.83267	1	8	4
US pint	0.47318	4.73176 E-4	0.10408	**0.125**	1	0.5
US quart	0.94635	9.46353 E-4	0.20817	**0.25**	2	1

Figure 4-1b
Volume-related dimensional equivalents
Quantities in any *row* are equivalent; numbers in [ ] are for notes in Appendix 4;
example (highlighted): 1 US gallon = 3.7854 liter

	kg	$\dfrac{\text{kgf·s}^2}{\text{m}}$	lb	slug	$\dfrac{\text{lbf·s}^2}{\text{in}}$
kg	1	0.10197	2.20462	6.85218 E-2	5.71015 E-3
$\dfrac{\text{kgf·s}^2}{\text{m}}$	**9.80665**	1	21.61996	0.67197	5.59974 E-2
lb [137]	0.45359	4.62535 E-2	1	3.10810 E-2	2.59008 E-3
$\dfrac{\text{lbf·s}^2}{\text{ft}}$ = slug	14.5939	1.48816	32.17405	1	8.33333 E-2
$\dfrac{\text{lbf·s}^2}{\text{in}}$	175.12684	17.85797	386.08858	**12**	1

Figure 4-1c
Mass-related dimensional equivalents
Quantities in any *row* are equivalent; numbers in [ ] are for notes in Appendix 4;
example (highlighted): 1 slug = 14.59390 kg

TRANSFORMATION OF DIMENSIONS

	dyne	newton	pound-force	poundal	kilogram-force
dyne [123]	1	E-5	2.24809 E-6	7.23301 E-5	1.01972 E-6
newton	E5	1	0.22481	7.23301	0.10197
pound-force [147]	4.44822 E5	4.44822	1	32.17405	0.45359
poundal [146]	1.38255 E4	0.13825	3.10810 E-2	1	1.49098 E-2
kilgram-force [118]	980,665	9.80665	2.20462	70.93164	1

Figure 4-1d
Force-related dimensional equivalents
Quantities in any *row* are equivalent; numbers in [ ] are for notes in Appendix 4; example (highlighted): 1 pound-force (lbf) = 4.44822 newton

	J	BTU	calorie	kW·h	lbf·ft	kgf·m	kg (relativistic)
joule (J) [8]	1	9.47817 E-4	0.23885	2.77777 E-7	0.73756	0.10197	1.11265 E-17
BTU [138]	1.05506 E3	1	251.99576	2.93071 E-4	778.16926	107.58576	1.17391 E-14
calorie [121]	4.1868	3.96832 E-3	1	1.163 E-6	3.08803	0.426935	4.65844 E-17
kW·h [139]	3.6 E6	3412.14163	8.59845 E5	1	2.65522 E6	3.67098 E5	4.00554 E-11
lbf·ft [147]	1.35582	1.28507 E-3	0.32383	3.76616 E-7	1	0.13825	1.50855 E-17
kgf·m [118]	9.80665	9.29491 E-3	2.34228	2.72407 E-6	7.23301	1	1.09114 E-16
kg [140] (relativistic)	8.98755 E16	8.51856 E13	2.14664 E16	2.49654 E10	6.62888 E16	9.16475 E15	1

Figure 4-1e
Energy-related dimensional equivalents
Quantities in any *row* are equivalent; numbers in [ ] are for notes in Appendix 4; examples (highlighted): 1 cal = 4.1868 joule; 1 kg mass converted to energy = 2.14664 × 10^{16} cal

	watt	HP (metric)	HP
watt [18]	1	1.35962 E-3	1.34102 E-3
horsepower (HP) (metric) [141]	735.49875	1	0.98632
horsepower (HP) [142]	745.69987	1.01387	1

Figure 4-1f
Power-related dimensional equivalents
Quantities in any *row* are equivalent; numbers in [ ] are for notes in Appendix 4; example (highlighted): 1 metric horse-power = 0.98632 (US) horse-power

	atm	pascal	psi	torr
standard atmosphere (atm)	1	**101325**	14.69595	**760**
pascal	9.86923 E-6	1	1.45038 E-4	7.50062 E-3
lbf/in² (psi)	6.80460 E-2	6894.75729	1	51.71493
torr [143]	1.31579 E-3	133.32237	1.93368 E-2	1

Figure 4-1g
Pressure-related dimensional equivalents
Quantities in any *row* are equivalent; numbers in [ ] are for notes in Appendix 4; example (highlighted): 1 psi = 6894.75729 pascal

	gauss	tesla	gamma	maxwell/cm²
gauss [128]	1	E-4	E-5	1
tesla [16]	E4	1	E9	E4
gamma	E-5	E-9	1	E-5
maxwell/cm² [127]	1	E-4	E5	1

Figure 4-1h
Magnetic flux-related dimensional equivalents
Quantities in any *row* are equivalent; numbers in [ ] are for notes in Appendix 4; example (highlighted): 1 tesla = 10^4 Mx/cm²; all quantities in this table are exact

	K	°C	°F	°R
kelvin (K)	K	°C + 273.15	$\frac{5}{9}\cdot$(°F + 459.67)	$\frac{5}{9}\cdot$°R
Celsius (°C)	K–273.15	°C	$\frac{5}{9}\cdot$(°F – 32)	$\frac{5}{9}\cdot$°R – 273.15
Fahrenheit (°F) [144]	$\frac{9}{5}\cdot$K – 459.67	32+1.8·°C	°F	°R – 459.67
Rankine (°R) [145]	$\frac{9}{5}\cdot$K	$\frac{9}{5}\cdot$°C+491.67	°F + 459.67	°R

Figure 4-2a
Relations among temperature scales
Quantities in any *row* are equivalent; numbers in [ ] are for notes in Appendix 4; example (highlighted): 300 Celsius deg = 573.15 K; all quantities in this table are exact

	kelvin	Celsius	Fahrenheit	Rankine
kelvin	1	1	9/5	9/5
Celsius	1	1	9/5	9/5
Fahrenheit	5/9	5/9	1	1
Rankine	5/9	5/9	1	1

Figure 4-2b
Relations among temperature intervals expressed in different temperature scales
Example (highlighted): temperature *interval* of 1 °R = temperature *interval* of 5/9 K; all quantities in this table are exact

TRANSFORMATION OF DIMENSIONS   73

	kelvin	Celsius	Fahrenheit	Rankine
absolute zero	0	−273.15	−459.67	0
ice point	273.15	0	32	491.67
triple point of distilled water	273.16	0.01	32.018	491.688
boiling point of distilled water at sea level	373.15	100	212	671.67

Figure 4-3
Principal temperature points on different scales
All values are exact; *triple point* is defined in Art. 3.3.2a

4.2. TECHNIQUE

Transformations between dimensional systems, or between units of the same system are rather straightforward. First, we present the problem and the solution in general terms, and then we demonstrate the procedure by way of several examples. Finally, a number of problems are posed for the enthusiastic reader.

The general problem is this: *Given a dimensional quantity Q in dimensional system 1, what is the corresponding numerical value of Q in dimensional system 2?*

To solve this problem, we write

$$Q \cdot d_1^{e_1} \cdot d_2^{e_2} \ldots d_n^{e_n} = x \cdot D_1^{e_1} \cdot D_2^{e_2} \ldots D_n^{e_n} \qquad (4\text{-}1)$$

where $d_1, d_2, \ldots$ = dimensions in system 1;
$D_1, D_2, \ldots$ = dimensions in system 2;
$e_1, e_2, \ldots$ = exponents of dimensions in both systems;
n = number of dimensions in each system;
x = a numerical factor.

Obviously, the aim is to find x. Dimensions $d_1, d_2, \ldots$, must of course correspond to dimensions $D_1, D_2, \ldots$. In other words, if d_1 represents mass, then D_1 must also represent mass; if d_2 represents temperature, then so must D_2, etc. Of course, the magnitudes of these dimensions may differ, but they have to represent the same type of physical dimension. Apart from this plain condition, the d's and D's can be either *fundamental* or *derived*.

Generally we now have

$$\left.\begin{array}{l} d_1 = k_1 \cdot D_1 \\ d_2 = k_2 \cdot D_2 \\ \vdots \\ d_n = k_n \cdot D_n \end{array}\right\} \qquad (4\text{-}2)$$

where factors $k_1, k_2, \ldots$ are absolute numbers. For example, if d_1 = foot and D_1 = inch, then k_1 = 12, since 1 foot = 12 inches. Similarly, if d_2 = mile and D_2 = foot, then k_2 = 5280, since 1 mile = 5280 feet.

That the exponents $e_1, e_2, \ldots$ of both sides of (4-1) must be identical to the re-

74   APPLIED DIMENSIONAL ANALYSIS AND MODELING

spective dimensions is mandated by the *Principle of Dimensional Homogeneity,* which is discussed in Chapter 6. Here we only mention briefly that if, say, the length is raised to a certain power on the left side of (4-1), then length must be raised to the *same* power on the right side, otherwise length raised to some power would remain either on just the left or the right side of the equation after simplification. Since this residual length could not then be equal to any other dimension, therefore dimensional equality between the sides of (4-1) could not be achieved.

Substituting (4-2) into (4-1), we obtain

$$Q \cdot (k_1 \cdot D_1)^{e_1} \cdot (k_2 \cdot D_2)^{e_2} \ldots (k_n \cdot D_n)^{e_n} = x \cdot D_1^{e_1} \cdot D_2^{e_2} \ldots D_n^{e_n} \qquad (4\text{-}3)$$

which yields, after elementary simplification

$$x = Q \cdot (k_1^{e_1} \cdot k_2^{e_2} \ldots k_n^{e_n}) \qquad (4\text{-}4)$$

Thus, this *is* the simple solution to *all* transformation problems. We now present some examples to show how easily this technique works.

4.3. EXAMPLES

Example 4-1

The pressure in a gas tank is 68 kgf/m². What is this pressure in units of N/in²?
We write 68 kgf·m^{-2} = x·N·in^{-2}, thus, by (4-1) and (4-2), $Q = 68$. Moreover

1 kgf = 9.80665 N (Fig. 4-1d), therefore $k_1 = 9.80665$, $e_1 = 1$
1 m = 39.37008 in (Fig. 4-1a), therefore $k_2 = 39.37008$, $e_2 = -2$

and hence, by (4-4),

$$x = 68 \times (9.80665^1 \times 39.37008^{-2}) = 0.43023$$

So that

$$68 \frac{\text{kgf}}{\text{m}^2} = 0.43023 \frac{\text{N}}{\text{in}^2}$$

⇑

Example 4-2

How many kg³/m$^{2.8}$ is 4 slug³/ft$^{2.8}$?
We write the problem such that

$$4 \frac{\text{slug}^3}{\text{ft}^{2.8}} = x \frac{\text{kg}^3}{\text{m}^{2.8}}, \quad \text{or} \quad 4 \cdot \text{slug}^3 \cdot \text{ft}^{-2.8} = x \cdot \text{kg}^3 \cdot \text{m}^{-2.8}$$

Thus, by (4-1) and (4-2),

$Q = 4$
1 slug = 14.5939 kg (Fig. 4-1c), therefore $k_1 = 14.5939$, $e_1 = 3$
1 ft = 0.3048 m (Fig. 4-1a), therefore $k_2 = 0.3048$, $e_2 = -2.8$

and by (4-4)

$$x = Q \cdot (k_1^{e_1} \cdot k_2^{e_2}) = 4 \times (14.5939^3 \times 0.3048^{-2.8}) = 346204.08984$$

This gives us the answer

$$4 \frac{\text{slug}^3}{\text{ft}^{2.8}} = 346204.08984 \text{ kg}^3/\text{m}^{2.8}$$

Example 4-3

How many slugs is 1 (kgf·s²)/m?

We write 1 kgf·s²/m = x slug. The slug is a *derived* unit in the American/British Force System (Art. 3.4.3) in which the *fundamental* dimensions are lbf, ft, s. Thus, we have

$$1 \frac{\text{kgf} \cdot \text{s}^2}{\text{m}} = x \frac{\text{lbf} \cdot \text{s}^2}{\text{ft}}$$

This can be written

$$1 \text{kgf} \cdot \text{s}^2 \cdot \text{m}^{-1} = x \cdot \text{lbf}^1 \cdot \text{s}^2 \cdot \text{ft}^{-1}$$

Hence, by (4-1) and (4-2),

$Q = 1$
1 kgf = 2.20462 lbf (Fig. 4-1d), therefore k_1 = 2.20462, e_1 = 1
1 s = 1 s, therefore k_2 = 1, e_2 = 2
1 m = 3.28084 ft (Fig. 4-1a), therefore k_3 = 3.28084, e_3 = −1

From which, by (4-4)

$$x = Q \cdot (k_1^{e_1} \cdot k_2^{e_2} \cdot k_3^{e_3}) = 1 \times (2.20462^1 \times 1^2 \times 3.28084^{-1}) = 0.67197$$

providing the answer

$$1 \frac{\text{kgf} \cdot \text{s}^2}{\text{m}} = 0.67197 \text{ slug}$$

This confirms the datum found in Fig. 4-1c.

Example 4-4

Newton's gravitational law states that

$$F = k \cdot \frac{M_1 \cdot M_2}{R^2}$$

where F = gravitational force attracting masses M_1, M_2 toward each other;
M_1, M_2 = masses of body 1 and body 2;
R = distance between centers of the masses;
k = universal gravitational constant.

The value of k in SI is $k_{SI} = 6.67259 \times 10^{-11} \cdot \text{m}^3/(\text{kg}\cdot\text{s}^2)$ (Appendix 2). Question: What is the value of this constant k_{US} in the *American/British Engineering System*? To answer, we must first establish the dimension of k_{US}. Using the Newton equation, and by considering that the fundamental dimensions of this system are ft, lbf, s,

$$[k_{US}] = \frac{[F]\cdot[R]^2}{[M_1]\cdot[M_2]} = \frac{\text{lbf}\cdot\text{ft}^2}{\left(\frac{\text{lbf}\cdot\text{s}^2}{\text{ft}}\right)^2} = \frac{\text{ft}^4}{\text{lbf}\cdot\text{s}^4}$$

Therefore, by (4-3), we now have to solve for x

$$6.67259 \times 10^{-11} \frac{\text{m}^3}{\text{kg}\cdot\text{s}^2} = x \cdot \frac{\text{ft}^4}{\text{lbf}\cdot\text{s}^4}$$

Rearrangement and simplification yield

$$6.67259 \times 10^{-11} \cdot \text{m}^3 \cdot \text{kg}^{-1} \cdot \text{lbf} = x \cdot \text{ft}^4 \cdot \text{s}^{-2}$$

But now we notice a disturbing fact: the number of dimensions on the left is *not* the same as on the right! Hence we cannot continue, since the number of dimensions on both sides of an equation *must* be the same, else we do not have dimensional homogeneity. The reason for this anomaly is that we are dealing with *both* mass *and* force dimensions in the same formula, which is always unhealthy. The cure lies in the elimination of one of these dimensions. Let us eliminate force. Thus, by Fig. 4-1d,

$$1 \text{ lbf} = 4.44822 \text{ m}\cdot\text{kg}\cdot\text{s}^{-2}$$

which we substitute into the above formula. This will yield

$$6.67259 \times 10^{-11} \cdot \text{m}^3 \cdot \text{kg}^{-1} \cdot (4.44822 \cdot \text{m}\cdot\text{kg}\cdot\text{s}^{-2}) = x \cdot \text{ft}^4 \cdot \text{s}^{-2}$$

which can be nicely simplified to read

$$2.96811 \times 10^{-10} \cdot \text{m}^4 = x \cdot \text{ft}^4$$

This now has the *same* number of dimensions on both sides, namely 1. By (4-1) and (4-2)

$Q = 2.96811 \times 10^{-10}$
$1 \text{ m} = 3.28084 \text{ ft}$ (Fig. 4-1a), therefore $k_1 = 3.28084$, $e_1 = 4$

Thus, by (4-4),

$$x = Q \cdot k_1^{e_1} = 2.96811 \times 10^{-10} \cdot (3.28084^4) = 3.43891 \times 10^{-8}$$

and therefore

$$6.67259 \times 10^{-11} \frac{\text{m}^3}{\text{kg}\cdot\text{s}^2} = 3.43891 \times 10^{-8} \frac{\text{ft}^4}{\text{lbf}\cdot\text{s}^4}$$

which gives us the final answer

$$k_{US} = 3.43891 \times 10^{-8} \frac{\text{ft}^4}{\text{lbf}\cdot\text{s}^4}$$

Example 4-5

What is the value of the gravitational constant k_{inch} in the in–lbf–s system? By the above example, we have

$$3.43891 \times 10^{-8} \cdot \text{ft}^4 \cdot \text{lbf}^{-1} \cdot \text{s}^{-4} = x \cdot \text{in}^4 \cdot \text{lbf}^{-1} \cdot \text{s}^{-4}$$

Simplification by $\text{lbf}^{-1} \cdot \text{s}^{-4}$, which appears on both sides, yields

$$3.43891 \times 10^{-8} \cdot \text{ft}^4 = x \cdot \text{in}^4$$

Hence, by (4-1), $Q = 3.43891 \times 10^{-8}$, and by (4-2)

$$1 \text{ ft} = 12 \text{ in, therefore } k_1 = 12, \ e_1 = 4$$

So that, by (4-4),

$$x = Q \cdot k_1^{e_1} = 3.43891 \times 10^{-8} \cdot (12^4) = 7.13093 \times 10^{-4}$$

and therefore, the answer is

$$k_{\text{inch}} = 7.13093 \times 10^{-4} \, \frac{\text{in}^4}{\text{lbf} \cdot \text{s}^4}$$

⇑

Example 4-6

A man's mass is M kg. What is his weight in *newton* and in kgf units?
By Newton's second law

$$G_N = M \cdot g$$

where G_N is the weight expressed in newtons. By definition, 1 kgf (Note 118 in Appendix 4) is the force necessary to accelerate 1 kg mass at the rate of g m/s². Thus, 1 kgf = g N, whence 1 N = $1/g$ kgf. It therefore follows that the weight in terms of kgf is

$$G_{\text{kgf}} = G_N \cdot \frac{1}{g} = (M \cdot g) \cdot \frac{1}{g} = M$$

Consequently, the important result is that the *weight* of a body expressed in *kgf* is numerically equal to the *mass* of that body expressed in *kg*. So, if our man has a mass of 75 kg, then he weighs exactly 75 kgf which in SI is $75 \cdot g = 75 \times 9.80665 = 735.49875$ newton.

⇑

Example 4-7

Express the quantity $608.4 \, \dfrac{\text{kg}^4 \cdot \text{ft}^{3.9}}{\text{slug}^{0.7} \cdot (\text{light-year})^{1/30}}$ in SI.

It is immediately apparent that here we have the dimensions of mass and length only, since "kg" and "slug" are mass dimensions, and "ft" and "light-year" are length dimensions. Therefore by (4-1)

$$608.4 \text{ kg}^4 \cdot \text{ft}^{3.9} \cdot \text{slug}^{-0.7} \cdot \text{ly}^{-1/30} = x \cdot \text{kg}^4 \cdot \text{m}^{3.9} \cdot \text{kg}^{-0.7} \cdot \text{m}^{-1/30}$$

Note that both dimensions "kg" and "m" occur twice on the right side. This is of course entirely permissible, although at the end (not now!) simplification can be done (if we sim-

78 APPLIED DIMENSIONAL ANALYSIS AND MODELING

plified the right side now, then the corresponding exponents would not be identical, and the number of dimensions on the left and right sides would be different). Following the established procedure, we write

Q = 608.4
1 kg = 1 kg, therefore $k_1 = 1$, $e_1 = 4$
1 ft = 0.3048 m, therefore $k_2 = 0.3048$, $e_2 = 3.9$
1 slug = 14.5939 kg (Fig. 4-1c), therefore $k_3 = 14.59399$, $e_3 = -0.7$
1 light-year = 9.46053 × 10^{15} m (Fig. 4-1a), therefore $k_4 = 9.46053 \times 10^{15}$, $e_4 = -1/30$

By (4-4) now,

$$x = Q \cdot (k_1^{e_1} \cdot k_2^{e_2} \cdot k_3^{e_3} \cdot k_4^{e_4}) = (608.4) \cdot ((1)^4 \cdot (0.3048)^{3.9} \cdot (14.5939)^{-0.7} \cdot (9.46053 \times 10^{15})^{-1/30})$$
$$= 0.26570$$

and hence, the solution for this problem is

$$608.4 \ \frac{\text{kg}^4 \cdot \text{ft}^{3.9}}{\text{slug}^{0.7} \cdot \text{ly}^{1/30}} = 0.26570 \ \frac{\text{kg}^4 \cdot \text{m}^{3.9}}{\text{kg}^{0.7} \text{m}^{1/30}} = 0.26570 \ \text{kg}^{3.3} \cdot \text{m}^{58/15}$$

where on the extreme right the expression is in the required SI.

⇑

Example 4-8. Strain Energy in a Cantilever Loaded Laterally by a Concentrated Force

A cantilever is loaded at its free end by a concentrated lateral force P (Fig. 4-4). The *stored energy U* in the deformed beam is equal to

$$U = 5.5556 \times 10^{-9} \cdot \frac{P^2 \cdot L^3}{I}$$

where L is the length, and I is the second moment of area of the cross-section about the centroidal axis $n-n$, as illustrated in the figure. In the formula, all dimensions are in the *American/British Engineering System* (Art. 3.4.3); thus the dimensions of P, L, I, and U

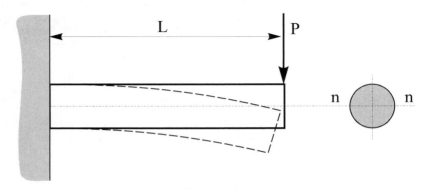

Figure 4-4
Cantilever loaded laterally at its free end by concentrated force

are lbf, in, in⁴ and lbf·in, respectively. It is now desired to modify the formula so that it yields correct results when all quantities are expressed in SI. In other words, we want the value of the dimensional factor in SI.

We *could* proceed by converting all the variables to SI dimensions and then determining the value of the dimensional factor. However, this would be a *laborious* route to follow; a better way is to use a neat "trick" that saves considerable effort in this type of problem. From the formula, the dimension of the factor is

$$[\text{factor}] = \frac{(\text{lbf·in})\cdot(\text{in})^4}{(\text{lbf})^2\cdot(\text{in})^3} = \frac{\text{in}^2}{\text{lbf}}$$

Therefore we can write

$$5.5556 \times 10^{-9} \text{ in}^2\cdot\text{lbf}^{-1} = x\cdot\text{m}^2\cdot\text{N}^{-1}$$

Thus, by (4-1) and (4-2),

$Q = 5.5556 \times 10^{-9}$
1 in $= 0.0254$ m (Fig. 4-1a), therefore $k_1 = 0.0254$, $e_1 = 2$
1 lbf $= 4.44822$ N (Fig. 4-1d), therefore $k_2 = 4.44822$, $e_2 = -1$

So, by (4-4),

$$x = Q\cdot k_1^{e_1}\cdot k_2^{e_2} = (5.5556 \times 10^{-9})\cdot(0.0254)^2\cdot(4.44822)^{-1} = 8.05772 \times 10^{-13}$$

Therefore our strain–energy formula in SI will be

$$U = (8.05772 \times 10^{-13})\cdot\frac{P^2\cdot L^3}{I}$$

in which U, P, L, and I are of the dimensions joule, newton, meter, and meter⁴, respectively. Note that this method obviated the need of converting *all* the variables appearing in the formula, for only *one* quantity (the dimensional factor) had to be converted.

Example 4-9

Suppose somebody derives an *empirical* formula for the deflection of the cantilever seen in Example 4-8. The formula is

$$f = k\cdot\frac{P\cdot L^3}{I\cdot E}$$

where the dimensions of the variables are as follows:

variable	symbol	dimension
deflection	f	in
force	P	lbf
length	L	in
Young's modulus	E	lbf/in²
second moment of cross-section	I	in⁴

The factor is *experimentally* found to be $k = 1/3$. Question: What is the numerical value of this factor, if all variables are expressed in SI? To answer, we express k from the above formula

$$k = \frac{f \cdot I \cdot E}{P \cdot L^3}$$

Therefore the dimension of k is

$$[k] = \frac{\text{in} \cdot \text{in}^4 \cdot \left(\frac{\text{lbf}}{\text{in}^2}\right)}{\text{lbf} \cdot \text{in}^3} = 1$$

i.e., k is a *dimensionless* number and hence its numerical value is *unaffected* by the change of dimensional system. Consequently, the formula expressing the deflection remains *unchanged*; in SI the same formula is valid with an identical numerical value for factor k. This is an important characteristic of a dimensionally correct formula; if a factor in it is dimensionless (has a dimension of 1), then its numerical value is *invariant* with regard to the dimensional system used.

⇑

Example 4-10. Fuel Consumption of Cars

An advertisement purports that a European car's fuel consumption is (only) $C_m = 6.2$ liter/(100·km), where the subscript "m" stands for "metric." What is this car's "mileage," C_{US}, expressed in the dimension "miles/USgallon"?

One immediately sees that the fuel consumption expressed in the European and American systems are *inversely* proportional; if one increases, the other decreases. To circumvent this "complication," let us temporarily consider the *reciprocal* of the dimension of the American fuel consumption. Accordingly, we set up our equation as follows:

$$C_m \frac{\text{liter}}{100 \cdot \text{km}} = x \frac{\text{gallon}}{\text{mile}}$$

where x is the *reciprocal* of fuel consumption in the American system. We write this formula in the standard form of (4-1)

$$C_m \text{ liter} \cdot (100 \cdot \text{km})^{-1} = x \text{ gallon} \cdot (\text{mile})^{-1}$$

Thus, by (4-2),

$Q \quad = C_m$

1 liter $= 0.26417$ gallon (Fig. 4-1b), therefore $k_1 = 0.26417$, $e_1 = 1$
100 km $= 62.13712$ mile (Fig. 4-1/a), therefore $k_2 = 62.13712$, $e_2 = -1$

So, by (4-4),

$$x = Q \cdot k_1^{e_1} \cdot k_2^{e_2} = C_m \cdot (0.26417)^1 \cdot (62.13712)^{-1} = C_m \cdot 4.25140 \times 10^{-3}$$

But x is the reciprocal of C_{US}, thus

$$C_{US} = \frac{1}{x} = \frac{1}{C_m \cdot (4.25140 \times 10^{-3})} = \frac{235.21663}{C_m} \qquad (a)$$

Therefore, if for our European car the fuel consumption is $C_m = 6.2$ liter/(100·km), then this car's "mileage" is

$$C_{US} = \frac{235.21663}{6.2} = 37.938 \frac{\text{mile}}{\text{gallon}}$$

a rather acceptable—and rare—figure.

⇑

Example 4-11

How would relation (a) of the previous example be modified if we use UK gallon, instead of US gallon (see Fig 4-1b)? In this case (4-1) will be written

$$C_{UK} \frac{\text{mile}}{\text{UKgallon}} = x \cdot \frac{\text{mile}}{\text{USgallon}}$$

or

$$C_{UK} \cdot \text{mile} \cdot \text{UKgallon}^{-1} = x \cdot \text{mile} \cdot \text{USgallon}^{-1}$$

Hence, by (4-2),

$$Q = C_{UK}$$

1 UKgallon = 1.20095 USgallon (Fig. 4-1b), therefore $k_1 = 1.20095$, $e_1 = -1$

Next, by (4-4),

$$x = Q \cdot k_1^{e_1} = C_{UK} \cdot (1.20095^{-1}) = C_{UK} \cdot (0.83267)$$

But x is clearly C_{US}, thus $C_{US} = 0.83267 \cdot C_{UK}$, whence

$$C_{UK} = \frac{1}{0.83267} \cdot C_{US} = 1.20095 \cdot C_{US} \qquad \text{(a)}$$

So, our 37.938 mile/USgallon "mileage" (see previous example) would become 45.562 mile/UKgallon in Canada, if Canada still used the Imperial system to measure fuel*—but she does not; Canada now uses the metric system for fuel.

Of course, the same result could have been obtained much more simply by a little "heuristic meditation": since the UK gallon is 1.20095 times the US gallon, therefore the distance traveled on a UK gallon must be 1.20095 times longer; hence the laboriously derived relation (a). The reasons why we followed the formal method were to illustrate the method, and also to show that the technique could be used rather automatically, thereby providing less chance for error.

⇑

Example 4-12a

The value of Young's modulus—also known as the *modulus of elasticity*—of a particular steel is $E = 3 \times 10^7$ lbf/in². What is the value of E in SI?

---

*The Canadian gallon is not *exactly* the same as the UK gallon; 1 UKgallon = 1.0000043994 CANgallon.

82 APPLIED DIMENSIONAL ANALYSIS AND MODELING

We set up our equation as per (4-1):

$$3 \times 10^7 \text{ lbf·in}^{-2} = x \cdot \text{N·m}^{-2}$$

where, by (4-2),

$Q = 3 \times 10^7$
1 lbf = 4.44822 N (Fig. 4-1d), therefore $k_1 = 4.44822$, $e_1 = 1$
1 in = 0.0254 m (Fig. 4-1a), therefore $k_2 = 0.0254$, $e_2 = -2$

So, by (4-4),

$$x = Q \cdot k_1^{e_1} \cdot k_2^{e_2} = (3 \times 10^7) \cdot (4.44822)^1 \cdot (0.0254)^{-2} = 2.06843 \times 10^{11}$$

Therefore

$$3 \times 10^7 \text{ lbf/in}^2 = 2.06843 \times 10^{11} \text{ Pa} = 206.843 \text{ GPa}$$

⇑

Example 4-12/b

A parameter frequently used in fluid dynamics is the *Prandtl number*, Pr. It is defined as

$$\text{Pr} = \frac{c \cdot \mu}{k} \qquad (a)$$

where the variables for water at 60 °F in the American/British Mass System are as follows:

Variable	Symbol	Value	Dimension
Specific heat capacity	c	1	BTU/(lb$_m$·°F)
Dynamic viscosity	μ	2.49	lb$_m$/(ft·h)
Thermal conductivity	k	343.5	BTU/(h·ft·°F)

Therefore Pr, by (a) and the data in the above table, is

$$\text{Pr} = \frac{(1) \cdot (2.49)}{343.5} \frac{(\text{BTU}/(\text{lb}_m \cdot °\text{F})) \cdot (\text{lb}_m/(\text{ft·h}))}{\text{BTU}/(\text{h·ft·°F})} = 0.00725$$

Notice the absence of a dimension on the extreme right of this expression. This is because all dimensions cancel out, so the Prandtl number is dimensionless, and as such—by this characteristic alone—its numerical value is independent of the dimensional system in which it is expressed. Let us verify that this is indeed so by expressing all these variables in another system—say SI—and then substituting those values back into the original defining formula (a).

Accordingly, using the format (4-1) for *specific heat capacity c* we write

$$1 \frac{\text{BTU}}{\text{lb}_m \cdot °\text{F}} = x \frac{\text{J}}{\text{kg·K}}$$

so that, by (4-1) and (4-2),

$Q = 1$
1 BTU = 1055.05585 J (Fig. 4-1e), therefore $k_1 = 1055.05585$, $e_1 = 1$

1 lb$_m$ = 0.45359 kg (Fig. 4-1c), therefore $k_2 = 0.45359$, $e_2 = -1$
1 °F (interval) = $\frac{5}{9}$ K (interval; Fig. 4-2b), therefore $k_3 = \frac{5}{9}$, $e_3 = -1$

Consequently, by (4-4),

$$x = Q \cdot k_1^{e_1} \cdot k_2^{e_2} \cdot k_3^{e_3} = (1055.05585)^1 \cdot (0.45359)^{-1} \cdot \left(\frac{5}{9}\right)^{-1} = 4186.8$$

yielding in SI

$$c = 4186.8 \, \frac{J}{kg \cdot K} \tag{b}$$

For the *dynamic viscosity* μ we have

$$2.49 \, \frac{lb_m}{ft \cdot h} = x \cdot \frac{kg}{m \cdot s}$$

so that, by (4-1) and (4-2)

Q = 2.49
1 lb$_m$ = 0.45359 kg (Fig. 4-1c), therefore $k_1 = 0.45359$, $e_1 = 1$
1 ft = 0.3048 m (Fig. 4-1a), therefore $k_2 = 0.3048$, $e_2 = -1$
1 h = 3600 s (Fig. 3-14), therefore $k_3 = 3600$, $e_3 = -1$

By (4-4), then

$$x = Q \cdot k_1^{e_1} \cdot k_2^{e_2} \cdot k_3^{e_3} = (2.49) \cdot (0.45359)^1 \cdot (0.3048)^{-1} \cdot (3600)^{-1} = 1.02931 \times 10^{-3}$$

and hence

$$\mu = 1.02931 \times 10^{-3} \, \frac{kg}{m \cdot s} \tag{c}$$

Finally, for *thermal conductivity k*, we write

$$343.5 \, \frac{BTU}{h \cdot ft \cdot °F} = x \cdot \frac{J}{s \cdot m \cdot K}$$

and hence, following our usual routine

Q = 343.5
1 BTU = 1055.05585 J (Fig. 4-1e), therefore $k_1 = 1055.05585$, $e_1 = 1$
1 h = 3600 s (Fig. 3-14), therefore $k_2 = 3600$, $e_2 = -1$
1 ft = 0.3048 m (Fig. 4-1a), therefore $k_3 = 0.3048$, $e_3 = -1$
1 °F(interval) = (5/9) K (interval), therefore $k_4 = 5/9$, $e_4 = -1$

for which we obtain, by (4-4),

$$x = Q \cdot k_1^{e_1} \cdot k_2^{e_2} \cdot k_3^{e_3} \cdot k_4^{e_4} = (343.5) \cdot (1055.05585)^1 \cdot (3600)^{-1} \cdot (0.3048)^{-1} \cdot \left(\frac{5}{9}\right)^{-1} = 594.50736$$

Therefore,

$$k = 594.50736 \, \frac{J}{s \cdot m \cdot K} \tag{d}$$

Finally, the above three quantities in (b), (c), and (d)—all expressed in SI—are substituted back into the original expression (a)

$$\Pr = \frac{c \cdot \mu}{k} = \frac{(4186.8) \cdot (1.02931 \times 10^{-3})}{594.50736} \cdot \frac{(J/(kg \cdot K)) \cdot (kg/(m \cdot s))}{J/(s \cdot m \cdot K)} = 0.00725$$

and again all dimensions cancel out (as expected) and the value is 0.00725, which is *identical* (as also expected) to the previously calculated one in the American/British Mass System.

As already mentioned, the fact that the value of a dimensionless variable (number) is independent of the dimensional system used is entirely general. For *all* dimensionless numbers, variables, and parameters have this property, and this is why these numbers are so useful.

A list of the more important dimensionless numbers and parameters is given in Appendix 3; a more comprehensive collection of these quantities can be found in Ref. 12.

⇑

Example 4-13. Lengths vs. Weights of Dolphins

A. V. Hill (Ref. 112, p. 216) published an experimentally determined relation between the lengths and weights of dolphins. His formula—in a slightly modified form—is

$$W = (3.613 \times 10^{-4}) \cdot L^3 \qquad (a)$$

where L is the *length* of the dolphin in *inches,* and W is the *weight* of the dolphin in *pounds.* What is the magnitude of the dimensional constant on the right-hand side of (a) in SI?

We first emphatically note that in the formula "pound" is really "pound-force" (lbf), since we are dealing with *weight,* which is *force.*

We have the dimension of the constant in (a)

$$[3.613 \times 10^{-4}] = \left[\frac{W}{L^3}\right] = \frac{[W]}{[L]^3} = \frac{\text{lbf}}{\text{in}^3}$$

Hence, by (4-1),

$$3.613 \times 10^{-4} \text{ lbf} \cdot \text{in}^{-3} = x \cdot \text{kgf} \cdot \text{m}^{-3}$$

and by (4-2) and (4-3)

1 lbf = 0.45359 kgf, therefore $k_1 = 0.45359$, $e_1 = 1$
1 in = 0,0254 m, therefore $k_2 = 0.0254$, $e_2 = -3$

So that, by (4-4),

$$x = (3.613 \times 10^{-4}) \cdot k_1^{e_1} \cdot k_2^{e_2} = (3.613 \times 10^{-4}) \cdot (0.45359)^1 \cdot (0.0254)^{-3} = 10$$

Therefore

$$W = 10 \cdot L^3 \qquad (b)$$

where W is the weight in kgf, and L is the length in m. Next, we consider that *numerically,* the weight expressed in kgf is the same as the mass expressed in kg (see Example 4-6). Consequently, (b) is really

$$M = 10 \cdot L^3 \qquad (c)$$

where M is the *mass* of the dolphin in kg, and L is its length in m. For example, an $L = 2$ m long dolphin's mass is about $M = 80$ kg, i.e., it is 80 kgf (= 176.4 lbf) heavy.

⇑

Example 4-14. End Slope of a Simply Supported Beam Subjected to Distributed Load

A simply supported beam of uniform circular cross-section loaded only by its own weight assumes the elastic shape shown in Fig 4-5. For a given material and terrestrial location, the slopes at the ends (identical because of symmetry) are determined experimentally and found to be in SI

$$\varphi = 2.56 \times 10^{-7} \cdot \frac{L^3}{D^2} \text{ radian} \quad \text{(a)}$$

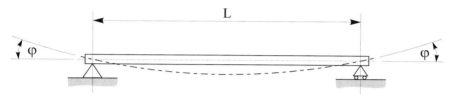

Figure 4-5
Deformed shape of a simply supported uniform beam loaded by its own weight

where D is the diameter of the uniform circular cross-section. Find the value of the dimensional factor on the right, if the system used has the following fundamental dimensions:

mass   = M_U, mass of the universe
length = m, meter
time   = T_U, age of the universe

Let us designate this system SU (System Universe). The dimension of the numeric factor in (a) is

$$[2.56 \times 10^{-7}] = \frac{[\varphi] \cdot [D]^2}{[L]^3} = \frac{1 \cdot m^2}{m^3} = \frac{1}{m} \quad \text{(b)}$$

We see that the dimension of this factor does *not* include either mass, or time, and therefore these dimensions have no influence on this number. Now since the number 2.56×10^{-7} as originally given in SI, in which the dimension of length is m, and in SU the dimension of length is *still* m, therefore our factor in SU is the *same* as it is in SI. This finding is mildly astonishing—almost contrary to intuition—since the ingredients of this factor must involve gravitational acceleration (hence time), and gravitational force (hence mass). So why do these fundamental dimensions not show up? The answer of course is that they cancel each other out.

⇑

Example 4-15. An Unusual Dimensional System (I)

Three of the universal constants of nature are the *universal gravitational constant*, k, the *speed of light* in vacuum, c, and the *Planck's constant*, h. Their numerical values rounded to five significant digits in SI are as follows (see also Appendix 2):

$$k = 6.67259 \times 10^{-11} \, \frac{m^3}{kg \cdot s^2} \quad \text{(universal gravitational constant)}$$

$$c = 2.99792 \times 10^{8} \, \frac{m}{s} \quad \text{(speed of light)}$$

$$h = 6.62608 \times 10^{-34} \, \frac{m^2 \cdot kg}{s} \quad \text{(Planck's constant)}$$

However, SI is not the only system possible. For example, we can concoct a system in which the numerical values of all of the above three universal constants are 1. The question is now: What will be the dimensions of mass, length, and time in this *new* system in terms of the dimensions of SI?

For the sake of expediency, let us denote these new dimensions of mass, length, and time by the symbols x, y, z, and the old magnitudes by μ_k, μ_c, μ_h (for example, $\mu_c = 2.99792 \times 10^8$). We can therefore write

$$\mu_k \cdot m^3 \cdot kg^{-1} \cdot s^{-2} = 1 \cdot x^3 \cdot y^{-1} \cdot z^{-2}$$

$$\mu_c \cdot m \cdot s^{-1} = 1 \cdot x \cdot z^{-1}$$

$$\mu_h \cdot m^2 \cdot kg \cdot s^{-1} = 1 \cdot x^2 \cdot y \cdot z^{-1}$$

where the coefficient 1 is written out on the right side of the formulas to emphasize the fact that the magnitude of these constants in the new system is uniformly 1. By the above, then

$$\mu_k = \left(\frac{x}{m}\right)^3 \cdot \left(\frac{y}{kg}\right)^{-1} \cdot \left(\frac{z}{s}\right)^{-2}$$

$$\mu_c = \left(\frac{x}{m}\right) \cdot \left(\frac{z}{s}\right)^{-1}$$

$$\mu_h = \left(\frac{x}{m}\right)^2 \cdot \left(\frac{y}{kg}\right) \cdot \left(\frac{z}{s}\right)^{-1}$$

This nonlinear simultaneous equation system can be easily reduced to a linear system by taking logarithms. Therefore

$$\ln \mu_k = 3 \cdot \ln \frac{x}{m} - \ln \frac{y}{kg} - 2 \cdot \ln \frac{z}{s}$$

$$\ln \mu_c = \ln \frac{x}{m} - \ln \frac{z}{s}$$

$$\ln \mu_h = 2 \cdot \ln \frac{x}{m} + \ln \frac{y}{kg} - \ln \frac{z}{s}$$

Now this can be conveniently written

$$\begin{bmatrix} \ln \mu_k \\ \ln \mu_c \\ \ln \mu_h \end{bmatrix} = \mathbf{T} \cdot \begin{bmatrix} \ln(x/m) \\ \ln(y/kg) \\ \ln(z/s) \end{bmatrix} \qquad (a)$$

where

$$\mathbf{T} = \begin{bmatrix} 3 & -1 & -2 \\ 1 & 0 & -1 \\ 2 & 1 & -1 \end{bmatrix} \quad (b)$$

is the *transformation matrix* between the two dimensional systems. From the above then

$$\begin{bmatrix} \ln(x/m) \\ \ln(y/kg) \\ \ln(z/s) \end{bmatrix} = \mathbf{T}^{-1} \cdot \begin{bmatrix} \ln \mu_k \\ \ln \mu_c \\ \ln \mu_h \end{bmatrix} \quad (c)$$

It is thus obvious that the new system—based on the fundamental dimensions x, y, z—can exist *only* if the transformation matrix $\mathbf{T}$ is *invertible*, i.e., nonsingular. This means that the determinant of matrix $\mathbf{T}$ must not be zero. In our case $|\mathbf{T}| = 2 \neq 0$, hence the desired new system is possible, and we can continue.

We now have, by (b),

$$\mathbf{T}^{-1} = \begin{bmatrix} 3 & -1 & -2 \\ 1 & 0 & -1 \\ 2 & 1 & -1 \end{bmatrix}^{-1} = \begin{bmatrix} 0.5 & -1.5 & 0.5 \\ -0.5 & 0.5 & 0.5 \\ 0.5 & -2.5 & 0.5 \end{bmatrix} \quad (d)$$

Taking (d) and substituting it into (c), together with the other relevant values, we obtain

$$\begin{bmatrix} \ln(x/m) \\ \ln(y/kg) \\ \ln(z/s) \end{bmatrix} = \begin{bmatrix} 0.5 & -1.5 & 0.5 \\ -0.5 & 0.5 & 0.5 \\ 0.5 & -2.5 & 0.5 \end{bmatrix} \cdot \begin{bmatrix} -23.43043 \\ 19.51860 \\ -76.39688 \end{bmatrix} = \begin{bmatrix} -79.19155 \\ -16.72393 \\ -98.71015 \end{bmatrix}$$

from which

$$\left. \begin{array}{l} x = e^{-79.19155} \text{ m} = 4.05084 \times 10^{-35} \text{ m} \\ y = e^{-16.72393} \text{ kg} = 5.45621 \times 10^{-8} \text{ kg} \\ z = e^{-98.71015} \text{ s} = 1.35122 \times 10^{-43} \text{ s} \end{array} \right\} \quad (e)$$

meaning, for example, the mass' new dimension y is 5.45621×10^{-8} kg, etc. While this new system might be popular with nuclear scientists and with people in the cosmology business, it would probably be less welcome among ordinary mortals. For example, Mrs. Smith would have to say to the grocery clerk "Give me 8,313,323.2 y bread"—if she wants to buy 1 lb of this commodity. Or one might hear the following pleasant exchange: "How tall are you, Sir?" "Oh, I am just over 4.70272×10^{34} x." And if he is 6 ft 3 in, he would be right.

⇑

Example 4-16. An Unusual Dimensional System (II)

We now plan to create a dimensional system in which the numerical values of standard atmospheric pressure p and gravitational constant k are unity. The values of these constants in SI are

$$p = 101325 \, \frac{\text{kg}}{\text{m·s}^2} \quad \text{(standard atmospheric pressure)}$$

$$k = 6.67259 \times 10^{-11} \frac{\text{m}^3}{\text{kg·s}^2} \quad \text{(gravitational constant)}$$

Following the process developed in Example 4-15, we write

$$\mu_p \cdot m^{-1} \cdot kg \cdot s^{-2} = 1 \cdot x^{-1} \cdot y \cdot z^{-2}$$

$$\mu_k \cdot m^3 \cdot kg^{-1} \cdot s^{-2} = 1 \cdot x^3 \cdot y^{-1} \cdot z^{-2}$$

where x, y, z are the *new* dimensions of length, mass and time, respectively, while μ_p and μ_k are the respective *old* dimensions. By the above, then

$$\mu_p = \left(\frac{x}{m}\right)^{-1} \cdot \left(\frac{y}{kg}\right) \cdot \left(\frac{z}{s}\right)^{-2}$$

$$\mu_k = \left(\frac{x}{m}\right)^{3} \cdot \left(\frac{y}{kg}\right)^{-1} \cdot \left(\frac{z}{s}\right)^{-2}$$

from which

$$\ln \mu_p = -\ln \frac{x}{m} + \ln \frac{y}{kg} - 2 \cdot \ln \frac{z}{s}$$

$$\ln \mu_k = 3 \cdot \ln \frac{x}{m} - \ln \frac{y}{kg} - 2 \cdot \ln \frac{z}{s}$$

which can be written

$$\begin{bmatrix} \ln \mu_p \\ \ln \mu_k \end{bmatrix} = \mathbf{T} \cdot \begin{bmatrix} \ln(x/m) \\ \ln(y/kg) \\ \ln(z/s) \end{bmatrix} \quad (a)$$

where

$$\mathbf{T} = \begin{bmatrix} -1 & 1 & -2 \\ 3 & -1 & -2 \end{bmatrix} \quad (b)$$

is the *transformation matrix* between the original and the new dimensional systems. We now observe that the *inverse* of **T** does not exist since **T** is not a square matrix. This always happens if the number of new dimensions (in this case three) exceeds the number of defining equations (in this case two). Therefore, in this instance, *one* of the new variables (i.e., one of x, y, z) can be freely chosen and then the other two can be expressed as a function of this preselected first one. We therefore proceed by writing our formula (a) as:

$$\begin{bmatrix} \ln \mu_p \\ \ln \mu_k \end{bmatrix} = \begin{bmatrix} -1 & 1 \\ 3 & -1 \end{bmatrix} \cdot \begin{bmatrix} \ln(x/m) \\ \ln(y/kg) \end{bmatrix} + \begin{bmatrix} -2 \\ -2 \end{bmatrix} \cdot \ln \frac{z}{s}$$

As the reader can easily verify, this form is equivalent to our original relation (a). Therefore

$$\begin{bmatrix} -1 & 1 \\ 3 & -1 \end{bmatrix} \cdot \begin{bmatrix} \ln(x/m) \\ \ln(y/kg) \end{bmatrix} = \begin{bmatrix} \ln \mu_p \\ \ln \mu_k \end{bmatrix} - \begin{bmatrix} -2 \\ -2 \end{bmatrix} \cdot \ln \frac{z}{s}$$

which immediately yields

$$\begin{bmatrix} \ln(x/m) \\ \ln(y/kg) \end{bmatrix} = \begin{bmatrix} -1 & 1 \\ 3 & -1 \end{bmatrix}^{-1} \cdot \left(\begin{bmatrix} \ln \mu_p \\ \ln \mu_k \end{bmatrix} - \begin{bmatrix} -2 \\ -2 \end{bmatrix} \cdot \ln \frac{z}{s} \right)$$

Inverting the matrix on the right, and substituting the relevant μ_p, μ_k numerical values, we obtain

$$\begin{bmatrix} \ln(x/\text{m}) \\ \ln(y/\text{kg}) \end{bmatrix} = \begin{bmatrix} -5.95217 \\ 5.57392 \end{bmatrix} - \begin{bmatrix} -2 \\ -4 \end{bmatrix} \cdot \ln\frac{z}{\text{s}}$$

from which

$$\frac{x}{\text{m}} = e^{-5.95217} \cdot \left(\frac{z}{\text{s}}\right)^2 = 2.60019 \times 10^{-3} \cdot \left(\frac{z}{\text{s}}\right)^2$$

$$\frac{y}{\text{kg}} = e^{5.57392} \cdot \left(\frac{z}{\text{s}}\right)^4 = (263.46486) \cdot \left(\frac{z}{\text{s}}\right)^4$$

We can now choose any value for z/s, and then calculate the corresponding ratios x/m and y/kg. It therefore follows that an *infinite* number of new systems can be created, all satisfying the stated criteria. For example, let the time dimension of the new system be 1 min. Then $z = 60$ s, and therefore $z/\text{s} = 60$. With this value now, the above two equations supply

$$\frac{x}{\text{m}} = 2.60019 \times 10^{-3} \cdot 60^2 = 9.36069$$

$$\frac{y}{\text{kg}} = (263.46486) \cdot 60^4 = 3.41450 \times 10^9$$

thus the *new* dimensions of length, mass and time will be

$x = 9.36069$ m   (length)
$y = 3.41450 \times 10^9$ kg   (mass)
$z = 60$ s   (time)

⇑

Example 4-17. An Impossible Dimensional System

We wish to create a dimensional system in which the numerical values of the universal gravitational constant k, the mass of electron m_e, the Planck's constant h, and the speed of light in vacuum are all unity. In SI these constants are

$k = 6.67259 \times 10^{-11} \dfrac{\text{m}^3}{\text{kg} \cdot \text{s}^2}$ (gravitational constant)

$m_e = 9.10939 \times 10^{-31}$ kg   (mass of electron)

$h = 6.62608 \times 10^{-3} \dfrac{\text{m}^2 \cdot \text{kg}}{\text{s}}$ (Planck's constant)

$c = 2.99792 \times 10^8 \dfrac{\text{m}}{\text{s}}$ (speed of light)

If the new dimensions of length, mass, and time are x, y, and z, respectively, then we can write

$$\left.\begin{array}{l}\mu_k \cdot m^3 \cdot kg^{-1} \cdot s^{-2} = 1 \cdot x^3 \cdot y^{-1} \cdot z^{-2} \\ \mu_{m_e} \cdot kg = 1 \cdot y \\ \mu_h \cdot m^2 \cdot kg \cdot s^{-1} = 1 \cdot x^2 \cdot y \cdot z^{-1} \\ \mu_c \cdot m \cdot s^{-1} = 1 \cdot x \cdot z^{-1}\end{array}\right\} \quad (a)$$

where μ_k, μ_{m_e}, μ_h, and μ_c are the *magnitudes* of the constants k, m_e, h, and c, respectively, in SI. By the above, then

$$\mu_k = \left(\frac{x}{m}\right)^3 \cdot \left(\frac{y}{kg}\right)^{-1} \cdot \left(\frac{z}{s}\right)^{-2}$$

$$\mu_{m_e} = \left(\frac{y}{kg}\right)$$

$$\mu_h = \left(\frac{x}{m}\right)^2 \cdot \left(\frac{y}{kg}\right) \cdot \left(\frac{z}{s}\right)^{-1}$$

$$\mu_c = \left(\frac{x}{m}\right) \cdot \left(\frac{z}{s}\right)^{-1}$$

Taking logarithms of both sides

$$\left.\begin{array}{l}\ln \mu_k = 3 \cdot \ln\dfrac{x}{m} - \ln\dfrac{y}{kg} - 2 \cdot \ln\dfrac{z}{s} \\[4pt] \ln \mu_{m_e} = \ln\dfrac{y}{kg} \\[4pt] \ln \mu_h = 2 \cdot \ln\dfrac{x}{m} + \ln\dfrac{y}{kg} - \ln\dfrac{z}{s} \\[4pt] \ln \mu_c = \ln\dfrac{x}{m} - \ln\dfrac{z}{s}\end{array}\right\} \quad (b)$$

which, following the process used in Examples 4-14 and 4-15, can be written

$$\begin{bmatrix}\ln\mu_k \\ \ln\mu_{m_e} \\ \ln\mu_h \\ \ln\mu_c\end{bmatrix} = \mathbf{T} \cdot \begin{bmatrix}\ln(x/m) \\ \ln(y/kg) \\ \ln(z/s)\end{bmatrix}$$

where

$$\mathbf{T} = \begin{bmatrix} 3 & -1 & -2 \\ 0 & 1 & 0 \\ 2 & 1 & -1 \\ 1 & 0 & -1 \end{bmatrix}$$

is the *transformation matrix*. But now we see that the *inverse* of this matrix does not exist because it is not a square matrix. Therefore the system has a solution only if the *augmented* matrix and matrix **T** have the same rank (see Art. 1.4.2). However, this is not the case in the present application since the rank of **T** is 3, whereas the augmented matrix has the rank of 4 [this finding is strictly concordant with the well-known rule that states that, generally, if the number of conditions (equations) in a linear system—such as (b) in our case—exceeds the number of unknowns, then there is no solution]. Hence system (a), and thus (b), have no solution for x, y, and z. Consequently, it is *impossible* to create a dimensional system in which the stated 4 constants of nature, viz. k, m_e, h, and c, would be unity.

⇑

4.4. PROBLEMS

Solutions are in Appendix 6.

4/1 The value of Planck's constant in SI is $h = 6.6260755 \times 10^{-34}$ J·s (Appendix 2). What is h in (a) the *American/British Mass System* (Art. 3.4.4); (b) the *American/British Force (Engineering) System* (Art. 3.4.3)?

4/2 The mechanical equivalent of heat is $q = 4.18680$ J/cal. What is the value of q if the dimensions of force, length and energy are lbf, in and BTU, respectively?

4/3 Writers' efficiency.

The efficiency η of a writer may be defined as the ratio of his word-count to his energy (food) intake, in units of m·kgf, needed to produce written material. On the other hand, he could be judged by his *z factor*, defined as the ratio of his physiologically expended energy, in units of calorie, to the average number of question marks per page in a manuscript. It is found that, on average, a writer produces 3.8 question marks and 662.5 words per page. Derive a relation between this writers' efficiency η and his z factor.

4/4 The dimension of *dynamic viscosity* in SI is kg/(m·s). What is its dimension in the MKS system (Art. 3.4.2)?

4/5 Quantity Z is defined as $Z = 40$ ft$^{3.3}$·s$^{2.2}$/lbf$^{1.1}$. Express Z in SI.

4/6 The mass of a body is $M = 1/250$ ton·year^2/parsec. Express M in SI.

4/7 Cost to build a medicinal bathhouse.

James B. S. Broggg (correctly spelled), the eminent balneologist, developed an *empirical* formula for the cost of building medicinal bathhouses. His formula is

$$C = \left(39 \cdot \left(\frac{A}{\lambda}\right)^{2.2} + T^{0.7} \cdot M^{1.9} \cdot \frac{37}{50} - \frac{10^{16}}{n^2 \cdot w^{1.1}}\right)^{0.52}$$

where C = cost in U.S. dollars to build the bathhouse;
A = area of the main basin, in ft^2;
T = time in days for the planning committee to decide whether to build the bathhouse;
M = total mass of bricks used in building the bathhouse, in pounds;
n = the numbers of lockers available for bathers;
w = the average weight of a *naked* customer, in lbf;
λ = the absolute value of the terrestrial longitude of the bathhouse, in decimal degrees.

Using SI dimensions, derive a formula that yields identical results for C. Check your formula by a numerical example.

4/8 The film heat transfer coefficient h for 120 °F, water flowing in a 0.62 in ID tube at 3 ft/s velocity is $h = 900$ BTU/(hour·ft^2·°F). What is the corresponding value of h in SI?

4/9 The value of quantity q is $c \cdot \text{lbf}^{0.7} \cdot \text{cm}^{-4.3} \cdot °F^{-2} \cdot h^{-2.8}$, where $c = 99$ and °F and h designate temperature difference and hour, respectively. What is the value of c, if q is expressed in SI?

4/10 In 1935 K. D. Wood published an *empirical* formula for the cost to build a wind tunnel:

$$C = A \cdot (150 + 0.04 \times v^2)$$

where C = cost in U.S. dollars;
v = velocity of air in miles per hour;
A = area of the "throat" of the wind tunnel in ft^2.

Modify this formula such that only SI dimensions are used and C remains unchanged (this problem was adapted from Ref. 19, p. 12).

4/11 In the American/British Engineering System (Art.3.4.3), the critical value of the compressive force P_{cr} for a strut of uniform circular cross-section and hinged ends was experimentally found to be

$$P_{cr} = 46300 \cdot I$$

where I is the cross-section's second moment of area. Derive the corresponding relation in SI.

4/12 Numerical measures of importance of public demonstrations.

It is widely recognized that the importance of a social, political, military, cultural, etc., event is reflected by the "air-time" TV networks spend to "cover" it. This fact gave public relations experts, "media scientists," and sociology gurus the idea that the *length* of dedicated TV time could be used to *quantify* the importance, or unimportance, of an event.

Accordingly, environmental activists joined forces with communication specialists and they composed a numerical measure, the *importance factor, i*, to *quantify* the significance of demonstrations in general, and environmental demonstrations in particular:

$$i = \left(\frac{2 \times 10^4 \cdot t_{TV}}{n \cdot t_D} \right)^{1.5}$$

where t_{TV} = TV coverage time in seconds;
n = number of participants in the demonstration;
t_D = duration of demonstration (including speeches) in minutes.

In contrast to this approach, a group of eminent communication scholars developed the *unimportance factor, u,* to apply to an environmental rally:

$$u = \left(\frac{n \cdot t_D}{1000 \cdot t_{TV}} \right)^{1.5}$$

where t_{TV} = TV coverage time in seconds;
n = number of participants in the demonstration;
t_D = duration of demonstration (including speeches) in hours.

Derive a relation between i and u.

4/13 Ferguson's lubrication problem.

A drop of oil is on the inside surface of a frustum of a cone (Fig. 4-6). The half apex angle of the cone is φ and its radius at the drop is r—as illustrated. The cone rotates at constant speed about its vertical axis of symmetry. The drop does not move *horizontally* with respect to the cone.

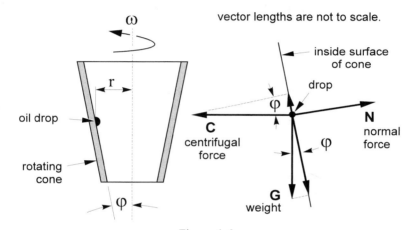

Figure 4-6
A drop of oil is in equilibrium on the inside surface of a rotating cone

Assuming no friction between the drop and the cone's surface, three forces act on the drop: gravitational force G, centrifugal force C, and normal force N. If the component of C along the cone's surface is larger than the similar component of G, then the drop moves upward, else it moves downward (N has no relevant component, and hence is ignored). In between, there is a *particular* rotational speed ω_e at which the drop is in *equilibrium*. It follows, therefore, that if $\omega > \omega_e$, then the cone acts as a pump, i.e., moves oil vertically upward. It can be shown that if r is in meters, then

$$\omega_e = \frac{3.132}{\sqrt{r \cdot \tan \varphi}} \; \frac{\text{rad}}{\text{s}}$$

What is the *constant* in the numerator, if (a) r is expressed in inches, and ω_e is expressed in revolutions per minute?; (b) r is expressed in parsecs, and ω_e is expressed in degrees per millennium?

Note: This problem was suggested to the author by Robert Ferguson, President of Taiga Engineering Group, Inc., Ontario, Canada.

4/14 An unusual dimensional system (III).

Create a dimensional system in which the *magnitudes* of the following constants of nature are unity:

Stefan–Boltzmann constant, σ

Speed of light in vacuum, c

Charge of electron (elementary charge), e

Universal gravitational constant, k

Gravitational acceleration at sea level on Earth, g

Verify your result for at least one constant. The SI values and dimensions of the first four constants are in Appendix 2. The standard value of g is 9.80665 m/s².

4/15 The Kapitza dimensional system.

In 1966 Pyotr Leonidovich Kapitza (1894–1984), a Russian physicist (famous for his work on high energy magnetism, a Nobel prize winner in 1978, and for a time assistant director of magnetic research at the Cavendish Laboratory) proposed a dimensional system (Ref. 1, p. 27) in which the electron's radius, relativistic rest-mass, and mass are the dimensions, respectively, of length, energy and mass. In other words, in the Kapitza system the magnitudes of these constants are unity. According to Barenblatt (Ref. 1, p. 27), the Kapitza system is "... extremely convenient to use in classical electrodynamics problems since it allows one to avoid very large or very small numerical values for all quantities of practical interest."

Express the length, mass and time fundamental dimensions of the Kapitza system in terms of SI. Verify your results for the three fundamental dimensions involved. Use the relevant numeric values (rounded to five decimals) found in Appendix 2.

4/16 Heart rate of mammals.

The *Clark formula,* quoted by A. V. Hill (Ref. 112, p. 219), gives the heartbeat frequency of mammals (including man) as a function of linear size L. The formula, in a modified form, is

$$f = 2273 \cdot L^{-0.81}$$

where f is the resting heartbeats per minute and L is in inches. Express the Clark formula in SI.

CHAPTER 5
ARITHMETIC OF DIMENSIONS

In this short chapter we state some of the theorems dealing with the arithmetic of dimensions. As the reader will see, these theorems are rather intuitive.

Theorem 5-1 (Theorem of products). *The product of the dimensions of two variables is the dimension of the product of two variables.* That is, if V_1 and V_2 are variables, then

$$[V_1] \cdot [V_2] = [V_1 \cdot V_2] \tag{5-1}$$

Proof. Using the format described in Art. 2.1, we can write

$$\left. \begin{array}{l} V_1 = m_1 \cdot d_1 \\ V_2 = m_2 \cdot d_2 \end{array} \right\} \tag{5-2}$$

where m_1 and m_2 are the *magnitudes,* and d_1 and d_2 are the *dimensions* of variables V_1 and V_2, respectively. Then by definition $[V_1] = d_1$ and $[V_2] = d_2$. Therefore

$$[V_1] \cdot [V_2] = d_1 \cdot d_2$$

On the other hand, obviously, $V_1 \cdot V_2 = m_1 \cdot m_2 \cdot d_1 \cdot d_2$, from which, by the above relation,

$$[V_1 \cdot V_2] = d_1 \cdot d_2 = [V_1] \cdot [V_2]$$

which is identical to (5-1). This proves the theorem.

Note that this theorem is general, inasmuch as it can be extended to any number of variables. Accordingly, we state the following corollary of Theorem 5-1 without formal proof.

Corollary of Theorem 5-1. *The product of the dimensions of a set of* n *variables is the dimension of the product of this set of* n *variables.* That is, if $V_1, V_2, \ldots, V_n$ is a set of n variables, then

$$[V_1] \cdot [V_2] \ldots [V_n] = [V_1 \cdot V_2 \ldots V_n] \tag{5-3}$$

Theorem 5-2 (Theorem of quotient). *The quotient of the dimensions of any two variables is the dimension of the quotient of these two variables.* That is, if V_1 and V_2 are variables, then

$$\frac{[V_1]}{[V_2]} = \left[\frac{V_1}{V_2}\right] \tag{5-4}$$

Proof. Using the format of (5-2), we have

$$\frac{[V_1]}{[V_2]} = \frac{d_1}{d_2} \tag{5-5}$$

But

$$\frac{V_1}{V_2} = \frac{m_1 \cdot d_1}{m_2 \cdot d_2} = \frac{m_1}{m_2} \cdot \frac{d_1}{d_2} \tag{5-6}$$

Therefore, by (5-5) and (5-6)

$$\left[\frac{V_1}{V_2}\right] = \frac{d_1}{d_2} = \frac{[V_1]}{[V_2]}$$

which was to be proven.

Theorem 5-3 (Theorem of association). *If d_1, d_2, and d_3 are the dimensions of V_1, V_2, and V_3, respectively, then $d_1 \cdot (d_2 \cdot d_3) = (d_1 \cdot d_2) \cdot d_3$.*

Proof. Let us temporarily define two new variables, $V_{12} \equiv V_1 \cdot V_2$ and $V_{23} \equiv V_2 \cdot V_3$. Therefore, by Theorem 5-1,

$$[V_{12}] \cdot [V_3] = (d_1 \cdot d_2) \cdot d_3 = [V_1 \cdot V_2] \cdot [V_3] = [V_1] \cdot [V_2] \cdot [V_3]$$

$$[V_1] \cdot [V_{23}] = d_1 \cdot (d_2 \cdot d_3) = [V_1] \cdot [V_2 \cdot V_3] = [V_1] \cdot [V_2] \cdot [V_3]$$

from which

$$d_1 \cdot (d_2 \cdot d_3) = (d_1 \cdot d_2) \cdot d_3. \tag{5-7}$$

This proves the theorem.

Theorem 5-4 (Theorem of exponents). *The dimension of a power of a variable is the power of the dimension of that variable.* In other words, $[V^n] = [V]^n$.

Proof. Again, let $V = m \cdot d$, where m is the magnitude and d is the dimension of variable V. Then, obviously, $[V] = d$. Therefore $[V]^n = d^n$. On the other hand, $V^n = (m \cdot d)^n = m^n \cdot d^n$, and hence

$$[V^n] = d^n = [V]^n \tag{5-8}$$

This proves the theorem.

Theorem 5-5 (Theorem of differentials). *The dimension of the quotient of differentials of variables* V_1 *and* V_2 *is the quotient of dimensions* V_1 *and* V_2. *That is*

$$\left[\frac{dV_1}{dV_2}\right] = \frac{[V_1]}{[V_2]} \tag{5-9}$$

Proof. We return to the definition of the derivative of a function. Accordingly, if V_1 and V_2 are variables, then

$$\frac{dV_1}{dV_2} = \lim_{\Delta V_2 \to 0} \frac{\Delta V_1}{\Delta V_2}$$

where ΔV_1 and ΔV_2 are both *finite*. Hence, by the above and Theorem 5-2 (Theorem of quotient)

$$\left[\frac{dV_1}{dV_2}\right] = \left[\frac{\Delta V_1}{\Delta V_2}\right] = \frac{[\Delta V_1]}{[\Delta V_2]} = \frac{[V_1]}{[V_2]}$$

This proves the theorem.

Theorem 5-6 (Theorem of nth order differential). *The dimension of the nth order differential of variable* V_1 *with respect to variable* V_2 *is the quotient of dimensions of* V_1 *and* V_2^n. *That is*

$$\left[\frac{d^n V_1}{dV_2^n}\right] = \frac{[V_1]}{[V_2^n]} \tag{5-10}$$

Proof. The dimension of the second derivative can be found as follows: Let us create a temporary variable $V_3 \equiv dV_1/dV_2$. Then, by Theorem 5-2 and Theorem 5-5,

$$\left[\frac{dV_3}{dV_2}\right] = \frac{[V_3]}{[V_2]} = \frac{[V_1]/[V_2]}{[V_2]} = \frac{[V_1]}{[V_2]^2} = \frac{[V_1]}{[V_2^2]}$$

Therefore, by substituting the above-defined V_3, we obtain

$$\left[\frac{d^2 V_1}{dV_2^2}\right] = \frac{[V_1]}{[V_2^2]}$$

This process can continue with progressively higher order differentials, yielding relation (5-10). This proves the theorem.

Example 5-1. Dimension of Jerk

The noun "jerk," contrary to popular opinion, is not necessarily a derogatory expression; for it also *may* mean the change of acceleration in unit time. Question: What is the dimension of jerk in SI?

If t is time, x is distance traveled, and q is acceleration, then

$$\text{jerk} = \frac{dq}{dt} = \frac{d}{dt}\left(\frac{d^2 x}{dt^2}\right) = \frac{d^3 x}{dt^3} \tag{a}$$

and therefore

$$[\text{jerk}] = \left[\frac{d^3 x}{dt^3}\right] = \left[\frac{x}{t^3}\right] = \frac{[x]}{[t^3]} = \frac{[x]}{[t]^3} = \frac{m}{s^3}$$

where m is meter, s is second.

⇑

Theorem 5-7 (Theorem of integrals). The dimension of an integral is the product of the dimension of the integrand and the dimension of the independent variable. That is if $I = \int y(x) \cdot dx$, then

$$[I] = [y(x) \cdot dx] = [y(x)] \cdot [x] \quad (5\text{-}11)$$

where $y(x)$ is the *integrand* and x is the *independent* variable.

Proof. The expression under the integral sign is really a *product*, to which, therefore, Theorem 5-1 applies. But this latter theorem was already proved; therefore the present theorem is also proved.

Theorem 5-8 (Theorem of multiple integrals). The dimension of a multiple integral is the product of the dimension of the integrand and the dimensions of all the independent variables. That is, if

$$I = \int_{(1)} \int_{(2)} \cdots \int_{(n)} y(x_1, x_2, \ldots, x_n) \cdot dx_1 \cdot dx_2 \cdots dx_n,$$

then

$$[I] = [y(x_1, x_2, \ldots, x_n)] \cdot [x_1] \cdot [x_2] \cdots [x_n] \quad (5\text{-}12)$$

Proof. The proof closely follows that of Theorem 5-6; therefore no separate proof of the present theorem is necessary.

CHAPTER 6
DIMENSIONAL HOMOGENEITY

6.1. EQUATIONS

Any equation, to meaningfully express properties of the physical world around us, must fulfill two criteria: (a) its two sides must have *numerical equality;* (b) its two sides must have *dimensional homogeneity*.

That both sides of an equation must be numerically equal in order to be meaningful (useful) is a self-evident truth, for otherwise we would not have an *equation*, but an *inequality*. For example, 6 = 6 is an equality (equation), but 7 = 8 is not, and therefore the latter should be written as an inequality 7 ≠ 8. Now consider the equation

$$5 \text{ elephants} = 3 \text{ elephants} + 2 \text{ elephants}$$

The sides of this equation have *numerical equality* since the *magnitudes* of both sides are the same, viz., 5. Moreover, since the dimensions of the left and the right are both "elephant," our relation is *dimensionally homogeneous*. Next, consider

$$5 \text{ elephants} = 1 \text{ zebra} + 4 \text{ antelopes}$$

Here we have numerical equality, but not dimensional homogeneity. Consequently, this relation is meaningless.

If a relation is dimensionally homogeneous and numerically correct in one dimensional system, then it is also correct in any other dimensional system consistently applied.

Example 6-1

Consider the well-known formula $h = \frac{1}{2} \cdot g \cdot t^2$ in which h is the distance traveled by a falling body in vacuum in t time starting from zero speed. The gravitational acceleration g = constant. This equation is true whether we use SI, or any other system. Take, for instance, SI in which $g = 9.81$ m/s^2 and, say, $t = 4$ s. Then the distance of free fall will be $h = 78.48$ m. Now, let us use a system in which the dimension of length is "foot," and the dimension of time is "year." In this system $g = 3.20511 \times 10^{16}$ ft/year^2 and $t = 1.26755 \times 10^{-7}$ year. Thus, we will have

$$h = \frac{1}{2} \cdot (3.20511 \times 10^{16}) \cdot (1.26755 \times 10^{-7})^2 \frac{\text{ft}}{\text{year}^2} \cdot \text{year}^2 = 257.47979 \text{ ft}$$

But 257.47979 ft = 78.48 m, therefore our formula (since it is dimensionally homogeneous), supplied identical results in both dimensional systems.

⇑

This important characteristic of dimensionally homogeneous relations can be formally proven as follows: For the sake of simplicity, we shall only consider three variables, V_1, V_2, and V_3 and two dimensions d_1 and d_2. However, the procedure can be extended to any number of variables and dimensions. Let us have a formula connecting the variables

$$V_1 = Q \cdot V_2^{q_2} \cdot V_3^{q_3} \tag{6-1}$$

where Q is a nonzero number and q_1 and q_2 are exponents which can be of any value, including zero. However, if $q_1 = q_2 = 0$, then $V_1 = Q$ = constant. In Example 6-1, $Q = \frac{1}{2}$, $V_1 = h$, $V_2 = g$, $V_3 = t$, $q_2 = 1$, and $q_3 = 2$. Now obviously, if m_1, m_2, and m_3 are the respective *magnitudes*, then

$$V_1 = m_1 \cdot d_1^{e_{11}} \cdot d_2^{e_{12}}; \qquad V_2 = m_2 \cdot d_1^{e_{21}} \cdot d_2^{e_{22}}; \qquad V_3 = m_3 \cdot d_1^{e_{31}} \cdot d_2^{e_{32}} \tag{6-2}$$

where $e_{11}, \ldots$, etc. are the appropriate exponents. Therefore, by (6-1) and (6-2)

$$V_1 = m_1 \cdot d_1^{e_{11}} \cdot d_2^{e_{12}} = Q \cdot (m_2 \cdot d_1^{e_{21}} \cdot d_2^{e_{22}})^{q_2} \cdot (m_3 \cdot d_1^{e_{31}} \cdot d_2^{e_{32}})^{q_3}$$

or

$$V_1 = m_1 \cdot d_1^{e_{11}} \cdot d_2^{e_{12}} = (Q \cdot m_2^{q_2} \cdot m_3^{q_3}) \cdot (d_1^{e_{21} \cdot q_2 + e_{31} \cdot q_3} \cdot d_2^{e_{22} \cdot q_2 + e_{32} \cdot q_3}) \tag{6-3}$$

Since this equation was derived from (6-1) which is both numerically correct and dimensionally homogeneous, therefore both the magnitude and the dimension on the *left side* of (6-3) must be the same as on the *right side*. Hence, from the condition of *numerical correctness* we have

$$m_1 = Q \cdot m_2^{q_2} \cdot m_3^{q_3} \tag{6-4}$$

and from the condition of *dimensional homogeneity*

$$e_{11} = e_{21} \cdot q_2 + e_{31} \cdot q_3; \qquad e_{12} = e_{22} \cdot q_2 + e_{32} \cdot q_3 \tag{6-5}$$

since the exponents of dimensions d_1 and d_2 on the left and right sides of (6-3) must be identical.

Now we change our dimensional system. Our new dimensions are D_1 and D_2, instead of d_1 and d_2. The old and new systems are connected by the relations

$$d_1 = k_1 \cdot D_1; \qquad d_2 = k_2 \cdot D_2 \tag{6-6}$$

where k_1 and k_2 are constants. Considering these latest two relations, we write (6-3) as

$$V_1 = (Q \cdot m_2^{q_2} \cdot m_3^{q_3}) \cdot (k_1 \cdot D_1)^{e_{21}.q_2 + e_{31}.q_3} \cdot (k_2 \cdot D_2)^{e_{22}.q_2 + e_{32}.q_3}$$

so that

$$V_1 = (Q \cdot m_2^{q_2} \cdot m_3^{q_3} \cdot k_1^{e_{21}.q_2 + e_{31}.q_3} \cdot k_2^{e_{22}.q_2 + e_{32}.q_3}) \cdot (D_1^{e_{21}.q_2 + e_{31}.q_3} \cdot D_2^{e_{22}.q_2 + e_{32}.q_3}) \qquad (6\text{-}7)$$

But by the first of (6-2) and (6-6), we have $V_1 = m_1 \cdot (k_1 \cdot D_1)^{e_{11}} \cdot (k_2 \cdot D_2)^{e_{12}}$, which can be written

$$V_1 = (m_1 \cdot k_1^{e_{11}} \cdot k_2^{e_{12}}) \cdot D_1^{e_{11}} \cdot D_2^{e_{12}} \qquad (6\text{-}8)$$

It is seen that, by (6-5), the exponents of k_1, k_2, D_1, and D_2 in this equation are, respectively, the same as the exponents of k_1, k_2, D_1, and D_2 in (6-7). Moreover, by (6-4), the magnitude m_1 is identical to $Q \cdot m_2^{q_2} \cdot m_3^{q_3}$ in (6-7). Thus, V_1 in the first formula of (6-2), in which the dimensions are d_1 and d_2, supplies the *same* result as does formula (6-8), in which the dimensions are D_1 and D_2. The next example demonstrates the above procedure.

Example 6-2

Consider again Example 6-1. We have the formula for free fall $h = \frac{g}{2} \cdot t^2$. Therefore, by definition, the variables are $V_1 \equiv h$ (falling height); $V_2 \equiv g$ (gravitational acceleration); and $V_3 \equiv t$ (time). Also in relation (6-1):

$$Q = \frac{1}{2}, \qquad q_2 = 1, \qquad q_3 = 2$$

in relation (6-2):

$$m_2 = 9.81, \ m_3 = 4, \ d_1 = m \text{ (meter)}, \ d_2 = s \text{ (sec)}, \ e_{21} = 1, \ e_{22} = -2, \ e_{31} = 0, \ e_{32} = 1$$

Therefore the magnitude of V_1 will be, by (6-4),

$$m_1 = \frac{1}{2} \cdot (9.81) \cdot (4^2) = 78.48$$

and, by (6-5), the exponents of the dimensions are $e_{11} = 1$ and $e_{12} = 0$. Therefore, by the first part of (6-2),

$$V_1 = h = 78.48 \ \text{m}^1 \cdot \text{s}^0$$

In the *second* dimensional system we have $D_1 \equiv$ ft, $D_2 \equiv$ yr (year), and in (6-6), $k_1 = 3.28084$, $k_2 = 3.16888 \times 10^{-8}$, since 1 m = 3.28084 ft, and 1 s = 3.16888 × 10⁻⁸ yr. Therefore, by (6-8),

$$V_1 \equiv h = (78.48) \cdot (3.28084)^1 \cdot (3.16888 \times 10^{-8})^0 \ \text{ft}^1 \cdot \text{s}^0 = 257.48032 \ \text{ft}$$

But this value is 78.48 m and therefore the results in both dimensional systems are identical—as expected (since we started from a dimensionally homogeneous relation).

The reader might ponder the relation of the type 1 ft = 0.3048 m which, *by definition*, is certainly true, yet neither the magnitudes nor the dimensions of its sides are identical. How can this be? The key is the phrase "by definition." The formula expresses *nothing* about our physical world; it is not influenced by any characteristic of nature. Rather, its veracity stems from an *agreement* by which an arbitrary length we *decided* to call "foot" is exactly 0.3048 times as long as another arbitrary length we *decided* to call "meter." An individual sitting in a room completely isolated from the outside world—windows closed, curtains drawn, telephone line cut—is capable of arriving at this relation. Notwithstanding the above, the formula *can* be rendered both numerically correct and dimensionally homogeneous. We know that 1 m = 3.28084 ft. If this is now substituted into our original relation, we will have

$$1 \text{ ft} = 0.3048 \text{ m} = (0.3048) \cdot (1 \text{ m}) = (0.3048) \cdot (3.28084 \text{ ft}) = 1 \text{ ft}$$

and the equation becomes both numerically correct and dimensionally homogeneous (albeit utterly useless).

It is interesting that, although the requirement for numerical equality of both sides of an equation was recognized from almost prehistoric times, the equally self-evident condition that dimensions must also match originates from not earlier than 1765. In that year Leonhard Euler (Swiss mathematician, 1707–1783), in his *Theoria motus corporum solidorum seu rigidorum,* discussed units and dimensional homogeneity. Rather astonishingly, for almost six decades nobody recognized his profound ideas on the subject. Finally, Jean Baptiste Joseph Fourier (French mathematician and physicist, 1768–1830), in the last of his three versions of *Analytic Theory of Heat,* dated 1822, firmly established the theory of dimensions and formulated the inescapable requirements of dimensional homogeneity. The reader's attention is directed to the excellent essay by Enzio Macagno (Ref. 61) dealing with the history of dimensions and some related topics.

To fulfill the requirement of *dimensional homogeneity,* the following *rules* must be obeyed:

Rule 1. *In every analytically derived equation, both sides of the equation must have identical dimensions. All numbers appearing in this equation must be dimensionless constants; i.e., they must have the dimension of 1.*

To illustrate the use of this rule, consider again the formula for free fall, $h = \frac{1}{2} g \cdot t^2$, which is an *analytically* derived equation. Therefore, the *number* $\frac{1}{2}$ on the right is a *dimensionless* constant. Moreover, the dimension on the left and on the right is "length." Hence the formula obeys Rule 1 of dimensional homogeneity.

Example 6-3

Someone derived the following formula for the speed v attained by a body falling h distance in vacuum under constant gravitational acceleration g:

$$v = \sqrt{3 \cdot g \cdot h^2}$$

In SI, the dimension on the left is m·s^{-1}, while on the right it is m$^{1.5}$·s^{-1}. These two dimensions are unequal, hence Rule 1 of dimensional homogeneity is violated. The given formula cannot be correct.

In the case of an *experimentally* derived formula, Rule 1 is modified to the extent that numbers appearing in it, including 1, may have dimensions assigned to them, i.e., they are allowed to be *dimensional* constants. In the case where any one number is *dimensional*, then the dimension of *all* the variables in the formula must be stated, else the formula is useless (see also Art. 2.1).

Example 6-4

By measurement somebody established the formula for the speed v attained by a falling body in vacuum after traveling h distance

$$v = 4.43 \cdot \sqrt{h}$$

Considering only the given variables v and h, the dimension on the left is different from that on the right, hence if the factor 4.43 is dimensionless, then the formula is *not* dimensionally homogeneous. To make it so, we must assign dimensions to both v and h. This means we must *know* the dimensional system the experimenter used to generate the formula. Without this information the formula is worthless. For suppose the experimenter used SI. Then after a, say, $h = 10$ m fall, the speed given by the formula is $v = 14$ m/s. On the other hand, if the experimenter used the inch–second system, then after a fall of $h = 393.7$ in (which is 10 m), the speed by the formula is $v = 87.9$ in/s, which is $v = 2.23$ m/s, *not* the 14 m/s previously obtained. Therefore *to state the dimensional system used is mandatory in experimentally obtained formulas* (see also Art. 2.1.3).

The dimension of the numeric factor can be obtained from the formula and is $[4.43] = [v]/[h]^{0.5}$, viz., $m^{0.5} \cdot s^{-1}$. Now our formula is dimensionally homogeneous, since both sides have the same dimension, m/s.

Rule 2. *If either one or both sides of a theoretically derived equation has/have more than 1 member (joined by addition or substraction), then all of these members must have identical dimensions.*

Example 6-5

The height h reached by an object with an initial upward velocity v_0 in t time (neglecting air resistance and assuming constant gravitational acceleration g) is

$$h = v_0 \cdot t - \frac{1}{2} \cdot g \cdot t^2$$

Here the right side has two members joined by substraction. Both of these members have the dimension m (in SI), and therefore Rule 2 is obeyed. Moreover, the left side also has the dimension m, hence Rule 1 is also satisfied. We can say therefore, that the given formula is dimensionally homogeneous.

Example 6-6

Someone derived an equation in terms of "head" in a horizontal conduit through which incompressible fluid flows:

$$\frac{p^2}{\rho \cdot g} + \frac{v^2}{g} = \text{const.}$$

104   APPLIED DIMENSIONAL ANALYSIS AND MODELING

where p is the pressure in kg/(m·s²), ρ is the density in kg/m³, g is the gravitational acceleration in m/s² and v is the velocity of the fluid in m/s. The dimension of the first member on the left is kg/s², while that of the second member is m. Thus Rule 2 is violated and therefore the derived relation is wrong, no matter what the constant on the right is.

⇑

Rule 3.   *The number 0 (zero) may have any dimension.*

This rule must exist in order to be consistent with Rule 2. For example, for variables V_1, V_2, V_3 let

$$V_1 = 3 \cdot V_2 \cdot V_3^2$$

be a dimensionally homogeneous equation. Obviously we can write

$$V_1 + 0 = 3 \cdot V_2 \cdot V_3^2$$

which must remain dimensionally homogeneous. By Rule 2, it is mandatory that $[V_1] = [0]$ since V_1 and 0 are two members on the left side joined by *addition.* But V_1 can assume any dimension, therefore 0 must do likewise. Note that, in this context, zero, strictly speaking, is not a number, but a *physical quantity of zero magnitude*. This semantic peculiarity however can be safely ignored.

Rule 4.   *Exponents and arguments of transcendental functions in a theoretically derived formula must be dimensionless.*

If the formula is an experimental one, then the dimension(s) of the numerical factor(s) appearing in it (including the factor 1) must be adjusted so that the said exponents and arguments will become dimensionless.

At this point the reader might legitimately ask: why is this rule necessary? For example, if t is time, why does $y = e^t$ make no sense? There are actually two closely connected reasons for this:

- The expression $y = e^t$ cannot be dimensionally homogeneous because the numerical value of e^t is *not independent* of the dimensional system used. For example, if we use *second* as the dimension of time, then with $t = 3$ s we have $e^3 = 20.08554$, but if we use *hour*, then, since 3 s = 1/1200 h, we get $e^{1/1200} = 1.00083368$, yet the value of time in both cases is the same.
- Consider the Taylor expansion of e^t

$$e^t = 1 + t + \frac{t^2}{2!} + \frac{t^3}{3!} + \ldots$$

The right side obviously violates Rule 2, unless the dimension of t is 1, i.e., t is dimensionless. Therefore the exponent in e^t *must* be dimensionless.

Notwithstanding the above restriction embodied in Rule 4, it is emphasized that this rule applies only to the *end result* of an analytically derived formula comprising transcendental functions. For it may happen that *during* the derivation some transcendental functions appear that *temporarily* have dimensional arguments. The following example demonstrates this occurrence.

Example 6-7. Heat Loss Through a Pipe Wall

We wish to determine the heat loss through the wall of a pipe of circular and uniform cross-section. Fig 6-1 shows the arrangement.

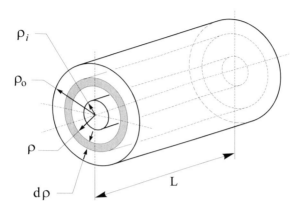

Figure 6-1
Heat loss through a pipe wall

Generally, the quantity of heat (energy) per unit time conducted through a material is

$$Q = \frac{k \cdot A}{\lambda} \cdot \Delta t \tag{a}$$

where k is the conductivity of the material, A is the area through which the heat is conducted, λ is the length of the conductor in the direction of heat flow, and Δt is the temperature difference between the end-surfaces of the heat conductor. In the case of a pipe, we consider the heat flowing through an annular element of thickness $d\rho$, radius ρ, and axial length L. Hence, we can write $A = 2 \cdot \rho \cdot \pi \cdot L$, $\lambda = d\rho$, $\Delta t = dt$, and, therefore

$$Q = \frac{k \cdot 2 \cdot \rho \cdot \pi \cdot L}{d\rho} \cdot dt$$

which can be written

$$\frac{d\rho}{\rho} = \frac{2 \cdot \pi \cdot L \cdot k}{Q} \cdot dt$$

Integration now yields

$$\ln \rho = \frac{2 \cdot \pi \cdot L \cdot k}{Q} \cdot t + C \tag{b}$$

where C is a constant.

But now we notice that the argument of the logarithm on the left is "length," which is not a dimensionless variable. What happened? How did (a), which is dimensionally homogeneous, became (b), which is not? Was an error made in the derivation? No. The simple fact is that we have *not yet finished* the derivation, and hence the "result" hitherto ob-

106 APPLIED DIMENSIONAL ANALYSIS AND MODELING

tained is not final. So, we can—at least for now—reduce our stress level and *continue* the derivative process.

If the inside and outside radii of the pipe are ρ_i and ρ_0 with corresponding temperatures t_i and t_0, then, by (b),

$$\ln \rho_i = \frac{2 \cdot \pi \cdot L \cdot k}{Q} \cdot t_i + C$$

from which

$$C = \ln \rho_i - \frac{2 \cdot \pi \cdot L \cdot k}{Q} \cdot t_i$$

This is now substituted into (b) yielding, after some simplification and rearrangement,

$$\ln \frac{\rho}{\rho_i} = \frac{2 \cdot \pi \cdot L \cdot k}{Q} \cdot (t - t_i)$$

Or, since at radius $\rho = \rho_0$, the temperature is $t = t_0$,

$$\ln \frac{\rho_0}{\rho_i} = \frac{2 \cdot \pi \cdot L \cdot k}{Q} \cdot (t_0 - t_i)$$

whence we finally get

$$Q = \frac{2 \cdot \pi \cdot L \cdot k}{\ln(\rho_0/\rho_i)} \cdot (t_0 - t_i)$$

which is the *end result* of the derivation. The mystery is solved! The argument of the transcendental function (logarithm) is now dimensionless, as it should be; Rule 4 is obeyed.

⇑

Example 6-8. Lengths of Dogs' Tails in Urban Areas

Someone *derived* a relation claiming it to be valid for any *urban* area situated north of latitude 34.6 deg. It is claimed that the formula contains only *three* physical variables:

$$L = (384) \cdot {_3}\log \frac{g}{7 \cdot H}$$

where L is the *arithmetic* average of lengths of dogs' tails, g is the gravitational acceleration at sea level, and H is the *geometric* average of heights of industrial chimneys. The base of logarithm is 3, as shown.

What is wrong with this formula? One is tempted to dismiss it outright because of its obviously absurd claims of functional relation among variables that are utterly unrelated. But, however unlikely it is that the claimed causal dependence of these variables exists, this is not the fault of our formula. For these grotesque claims only make it *unlikely* that the relation is true—they do not render it *impossible*. Nonetheless, there are other characteristics of the formula that *do* decide its sad fate—with *certainty*.

First, the number 7 in the denominator cannot be dimensionless, for otherwise the argument of the logarithm would be dimensional, which is a violation of Rule 4. Therefore 7 must be dimensional, viz., $[7] = s^{-2}$. But, in this case, the formula either cannot be a derivable one (numbers in derived formulas must be dimensionless, Rule 1), or 7 is the *magnitude* of a variable of dimension s^{-2}. However, this latter alternative cannot be the case since a fourth variable would then be involved, which is contrary to the claim that there are only three variables in the formula.

Second, the number 384 on the right cannot be dimensionless, or Rule 1 is violated. But it must be dimensionless since it appears in a theoretically derived formula, unless the number 384 is the *magnitude* of yet another variable. This alternative, as explained above, cannot be the case, either.

In conclusion, we are forced to accept the fact, which we suspected at the outset—but for a different reason—that the formula is false.

⇑

Example 6-9. Variation of Atmospheric Pressure with Altitude (adapted from Ref. 22, p. 22)

Atmospheric pressure is a function of altitude. Let us derive a simple formula relating pressure p to altitude h. Consider an air "column" of elementary thickness dh and cross-section A (Fig. 6-2). Obviously, the pressure difference Δp between the bottom and the top surfaces must be equal to the weight of the column. Therefore, if ρ is the density of air and g is the gravitational acceleration, then

$$dp = -\frac{d(\text{weight})}{\text{area}} = -\frac{A \cdot dh \cdot \rho \cdot g}{A} = -\rho \cdot g \cdot dh$$

Here the minus sign is used because the pressure *decreases* as the altitude *increases*.

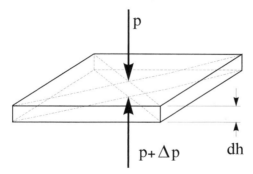

Figure 6-2
Illustration for the derivation of relation of air pressure versus altitude

As a first approximation, with constant air temperature, we can write $\rho = k \cdot p$, i.e., density is proportional to pressure with the factor of proportionality $k \cong 1.214 \times 10^{-5}$ m^{-2}·s^2 for air. Therefore, $dp = -k \cdot p \cdot g \cdot dh$, or

$$\frac{dp}{p} = -k \cdot g \cdot dh$$

By integrating, we obtain, by assuming constant g*

$$\ln p = -k \cdot g \cdot h + C$$

---

*By the inverse square law, it can be shown that below 32000 m altitude the error committed by assuming constant g is less than 1%.

where C is a constant. Here we see that the argument of the logarithm on the left is pressure, which is certainly not dimensionless, an apparent violation of Rule 4. However, we have *not yet* finished our derivation of the sought-after formula!

Now we determine the constant (parameter) C. At sea level $h = 0$, and $p = p_0$. Hence, by the above formula, $\ln p_0 = C$. This yields, after some rearrangement,

$$p = p_0 \cdot e^{-k \cdot g \cdot h}$$

where the exponent (to our relief) is dimensionless, as the reader should ascertain. Incidentally, this formula gives a rather good approximation (within about 3%) of air pressure up to 6000 m altitude. Beyond this, temperature and other effects render the relation inaccurate; for example, at $h = 10$ km, the relation is in error about 25%.

⇑

Certain arithmetical operations, which are entirely valid in mathematical formulas involving *dimensionless* numbers, are forbidden in equations connecting *dimensional* variables. For example, if a, b, c, and d are *dimensionless numbers*, and if $a = b$ and $c = d$, then the following operations are all valid:

$$a + c = b + d \; ; \; a^c = b^d \; ; \; \sqrt[c]{a} = \sqrt[d]{b} \; ; \; c^a = b^d \; ; \; \sqrt[a]{c} = \sqrt[b]{d} \; ; \; \ln a = \ln b \; ; \; \ln c = \ln d$$

However, if a, b, c, and d denote *dimensional* variables, then in each case it must be ascertained that all the rules of dimensional homogeneity hitherto defined have been adhered to. As an example, consider free-fall without air resistance, $v = \sqrt{2 \cdot g \cdot h}$, $g = v/t$, where g is gravitational acceleration, v is speed, h is height, and t is time. Both of these equations are true, and yet, none of the following relations makes sense:

$g + v = v/t + \sqrt{2 \cdot g \cdot h}$	(violation of Rule 2)
$v^g = (\sqrt{2 \cdot g \cdot h})^{v/t}$	(violation of Rule 4)
$\ln g = \ln v - \ln t$	(violation of Rules 2 and 4)
$\sin g = \sin \dfrac{v}{t}$	(violation of Rule 4)
$N^v = N^{\sqrt{2 \cdot g \cdot h}}$	(violation of Rule 4; N is any number)

Based on the above, the following important advice is given:

Important advice 1. *When combining formulas involving additions, substractions, or transcendental functions, always verify that no rule of dimensional homogeneity is violated.*

Note that relational operations involving multiplication or division are permitted—i.e., if the original relations are valid, so will the derived ones—at least dimensionally. In these cases, no additional dimensional checking is required. To illustrate, by quoting the above example, if relations

$$v = \sqrt{2 \cdot g \cdot h} \quad \text{and} \quad g = \frac{v}{t} \tag{6-9}$$

are true, so are

$$v \cdot g = \sqrt{2 \cdot g \cdot h} \cdot \frac{v}{t} \quad \text{and} \quad \frac{v}{g} = \frac{\sqrt{2 \cdot g \cdot h} \cdot t}{v} \tag{6-10}$$

The first formula of (6-10) can be simplified to

$$h = \frac{1}{2} \cdot g \cdot t^2 \tag{6-11}$$

while the second formula becomes

$$v = \sqrt[4]{2 \cdot g^3 \cdot t^2 \cdot h} \tag{6-12}$$

Both of these formulas are dimensionally correct, although numerically not necessarily so. For example, let $g = 9.81$ m/s², $h = 200$ m, $t = 6.385$ s. Then (6-12) supplies $v = 62.64$ m/s, which is correct, since it satisfies the pair in (6-9). But now let us choose $g = 9.81$ m/s², $h = 3.8$ m, $t = 25$ s. This case (6-12) yields $v = 46.02$ m/s. However, $v_1 = \sqrt{2 \cdot g \cdot h}$ with the given g and h values yields $v_1 = 8.63$ m/s and $v_2 = g \cdot t$ provides $v_2 = 245.25$ m/s. So, we observe that $v \neq v_1 \neq v_2$.

What happened? The sad fact is that only two of the three variables on the right side of (6-12) may be freely chosen; i.e., the three variables in question are *not independent*. We can choose g and t, or g and h, or t and h, but *not* g, t, and h. This finding prompts us to offer another important piece of advice.

Important advice 2. When combining dimensionally and numerically correct relationships by (maybe repeated) multiplications or divisions, verify that the resulting relation contains only one dependent variable, or the value furnished by the formula may be (and most likely will be) wrong.

From this advice, our next rule can be formulated.

Rule 5. Any meaningful (i.e., dimensionally and numerically correct) physical relationship must contain only one dependent variable.

Example 6-10. Quantity of Paint on a Sphere

We wish to determine the quantity (mass) of paint necessary to cover the outer surface of a sphere of radius R. The paint's constant thickness is t, its density is ρ. The surface area is

$$A = 4 \cdot R^2 \cdot \pi \tag{a}$$

and therefore the mass of paint is

$$M = A \cdot t \cdot \rho \tag{b}$$

For example, if $R = 4.5$ m, $t = 0.0005$ m, $\rho = 920$ kg/m³, then $A = 254.469$ m² and $M = 117.06$ kg. Dividing (a) by (b), we get

$$\frac{A}{M} = \frac{4 \cdot R^2 \cdot \pi}{A \cdot t \cdot \rho}$$

which can be written

$$M = \frac{A^2 \cdot t \cdot \rho}{4 \cdot R^2 \cdot \pi} \tag{c}$$

This equation is dimensionally homogeneous, but numerically meaningless; it violates Rule 5 inasmuch as it contains more than one dependent variable. In particular, on the right side, A, t, ρ, and R cannot all be independent since A is set by R, and therefore t, ρ, and R define *both* A and M. So, we have more than one independent variable in (c).

To show that (c) can yield an erroneous result for M, we *select* $R = 1.9$ m, $A = 188.7$ m², $t = 0.00065$ m, and $\rho = 874$ kg/m³. By these values, (c) provides $M = 445.914$ kg. This value, however, cannot be correct since by (b) the mass of paint is $M = 117.06$ kg.

Nevertheless, (c) is not always wrong, just most of the time! For example, if we select $R = 4.5$ m, $t = 0.0005$ m, $\rho = 920$ kg/m³, and $A = 254.469$ m², then we get the correct $M = 117.06$ kg.

⇑

The procedure to follow when there is more than one dependent variable in a relation is to break down (resolve) the formula into its constituent equations such that each of them would contain only *one* dependent variable. The following example illustrates this technique.

Example 6-11

Suppose one is presented with formula (c) of Example 6-10, regarding the mass of paint to cover a sphere of radius R. As we saw, this formula contains more than one dependent variable. Our task here is to resolve this relation into two formulas, such that each would contain only *one* dependent variable. Accordingly, by (c) of the previous example (repeated here for convenience)

$$M = \frac{A^2 \cdot t \cdot \rho}{4 \cdot R^2 \cdot \pi}$$

from which

$$A^2 = \frac{M \cdot 4 \cdot R^2 \cdot \pi}{t \cdot \rho} \tag{a}$$

which can be written

$$A \cdot A = \left(\frac{M}{t \cdot \rho}\right) \cdot (4 \cdot R^2 \cdot \pi) \tag{b}$$

Hence, we now have two equations:

$$A = \frac{M}{t \cdot \rho} \quad (\text{or } M = A \cdot t \cdot \rho) \tag{c}$$

and

$$A = 4 \cdot R^2 \cdot \pi \tag{d}$$

Each of these formulas contains only *one* dependent variable, namely M in (c), and A in (d).

⇑

6.2. GRAPHS

As with equations (discussed in Art. 6.1), graphs, diagrams, charts, etc. can also be either homogeneous or nonhomogeneous. A graph, by definition, is dimensionally homogeneous if, and only if, its shape is invariant upon the dimensions of both the abscissa and the ordinate.

DIMENSIONAL HOMOGENEITY

If there is a third variable, i.e., a *parameter*, then instead of a single curve, we have a family of curves, each of which is characterized by a distinct value of that parameter. This family of curves is called a *chart*. If the chart is homogeneous, then its every curve must be invariant upon the dimensional system used and, further, the numerical value of the parameter must also be independent of the dimensional system.

A graph can be still useful (i.e., informative, correct, etc.) even though it is not dimensionally homogeneous. Note that this is in sharp contrast to a formula, which is totally useless unless it is dimensionally homogeneous (as we have seen in Art. 6.1).

Now let us explore the world of dimensional homogeneity for graphs. Generally, a function can be written as

$$W = p_1 \cdot V^{q_1} + p_2 \cdot V^{q_2} + \ldots \quad (6\text{-}13)$$

where V is the independent variable, W is the dependent variable, the p's are the dimensional or dimensionless parameters, and the q exponents are dimensionless numbers.

In any graph in a rectangular coordinate system, the independent and dependent variables appear as *distances* x and y on the abscissa and ordinate, i.e., on the "horizontal" and "vertical," axes. Therefore, we can write

$$V = S_V \cdot x \, ; \qquad W = S_W \cdot y \quad (6\text{-}14)$$

where S_V and S_W are the *scales* of the independent and dependent variables on their coordinate axes. Note that, generally, the scales are dimensional constants, for if we measure the distance on the axes in, for example, units of mm, then

$$[S_V] = \frac{[V]}{\text{mm}} \, ; \qquad [S_W] = \frac{[W]}{\text{mm}} \quad (6\text{-}15)$$

Combining (6-13) with (6-14), we get

$$y = \frac{p_1 \cdot S_V^{q_1}}{S_W} \cdot x^{q_1} + \frac{p_2 \cdot S_V^{q_2}}{S_W} \cdot x^{q_2} + \ldots \quad (6\text{-}16)$$

or

$$y = z_1 \cdot x^{q_1} + z_2 \cdot x^{q_2} + \ldots \quad (6\text{-}17)$$

where we defined the dimensional coefficients as

$$z_1 = \frac{p_1 \cdot S_V^{q_1}}{S_W} \, ; \qquad z_2 = \frac{p_2 \cdot S_V^{q_2}}{S_W}, \ldots \quad (6\text{-}18)$$

If we now select another dimensional system for our curve(s), then, generally and analogously to (6-18), we will have

$$z_1' = \frac{p_1' \cdot (S_V')^{q_1}}{S_W'} \, ; \qquad z_2' = \frac{p_2' \cdot (S_V')^{q_2}}{S_W'}, \ldots \quad (6\text{-}19)$$

Obviously, the curves in the two dimensional systems are identical if—and only if—the respective z coefficients in (6-17) remain unchanged. That is, if for all x

$$z_1 = z_1' \, ; \qquad z_2 = z_2', \ldots \quad (6\text{-}20)$$

Combining (6-18) and (6-19) with (6-20), we get the important relations

$$\frac{p_1'}{p_1} = \frac{S_W'}{S_W} \cdot \left(\frac{S_V'}{S_V}\right)^{-q_1} ; \qquad \frac{p_2'}{p_2} = \frac{S_W'}{S_W} \cdot \left(\frac{S_V'}{S_V}\right)^{-q_2}, \ldots \qquad (6\text{-}21)$$

Next, we observe that, if the abscissa and ordinate scales do not change between two dimensional systems, then

$$\frac{S_W'}{S_W} = \frac{S_V'}{S_V} = 1 \qquad (6\text{-}22)$$

which, considering (6-21), yields

$$\frac{p_1'}{p_1} = \frac{p_2'}{p_2} = \ldots = 1 \qquad (6\text{-}23)$$

or

$$p_1' = p_1 ; \qquad p_2' = p_2, \ldots \qquad (6\text{-}24)$$

The following example illustrates this process.

Example 6-12

On the curve of Fig. 6-3, point Q represents specific values of independent variable $V = 6$ s (abscissa) and dependent variable $W = 70$ m (ordinate). The respective coordinate distances

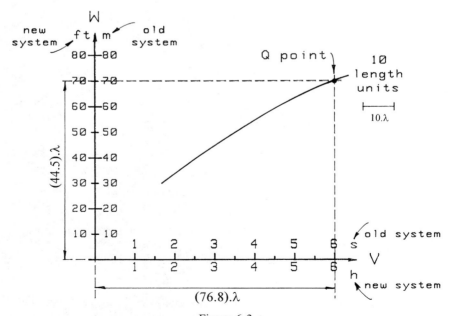

Figure 6-3
Coordinate scale transformation to accommodate dimensional system change

from origin 0 are $x = 76.8\,\lambda$ and $y = 44.5\,\lambda$, where λ is the "length unit." The length unit can be anything practical, e.g., inch, mm, etc. In the figure this unit is *defined* by length $10\cdot\lambda$, as shown.

In this figure, the *scales* in the m–s system (designated "old") will be, by (6-14),

$$S_V = \frac{6}{76.8} = 0.078125 \; \frac{s}{\lambda} \qquad \text{(abscissa scale)}$$

$$S_W = \frac{70}{44.5} = 1.57303 \; \frac{m}{\lambda} \qquad \text{(ordinate scale)}$$

Now we change the dimensional system to the one which uses "hour" for time and "ft" for length. So, to preserve the numerical values of the scales, in the "new" system we must have

$$S'_V = 0.078125 \; \frac{\text{hour}}{\lambda} \qquad \text{(abscissa scale)}$$

$$S'_W = 1.57303 \; \frac{\text{ft}}{\lambda} \qquad \text{(ordinate scale)}$$

⇑

Recall that relation (6-24) means that the numerical values of parameters must not change when we switch dimensional systems. But the only type of quantity impervious to any change of the dimensional system (in which they are expressed) is a dimensionless quantity (variable or constant). In short, all *p* parameters must be dimensionless. This fact now allows us to formulate an appropriate rule as follows:

Rule 6 (homogeneity of a graph). *A graph is homogeneous, i.e., valid in any dimensional system without change of coordinate scales, only if the parameter(s) associated with the variables is (are) dimensionless.*

Example 6-13

Consider a sphere of radius R and surface area A. Then $A = 4\cdot\pi\cdot R^2$, where, following the format of (6-13)

$R \equiv V$ is the independent variable (abscissa)
$W \equiv A$ is the dependent variable (ordinate)
$p \equiv 4\cdot\pi$ is the parameter
$q = 2$ is the exponent of the independent variable R

The parameter $p = 4\cdot\pi$ is a dimensionless constant and hence the graph in Fig. 6-4 is homogeneous. This means that the designations of the coordinate axes can be changed from one dimensional system to any other, without any alteration of the curve. For example, if we switch from the *meter* based system to the *inch* based system, then the numbers on the axes remain unchanged, the curve remains unchanged, and yet the read-off data from the curve remain correct.

For example, if $R = 4$ m, then by the curve, the surface area is $A = 201.06$ m². If now we change the dimension of length to "inch," then for $R = 4$ in the same curve yields $A = 201.06$ in².

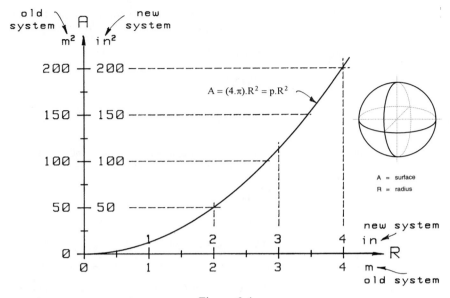

Figure 6-4
A dimensionally homogeneous graph for the surface area of a sphere
(The parameter $p = 4 \cdot \pi$ is dimensionless)

⇑

Example 6-14

Within elastic limit the stress–strain relationship is described by Hooke's law, named after Robert Hooke (English physicist, 1635–1703). By this law

$$\sigma = E \cdot \epsilon \qquad (a)$$

where σ is the stress in the material, normal to the cross-section
  ϵ is the elongation per unit length, called "strain"
  E is a proportionality constant, called Young's modulus, after Thomas Young (English physicist and physician, 1773–1829).

The dimension of strain is $[\epsilon] = [\text{length}]/[\text{length}] = 1$, i.e., strain is dimensionless. Now by (a) and Rule 1 of dimensional homogeneity (both sides of an equation must have the same dimension),

$$[\sigma] = [E] \cdot [\epsilon] = [E] \cdot 1 = [E]$$

This means that the dimension of the Young's modulus is the same as that of stress. If we now plot the stress vs. strain curve, then by (a) we get a straight line through the origin of the coordinate system. The tangent of the angle between this line and the abscissa axis is E. That is the parameter is E, a *dimensional* (i.e., not dimensionless) parameter. Therefore, by Rule 6, the plot shown in Fig. 6-5 is dimensionally *not* homogeneous.

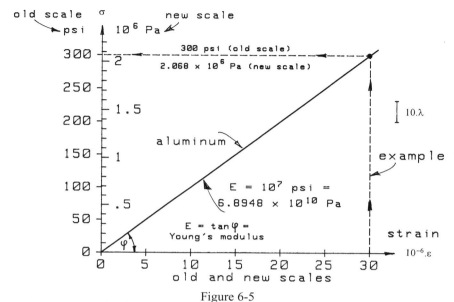

Figure 6-5
A stress vs. strain dimensionally nonhomogeneous plot of an elastic material
A change of dimensional system alters the coordinate scales
even though parameter E and the curve are left intact

This means that, if we wish to change the dimensional system, we have the following three variations to go by:

Variation 1: we leave the curve and the parameter intact, and change the coordinate scales;

Variation 2: we leave the parameters and the coordinate scales intact, and change the curve;

Variation 3: we leave the curve and coordinate scales intact, and change the parameter

In this example, we select *Variation 1*; in the following two examples the remaining two variations will be dealt with.

Let us then select the in–lbf–s dimensional system; this will be referred as the "old" system. By (6-13) and (6-14) now

$$\epsilon = S_\epsilon x \, ; \qquad \sigma = S_\sigma y \qquad (b)$$

where S_ϵ and S_σ are the coordinate scales for strain and stress, respectively. The right sides of the pair of relations in (b) are now substituted into (a). This will yield

$$y = \left(\frac{E \cdot S_\epsilon}{S_\sigma}\right) \cdot x$$

116 APPLIED DIMENSIONAL ANALYSIS AND MODELING

If we now select another dimensional system, say SI, then in this "new" system

$$y = \left(\frac{E' \cdot S'_\epsilon}{S'_\sigma}\right) \cdot x$$

But we insist that the curve (straight line) not change. Therefore, by the last two formulas,

$$\frac{E \cdot S_\epsilon}{S_\sigma} = \frac{E' \cdot S'_\epsilon}{S'_\sigma}$$

yielding

$$S'_\sigma = \left(\frac{E'}{E}\right)\left(\frac{S'_\epsilon}{S_\epsilon}\right) \cdot S_\sigma \qquad (c)$$

which is of course relation (6-21). Next, we must evaluate the right side of (c). To this end, we first observe that $S'_\epsilon = S_\epsilon$, since ϵ is dimensionless and therefore its scale is not affected by changes in the dimensional system. Next, considering the "length unit" λ in Fig. 6-5

$$S_\sigma = \frac{300}{95} \frac{\text{psi}}{\lambda} = 3.15789 \frac{\text{psi}}{\lambda}$$

and also 1 psi = 6894.75729 Pa (Fig. 4-1g). Therefore

$$\frac{E'}{E} = 6894.75729$$

By these values, (c) supplies

$$S'_\sigma = (6894.75729) \cdot (3.15789) = 21772.91777 \frac{\text{Pa}}{\lambda}$$

This means that 10^6 Pa is 45.92862 λ long on the σ axis (at this point we reiterate that λ is the "length unit" on the abscissa; it acts like meter or inch, etc.). The new scale, shown in Fig. 6-5, can be confirmed by checking *one* of its points. Say we select $\epsilon = 30 \times 10^{-6}$ (30 "microstrain") and $E = 10^7$ psi (aluminum). This pair gives us in the old system

$$\sigma = E \cdot \epsilon = (10^7) \cdot (30 \times 10^{-6}) = 300 \text{ psi}$$

which can be seen on the *outer* scale on the ordinate axis. Next, we consider that

$$10^7 \text{ psi} = (10^7) \cdot (6894.75729) = 6.89476 \times 10^{10} \text{ Pa}$$

Therefore in SI (new scale) the stress is

$$\sigma = (6.89476 \times 10^{10}) \cdot (30 \times 10^{-6}) = 2.068 \times 10^6 \text{ Pa} = 2.068 \text{ MPa}$$

The correctness of this value can be easily verified on the *inner* scale of the ordinate axis.
⇑

The next example deals with *Variation 2* in a dimensional system change. In this variation the coordinate scales and the parameter remain the same, but the graph changes, since it is non-homogeneous.

Example 6-15. Deflection of a Cantilever Upon a Concentrated Lateral Load (I)

Consider a cantilever loaded at its free end by a lateral force F. The length of the beam is L, its cross-section's second moment of area is I, and its material's Young's modulus is E. The free end is deflected U distance upon the applied load F. Fig 6-6 shows a sketch of the configuration.

DIMENSIONAL HOMOGENEITY **117**

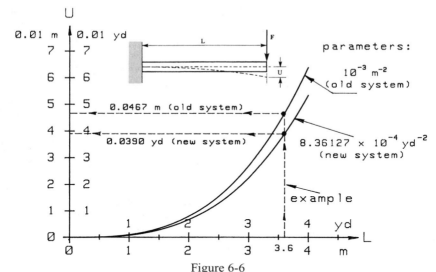

Figure 6-6
**Dimensionally nonhomogeneous plot for the
deflection of a laterally loaded cantilever**
A change of dimensional system alters the curve even though
parameter p and the coordinate scales are left intact

The deflection U can be expressed by the analytically derived formula

$$U = \frac{F}{3 \cdot I \cdot E} \cdot L^3 = p \cdot L^3 \tag{a}$$

in which L is the independent variable, U is the dependent variable, and $p \equiv F/(3 \cdot I \cdot E)$ is the parameter. We note with delight that the dimension of the parameter is (length)$^{-2}$, therefore when we switch dimensional systems we only have to deal with *one* dimension.

The graph of function (a) can be written as $y = z \cdot x^3$, where x is the abscissa and y is the ordinate. The coefficient z is by (6-18)

$$z = \frac{p \cdot S_L^3}{S_U}$$

where p is the parameter defined in (a) and S_L and S_U are the scale factors for length and deflection. Let us now designate SI as the "old" system and select a parameter $p = 10^{-3}$ 1/m². The graph of this relation (a cubical parabola) is shown in Fig. 6-6 (the upper curve). Now we switch to the "new" dimensional system. This system uses the dimension "yard" (yd) for length. We write

$$z' = \frac{p' \cdot (S_L')^3}{S_U'}$$

where p', S_L', and S_U' are the corresponding quantities in the new system. By the above two formulas

$$\frac{z'}{z} = \frac{p'}{p} \cdot \left(\frac{S_L'}{S_L}\right)^3 \cdot \left(\frac{S_U'}{S_U}\right)^{-1}$$

We do not change the scales on the coordinate axes, so $S'_L = S_L$ and $S'_U = S_U$. Hence, by this last relation

$$\frac{z'}{z} = \frac{p'}{p}$$

Furthermore, since $p = 10^{-3}$ 1/m² and 1 m = 1.09361 yd, therefore

$$p' = 8.36127 \times 10^{-4} \frac{1}{\text{yd}^2} \quad \text{(b)}$$

Consequently,

$$\frac{z'}{z} = \frac{p'}{p} = \frac{8.36127 \times 10^{-4}}{10^{-3}} = 0.83613$$

which means that the ordinate distances of our curve in the *yd* system (the "new" system) will be 0.83613 times as much as their values in the "old" system (i.e., in SI). To verify this, let us select the length $L = 3.6$ m, shown in the figure. In SI we have $p = 10^{-3}$ 1/m², thus the deflection U_m will be

$$U_m = p \cdot L^3 = (10^{-3}) \times 3.6^3 = 0.0467 \text{ m}$$

In the new system, with $L = 3.6$ yd, we have the deflection U_{yd} with the parameter from (b)

$$U_{yd} = p' \cdot L^3 = (8.36127 \times 10^{-4}) \cdot (3.6)^3 = 0.0390 \text{ yd}$$

Let us now see how large U_{yd} would be if the equivalent number of yards equal to 3.6 m were used. We have 3.6 m = 3.93701 yd, therefore

$$U_{yd} = (8.36127 \times 10^{-4}) \cdot (3.93701)^3 = 0.05102 \text{ yd} = 0.0467 \text{ m}$$

which agrees with the previous figure obtained directly.

We wish to emphasize that the two parameters characterizing the two curves in the two different dimensional systems are only *numerically* different; *physically* they are the same. A simplistic analogy is: 1 ft and 12 inches are only numerically different; since they both represent the same length, they are physically identical.

⇑

The next example deals with *Variation 3*, in which both the scales and the curve remain intact, but since the graph is nonhomogeneous, a change of dimensional system causes a change of parameter.

Example 6-16

Again we consider the cantilever, the subject of the previous example. On this beam the concentrated force F, acting at the free end, produces a deflection U expressed by

$$U = \frac{F}{3 \cdot I \cdot E} \cdot L^3 = p \cdot L^3 \quad \text{(a)}$$

where the variables are as defined in Example 6-15, and p is the parameter. The sketch in Fig. 6-7 shows the arrangement. According to (6-21), if the curve remains unchanged, then for two consistently applied dimensional systems

$$\frac{p'}{p} = \frac{S'_W}{S_W} \cdot \left(\frac{S'_V}{S_V}\right)^{-q_1} \tag{b}$$

where p and p' are the numerical values of the parameters in the old and new dimensional systems;
S_V and S'_V are the coordinate scales for the independent variable L in the old and new systems;
S_W and S'_W are the coordinate scales for the dependent variable U in the old and new systems;
q_1 is the exponent of the dependent variable. In this case $q_1 = 3$.

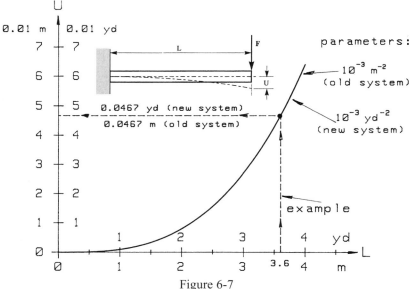

Figure 6-7
Dimensionally nonhomogeneous plot for the deflection of a laterally loaded cantilever
Change of dimensions alters the physical size of parameter p
even though curve and coordinate scales are left intact

Since, by definition, $S_V = S'_V$ and $S_W = S'_W$ (coordinate scales do not change), therefore, by relation (b) we have $p' = p$. In other words, the *numerical* values of parameters do not change. Consequently, if the parameter in the "old" system (using "meter") is, say, $p = 10^{-3}$ 1/m², then for the same curve and with the same coordinate scales, the parameter in the "new" system (using "yard") will be $p' = 10^{-3}$ 1/yd². The curve, with parameter, is shown in Fig. 6-7.

To verify this curve, let us select $L = 3.6$ m. With parameter $p = 10^{-3}$ 1/m² this L value is substituted into (a) to yield

$$U = (10^{-3}) \cdot (3.6)^3 = 0.0467 \text{ m}$$

If we now switch to the yard-based system, then $L = 3.6$ yd (abscissa scale unaltered). So, again, with parameter $p' = 10^{-3}$ 1/yd²

$$U = (10^{-3}) \cdot (3.6)^3 = 0.0467 \text{ yd}$$

as the plot correctly shows.

⇑

Let us again consider (6-13), which is repeated here for convenience

$$W = p_1 \cdot V^{q_1} + p_2 \cdot V^{q_2} + \ldots \qquad \text{repeated (6-13)}$$

where V is the dependent variable, W is the independent variable, $p_1, p_2, \ldots$ are the parameters, and $q_1, q_2, \ldots$ are dimensionless exponents. If this equation is homogeneous and if both V and W are dimensionless variables, then by Rule 1 and Rule 2, all the parameters must also be dimensionless. To prove this assertion, assume that p_1 is not dimensionless. Then, by Rule 2, none of the other p parameters can be dimensionless, and therefore they are all dimensional. Hence all members on the right side of (6-13) are dimensional. But, by assumption, the left side is dimensionless. Therefore Rule 1 is violated, and hence the equation is dimensionally nonhomogeneous. But this contradicts our original assumption. Thus the assertion is proved. In view of the above, the following rule can be formulated:

Rule 7. *If, in a dimensionally homogeneous equation, the variables are all dimensionless, then all the parameters in it must also be dimensionless.*

Now we consider a dimensionless quantity whose magnitude is calculated in a dimensional system of fundamental dimensions $d_1, d_2, \ldots$. Obviously, if we use the symbol* π_1 for this quantity, then

$$\pi_1 = Q \cdot d_1^0 \cdot d_2^0 \ldots \qquad (6\text{-}25)$$

where Q is the magnitude of the variable. Next, we switch to a dimensional system whose fundamental dimensions are—correspondingly—$D_1, D_2, \ldots$. Then we have the relations

$$\left. \begin{aligned} d_1 &= k_1 \cdot D_1^{e_{11}} \cdot D_2^{e_{12}} \ldots \\ d_2 &= k_2 \cdot D_1^{e_{21}} \cdot D_2^{e_{22}} \ldots \end{aligned} \right\} \qquad (6\text{-}26)$$

where $k_1, k_2, \ldots, e_{11}, e_{12}, \ldots$ are numeric constants. Substitution of (6-26) into (6-25) yields

$$\pi_1' = Q \cdot (k_1^0 \cdot k_2^0 \ldots) \cdot (D_1^0 \cdot D_2^0 \ldots) \qquad (6\text{-}27)$$

---

*Use of the symbol π, *always with subscript*, to denote a dimensionless variable is traditional. If there is only one π, then the subscript is 1, in which case we use the symbol π_1. In this context, of course, π has nothing to do with the number 3.14159....

DIMENSIONAL HOMOGENEITY   **121**

The expression in the first parentheses is a *number* whose value is 1, while that in the second is a *dimension* which is 1. Hence, by (6-27) and (6-25), $\pi_1 = \pi_1'$. Thus, we can formulate the following rules:

Rule 8.   *The magnitude of a dimensionless variable is invariant upon the dimensional system in which the constituents of that variable are expressed.*

Rule 9.   *If a quantity is dimensionless in a dimensional system, then it is also dimensionless in all other dimensional systems.*

The following two examples illustrate.

Example 6-17

We again have our cantilever discussed in previous examples. If E is the Young's modulus, F is the lateral deflecting force at the free end of the cantilever, and L is the length, then the quantity

$$\pi_1 = \frac{F}{L^2 \cdot E} \tag{a}$$

is dimensionless. Let us calculate the value of this quantity in SI if

$$F = 1000 \; \frac{\text{m} \cdot \text{kg}}{\text{s}^2} \; (\text{newton}) \; ; \quad L = 20 \text{ m} \; ; \quad E = 2 \times 10^{11} \; \frac{\text{kg}}{\text{m} \cdot \text{s}^2} \; (\text{pascal})$$

In SI, relation (a) provides

$$\pi_1 = \frac{1000}{20^2 \cdot (2 \times 10^{11})} = 1.25 \times 10^{-11}$$

Next, we switch to the in–lbf–s system. So now we have the equivalent values of

$$F' = 224.80894 \text{ lbf} \; ; \quad L' = 787.40157 \text{ in} \; ; \quad E' = 2.90075 \times 10^7 \text{ lbf/in}^2 \text{ (psi)}$$

which supplies, again by (a),

$$\pi_1' = \frac{224.80894}{(787.40157)^2 \cdot (2.90075 \times 10^7)} = 1.25 \times 10^{-11}$$

We see that indeed $\pi_1 = \pi_1'$, as expected.

⇑

Example 6-18

Using again our cantilever model of previous examples, we express the deflection U of the beam by

$$U = \frac{F \cdot L^3}{3 \cdot I \cdot E} \tag{a}$$

This formula can be transformed into the form

$$\left(\frac{U}{L} \right) = \left(\frac{F}{3 \cdot L^2 \cdot E} \right) \cdot \left(\frac{L^4}{I} \right) \tag{b}$$

using the method set forth in Chapter 7. Here we only mention in passing that all three quantities appearing in parentheses are dimensionless, as the reader can verify. The reader can also easily verify that formulas (a) and (b) are indeed equivalent.

Let us denote the above three dimensionless variables thus

$$\pi_1 = \frac{F}{3 \cdot L^2 \cdot E} \;;\quad \pi_2 = \frac{U}{L} \;;\quad \pi_3 = \frac{L^4}{I} \tag{c}$$

Therefore by (c), relation (b) can be written

$$\pi_2 = \pi_1 \cdot \pi_3 \tag{d}$$

If we now choose π_2 and π_3 as the *dependent* and *independent* variables, respectively, then π_1 becomes the *parameter*. But π_1 is dimensionless, so, by Rule 6, the graph of relation (b)—or alternatively (d)—is dimensionally homogeneous, i.e., it is valid without any modification in any dimensional system. Fig 6-8 presents such a graph.

For example, let us select

force $\qquad\qquad\qquad\qquad\qquad\qquad F = 1000$ m·kg/s² (newton)
length $\qquad\qquad\qquad\qquad\qquad\quad\; L = 15$ m
Young's modulus $\qquad\qquad\qquad\;\, E = 2 \times 10^{11}$ kg/(m·s²) (pascal)
second moment of area of cross-section $\quad I = 3 \times 10^{-5}$ m⁴

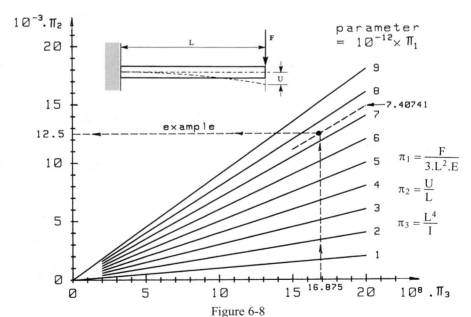

Figure 6-8
**Dimensionally homogeneous plot for the deflection
of a laterally loaded cantilever**
A change of dimensional system has no effect on the coordinate scales,
on the parameter, or on the curves

By (c), the parameter is

$$\pi_1 = \frac{F}{3 \cdot L^2 \cdot E} = 7.407407 \times 10^{-12}$$

and the independent variable

$$\pi_3 = \frac{L^4}{I} = 1.6875 \times 10^9$$

With these data, the chart on Fig. 6-8 provides the dependent variable

$$\pi_2 = 0.0125$$

but, by (c), $\pi_2 = U/L$, therefore the deflection U will be

$$U = \pi_2 \cdot L = (0.0125) \cdot 15 = 0.1875 \text{ m}$$

If we now switch to the in–lbf–s dimensional system, then we have the equivalent values

$F' = 224.80894$ lbf
$L' = 590.55118$ in
$E' = 2.9075 \times 10^7$ lbf/in^2
$I' = 72.07529$ in^4

This set will yield

$$\pi_1' = \frac{F'}{3 \cdot (L')^2 \cdot E'} = 7.40741 \times 10^{-12} = \pi_1$$

$$\pi_3' = \frac{(L_1)^4}{I'} = 1.6875 \times 10^9 = \pi_3$$

as expected and required. Moreover, by the figure, necessarily

$$\pi_2' = \pi_2 = 0.0125$$

So that $\pi_2' = U'/L' = 0.0125$, from which

$$U' = \pi_2' \cdot L' = (0.0125) \cdot (590.55118) = 7.38189 \text{ in}$$

But 7.38189 in = 0.1875 m, the very deflection obtained previously. Therefore it has been demonstrated that a change in the dimensional system has no effect on the dimensionally homogenous plot.

Finally, we note that the same result could have been obtained *without* any calculation by merely invoking Rules 9 and 8. For the former assured us that π_1', π_2', π_3' are dimensionless, while the latter mandated that the two sets of dimensionless variables (belonging to the two different dimensional systems) are numerically equal.

⇑

Incidentally, the method just expounded also shows the great advantage of *transforming* a relation into a dimensionless form, because the number of dimensionless variables completely describing a physical system is *always* less than the number of physical variables involved. A very pleasing consequence of this is that the logistics of presenting the results is *significantly* improved. To illustrate: in Example 6-18

124   APPLIED DIMENSIONAL ANALYSIS AND MODELING

we had five *physical* variables (F, U, I, L, E), but only three *dimensionless* variables. Thus, by using dimensionless variables, *one* chart was sufficient to describe the relation, but with the five physical variables a *set of set* of charts would have been needed (see also Chapter 12).

Sometimes by considering dimensional homogeneity alone, rather astonishing results can be easily derived. To be sure, some "heuristic reasoning" might be necessary but, as the following example shows, the process is quite simple.

Example 6-19. Jumping Heights of Animals (I)

Using the criterion of dimensional homogeneity, we will now show that all similarly constructed animals (including man) can jump from a resting position on the ground to (approximately) the same height. That is, the height of the jump is independent of the size of the animal.

Assuming that the proportion of muscle mass to body mass is the same for all animals, it is evident that the maximum force F is proportional to the cross-section of the muscle, which, therefore, varies as the square of the animal's linear size L. Therefore we can write

$$F = k_1 \cdot L^2$$

where k_1 is a dimensional constant having the dimension $m^{-1} \cdot kg \cdot s^{-2}$. The distance through which this force acts is λ (see Fig. 6-9), which is therefore proportional to L. Hence,

$$\lambda = k_2 \cdot L$$

where k_2 is a dimensionless constant.

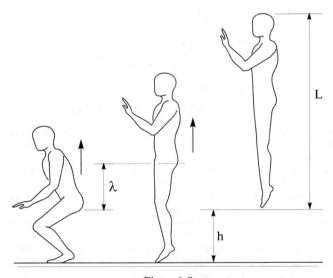

Figure 6-9
A man of size L jumps from a stationary position on the ground to height h
He accelerates through distance λ. The center figure shows
the configuration as his feet just leave the ground

DIMENSIONAL HOMOGENEITY

The energy imparted by the muscles is

$$E_m = F \cdot \lambda = (k_1 \cdot L^2) \cdot (k_2 \cdot L) = k_1 \cdot k_2 \cdot L^3 \tag{a}$$

The potential energy at height h is $E_p = M \cdot g \cdot h$, where M is the mass of the animal, and g is the gravitational acceleration. Now since M is proportional to L^3 and g is constant, we write

$$E_p = k_3 \cdot L^3 \cdot h \tag{b}$$

where k_3 is a dimensional constant of dimension $m^{-2} \cdot kg \cdot s^{-2}$.

Since the energy imparted by the muscles must be equal to potential energy, therefore $E_m = E_p$, from which, by (a) and (b),

$$k_1 \cdot k_2 \cdot L^3 = k_3 \cdot L^3 \cdot h$$

and, by elementary simplification,

$$h = \frac{k_1 \cdot k_2}{k_3} = \text{const.} \tag{c}$$

which, of course, must be dimensionally homogeneous, since it is a theoretically derived formula. Checking, we find that

$$\left[\frac{k_1 \cdot k_2}{k_3}\right] = \frac{[k_1] \cdot [k_2]}{[k_3]} = \frac{(m^{-1} \cdot kg \cdot s^{-2}) \cdot 1}{(m^{-2} \cdot kg \cdot s^{-2})} = m$$

which is identical to the dimension of h, thereby confirming the dimensionally homogeneous nature of (c).

Our equation (c) now says that all animals—including man—can jump to the same height.* Observation shows this to be approximately true, as dogs, men, and horses all jump to about the same height of 2 m. We shall revisit this problem in Art 11.2 (Example 11-16) and obtain the same result (but much more quickly) by a purely dimensional method.

⇑

Example 6-20. Urine Secretion Rate of Mammals

N. Edwards (Ref. 74) quotes other sources for an experimentally derived formula giving the urine secretion rate of mammals. The referred experiment encompassed 30 species and the correlation obtained was 0.95. A slightly modified version of the formula is

$$V = p \cdot M^q \tag{a}$$

where $V =$ the urine secretion rate, ml/day;
$M =$ the mass of the animal, lb;
$p = 33.63$, a dimensional constant (see below for its dimension);
$q = 0.75$, a dimensionless constant.

What is the value of parameter p, if we use SI throughout?

From the information given and by the condition of dimensional homogeneity, we can determine the dimension of p as follows: by (a) and by the given value for q

$$p = \frac{V}{M^{0.75}}$$

---
*A detailed treatment of this interesting biomechanical phenomenon can be found in Ref. 43, p. 155 and Ref. 122, p. 13.

Hence

$$[p] = \frac{[V]}{[M]^{0.75}} = \frac{\text{ml/day}}{\text{lb}^{0.75}} = \text{ml·day}^{-1}\cdot\text{lb}^{-0.75}$$

Thus, following the method outlined in Chapter 4, we can write

$$33.63 \text{ ml·day}^{-1}\cdot\text{lb}^{-0.75} = x \cdot \text{m}^3 \cdot \text{s}^{-1} \cdot \text{kg}^{-0.75}$$

and we have the following equivalences: 1 ml = 10^{-6} m^3, 1 day = 86400 s, 1 lb = 0.4536 kg. Hence, by relations (4-3) and (4-4)

$$x = 33.63 \cdot (10^{-6})^1 \cdot (86400)^{-1} \cdot (0.4536)^{-0.75} = 7.0422 \times 10^{-10}$$

Therefore the parameter in SI is

$$p_{SI} = 7.0422 \times 10^{-10} \frac{\text{m}^3}{\text{s·kg}^{-0.75}}$$

and in SI formula (a) for the urine secretion rate of mammals becomes

$$V = 7.0422 \times 10^{-10} \cdot M^{0.75} \frac{\text{m}^3}{\text{s}}$$

where M is the mass of the animal in kg. For example, a 20 kg dog produces about 6.66×10^{-9} m^3/s = 0.58 liters of urine per day.

⇑

6.3. PROBLEMS (Solutions are in Appendix 6)

6/1 P. Bridgeman mentions (Ref. 44, p. 74) that at a Yale University lecture entitled *Application of thermodynamics to chemistry,* Prof. H. Nernst (German physical chemist, 1864–1941) announced the following equation:

$$\ln C = -\frac{H}{R\cdot T} + \frac{k_1}{R}\cdot \ln T + \frac{k_2}{R}\cdot T + \frac{k_3}{2\cdot R}\cdot T^2 + q' \quad \text{(a)}$$

in which C = gas concentration in terms of mole per volume;
H = heat content per mole;
R = universal gas constant;
T = absolute temperature;
$\left.\begin{array}{l} k_1 \\ k_2 \\ k_3 \end{array}\right\}$ = constants of appropriate dimensions, in particular $[k_1] = [R]$;
q' = a constant of integration.

Obviously relation (a) is dimensionally nonhomogeneous, since it violates Rule 4 inasmuch as C is not dimensionless.

Modify (a) such that it becomes dimensionally homogeneous, and determine the dimensions of the k constants.

6/2 Express the surface area of a right circular cylinder in terms of its diameter and length, and then construct (a) a dimensionally non-homogeneous chart,

DIMENSIONAL HOMOGENEITY

and (b) a dimensionally homogeneous chart from which this area could be determined.

6/3 Somebody *experimentally* determined the following relationship for a worker's yearly income:

$$D = \frac{46800}{e^{(0.2) \cdot a}} + 6000 \cdot e$$

where a = the worker's age in years;
e = his education beyond secondary level in years;
D = his yearly income in dollars.

It is evident that if the numbers appearing are dimensionless, then the equation is dimensionally nonhomogeneous. Assign dimensions to (some of) the numbers to make the equation homogeneous.

6/4 One of the exponents of x in the following theoretically derived equation is in error

$$\int x^2 \cdot \sin(a \cdot x) \cdot dx = \frac{2 \cdot x}{a^2} \cdot \sin(a \cdot x) + \frac{2 \cdot x}{a^3} \cdot \cos(a \cdot x) - \frac{x^2}{a} \cdot \cos(a \cdot x)$$

Using only dimensional considerations, find and correct this erroneous exponent.

6/5 A ball dropped in still air achieves a terminal velocity v, which is a function of the ball's radius R, density ρ, and gravitational acceleration g (considered constant). Accordingly, the following formula is valid

$$v = k \cdot \sqrt{R^{n_1} \cdot \rho^{n_2} \cdot g^{n_3}}$$

where the dimension of k is $m^{1.5} \cdot kg^{-1.5}$ in SI. Assuming the dimensional homogeneity of this formula, determine the n exponents.

6/6 Pankhurst (Ref. 28, p.54) mentions an equation used in wave mechanics. The equation (in a slightly modified form) can be written as follows:

$$\frac{\partial^2 \Psi}{\partial x^2} - \frac{4 \cdot \pi \cdot M_0}{h} \cdot \frac{\partial \Psi}{\partial t} - \frac{8 \cdot \pi^2 \cdot M_0}{h^n} \cdot (E + M_0 \cdot c^2) \cdot \Psi = 0$$

where M_0 is the rest-mass of the particle;
x is the distance traveled;
h is Planck's constant;
E is the potential energy;
n is a number;
Ψ is a variable of unknown dimension;
t is time.

Determine n from the assumed dimensional homogeneity of the formula.

6/7 Volume of a barrel.
A barrel is defined by lengths R, D, and h, as shown in Fig. 6-10. The volume of this barrel is given by the formula

$$V = \frac{\pi}{6} \cdot \left[3 \cdot h \cdot (8 \cdot R^2 - 4 \cdot R \cdot D + D^{n_1}) - 4 \cdot h^{n_2} - 6 \cdot (2 \cdot R - D) \cdot \left(h \cdot \sqrt{R^2 - h^2} + R^{n_3} \cdot \arcsin \frac{h}{R} \right) \right]$$

128   APPLIED DIMENSIONAL ANALYSIS AND MODELING

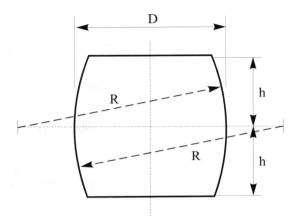

Figure 6-10
Definition of a barrel by three lengths

(a) By the criteria of dimensional homogeneity, determine the n exponents.
(b) Express the formula in a dimensionless form.
(c) Using the result of (b), construct a dimensionless chart to determine the volume.
(d) Prove the correctness of (b) in the degenerated cases: the barrel becomes (i) a cylinder; (ii) a sphere.

6/8 One of the *Navier–Stokes* equations (dealing with a steady two-dimensional flow of incompressible fluid) is

$$U \cdot \frac{\partial U}{\partial x} + V \cdot \frac{\partial U}{\partial y} = -\frac{1}{\rho^{n_1}} \cdot \frac{\partial p}{\partial x} + v \cdot g^{n_2} \cdot \left(\frac{\partial^2 U}{\partial x^2} + \frac{\partial^2 U}{\partial y^2} \right)$$

where U and V are the x, y directional velocity components, ρ and v are the density and kinematic viscosity of the fluid, and g is the gravitational acceleration. Assuming that the equation is dimensionally homogeneous, determine the n exponents.

6/9 Circulation velocity of money.
De Jong (Ref. 36, p. 23) mentions the *Fisher formula* for monetary exchange. This formula, with some modifications, is as follows:

$$V_m = \frac{c \cdot P \cdot V_\gamma}{M}$$

where V_m = the circulation velocity of money;
M = the quantity of money in circulation at any one time in a country;
P = the price of unit quantity of goods;
V_γ = velocity of mass flow of goods;
c = a dimensional constant.

Based on the above description,

(a) establish a dimensional system concordant with the Fisher equation;
(b) in system (a), establish the dimensions of all the *variables* of the Fisher formula;
(c) establish the dimension of constant c.

6/10 De Jong (Ref. 36, p. 35) gives the *Cobb–Douglas production function*, which, in a slightly modified form, is as follows:

$$U = c \cdot (n \cdot p)^\alpha \cdot k^{1-\alpha}$$

where U = the rate of production;
n = the number of persons working;
p = the average productivity of a person;
k = capital in use;
α = a constant between 0 and 1;
c = is a dimensional constant characterizing the *level* of technical knowledge. It is a function of such elusive parameters as "organization," "quality of capital goods," "skill of labor," etc. For this reason c is called *technology parameter*.

(a) By the definitions given above, determine the dimensions of variables. Use the dimensional system established in Problem 6.9 and augment this system with additional dimension(s) as necessary;
(b) By the conditions for dimensional homogeneity, determine the dimension of technology parameter c as a function of constant α.

6/11 Bending of light in a gravitational field.
According to Einstein's Theory of Relativity, light rays are bent in a gravitational field toward the mass creating that field. In particular, this deflection can be measured as light emanating from star A grazing a *massive* star B (Fig. 6-11). The deflection φ (see Fig. 6-11) is proportional to the mass M of star B, and is affected by B's radius R, the speed of light c, and the universal gravitational constant k.

(a) By the above given details, derive a formula for the relativistic deflection φ of light.
(b) Using the formula derived in (a), and the actually measured relativistic deflection of light $\varphi = 1.75$ arcsec by the Sun, what deflection does the planet Jupiter cause on a ray emitted by a distant star just *grazing* Jupiter's surface? Use the following data:

radius of Sun	$R_S = 6.955 \times 10^8$ m
mass of Sun	$M_S = 1.987 \times 10^{30}$ kg
radius of Jupiter	$R_J = 6.898 \times 10^7$ m
mass of Jupiter	$M_J = 1.897 \times 10^{27}$ kg
speed of light	$c = 2.998 \times 10^8$ m·s^{-1}
universal gravitational constant	$k = 6.673 \times 10^{-11}$ m^3·kg^{-1}·s^{-2}

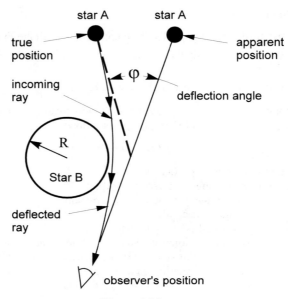

Figure 6-11
Gravitational field of massive star B bends a ray of light emitted by star A

6/12 Somebody composed the following *empirical* relationship for a city of 434,260 inhabitants:

$$\ln R = k_1 \cdot T^{-0.8} + (0.8) \cdot \ln((43.8) \cdot \rho + 1)$$

in which R = total revenue generated by taxi trips per day; thus $[R]$ = \$/day
T = ambient average absolute temperature for the day; thus $[T] = K$
ρ = rainfall for the day in meters; thus $[\rho]$ = m/day
k_1 = 1270, a dimensional constant called *affluence parameter*.

Obviously, and for several reasons, this equation is dimensionally nonhomogeneous.

(a) Determine the dimension of k_1.
(b) To make the equation dimensionally homogeneous, introduce dimensional constants as necessary, and state their dimensions and magnitudes.
(c) Make some (simple) qualitative observations regarding the nature and general usefulness of the given relation.

6/13 Given for free fall

$$v = \sqrt{3 \cdot g \cdot h}$$

where g = constant gravitational acceleration (m/s²);
h = height of the fall (m);
v = attained speed upon falling h distance (m/s).

(a) By the conditions of dimensional homogeneity, verify or refute the correctness of the formula.

(b) If, and only if, the formula is dimensionally homogeneous, comment on its *numerical* correctness from the viewpoint of its dimensional correctness (homogeneity).

6/14 Bending a uniform cross-section cantilever.

A uniform solid-circle cross-section cantilever is loaded at its free end with a bending moment M whose axis is perpendicular to the plane of Fig. 6-12, as shown. The maximum normal stress σ generated by M is obviously a function of M and diameter D of the cross-section. By relying entirely on the fulfillment of the condition of dimensional homogeneity, derive a formula giving σ in terms of D and M.

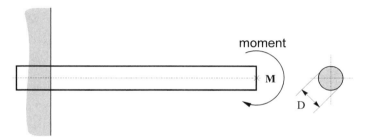

Figure 6-12
Uniform cross-section cantilever is loaded with bending moment M

6/15 Given the partial differential equation

$$F \cdot x^2 \cdot \frac{\partial^2 F}{\partial x \partial y} = K \cdot \left(\left(x \cdot \frac{\partial F}{\partial x} \right)^2 - \left(y \cdot \frac{\partial F}{\partial y} \right)^2 \right)$$

where F is an unknown function of x and y,
K is a dimensional constant,
x and y are dimensional variables.

By the conditions of dimensional homogeneity, determine the dimension of U, if $U = K \cdot y / x$.

6/16 Body surface area of slim individuals.

Kleiber (Ref. 76, p. 514) quotes the *DuBois formula* for the surface area of *ectomorphic* (stout and slim) individuals, as a function of their weights and heights. In a slightly modified form the formula is

$$A = k \cdot W^{0.425} \cdot L^{0.725}$$

where A = surface area, ft^2
W = weight, lbf
L = height, ft
k = 0.1086, a dimensional constant

(a) Determine the dimension of k.

(b) Determine the dimension and magnitude of k in SI.

6/17 Frequency of feeding versus mass of warm-blooded animals.

Assume that under stable conditions the metabolic rate (i.e., expended energy per time) of a warm-blooded animal (a homoiotherm) varies as the 0.75 power of its mass, and that the quantity of food per *one* meal is proportional to the animal's mass.

Using only dimensional considerations, derive a formula giving the frequency of feeding (number of meals per day) as a function of the mass of the animal.

CHAPTER 7
STRUCTURE OF PHYSICAL RELATIONS

7.1. MONOMIAL POWER FORM

Suppose we have a relation among variables

$$y = \Psi\{x_1, x_2, \ldots, x_n\} \tag{7-1}$$

where Ψ symbolizes a function. Then the approximation theory in mathematics (Ref. 27, p. 4) states that if Ψ extends within a certain range, then it can be approximated to any degree of accuracy by the series

$$y = k_1 \cdot (x_1^{a_1} \cdot x_2^{b_1} \ldots) + k_2 \cdot (x_1^{a_2} \cdot x_2^{b_2} \ldots) + \ldots \tag{7-2}$$

where $k_1, k_2, \ldots$ are dimensionless constant coefficients and $a_1, a_2, \ldots, b_1, b_2, \ldots$ are dimensionless exponents. By Rule 2 of Dimensional Homogeneity (Art. 6), (7-2) is dimensionally satisfied only if

$$\left. \begin{array}{c} a_1 = a_2 = \ldots = a \\ b_1 = b_2 = \ldots = b \\ \text{etc.} \end{array} \right\} \tag{7-3}$$

Therefore (7-2) can be written

$$y = (k_1 + k_2 + \ldots) \cdot (x_1^a \cdot x_2^b \ldots) = k \cdot x_1^a \cdot x_2^b \ldots \tag{7-4}$$

where of course

$$k = k_1 + k_2 + \ldots \tag{7-5}$$

is a dimensionless constant.

Next, by Rule 1 of Dimensional Homogeneity (Art.6.1), the dimension of the left-hand and right-hand sides of a valid physical relation (equation) must be the same. Consequently, if $[U]$ denotes the dimension of any quantity U, then, by (7-4),

$$[y] = [x_1]^a \cdot [x_2]^b \ldots \tag{7-6}$$

That is, as far as dimensions are concerned, the *monomial power* form is always applicable.

7.2. THE DIMENSIONAL MATRIX

In a physical relation the dimensions $d_1, d_2 \ldots$ of each variable $V_1, V_2, \ldots$ are best given and displayed by the *dimensional matrix* in which the *rows* are the dimensions, and the *columns* are the variables. A typical *element* of this matrix is the *exponent* to which the particular dimension (row) is raised in the particular variable (column). Fig 7-1 shows a dimensional matrix in which the elements are of course fictitious, as they are chosen for illustration only. For example, the dimension of variable V_2 is $d_1^2 \cdot d_2^4 \cdot d_3^4$ and of V_4 is $d_1^4 \cdot d_2^0 \cdot d_3^2 = d_1^4 \cdot d_3^2$.

	V_1	V_2	V_3	V_4	V_5
d_1	1	2	3	4	5
d_2	2	4	3	0	2
d_3	3	4	3	2	1

Figure 7-1
A dimensional matrix

Example 7-1

Construct the dimensional matrix for a relation involving the following physical variables:

Variable	Symbol	Dimension
pressure	p	$m^{-1} \cdot kg \cdot s^{-2}$
torque	T	$m^2 \cdot kg \cdot s^{-2}$
mass	M	kg
density	ρ	$m^{-3} \cdot kg$
angle of rotation	φ	1 (rad)
universal gravitational const.	k	$m^3 \cdot kg^{-1} \cdot s^{-2}$

Hence the dimensional matrix is

	p	T	M	ρ	φ	k
m	−1	2	0	−3	0	3
kg	1	1	1	1	0	−1
s	−2	−2	0	0	0	−2

Note that column φ contains all zeros, since it is measured in dimensionless radians.

⇑

The question now naturally emerges: what is the sequential order of variables in a dimensional matrix? For example, in the above matrix we wrote $p, T, M, \rho, \varphi, k$—why not $M, p, T, k, \varphi, \rho$? This question will be dealt with in some detail in Chapter 14; for now we only mention that it is *usually* advantageous to place the dependent variable in the leftmost position. Exceptions do occur, however, as will be shown later.

7.3. GENERATING PRODUCTS OF VARIABLES OF DESIRED DIMENSIONS

Assume we have a relation among N_v variables $V_1, V_2, \ldots$ and N_d dimensions $d_1, d_2, \ldots$. The task is to determine those particular combinations (groupings) of variables raised to certain powers (to be determined) that possess a preselected (i.e., desired) dimensional composition. To see this problem more concretely, we consider a relation involving five variables $V_1, V_2, \ldots, V_5$ and three dimensions d_1, d_2, d_3. We now seek those exponents $\epsilon_1, \epsilon_2, \ldots, \epsilon_5$ that satisfy

$$[V_1^{\epsilon_1} \cdot V_2^{\epsilon_2} \cdot V_3^{\epsilon_3} \cdot V_4^{\epsilon_4} \cdot V_5^{\epsilon_5}] = d_1^{q_1} \cdot d_2^{q_2} \cdot d_3^{q_3} \quad (7\text{-}7)$$

where q_1, q_2, q_3 in general are the desired exponents of the dimensions, including zero. To solve (7-7) for q_1, q_2, q_3 we of course must know the dimensions of every variable occurring in (7-7). Thus,

we know:	dimensions d_1, d_2, d_3
we know:	exponents of dimensions q_1, q_2, q_3
we know:	dimensions of each variable (from the dimensional matrix)
we do not know, but want to know:	exponents $\epsilon_1, \epsilon_2, \ldots, \epsilon_5$ of variables $V_1, \ldots, V_5$

The dimension of each variable is given by the dimensional matrix as presented in Fig. 7-1. By using the elements of this matrix, (7-7) can be written

$$(d_1^1 \cdot d_2^2 \cdot d_3^3)^{\epsilon_1} \cdot (d_1^2 \cdot d_2^4 \cdot d_3^4)^{\epsilon_2} \cdot (d_1^3 \cdot d_2^3 \cdot d_3^3)^{\epsilon_3} \cdot (d_1^4 \cdot d_2^0 \cdot d_3^2)^{\epsilon_4} \cdot (d_1^5 \cdot d_2^2 \cdot d_3^1)^{\epsilon_5} = d_1^{q_1} \cdot d_2^{q_2} \cdot d_3^{q_3}$$

from which, by Rule 2 of dimensional homogeneity (exponents of respective dimensions of both sides of an equation must be identical), we can compose the relations:

$$\left.\begin{array}{l} 1 \cdot \epsilon_1 + 2 \cdot \epsilon_2 + 3 \cdot \epsilon_3 + 4 \cdot \epsilon_4 + 5 \cdot \epsilon_5 = q_1 \\ 2 \cdot \epsilon_1 + 4 \cdot \epsilon_2 + 3 \cdot \epsilon_3 + 0 \cdot \epsilon_4 + 2 \cdot \epsilon_5 = q_2 \\ 3 \cdot \epsilon_1 + 4 \cdot \epsilon_2 + 3 \cdot \epsilon_3 + 2 \cdot \epsilon_4 + 1 \cdot \epsilon_5 = q_3 \end{array}\right\} \quad (7\text{-}8)$$

or, more elegantly

$$\begin{bmatrix} 1 & 2 & 3 & 4 & 5 \\ 2 & 4 & 3 & 0 & 2 \\ 3 & 4 & 3 & 2 & 1 \end{bmatrix} \cdot \begin{bmatrix} \epsilon_1 \\ \epsilon_2 \\ \epsilon_3 \\ \epsilon_4 \\ \epsilon_5 \end{bmatrix} = \begin{bmatrix} q_1 \\ q_2 \\ q_3 \end{bmatrix} \quad (7\text{-}9)$$

We now immediately notice—with anticipatory delight—that the coefficient matrix on the left *is* the dimensional matrix defined in Fig. 7-1.

In (7-9) we have five unknowns (exponents $\epsilon_1, \ldots, \epsilon_5$), but only three independent conditions (equations), the latter equals the rank of the dimensional matrix R_{DM}, as can be easily verified (there is at least one nonzero third-order determi-

136 APPLIED DIMENSIONAL ANALYSIS AND MODELING

nant). Thus we can determine only three (*any three*) of the five unknowns in terms of the remaining two. If, as in the present case, the third, fourth, and fifth columns of the dimensional matrix are independent, then the unknowns to be found are ϵ_3, ϵ_4, ϵ_5, and the arbitrary chosen "defining" two unknowns are ϵ_1 and ϵ_2. Accordingly, we partition the dimensional matrix into two submatrices as shown in Fig. 7-2.

$$\mathbf{B}\text{ matrix} \longrightarrow \begin{bmatrix} 1 & 2 \\ 2 & 4 \\ 3 & 4 \end{bmatrix} \begin{bmatrix} 3 & 4 & 5 \\ 3 & 0 & 2 \\ 3 & 2 & 1 \end{bmatrix} \longleftarrow \mathbf{A}\text{ matrix}$$

Dimensional Matrix

Figure 7-2
Partitioning the dimensional matrix into submatrices A and B

We note that **A** is a square matrix whose order is the number of dimensions $N_d = 3$, which (in this case) also equals the rank R_{DM}. An important criterion for **A** is now stated: matrix **A** must *not* be singular; i.e., its determinant must not be zero. The justification that this is an essential condition will follow later. Note also that **A** is formed by the *rightmost* determinant of the dimensional matrix.

By the partition defined in Fig. 7-2, relation (7-9) can now be written

$$[\mathbf{B} \quad \mathbf{A}] \cdot \begin{bmatrix} \begin{bmatrix} \epsilon_1 \\ \epsilon_2 \end{bmatrix} \\ \begin{bmatrix} \epsilon_3 \\ \epsilon_4 \\ \epsilon_5 \end{bmatrix} \end{bmatrix} = \begin{bmatrix} q_1 \\ q_2 \\ q_3 \end{bmatrix} \quad (7\text{-}10)$$

Now, obviously,

$$\begin{bmatrix} \epsilon_1 \\ \epsilon_2 \end{bmatrix} = \begin{bmatrix} 1 & 0 \\ 0 & 1 \end{bmatrix} \cdot \begin{bmatrix} \epsilon_1 \\ \epsilon_2 \end{bmatrix} = \mathbf{I} \cdot \begin{bmatrix} \epsilon_1 \\ \epsilon_2 \end{bmatrix} \quad (7\text{-}11)$$

where **I** is the *identity* matrix. By (7-10) and (7-11),

$$\begin{bmatrix} \mathbf{I} & \mathbf{0} \\ \mathbf{B} & \mathbf{A} \end{bmatrix} \cdot \begin{bmatrix} \begin{bmatrix} \epsilon_1 \\ \epsilon_2 \end{bmatrix} \\ \begin{bmatrix} \epsilon_3 \\ \epsilon_4 \\ \epsilon_5 \end{bmatrix} \end{bmatrix} = \begin{bmatrix} \begin{bmatrix} \epsilon_1 \\ \epsilon_2 \end{bmatrix} \\ \begin{bmatrix} q_1 \\ q_2 \\ q_3 \end{bmatrix} \end{bmatrix} \quad (7\text{-}12)$$

By premultiplying both sides of (7-12) by the inverse of the 2 × 2 matrix on the left, we obtain

$$\begin{bmatrix} \epsilon_1 \\ \epsilon_2 \\ \epsilon_3 \\ \epsilon_4 \\ \epsilon_5 \end{bmatrix} = \begin{bmatrix} \mathbf{I} & \mathbf{0} \\ \mathbf{B} & \mathbf{A} \end{bmatrix}^{-1} \cdot \begin{bmatrix} \epsilon_1 \\ \epsilon_2 \\ q_1 \\ q_2 \\ q_3 \end{bmatrix} = \begin{bmatrix} \mathbf{I} & \mathbf{0} \\ -\mathbf{A}^{-1} \cdot \mathbf{B} & \mathbf{A}^{-1} \end{bmatrix} \cdot \begin{bmatrix} \epsilon_1 \\ \epsilon_2 \\ q_1 \\ q_2 \\ q_3 \end{bmatrix} \quad (7\text{-}13)$$

STRUCTURE OF PHYSICAL RELATIONS   **137**

as can be easily verified by direct multiplication. If now we *define*

$$\mathbf{E} = \begin{bmatrix} \mathbf{I} & \mathbf{0} \\ -\mathbf{A}^{-1}\mathbf{B} & \mathbf{A}^{-1} \end{bmatrix} \quad (7\text{-}14)$$

then (7-13) becomes simply

$$\begin{bmatrix} \epsilon_1 \\ \epsilon_2 \\ \epsilon_3 \\ \epsilon_4 \\ \epsilon_5 \end{bmatrix} = \mathbf{E} \cdot \begin{bmatrix} \epsilon_1 \\ \epsilon_2 \\ q_1 \\ q_2 \\ q_3 \end{bmatrix} \quad (7\text{-}15)$$

In (7-14), **E** is the *exponent matrix* because it determines the sought exponents.

For convenience the characteristics of the matrices hitherto introduced are now summarized. If N_v and N_d are the numbers of variables and dimensions, respectively, then

I   is a $(N_v - N_d) \times (N_v - N_d)$ square identity matrix whose elements are all zeros, except in the main diagonal, where they are all 1

0   is a $(N_v - N_d) \times N_d$ null matrix, whose elements are all zeros

A   is a $N_d \times N_d$ square matrix defined in Fig. 7-2; it is formed by the non-zero *rightmost* determinant of the dimensional matrix defined in Fig 7-1

B   is a $N_d \times (N_v - N_d)$ matrix defined in Fig 7-2

E   is a $N_v \times N_v$ square matrix defined by (7-14)

A⁻¹·**B** is a $N_d \times (N_V - N_d)$ matrix (this follows from the above definitions).

Example 7-2

Given the dimensional matrix (same as in Fig. 7-1)

$$\begin{bmatrix} 1 & 2 & 3 & 4 & 5 \\ 2 & 4 & 3 & 0 & 2 \\ 3 & 4 & 3 & 2 & 1 \end{bmatrix}$$

then $N_V = 5$ and $N_d = 3$. Hence **I** is a 2 × 2 matrix, **0** is a 2 × 3 matrix, **A** is a 3 × 3 matrix, **B** is a 3 × 2 matrix, **E** is a 5 × 5 square matrix, **A**⁻¹·**B** is a 3 × 2 matrix and, therefore, in this case,

$$\mathbf{A} = \begin{bmatrix} 3 & 4 & 5 \\ 3 & 0 & 2 \\ 3 & 2 & 1 \end{bmatrix}; \qquad \mathbf{A}^{-1} = \frac{1}{30}\begin{bmatrix} -4 & 6 & 8 \\ 3 & -12 & 9 \\ 6 & 6 & -12 \end{bmatrix}$$

$$\mathbf{B} = \begin{bmatrix} 1 & 2 \\ 2 & 4 \\ 3 & 4 \end{bmatrix}; \qquad -\mathbf{A}^{-1}\cdot\mathbf{B} = \frac{1}{30}\begin{bmatrix} -32 & -48 \\ -6 & 6 \\ 18 & 12 \end{bmatrix}$$

$$\mathbf{I} = \begin{bmatrix} 1 & 0 \\ 0 & 1 \end{bmatrix} = \frac{1}{30}\begin{bmatrix} 30 & 0 \\ 0 & 30 \end{bmatrix}; \quad \mathbf{0} = \begin{bmatrix} 0 & 0 & 0 \\ 0 & 0 & 0 \end{bmatrix}$$

By these particulars, and (7-14)

$$E = \frac{1}{30} \left[\begin{array}{ccccc} 30 & 0 & 0 & 0 & 0 \\ 0 & 30 & 0 & 0 & 0 \\ -32 & -48 & -4 & 6 & 8 \\ -6 & 6 & 3 & -12 & 9 \\ 18 & 12 & 6 & 6 & -12 \end{array} \right]$$

with the first two columns labeled $\mathbf{I}$, the last three labeled $\mathbf{0}$, and the lower-left block $-\mathbf{A}^{-1}\cdot\mathbf{B}$, lower-right block $\mathbf{A}^{-1}$.

Therefore, by (7-15), the exponents will be supplied by

$$\left[\begin{array}{c} \epsilon_1 \\ \epsilon_2 \\ \epsilon_3 \\ \epsilon_4 \\ \epsilon_5 \end{array} \right] = \frac{1}{30} \left[\begin{array}{ccccc} 30 & 0 & 0 & 0 & 0 \\ 0 & 30 & 0 & 0 & 0 \\ -32 & -48 & -4 & 6 & 8 \\ -6 & 6 & 3 & -12 & 9 \\ 18 & 12 & 6 & 6 & -12 \end{array} \right] \cdot \left[\begin{array}{c} \epsilon_1 \\ \epsilon_2 \\ q_1 \\ q_2 \\ q_3 \end{array} \right]$$

where $\epsilon_1, \epsilon_2, q_1, q_2, q_3$ can be arbitrary numbers including zero.

Now suppose we want to create products of variables whose dimensions are, say, $d_1^3 \cdot d_2^5 \cdot d_3^7$. Hence $q_1 = 3$, $q_2 = 5$, $q_3 = 7$. If, $\epsilon_1 = 1$, $\epsilon_2 = 2$, then, by (7-15) and using the numerically given Exponent Matrix $\mathbf{E}$, we write

$$\left[\begin{array}{c} \epsilon_1 \\ \epsilon_2 \\ \epsilon_3 \\ \epsilon_4 \\ \epsilon_5 \end{array} \right] = \mathbf{E} \cdot \left[\begin{array}{c} 1 \\ 2 \\ 3 \\ 5 \\ 7 \end{array} \right] = \left[\begin{array}{c} 1 \\ 2 \\ -1.8 \\ 0.6 \\ 0.2 \end{array} \right]$$

Thus, by (7-7) and the dimensional matrix given in Fig. 7-1,

$$[V_1^1 \cdot V_2^2 \cdot V_3^{-1.8} \cdot V_4^{0.6} \cdot V_5^{0.2}] = \underbrace{(d_1^1 \cdot d_2^2 \cdot d_3^3)^1}_{V_1} \cdot \underbrace{(d_1^2 \cdot d_2^4 \cdot d_3^4)^2}_{V_2} \cdot \underbrace{(d_1^3 \cdot d_2^3 \cdot d_3^3)^{-1.8}}_{V_3} \cdot \underbrace{(d_1^4 \cdot d_2^0 \cdot d_3^2)^{0.6}}_{V_4} \cdot \underbrace{(d_1^5 \cdot d_2^2 \cdot d_3^1)^{0.2}}_{V_5}$$

$$= d_1^3 \cdot d_2^5 \cdot d_3^7$$

as expected and required. Similarly, if, say, $\epsilon_1 = 0$, $\epsilon_2 = 0$, then

$$\left[\begin{array}{c} \epsilon_1 \\ \epsilon_2 \\ \epsilon_3 \\ \epsilon_4 \\ \epsilon_5 \end{array} \right] = \mathbf{E} \cdot \left[\begin{array}{c} 0 \\ 0 \\ 3 \\ 5 \\ 7 \end{array} \right] = \frac{1}{15} \left[\begin{array}{c} 0 \\ 0 \\ 37 \\ 6 \\ -18 \end{array} \right]$$

So, similarly to the above expression,

$$[V_1^0 \cdot V_2^0 \cdot V_3^{37/15} \cdot V_4^{6/15} \cdot V_5^{-18/15}] = d_1^3 \cdot d_2^5 \cdot d_3^7$$

again, as expected and required.

7.4. NUMBER OF INDEPENDENT SETS OF PRODUCTS OF GIVEN DIMENSIONS (I)

Next, we pose the important question: given the values of $q_1, q_2, \ldots$, how many *independent* sets of exponents $\epsilon_1, \epsilon_2, \ldots$ can exist? In other words, how many different sets of exponents can there be such that none would be a *linear combination* of the others? Or, still saying the same thing, how many sets of exponents are there such that none of them is expressible—using only linear functions—by the rest?

These questions can be easily answered by considering that the rank of the dimensional matrix is equal to the number of dimensions, i.e., the *rows* of the dimensional matrix are linearly independent. We now introduce the following notation:

- N_V is the number of variables
- N_d is the number of dimensions
- $N_{q \neq 0}$ is the number of q values (i.e., given exponents of dimensions) different from zero
- N_P is the number of independent dimensional or dimensionless groups which can be formed by the ascribed (i.e., imposed) dimensions
- R_{DM} is the rank of the dimensional matrix (in our case $R_{DM} = N_d$)

There are two cases to be distinguished:

- If $N_{q \neq 0} = 0$, i.e., all q values are 0, then the formula corresponding to (7-9) represents a homogeneous system of equations. The number of linearly independent solutions of such a system is equal to the difference between the number of unknowns and the rank of the coefficient matrix (dimensional matrix). Hence we can write

$$N_P = N_V - N_d \qquad (\text{if } N_{q \neq 0} = 0)$$

- If $N_{q \neq 0} > 0$, i.e., at least one of the q elements is different from zero, then the formula corresponding to (7-9) represents a nonhomogeneous linear system of equations. Since the rank of the coefficient matrix (dimensional matrix) enlarged by the q elements cannot be greater than R_{DM}, therefore the condition of solvability is automatically satisfied and the number of linearly independent solutions will be increased by 1 (see Chapter 1). Namely: when taking the *particular* solution of the system in which the arbitrarily chosen unknowns are all equal to zero, we obviously get one additional solution. Hence we have in this case

$$N_P = N_V - N_d + 1 \qquad (\text{if } N_{q \neq 0} > 0)$$

Again assuming five variables and three dimensions, $N_P = 5 - 3 + 1 = 3$ and by (7-15) we write

$$\begin{bmatrix} \epsilon_{11} \\ \epsilon_{21} \\ \epsilon_{31} \\ \epsilon_{41} \\ \epsilon_{51} \end{bmatrix} = \mathbf{E} \cdot \begin{bmatrix} \epsilon_{11} \\ \epsilon_{21} \\ q_1 \\ q_2 \\ q_3 \end{bmatrix} \;;\; \begin{bmatrix} \epsilon_{12} \\ \epsilon_{22} \\ \epsilon_{32} \\ \epsilon_{42} \\ \epsilon_{52} \end{bmatrix} = \mathbf{E} \cdot \begin{bmatrix} \epsilon_{12} \\ \epsilon_{22} \\ q_1 \\ q_2 \\ q_3 \end{bmatrix} \;;\; \begin{bmatrix} \epsilon_{13} \\ \epsilon_{23} \\ \epsilon_{33} \\ \epsilon_{43} \\ \epsilon_{53} \end{bmatrix} = \mathbf{E} \cdot \begin{bmatrix} \epsilon_{13} \\ \epsilon_{23} \\ q_1 \\ q_2 \\ q_3 \end{bmatrix} \qquad (7\text{-}16)$$

This can be written more compactly

$$\begin{bmatrix} \epsilon_{11} & \epsilon_{12} & \epsilon_{13} \\ \epsilon_{21} & \epsilon_{22} & \epsilon_{23} \\ \epsilon_{31} & \epsilon_{32} & \epsilon_{33} \\ \epsilon_{41} & \epsilon_{42} & \epsilon_{43} \\ \epsilon_{51} & \epsilon_{52} & \epsilon_{53} \end{bmatrix} = \mathbf{E} \cdot \begin{bmatrix} \epsilon_{11} & \epsilon_{12} & \epsilon_{13} \\ \epsilon_{21} & \epsilon_{22} & \epsilon_{23} \\ q_1 & q_1 & q_1 \\ q_2 & q_2 & q_2 \\ q_3 & q_3 & q_3 \end{bmatrix} \quad (7\text{-}17)$$

$$\underbrace{}_{\mathbf{P}} \qquad \underbrace{}_{\mathbf{Z}}$$

where matrices **P** and **Z** are as defined here. We see that both **P** and **Z** are $N_V \times N_P$ matrices and their rank must be equal since **E** is nonsingular.

It should be pointed out that the elements in the last N_d rows of **Z** are equal. Therefore, if $N_{q \neq 0} = 0$, then the first $N_V - N_d$ rows augmented with a row containing only 1's must be linearly independent. Since the first $N_V - N_d$ rows of **Z** are formed by the values of unknowns, which can be chosen arbitrarily, the above condition can be easily satisfied.

In summary, we formulate the following important relations:

$$N_P = N_V - N_d \qquad \text{if } N_{q \neq 0} = 0 \quad (7\text{-}18)$$

$$N_P = N_V - N_d + 1 \qquad \text{if } N_{q \neq 0} > 0 \quad (7\text{-}19)$$

where the symbols are as defined above.

Example 7-3

Using the data introduced in Example 7-2, we have:

Five variables, therefore $N_V = 5$
Three dimensions, therefore $N_d = 3$
Three nonzero q (viz., $q_1 = 3$, $q_2 = 5$, $q_3 = 7$), therefore $N_{q \neq 0} = 3$

Consequently, (7-19) is applicable:

$$N_P = N_V - N_d + 1 = 5 - 3 + 1 = 3$$

and hence only $N_P = 3$ independent products of the variables $V_1, V_2, \ldots, V_5$ can be formed (each of whose dimension is $d_1^3 \cdot d_2^5 \cdot d_3^7$).
⇑

The next example shows how to form *independent* products of variables fulfilling the required dimensional requirements.

Example 7-4

Using the input data presented in Example 7-3, we have five variables $V_1, V_2, \ldots V_5$, three dimensions d_1, d_2, d_3, and the dimensional matrix (defined in Fig. 7-1) providing the dimensions of the five variables. Moreover, as before, $q_1 = 3$, $q_2 = 5$, and $q_3 = 7$; i.e., we are looking for the products (of variables) whose dimensions are uniformly $d_1^3 \cdot d_2^5 \cdot d_3^7$. In Example 7-3 the number of such products is $N_P = 3$. Therefore the **Z** matrix contains 3 columns and is

$$\mathbf{Z} = \begin{bmatrix} 1 & 1 & 0 \\ 0 & 1 & 1 \\ 3 & 3 & 3 \\ 5 & 5 & 5 \\ 7 & 7 & 7 \end{bmatrix}$$

Consequently, with the **E** matrix given in Example 7-2, and by (2-17),

$$\mathbf{P} = \mathbf{E} \cdot \mathbf{Z} = \begin{bmatrix} 1 & 1 & 0 \\ 0 & 1 & 1 \\ 1.4 & -0.2 & \dfrac{13}{15} \\ 0.2 & 0.4 & 0.6 \\ -0.6 & -0.2 & -0.8 \end{bmatrix}$$

and hence the sought after $N_P = 3$ dimensional products are

$$\tau_1 = \frac{V_1 \cdot V_3^{1.4} \cdot V_4^{0.2}}{V_5^{0.6}}; \quad \tau_2 = \frac{V_1 \cdot V_2 \cdot V_4^{0.4}}{V_3^{0.2} \cdot V_5^{0.2}}; \quad \tau_3 = \frac{V_2 \cdot V_3^{13/15} \cdot V_4^{0.6}}{V_5^{0.8}}$$

where τ_1, τ_2, τ_3 designate the respective *dimensional* products of variables.

To check, let us verify that τ_3 indeed fulfills the dimensional requirements. So, we write for τ_3

$$[V_2^1 \cdot V_3^{13/15} \cdot V_4^{0.6} \cdot V_5^{-0.8}] = (d_1^2 \cdot d_2^4 \cdot d_3^4) \cdot (d_1^3 \cdot d_2^3 \cdot d_3^3)^{13/15} \cdot (d_1^4 \cdot d_2^0 \cdot d_3^2)^{0.6} \cdot (d_1^5 \cdot d_2^2 \cdot d_3^1)^{-0.8} = d_1^3 \cdot d_2^5 \cdot d_3^7$$

as expected and required. The reader should verify that the same result is obtained for τ_1 and τ_2, as well.

⇑

The next example—which is somewhat less abstract—further illustrates this process.

Example 7-5. Deflection of a Cantilever Upon a Concentrated Lateral Load (II)

The cantilever is of uniform cross-section; its deformation under the load is considered strictly lateral. Fig 7-3 shows the schematic of the arrangement. The relevant variables, their symbols and dimensions are listed as follows:

Variable	Symbol	Dimension	Remark
deflection	U	m	lateral
load	F	N	lateral
length	L	m	
Young's modulus	E	$m^{-2} \cdot N$	
diameter of cross-section	D	m	uniform, solid circle

Therefore the dimensional matrix is:

	U	F	L	E	D
m	1	0	1	-2	1
N	0	1	0	1	0

A matrix (columns E, D); **B** matrix (columns U, F, L)

142 APPLIED DIMENSIONAL ANALYSIS AND MODELING

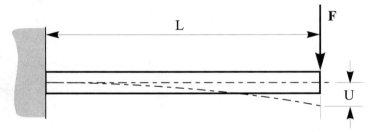

Figure 7-3
A cantilever loaded laterally by a concentrated force

As before (see Fig. 7-2), the dimensional matrix is partitioned into matrices **A** and **B**. The former is formed by the *rightmost determinant* of the dimensional matrix, the latter is formed by the *rest* of the dimensional matrix. Accordingly

$$\mathbf{A} = \begin{bmatrix} -2 & 1 \\ 1 & 0 \end{bmatrix}; \quad \mathbf{B} = \begin{bmatrix} 1 & 0 & 1 \\ 0 & 1 & 0 \end{bmatrix}$$

By the dimensional matrix $N_V = 5$, $N_d = 2$ and therefore **I** is a 3×3 matrix, **0** is a 3×2 matrix, and further, by the above given **A** and **B** matrices

$$\mathbf{A}^{-1} = \begin{bmatrix} 0 & 1 \\ 1 & 2 \end{bmatrix}; \quad -\mathbf{A}^{-1} \cdot \mathbf{B} = \begin{bmatrix} 0 & -1 & 0 \\ -1 & -2 & -1 \end{bmatrix}$$

Therefore, by (7-14), the exponent matrix **E** is

$$\mathbf{E} = \begin{bmatrix} 1 & 0 & 0 & 0 & 0 \\ 0 & 1 & 0 & 0 & 0 \\ 0 & 0 & 1 & 0 & 0 \\ 0 & -1 & 0 & 0 & 1 \\ -1 & -2 & -1 & 1 & 2 \end{bmatrix}$$

Now let us form products of variables whose dimension is, say, m³·N⁴; thus $q_1 = 3$, $q_2 = 4$. Hence, $N_{q \neq 0} = 2 > 0$ and therefore relation (7-19) is applicable

$$N_P = N_V - N_d + 1 = 5 - 2 + 1 = 4$$

i.e., four independent dimensional products of variables can be formed, each of whose (the products) dimensions is m³·N⁴. Consequently, the **Z** matrix, defined in (7-17), will have $N_V = 5$ rows and $N_P = 4$ columns:

$$\mathbf{Z} = \begin{bmatrix} 1 & 0 & 0 & 1 \\ 0 & 1 & 0 & 1 \\ 0 & 0 & 1 & 0 \\ 3 & 3 & 3 & 3 \\ 4 & 4 & 4 & 4 \end{bmatrix}$$

Note that the *top* $N_V - N_d = 3$ rows of this matrix are *selected* arbitrarily, satisfying the only condition that these three rows augmented with any other nonzero row of **Z** must not generate a singular matrix.

Now, by the given **E** and **Z** matrices, (7-17) furnishes the required exponents. Thus

$$\mathbf{P} = \mathbf{E} \cdot \mathbf{Z} = \begin{bmatrix} 1 & 0 & 0 & 1 \\ 0 & 1 & 0 & 1 \\ 0 & 0 & 1 & 0 \\ 4 & 3 & 4 & 3 \\ 10 & 9 & 10 & 8 \end{bmatrix} \begin{matrix} \leftarrow U \\ \leftarrow F \\ \leftarrow L \\ \leftarrow E \\ \leftarrow D \end{matrix}$$

As said before, each *row* of this matrix corresponds to a particular physical variable forming the respective *column* of the dimensional matrix. Thus the first row corresponds to U (deflection), the second row to F (load), etc. For convenience these variables were written out besides the relevant rows of the **P** matrix.

By the above relation, the four *independent* products of variables are:

$$\tau_1 = U \cdot E^4 \cdot D^{10}; \quad \tau_2 = F \cdot E^3 \cdot D^9; \quad \tau_3 = L \cdot E^4 \cdot D^{10}; \quad \tau_4 = U \cdot F \cdot E^3 \cdot D^8$$

The dimension of each of these products is the required m$^3 \cdot$N^4—as the reader can easily verify—and is urged to do so.

⇑

7.5. COMPLETENESS OF THE SET OF PRODUCTS OF VARIABLES

A set of N_P products of variables, possessing the given (i.e., required) dimension, *uniquely* defines the relation among the relevant physical variables [N_P is defined in (7-18) and (7-19)]. Therefore the set so obtained is said to be (and is defined as) *complete*. That is, if there is a relation among N_V physical variables $V_1, V_2 \ldots$

$$\Psi_1 \{V_1, V_2, \ldots, V_{N_V}\} = \text{const} \tag{7-20}$$

then it is also true that

$$\Psi_2 \{\tau_1, \tau_2, \ldots, \tau_{N_P}\} = \text{const} \tag{7-21}$$

where $\tau_1, \ldots, \tau_{N_P}$ is the complete set of products of physical variables, N_P is as defined in (7-18) and (7-19), and Ψ_1, Ψ_2 symbolize functions.

It is emphasized that the information contained in (7-21) is the same as in (7-20), yet the number of variables in the former is N_V, whereas in the latter it is N_P, and, as relations (7-18) and (7-19) tell us, $N_P < N_V$. That is, we deal with *fewer* variables if we use dimensional—or dimensionless—products of variables.

Example 7-6. Deflection of a Cantilever Upon a Concentrated Lateral Load (III)

This is a continuation of Example 7-5. In that example we derived four independent products of variables, each of whose dimensions was m$^3 \cdot$N^4. For convenience, they are repeated here

$$\tau_1 = U \cdot E^4 \cdot D^{10}; \quad \tau_2 = F \cdot E^3 \cdot D^9; \quad \tau_3 = L \cdot E^4 \cdot D^{10}; \quad \tau_4 = U \cdot F \cdot E^3 \cdot D^8$$

where the physical variables are as defined in Fig. 7-3 and in the table in Example 7-5. By simple analytic derivation we know that the deflection U of a cantilever upon a concentrated lateral load F acting at the free end is

$$U = \frac{64 \cdot F \cdot L^3}{3 \cdot \pi \cdot D^4 \cdot E} \tag{a}$$

Hence, following the notation introduced in (7-20), we write

$$\Psi_1\{U, D, E, F, L\} = U \cdot D^4 \cdot E \cdot F^{-1} \cdot L^{-3} = \frac{64}{3 \cdot \pi} = \text{const} \tag{b}$$

As we can see, the function Ψ_1 is defined by five variables. In contrast, if we now employ the *products* of variables, as defined above, then our analytical formula (a) assumes the form defined in (7-21)

$$\Psi_2\{\tau_1, \tau_2, \tau_3, \tau_4\} = \tau_1^4 \cdot \tau_2^2 \cdot \tau_3^{-3} \cdot \tau_4^{-3} = \frac{64}{3 \cdot \pi} = \text{const} \tag{c}$$

in which there are only four variables—a modest, but noteworthy improvement. To inject some concrete numbers into our deliberations, let us say our cantilever has

length	$L = 1.8$ m
diameter of cross section	$D = 0.080$ m (solid circle)
Young's modulus	$E = 2.05 \times 10^{11}$ N/m^2
load	$F = 4650$ N
deflection	$U = 0.02193$ m

Then, by (b),

$$\Psi_1\{\ \} = (0.02193) \cdot (0.08)^4 \cdot (2.05 \times 10^{11}) \cdot (4650)^{-1} \cdot (1.8)^{-3} = 6.79019$$

The constant at the extreme right is $64/(3 \cdot \pi)$ (within 0.006%), as required and expected.
Next, the defined products of variables are:

$\tau_1 = 4.15867 \times 10^{32}$ m$^3 \cdot$N^4
$\tau_2 = 5.37681 \times 10^{27}$ m$^3 \cdot$N^4
$\tau_3 = 3.41340 \times 10^{34}$ m$^3 \cdot$N^4
$\tau_4 = 1.47392 \times 10^{27}$ m$^3 \cdot$N^4

Note the identical dimensions of all of the above dimensional products of variables. Therefore by (c), with 5 decimal accuracy

$$\Psi_2\{\ \} = (4.15867 \times 10^{32})^4 \cdot (5.37681 \times 10^{27})^2 \cdot (3.4134 \times 10^{34})^{-3} \cdot (1.47392 \times 10^{27})^{-3} = 6.79022$$

The constant on the extreme right is again $64/(3 \cdot \pi)$ (within 0.006%), as required and expected.

⇑

7.6. SPECIAL CASE: MATRIX A IS SINGULAR

We saw in Art. 7.3 that in the process of determining the dimensional products, the inversion of the **A** matrix was necessary [the reader should recall that matrix **A** is formed by the elements of the *rightmost* determinant of the dimensional matrix (see Fig 7-2)]. Therefore **A** cannot be singular, i.e., its determinant must not be zero. If it is zero, we should make it nonzero by either of the two methods to be described presently. But first we define the quantity

$$\Delta = N_d - R_{DM} \tag{7-22}$$

where N_d is the number of dimensions in the dimensional matrix;
R_{DM} is the rank of the dimensional matrix.

Now, obviously, Δ cannot be negative, because in this case the rank of the dimensional matrix would exceed the number of its rows N_d, which is impossible. Therefore Δ must be either zero or larger than zero. If $\Delta = 0$, we use *Method 1;* if $\Delta > 0$, we use *Method 2*.

Method 1. In this case $\Delta = 0$, i.e., the rank of the dimensional matrix equals the number of its rows. Therefore a suitable *interchange* of two *columns* (i.e., two variables) of this matrix will *always* result in a nonsingular **A** matrix, i.e., a nonzero *rightmost* determinant. The example below illustrates this case:

Example 7-7

We have five variables $V_1, V_2, \ldots, V_5$ and three dimensions d_1, d_2, d_3—as given in the dimensional matrix

	V_1	V_2	V_3	V_4	V_5
d_1	2	1	3	4	−2
d_2	5	4	2	1	0
d_3	2	1	−1	2	−2

A matrix (rightmost 3 columns), **B** matrix (leftmost 2 columns)

Here the **A** matrix is singular since its determinant (the rightmost determinant of the dimensional matrix) is zero. However, the rank of the dimensional matrix is $R_{DM} = 3$. To demonstrate this, we form the *leftmost* determinant

$$\begin{vmatrix} 2 & 1 & 3 \\ 5 & 4 & 2 \\ 2 & 1 & -1 \end{vmatrix} = -12 \neq 0$$

Therefore it must be possible to *interchange* two columns in the dimensional matrix such that the *rightmost* determinant becomes nonzero. To this end, let us interchange columns 1 and 3 (variables V_1, V_3). Thus we obtain the (modified) dimensional matrix

	V_3	V_2	V_1	V_4	V_5
d_1	3	1	2	4	−2
d_2	2	4	5	1	0
d_3	−1	1	2	2	−2

A matrix (rightmost 3 columns), **B** matrix (leftmost 2 columns)

in which the value of the *rightmost* determinant is $20 \neq 0$. Therefore we can now proceed, using this new dimensional matrix, since its **A** matrix became nonsingular.

Method 2. In this case $\Delta > 0$, i.e., the number of dimensions N_d exceeds the rank R_{DM} of the dimensional matrix. It follows, therefore, that no exchange of any two (or more) columns of this matrix could make its rightmost determinant nonzero. The reason is simple: by definition, this determinant is of the order N_d, therefore if it were nonzero, then R_{DM} would be equal to N_d, which is impossible if $\Delta = N_d - R_{DM} > 0$. The only thing we can do is discard Δ number of dimensions (rows) from the dimensional matrix. This will necessarily make $\Delta = 0$. Hence matrix **A** will be either nonsingular in which case we can proceed since everything is as it should be—or if **A** is still singular, then we should follow the process described in *Method 1* above.

The example below illustrates *Method 2*.

Example 7-8. Deflection of a Cantilever Upon a Concentrated Lateral Load (IV)

We again deal with the cantilever described and shown in Example 7-5. But now, instead of using two dimensions, viz., metre "m" for length, and newton "N" for force, we use three dimensions, viz., meter "m" for length, kilogram "kg" for mass, and second "s" for time. Accordingly, the variables will be:

Variable	Symbol	Dimension	Remark
deflection	U	m	lateral
load	F	m·kg/s²	lateral
length	L	m	
Young's modulus	E	kg/(m·s²)	
diameter of cross-section	D	m	uniform, solid circle

yielding the dimensional matrix

$$\begin{array}{c|ccccc} & U & F & L & E & D \\ \hline m & 1 & 1 & 1 & -1 & 1 \\ kg & 0 & 1 & 0 & 1 & 0 \\ s & 0 & -2 & 0 & -2 & 0 \end{array}$$

with the boxed submatrix on columns L, E, D being the **A** matrix, and the bottom row (s) deleted.

We immediately see that matrix **A** is singular, since two of its columns are identical. If we try to make it nonsingular by interchanging any of its three columns L, E, D with the other two viz., U or F, we will fail. For no matter which two columns (variables) of this matrix are interchanged, elements of the third row will remain two times that of the second row. Therefore, no third-order nonzero determinant can be formed. Consequently the rank of the dimensional matrix must be less than 3. That the rank is 2 is easily ascertained by observing that the 2×2 determinant in the upper right corner is not zero. Therefore we have, by (7-22)

$$\Delta = N_d - R_{DM} = 3 - 2 = 1 \neq 0$$

and hence we must use *Method 2*, i.e., eliminate 1 row. Now the obvious question is: which one?

Sometimes it is easier to answer the complementary question: which row should *not* be discarded? In our case, obviously, if the first row "m" were eliminated, then the rank would become 1, since the *remaining* "s" row is twice the "kg" row. Therefore we cannot eliminate "m," and thus we have to eliminate either "kg" or "s". Let us select "s" (third row) for elimination. Thus the modified dimensional matrix will be

STRUCTURE OF PHYSICAL RELATIONS

$$\begin{array}{c|ccccc} & U & F & L & E & D \\ \hline m & 1 & 1 & 1 & -1 & 1 \\ kg & 0 & 1 & 0 & 1 & 0 \end{array} \quad \text{A matrix}$$

in which the rightmost determinant is not zero, and **A** is nonsingular. Note that effectively we are dealing with two dimensions, rather than the originally contemplated three.

In this example we encountered two dimensional matrices, the 3×5 "original," and the 2×5 "new" one. We call the latter one the *reduced dimensional matrix* (RDM), because in it the rightmost determinant is surely nonzero. Now from the construction of the dimensional matrix DM and the RDM it is obvious that:

- the ranks of DM and RDM are identical, i.e.,

$$R_{DM} = R_{RDM} \tag{7-23}$$

- the number of dimensions in the RDM is equal to its rank, i.e.,

$$N_d = R_{RDM} \tag{7-24}$$

From the above two relations it follows that if the rank of the dimensional matrix is equal to the number of dimensions, then the dimensional matrix and the reduced dimensional matrix are identical; i.e.,

$$\text{if} \quad R_{DM} = N_d, \quad \text{then} \quad (DM) = (RDM) \tag{7-25}$$

From the above presented material it is also obvious that only the RDM can serve as the basis for determining (creating) products of dimensional or dimensionless

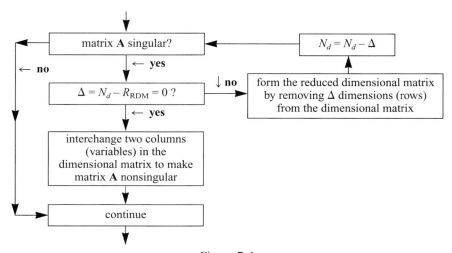

Figure 7-4
Process to make submatrix A of the dimensional matrix nonsingular
Dimensional matrix and matrix **A** are defined in Fig. 7-1 and Fig. 7-2.
N_d = number of dimensions, R_{DM} = rank of dimensional matrix.

148 APPLIED DIMENSIONAL ANALYSIS AND MODELING

variables. However, in the great majority of cases, these two matrices are identical so, for economy, and if there is no danger of confusion, the term *dimensional matrix* will be used. In these cases, unless stated otherwise, the **A** matrix is always nonsingular, i.e., the rightmost determinant of the dimensional matrix is not zero.

The above protocol is conveniently displayed in the self-explanatory schematic of Fig. 7-4.

7.7. NUMBER OF INDEPENDENT SETS OF PRODUCTS OF A GIVEN DIMENSION (II); BUCKINGHAM'S THEOREM

It is evident that because of the alteration of the dimensional matrix, which was manifested in the reduction of the number of dimensions N_d, we must modify relations (7-18) and (7-19). We see from (7-23) and (7-24) that the number of dimensions N_d in the reduced dimensional matrix is the same as the rank of the dimensional matrix. Hence it follows that the number of independent dimensional products N_P which are composed by the given number of variables is

$$\left. \begin{array}{l} N_P = N_V - R_{DM} \quad \text{if} \quad N_{q \neq 0} = 0 \\ N_P = N_V - R_{DM} + 1 \quad \text{if} \quad 0 < N_{q \neq 0} \leq R_{DM} \end{array} \right\} \quad (7\text{-}26)$$

Accordingly, the constituents of exponent matrix **E** [defined in (7-14)] are of the following sizes:

A	is a $R_{DM} \times R_{DM}$	square matrix
B	is a $R_{DM} \times (N_V - R_{DM})$	matrix
I	is a $(N_V - R_{DM}) \times (N_V - R_{DM})$	square unit matrix
0	is a $(N_V - R_{DM}) \times R_{DM}$	null matrix
E	is a $N_V \times N_V$	square matrix
A^{-1}·B	is a $R_{DM} \times (N_V - R_{DM})$	matrix

In the above, N_V is the number of variables, and R_{DM} is the rank of the dimensional matrix.

The 1st of the relations (7-26) is known as *Buckingham's Theorem* (Ref. 70, p. 345), dating from 1914. However, according to Macagno (Ref. 61, p. 397), Vaschy already stated (without proof) the same thing in a paper published in French in 1892. Baker (Ref. 34, p. 21) mentions that the proof of the theorem has been provided by many authors, notably Buckingham himself, as well as K. Brenkert, P. Bridgeman, H. Langhaar, G. Birkhoff, W. Durand, S. Drobot, L. Brand, and L. Sedov. However, not all of these proofs are easily "accessible" to mortals of only finite mathematical sophistication. The present author considers himself a member of this latter very populous group.

STRUCTURE OF PHYSICAL RELATIONS **149**

A remark on notation

We used the Greek letter τ *always with a subscript* to denote a product of variables whose dimension is different from 1, i.e., a *dimensional* product. Thus if $\tau_1, \tau_2, \ldots$ denote products of variables in a particular relation $\Psi_1\{V_1, V_2, \ldots\} = \text{const}$, then $[\tau_1] = [\tau_2] \ldots \neq 1$, i.e., *none* of the quantities $\tau_1, \tau_2, \ldots$ are dimensionless. On the other hand, if the dimensions of the products of variables are 1, i.e., they are *dimensionless*, then they are denoted by $\pi_1, \pi_2, \ldots$. Here, of course, the Greek letter π has nothing to do with the dimensionless constant 3.1415.... Again, a *subscript is always used,* even if there is but a single dimensionless product in a relation. Thus, if $\pi_1, \pi_2, \ldots$ denote different products of variables in a particular relationship of (7-20), then $[\pi_1] = [\pi_2] = \ldots = 1$; i.e., these quantities are *all* dimensionless.

The following example illustrates a case when $N_{q \neq 0} > 0$, i.e., when the second relation in (7-26) is applicable.

Example 7-9. Deflection of a Cantilever Upon a Concentrated Lateral Load (V)

We again deal with the cantilever shown in Example 7-5. We now are required to form dimensional products of variables U, F, L, E, D (defined in the table in Example 7-8) such that this dimension is m². The reduced dimensional matrix is

	U	F	L	E	D
m	1	1	1	−1	1
kg	0	1	0	1	0

⬅ A Matrix

therefore $R_{DM} = R_{RDM} = 2$ rank of dimensional matrix
$N_V = 5$ number of variables
$N_d = 2$ number of dimensions
$q_1 = 2$ exponent of "m" in the to-be-created dimensionless products
$q_2 = 0$ exponent of "kg" in the to-be-created dimensionless products

Consequently, $N_{q \neq 0} = 1 > 0$, so by (7-26), $N_P = N_V - R_{DM} + 1 = 5 - 2 + 1 = 4$, i.e., there are four independent products of variables, each having the dimension m². In order to get these products, we construct the exponent matrix **E**, which, by (7-14), comprises

$$\mathbf{A} = \begin{bmatrix} -1 & 1 \\ 1 & 0 \end{bmatrix}; \quad \mathbf{B} = \begin{bmatrix} 1 & 1 & 1 \\ 0 & 1 & 0 \end{bmatrix}; \quad \mathbf{A}^{-1} = \begin{bmatrix} 0 & 1 \\ 1 & 1 \end{bmatrix}$$

$$-\mathbf{A}^{-1} \cdot \mathbf{B} = \begin{bmatrix} 0 & -1 & 0 \\ -1 & -2 & -1 \end{bmatrix}; \quad \mathbf{0} = \begin{bmatrix} 0 & 0 \\ 0 & 0 \\ 0 & 0 \end{bmatrix}; \quad \mathbf{I} = \begin{bmatrix} 1 & 0 & 0 \\ 0 & 1 & 0 \\ 0 & 0 & 1 \end{bmatrix}$$

Therefore, by (7-14) and (7-17),

$$\mathbf{E} = \begin{bmatrix} 1 & 0 & 0 & 0 & 0 \\ 0 & 1 & 0 & 0 & 0 \\ 0 & 0 & 1 & 0 & 0 \\ 0 & -1 & 0 & 0 & 1 \\ -1 & -2 & -1 & 1 & 1 \end{bmatrix} \quad \mathbf{Z} = \begin{bmatrix} 1 & 0 & 0 & 1 \\ 0 & 1 & 0 & 1 \\ 0 & 0 & 1 & 0 \\ 2 & 2 & 2 & 2 \\ 0 & 0 & 0 & 0 \end{bmatrix}$$

Note that the elements of the top three rows of **Z** can be any numbers, as long as the rows remain independent. We selected numbers 1 and 0 to yield the *simplest* result. The exponents are provided by (7-17); thus

$$P = E \cdot Z = \begin{array}{c} \phantom{\begin{bmatrix}} \tau_1 \tau_2 \tau_3 \tau_4 \\ \phantom{\begin{bmatrix}} \downarrow \downarrow \downarrow \downarrow \\ \begin{bmatrix} 1 & 0 & 0 & 1 \\ 0 & 1 & 0 & 1 \\ 0 & 0 & 1 & 0 \\ 0 & -1 & 0 & -1 \\ 1 & 0 & 1 & -1 \end{bmatrix} \begin{array}{l} \leftarrow U \\ \leftarrow F \\ \leftarrow L \\ \leftarrow E \\ \leftarrow D \end{array} \end{array}$$

Since the rows of **P** correspond to the variables U, F, L, E, D, and the columns to the dimensional products $\tau_1, \tau_2, \tau_3, \tau_4, \tau_5$, we can write

$$\tau_1 = U \cdot D; \qquad \tau_2 = \frac{F}{E}; \qquad \tau_3 = L \cdot D; \qquad \tau_4 = \frac{U \cdot F}{E \cdot D}$$

The dimensions of all of these products, as the reader is advised to verify, is m², as expected and required.

⇑

The next example again shows the process where the rank R_{DM} of the dimensional matrix is less than the number of dimensions N_d.

Example 7-10. Deflection of a Cantilever Upon a Concentrated Lateral Load (VI)

We repeat the previous example with the difference that the required dimension of the generated products should now be m⁻¹·kg³; thus we have in this application $q_1 = -1$, $q_2 = 3$. For the purpose which will be clear presently, we repeat the dimensional matrix

	U	F	L	E	D
m	1	1	1	-1	1
kg	0	1	0	1	0
s	0	-2	0	-2	0	← row deleted

whose rank is $R_{DM} = 2$. However, the number of dimensions is $N_d = 3$. Therefore, by (7-22), $\Delta = N_d - R_{DM} = 3 - 2 = 1$, and hence we have to delete one row. As a consequence, the reduced dimensional matrix will be the same as was determined in Example 7-9. From this it follows that the **E** matrix will also be the same, and the **Z** matrix will be

$$Z = \begin{bmatrix} 1 & 0 & 0 & 1 \\ 0 & 1 & 0 & 1 \\ 0 & 0 & 1 & 0 \\ -1 & -1 & -1 & -1 \\ 3 & 3 & 3 & 3 \end{bmatrix}$$

where the top three rows are as before, and the bottom two rows are in accordance with the given (required) exponents $q_1 = -1, q_2 = 3$. Thus

STRUCTURE OF PHYSICAL RELATIONS   151

$$P = E \cdot Z = \begin{bmatrix} \tau_1 & \tau_2 & \tau_3 & \tau_4 \\ \downarrow & \downarrow & \downarrow & \downarrow \\ 1 & 0 & 0 & 1 \\ 0 & 1 & 0 & 1 \\ 0 & 0 & 1 & 0 \\ 3 & 2 & 3 & 2 \\ 1 & 0 & 1 & -1 \end{bmatrix} \begin{matrix} \leftarrow U \\ \leftarrow F \\ \leftarrow L \\ \leftarrow E \\ \leftarrow D \end{matrix}$$

yielding the sought-after 4 products of variables:

$$\tau_1 = U \cdot E^3 \cdot D; \qquad \tau_2 = F \cdot E^2; \qquad \tau_3 = L \cdot E^3 \cdot D; \qquad \tau_4 = \frac{U \cdot F \cdot E^2}{D}$$

The dimensions of all of these products are the required $m^{-1} \cdot kg^3 \cdot s^{-6}$—as the meticulous reader can easily verify.

⇑

7.8. SELECTABLE AND NONSELECTABLE DIMENSIONS IN A PRODUCT OF VARIABLES

In Example 7-10, the exponent of "s" is –6. This stems from the fact that the *matrix product* of the "s" row (the deleted row) in the dimensional matrix *and* any of the columns of the **P** matrix is –6. For example, the product of the last (third) row of the dimensional matrix and the second column of the **P** matrix is

$$[0 \quad -2 \quad 0 \quad -2 \quad 0] \cdot \begin{bmatrix} 0 \\ 1 \\ 0 \\ 2 \\ 0 \end{bmatrix} = -6$$

It follows therefore that the dimension represented by a deleted row(s) cannot be freely chosen, since its numerical value is set by the *selectable* exponents. To illustrate, in the last example the row "s" was *deleted* from the dimensional matrix, therefore the exponent of "s" in the derived product of variables is *nonselectable* (i.e., it must be –6, as in the example). However, the exponents of the *remaining* dimensions in the dimensional matrix are freely selectable (in the example: $q_1 = -1, q_2 = 3$). Since the number of the remaining dimensions is the rank of the dimensional matrix, therefore we can state that the number n_s of *selectable* exponents is

$$n_s = R_{DM} \qquad (7\text{-}27)$$

and the number n_{ns} of *nonselectable* exponents is

$$n_{ns} = N_d - R_{DM} \qquad (7\text{-}28)$$

However, by (7-22) and (7-27),

$$n_{ns} = N_d - R_{DM} = \Delta \qquad (7\text{-}29)$$

i.e., the number of *nonselectable* exponents is equal to the number of rows (dimensions) one must discard from the dimensional matrix in order to make its *rightmost* determinant nonzero (i.e., to make the submatrix **A** nonsingular).

To illustrate the employment of these simple relations, recall Example 7-5. There $N_d = 2$ and $R_{DM} = 2$. Hence the number of selectable exponents, by (7-27), is $n_s = R_{DM} = 2$ and the number of nonselectable ones, by (7-29), is $n_{ns} = N_d - R_{DM} = 2 - 2 = 0$. This latter result of course is hardly astonishing; we only have two dimensions, so if two are selectable, then zero of them is nonselectable.

However, in Example 7-7 we have $R_{DM} = 2$, $N_d = 3$; therefore, by (7-27), the number of selectable exponents is $n_s = R_{DM} = 2$, and the number of nonselectable exponents, by (7-29), is $n_s = N_d - R_{DM} = 3 - 2 = 1$, i.e., two exponents are selectable, 1 is not.

Note that when we say two exponents are selectable, it does not necessarily mean *any* two! Observation of the dimensional matrix always tells us which dimension (row) cannot be eliminated. The dimensions *retained* (i.e., dimensions making up the reduced dimensional matrix) are those whose exponents are *selectable;* dimensions *deleted* are those whose exponents are *nonselectable*. To illustrate, in Example 7-8 we eliminated "s" from the dimensional matrix, therefore the exponent of "s" occurring in the dimensional products $\tau_1, \ldots, \tau_4$ is not selectable, but of the other two, viz., of "m" and "kg" are. In this example it is obvious that the exponents of "kg" and "s" cannot be the selectable pair, since the remaining "m" cannot be nonselectable because it cannot be deleted, as pointed out earlier.

7.9. MINIMUM NUMBER OF INDEPENDENT PRODUCTS OF VARIABLES OF GIVEN DIMENSION

It is obvious that the smaller the number N_P of independent products of variables (of a prescribed dimension)—i.e., the membership of the *complete set* (Art. 7.5.)—the more useful the given relationship is. The reason is simple: the "membership" is equal to the number of variables *uniquely* defining the subject function, and hence the smaller this number, the better. A three-variable function can be presented in *one* chart, but for a 4-variable relation, a *set* of charts is needed. Therefore the goal is to have as few variables as possible; i.e., N_P should be minimum. The question now is: what is the this *minimum* value of N_P? Before this question is answered we must state and prove two important theorems.

Theorem 7-1. *N_P cannot be negative.*

Proof. Assume the contrary, i.e., that N_P is negative. Then, by (7-26),

$$\left. \begin{array}{ll} \text{either} & R_{DM} > N_V \\ \text{or} & R_{DM} > N_V + 1 \end{array} \right\} \quad (7\text{-}30)$$

Since N_V is the number of columns of the dimensional matrix, therefore both versions of (7-30) purport that the rank of this matrix exceeds its number of columns, which is impossible. Hence N_P cannot be negative. This proves the theorem.

Theorem 7-2. *N_P cannot be zero.*

Proof. Obviously every relation among variables $V_1, V_2, \ldots$ can be written as

$$V_1 = \Psi\{V_1, V_2, \ldots\} \qquad (7\text{-}31)$$

where V_1 on the right may be absent and Ψ is the symbol for function. For example

$$V_1 \cdot V_2 \cdot \sin(V_1 \cdot V_2) = 5$$

can be written

$$V_1 = \frac{5}{V_2 \cdot \sin(V_1 \cdot V_2)}$$

in which case

$$\Psi\{\ \} = \frac{5}{V_2 \cdot \sin(V_1 \cdot V_2)}$$

Now by Rule 1 of dimensional homogeneity (Art. 6.1), the dimensions of the left and right sides of (7-31) must be identical. That is $[V_1] = [\Psi]$ from which

$$\frac{[V_1]}{[\Psi]} = \left[\frac{V_1}{\Psi}\right] = 1$$

Therefore V_1/Ψ is a dimensionless product, and hence *at least one* dimensionless product of variables in any physical relation must exist. But, according to (7-26), the number of *dimensional* products of variables exceeds the number of *dimensionless* product of variables. The former's minimum value is 1, therefore the latter's must be *at least* 1, and hence neither of them can be less than 1, and hence neither of them can be zero. This proves the theorem.

Now we are ready to answer the question posed earlier: what is the minimum value of N_P? By (7-26) and Theorem 7-2, the minimum number of dimensionless and dimensional products of variables is

$$\left.\begin{array}{ll} (N_P)_{\min} = 1 & \text{if } N_{q \neq 0} = 0 \\ (N_P)_{\min} = 2 & \text{if } N_{q \neq 0} > 0 \end{array}\right\} \qquad (7\text{-}32)$$

In other words, if the products are dimensional, their minimum number is 2; if they are dimensionless, their minimum number is 1.

7.10. CONSTANCY OF THE SOLE DIMENSIONLESS PRODUCT

We saw in Art. 7.5 that a relation among physical variables can always be expressed by

$$\Psi\{\tau_1, \tau_2, \ldots, \tau_{N_P}\} = \text{const} \qquad (7\text{-}33)$$

in the case of *dimensional* products $\tau_1, \tau_2, \ldots$ of variables, and

$$\Psi\{\pi_1, \pi_2, \ldots, \pi_{N_P}\} = \text{const} \tag{7-34}$$

in the case of *dimensionless* products $\pi_1, \pi_2, \ldots$ of variables, where N_P is the respective number of such products defined in (7-32).

We now state and prove the following theorems.

Theorem 7-3. *If $N_P = 1$, then the generated product of variables must be dimensionless.*

Proof. By (7-32), if the subject product is dimensional, then N_P must be at least 2. But here we have $N_P = 1$, hence the product cannot be dimensional and therefore it must be dimensionless. This proves the theorem.

Theorem 7-4. *If $N_P = 1$, then this sole dimensionless product must be a constant.*

Proof. By (7-34), if for *all* possible values of its argument $\Psi\{\pi_1\} = \text{const}$, then necessarily $\pi_1 = \text{const}$. This proves the theorem.

The following example illustrates the great usefulness of this theorem.

Example 7-11. Propagation of the Wavefront in an Atomic Explosion

In an atomic (more accurately "nuclear") explosion, very large amount of energy is released within a very short time (one might say "instantaneously") and within a very small volume (one might say "at a point"). The pressure buildup can be considered spherically symmetric within the confines of a half-sphere, the center of which is the point of explosion. By definition, within this semisphere the pressure is very large, outside of it, it is normal atmospheric. The surface of this hemisphere is called the *wavefront*.

Now we ask the question: For a given amount of released energy, how does the radius of the wavefront vary with time? Or, given the radius versus time relation, what is the released energy?

This problem is treated by Barenblatt (Ref. 1, p.41), who employs dimensional methods. We shall discuss the case in a more streamlined fashion using the techniques developed in the present book. The first task, always, is to list the variables assumed relevant:

Variable	Symbol	Dimension
radius of wavefront	R	m
time	t	s
initial air density	ρ_0	kg/m³
released energy	Q	m²·kg/s²

The dimensional matrix therefore is

	R	t	ρ_0	Q
m	1	0	-3	2
kg	0	0	1	1
s	0	1	0	-2

The rank of this matrix is $R_{DM} = 3$ and the number of variables is $N_V = 4$. Therefore the number of independent dimensionless products of variables, by (7-26), is $N_P = N_V - R_{DM}$

STRUCTURE OF PHYSICAL RELATIONS

= 4 − 3 = 1 (since $N_{q\neq 0} = 0$). Consequently, we have a *single* such product, which we will now determine. By the given dimensional matrix

$$\mathbf{A} = \begin{bmatrix} 0 & -3 & 2 \\ 0 & 1 & 1 \\ 1 & 0 & -2 \end{bmatrix}; \quad \mathbf{B} = \begin{bmatrix} 1 \\ 0 \\ 0 \end{bmatrix}$$

and therefore the submatrices of matrix **E**, defined in (7-14), are (with $N_d = 3$)

$$\mathbf{I} = [1] \quad \mathbf{0} = [0 \ 0 \ 0]$$

$$-\mathbf{A}^{-1}\cdot\mathbf{B} = \begin{bmatrix} -0.4 \\ 0.2 \\ -0.2 \end{bmatrix}; \quad \mathbf{A}^{-1} = \begin{bmatrix} 0.4 & 1.2 & 1 \\ -0.2 & 0.4 & 0 \\ 0.2 & 0.6 & 0 \end{bmatrix}$$

Therefore

$$\mathbf{E} = \begin{bmatrix} 1 & 0 & 0 & 0 \\ -0.4 & 0.4 & 1.2 & 1 \\ 0.2 & -0.2 & 0.4 & 0 \\ -0.2 & 0.2 & 0.6 & 0 \end{bmatrix} \quad \text{and} \quad \mathbf{Z} = \begin{bmatrix} 1 \\ 0 \\ 0 \\ 0 \end{bmatrix}$$

Thus, by (7-17),

$$\mathbf{P} = \mathbf{E}\cdot\mathbf{Z} = \begin{bmatrix} 1 \\ -0.4 \\ 0.2 \\ -0.2 \end{bmatrix} \begin{matrix} \leftarrow R \\ \leftarrow t \\ \leftarrow \rho_0 \\ \leftarrow Q \end{matrix}$$

We recall that every *row* in the **P** matrix corresponds to a *variable* in the dimensional matrix in the same sequential order. Therefore the sole dimensionless variable will be

$$\pi_1 = R\cdot t^{-0.4}\cdot\rho_0^{0.2}\cdot Q^{-0.2} \tag{a}$$

By Theorem 7-4, π_1 = const, therefore (a) can be written

$$\pi_1 = R\cdot\sqrt[5]{\frac{\rho_0}{t^2\cdot Q}} = \text{const} \tag{b}$$

Based on theoretical considerations (Ref. 1, p. 44), the constant in (b) is close to unity. Therefore (b) can be written using 10 based logarithm

$$\log R = \frac{1}{5}\log\frac{Q}{\rho_0} + \frac{2}{5}\log t \tag{c}$$

i.e., in a log-log coordinate system the plot of (c) is a straight line (Fig. 7-5). Indeed, by Barenblatt (Ref. 1, p. 44), photographs of an atomic explosion confirm this relation.

By the plot, the ordinate intercept of the curve is at $\log R = 2.781$; i.e., at $t = 1$ s (log 1 = 0) the radius of the wavefront is $R = 603.95$ m. Moreover, by (c), if $t = 1$ s, then

$$\log R = 2.781 = \frac{1}{5}\log\frac{Q}{\rho_0} \tag{d}$$

from which the released energy, considering $\rho_0 = 1.25$ kg/m³ for air, is

$$Q = (10^{5\cdot(2.781)})\cdot\rho_0 = 10^{14} \text{ J} (= 7.4 \times 10^{13} \text{ lbf·ft})$$

156 APPLIED DIMENSIONAL ANALYSIS AND MODELING

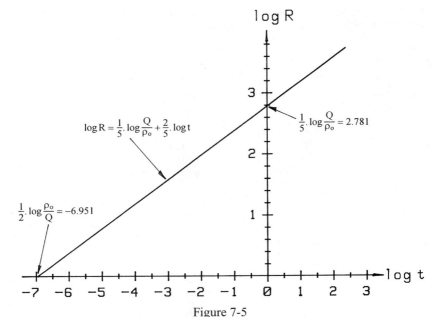

**Figure 7-5
Radius of the hemispherical wavefront surface
versus time in an atomic explosion**
The plot was constructed by the author from data published by Barenblatt (Ref. 1, p. 44).
See table in Example 7-11 for definition of symbols; logarithms are of base 10.

Thus, astonishingly, the energy yield of the atom bomb can be derived from a pair of numbers easily read off a single time fixed photograph of the "mushroom."

From (c) the radial speed or propagation of the wavefront $dR/dt = \dot{R}$ is

$$\log \dot{R} = \log \frac{2}{5} + \frac{1}{5} \log \frac{Q}{\rho_0} - \frac{3}{5} \log t \qquad (e)$$

which, since $Q = 10^{14}$ J,

$$\log \dot{R} = 2.3827 - \frac{3}{5} \log t \qquad (f)$$

For example, at $t = 10^{-3}$ s, the speed of propagation of the wavefront is $\dot{R} = 15230$ m/s, and at $t = 1$ s, $\dot{R} = 241.4$ m/s. At the instant of explosion ($t = 0$), the speed of propagation is very large ($\approx \infty$).

Expressing $\log t$ from (c), and substituting it into (e), we obtain, after some simple rearrangement,

$$\dot{R} = \frac{2}{5} \sqrt{\frac{Q}{\rho_0 \cdot R^3}}$$

For example, when the radius of the wavefront is just $R = 1$ m, the speed is $\dot{R} = 3.57 \times 10^6$ m/s, but at $R = 1000$ m the speed of the wavefront propagation is only $\dot{R} = 113.14$ m/s. This example thus shows the significant practical benefit of dimensional considerations.

⇑

7.11. NUMBER OF DIMENSIONS EQUALS OR EXCEEDS THE NUMBER OF VARIABLES

There are two cases, which are discussed separately.

7.11.1. Number of Dimensions Equals the Number of Variables

Since the dimensions and variables constitute the rows and columns of a dimensional matrix (Fig. 7-1), therefore (in this case) it is a square matrix of size $N_d \times N_d$ (or, equivalently $N_V \times N_V$), where N_d and N_V are the numbers of dimensions and variables, respectively. We now state and prove the following theorem:

Theorem 7-5. *If the dimensional matrix is a square matrix, then it must be singular—otherwise no relation among the stated variables is possible.*

Proof. Consider the contrary, i.e., that the dimensional matrix is not singular. Then its rank R_{DM} must equal its number of variables, viz., $R_{DM} = N_V$, or $N_V - R_{DM} = 0$. But, by (7-26),

$$N_P = N_V - R_{DM} \qquad \text{repeated (7-26)}$$

hence $N_P = 0$. This, however, violates Theorem 7-2 which states that N_P cannot be zero. Therefore the dimensional matrix—if it is square—must be singular. This proves the theorem. The following example illustrates.

Example 7-12. Relativistic Energy-Mass Equivalence

According to Einstein, under suitable conditions, mass and energy are convertible. Let us try to derive a relation between the relevant variables and dimensional constants. But first, as always, we list the relevant entries.

Variable	Symbol	Dimension	Remark
energy	Q	m²·kg/s²	
mass	M	kg	
speed of light	c	m/s	in vacuum

Accordingly, the dimensional matrix is:

	Q	M	c
m	2	0	1
kg	1	1	0
s	−2	0	−1

This is a square matrix, therefore, by Theorem 7-12, to have any relation among the incumbent variables at all, the matrix must be singular, i.e., its determinant must be zero. Indeed

$$\begin{vmatrix} 2 & 0 & 1 \\ 1 & 1 & 0 \\ -2 & 0 & -1 \end{vmatrix} = 0$$

158   APPLIED DIMENSIONAL ANALYSIS AND MODELING

and hence it *is* possible to connect the listed variables. Since the rank of this matrix is $R_{DM} = 2$, therefore, by (7-22), $\Delta = N_d - R_{DM} = 3 - 2 = 1$. Thus 1 row (dimension) has to go. Let this sacrificial dimension be "s." Hence the reduced dimensional matrix will be

	Q	M	c
m	2	0	1
kg	1	1	0

with the **A** matrix being the right 2×2 block and the **B** matrix being the left 2×1 column.

So that the **A** and **B** matrices are

$$\mathbf{A} = \begin{bmatrix} 0 & 1 \\ 1 & 0 \end{bmatrix}; \quad \mathbf{B} = \begin{bmatrix} 2 \\ 1 \end{bmatrix}$$

Let us now determine the complete set of products of variables whose dimension is 1, i.e., the *dimensionless* products. By the 1st relation of (7-26), $N_P = N_V - R_{DM} = 3 - 2 = 1$, since $N_{q \neq 0} = 0$. That is, we have just *one* such product, and hence this product must be a *constant* (Theorem 7-4). Next, we construct the **E** and **Z** matrices—in accordance with the scheme described in (7-14) and (7-17). For these, we have the constituents

$$-\mathbf{A}^{-1} \cdot \mathbf{B} = \begin{bmatrix} -1 \\ -2 \end{bmatrix}; \quad \mathbf{A}^{-1} = \begin{bmatrix} 0 & 1 \\ 1 & 0 \end{bmatrix}; \quad \mathbf{I} = [1]; \quad \mathbf{0} = [0 \ 0]$$

hence

$$\mathbf{E} = \begin{bmatrix} 1 & 0 & 0 \\ -1 & 0 & 1 \\ -2 & 1 & 0 \end{bmatrix}; \quad \mathbf{Z} = \begin{bmatrix} \epsilon_1 \\ q_1 \\ q_2 \end{bmatrix} = \begin{bmatrix} 1 \\ 0 \\ 0 \end{bmatrix}$$

since $q_1 = q_2 = 0$ (the products have to be dimensionless). Thus, by (7-17)

$$\mathbf{P} = \mathbf{E} \cdot \mathbf{Z} = \begin{bmatrix} 1 \\ -1 \\ -2 \end{bmatrix} \begin{matrix} \leftarrow Q \\ \leftarrow M \\ \leftarrow c \end{matrix}$$

yielding the *one* only (hence constant) dimensionless product

$$\pi_1 = \text{const} \cdot \frac{Q}{M \cdot c^2} \quad \text{or} \quad Q = \text{const} \cdot M \cdot c^2$$

which is, of course, the famous Einstein equation with const = 1.

Let us now check the dimension of π_1. By the given relation

$$[\pi_1] = \left[\frac{Q}{M \cdot c^2}\right] = \frac{[Q]}{[M] \cdot [c^2]} = \frac{m^2 \cdot kg/s^2}{kg \cdot (m/s)^2} = 1$$

i.e., π_1 is indeed dimensionless, as required and expected.

7.11.2. Number of Dimensions Exceeds the Number of Variables

Since the number of rows and columns in a dimensional matrix equals the number of dimensions and variables, respectively, we have a dimensional matrix whose number of rows exceeds its number of columns. We now state and prove the following theorem.

Theorem 7-6. *If the dimensional matrix has more dimensions (rows) than variables (columns), then its rank must be equal to or less than the number of variables diminished by one, viz., $R_{DM} \leq N_V - 1$, else no relation among the constituting variables is possible.*

Proof. By relation (7-32)

$$N_P \geq 1 \qquad \text{repeated (7-32)}$$

and by (7-26)

$$N_P = N_V - R_{DM} \qquad \text{repeated (7-26)}$$

Therefore, by these two formulas $N_V - R_{DM} \geq 1$ from which

$$R_{DM} \leq N_V - 1 \qquad (7\text{-}35)$$

This proves the theorem. The following example illustrates:

Example 7-13.

Given the dimensional matrix

	V_1	V_2	V_3
d_1	7	−5	6
d_2	0	−3	5
d_3	3	0	−1
d_4	−1	−4	7
d_5	8	−10	14

The number of variables is $N_V = 3$, the number of dimensions is $N_d = 5$, hence $N_d > N_V$. To see if *any* relation among the variables is possible, we have to determine the rank of this matrix. By the method of *minimal dyadic decomposition* (Art. 1.3), it is found that the rank is $R_{DM} = 2$. So $R_{DM} = N_V - 1$, hence Theorem 7-6 [relation (7-33)] is fulfilled, and therefore a relation among the stated variables *can* exist. Indeed, as the indefatigable reader can verify, $V_1 = \text{const}/(V_2^5 \cdot V_3^3)$ is such a relation—as it fulfils the requirements of the dimensional matrix.

⇑

Example 7-14.

Consider the following three universal dimensional constants. For the present purpose their numerical values are unimportant.

Universal Constant	Symbol	Dimension
speed of light in vacuum	c	$m \cdot s^{-1}$
Stefan-Boltzmann constant	σ	$kg \cdot s^{-3} \cdot K^{-4}$
universal gravitational const.	k	$m^3 \cdot kg^{-1} \cdot s^{-2}$

Question: Can these three constants be related by a formula? *Answer:* The dimensional matrix of the problem is:

	c	σ	k
m	1	0	3
kg	0	1	−1
s	−1	−3	−2
K	0	−4	0

We see that the number of "variables" is $N_V = 3$, and the number of dimensions is $N_d = 4$. Therefore, by Theorem 7-6, a relation among c, σ, and k is possible *only* if the rank of this matrix is less than 3, since $R_{DM} \leq N_V - 1 = 3 - 1 = 2$. We see that R_{DM} is in fact 3 (the determinant formed by the top three rows is $-2 \neq 0$). Thus, by Theorem 7-6, no relation among the 3 stated universal constants can exist. Note that in this example c, σ, and k are constants, not variables. However, from a dimensional point of view, this difference has no bearing upon the conclusion.

⇑

Finally, it is emphasized that the fact that a relation among some variables is *dimensionally possible,* does not guarantee the *existence* of such a relation. The next example illustrates:

Example 7-15.

Consider a mass dropped from a given height. We assume the following variables relevant:

Variable	Symbol	Dimension	Remark
time duration of free fall	t	s	
gravitational acceleration	g	m/s²	constant
distance of fall	L	m	zero starting speed
experimenter's height	h	m	without shoes on

Now the relation $t = \sqrt{L/g} \cdot (h/L)^{3.5}$ is *possible*—as far as dimensions are concerned—yet it is patently false since, obviously, in real life no such function can exist.

⇑

7.12. PROBLEMS

The problems below deal with the interdependence of universal constants. Some of these constants, with their symbols, dimensions (in SI) and magnitudes are given in the table of Fig. 7-6 below. Use this information for solving the problems. Solutions are in Appendix 6.

STRUCTURE OF PHYSICAL RELATIONS 161

##	Constant Name	Symbol	Mantissa	Exponent	m	kg	s	K	A	mol	cd
1	Wien's displacement constant	c_1	2.897756	−3	1			1			
2	nuclear magneton	μ_N	5.0507866	−27	2				1		
3	Bohr magneton	μ_B	9.2740154	−24	2				1		
4	Bohr radius	a_0	5.29177249	−11	1						
5	Rydberg constant	R_∞	1.09737315	7	−1						
6	Faraday constant	F	9.6485309	4			1		1	−1	
7	magnetic flux quantum	ϕ	2.06783461	−15	2	1	−2		−1		
8	fine structure constant	α	7.29735308	−3							
9	proton-electron mass ratio	q_1	1.836152701	3							
10	proton rest mass	m_p	1.67262305	−27		1					
11	electron charge/mass ratio	q_2	1.75881961	11		−1	1		1		
12	electron rest mass	m_e	9.1093897	−31		1					
13	electron charge	e	1.60217733	−19			1		1		
14	Dirac constant	$\bar{h}$	1.05457266	−34	2	1	−1				
15	Planck constant	h	6.6260755	−34	2	1	−1				
16	universal gravitational constant	k	6.67259	−11	3	−1	−2				
17	gravitational acceleration (Earth)	g	9.80665	0	1		−2				
18	permeability of vacuum	μ_0	1.25663706	−6	1	1	−2		−2		
19	permittivity of vacuum	ϵ_0	8.85418781	−12	−3	−1	4		2		
20	speed of light (in vacuum)	c	2.99792458	8	1		−1				
21	Stefan–Boltzmann constant	σ	5.67051	−8		1	−3	−4			
22	universal gas constant	R	8.31451	0	2	1	−2	−1		−1	
23	molar volume	V_m	2.24141	−2	3					−1	
24	Boltzmann constant	k_B	1.380658	−23	2	1	−2	−1			
25	Avogadro number	N_A	6.0221367	23						−1	

Figure 7-6
Some universal constants
Dimensions are in SI; data are mainly from Ref. 8, p. 63; the base of exponents is 10

7/1 Are k, c, and m_e independent? If yes, why? If no, express k in terms of c and m_e.

7/2 If possible express h in terms of k, c, and m_e. If this cannot be done, give reason.

7/3 Can m_e be expressed by ϵ_0, e, c, and h?

7/4 Are ϵ_0, e, c, and h dependent? That is, can any one of these four constants be expressed by the other three?

7/5 Given the following six constants: N_A, k_B, V_m, R, h, F. How many of these are independent?

7/6 Given the following five constants: c, g, k, h, m_e. How many of these are independent?

CHAPTER 8
SYSTEMATIC DETERMINATION OF COMPLETE SET OF PRODUCTS OF VARIABLES

8.1. DIMENSIONAL SET; DERIVATION OF PRODUCTS OF VARIABLES OF A GIVEN DIMENSION

The process yielding a complete set of products of variables—of assigned dimensions—was outlined in Chapter 7. Although this process is always successful (it results in the required number of products of prescribed dimensions), it is somewhat cumbersome. Moreover, it requires a rather high level of attention to detail, with the associated significant risk of committing errors in the required manipulation of matrices. Therefore it is highly desired that this process be "streamlined" to allow a more compact and economical treatment, and—most importantly—less matrix arithmetic. We shall now create such a simple and highly practical method.

First, we recall the two important characteristics of the **P** matrix: each of its *rows* corresponds to a physical *variable,* and each of its *columns* corresponds to a particular *product* of variables (dimensional or dimensionless). For example, in Example 7-10 dealing with the deflection of a cantilever under a force F, the **P** matrix is

$$\mathbf{P} = \begin{bmatrix} 1 & 0 & 0 & 1 \\ 0 & 1 & 0 & 1 \\ 0 & 0 & 1 & 0 \\ 3 & 2 & 3 & 2 \\ 1 & 0 & 1 & -1 \end{bmatrix} \begin{matrix} \leftarrow U \\ \leftarrow F \\ \leftarrow L \\ \leftarrow E \\ \leftarrow D \end{matrix} \Bigg\} \text{variables} \qquad (8\text{-}1)$$

$$\underbrace{\begin{matrix} \uparrow & \uparrow & \uparrow & \uparrow \\ \tau_1 & \tau_2 & \tau_3 & \tau_4 \end{matrix}}_{\text{products of variables}}$$

in which the first *row* corresponds to variable U, the second to F, and so on. Similarly, the first column defines the product of variables τ_1, the second column defines the product of variables τ_2, etc. Thus, for example, in the first column the exponent of variable U is 1, of F it is 0, of L it is 0, of E it is 3, and of D it is 1. Therefore $\tau_1 = U^1 \cdot F^0 \cdot L^0 \cdot E^3 \cdot D^1 = U \cdot E^3 \cdot D$. These characteristics of the **P** matrix suggest that if we form the *transpose* of it—i.e., interchange its rows and columns—then it would fit nicely under the dimensional matrix whose *columns* represent variables.

For example, if we put $\mathbf{P}^T$ under the dimensional matrix shown in Fig. 8-1, then we have the scheme conveniently yielding the required products of variables presented in Example 7-10. The specific array so generated is called the *Dimensional Set* (Fig. 8-1). To demonstrate the facility accorded by this technique, let us construct τ_2. In the Dimensional Set we go along the second row (τ_2) and read off the respective *exponents* of variables placed at the top of the columns. Thus we will find that $\tau_2 = U^0 \cdot F^1 \cdot L^0 \cdot E^2 \cdot D^0 = F \cdot E^2$, exactly as was established in Example 7-10.

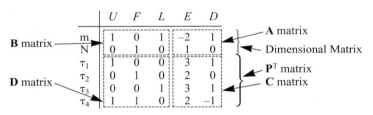

Figure 8-1
An example of the Dimensional Set (matrix A must not be singular)

The dimensional set also allows the easy determination of the exponent in any product of variables. For example, let us find the exponent of "m" in any of the product τ_x of variables—where x is any integer between 1 and 4. Since all τ_x products have identical dimensions by definition, it does not matter what x value we choose. Let us choose $x = 1$, i.e., τ_1. Hence, by multiplying the elements of row "m" by the elements of row "τ_1"—column by column—and adding the results, we obtain

$$\text{exponent of ``m''} = (1)\cdot(1) + (0)\cdot(0) + (1)\cdot(0) + (-2)\cdot(3) + (1)\cdot(1) = -4$$

$$\text{exponent of ``N''} = (0)\cdot(1) + (1)\cdot(0) + (0)\cdot(0) + (1)\cdot(3) + (0)\cdot(1) = 3$$

Therefore every τ_x has the dimension $m^{-4} \cdot N^3$.

The next task is to "streamline" the construction of $\mathbf{P}^T$ as it appears in Fig. 8-1. To do this, we *temporarily* complicate the matter a little. In (7-17) we established the relation $\mathbf{P} = \mathbf{E} \cdot \mathbf{Z}$ where **E** is as defined in (7-14)

$$\mathbf{E} = \begin{bmatrix} \mathbf{I} & \mathbf{0} \\ -\mathbf{A}^{-1}\mathbf{B} & \mathbf{A}^{-1} \end{bmatrix} \qquad \text{repeated (7-14)}$$

SYSTEMATIC DETERMINATION OF COMPLETE SET OF PRODUCTS OF VARIABLES **165**

The **Z** matrix is defined in (7-17)

$$Z = \begin{bmatrix} \epsilon_{11} & \epsilon_{12} & \cdots & \epsilon_{1N_P} \\ \epsilon_{21} & \epsilon_{22} & \cdots & \epsilon_{2N_P} \\ \vdots & \vdots & & \vdots \\ q_1 & q_1 & & q_1 \\ q_2 & q_2 & & q_2 \\ \vdots & \vdots & & \vdots \end{bmatrix} \qquad \text{repeated (7-17)}$$

This latter can be written

$$Z = \begin{bmatrix} \epsilon \\ q \end{bmatrix} \qquad (8\text{-}2)$$

where ϵ and q are submatrices. In particular ϵ is an $(N_V - R_{DM}) \times N_P$ matrix, and q is an $R_{DM} \times N_P$ matrix, where N_P is defined in (7-26), and R_{DM} is the rank of the dimensional matrix. To illustrate, in Example 7-9, $N_V = 5$, $R_{DM} = 2$, $N_P = 4$, therefore ϵ is a 3×4 matrix, and q is a 2×4 matrix. Thus—in that example—

$$\epsilon = \begin{bmatrix} 1 & 0 & 0 & 1 \\ 0 & 1 & 0 & 1 \\ 0 & 0 & 1 & 0 \end{bmatrix}; \qquad q = \begin{bmatrix} 2 & 2 & 2 & 2 \\ 0 & 0 & 0 & 0 \end{bmatrix}$$

so that

$$Z = \begin{bmatrix} \epsilon \\ q \end{bmatrix} = \begin{bmatrix} 1 & 0 & 0 & 1 \\ 0 & 1 & 0 & 1 \\ 0 & 0 & 1 & 0 \\ 2 & 2 & 2 & 2 \\ 0 & 0 & 0 & 0 \end{bmatrix}$$

as given in that example.

Next, considering (7-17), (7-14), and (8-2)

$$P = \begin{bmatrix} I & 0 \\ -A^{-1}\cdot B & A^{-1} \end{bmatrix} \cdot \begin{bmatrix} \epsilon \\ q \end{bmatrix} \qquad (8\text{-}3)$$

which yields

$$P = \begin{bmatrix} I\cdot\epsilon + 0\cdot q \\ -A^{-1}\cdot B\cdot\epsilon + A^{-1}\cdot q \end{bmatrix} = \begin{bmatrix} \epsilon \\ -A^{-1}\cdot B\cdot\epsilon + A^{-1}\cdot q \end{bmatrix} \qquad (8\text{-}4)$$

and hence its transpose is

$$P^T = [\epsilon^T \quad (-A^{-1}\cdot B\cdot\epsilon)^T + (A^{-1}\cdot q)^T] \qquad (8\text{-}5)$$

Let us now—for the sake of expediency and brevity—introduce the notation

$$\left. \begin{array}{l} D = \epsilon^T \\ C = \epsilon^T\cdot(-A^{-1}\cdot B)^T + (A^{-1}\cdot q)^T \end{array} \right\} \qquad (8\text{-}6)$$

by which (8-5) can be written

$$\mathbf{P}^T = [\mathbf{D} \quad \mathbf{C}] \tag{8-7}$$

Therefore, by virtue of the first part of (8-6), we have the formula

$$\mathbf{C} = -\mathbf{D} \cdot (\mathbf{A}^{-1} \cdot \mathbf{B})^T + (\mathbf{A}^{-1} \cdot \mathbf{q})^T \tag{8-8}$$

This is an important relation since it allows us to almost effortlessly construct the Dimensional Set, which, in turn, immediately yields the desired products of variables.

At this stage it may be useful if the sizes of matrices hitherto introduced are collected and listed. The table in Fig 8-2 provides such a list. This table shows that

- matrices **D** and **B** have the same number of columns;
- matrices **D** and **C** have the same number of rows;
- matrices **C** and **A** have the same number of columns.

Therefore in the Dimensional Set **D** fits nicely under **B**, and **C** under **A**—as seen in Fig. 8-1. The same structure is schematically shown in Fig. 8-3.

Matrix	Relation or figure #	Number of rows	Number of columns	Remark
A	Fig. 7-2	R_{DM}	R_{DM}	square
B	Fig. 7-2	R_{DM}	$N_V - R_{DM}$	
I	(7-13)	$N_V - R_{DM}$	$N_V - R_{DM}$	unit square
0	(7-13)	$N_V - R_{DM}$	R_{DM}	null
ϵ	(8-2)	$N_V - R_{DM}$	N_P	
q	(8-2)	R_{DM}	N_P	null if $N_{q \neq 0} = 0$
E	(7-14)	N_V	N_V	square
Z	(7-17)	N_V	N_P	
P	(7-17)	N_V	N_P	
D	(8-6)	N_P	$N_V - R_{DM}$	square if $N_{q \neq 0} = 0$
C	(8-8)	N_P	R_{DM}	
Reduced dimensional	Fig. 7-4	R_{DM}	N_V	
Dimensional Set	Fig 8-1	$N_P + R_{DM}$	N_V	square if $N_{q \neq 0} = 0$

Figure 8-2
Sizes and reference data of principal matrices
R_{DM} = rank of dimensional matrix, N_V = number of variables, $N_{q \neq 0}$ = number of products of variables that are not dimensionless, N_P = as defined in relation (7-26)

SYSTEMATIC DETERMINATION OF COMPLETE SET OF PRODUCTS OF VARIABLES 167

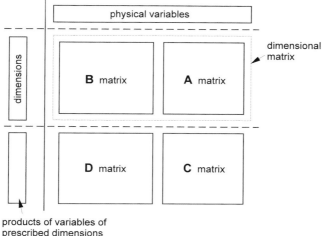

Figure 8-3
Structure of the Dimensional Set
Matrix **A** must be nonsingular; see Fig 8-2 for the characteristics of submatrices

Since, as we have seen, the *rows* of the ϵ matrix must be independent, therefore—since $\mathbf{D} = \epsilon^T$—exactly $N_V - R_{DM}$ number of *columns* of matrix **D** must also be independent; otherwise it can be arbitrary. So, to keep everything as simple as possible, it is wise to compose **D** with as many zeros and 1's as possible (although, of course, the scheme yields a correct result no matter how **D** is composed, as long as its columns are independent). Example 8-1 illustrates.

Example 8-1

We have a system of $N_V = 6$ variables and $N_d = 4$ dimensions. The dimensional matrix is shown below. The rank of this matrix is $R_{DM} = 2$, as seen easily since the third row is the sum of the first and second rows, and the fourth row is the sum of second and third rows. Therefore the number of rows (dimensions) to be discarded Δ will be, by (7-22) and Fig. 7-4, $\Delta = N_d - R_{DM} = 4 - 2 = 2$. Let these sacrificial dimensions be d_3 and d_4.

		V_1	V_2	V_3	V_4	V_5	V_6		
	d_1	1	2	3	5	2	3	← A matrix	
B matrix	d_2	4	1	2	1	2	4		(a)
	d_3	5	3	5	6	4	7	← deleted	
	d_4	9	4	7	7	6	11	← dimensions	

The reduced dimensional matrix therefore will be

		V_1	V_2	V_3	V_4	V_5	V_6		
B matrix	d_1	1	2	3	5	2	3	← A matrix	(b)
	d_2	4	1	2	1	2	4		

By (7-27) now, the selectable number of exponents for variables is $n_s = R_{DM} = 2$, and for the nonselectable ones, by (7-29), it is $n_{ns} = N_d - R_{DM} = \Delta = 2$. Accordingly, let us select $q_1 = 2$ and $q_2 = 3$, so that 2 will be the exponent of d_1 and 3 of d_2. Now since $N_{q \neq 0} = 2 \neq 0$, therefore, by (7-26)

$$N_P = N_V - R_{DM} + 1 = 6 - 2 + 1 = 5 \tag{c}$$

and hence, by Fig. 8-2, **q** is a 2×5 matrix

$$\mathbf{q} = \begin{bmatrix} q_1 & q_1 & q_1 & q_1 & q_1 \\ q_2 & q_2 & q_2 & q_2 & q_2 \end{bmatrix} = \begin{bmatrix} 2 & 2 & 2 & 2 & 2 \\ 3 & 3 & 3 & 3 & 3 \end{bmatrix} \tag{d}$$

Next, we must compose the **D** matrix which, by Fig. 8-2 and (c), must have $N_P = 5$ rows and $N_V - R_{DM} = 6 - 2 = 4$ columns. Thus we write, for example,

$$\mathbf{D} = \begin{bmatrix} 1 & 0 & 0 & 0 \\ 0 & 1 & 0 & 0 \\ 0 & 0 & 1 & 0 \\ 0 & 0 & 0 & 1 \\ 1 & 1 & 0 & 0 \end{bmatrix}$$

in which—as the reader should carefully observe—the columns are independent, as required. By the presented reduced dimensional matrix

$$\mathbf{A} = \begin{bmatrix} 2 & 3 \\ 2 & 4 \end{bmatrix}; \quad \mathbf{B} = \begin{bmatrix} 1 & 2 & 3 & 5 \\ 4 & 1 & 2 & 1 \end{bmatrix}$$

and hence, by (8-8),

$$\mathbf{C} = \begin{bmatrix} 3.5 & -2 \\ -3 & 2 \\ -3.5 & 2 \\ -9 & 5 \\ 1 & -1 \end{bmatrix}$$

Therefore the Dimensional Set, by Fig. 8-3, will be

		V_1	V_2	V_3	V_4	V_5	V_6	
B matrix	d_1	1	2	3	5	2	3	
	d_2	4	1	2	1	2	4	← **A** matrix
	τ_1	1	0	0	0	3.5	-2	
	τ_2	0	1	0	0	-3	2	
D matrix	τ_3	0	0	1	0	-3.5	2	← **C** matrix
	τ_4	0	0	0	1	-9	5	
	τ_5	1	1	0	0	1	-1	

from which the five products of variables of the required dimensions can be read off as

$$\tau_1 = \frac{V_1 \cdot V_5^{3.5}}{V_6^2}; \quad \tau_2 = \frac{V_2 \cdot V_6^2}{V_5^3}; \quad \tau_3 = \frac{V_3 \cdot V_6^2}{V_5^{3.5}}; \quad \tau_4 = \frac{V_4 \cdot V_6^5}{V_5^9}; \quad \tau_5 = \frac{V_1 \cdot V_2 \cdot V_5}{V_6} \tag{e}$$

All these products now have the required dimensions, viz., the exponent of d_1 is 2, and of d_2 is 3. Let us check one of them, say τ_3; the rest can be done the same way. Thus, by (a),

$$[\tau_3] = [V_3 \cdot V_5^{-3.5} \cdot V_6^2] = (d_1^3 \cdot d_2^2 \cdot d_3^5 \cdot d_4^7)^1 \cdot (d_1^2 \cdot d_2^2 \cdot d_3^4 \cdot d_4^6)^{-3.5} \cdot (d_1^3 \cdot d_2^4 \cdot d_3^7 \cdot d_4^{11})^2 = d_1^2 \cdot d_2^3 \cdot d_3^5 \cdot d_4^8$$

SYSTEMATIC DETERMINATION OF COMPLETE SET OF PRODUCTS OF VARIABLES **169**

We see that the exponents of d_1 and d_2 are indeed the required $q_1 = 2$, and $q_2 = 3$. We also observe that there are two *nonselectable* exponents 5 and 8 for dimensions d_3 and d_4. Of course, these two exponents are also constants for all τ_x values—as easily confirmed by the dimensional set given in (a). For example, what is the exponent of d_4 in τ_5? All we have to do is to add the "vertically pairwise" products of rows d_4 and τ_5. Thus the exponent of d_4 in τ_5 is $(9)\cdot(1) + (4)\cdot(1) + (7)\cdot(0) + (7)\cdot(0) + (6)\cdot(1) + (11)\cdot(-1) = 8$, as expected.

⇑

Example 8-2

Here we have the same system as we had in Example 8-1, but now we select a more complex **D** matrix, as in the Dimensional Set below

		V_1	V_2	V_3	V_4	V_5	V_6	
B matrix	d_1	1	2	3	5	2	3	
	d_2	4	1	2	1	2	4	
	τ_1	1	0	1	3	−25	11	**A** matrix
D matrix	τ_2	2	1	0	5	−37.5	16	(a)
	τ_3	0	2	4	0	−17.5	7	← **C** matrix
	τ_4	2	0	3	9	−78	34	
	τ_5	1	1	0	2	−16	7	

where matrix **C** is as defined in (8-8), with the **A**, **B**, and **q** matrices identical to those in Example 8-1. By the Dimensional Set, now

$$\left. \begin{array}{c} \tau_1 = \dfrac{V_1 \cdot V_3 \cdot V_4^3 \cdot V_6^{11}}{V_5^{25}}; \qquad \tau_2 = \dfrac{V_1^2 \cdot V_2 \cdot V_4^5 \cdot V_6^{16}}{V_5^{-37.5}}; \qquad \tau_3 = \dfrac{V_2^2 \cdot V_3^4 \cdot V_6^6}{V_5^{17.5}}; \\[1em] \tau_4 = \dfrac{V_1^2 \cdot V_3^3 \cdot V_4^9 \cdot V_6^{34}}{V_5^{78}}; \qquad \tau_5 = \dfrac{V_1 \cdot V_2 \cdot V_4^2 \cdot V_6^7}{V_5^{16}} \end{array} \right\} \quad (b)$$

Let us check the dimension of one of these products, say τ_4; we use the data of the dimensional matrix given in (a) of Example 8-1,

$$[\tau_4] = [V_1^2 \cdot V_3^3 \cdot V_4^9 \cdot V_5^{-78} \cdot V_6^{34}] = (d_1^1 \cdot d_2^4 \cdot d_3^5 \cdot d_4^9)^2 \cdot (d_1^3 \cdot d_2^2 \cdot d_3^5 \cdot d_4^7)^3 \cdot (d_1^5 \cdot d_2^1 \cdot d_3^6 \cdot d_4^7)^9 \cdot$$
$$\cdot (d_1^2 \cdot d_2^2 \cdot d_3^4 \cdot d_4^6)^{-78} \cdot (d_1^3 \cdot d_2^4 \cdot d_3^7 \cdot d_4^{11})^{34} = d_1^2 \cdot d_2^3 \cdot d_3^5 \cdot d_4^8$$

which is identical to the value obtained in Example 8-1—as expected and required.

That the products $\tau_1, \ldots, \tau_5$ in (b) of Example 8-2 are independent can be ascertained easily by observing that the rank of **D** is not less than the number of its columns. In our case here, both the rank and the number of columns is 4. Hence the products of variables, calculated as in (b) above, are indeed independent.

Although each of the products in (b) of Example 8-2 is independent of the others, the *set* in (b) is *not* independent from any other set similarly derived. This means that any member of one set can be expressed as a *monomial* function of the members of any other set. For example, if we temporarily denote τ_1 in (b) as $\hat{\tau}_1$, and consider $\tau_1, \ldots, \tau_5$ as given in (e) in Example 8-1, then, we can easily verify that

$$\hat{\tau}_1 = \dfrac{\tau_3 \cdot \tau_4^3 \cdot \tau_5^4}{\tau_1^3 \cdot \tau_2^4} \qquad (c)$$

Similar expressions can be found for $\hat{\tau}_2 \ldots \hat{\tau}_5$. In Chapter 9 a systematic and general method will be presented by which a transformation between product sets can be made.

⇑

The next example deals with a case in which the **D** matrix is *singular*. As previously explained, the singularity of **D** causes *dependence* of the derived products of variables (i.e., they are not independent).

Example 8-3

Suppose we select—by oversight—a **D** matrix in which the columns are not independent. In other words, at least one column of **D** is a linear combination of the other columns. We use the same dimensional matrix as shown in Example 8-1. The **D** matrix selected is

$$\mathbf{D} = \begin{bmatrix} 1 & 2 & 3 & 2 \\ 2 & 1 & 1 & 4 \\ -1 & 2 & 1 & -2 \\ 3 & 1 & 1 & 6 \\ 4 & 2 & 4 & 8 \end{bmatrix} \quad \text{(a)}$$

in which—as seen—the fourth column is twice the first column. The Dimensional Set therefore is as shown in (b), where the **C** matrix of course is as determined by relation (8-8), in which the **q** matrix is as given in (d) of Example 8-1. Hence, by (b), we have the set shown in (c).

		V_1	V_2	V_3	V_4	V_5	V_6
B matrix	d_1	1	2	3	5	2	3
	d_2	4	1	2	1	2	4
D matrix	τ_1	1	2	3	2	−27.5	11
	τ_2	2	1	1	4	−32	13
	τ_3	−1	2	1	−2	4.5	−1
	τ_4	3	1	1	6	−45	18
	τ_5	4	2	4	8	−69.5	27

(b), with **A** matrix and **C** matrix labels on the right

$$\left. \begin{array}{c} \tau_1 = \dfrac{V_1 \cdot V_2^2 \cdot V_3^3 \cdot V_4^2 \cdot V_6^{11}}{V_5^{27.5}}; \quad \tau_2 = \dfrac{V_1^2 \cdot V_2 \cdot V_3 \cdot V_4^4 \cdot V_6^{13}}{V_5^{32}}; \quad \tau_3 = \dfrac{V_2^2 \cdot V_3 \cdot V_5^{4.5}}{V_1 \cdot V_4^2 \cdot V_6} \\ \tau_4 = \dfrac{V_1^3 \cdot V_2 \cdot V_3 \cdot V_4^6 \cdot V_6^{18}}{V_5^{45}}; \quad \tau_5 = \dfrac{V_1^4 \cdot V_2^2 \cdot V_3^4 \cdot V_4^8 \cdot V_6^{27}}{V_5^{69.5}} \end{array} \right\} \quad \text{(c)}$$

The dimension of all five of these products is uniformly $d_1^2 \cdot d_2^3 \cdot d_3^5 \cdot d_4^8$, as in the previous two examples. The question is now: Are the products in (c) independent? If yes, the $\Psi_1\{V_1, \ldots, V_6\} = \text{const}$ relation can be uniquely represented by the $\Psi_2\{\tau_1 \ldots \tau_5\} = \text{const}$ relation, otherwise no such representation is possible—since the obtained set of τ_x variables in (c) is not *complete* (Art. 7.5). So, to answer the question all we have to do is to determine the rank of matrix **D**. If the rank of **D** is the same as the number of its columns, then the products in (c) are independent and therefore the set is *complete*; otherwise the products are not independent and the set is *incomplete*.

In this example, the rank of matrix **D** in (a) is 3—which can be easily verified—although the number of its columns is 4. Therefore the rank is less than the number of columns, and hence the τ_x products in (c) are not independent and thus they do *not* form a complete set.

Indeed, as the motivated reader can easily verify,

$$\frac{\tau_2^4 \cdot \tau_3 \cdot \tau_5^2}{\tau_1^3 \cdot \tau_4^4} = 1 \quad \text{(d)}$$

So, any one of the products in (c) can be expressed as a function of the other four.

8.2. CHECKING THE RESULTS

A very convenient and efficient verification of the foregoing matrix arithmetic can be carried out as follows: The dimensional matrix **M** can be written (Fig. 8-3)

$$\mathbf{M} = [\mathbf{B} \quad \mathbf{A}] \tag{8-9}$$

Let us temporarily call the matrix under **M** the **U** matrix (see Fig. 8-3). Then

$$\mathbf{U} = [\mathbf{D} \quad \mathbf{C}] \tag{8-10}$$

By this relation and the data found in Fig. 8-2 it is evident that **U** is an $N_P \times N_V$ matrix, where N_P is as defined in (7-26), and N_V is the number of variables. Next, by a good hunch, we determine the matrix product $\mathbf{M} \cdot \mathbf{U}^T$. Accordingly, by (8-9) and (8-10),

$$\mathbf{M} \cdot \mathbf{U}^T = [\mathbf{B} \quad \mathbf{A}] \cdot \begin{bmatrix} \mathbf{D}^T \\ \mathbf{C}^T \end{bmatrix} \tag{8-11}$$

At this point we have to see whether the matrix products $\mathbf{B} \cdot \mathbf{D}^T$ and $\mathbf{A} \cdot \mathbf{C}^T$ implied by this relation are possible. From the information given above for **U**, plus the data contained in Fig. 8-2, the reader can verify that the matrix products under scrutiny are indeed possible. Hence, by (8-11), we can write

$$\mathbf{M} \cdot \mathbf{U}^T = \mathbf{B} \cdot \mathbf{D}^T + \mathbf{A} \cdot \mathbf{C}^T \tag{8-12}$$

We now consider that $\mathbf{C}^T$, by (8-8), can be written as

$$\mathbf{C}^T = \mathbf{A}^{-1} \cdot \mathbf{q} - \mathbf{A}^{-1} \cdot \mathbf{B} \cdot \mathbf{D}^T \tag{8-13}$$

which we substitute into (8-12). This yields

$$\mathbf{M} \cdot \mathbf{U}^T = \mathbf{B} \cdot \mathbf{D}^T + \mathbf{A} \cdot (\mathbf{A}^{-1} \cdot \mathbf{q} - \mathbf{A}^{-1} \cdot \mathbf{B} \cdot \mathbf{D}^T) = \mathbf{B} \cdot \mathbf{D}^T + \mathbf{q} - \mathbf{B} \cdot \mathbf{D}^T = \mathbf{q} \tag{8-14}$$

where **q** is a matrix defined in Art. 8.1. in relation (8-2). Recall that **q** is composed of the assigned exponents to dimensions. Now relation (8-14) purports simply that the product of the dimensional matrix and the *transpose* of the matrix under it must be identical to the **q** matrix.

Note that if the dimensional matrix and the reduced dimensional matrix are not equal, then (8-14) will yield the **q** matrix comprising both the *selectable* and the *nonselectable* exponents of dimensions. If we want to have a **q** matrix of only selectable values, then **M** in (8-14) must be interpreted as a reduced dimensional matrix. It is advisable for this checking process to consider the dimensional matrix, rather than only the reduced dimensional matrix, because the latter does not include the nonselectable exponents. The following example demonstrates this process.

Example 8-4

We use the data derived in Example 8-1. The dimensional matrix [relation (a) in Example 8-1] is repeated here for convenience

$$\mathbf{M} = \begin{bmatrix} 1 & 2 & 3 & 5 & 2 & 3 \\ 4 & 1 & 2 & 1 & 2 & 4 \\ 5 & 3 & 5 & 6 & 4 & 7 \\ 9 & 4 & 7 & 7 & 6 & 11 \end{bmatrix} \tag{a}$$

172 APPLIED DIMENSIONAL ANALYSIS AND MODELING

and by (8-10) the **U** matrix is

$$\mathbf{U} = \begin{bmatrix} 1 & 0 & 0 & 0 & 3.5 & -2 \\ 0 & 1 & 0 & 0 & -3 & 2 \\ 0 & 0 & 1 & 0 & -3.5 & 2 \\ 0 & 0 & 0 & 1 & -9 & 5 \\ 1 & 1 & 0 & 0 & 1 & -1 \end{bmatrix} \quad \text{(b)}$$

Hence, by (8-14)

$$\mathbf{M} \cdot \mathbf{U}^T = \begin{bmatrix} 2 & 2 & 2 & 2 & 2 \\ 3 & 3 & 3 & 3 & 3 \\ 5 & 5 & 5 & 5 & 5 \\ 8 & 8 & 8 & 8 & 8 \end{bmatrix} = \mathbf{q} \quad \text{(c)}$$

and we see that the dimensions of all τ_x products in the example are $[\tau_x] = d_1^2 \cdot d_2^3 \cdot d_3^5 \cdot d_4^8$ for $x = 1,2,3,4$—as expected and required. Note that here both the selectable and the nonselectable exponents are involved. If we had used the reduced dimensional matrix—as in (b) in Example 8-1—then only the veracity of selectable exponents, viz., 2 and 3 could have been confirmed.

⇑

Example 8-5. Power Requirement of a Rotating-Blade Mixer

Consider the mixer as schematically illustrated in Fig 8-4. The blades are rotated *slowly* by an external motor. The power required is obviously a function of the rotational speed, the diameter of the blades and the dynamic viscosity of the medium (a liquid). Accordingly, the relevant variables, their symbols, and dimensions are as follows:

Variable	Symbol	Dimension	Remark
power required	P	$m^2 kg/s^3$	
rotational speed	ω	$1/s$	
blade diameter	D	m	characteristic value
viscosity of medium	μ	$kg/(m \cdot s)$	dynamic, constant

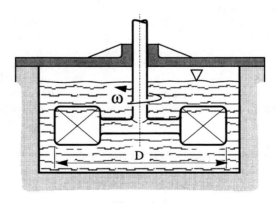

Figure 8-4
A rotating-blade mixer

SYSTEMATIC DETERMINATION OF COMPLETE SET OF PRODUCTS OF VARIABLES 173

Because of the slow rotational speed of the blades, the inertial resistance offered by the liquid can be ignored. Hence the density of the medium is not considered.

We now wish to form products of variables whose dimensions are uniformly $m^2 = m^2 \cdot kg^0 \cdot s^0$. Therefore $q_1 = 2$, $q_2 = q_3 = 0$. The dimensional matrix **M** is

$$\textbf{B matrix} \longrightarrow \begin{array}{c|cccc} & P & \omega & D & \mu \\ \hline m & 2 & 0 & 1 & -1 \\ kg & 1 & 0 & 0 & 1 \\ s & -3 & -1 & 0 & -1 \end{array} \begin{array}{l} \text{dimensional matrix} \\ \\ \textbf{A matrix} \end{array}$$

whose rank is $R_{DM} = 3$, the number of dimension is $N_d = 3$, and its rightmost determinant is not zero. Thus the reduced dimensional matrix is identical to **M**, so we do not have to jettison any rows (dimensions). We have, therefore

$$\mathbf{A} = \begin{bmatrix} 0 & 1 & -1 \\ 0 & 0 & 1 \\ -1 & 0 & -1 \end{bmatrix}; \quad \mathbf{B} = \begin{bmatrix} 2 \\ 1 \\ -3 \end{bmatrix} \tag{a}$$

The number of dimensions whose exponents is different from zero is $N_{q \neq 0} = 1 > 0$, therefore, by (7-26), the number of independent products of variables is

$$N_P = N_V - R_{DM} + 1 = 4 - 3 + 1 = 2 \tag{b}$$

Consequently, by Fig. 8-2, **D** is a 2×1 matrix, i.e., it has only 1 column, so, by its uniqueness, it is "independent." For its two rows we now select the simplest pair of numbers (not both zeros)

$$\mathbf{D} = \begin{bmatrix} 1 \\ 0 \end{bmatrix} \tag{c}$$

The **q** matrix—by Fig. 8-2—is of the size 3×2

$$\mathbf{q} = \begin{bmatrix} q_1 & q_1 \\ q_2 & q_2 \\ q_3 & q_3 \end{bmatrix} = \begin{bmatrix} 2 & 2 \\ 0 & 0 \\ 0 & 0 \end{bmatrix} \tag{d}$$

Therefore the **C** matrix, by (8-8), is

$$\mathbf{C} = -\mathbf{D} \cdot (\mathbf{A}^{-1} \cdot \mathbf{B})^T + (\mathbf{A}^{-1} \cdot \mathbf{q})^T = \begin{bmatrix} -2 & -1 & -1 \\ 0 & 2 & 0 \end{bmatrix}$$

by which, following the scheme on Fig. 8-3, the Dimensional Set will be

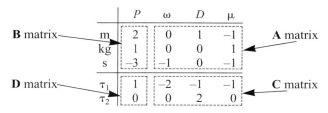

yielding, by (b), $N_P = 2$, independent products of variables

$$\tau_1 = \frac{P}{\omega^2 \cdot D \cdot \mu}; \qquad \tau_2 = D^2 \tag{e}$$

Since the dimension of τ_2 is obviously m², let us check τ_1 for dimension

$$[\tau_1] = [P \cdot \omega^{-2} \cdot D^{-1} \cdot \mu^{-1}] = (m^2 \cdot kg \cdot s^{-3}) \cdot (s^{-1})^{-2} \cdot (m)^{-1} \cdot (m^{-1} \cdot kg \cdot s^{-1})^{-1} = m^2 \cdot kg^0 \cdot s^0 = m^2$$

as required and expected.

Let us now determine $\mathbf{M} \cdot \mathbf{U}^T$. From the above data

$$\mathbf{M} = \begin{bmatrix} 2 & 0 & 1 & -1 \\ 1 & 0 & 0 & 1 \\ -3 & -1 & 0 & -1 \end{bmatrix}; \quad \mathbf{U} = \begin{bmatrix} 1 & -2 & -1 & -1 \\ 0 & 0 & 2 & 0 \end{bmatrix}$$

and hence, by (8-14)

$$\mathbf{M} \cdot \mathbf{U}^T = \begin{bmatrix} 2 & 2 \\ 0 & 0 \\ 0 & 0 \end{bmatrix} = \mathbf{q}$$

as given by (d)—as expected and required. The *two* products, τ_1 and τ_2 by (e), uniquely describe the relation among the *four* variables P, ω, D, μ. Accordingly, we can write

$$\Psi_1\{\tau_1, \tau_2\} = \text{const} \tag{f}$$

from which

$$\tau_1 = \text{const} \cdot \Psi_2\{\tau_2\} \tag{g}$$

and hence, by (e)

$$\frac{P}{\omega^2 \cdot D \cdot \mu} = \text{const} \cdot \Psi_2\{\tau_2\} \tag{h}$$

yielding

$$P = \text{const} \cdot \omega^2 \cdot D \cdot \mu \cdot \Psi_2\{\tau_2\} \tag{i}$$

where the constant and the function Ψ_2 should be determined by experiments.

Note that, by dealing with dimensional products of variables τ_1 and τ_2, instead of the variables themselves, the four-variable (original) problem is reduced to a two-variable one, already a significant improvement. Later, we will increase this advantage by reducing the number of products of variables even more!

⇑

In the special case when all products of variables are dimensionless, i.e., $N_{q \neq 0} = 0$, then $\mathbf{q}$, by Fig. 8-2, is a *null matrix*, so, by (8-12) and (8-14)

$$\mathbf{M} \cdot \mathbf{U}^T = \mathbf{B} \cdot \mathbf{D}^T + \mathbf{A} \cdot \mathbf{C}^T = \mathbf{0} \tag{8-15}$$

We now note that, by Fig. 8-2, in this case $\mathbf{D}$ is a square matrix. Also, as pointed out previously, $\mathbf{D}$ should be as simple as possible, thus—since it is square—it should be an *identity* matrix, $\mathbf{I}$ (all elements are zero, except in the main diagonal, where they are all 1). If $\mathbf{D}$ is an identity matrix, then of course

$$\mathbf{D} = \mathbf{D}^T = \mathbf{I} \tag{8-16}$$

and (8-15) is further simplified to

$$\mathbf{B} + \mathbf{A} \cdot \mathbf{C}^T = \mathbf{0} \tag{8-17}$$

SYSTEMATIC DETERMINATION OF COMPLETE SET OF PRODUCTS OF VARIABLES

This formula will be frequently used to test the arithmetic of our matrix operations. To illustrate the application of this formula, consider the following example:

Example 8-6. Resistance of a Flat Surface Moving Tangentially on Water (adapted from Ref. 104, p. 66)

A flat body is sliding on water at a constant speed. We wish to find the horizontal force resisting such motion (Fig. 8-5). It is assumed that the body does *not* break the water's surface; it merely "floats," but is not immersed.

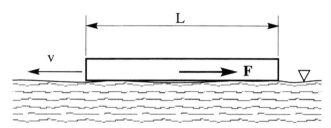

Figure 8-5
Resistance of a flat surface moving on water

As a first step, we list the variables and dimensional constants (if any) assumed to be relevant.

Variable	Symbol	Dimension	Remark
resistance	F	$m \cdot kg \cdot s^{-2}$	horizontal
size of body	L	m	representative value
water density	ρ	kg/m^3	
body velocity	v	m/s^2	constant
water viscosity	ν	m^2/s	kinematic

Hence the dimensional matrix is

$$\begin{array}{c|ccccc} & F & L & \rho & v & \nu \\ \hline m & 1 & 1 & -3 & 1 & 2 \\ kg & 1 & 0 & 1 & 0 & 0 \\ s & -2 & 0 & 0 & -1 & -1 \end{array}$$

B matrix ← | → **A** matrix (a)

We see that the rank of this matrix is $R_{DM} = 3$, matrix **A** is not singular, and the number of variables is $N_V = 5$. Let us now construct products of variables whose dimensions are $m^0 kg^0 s^0$, i.e., the products are *dimensionless*, hence $N_{q \neq 0} = 0$ since $\mathbf{q} = \mathbf{0}$. By Fig. 8-2 now, **D** is a 2×2 matrix, and **C** is a 2×3 matrix. For simplicity, we choose a unit matrix for **D**, so

$$\mathbf{D} = \begin{bmatrix} 1 & 0 \\ 0 & 1 \end{bmatrix} = \mathbf{I} \tag{b}$$

and by (8-8), (a) and (b)

$$\mathbf{C} = -\mathbf{D} \cdot (\mathbf{A}^{-1} \cdot \mathbf{B})^T = -(\mathbf{A}^{-1} \cdot \mathbf{B})^T = \begin{bmatrix} -1 & 0 & -2 \\ 0 & 1 & -1 \end{bmatrix} \tag{c}$$

176   APPLIED DIMENSIONAL ANALYSIS AND MODELING

So the Dimensional Set, by Fig. 8-3, will be

$$
\begin{array}{c|ccc:cc}
 & F & L & \rho & v & \nu \\
\hline
m & 1 & 1 & -3 & 1 & 2 \\
kg & 1 & 0 & 1 & 0 & 0 \\
s & -2 & 0 & 0 & -1 & -1 \\
\hdashline
\pi_1 & 1 & 0 & -1 & 0 & -2 \\
\pi_2 & 0 & 1 & 0 & 1 & -1
\end{array}
\quad (d)
$$

providing two dimensionless variables (products) as follows:

$$\pi_1 = \frac{F}{\rho \cdot v^2}; \qquad \pi_2 = \frac{L \cdot v}{\nu} \quad (e)$$

Now we check the arithmetic. By (8-17)

$$\mathbf{B} + \mathbf{A} \cdot \mathbf{C}^T = \begin{bmatrix} 1 & 1 \\ 1 & 0 \\ -2 & 0 \end{bmatrix} + \begin{bmatrix} -3 & 1 & 2 \\ 1 & 0 & 0 \\ 0 & -1 & -1 \end{bmatrix} \cdot \begin{bmatrix} -1 & 0 \\ -1 & 0 \\ -2 & -1 \end{bmatrix} = \begin{bmatrix} 0 & 0 \\ 0 & 0 \\ 0 & 0 \end{bmatrix}$$

as required and expected. The reader can also easily verify that both π_1 and π_2 in (e) are indeed dimensionless, i.e., their dimensions are 1.

We then have

$$\pi_1 = \Psi\{\pi_2\} \quad (f)$$

which can be represented by a *single* graph of two variables, as opposed to the original five-variable case for which a *set of sets* of charts would be needed!

⇑

8.3.   THE FUNDAMENTAL FORMULA

As pointed out in Art. 7.9, it is highly desirable to have as few dimensionless or dimensional products as possible, i.e., N_p [defined in (7-26)] should be minimum. It is now obvious from (7-26) that to achieve this objective, we should deal with *dimensionless*, rather than *dimensional*, products, for the number of the former is always 1 less than that of the latter. It follows, therefore, that the **q** matrix (Art. 8.1) must be a *null* matrix (see also Fig. 8-2) in which case, by Fig 8-2, **D** is a square matrix, and since it should be simple and its determinant should be the smallest nonzero positive integer, viz., 1, therefore it should be an *identity* matrix. This was also pointed out in Art. 8.2.

The end result is that relation (8-8) for the **C** matrix is simplified to

$$\boxed{\mathbf{C} = -(\mathbf{A}^{-1} \cdot \mathbf{B})^T} \quad (8\text{-}18)$$

where matrices **A** and **B** are defined in Fig. 7-2, and the "location" of the **C** matrix in the Dimensional Set is given in Fig 8-3. On rare occasions we deal with a **D** ma-

SYSTEMATIC DETERMINATION OF COMPLETE SET OF PRODUCTS OF VARIABLES 177

trix which, although a square matrix, is not an identity matrix. In these cases we shall use

$$\mathbf{C} = -\mathbf{D} \cdot (\mathbf{A}^{-1} \cdot \mathbf{B})^T \tag{8-19}$$

which of course is also derivable from (8-8) by the substitution of $\mathbf{q} = \mathbf{0}$.

Since formula (8-18) [or its variation (8-19)] is of great importance due to its facility and usefulness, as will be demonstrated throughout this book, we call it the *Fundamental Formula,* and shall always refer to it as such.

In viewing this formula, the reader's attention is directed to its *simplicity* and *brevity*—noteworthy and frequently occurring characteristics of most of the important formulas in engineering, mathematics, and physics.

The following example illustrates the convenience of the Fundamental Formula.

Example 8-7.   Radiation Pressure on Satellites

Radiation pressure by the Sun on space vehicles can have a substantial effect on their orbital behavior. The effect is especially pronounced if the surface-to-mass ratio is large—e.g., balloons and satellites with large solar arrays.

Let us derive a formula for radiation pressure. Obviously the variables influencing this pressure must include the total radiating power of the Sun, the Sun–satellite distance, and the speed of light. The following table lists these quantities, their symbols, and dimensions.

Variable	Symbol	Dimension
radiation pressure	p	$m^{-1} \cdot kg \cdot s^{-2}$
radiating power by Sun	Q	$m^2 \cdot kg \cdot s^{-3}$
speed of light	c	$m \cdot s^{-1}$
Sun-Earth distance	R	m

Here Q and c are *dimensional constants* whose values are

$Q = 3.86 \times 10^{26}$ W (Ref. 42, p. 225)
$c = 3 \times 10^8$ m/s

The dimensional matrix is

$$\mathbf{B} \text{ matrix} \quad \begin{array}{c|cccc} & p & Q & c & R \\ \hline m & -1 & 2 & 1 & 1 \\ kg & 1 & 1 & 0 & 0 \\ s & -2 & -3 & -1 & 0 \end{array} \quad \mathbf{A} \text{ matrix} \tag{a}$$

where the submatrices **A** and **B** are as indicated. Hence, by the Fundamental Formula (8-18), the **C** matrix is

$$\mathbf{C} = -(\mathbf{A}^{-1} \cdot \mathbf{B})^T = [-1 \quad 1 \quad 2] \tag{b}$$

and of course the **D** matrix, by Fig. 8-2, has but one element: $\mathbf{D} = [1]$. The number of variables is $N_V = 4$, and the rank of the dimensional matrix is $R_{DM} = 3$. Consequently, by (7-26), the number of dimensionless variables is $N_P = N_v - R_{DM} = 4 - 3 = 1$, thus this

178 APPLIED DIMENSIONAL ANALYSIS AND MODELING

sole dimensionless "variable" must be a constant (Theorem 7-4); it is supplied by the Dimensional Set

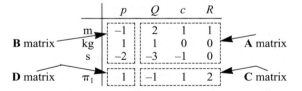

from which

$$\pi_1 = \frac{p \cdot c \cdot R^2}{Q} = \text{const} \qquad (c)$$

or

$$p = \text{const} \; \frac{Q}{c \cdot R^2} \qquad (d)$$

where the const is $1/(4 \cdot \pi)$. If now we substitute the Sun–Earth (average) distance $R = 1.5 \times 10^{11}$ m into (d), we obtain the solar pressure experienced by all objects in the vicinity of Earth. Therefore

$$p = \frac{1}{4\pi} \cdot \frac{3.86 \times 10^{26}}{(3 \times 10^8) \cdot (1.5 \times 10^{11})^2} = 4.55 \times 10^{-6} \text{ Pa} \; (= 6.6 \times 10^{-10} \text{ psi})$$

which is a rather small pressure; on a square kilometer area it generates a mere 4.55 N (= 1.03 lbf) force. Nevertheless, over the entire Earth it means a 5.81×10^8 N (= 1.31×10^8 lbf) force pointing away from the Sun.

⇑

From the above example it is evident that we really do not have to write down the dimensional matrix at all, since the Dimensional Set already includes it. Therefore—unless warranted for some special reason—in the rest of this book only the Dimensional Set will be constructed. We will see that this labor-saving technique in no way hinders our work.

The following example illustrates this point.

Example 8-8. Mass of a Drop of Liquid Slowly Emerging from a Pipe

Consider the arrangement shown Fig. 8-6. We let water flow *slowly* through a pipe, so that individual drops will be formed at the mouth of the pipe. Question: What will be the mass of the individual drops of water?

We deal with this problem using the dimensional method. First, we list the relevant variables.

Variable	Symbol	Dimension
mass of drop	M	kg
liquid density	ρ	$m^{-3} \cdot kg$
surface tension	σ	$kg \cdot s^{-2}$
inside diameter of pipe	D	m
gravitational acceleration	g	$m \cdot s^{-2}$

SYSTEMATIC DETERMINATION OF COMPLETE SET OF PRODUCTS OF VARIABLES 179

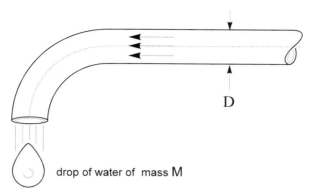

Figure 8-6
A drop of liquid slowly emerging from a pipe

We have five variables and three dimensions. Accordingly, the Dimensional Set is

		M	ρ	σ	D	g	
B matrix	m	0	−3	0	1	1	
	kg	1	1	1	0	0	**A matrix**
	s	0	0	−2	0	−2	
D matrix	π_1	1	0	−1	−1	1	
	π_2	0	1	−1	2	1	**C matrix**

The **A** matrix is nonsingular, hence the rank of the dimensional matrix $R_{DM} = N_d = 3$. The **C** matrix is obtained by the Fundamental Formula (8-18). Thus, by relation (7-26),

$$N_P = N_V - R_{DM} = 5 - 3 = 2$$

i.e., we have two dimensionless variables, supplied by the above dimensional set:

$$\pi_1 = \frac{M \cdot g}{\sigma \cdot D}; \quad \pi_2 = \frac{\rho \cdot D^2 \cdot g}{\sigma} \quad \text{(a)}$$

from which the mass of the drop

$$M = \text{const} \frac{\sigma \cdot D}{g} \cdot \Psi \left\{ \frac{\rho \cdot D^2 \cdot g}{\sigma} \right\} \quad \text{(b)}$$

where the constant and the function Ψ must be determined by other means.

According to Rayleigh (Ref. 108, p. 326), "for many purposes it may be suffice to treat $M \cdot g/(\sigma \cdot D)$ as a constant, say 3.8." From this it follows that the *weight* of the drop is simply

$$M \cdot g = 3.8 \times \sigma \cdot D \quad \text{(c)}$$

Note that here 3.8 is a dimensionless constant; therefore formula (c) can be applied in any dimensional system consistently used.

CHAPTER 9
TRANSFORMATIONS

In this chapter we describe a number of useful methods to transform one Dimensional Set to another. Of course, the basic condition of all these transformations is that both sets must be based on the same collection of variables, and that the dimensions of these variables do not change. For example, if Set 1 comprises variables V_1, V_2, V_3, V_4, then Set 2 must also comprise variables V_1, V_2, V_3, V_4, although of course they can appear in a different sequential order. Similarly, if, say, variable V_2 has a dimension in Set 1, then V_2 must have the *same* dimension in Set 2. These conditions essentially express the basic idea that we have to describe the same physical systems in both sets. It is also understood that we use the *same* dimensional system for both sets. For imagine that one set uses the *force*-based, and the other the *mass*-based dimensional system. Obviously the same physical variable will assume different dimensions in these two systems (force has a dimension m·kg/s² in the mass based, but kg in the force based metric system).

We will first deal with theorems facilitating some specific transformations, then compare systems with different **D** matrices, and finally present a technique encompassing the general case.

9.1. THEOREMS RELATED TO SOME SPECIFIC TRANSFORMATIONS

Theorem 9-1.   If in Dimensional Sets 1, 2 $\mathbf{A}_2 = \mathbf{B}_1$ and $\mathbf{B}_2 = \mathbf{A}_1$, then

$$\mathbf{C}_2 = \mathbf{D}_2 \cdot \mathbf{C}_1^{-1} \cdot \mathbf{D}_1 \tag{9-1}$$

In other words, this theorem deals with a case in which matrices **A** and **B** are interchanged between sets of (maybe) different **D** matrices. Of course we assume that **A** and **B** are both nonsingular and of the same size. In the special case when $\mathbf{D}_1 = \mathbf{D}_2 = \mathbf{I}$ (i.e., identity matrix), then (9-1) is reduced to

$$\mathbf{C}_2 = \mathbf{C}_1^{-1} \tag{9-2}$$

Proof.   By the Fundamental Formula (8-19)

$$\mathbf{C}_1 = -\mathbf{D}_1 \cdot (\mathbf{A}_1^{-1} \cdot \mathbf{B}_1)^{\mathrm{T}}; \qquad \mathbf{C}_2 = -\mathbf{D}_2 \cdot (\mathbf{A}_2^{-1} \cdot \mathbf{B}_2)^{\mathrm{T}} \tag{9-3}$$

182 APPLIED DIMENSIONAL ANALYSIS AND MODELING

from which, since by assumption $\mathbf{A}_2 = \mathbf{B}_1$ and $\mathbf{B}_2 = \mathbf{A}_1$

$$-(\mathbf{D}_1^{-1}\cdot\mathbf{C}_1)^T = \mathbf{A}_1^{-1}\cdot\mathbf{B}_1; \qquad -(\mathbf{D}_2^{-1}\cdot\mathbf{C}_2)^T = \mathbf{A}_2^{-1}\cdot\mathbf{B}_2 = \mathbf{B}_1^{-1}\cdot\mathbf{A}_1 \qquad (9\text{-}4)$$

Postmultiplying the first of (9-4) by the second, we can write

$$(\mathbf{D}_1^{-1}\cdot\mathbf{C}_1)^T\cdot(\mathbf{D}_2^{-1}\cdot\mathbf{C}_2)^T = (\mathbf{A}_1^{-1}\cdot\mathbf{B}_1)\cdot(\mathbf{B}_1^{-1}\cdot\mathbf{A}_1) \qquad (9\text{-}5)$$

The right side of this relation is really the unit matrix, for

$$(\mathbf{A}_1^{-1}\cdot\mathbf{B}_1)\cdot(\mathbf{B}_1^{-1}\cdot\mathbf{A}_1) = \mathbf{A}_1^{-1}\cdot(\mathbf{B}_1\cdot\mathbf{B}_1^{-1})\cdot\mathbf{A}_1 = \mathbf{A}^{-1}\cdot\mathbf{I}\cdot\mathbf{A}_1 = \mathbf{A}^{-1}\cdot\mathbf{A} = \mathbf{I} \qquad (9\text{-}6)$$

By (9-5) and (9-6) now

$$(\mathbf{D}_2^{-1}\cdot\mathbf{C}_2)\cdot(\mathbf{D}_1^{-1}\cdot\mathbf{C}_1) = \mathbf{I} \qquad (9\text{-}7)$$

Both sides of this equation are now premultiplied by $\mathbf{D}_2$, and then postmultiplied by $(\mathbf{D}_1^{-1}\cdot\mathbf{C}_1)^{-1}$. This yields

$$\mathbf{C}_2 = \mathbf{D}_2\cdot\mathbf{I}\cdot(\mathbf{D}_1^{-1}\cdot\mathbf{C}_1)^{-1} = \mathbf{D}_2\cdot(\mathbf{C}_1^{-1}\cdot\mathbf{D}_1) = \mathbf{D}_2\cdot\mathbf{C}_1^{-1}\cdot\mathbf{D}_1$$

concluding the proof, since this relation is identical to (9-1).

Example 9-1

Consider the following two Dimensional Sets

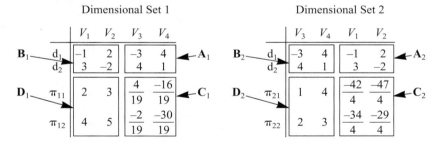

As the reader can observe in the above sets $\mathbf{A}_2 = \mathbf{B}_1$ and $\mathbf{B}_2 = \mathbf{A}_1$. By (9-1) now

$$\mathbf{C}_2 = \mathbf{D}_2\cdot\mathbf{C}_1^{-1}\cdot\mathbf{D}_1 = \begin{bmatrix} 1 & 4 \\ 2 & 3 \end{bmatrix} \cdot \begin{bmatrix} \frac{4}{19} & \frac{-16}{19} \\ \frac{-2}{19} & \frac{-30}{19} \end{bmatrix}^{-1} \cdot \begin{bmatrix} 2 & 3 \\ 4 & 5 \end{bmatrix} = \frac{1}{4}\cdot\begin{bmatrix} -42 & -47 \\ -34 & -29 \end{bmatrix}$$

confirming the $\mathbf{C}_2$ matrix in set 2.

⇑

Next, we wish to derive the actual set of dimensionless products if matrices $\mathbf{A}$ and $\mathbf{B}$ are interchanged. Suppose we have—as before—systems 1 and 2. Further, suppose each set comprises two dimensionless variables. Then we have π_{11}, π_{12} in system 1, and π_{21}, π_{22} in system 2. Since the two systems are interrelated, we can write

$$\pi_{21} = \pi_{11}^{a_1} \cdot \pi_{12}^{b_1}; \qquad \pi_{22} = \pi_{11}^{a_2} \cdot \pi_{12}^{b_2} \qquad (9\text{-}8)$$

Our task now is to determine the exponents in these equations.

Relations (9-8) can be symbolically written

$$\begin{bmatrix} \ln \pi_{21} \\ \ln \pi_{22} \end{bmatrix} = \mathbf{T} \cdot \begin{bmatrix} \ln \pi_{11} \\ \ln \pi_{12} \end{bmatrix} \qquad (9\text{-}9)$$

where **T** is the *transformation matrix* defined as

$$\mathbf{T} = \begin{bmatrix} a_1 & b_1 \\ a_2 & b_2 \end{bmatrix} \qquad (9\text{-}10)$$

Now we can formulate a theorem.

Theorem 9-2. *If we have Dimensional Sets 1 and 2, in which the **A** and **B** matrices are interchanged, then the transformation matrix defined in (9-9) is*

$$\mathbf{T} = \mathbf{C}_2 \cdot \mathbf{D}_1^{-1} \qquad (9\text{-}11\text{a})$$

or, equivalently

$$\mathbf{T} = \mathbf{D}_2 \cdot \mathbf{C}_1^{-1} \qquad (9\text{-}11\text{b})$$

Proof. The proof is offered for a case of four variables and two dimensions, but of course it can be applied to any duo of Dimensional Sets in which matrices **A**, **B**, **D**$_1$, and **D**$_2$ are nonsingular and of the same size. We have, then, 2 sets as shown below

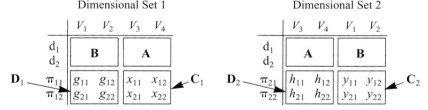

Therefore, by definition,

$$\pi_{11} = V_1^{g_{11}} \cdot V_2^{g_{12}} \cdot V_3^{x_{11}} \cdot V_4^{x_{12}}; \qquad \pi_{21} = V_1^{y_{11}} \cdot V_2^{y_{12}} \cdot V_3^{h_{11}} \cdot V_4^{h_{12}}$$

$$\pi_{12} = V_1^{g_{21}} \cdot V_2^{g_{22}} \cdot V_3^{x_{21}} \cdot V_4^{x_{22}}; \qquad \pi_{22} = V_1^{y_{21}} \cdot V_2^{y_{22}} \cdot V_3^{h_{21}} \cdot V_4^{h_{22}}$$

which can be written more compactly

$$\left. \begin{array}{c} \begin{bmatrix} \ln \pi_{11} \\ \ln \pi_{12} \end{bmatrix} = \mathbf{D}_1 \cdot \begin{bmatrix} \ln V_1 \\ \ln V_2 \end{bmatrix} + \mathbf{C}_1 \cdot \begin{bmatrix} \ln V_3 \\ \ln V_4 \end{bmatrix} \\[2ex] \begin{bmatrix} \ln \pi_{21} \\ \ln \pi_{22} \end{bmatrix} = \mathbf{C}_2 \cdot \begin{bmatrix} \ln V_1 \\ \ln V_2 \end{bmatrix} + \mathbf{D}_2 \cdot \begin{bmatrix} \ln V_3 \\ \ln V_4 \end{bmatrix} \end{array} \right\} \qquad (9\text{-}12)$$

Now, premultiplying the first of (9-12) by $\mathbf{D}_1^{-1}$, and the second by $\mathbf{C}_2^{-1}$, we obtain

$$\left.\begin{array}{c} \mathbf{D}_1^{-1} \cdot \begin{bmatrix} \ln \pi_{11} \\ \ln \pi_{12} \end{bmatrix} = \begin{bmatrix} \ln V_1 \\ \ln V_2 \end{bmatrix} + \mathbf{D}_1^{-1} \cdot \mathbf{C}_1 \cdot \begin{bmatrix} \ln V_3 \\ \ln V_4 \end{bmatrix} \\ \mathbf{C}_2^{-1} \cdot \begin{bmatrix} \ln \pi_{21} \\ \ln \pi_{22} \end{bmatrix} = \begin{bmatrix} \ln V_1 \\ \ln V_2 \end{bmatrix} + \mathbf{C}_2^{-1} \cdot \mathbf{D}_2 \cdot \begin{bmatrix} \ln V_3 \\ \ln V_4 \end{bmatrix} \end{array}\right\} \quad (9\text{-}13)$$

We now premultiply both sides of (9-7) by $(\mathbf{D}_2^{-1} \cdot \mathbf{C}_2)^{-1}$. This simple operation will yield

$$\mathbf{D}_1^{-1} \cdot \mathbf{C}_1 = (\mathbf{D}_2^{-1} \cdot \mathbf{C}_2)^{-1} \cdot \mathbf{I} = \mathbf{C}_2^{-1} \cdot \mathbf{D}_2$$

Therefore the right-hand sides of both relations of (9-13) are equal; and hence their left-hand sides must also be equal. Thus

$$\mathbf{D}_1^{-1} \cdot \begin{bmatrix} \ln \pi_{11} \\ \ln \pi_{12} \end{bmatrix} = \mathbf{C}_2^{-1} \cdot \begin{bmatrix} \ln \pi_{21} \\ \ln \pi_{22} \end{bmatrix}$$

Premultiplying both sides now by $\mathbf{C}_2$, we get

$$\begin{bmatrix} \ln \pi_{21} \\ \ln \pi_{11} \end{bmatrix} = \mathbf{C}_2 \cdot \mathbf{D}_1^{-1} \cdot \begin{bmatrix} \ln \pi_{22} \\ \ln \pi_{12} \end{bmatrix}$$

which, if compared with (9-9), yields $\boldsymbol{\tau} = \mathbf{C}_2 \cdot \mathbf{D}_1^{-1}$, identical to (9-11a).

Next, we premultiply both sides of (9-7) by $\mathbf{D}_2$. This will yield $\mathbf{C}_2 \cdot \mathbf{D}_1^{-1} \cdot \mathbf{C}_1 = \mathbf{D}_2$. Then both sides of this relation are postmultiplied by $\mathbf{C}_1^{-1}$ resulting in $\mathbf{C}_2 \cdot \mathbf{D}_1^{-1} = \mathbf{D}_2 \cdot \mathbf{C}_1^{-1}$, thereby verifying the equivalence of relations (9-11a) and (9-11b). This proves the theorem.

If, as usual, $\mathbf{D}_1 = \mathbf{D}_2 = \mathbf{I}$, then (9-11a) and (9-11b) are reduced to the simpler form of

$$\boldsymbol{\tau} = \mathbf{C}_2 = \mathbf{C}_1^{-1} \qquad (9\text{-}14)$$

An alternative and more "elegant" proof of this theorem will be offered as a special case of Theorem 9-17.

Example 9-2

In Example 9-1 we had

$$\mathbf{C}_1 = \frac{1}{19} \cdot \begin{bmatrix} 4 & -16 \\ -2 & -30 \end{bmatrix}; \quad \mathbf{C}_2 = \frac{1}{4} \cdot \begin{bmatrix} -42 & -47 \\ -34 & -29 \end{bmatrix}; \quad \mathbf{D}_1 = \begin{bmatrix} 2 & 3 \\ 4 & 5 \end{bmatrix}; \quad \mathbf{D}_2 = \begin{bmatrix} 1 & 4 \\ 2 & 3 \end{bmatrix} \quad (a)$$

Therefore, by (9-11a),

$$\boldsymbol{\tau} = \frac{1}{4} \cdot \begin{bmatrix} -42 & -47 \\ -34 & -29 \end{bmatrix} \cdot \begin{bmatrix} 2 & 3 \\ 4 & 5 \end{bmatrix}^{-1} = \frac{1}{4} \cdot \begin{bmatrix} 11 & -16 \\ 27 & -22 \end{bmatrix} \quad (b)$$

and by (9-11b)

$$\mathbf{T} = \begin{bmatrix} 1 & 4 \\ 2 & 3 \end{bmatrix} \cdot \begin{bmatrix} \frac{4}{19} & \frac{-16}{19} \\ \frac{-2}{19} & \frac{-30}{19} \end{bmatrix}^{-1} = \frac{1}{4} \begin{bmatrix} 11 & -16 \\ 27 & -22 \end{bmatrix} \quad (c)$$

as expected and required. Next, by (9-9) and (b)

$$\begin{bmatrix} \ln \pi_{21} \\ \ln \pi_{22} \end{bmatrix} = \mathbf{T} \cdot \begin{bmatrix} \ln \pi_{11} \\ \ln \pi_{12} \end{bmatrix} = \frac{1}{4} \begin{bmatrix} 11 & -16 \\ 27 & -22 \end{bmatrix} \cdot \begin{bmatrix} \ln \pi_{11} \\ \ln \pi_{12} \end{bmatrix} \quad (d)$$

from which

$$\pi_{21} = \sqrt[4]{\pi_{11}^{11} \cdot \pi_{12}^{-16}}; \qquad \pi_{22} = \sqrt[4]{\pi_{11}^{27} \cdot \pi_{12}^{-22}} \quad (e)$$

If we now want to express π_{11} and π_{12} in terms of π_{21} and π_{22}, then we obtain, by (c)

$$\mathbf{T}^{-1} = \begin{bmatrix} \frac{11}{4} & \frac{-16}{4} \\ \frac{27}{4} & \frac{-22}{4} \end{bmatrix}^{-1} = \frac{1}{95} \cdot \begin{bmatrix} -44 & 32 \\ -54 & 22 \end{bmatrix} \quad (f)$$

thus, again by (9-9),

$$\begin{bmatrix} \ln \pi_{11} \\ \ln \pi_{12} \end{bmatrix} = \mathbf{T}^{-1} \cdot \begin{bmatrix} \ln \pi_{21} \\ \ln \pi_{22} \end{bmatrix} = \frac{1}{95} \begin{bmatrix} -44 & 32 \\ -54 & 22 \end{bmatrix} \cdot \begin{bmatrix} \ln \pi_{21} \\ \ln \pi_{22} \end{bmatrix} \quad (g)$$

yielding

$$\pi_{11} = (\pi_{21}^{-44} \cdot \pi_{22}^{32})^{1/95}; \qquad \pi_{12} = (\pi_{21}^{-54} \cdot \pi_{22}^{22})^{1/95} \quad (h)$$

and of course (e) and (h) are fully compatible, as easily proven.

⇑

Theorem 9-3. *Premultiplication of matrices* **A** *and* **B** *in a Dimensional Set by the same compatible, nonsingular (but otherwise arbitrary) matrix, does not change the* **C** *matrix.*

In other words, the **C** matrix is *invariant* upon the premultiplication of both **A** and **B** by a matrix. Let this multiplier matrix be denoted by **Z**. Then, for **Z** to be *compatible* it

- must be a square matrix (otherwise it could not be nonsingular);
- must have as many rows (and columns) as **A** (and **B**).

From this theorem it also follows that premultiplication of both **A** and **B** by an identical (and compatible) matrix does not alter the composition of the resulting dimensionless products.

Proof. Let us have two Sets, 1 and 2. For the first we have—by the Fundamental Formula—

$$\mathbf{C}_1 = -\mathbf{D} \cdot (\mathbf{A}^{-1} \cdot \mathbf{B})^{\mathrm{T}} \quad (9\text{-}15)$$

and, similarly, for the second, multiplied by **Z**

$$\mathbf{C}_2 = -\mathbf{D} \cdot ((\mathbf{Z} \cdot \mathbf{A})^{-1} \cdot (\mathbf{Z} \cdot \mathbf{B}))^T \qquad (9\text{-}16)$$

where **Z** is an arbitrary *compatible* matrix. Then, by (9-16)

$$\mathbf{C}_2 = -\mathbf{D} \cdot ((\mathbf{A}^{-1} \cdot \mathbf{Z}^{-1}) \cdot (\mathbf{Z} \cdot \mathbf{B}))^T = -\mathbf{D} \cdot (\mathbf{A}^{-1} \cdot (\mathbf{Z}^{-1} \cdot \mathbf{Z}) \cdot \mathbf{B})^T = -\mathbf{D} \cdot (\mathbf{A}^{-1} \cdot \mathbf{B})^T \qquad (9\text{-}17)$$

Comparison of (9-15) with (9-17) yields $\mathbf{C}_1 = \mathbf{C}_2$, which was to be proved.

Example 9-3

Given Dimensional Set 1

Dimensional Set 1

	V_1	V_2	V_3	V_4	V_5	V_6
d_1	−5	0	−3	2	0	1
d_2	−5	5	−5	3	−2	1
d_3	−8	−6	0	2	0	4
π_{11}	2	0	4	12	2	−2
π_{12}	3	2	1	6	7	6
π_{13}	−1	2	2	0	3	1

in which

$$\mathbf{A}_1 = \begin{bmatrix} 2 & 0 & 1 \\ 3 & -2 & 1 \\ 2 & 0 & 4 \end{bmatrix}; \quad \mathbf{B}_1 = \begin{bmatrix} -5 & 0 & -3 \\ -5 & 5 & -5 \\ -8 & -6 & 0 \end{bmatrix}$$

Now we premultiply both above matrices by

$$\mathbf{Z} = \begin{bmatrix} 2 & 1 & 4 \\ -3 & 2 & 1 \\ 5 & 6 & 2 \end{bmatrix}$$

We see that **Z** is *compatible* since it fulfills the stated two conditions. Hence,

$$\mathbf{A}_2 = \mathbf{Z} \cdot \mathbf{A}_1 = \begin{bmatrix} 15 & -2 & 19 \\ 2 & -4 & 3 \\ 32 & -12 & 19 \end{bmatrix}; \quad \mathbf{B}_2 = \mathbf{Z} \cdot \mathbf{B}_1 = \begin{bmatrix} -47 & -19 & -11 \\ -3 & 4 & -1 \\ -71 & 18 & -45 \end{bmatrix}$$

and therefore the new Dimensional Set, i.e., Set 2, will be

Dimensional Set 2

	V_1	V_2	V_3	V_4	V_5	V_6
d_1	−47	−19	−11	15	−2	19
d_2	−3	4	−1	2	−4	3
d_3	−71	18	−45	32	−12	19
π_1	2	0	4	12	2	−2
π_2	3	2	1	6	7	6
π_3	−1	2	2	0	3	1

in which—as the reader can readily see—matrix $\mathbf{C}_2$ in Set 2 is identical to $\mathbf{C}_1$ in Set 1.

Corollary 1 of Theorem 9-3. *If $\mathbf{Z}$ is a diagonal matrix having a scalar q in its main diagonal, then this is equivalent to multiplying both $\mathbf{A}$ and $\mathbf{B}$ matrices by q.*

Proof. Let $\mathbf{Z}$ be a 3×3 diagonal matrix, and $\mathbf{I}$ a 3×3 identity matrix. Then

$$\mathbf{Z} = \begin{bmatrix} q & 0 & 0 \\ 0 & q & 0 \\ 0 & 0 & q \end{bmatrix} = q \cdot \begin{bmatrix} 1 & 0 & 0 \\ 0 & 1 & 0 \\ 0 & 0 & 1 \end{bmatrix} = q \cdot \mathbf{I}$$

and hence

$$\left. \begin{array}{l} \mathbf{A}_2 = \mathbf{Z} \cdot \mathbf{A}_1 = (q \cdot \mathbf{I}) \cdot \mathbf{A}_1 = q \cdot (\mathbf{I} \cdot \mathbf{A}_1) = q \cdot \mathbf{A}_1 \\ \mathbf{B}_2 = \mathbf{Z} \cdot \mathbf{B}_1 = (q \cdot \mathbf{I}) \cdot \mathbf{B}_1 = q \cdot (\mathbf{I} \cdot \mathbf{B}_1) = q \cdot \mathbf{B}_1 \end{array} \right\} \quad (9\text{-}18)$$

This proves the corollary.

Example 9-4

Consider the Dimensional Set 1 in Example 9-3, and let

$$\mathbf{Z} = \begin{bmatrix} 4 & 0 & 0 \\ 0 & 4 & 0 \\ 0 & 0 & 4 \end{bmatrix}$$

then

$$\mathbf{A}_2 = \mathbf{Z} \cdot \mathbf{A}_1 = \begin{bmatrix} 4 & 0 & 0 \\ 0 & 4 & 0 \\ 0 & 0 & 4 \end{bmatrix} \cdot \begin{bmatrix} 2 & 0 & 1 \\ 3 & -2 & 1 \\ 2 & 0 & 4 \end{bmatrix} = 4 \cdot \begin{bmatrix} 2 & 0 & 1 \\ 3 & -2 & 1 \\ 2 & 0 & 4 \end{bmatrix} = \begin{bmatrix} 8 & 0 & 4 \\ 12 & -8 & 4 \\ 8 & 0 & 16 \end{bmatrix}$$

$$\mathbf{B}_2 = \mathbf{Z} \cdot \mathbf{B}_1 = \begin{bmatrix} 4 & 0 & 0 \\ 0 & 4 & 0 \\ 0 & 0 & 4 \end{bmatrix} \cdot \begin{bmatrix} -5 & 0 & -3 \\ -5 & 5 & -5 \\ -8 & -6 & 0 \end{bmatrix} = 4 \cdot \begin{bmatrix} -5 & 0 & -3 \\ -5 & 5 & -5 \\ -8 & -6 & 0 \end{bmatrix} = \begin{bmatrix} -20 & 0 & -12 \\ -20 & 20 & -20 \\ -32 & -24 & 0 \end{bmatrix}$$

With these data we now have

Dimensional Set 2

	V_1	V_2	V_3	V_4	V_5	V_6
d_1	-20	0	-12	8	0	4
d_2	-20	20	-20	12	-8	4
d_3	-32	20	0	8	0	16
π_1	2	0	4	12	2	-2
π_2	3	2	1	6	7	6
π_3	-1	2	2	0	3	1

in which, compared with Dimensional Set 1, the $\mathbf{C}$ matrix remained unchanged.

⇑

Corollary 2 of Theorem 9-3. *In the special case when the multiplier matrix is $\mathbf{A}^{-1}$, and $\mathbf{D} = \mathbf{I}$, then the new $\mathbf{A}$ becomes $\mathbf{I}$, and the new $\mathbf{B}$ becomes the negative transpose of the (unchanging) matrix $\mathbf{C}$.*

188 APPLIED DIMENSIONAL ANALYSIS AND MODELING

That is, if $\mathbf{A}_1$, $\mathbf{B}_1$ and $\mathbf{A}_2$, $\mathbf{B}_2$ are the respective matrices in Sets 1 and 2, and if

$$\mathbf{Z} = \mathbf{A}_1^{-1} \tag{9-19}$$

then

$$\mathbf{A}_2 = \mathbf{I}; \qquad \mathbf{B}_2 = -\mathbf{C}_1^T = -\mathbf{C}_2^T \tag{9-20}$$

This can be schematically represented:

Dimensional Set 1			Dimensional Set 2		
	$V_1\ V_2\ V_3$	$V_4\ V_5\ V_6$		$V_1\ V_2\ V_3$	$V_4\ V_5\ V_6$
d_1 d_2 d_3	**B**	**A**	d_1 d_2 d_3	$-\mathbf{C}_1^T$	**I**
π_1 π_2 π_3	**I**	$\mathbf{C}_1$	π_1 π_2 π_3	**I**	$\mathbf{C}_1$

Proof. By definition, $\mathbf{A}_2 = \mathbf{Z} \cdot \mathbf{A}_1$, and hence, by this and (9-19), $\mathbf{A}_2 = \mathbf{A}_1^{-1} \cdot \mathbf{A}_1 = \mathbf{I}$, proving the first part of (9-20). Moreover, by the Fundamental Formula, $\mathbf{C}_2 = -(\mathbf{A}_2 \cdot \mathbf{B}_2)^T$. But now $\mathbf{A}_2 = \mathbf{I}$, therefore $\mathbf{C}_2 = -(\mathbf{A}_2^{-1} \cdot \mathbf{B}_2)^T = -\mathbf{B}_2^T$, yielding $\mathbf{B}_2 = -\mathbf{C}_2^T$, which is the second part of (9-20). Now since $\mathbf{C}_1 = \mathbf{C}_2$, therefore $\mathbf{B}_2 = -\mathbf{C}_1^T$, and thus the new $\mathbf{B}$ (i.e., $\mathbf{B}_2$) can be determined by inspection.

Example 9-5

Given the Dimensional Set 1

Dimensional Set 1

	V_1	V_2	V_3	V_4	V_5	V_6
d_1	−5	0	−3	2	0	1
d_2	−5	5	−5	3	−2	1
d_3	−8	−6	0	2	0	4
π_1	1	0	0	2	1	1
π_2	0	1	0	−1	2	2
π_3	0	0	1	2	0	−1

in which

$$\mathbf{A}_1 = \begin{bmatrix} 2 & 0 & 1 \\ 3 & -2 & 1 \\ 2 & 0 & 4 \end{bmatrix}; \quad \mathbf{B}_1 = \begin{bmatrix} -5 & 0 & -3 \\ -5 & 5 & -5 \\ -8 & -6 & 0 \end{bmatrix}; \quad \mathbf{C}_1 = \begin{bmatrix} 2 & 1 & 1 \\ -1 & 2 & 2 \\ 2 & 0 & -1 \end{bmatrix}$$

We now have the multiplier $\mathbf{Z}$

$$\mathbf{Z} = \mathbf{A}_1^{-1} = \frac{1}{12} \cdot \begin{bmatrix} 8 & 0 & -2 \\ 10 & -6 & -1 \\ -4 & 0 & 4 \end{bmatrix}$$

Therefore $A_2 = Z \cdot A_1^{-1} = A_1^{-1} \cdot A_1 = I$, thus

$$B_2 = Z \cdot B_1 = \frac{1}{12} \cdot \begin{bmatrix} 8 & 0 & -2 \\ 10 & -6 & -1 \\ -4 & 0 & 4 \end{bmatrix} \begin{bmatrix} -5 & 0 & -3 \\ -5 & 5 & -5 \\ -8 & -6 & 0 \end{bmatrix} = \begin{bmatrix} -2 & 1 & -2 \\ -1 & -2 & 0 \\ -1 & -2 & 1 \end{bmatrix}$$

as expected and required. Therefore Dimensional Set 2 will be

Dimensional Set 2

	V_1	V_2	V_3	V_4	V_5	V_6
d_1	-2	1	-2	1	0	0
d_2	-1	-2	0	0	1	0
d_3	-1	-2	1	0	0	1
π_1	1	0	0	2	1	1
π_2	0	1	0	-1	2	2
π_3	0	0	1	2	0	-1

where we see that the **B** matrix is the negative transpose of matrix **C**, as required and expected.

⇑

Theorem 9-4. *Interchange of any two rows of the* **A** *matrix, and the interchange of the same two rows of the* **B** *matrix, do not change the* **C** *matrix.*

Proof. Since the interchange of any two rows in any matrix can be always effected by the premultiplication of that matrix by an appropriate *permutation* matrix, therefore this theorem is a special case of Theorem 9-3, which was already proved.

Example 9-6

Given Dimensional Set 1

Dimensional Set 1

	V_1	V_2	V_3	V_4	V_5	V_6
d_1	3	2	1	4	2	2
d_2	3	2	2	1	0	4
d_3	1	4	0	2	1	0
π_1	1	0	0	-1	1	$-\frac{1}{2}$
π_2	0	1	0	-14	24	3
π_3	0	0	1	0	0	$-\frac{1}{2}$

in which

$$A_1 = \begin{bmatrix} 4 & 2 & 2 \\ 1 & 0 & 4 \\ 2 & 1 & 0 \end{bmatrix}; \quad B_1 = \begin{bmatrix} 3 & 2 & 1 \\ 3 & 2 & 2 \\ 1 & 4 & 0 \end{bmatrix}; \quad C_1 = \begin{bmatrix} -1 & 1 & -\frac{1}{2} \\ -14 & 24 & 3 \\ 0 & 0 & -\frac{1}{2} \end{bmatrix}$$

Now we interchange the first and third rows in both $\mathbf{A}_1$ and $\mathbf{B}_1$. Thus the new $\mathbf{A}_2$ and $\mathbf{B}_2$ are

$$\mathbf{A}_2 = \begin{bmatrix} 2 & 1 & 0 \\ 1 & 0 & 4 \\ 4 & 2 & 2 \end{bmatrix}; \quad \mathbf{B}_2 = \begin{bmatrix} 1 & 4 & 0 \\ 3 & 2 & 2 \\ 3 & 2 & 1 \end{bmatrix}$$

by which Dimensional Set 2 will be

Dimensional Set 2

	V_1	V_2	V_3	V_4	V_5	V_6
d_1	1	4	0	2	1	0
d_2	3	2	2	1	0	4
d_3	3	2	1	4	2	2
π_1	1	0	0	-1	1	$-\frac{1}{2}$
π_2	0	1	0	-14	24	3
π_3	0	0	1	0	0	$-\frac{1}{2}$

As we see, the new $\mathbf{C}$ matrix

$$\mathbf{C}_2 = \begin{bmatrix} -1 & 1 & -.5 \\ -14 & 24 & 3 \\ 0 & 0 & -.5 \end{bmatrix}$$

is identical to $\mathbf{C}_1$, as required and expected.

⇑

Corollary of Theorem 9-4. *Sequential order of rows (i.e., dimensions) in the dimensional matrix does not affect the $\mathbf{C}$ matrix, nor are the compositions of the derived dimensionless variables influenced.*

Proof. It is evident that any row order can be achieved by successive interchanges of appropriate row-pairs, which, individually (by Theorem 9-4) have no effect on the resulting $\mathbf{C}$ matrix. If, therefore, $\mathbf{C}$ is the same, and (by definition) the sequential order of variables is left intact, then the compositions of dimensionless variables must also remain unaltered. This proves the corollary.

Theorem 9-5. *Postmultiplication of the $\mathbf{B}$ matrix by any compatible $\mathbf{Z}$ matrix causes the premultiplication of the $\mathbf{C}$ matrix by the transpose of $\mathbf{Z}$.*

That is, if in a system we have matrices $\mathbf{A}_1$, $\mathbf{B}_1$, and $\mathbf{C}_1$, and if we set

$$\mathbf{B}_2 = \mathbf{B}_1 \cdot \mathbf{Z} \tag{9-21}$$

then, in the new (changed) system

$$\mathbf{C}_2 = \mathbf{Z}^T \cdot \mathbf{C}_1 \tag{9-22}$$

TRANSFORMATIONS

Proof. By the Fundamental Formula

$$C_2 = -(A^{-1} \cdot B_2)^T \qquad (9\text{-}23)$$

into which B_2 from (9-21) is now substituted. This yields

$$C_2 = -(A^{-1} \cdot (B_1 \cdot Z))^T = -((A^{-1} \cdot B_1) \cdot Z)^T = -(Z^T \cdot (A^{-1} \cdot B_1)^T) \qquad (9\text{-}24)$$

But, analogously to (9-23), we can write $C_1 = -(A^{-1} \cdot B_1)^T$ or $(A^{-1} \cdot B_1)^T = -C_1$. This last expression is now inserted into (9-24), yielding $C_2 = -(Z^T \cdot (-C_1)) = Z^T \cdot C_1$, identical to (9-22), which was to be proved.

Example 9-7

Suppose a given Dimensional Set 1 (not shown here) yields the following submatrices

$$A_1 = \begin{bmatrix} 2 & 1 & 0 \\ 1 & 0 & 4 \\ 4 & 2 & 2 \end{bmatrix}; \quad B_1 = \begin{bmatrix} 1 & 4 & 0 \\ 3 & 2 & 2 \\ 3 & 2 & 1 \end{bmatrix}; \quad C_1 = \begin{bmatrix} -1 & 1 & -.5 \\ -14 & 24 & 3 \\ 0 & 0 & -.5 \end{bmatrix}$$

Let us postmultiply B_1 by $Z = \begin{bmatrix} 2 & 1 & 0 \\ 2 & -2 & 4 \\ 3 & 2 & 1 \end{bmatrix}$ so that $B_2 = B_1 \cdot Z = \begin{bmatrix} 10 & -7 & 16 \\ 16 & 3 & 10 \\ 13 & 1 & 9 \end{bmatrix}$

and therefore, since $A_2 = A_1$

$$C_2 = -(A_1^{-1} \cdot B_2)^T = \begin{bmatrix} -30 & 50 & 3.5 \\ 27 & -47 & -7.5 \\ -56 & 96 & 11.5 \end{bmatrix}$$

Now by (9-22),

$$C_2 = Z^T \cdot C_1 = \begin{bmatrix} 2 & 1 & 0 \\ 2 & -2 & 4 \\ 3 & 2 & 1 \end{bmatrix}^T \cdot \begin{bmatrix} -1 & 1 & -.5 \\ -14 & 24 & 3 \\ 0 & 0 & -.5 \end{bmatrix} = \begin{bmatrix} -30 & 50 & 3.5 \\ 27 & -47 & -7.5 \\ -56 & 96 & 11.5 \end{bmatrix}$$

The end results of the above two relations are identical, as required and expected. ⇑

Theorem 9-6. *Interchange of any two columns of the **B** matrix causes the interchange of the corresponding rows of the **C** matrix.*

Proof. The interchange of any two *columns* of any matrix **U** can always be effected by the postmultiplication of **U** by the appropriate elementary *permutation* matrix. This matrix—say **Z**—can be generated from the *identity* matrix whose *columns* are interchanged in unison with that of **U**. For example, in the following relation

$$\underbrace{\begin{bmatrix} 2 & 4 & 5 \\ 9 & 10 & 6 \\ 7 & 8 & 3 \end{bmatrix}}_{U_1} \cdot \underbrace{\begin{bmatrix} 0 & 1 & 0 \\ 1 & 0 & 0 \\ 0 & 0 & 1 \end{bmatrix}}_{Z} = \underbrace{\begin{bmatrix} 4 & 2 & 5 \\ 10 & 9 & 6 \\ 8 & 7 & 3 \end{bmatrix}}_{U_2}$$

192   APPLIED DIMENSIONAL ANALYSIS AND MODELING

Z is the permutation matrix which, in this case, can be derived from an identity matrix by interchanging its first and second *columns*. Hence, if we interchange the *same* two columns in $\mathbf{U}_1$, we get $\mathbf{U}_2$. This fact is verified in the above illustration.

Similarly, the interchange of any two *rows* of any matrix **U** can always be effected by the *premultiplication* of that matrix by an appropriate permutation matrix **Z**. This matrix is derived from the identity matrix with two of its *rows* interchanged in unison with that of **U**. For example, in the following relation

$$\underbrace{\begin{bmatrix} 0 & 1 & 0 \\ 1 & 0 & 0 \\ 0 & 0 & 1 \end{bmatrix}}_{\mathbf{Z}} \cdot \underbrace{\begin{bmatrix} 2 & 4 & 5 \\ 9 & 10 & 6 \\ 7 & 8 & 3 \end{bmatrix}}_{\mathbf{U}_1} = \underbrace{\begin{bmatrix} 9 & 10 & 6 \\ 2 & 4 & 5 \\ 7 & 8 & 3 \end{bmatrix}}_{\mathbf{U}_2}$$

Z is the permutation matrix, which is an identity matrix whose first and second *rows* are interchanged. Hence, if we interchange the same two rows in $\mathbf{U}_1$, we get $\mathbf{U}_2$. This is verified above. Now, by its construction, **Z** always equals its own transpose, i.e., $\mathbf{Z} = \mathbf{Z}^T$ and hence if **Z** is generated by the interchange of any two *columns* of an identity matrix, then it also can be generated by the interchange of the corresponding *rows* of the identity matrix. That is, schematically

$$\left. \begin{array}{l} \text{columns } i \text{ and } j \text{ of } \mathbf{I} \text{ are interchanged} \Rightarrow \mathbf{Z}_1 \\ \text{rows } i \text{ and } j \text{ of } \mathbf{I} \text{ are interchanged} \Rightarrow \mathbf{Z}_2 \end{array} \right\} \text{ then } \mathbf{Z}_1 = \mathbf{Z}_2 = \mathbf{Z} = \mathbf{Z}^T$$

But now we have, by assumption, $\mathbf{B}_2 = \mathbf{B}_1 \cdot \mathbf{Z}$, where **Z** is the permutation matrix, to yield the defined interchanges of two *columns* in $\mathbf{B}_1$. Therefore, by Theorem 9-5, $\mathbf{C}_2 = \mathbf{Z}^T \cdot \mathbf{C}_1 = \mathbf{Z} \cdot \mathbf{C}_1$, thus $\mathbf{C}_2$ is generated by the interchange of the corresponding *rows* in $\mathbf{C}_1$. This proves the theorem.

Example 9-8

Given

Dimensional Set 1

	V_1	V_2	V_3	V_4	V_5	V_6
d_1	−8	−40	−16	4	6	8
d_2	−7	−18	−7	2	4	3
d_3	0	−16	−8	2	1	4
π_1	1	0	0	1	2	−1
π_2	0	1	0	−1	2	4
π_3	0	0	1	2	0	1

in which $\mathbf{A}_1 = \begin{bmatrix} 4 & 6 & 8 \\ 2 & 4 & 3 \\ 2 & 1 & 4 \end{bmatrix}$; $\mathbf{B}_1 = \begin{bmatrix} -8 & -40 & -16 \\ -7 & -18 & -7 \\ 0 & -16 & -8 \end{bmatrix}$

and therefore, by the Fundamental Formula

$$\mathbf{C}_1 = \begin{bmatrix} 1 & 2 & -1 \\ -1 & 2 & 4 \\ 2 & 0 & 1 \end{bmatrix}$$

Now we interchange the first and third *columns* in $\mathbf{B}_1$. This yields

$$\mathbf{B}_2 = \begin{bmatrix} -16 & -40 & -8 \\ -7 & -18 & -7 \\ -8 & -16 & 0 \end{bmatrix}$$

and hence by the Fundamental Formula, since $\mathbf{A}_2 = \mathbf{A}_1$,

$$\mathbf{C}_2 = \begin{bmatrix} 2 & 0 & 1 \\ -1 & 2 & 4 \\ 1 & 2 & -1 \end{bmatrix}$$

in which, as seen, the first and third *rows* are interchanged with respect to $\mathbf{C}_1$.

⇑

Theorem 9-7. *The interchange of any two columns of* $\mathbf{A}$ *causes the interchange of the same two columns in* $\mathbf{C}$.

Proof. Suppose we have two Dimensional Sets. The first has submatrices $\mathbf{A}_1$, $\mathbf{B}_1$, $\mathbf{C}_1$, and the second $\mathbf{A}_2$, $\mathbf{B}_2$, $\mathbf{C}_2$. Then, by assumption, $\mathbf{B}_1 = \mathbf{B}_2$, and

$$\mathbf{A}_2 = \mathbf{A}_1 \cdot \mathbf{Z} \tag{9-25}$$

where $\mathbf{Z}$ is the appropriate permutation matrix causing the interchange of the two designated columns in $\mathbf{A}_1$. Now by (9-25) and the Fundamental Formula

$$\mathbf{C}_2 = -(\mathbf{A}_2^{-1} \cdot \mathbf{B}_2)^{\mathrm{T}} = -((\mathbf{A}_1 \cdot \mathbf{Z})^{-1} \cdot \mathbf{B}_2)^{\mathrm{T}} = -(\mathbf{Z}^{-1} \cdot \mathbf{A}_1^{-1} \cdot \mathbf{B}_2)^{\mathrm{T}}$$

from which, since $\mathbf{B}_1 = \mathbf{B}_2$,

$$\mathbf{C}_2 = -(\mathbf{A}_1^{-1} \cdot \mathbf{B}_2)^{\mathrm{T}} \cdot (\mathbf{Z}^{-1})^{\mathrm{T}} = \mathbf{C}_1 (\mathbf{Z}^{-1})^{\mathrm{T}} \tag{9-26}$$

But $\mathbf{Z}$ is orthogonal, thus its transpose is equal to its inverse. Hence $\mathbf{Z}^{-1} = \mathbf{Z}^{\mathrm{T}}$, so that $(\mathbf{Z}^{-1})^{\mathrm{T}} = \mathbf{Z}$. Substituting this result into (9-26), we obtain

$$\mathbf{C}_2 = \mathbf{C}_1 \cdot \mathbf{Z} \tag{9-27}$$

Comparing (9-25) with (9-27), we see that $\mathbf{A}_1$ and $\mathbf{C}_1$ undergo the same transformation, and therefore if $\mathbf{Z}$ causes the interchange of any particular two columns in $\mathbf{A}_1$, it also causes the interchange of the *same* two columns in $\mathbf{C}_1$. This proves the theorem.

Example 9-9

Consider the following two Dimensional Sets:

Dimensional Set 1

	V_1	V_2	V_3	V_4	V_5	V_6
d_1	3	1	2	4	6	8
d_2	1	2	3	2	4	3
d_3	1	2	5	2	1	4
π_{11}	1	0	0	$\dfrac{9}{8}$	$\dfrac{-2}{8}$	$\dfrac{-6}{8}$
π_{12}	0	1	0	$\dfrac{-47}{8}$	$\dfrac{6}{8}$	$\dfrac{18}{8}$
π_{13}	0	0	1	$\dfrac{-92}{8}$	$\dfrac{16}{8}$	$\dfrac{32}{8}$

Dimensional Set 2

	V_1	V_2	V_3	V_4	V_5	V_6
d_1	3	1	2	8	6	4
d_2	1	2	3	3	4	2
d_3	1	2	5	4	1	2
π_{21}	1	0	0	$\dfrac{-6}{8}$	$\dfrac{-2}{8}$	$\dfrac{9}{8}$
π_{22}	0	1	0	$\dfrac{18}{8}$	$\dfrac{6}{8}$	$\dfrac{-47}{8}$
π_{23}	0	0	1	$\dfrac{32}{8}$	$\dfrac{16}{8}$	$\dfrac{-92}{8}$

In Set 2 the first and third columns of $\mathbf{A}$ are interchanged with respect to Set 1. We see that the *same* two columns are interchanged in the $\mathbf{C}$ matrix, as Theorem 9-7 prescribes.

⇑

Corollary of Theorem 9-7. *The interchange of any 2 columns in the* **A** *matrix has no effect on the composition of dimensionless variables.*

Proof. Since the columns of **A** and **C** are changed in unison, this change affects only the sequential order in which the variables are written in any particular dimensionless variable which, being a scalar product, is impervious to the sequential order. This proves the corollary.

Example 9-10

Consider again the Dimensional Sets in Example 9-9. In these Sets the only difference is that in Set 2 columns V_4 and V_6 are interchanged with respect to Set 1. As we can see

$$\pi_{11} = V_1^1 \cdot V_2^0 \cdot V_3^0 \cdot (V_4^9 \cdot V_5^{-2} \cdot V_6^{-6})^{1/8} = \pi_{21}$$

$$\pi_{12} = V_1^0 \cdot V_2^1 \cdot V_3^0 \cdot (V_4^{-47} \cdot V_5^6 \cdot V_6^{18})^{1/8} = \pi_{22}$$

$$\pi_{13} = V_1^0 \cdot V_2^0 \cdot V_3^1 \cdot (V_4^{-92} \cdot V_5^{16} \cdot V_6^{32})^{1/8} = \pi_{23}$$

The two sets of dimensionless variables are indeed identical, as expected and required.

⇑

Theorem 9-8. *If the element* c_{ij} *at the intersection of the* i*th row and* j*th column of the* **C** *matrix is zero, then the* i*th column of the* **B** *matrix cannot be exchanged with the* j*th column of the* **A** *matrix.*

For if such an exchange took place, then it would result in a singular **A** matrix which, of course, cannot be allowed to happen.

Proof. By Corollary 2 of Theorem 9-3, if both **A** and **B** matrices are premultiplied by $\mathbf{A}^{-1}$, then the new **A** will be an identity matrix, and the new **B** will be the negative transpose of matrix **C**. Thus, if $c_{ij} = 0$, then the b_{ji} element of the new **B** matrix is zero. Consequently, if the ith column of the new **B** is exchanged with the jth column of the new **A**, then the jth column of the new **A** will be all zeros, since the only nonzero member in the new **A** occurs in its ith (= jth) row (or column). Hence the new **A** will be singular, and therefore the exchange cannot take place. This proves the theorem (the example below will help the reader to understand this rather wordy proof).

Example 9-11

In the given Dimensional Set 1 the **C** matrix has zero elements in the intersection of its

- $i = 1$ row and $j = 3$ column; hence variables V_1 and V_6 (first column of **B**, and third column of **A**) cannot be interchanged;
- $i = 2$ row and $j = 2$ column; hence variables V_2 and V_5 cannot be interchanged;
- $i = 3$ row and $j = 3$ column; hence variables V_3 and V_6 cannot be interchanged.

Dimensional Set 1

	V_1	V_2	V_3	V_4	V_5	V_6
d_1	4	−10	−8	1	−2	3
d_2	7	−4	−14	4	1	0
d_3	−2	−5	4	−1	0	2
π_1	1	0	0	−2	1	0
π_2	0	1	0	1	0	3
π_3	0	0	1	4	−2	0

Now we multiply both **A** and **B** by $\mathbf{A}^{-1}$. This operation will produce Dimensional Set 2, wherein $\mathbf{B}_2 = \mathbf{A}^{-1} \cdot \mathbf{B}_1$, $\mathbf{A}_2 = \mathbf{A}_1^{-1} \cdot \mathbf{A}_1 = \mathbf{I}$ and of course $\mathbf{C}_2 = \mathbf{C}_1$.

Dimensional Set 2

	V_1	V_2	V_3	V_4	V_5	V_6
d_1	2	−1	−4	1	0	0
d_2	−1	⓪	2	0	①	0
d_3	0	−3	0	0	0	1
π_1	1	0	0	−2	1	0
π_2	0	1	0	1	0	3
π_3	0	0	1	4	−2	0

Now we notice that if we interchanged variables V_2 and V_5 in Set 2, we would get an **A** matrix whose second row would be all zeros; hence **A** would become singular—which is strictly forbidden. Therefore, variables V_2 and V_5 cannot be interchanged. The same thing would happen if we interchanged variables V_1 and V_6, in which case the third row of **A** would become all zeros, and again **A** would be singular. Thus variables V_1 and V_6 cannot be exchanged, either. Finally, **A** would become singular if V_3 and V_6 were interchanged—and hence this exchange is also forbidden.

⇑

A common characteristic of these proscribed transfers that they always involve a zero in the $\mathbf{B}_2$ matrix. This zero then replaces the solitary nonzero in the particular row in $\mathbf{A}_2$. Therefore this row in $\mathbf{A}_2$ becomes all zeros. The attentive reader will also notice that $\mathbf{C}_2 = -\mathbf{B}_2^T$ in Set 2; this, of course, follows directly from the second relation of (9-20).

Theorem 9-9. *If a column in the* **D** *matrix is identical to any column in the* **C** *matrix, then the respective variables in the* **B** *and* **A** *matrices can be interchanged without affecting* **C**.

Proof. By assumption the indicated transposition does not change the composition of the dimensionless variables. Nor are the dimensional composition of the variables affected. Thus the **C** matrix, which is defined solely by these two characteristics, must also remain unchanged. This proves the theorem.

Example 9-12

Consider the following two Dimensional Sets:

Dimensional Set 1   Dimensional Set 2

interchanged

	V_1	V_2	V_3	V_4	V_5	V_6
d_1	−2	3	−4	2	−1	1
d_2	−2	2	−4	2	−1	2
d_3	−4	−2	1	1	1	1
π_{11}	0	1	2	1	−2	1
π_{12}	1	1	1	2	2	1
π_{13}	1	2	2	2	2	2

	V_1	V_2	V_3	V_4	V_5	V_6
d_1	−2	1	−4	2	−1	3
d_2	−2	2	−4	2	−1	2
d_3	−4	1	1	1	1	−2
π_{21}	0	1	2	1	−2	1
π_{22}	1	1	1	2	2	1
π_{23}	1	2	2	2	2	2

columns are identical

As we see in Set 1, the second column of the **D** matrix and the third column of the **C** matrix are identical. Therefore variables V_2 and V_6 can be interchanged without affecting **C**. Indeed, the **C** matrix in Set 2 is identical to that of Set 1, as can be observed. Also, by the given sets

$$\pi_{11} = V_1^0 \cdot V_2^1 \cdot V_3^2 \cdot V_4^1 \cdot V_5^{-2} \cdot V_6^1 = \pi_{21}$$

$$\pi_{12} = V_1^1 \cdot V_2^1 \cdot V_3^1 \cdot V_4^2 \cdot V_5^2 \cdot V_6^1 = \pi_{22}$$

$$\pi_{13} = V_1^1 \cdot V_2^2 \cdot V_3^2 \cdot V_4^2 \cdot V_5^2 \cdot V_6^2 = \pi_{23}$$

as expected and required. Note that **C** did not change, despite the fact that the dimensions of the interchanged variables V_2 and V_6 are *different*.

⇑

Corollary of Theorem 9-9. *If* **D** *is an identity matrix, and if the* j*th column of* **C** *contains all zeros except one "1" in the* i*th row, then the variables in the* i*th column of the* **B** *matrix and the* j*th column of the* **A** *matrix can be interchanged without altering the* **C** *matrix.*

Proof. If **D** is an identity matrix, then its *i*th row contains all zeros, except one "1" which is in its *i*th (= *j*th) column. Therefore the *i*th column of **D** is identical to the *j*th column of **C**, and thus their interchange, by Theorem 9-9, will not affect **C**. This proves the corollary.

Example 9-13

In Dimensional Set 1 the third ($j = 3$) column of **C** is all zeros, except in its second ($i = 2$) row which is a "1." Hence the second column of **B** (variable V_2) can be interchanged with the third column of **A** (variable V_6) without any effect on the **C** matrix, as Dimensional Set 2 shows. Indeed, as seen $\pi_{11} = \pi_{21}$; $\pi_{12} = \pi_{22}$; $\pi_{13} = \pi_{23}$, as expected and required.

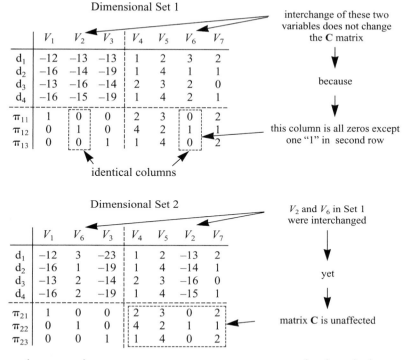

As can be seen again, $\pi_{11} = \pi_{21}$, $\pi_{12} = \pi_{22}$, $\pi_{13} = \pi_{23}$, as expected and required.

Example 9-14

Consider the Dimensional Set

	V_1	V_2	V_3	V_4	V_5	V_6	V_7
d_1	−28	40	−12	4	6	2	6
d_2	−19	30	−8	3	4	2	8
d_3	−7	11	−4	−1	2	−3	−4
d_4	−8	11	−4	0	2	−1	−2
π_1	1	0	0	1	4	0	0
π_2	0	1	0	0	−6	1	−1
π_3	0	0	1	0	2	0	0

We see here that variables V_1 and V_4 can be interchanged with impunity (i.e., without affecting the **C** matrix) as can V_2 and V_6. However, V_2 and V_7 cannot be interchanged without affecting **C**, because the intersection of the fourth column and the second row in **C** is not "1," but "−1."

Theorem 9-10. *If a column in matrix* **A** *is identical to a column in matrix* **B**, *then these columns (i.e., variables) can be interchanged with no effect on matrix* **C**.

Proof. Since the construction of the **C** matrix—by virtue of the Fundamental Formula—is solely defined by matrices **A** and **B**, and since these matrices are not changed by the subject interchange of columns (since they are identical), therefore **C** cannot change. This proves the theorem.

Example 9-15

Consider the following Dimensional Sets:

Dimensional Set 1

	V_1	V_2	V_3	V_4	V_5
d_1	1	2	1	1	−5
d_2	3	2	1	1	−6
π_{11}	1	0	0	9	2
π_{12}	0	1	0	−2	0
π_{13}	0	0	1	−1	0

Dimensional Set 2

	V_1	V_2	V_4	V_3	V_5
d_1	1	2	1	1	5
d_2	3	2	1	1	−6
π_{21}	1	0	0	9	2
π_{22}	0	1	0	−2	0
π_{23}	0	0	1	−1	0

In Set 1 the third column of **B** (variable V_3) and the first column of **A** (variable V_4) are identical. Therefore their interchange does not affect the **C** matrix, as Set 2 shows.

⇑

Theorem 9-11. *If the* ith *row of the* **C** *matrix consists of all zeros, except a nonzero element q in its* jth *column, then the* ith *column of the* **B** *matrix is* −q *times the* jth *column of the* **A** *matrix.*

Proof. Let the involved variables in **A** and **B** matrices be denoted by V_A, V_B, respectively. Then obviously the quantity $V_B \cdot V_A^q$ must be dimensionless. Therefore the exponents of all dimensions in V_B (i.e., the *i*th column in **B**) must be —*q* times as much as that of V_A (the *j*th column in **A**). Therefore the *sum* of exponents of every dimension will be zero. This proves the theorem.

The following example demonstrates the application of this theorem.

Example 9-16

Given the Dimensional Set:

	V_1	V_2	V_3	V_4	V_5
d_1	−20	18	1	2	3
d_2	3	6	4	−5	1
d_3	−23	24	2	1	4
π_1	1	0	2	3	4
π_2	0	1	0	0	−6

In this Set the second ($i = 2$) row of the **C** matrix is all zeros, except the third ($j = 3$) column, which is $q = -6$. Therefore the second column of the **B** matrix is $-q = 6$ times the third column of the **A** matrix. Indeed, as seen

$$\begin{bmatrix} 18 \\ 6 \\ 24 \end{bmatrix} = 6 \cdot \begin{bmatrix} 3 \\ 1 \\ 4 \end{bmatrix}$$

⇑

Collorary of Theorem 9-11. *If the* ith *row of the* **C** *matrix consists of all zeros except one of its elements,* -1 *in its* jth *column, then the* ith *column of the* **B** *matrix is identical to the* jth *column of the* **A** *matrix.*

Proof. This is a special case of Theorem 9-11 when $q = -1$. This proves this corollary.

Example 9-17

Given the dimensional set

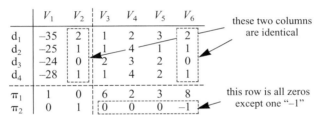

In the second row of **C**, every element is zero, except one element, which is -1. This element occurs in the fourth column of **C**. Hence the second column of **B** and the fourth column of **A** must be identical, and indeed they are, as seen above.

Note that by inspection we are certain that V_2 cannot be interchanged with either V_3, V_4, or V_5. For in the **C** matrix we have zeros in the relevant locations, which—by Theorem 9-8—preclude these interchanges.

⇑

Theorem 9-12. *In a dimensional set there exists a matrix* **Z** *such that*

$$\mathbf{B} = \mathbf{A} \cdot \mathbf{Z} \tag{9-28}$$

where the size of **Z** *is identical to that of* **B**.

Proof. By (9-28),

$$\mathbf{Z} = \mathbf{A}^{-1} \cdot \mathbf{B} \tag{9-29}$$

and by the Fundamental Formula (8-18) $\mathbf{C} = -(\mathbf{A}^{-1} \cdot \mathbf{B})^T$, from which

$$-\mathbf{C}^T = \mathbf{A}^{-1} \cdot \mathbf{B} \tag{9-30}$$

and hence, by (9-29) and (9-30)

$$Z = -C^T \tag{9-31}$$

proving the theorem.

Example 9-18

Given the Dimensional Set

	V_1	V_2	V_3	V_4	V_5	V_6
d_1	-10	73	1	3	-2	3
d_2	26	-5	2	-1	4	2
d_3	14	7	3	-2	1	1
d_4	24	18	3	0	2	1
π_1	1	0	-5	-2	-6	3
π_2	0	1	-8	-9	7	-8

in which

$$\mathbf{A} = \begin{bmatrix} 1 & 3 & -2 & 3 \\ 2 & -1 & 4 & 2 \\ 3 & -2 & 1 & 1 \\ 3 & 0 & 2 & 1 \end{bmatrix}; \quad \mathbf{B} = \begin{bmatrix} -10 & 73 \\ 26 & -5 \\ 14 & 7 \\ 24 & 18 \end{bmatrix}; \quad \mathbf{C} = \begin{bmatrix} -5 & -2 & -6 & 3 \\ -8 & -9 & 7 & -8 \end{bmatrix}$$

and therefore, by (9-31)

$$\mathbf{Z} = -\mathbf{C}^T = \begin{bmatrix} 5 & 8 \\ 2 & 9 \\ 6 & -7 \\ -3 & 8 \end{bmatrix}$$

and indeed, by (9-28),

$$\mathbf{B} = \mathbf{A} \cdot \mathbf{Z} = \begin{bmatrix} 1 & 3 & -2 & 3 \\ 2 & -1 & 4 & 2 \\ 3 & -2 & 1 & 1 \\ 3 & 0 & 2 & 1 \end{bmatrix} \cdot \begin{bmatrix} 5 & 8 \\ 2 & 9 \\ 6 & -7 \\ -3 & 8 \end{bmatrix} = \begin{bmatrix} -10 & 73 \\ 26 & -5 \\ 14 & 7 \\ 24 & 18 \end{bmatrix}$$

as expected and required.

⇑

Theorem 9-13. *If there are two Dimensional Sets 1 and 2 such that*

$$\mathbf{A}_2 = \mathbf{A}_1 \cdot \mathbf{Z}_A; \qquad \mathbf{B}_2 = \mathbf{B}_1 \cdot \mathbf{Z}_B \tag{9-32}$$

then

$$\mathbf{C}_2 = \mathbf{Z}_B^T \cdot \mathbf{C}_1 \cdot (\mathbf{Z}_A^{-1})^T \tag{9-33}$$

Proof. By the Fundamental Formula (8-18), $\mathbf{C}_2 = -(\mathbf{A}_2^{-1} \cdot \mathbf{B}_2)^T$ into which (9-32) is substituted. This will yield, in order,

TRANSFORMATIONS

$$\mathbf{C}_2 = -[(\mathbf{A}_1 \cdot \mathbf{Z}_A)^{-1} \cdot (\mathbf{B}_1 \cdot \mathbf{Z}_B)]^T = -[\mathbf{Z}_A^{-1} \cdot \mathbf{A}_1^{-1} \cdot \mathbf{B}_1 \cdot \mathbf{Z}_B]^T = [\mathbf{Z}_A^{-1} \cdot \mathbf{C}_1^T \cdot \mathbf{Z}_B]^T$$

or

$$\mathbf{C}_2 = \mathbf{Z}_B^T \cdot \mathbf{C}_1 \cdot (\mathbf{Z}_A^{-1})^T$$

which was to be proven. Note that the nature of this relation makes the nonsingularity of $\mathbf{Z}_A$ and $\mathbf{Z}_B$ mandatory, i.e., they have to be nonsingular square matrices.

There are two special cases:

(a) **B** matrices are identical; i.e., $\mathbf{B}_2 = \mathbf{B}_1$, hence, by (9-32), $\mathbf{Z}_B = \mathbf{I}$, thus (9-33) is reduced to

$$\mathbf{C}_2 = \mathbf{C}_1 \cdot (\mathbf{Z}_A^{-1})^T \qquad (9\text{-}34)$$

(b) **A** matrices are identical; i.e., $\mathbf{A}_2 = \mathbf{A}_1$, hence, by (9-32), $\mathbf{Z}_A = \mathbf{I}$, thus (9-33) is reduced to

$$\mathbf{C}_2 = \mathbf{Z}_B^T \cdot \mathbf{C}_1 \qquad (9\text{-}35)$$

Example 9-19

Given the Dimensional Set

	V_1	V_2	V_3	V_4	V_5	V_6
d_1	2	3	8	2	2	−1
d_2	0	−4	−4	−2	−1	0
d_3	3	−4	−1	1	2	3
π_{11}	1	0	0	1	−2	0
π_{12}	0	1	0	−3	2	1
π_{13}	0	0	1	−1	−2	2

in which

$$\mathbf{B}_1 = \begin{bmatrix} 2 & 3 & 8 \\ 0 & -4 & -4 \\ 3 & -4 & -1 \end{bmatrix}; \quad \mathbf{A}_1 = \begin{bmatrix} 2 & 2 & -1 \\ -2 & -1 & 0 \\ 1 & 2 & 3 \end{bmatrix}; \quad \mathbf{C}_1 = \begin{bmatrix} 1 & -2 & 0 \\ -3 & 2 & 1 \\ -1 & -2 & 2 \end{bmatrix}$$

Now we define the multiplier matrices $\mathbf{Z}_A$ and $\mathbf{Z}_B$

$$\mathbf{Z}_A = \begin{bmatrix} 1 & 2 & -1 \\ 2 & 0 & 1 \\ 3 & 2 & -2 \end{bmatrix}; \quad \mathbf{Z}_B = \begin{bmatrix} 1 & -2 & 1 \\ -2 & 1 & -4 \\ 2 & -1 & -2 \end{bmatrix}$$

so that $|\mathbf{Z}_A| = 8 \neq 0$. Hence, by (9-32)

$$\mathbf{A}_2 = \mathbf{A}_1 \cdot \mathbf{Z}_A = \begin{bmatrix} 3 & 2 & 2 \\ -4 & -4 & 1 \\ 14 & 8 & -5 \end{bmatrix}; \quad \mathbf{B}_2 = \mathbf{B}_1 \cdot \mathbf{Z}_B = \begin{bmatrix} 12 & -9 & -26 \\ 0 & 0 & 24 \\ 9 & -9 & 21 \end{bmatrix}$$

from which, by (9-33)

$$\mathbf{C}_2 = \mathbf{Z}_B^T \cdot \mathbf{C}_1 \cdot (\mathbf{Z}_A^{-1})^T = \frac{1}{8} \cdot \begin{bmatrix} -26 & 19 & -28 \\ 22 & -17 & 20 \\ -58 & 123 & 68 \end{bmatrix}$$

To confirm this, we determine $\mathbf{C}_2$ directly by the Fundamental Formula

$$\mathbf{C}_2 = -(\mathbf{A}_2^{-1} \cdot \mathbf{B}_2)^T = -\left[\begin{bmatrix} 3 & 2 & 2 \\ -4 & -4 & 1 \\ 14 & 8 & -5 \end{bmatrix}^{-1} \cdot \begin{bmatrix} 4 & -17 & -26 \\ 4 & 4 & 24 \\ 10 & -8 & 21 \end{bmatrix} \right]^T = \frac{1}{8} \cdot \begin{bmatrix} -26 & 19 & -28 \\ 22 & -17 & 20 \\ -58 & 123 & 68 \end{bmatrix}$$

The above two $\mathbf{C}_2$ matrices are identical, as required and expected.

⇑

Theorem 9-14. *If in a Dimensional Set* $\mathbf{A} = \mathbf{B}$, *then* $\mathbf{C} = -\mathbf{I}$; *i.e.,* $\mathbf{C}$ *is a negative identity matrix.*

Proof. By the Fundamental Formula (8-18) we have $\mathbf{C} = -(\mathbf{A}^{-1} \cdot \mathbf{B})^T$. If now $\mathbf{A} = \mathbf{B}$, then

$$\mathbf{C} = -(\mathbf{A}^{-1} \cdot \mathbf{A})^T = -\mathbf{I}^T = -\mathbf{I}$$

which was to be proven.

Example 9-20

Given the Dimensional Set

	V_1	V_2	V_3	V_4	V_5	V_6
d_1	1	2	3	1	2	3
d_2	-2	1	0	-2	1	0
d_3	2	1	-4	2	1	-4
π_1	1	0	0	-1	0	0
π_2	0	1	0	0	-1	0
π_3	0	0	1	0	0	-1

in which $\mathbf{A} = \mathbf{B}$; hence $\mathbf{C} = \begin{bmatrix} -1 & 0 & 0 \\ 0 & -1 & 0 \\ 0 & 0 & -1 \end{bmatrix} = -\mathbf{I}$

as expected and required.

⇑

9.2. TRANSFORMATION BETWEEN SYSTEMS OF DIFFERENT D MATRICES

As mentioned in Art. 8.3, on occasion in a Dimensional Set (Fig. 8-3) we have to deal with a $\mathbf{D}$ matrix which, although a square matrix, is not an identity matrix. Therefore it may be useful to have a formula handy to relate systems having—in general—different $\mathbf{D}$ matrices.

First, we state and then prove a theorem dealing with $\mathbf{C}$ and $\mathbf{D}$ matrices.

Theorem 9-15. *If, in Dimensional Sets 1 and 2,* $\mathbf{A}_1 = \mathbf{A}_2$ *and* $\mathbf{B}_1 = \mathbf{B}_2$ *(i.e., the dimensional matrices are identical), but the* $\mathbf{D}$ *matrices are different (and nonsingular), then*

$$\mathbf{C}_1 \cdot \mathbf{C}_2^{-1} = \mathbf{D}_1 \cdot \mathbf{D}_2^{-1} \tag{9-36}$$

Proof. By the Fundamental Formula (8-19)

$$\mathbf{C}_1 = -\mathbf{D}_1 \cdot (\mathbf{A}^{-1} \cdot \mathbf{B})^T \; ; \qquad \mathbf{C}_2 = -\mathbf{D}_2 \cdot (\mathbf{A}^{-1} \cdot \mathbf{B})^T$$

From the first and second of these, in order

$$-(\mathbf{A}^{-1} \cdot \mathbf{B})^T = \mathbf{D}_1^{-1} \cdot \mathbf{C}_1 \; ; \qquad -(\mathbf{A}^{-1} \cdot \mathbf{B})^T = \mathbf{D}_2^{-1} \cdot \mathbf{C}_2$$

Hence

$$\mathbf{D}_1^{-1} \cdot \mathbf{C}_1 = \mathbf{D}_2^{-1} \cdot \mathbf{C}_2$$

Premultiplication of both sides of this relation by $\mathbf{D}_1$, followed by postmultiplication of both sides by $\mathbf{C}_2^{-1}$ yield $\mathbf{C}_1 \cdot \mathbf{C}_2^{-1} = \mathbf{D}_1 \cdot \mathbf{D}_2^{-1}$, which is (9-36). This proves the theorem.

Example 9-21

Consider Dimensional Sets 1 and 2

Dimensional Set 1

		V_1	V_2	V_3	V_4	V_5	V_6	
B	d_1	1	2	3	4	5	6	
	d_2	4	1	3	2	6	5	**A**
	d_3	2	1	3	5	4	6	
$\mathbf{D}_1$	π_{11}	2	1	3	−18	−17	24	
	π_{12}	5	4	2	−32	−31	44	$\mathbf{C}_1$
	π_{13}	1	6	3	17	12	−25	

Dimensional Set 2

		V_1	V_2	V_3	V_4	V_5	V_6	
B	d_1	1	2	3	4	5	6	
	d_2	4	1	3	2	6	5	**A**
	d_3	2	1	3	5	4	6	
$\mathbf{D}_2$	π_{21}	2	1	1	−16	−15	22	
	π_{22}	6	2	5	−55	−51	75	$\mathbf{C}_2$
	π_{23}	4	3	3	−28	−27	38	

We see that **A** and **B** are identical for both Sets, but $\mathbf{D}_1$ is different from $\mathbf{D}_2$, and correspondingly $\mathbf{C}_1$ differs from $\mathbf{C}_2$. By these given data and (9-36), then

$$\mathbf{C}_2 = \underbrace{\begin{bmatrix} 2 & 1 & 1 \\ 6 & 2 & 5 \\ 4 & 3 & 3 \end{bmatrix}}_{\mathbf{D}_2} \cdot \underbrace{\begin{bmatrix} 2 & 1 & 3 \\ 5 & 4 & 2 \\ 1 & 6 & 3 \end{bmatrix}^{-1}}_{\mathbf{D}_1^{-1}} \cdot \underbrace{\begin{bmatrix} -18 & -17 & 24 \\ -32 & -31 & 44 \\ 17 & 12 & -25 \end{bmatrix}}_{\mathbf{C}_1} = \underbrace{\begin{bmatrix} -16 & -15 & 22 \\ -55 & -51 & 75 \\ -28 & -27 & 38 \end{bmatrix}}_{\mathbf{C}_2}$$

which confirms, of course, the directly determined $\mathbf{C}_2$, as Set 2 shows.

⇑

Corollary of Theorem 9-15. If $\mathbf{D}_1 = \mathbf{I}$, or $\mathbf{D}_2 = \mathbf{I}$, or $\mathbf{D}_1 = \mathbf{D}_2$, then

$$\left. \begin{array}{ll} \mathbf{C}_2 = \mathbf{D}_2 \cdot \mathbf{C}_1 & \text{if } \mathbf{D}_1 = \mathbf{I} \\ \mathbf{C}_2 = \mathbf{D}_1^{-1} \cdot \mathbf{C}_1 & \text{if } \mathbf{D}_2 = \mathbf{I} \\ \mathbf{C}_2 = \mathbf{C}_1 & \text{if } \mathbf{D}_1 = \mathbf{D}_2 \end{array} \right\} \qquad (9\text{-}37)$$

Proof. All of the above equations can be obtained by direct substitution of the respective $\mathbf{D}_1$ and $\mathbf{D}_2$ values into relation (9-36). This proves the corollary.

Example 9-22

Consider Dimensional Set 1

	V_1	V_2	V_3	V_4	V_5
d_1	2	0	-3	1	4
d_2	4	2	-2	5	2
π_{11}	4	6	2	$\dfrac{-92}{18}$	$\dfrac{14}{18}$
π_{12}	3	1	-3	$\dfrac{-50}{18}$	$\dfrac{-55}{18}$
π_{13}	2	1	4	$\dfrac{-24}{18}$	$\dfrac{42}{18}$

$\qquad\qquad\qquad\qquad\quad\uparrow\qquad\qquad\uparrow$
$\qquad\qquad\qquad\quad\mathbf{D}_1$ matrix $\quad\mathbf{C}_1$ matrix

Question: What would be the **C** matrix (i.e., $\mathbf{C}_2$), if **D** (i.e., $\mathbf{D}_2$) were the unit matrix **I**? *Answer:* Using the $\mathbf{C}_1$ and $\mathbf{D}_1$ matrices given above, and employing the relevant (i.e., the second) relation of (9-37), we obtain

$$\mathbf{C}_2 = \mathbf{D}_1^{-1} \cdot \mathbf{C}_1 = \left(\frac{1}{78} \cdot \begin{bmatrix} -7 & 22 & 20 \\ 18 & -12 & -18 \\ -1 & -8 & 14 \end{bmatrix}\right) \cdot \left(\frac{1}{18} \cdot \begin{bmatrix} -92 & 14 \\ -50 & -55 \\ -24 & 42 \end{bmatrix}\right) = \frac{1}{18} \cdot \begin{bmatrix} -12 & -6 \\ -8 & 2 \\ 2 & 13 \end{bmatrix} \quad \text{(a)}$$

Indeed, if we now have the new Dimensional Set

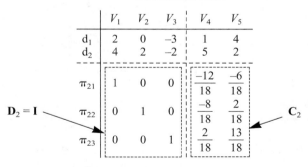

Dimensional Set 2

we see that the $\mathbf{C}_2$ matrix in it is as determined by (a) above.

⇑

Theorem 9-16. *If two Dimensional Sets, 1 and 2, have identical* **A, B** *matrices (i.e.,* $\mathbf{A}_1 = \mathbf{A}_2$ *and* $\mathbf{B}_1 = \mathbf{B}_2$*), but different* $\mathbf{D}_1$, $\mathbf{D}_2$ *matrices, then the generated two sets of dimensionless variables* $\pi_{11}, \pi_{12}, \ldots$ *and* $\pi_{21}, \pi_{22}, \ldots$ *are connected by*

TRANSFORMATIONS

$$\begin{bmatrix} \ln \pi_{11} \\ \ln \pi_{12} \\ \vdots \\ \ln \pi_{1N_P} \end{bmatrix} = \mathbf{T}^{-1} \cdot \begin{bmatrix} \ln \pi_{21} \\ \ln \pi_{22} \\ \vdots \\ \ln \pi_{2N_P} \end{bmatrix} \quad (9\text{-}38)$$

or, correspondingly

$$\begin{bmatrix} \ln \pi_{21} \\ \ln \pi_{22} \\ \vdots \\ \ln \pi_{2N_P} \end{bmatrix} = \mathbf{T} \cdot \begin{bmatrix} \ln \pi_{11} \\ \ln \pi_{12} \\ \vdots \\ \ln \pi_{1N_P} \end{bmatrix} \quad (9\text{-}39)$$

where $\mathbf{T}$ is the transformation matrix defined as

$$\mathbf{T} = \mathbf{D}_2 \cdot \mathbf{D}_1^{-1} \quad (9\text{-}40)$$

In the above, N_P is the number of dimensionless variables defined by relation (7-26); in this case N_P is the difference between the number of variables and the rank of the dimensional matrix.

Proof. We prove this important set of relations on a system of five variables and three dimensions, but, of course, the process can be applied (extended) to systems of arbitrary numbers of variables and dimensions.

In general, consider the Dimensional Set

$$\begin{array}{c|ccccc} & V_1 & V_2 & V_3 & V_4 & V_5 \\ \hline d_1 & & & & & \\ d_2 & & & & & \\ d_3 & & & & & \\ \hline \pi_1 & g_{11} & g_{12} & c_{11} & c_{12} & c_{13} \\ \pi_2 & g_{21} & g_{22} & c_{21} & c_{22} & c_{23} \end{array} \quad (9\text{-}41)$$

D matrix ← ... → **C** matrix

by which we can write

$$\left. \begin{array}{l} \pi_1 = V_1^{g_{11}} \cdot V_2^{g_{12}} \cdot V_3^{c_{11}} \cdot V_4^{c_{12}} \cdot V_5^{c_{13}} \\ \pi_2 = V_1^{g_{21}} \cdot V_2^{g_{22}} \cdot V_3^{c_{21}} \cdot V_4^{c_{22}} \cdot V_5^{c_{23}} \end{array} \right\} \quad (9\text{-}42)$$

or taking logarithms of both sides

$$\left. \begin{array}{l} \ln \pi_1 = g_{11} \cdot \ln V_1 + g_{12} \cdot \ln V_2 + c_{11} \cdot \ln V_3 + c_{12} \cdot \ln V_4 + c_{13} \cdot \ln V_5 \\ \ln \pi_2 = g_{21} \cdot \ln V_1 + g_{22} \cdot \ln V_2 + c_{21} \cdot \ln V_3 + c_{22} \cdot \ln V_4 + c_{23} \cdot \ln V_5 \end{array} \right\} \quad (9\text{-}43)$$

These last two relations can be conveniently written in matrix form

$$\begin{bmatrix} \ln \pi_1 \\ \ln \pi_2 \end{bmatrix} = \begin{bmatrix} g_{11} & g_{12} \\ g_{21} & g_{22} \end{bmatrix} \cdot \begin{bmatrix} \ln V_1 \\ \ln V_2 \end{bmatrix} + \begin{bmatrix} c_{11} & c_{12} & c_{13} \\ c_{21} & c_{22} & c_{23} \end{bmatrix} \cdot \begin{bmatrix} \ln V_3 \\ \ln V_4 \\ \ln V_5 \end{bmatrix}$$

which, by using the construction of the Dimensional Set (9-41), can assume the compact form

$$\begin{bmatrix} \ln \pi_1 \\ \ln \pi_2 \end{bmatrix} = \mathbf{D} \cdot \begin{bmatrix} \ln V_1 \\ \ln V_2 \end{bmatrix} + \mathbf{C} \cdot \begin{bmatrix} \ln V_3 \\ \ln V_4 \\ \ln V_5 \end{bmatrix} \qquad (9\text{-}44)$$

If now there are *two* sets, then *analogously* we have

$$\left. \begin{aligned} \begin{bmatrix} \ln \pi_{11} \\ \ln \pi_{12} \end{bmatrix} &= \mathbf{D}_1 \cdot \begin{bmatrix} \ln V_1 \\ \ln V_2 \end{bmatrix} + \mathbf{C}_1 \cdot \begin{bmatrix} \ln V_3 \\ \ln V_4 \\ \ln V_5 \end{bmatrix} \\ \begin{bmatrix} \ln \pi_{21} \\ \ln \pi_{22} \end{bmatrix} &= \mathbf{D}_2 \cdot \begin{bmatrix} \ln V_1 \\ \ln V_2 \end{bmatrix} + \mathbf{C}_2 \cdot \begin{bmatrix} \ln V_3 \\ \ln V_4 \\ \ln V_5 \end{bmatrix} \end{aligned} \right\} \qquad (9\text{-}45)$$

But, by (9-36), $\mathbf{C}_2 = (\mathbf{D}_2 \cdot \mathbf{D}_1^{-1}) \cdot \mathbf{C}_1$, hence the second of (9-45) can be written

$$\begin{bmatrix} \ln \pi_{21} \\ \ln \pi_{22} \end{bmatrix} = \mathbf{D}_2 \cdot \begin{bmatrix} \ln V_1 \\ \ln V_2 \end{bmatrix} + (\mathbf{D}_2 \cdot \mathbf{D}_1^{-1}) \cdot \mathbf{C}_1 \cdot \begin{bmatrix} \ln V_3 \\ \ln V_4 \\ \ln V_5 \end{bmatrix}$$

Multiplying both sides by $\mathbf{D}_2^{-1}$ and rearranging, we get

$$\begin{bmatrix} \ln V_1 \\ \ln V_2 \end{bmatrix} = \mathbf{D}_2^{-1} \cdot \begin{bmatrix} \ln \pi_{21} \\ \ln \pi_{22} \end{bmatrix} - \mathbf{D}_1^{-1} \cdot \mathbf{C}_1 \cdot \begin{bmatrix} \ln V_3 \\ \ln V_4 \\ \ln V_5 \end{bmatrix} \qquad (9\text{-}46)$$

Similarly, if we multiply both sides of the first of (9-45) by $\mathbf{D}_1^{-1}$, then after some rearrangement we obtain

$$\begin{bmatrix} \ln V_1 \\ \ln V_2 \end{bmatrix} = \mathbf{D}_1^{-1} \cdot \begin{bmatrix} \ln \pi_{11} \\ \ln \pi_{12} \end{bmatrix} - \mathbf{D}_1^{-1} \cdot \mathbf{C}_1 \cdot \begin{bmatrix} \ln V_3 \\ \ln V_4 \\ \ln V_5 \end{bmatrix} \qquad (9\text{-}47)$$

Now we observe—with delight—that the left-hand sides of (9-46) and (9-47) are identical. It follows, then, that their right-hand sides must also be identical. Thus, by equating the right sides and eliminating the common member, we get

$$\mathbf{D}_1^{-1} \cdot \begin{bmatrix} \ln \pi_{11} \\ \ln \pi_{12} \end{bmatrix} = \mathbf{D}_2^{-1} \cdot \begin{bmatrix} \ln \pi_{21} \\ \ln \pi_{22} \end{bmatrix} \qquad (9\text{-}48)$$

from which, in view of (9-40), both (9-38) and (9-39) follow at once. This proves the theorem.

Example 9-23

Suppose we have the following two dimensionless variables:

$$\pi_{11} = V_1^4 \cdot V_2^2 \cdot V_3^2 \cdot V_4^1 \cdot V_5^4; \qquad \pi_{12} = V_1^1 \cdot V_2^{-1} \cdot V_3^{-2} \cdot V_4^0 \cdot V_5^2 \qquad (a)$$

TRANSFORMATIONS

generated by the $\mathbf{D}_1$ matrix

$$\mathbf{D}_1 = \begin{bmatrix} 4 & 2 \\ 1 & -1 \end{bmatrix} \tag{b}$$

If now the **D** matrix is changed to

$$\mathbf{D}_2 = \begin{bmatrix} 6 & 4 \\ 1 & 9 \end{bmatrix} \tag{c}$$

what would the new set of dimensionless variables be, and what would be their composition in terms of variables $V_1, V_2, \ldots, V_5$?

By the given $\mathbf{D}_1$ and $\mathbf{D}_2$ matrices in (b) and (c), and the definition of the transformation matrix in (9-40), we can write

$$\mathbf{T} = \mathbf{D}_2 \cdot \mathbf{D}_1^{-1} = \frac{1}{3} \cdot \begin{bmatrix} 5 & -2 \\ 5 & -17 \end{bmatrix} \tag{d}$$

Therefore, by (9-39),

$$\begin{bmatrix} \ln \pi_{21} \\ \ln \pi_{22} \end{bmatrix} = \frac{1}{3} \begin{bmatrix} 5 & -2 \\ 5 & -17 \end{bmatrix} \cdot \begin{bmatrix} \ln \pi_{11} \\ \ln \pi_{12} \end{bmatrix} \tag{e}$$

from which the sought-after connecting formulas are

$$\pi_{21} = \left(\frac{\pi_{11}^5}{\pi_{12}^2} \right)^{1/3}; \qquad \pi_{22} = \left(\frac{\pi_{11}^5}{\pi_{12}^{17}} \right)^{1/3} \tag{f}$$

Now by (a)

$$\mathbf{D}_1 \text{ matrix} \longrightarrow \begin{array}{c|cc|ccc} & V_1 & V_2 & V_3 & V_4 & V_5 \\ \hline \pi_{21} & 4 & 2 & 2 & 1 & 4 \\ \pi_{22} & 1 & -1 & -2 & 0 & 2 \end{array} \longleftarrow \mathbf{C}_1 \text{ matrix} \tag{g}$$

from which the $\mathbf{C}_1$ matrix is

$$\mathbf{C}_1 = \begin{bmatrix} 2 & 1 & 4 \\ -2 & 0 & 2 \end{bmatrix} \tag{h}$$

and therefore, by (9-36), (d) and (h),

$$\mathbf{C}_2 = (\mathbf{D}_2 \cdot \mathbf{D}_1^{-1}) \cdot \mathbf{C}_1 = \mathbf{T} \cdot \mathbf{C}_1 = \frac{1}{3} \begin{bmatrix} 5 & -2 \\ 5 & -17 \end{bmatrix} \cdot \begin{bmatrix} 2 & 1 & 4 \\ -2 & 0 & 2 \end{bmatrix} = \frac{1}{3} \begin{bmatrix} 14 & 5 & 16 \\ 44 & 5 & -14 \end{bmatrix} \tag{i}$$

Hence, we can now write, by (c) and (i),

$$\mathbf{D}_2 \text{ matrix} \longrightarrow \begin{array}{c|cc|ccc} & V_1 & V_2 & V_3 & V_4 & V_5 \\ \hline \pi_{21} & 6 & 4 & \frac{14}{3} & \frac{5}{3} & \frac{16}{3} \\ \pi_{22} & 1 & 9 & \frac{44}{3} & \frac{5}{3} & \frac{-14}{3} \end{array} \longleftarrow \mathbf{C}_2 \text{ matrix}$$

which yields

$$\pi_{21} = V_1^6 \cdot V_2^4 \cdot (V_3^{14} \cdot V_4^5 \cdot V_5^{16})^{1/3}; \qquad \pi_{22} = V_1^1 \cdot V_2^9 \cdot (V_3^{44} \cdot V_4^5 \cdot V_5^{-14})^{1/3}$$

Note that for none of the above operations was it necessary to know the dimensional compositions of variables $V_1, V_2, \ldots, V_5$.

⇑

Example 9-24. Beach Profile Characteristics

Consider a beach subjected to the action of waves propagating towards and in the direction perpendicular to the shore line. The amount of sediment, moving towards and along the coast, is a function of the *slope of the be*d in the vicinity of the shore—as Fig. 9-1 illustrates.

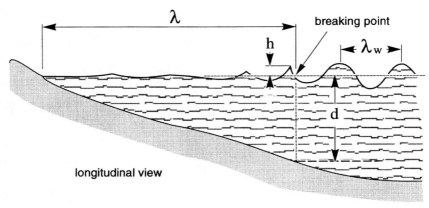

Figure 9-1
Geometric characteristics of waves near shore
For definition of symbols see table below

The slope φ—using the symbols of the figure—is given by

$$\tan \varphi = \frac{d}{\lambda} \qquad (a)$$

According to Sayao (Ref. 92, p. 242), the following are the relevant variables for this angle:

Variable	Symbol	Dimension	Remark
water depth at breaking point	d	m	see Fig. 9-1
water density	ρ	$m^{-3} \cdot kg$	assumed constant
gravitational acceleration	g	$m \cdot s^{-2}$	
water dynamic viscosity	μ	$m^{-1} \cdot kg \cdot s^{-1}$	assumed constant
breaker height	h	m	see Fig. 9-1
wave period	T	s	time to travel λ_W distance; see Fig. 9-1
breaker-point's distance	λ	m	from shore; see Fig. 9-1
sediment diameter	D	m	no cohesion among particles
sediment density	ρ_S	$m^{-3} \cdot kg$	

We have nine variables and three dimensions. Therefore, since the rank of the dimensional matrix is also 3, we have 9 – 3 = 6 dimensionless variables, supplied by the Dimensional Set, as follows:

TRANSFORMATIONS

Dimensional Set 1

		d	μ	g	h	D	ρ_S	ρ	λ	T
B matrix	m	1	-1	1	1	1	-3	-3	1	0
	kg	0	1	0	0	0	1	1	0	0
	s	0	-1	-2	0	0	0	0	0	1
D₁ matrix	π_{11}	1	0	0	0	0	0	0	-1	0
	π_{12}	0	1	0	0	0	0	-1	-2	1
	π_{13}	0	0	1	0	0	0	0	-1	2
	π_{14}	0	0	0	1	0	0	0	-1	0
	π_{15}	0	0	0	0	1	0	0	-1	0
	π_{16}	0	0	0	0	0	1	-1	0	0

A matrix (right block); C₁ matrix (right block of D₁).

which yields

$$\pi_{11} = \frac{d}{\lambda}; \quad \pi_{12} = \frac{\mu \cdot T}{\rho \cdot \lambda^2}; \quad \pi_{13} = \frac{g \cdot T^2}{\lambda}; \quad \pi_{14} = \frac{h}{\lambda}; \quad \pi_{15} = \frac{D}{\lambda}; \quad \pi_{16} = \frac{\rho_S}{\rho} \quad (b)$$

and therefore we can write

$$\pi_{11} = \Psi_1\{\pi_{12}, \pi_{13}, \pi_{14}, \pi_{15}, \pi_{16}\} \quad (c)$$

or by (a) and (b)

$$\frac{d}{\lambda} = \Psi_1\left\{\frac{\mu \cdot T}{\rho \cdot \lambda^2}, \frac{g \cdot T^2}{\lambda}, \frac{h}{\lambda}, \frac{D}{\lambda}, \frac{\rho_S}{\rho}\right\} \quad (d)$$

where Ψ_1 designates a function to be determined by measurements, modeling, etc.

Now Sayao (op. cit.) gives

$$\frac{d}{\lambda} = \Psi_2\left\{\frac{\rho \cdot \lambda^2}{\mu \cdot T}, \frac{g \cdot T^2}{\lambda}, \frac{h}{d}, \frac{\lambda}{D}, \frac{\rho_S}{\rho}\right\} \quad (e)$$

from which the dimensionless variables are

$$\pi_{21} = \frac{d}{\lambda}; \quad \pi_{22} = \frac{\rho \cdot \lambda^2}{\mu \cdot T}; \quad \pi_{23} = \frac{g \cdot T^2}{\lambda}; \quad \pi_{24} = \frac{h}{d}; \quad \pi_{25} = \frac{\lambda}{D}; \quad \pi_{26} = \frac{\rho_S}{\rho} \quad (f)$$

Question: What is the relation between the sets of dimensionless variables (b) and (f)? To answer, all we have to do is construct the **D₂** matrix concordant with Sayao's dimensionless variables in (f), and proceed to determine the transformation matrix **T** by (9-40).

Accordingly, we construct Dimensional Set 2

Dimensional Set 2

		d	μ	g	h	D	ρ_S	ρ	λ	T
B matrix	m	1	-1	1	1	1	-3	-3	1	0
	kg	0	1	0	0	0	1	1	0	0
	s	0	-1	-2	0	0	0	0	0	1
D₂ matrix	π_{21}	1	0	0	0	0	0	0	-1	0
	π_{22}	0	-1	0	0	0	0	1	2	-1
	π_{23}	0	0	1	0	0	0	0	-1	2
	π_{24}	-1	0	0	1	0	0	0	0	0
	π_{25}	0	0	0	0	-1	0	0	1	0
	π_{26}	0	0	0	0	0	1	-1	0	0

A matrix (right block); C₂ matrix (right block of D₂).

Note that the **A** and **B** matrices are identical to that of Set 1, and—importantly—**D**$_2$ is no longer a unit matrix, as it was in Set 1.

By (9-40), and by the above given **D**$_1$ and **D**$_2$ matrices,

$$\boldsymbol{\tau} = \mathbf{D}_2 \cdot \mathbf{D}_1^{-1} = \mathbf{D}_2 \cdot \mathbf{I}^{-1} = \mathbf{D}_2 \tag{g}$$

i.e., the transformation matrix is simply **D**$_2$. Hence, we can write, by (9-39),

$$\begin{bmatrix} \ln \pi_{21} \\ \ln \pi_{22} \\ \ln \pi_{23} \\ \ln \pi_{24} \\ \ln \pi_{25} \\ \ln \pi_{26} \end{bmatrix} = \mathbf{D}_2 \cdot \begin{bmatrix} \ln \pi_{11} \\ \ln \pi_{12} \\ \ln \pi_{13} \\ \ln \pi_{14} \\ \ln \pi_{15} \\ \ln \pi_{16} \end{bmatrix} = \underbrace{\begin{bmatrix} 1 & 0 & 0 & 0 & 0 & 0 \\ 0 & -1 & 0 & 0 & 0 & 0 \\ 0 & 0 & 1 & 0 & 0 & 0 \\ -1 & 0 & 0 & 1 & 0 & 0 \\ 0 & 0 & 0 & 0 & -1 & 0 \\ 0 & 0 & 0 & 0 & 0 & 1 \end{bmatrix}}_{\boldsymbol{\tau}} \cdot \begin{bmatrix} \ln \pi_{11} \\ \ln \pi_{12} \\ \ln \pi_{13} \\ \ln \pi_{14} \\ \ln \pi_{15} \\ \ln \pi_{16} \end{bmatrix} \tag{h}$$

from which the desired relations are:

$$\pi_{21} = \pi_{11}; \quad \pi_{22} = \frac{1}{\pi_{12}}; \quad \pi_{23} = \pi_{13}; \quad \pi_{24} = \frac{\pi_{14}}{\pi_{11}}; \quad \pi_{25} = \frac{1}{\pi_{15}}; \quad \pi_{26} = \pi_{16} \tag{i}$$

The reader can easily verify that these relations do indeed conform to the defined constructions of dimensionless variables given in (b) and (f).

The reader might also wonder whether it would be *possible* to rearrange the variables in a different sequence yielding the desired Sayao group, but which is based on *unity* **D**. The answer is *no*. Our **D**$_2$ is a 6 × 6 matrix, hence if it is a unit matrix, then there must be *at least* 6 variables which appear only *once* in the group. The reason for this is that if a variable appears more than once, then it cannot be represented in a column of unit **D**, because in a unit matrix there is only one element in a column and that element must be a "1." For this reason, we have the additional requirement that each of the six variables' exponent be 1.

In our present case, we have the following statistics for the composition of Sayao's dimensionless variables:

physical variable	d	ρ	μ	g	h	T	λ	D	ρ_S
occurrence number	2	2	1	1	1	2	4	1	1
exponent(s)	1, –1	1, –1	–1	1	1	–1, 2	–1, 2, –1, 1	–1	1

As we can see, there are only five variables which appear only once, and only three of them, viz., g, h, and ρ_S have the exponent 1. Therefore it is impossible to arrive at the Sayao collection of dimensionless variables using a unity **D** matrix.

⇑

It has been mentioned repeatedly that the principal advantage of using a unity **D** matrix is that it provides the simplest set of dimensionless variables, i.e., a set in which—in general—the physical variables possess the smallest exponents. The following example illustrates this important advantage.

Example 9-25

Suppose we have the following two dimensionless variables:

$$\left. \begin{array}{l} \pi_{11} = V_1^6 \cdot V_2^4 \cdot V_3^{28} \cdot V_4^{14} \cdot V_5^{22} \\ \pi_{12} = V_1^1 \cdot V_2^9 \cdot V_3^{38} \cdot V_4^{19} \cdot V_5^{12} \end{array} \right\} \tag{a}$$

where $V_1, V_2, \ldots$ are some physical variables. We see that the exponents of these variables are *suspiciously large;* the *sums* of these exponents—used as an arbitrary measure—in the two groups are 74 and 79, respectively. *Question:* Could we reformulate these variables to get some simpler expressions for the two dimensionless variables?

The **D** and **C** matrices for (a) are obviously

$$\mathbf{D}_1 = \begin{bmatrix} 6 & 4 \\ 1 & 9 \end{bmatrix}; \quad \mathbf{C}_1 = \begin{bmatrix} 28 & 14 & 22 \\ 38 & 19 & 12 \end{bmatrix} \quad (b)$$

Now we select $\mathbf{D}_2 = \mathbf{I}$. Hence, by (9-36),

$$\mathbf{C}_2 = \mathbf{D}_2 \cdot \mathbf{D}_1^{-1} \cdot \mathbf{C}_1 = \mathbf{D}_1^{-1} \cdot \mathbf{C}_1 = \begin{bmatrix} 6 & 4 \\ 1 & 9 \end{bmatrix}^{-1} \cdot \begin{bmatrix} 28 & 14 & 22 \\ 38 & 19 & 12 \end{bmatrix} = \begin{bmatrix} 2 & 1 & 3 \\ 4 & 2 & 1 \end{bmatrix} \quad (c)$$

Therefore

$$\begin{aligned} \pi_{21} &= V_1^1 \cdot V_2^0 \cdot V_3^2 \cdot V_4^1 \cdot V_5^3 \\ \pi_{22} &= V_1^0 \cdot V_2^1 \cdot V_3^4 \cdot V_4^2 \cdot V_5^1 \end{aligned} \quad (d)$$

This collection is *significantly* simpler than the set in (a)—as the sums of the exponents are only 7 (instead of 74) and 8 (instead of 79), respectively.

The transformation matrix, by (9-40) is now

$$\mathbf{T} = \mathbf{D}_2 \cdot \mathbf{D}_1^{-1} = \frac{1}{50} \cdot \begin{bmatrix} 9 & -4 \\ -1 & 6 \end{bmatrix} \quad (e)$$

hence, by (9-39),

$$\begin{bmatrix} \ln \pi_{21} \\ \ln \pi_{22} \end{bmatrix} = \mathbf{T} \cdot \begin{bmatrix} \ln \pi_{11} \\ \ln \pi_{12} \end{bmatrix} = \frac{1}{50} \cdot \begin{bmatrix} 9 & -4 \\ -1 & 6 \end{bmatrix} \begin{bmatrix} \ln \pi_{11} \\ \ln \pi_{12} \end{bmatrix} \quad (f)$$

from which the relation between sets (a) and (d) is

$$\pi_{21} = \sqrt[50]{\frac{\pi_{11}^9}{\pi_{12}^4}}; \quad \pi_{22} = \sqrt[50]{\frac{\pi_{12}^6}{\pi_{11}}} \quad (g)$$

or correspondingly, by (9-38) and (e)

$$\begin{bmatrix} \ln \pi_{11} \\ \ln \pi_{12} \end{bmatrix} = \mathbf{T}^{-1} \cdot \begin{bmatrix} \ln \pi_{21} \\ \ln \pi_{22} \end{bmatrix} = \begin{bmatrix} \frac{9}{50} & \frac{-4}{50} \\ \frac{-1}{50} & \frac{6}{50} \end{bmatrix}^{-1} \cdot \begin{bmatrix} \ln \pi_{21} \\ \ln \pi_{22} \end{bmatrix} = \begin{bmatrix} 6 & 4 \\ 1 & 9 \end{bmatrix} \cdot \begin{bmatrix} \ln \pi_{21} \\ \ln \pi_{22} \end{bmatrix} \quad (h)$$

from which

$$\pi_{11} = \pi_{21}^6 \cdot \pi_{22}^4; \quad \pi_{12} = \pi_{21} \cdot \pi_{22}^9 \quad (i)$$

⇑

9.3. TRANSFORMATION BETWEEN DIMENSIONAL SETS

We saw in the preceding theorems in Art. 9.1 and Art. 9.2 how the **C** matrix varies with different exchanges of variables within the **A** matrix, and within the **B** matrix, and between the **A** and **B** matrices. Now we will find the effects of these changes upon the resulting dimensionless variables themselves.

212 APPLIED DIMENSIONAL ANALYSIS AND MODELING

Since, by assumption, neither the variables nor the dimensions change between Dimensional Sets, therefore the ensuing sets of dimensionless variables will not be independent from each other. That is, any one set will always be expressible by another. To illustrate, assume we have two sets of dimensionless variables based on the same physical variables and dimensions

$$\text{set } 1 \Rightarrow \pi_{11}, \pi_{12}, \pi_{13}$$

$$\text{set } 2 \Rightarrow \pi_{21}, \pi_{22}, \pi_{23}$$

Then these sets are connected by relations

$$\pi_{21} = \pi_{11}^{\alpha_1} \cdot \pi_{12}^{\beta_1} \cdot \pi_{13}^{\gamma_1}; \quad \pi_{22} = \pi_{11}^{\alpha_2} \cdot \pi_{12}^{\beta_2} \cdot \pi_{13}^{\gamma_2}; \quad \pi_{23} = \pi_{11}^{\alpha_3} \cdot \pi_{12}^{\beta_3} \cdot \pi_{13}^{\gamma_3} \quad (9\text{-}49)$$

where the exponents are some real numbers, including zero. By taking the logarithm of both sides, these relations can be written symbolically in a compact matrix form

$$\begin{bmatrix} \ln \pi_{21} \\ \ln \pi_{22} \\ \ln \pi_{23} \end{bmatrix} = \mathbf{T} \cdot \begin{bmatrix} \ln \pi_{11} \\ \ln \pi_{12} \\ \ln \pi_{13} \end{bmatrix} \quad (9\text{-}50)$$

where, by definition

$$\mathbf{T} = \begin{bmatrix} \alpha_1 & \beta_1 & \gamma_1 \\ \alpha_2 & \beta_2 & \gamma_2 \\ \alpha_3 & \beta_3 & \gamma_3 \end{bmatrix} \quad (9\text{-}51)$$

is the *transformation matrix* between the two given sets of dimensionless variables. The attentive reader will notice that (9-50) is identical in appearance to (9-39). However, relation (9-50) is much more general, as will be seen presently. Our aim now is to determine **T** as in (9-50). Note that **T** cannot be singular, else the transformation as in (9-50) produces a set of dimensionless variables which are not independent.

By premultiplying both sides of (9-50) by the inverse of **T**, we immediately obtain

$$\begin{bmatrix} \ln \pi_{11} \\ \ln \pi_{12} \\ \ln \pi_{13} \end{bmatrix} = \mathbf{T}^{-1} \cdot \begin{bmatrix} \ln \pi_{21} \\ \ln \pi_{22} \\ \ln \pi_{23} \end{bmatrix} \quad (9\text{-}52)$$

Now we consider two Dimensional Sets of the same five variables $V_1, V_2, \ldots, V_5$, and two dimensions d_1, d_2. The two sets differ in their *sequence* of variables placed at the top, so that the construction of the two sets will also differ. We therefore have, for example,

Dimensional Set 1

	V_1	V_2	V_3	V_4	V_5
d_1	7	−42	−50	−8	9
d_2	−19	46	58	12	−5
π_{11}	1	0	0	2	1
π_{12}	0	1	0	−3	2
π_{13}	0	0	1	−4	2

Dimensional Set 2

	V_4	V_5	V_1	V_3	V_2
d_1	−8	9	7	−50	−42
d_2	12	−5	−19	58	46
π_{21}	1	0	0	−1	1
π_{22}	0	1	0	−1.5	2
π_{23}	0	0	1	3.5	−4

Note that the sequences of variables at the top of the sets are different. This difference, of course, results in different **A**, **B**, and **C** matrices in the sets.

We now define the *Shift Matrix* **S**—shown in Fig. 9-2—in which the rows and columns are defined by the sequential order of variables in Set 1 and Set 2, respectively. The elements of **S** are zeros, except those which are at the intersection of a row and a column marked by an *identical* variable. Thus, the element at the intersection of V_3 and V_4 is "0," since $V_3 \neq V_4$, but the element at the intersection of V_4 and V_4 is "1," because $V_4 = V_4$. By this construction, therefore, every row and every column of **S** has exactly one "1," and, as a consequence, the sumsquares of its elements in any row or column is 1 (by these characteristics **S** can also be defined as a *permutation* matrix). Moreover, the sum of products of the respective elements of any two columns or any two rows is zero. By these properties, **S** is an *orthogonal* matrix, so that its inverse equals its transpose, viz., $\mathbf{S}^{-1} = \mathbf{S}^T$. It is also evident that **S** must be a square matrix whose number of columns, and rows, is the number of variables N_V.

sequence of variables in Set 1 (top-to-bottom)

sequence of variables in Set 2 (left-to-right)

	V_4	V_5	V_1	V_3	V_2
V_1	0	0	1	0	0
V_2	0	0	0	0	1
V_3	0	0	0	1	0
V_4	1	0	0	0	0
V_5	0	1	0	0	0

Figure 9-2
Construction of the Shift Matrix
Elements having identical row and column headings are "1"; all other elements are zero

Now we partition the **S** matrix into four submatrices, as Fig. 9-3 illustrates. Submatrices $\mathbf{S}_1$, $\mathbf{S}_2$, $\mathbf{S}_3$, $\mathbf{S}_4$ are defined as follows:

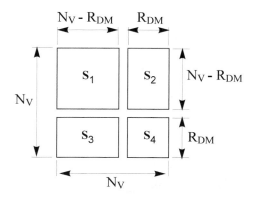

Figure 9-3
Partitioning the S matrix
N_V = number of variables; R_{DM} = rank of dimensional matrix

- Submatrix S_1 is at the *top left* corner of S; it has $N_V - R_{DM}$ rows and the same number of columns, where N_V is the number of variables, and R_{DM} is the rank of the dimensional matrix. Therefore, S_1 is always a square matrix, equal in size to D.
- Submatrix S_2 is at the *top right* corner of S; it has $N_V - R_{DM}$ rows and R_{DM} columns. Therefore, S_2 is a square matrix only if $N_V = 2 \cdot R_{DM}$, in which case matrices A and B in the Dimensional Set are of equal size.
- Submatrix S_3 is at the *bottom left* corner of S; it has R_{DM} rows and $N_V - R_{DM}$ columns. Therefore, S_3 is a square matrix only if $N_V = 2 \cdot R_{DM}$, in which case matrices A and B are of equal size.
- Submatrix S_4 is at the *bottom right* corner of S; it has R_{DM} rows and the same number of columns. Therefore, S_4 is always a square matrix.

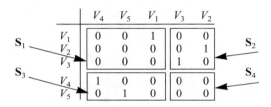

Figure 9-4
Partitioning of the Shift Matrix into submatrices in a case shown in Fig. 9-2

For example, in the numerical case of Fig. 9-2, the S matrix is partitioned as shown in Fig. 9-4. In this case, since $N_V = 5$, $R_{DM} = 2$, therefore, as seen above

S_1 is a 3 × 3 square matrix;
S_2 is a 3 × 2 matrix;
S_3 is a 2 × 3 matrix;
S_4 is a 2 × 2 square matrix.

Now we are ready to formulate a theorem defining the transformation matrix **T** as used in formulas (9-50) and (9-51).

Theorem 9-17. *The transformation matrix* **T**—*as defined in formulas (9-50) and (9-51)—between two dimensional sets of identical variables is*

$$\left. \begin{array}{l} \mathbf{T} = \mathbf{D}_2 \cdot (\mathbf{D}_1 \cdot \mathbf{S}_1 + \mathbf{C}_1 \cdot \mathbf{S}_3)^{-1} \\ \mathbf{T} = (\mathbf{D}_2 \cdot \mathbf{S}_1^T + \mathbf{C}_2 \cdot \mathbf{S}_2^T) \cdot \mathbf{D}_1^{-1} \end{array} \right\} \quad (9\text{-}53)$$

In these two formulas (both valid at all times):

S_1, S_2, S_3 are submatrices of the Shift Matrix S, as defined in Fig. 9-3;
D_1, D_2 are submatrices of the respective Dimensional Sets, as defined in Fig. 8-3;
C_1, C_2 are the submatrices, as determined by the Fundamental Formula (8-18).

TRANSFORMATIONS

If $\mathbf{D}_1 = \mathbf{D}_2 = \mathbf{I}$—which is usually the case—then relations (9-53) assume the simpler forms

$$\left. \begin{array}{l} \boldsymbol{\tau} = (\mathbf{S}_1 + \mathbf{C}_1 \cdot \mathbf{S}_3)^{-1} \\ \boldsymbol{\tau} = \mathbf{S}_1^T + \mathbf{C}_2 \cdot \mathbf{S}_2^T \end{array} \right\} \quad (9\text{-}54)$$

The proof of this important theorem is presented in Appendix 7; we now offer several examples of its use.

Example 9-26

We have, by the Dimensional Set 1 presented above,

$$\pi_{11} = V_1 \cdot V_4^2 \cdot V_5 \ ; \qquad \pi_{12} = V_2 \cdot V_4^{-3} \cdot V_5^2 \ ; \qquad \pi_{13} = V_3 \cdot V_4^{-4} \cdot V_5^2 \qquad (a)$$

and

$$\mathbf{C}_1 = \begin{bmatrix} 2 & 1 \\ -3 & 2 \\ -4 & 2 \end{bmatrix} ; \quad \mathbf{C}_2 = \begin{bmatrix} -1 & 1 \\ -1.5 & 2 \\ 3.5 & -4 \end{bmatrix} ; \quad \mathbf{D}_1 = \mathbf{D}_2 = \mathbf{I} \qquad (b)$$

Moreover, by Fig 9-4,

$$\mathbf{S}_1 = \begin{bmatrix} 0 & 0 & 1 \\ 0 & 0 & 0 \\ 0 & 0 & 0 \end{bmatrix} ; \quad \mathbf{S}_2 = \begin{bmatrix} 0 & 0 \\ 0 & 1 \\ 1 & 0 \end{bmatrix} ; \quad \mathbf{S}_3 = \begin{bmatrix} 1 & 0 & 0 \\ 0 & 1 & 0 \end{bmatrix} \qquad (c)$$

hence, by the first part of (9-54),

$$\boldsymbol{\tau} = (\mathbf{S}_1 + \mathbf{C}_1 \cdot \mathbf{S}_3)^{-1} = \frac{1}{2} \cdot \begin{bmatrix} 0 & 2 & -2 \\ 0 & 4 & -3 \\ 2 & -8 & 7 \end{bmatrix} \qquad (d)$$

and by the second part,

$$\boldsymbol{\tau} = (\mathbf{S}_1^T + \mathbf{C}_2 \cdot \mathbf{S}_2^T)^{-1} = \frac{1}{2} \cdot \begin{bmatrix} 0 & 2 & -2 \\ 0 & 4 & -3 \\ 2 & -8 & 7 \end{bmatrix}$$

which is identical to (d), as expected and required.

Thus, by (9-50),

$$\begin{bmatrix} \ln \pi_{21} \\ \ln \pi_{22} \\ \ln \pi_{23} \end{bmatrix} = \frac{1}{2} \cdot \begin{bmatrix} 0 & 2 & -2 \\ 0 & 4 & -3 \\ 2 & -8 & 7 \end{bmatrix} \cdot \begin{bmatrix} \ln \pi_{11} \\ \ln \pi_{12} \\ \ln \pi_{13} \end{bmatrix} \qquad (e)$$

which yields

$$\pi_{21} = \pi_{12} \cdot \pi_{13}^{-1} \ ; \qquad \pi_{22} = \pi_{12}^2 \cdot \pi_{13}^{-1.5} \ ; \qquad \pi_{23} = \pi_{11} \cdot \pi_{12}^{-4} \cdot \pi_{13}^{3.5} \qquad (f)$$

This result can be easily verified by direct enumeration. As an illustration, let us check π_{21}.

$$\underbrace{(V_1^0 \cdot V_2^1 \cdot V_3^0 \cdot V_4^{-3} \cdot V_5^2)^1}_{\pi_{12}} \cdot \underbrace{(V_1^0 \cdot V_2^0 \cdot V_3^1 \cdot V_4^{-4} \cdot V_5^2)^{-1}}_{\pi_{13}} = \underbrace{V_1^0 \cdot V_2^1 \cdot V_3^{-1} \cdot V_4^1 \cdot V_5^0}_{\pi_{21}}$$

which checks with π_{21} found in Dimensional Set 2.

216 APPLIED DIMENSIONAL ANALYSIS AND MODELING

If we now wish to do the *reverse* process, i.e., to express π_{11}, π_{12}, and π_{13} in terms of π_{21}, π_{22}, and π_{23}, then we use (9-52). Thus, in view of (d),

$$\begin{bmatrix} \ln \pi_{11} \\ \ln \pi_{12} \\ \ln \pi_{13} \end{bmatrix} = \mathbf{T}^{-1} \cdot \begin{bmatrix} \ln \pi_{21} \\ \ln \pi_{22} \\ \ln \pi_{23} \end{bmatrix} = \begin{bmatrix} 2 & 1 & 1 \\ -3 & 2 & 0 \\ -4 & 2 & 0 \end{bmatrix} \cdot \begin{bmatrix} \ln \pi_{21} \\ \ln \pi_{22} \\ \ln \pi_{23} \end{bmatrix} \quad (g)$$

by which

$$\pi_{11} = \pi_{21}^{2} \cdot \pi_{22}^{1} \cdot \pi_{23}^{1} \; ; \qquad \pi_{12} = \pi_{21}^{-3} \cdot \pi_{22}^{2} \cdot \pi_{23}^{0} \; ; \qquad \pi_{13} = \pi_{21}^{-4} \cdot \pi_{22}^{2} \cdot \pi_{23}^{0} \quad (h)$$

These latter results can be also easily verified by the method just described.

⇑

The next example deals with a case in which the **D** matrices are not unity.

Example 9-27

Given the following two Dimensional Sets:

Dimensional Set 1

	V_1	V_2	V_3	V_4	V_5
d_1	-1	-3.5	2	1	-1
d_2	1.2	1.6	1	-0.4	2
d_3	2.5	3.25	2	-1	4
π_{11}	1	2	2	1	-3
π_{12}	3	-2	-2	1	1

Dimensional Set 2

	V_3	V_1	V_4	V_2	V_5
d_1	2	-1	1	-3.5	-1
d_2	1	1.2	-0.4	1.6	2
d_3	2	2.5	-1	3.25	4
π_{21}	5	4	3.25	5	-8.25
π_{22}	3	2	1.75	3	-4.75

yielding two sets of dimensionless variables

$$\left. \begin{array}{l} \pi_{11} = V_1^1 \cdot V_2^2 \cdot V_3^2 \cdot V_4^1 \cdot V_5^{-3} \\ \pi_{12} = V_1^3 \cdot V_2^{-2} \cdot V_3^{-2} \cdot V_4^1 \cdot V_5^1 \end{array} \right\} \quad (a)$$

$$\left. \begin{array}{l} \pi_{21} = V_1^4 \cdot V_2^5 \cdot V_3^5 \cdot V_4^{3.25} \cdot V_5^{-8.25} \\ \pi_{22} = V_1^2 \cdot V_2^3 \cdot V_3^3 \cdot V_4^{1.75} \cdot V_5^{-4.75} \end{array} \right\} \quad (b)$$

For the relation between sets (a) and (b), we compose the Shift Matrix **S** (Fig. 9-2) and its partitioning (Fig. 9-3).

		V_3	V_1	V_4	V_2	V_5	
S_1	V_1	0	1	0	0	0	S_2
	V_2	0	0	0	1	0	
S_3	V_3	1	0	0	0	0	S_4
	V_4	0	0	1	0	0	
	V_5	0	0	0	0	1	

Hence we have

$$\mathbf{S}_1 = \begin{bmatrix} 0 & 1 \\ 0 & 0 \end{bmatrix} ; \qquad \mathbf{S}_3 = \begin{bmatrix} 1 & 0 \\ 0 & 0 \\ 0 & 0 \end{bmatrix} \quad (c)$$

and by the Dimensional Sets

$$\mathbf{D}_1 = \begin{bmatrix} 1 & 2 \\ 3 & -2 \end{bmatrix}; \quad \mathbf{D}_2 = \begin{bmatrix} 5 & 4 \\ 3 & 2 \end{bmatrix}; \quad \mathbf{C}_1 = \begin{bmatrix} 2 & 1 & -3 \\ -2 & 1 & 1 \end{bmatrix} \quad (d)$$

Therefore, the transformation matrix, by (9-53), (c) and (d), is

$$\mathbf{T} = \mathbf{D}_2 \cdot (\mathbf{D}_1 \cdot \mathbf{S}_1 + \mathbf{C}_1 \cdot \mathbf{S}_3)^{-1} = \frac{1}{8} \begin{bmatrix} 23 & 3 \\ 13 & 1 \end{bmatrix} \quad (e)$$

Thus, by (9-50),

$$\begin{bmatrix} \ln \pi_{21} \\ \ln \pi_{22} \end{bmatrix} = \mathbf{T} \cdot \begin{bmatrix} \ln \pi_{11} \\ \ln \pi_{12} \end{bmatrix} = \frac{1}{8} \begin{bmatrix} 23 & 3 \\ 13 & 1 \end{bmatrix} \cdot \begin{bmatrix} \ln \pi_{11} \\ \ln \pi_{12} \end{bmatrix} \quad (f)$$

from which we obtain

$$\pi_{21} = (\pi_{11}^{23} \cdot \pi_{12}^{3})^{1/8}; \quad \pi_{22} = (\pi_{11}^{13} \cdot \pi_{12}^{1})^{1/8} \quad (g)$$

Alternatively, since by (e)

$$\mathbf{T}^{-1} = \begin{bmatrix} \frac{23}{8} & \frac{3}{8} \\ \frac{13}{8} & \frac{1}{8} \end{bmatrix}^{-1} = \frac{1}{2} \begin{bmatrix} -1 & 3 \\ 13 & -23 \end{bmatrix} \quad (h)$$

therefore, by (9-52)

$$\begin{bmatrix} \ln \pi_{11} \\ \ln \pi_{12} \end{bmatrix} = \mathbf{T}^{-1} \cdot \begin{bmatrix} \ln \pi_{21} \\ \ln \pi_{22} \end{bmatrix} = \frac{1}{2} \begin{bmatrix} -1 & 3 \\ 13 & -23 \end{bmatrix} \cdot \begin{bmatrix} \ln \pi_{21} \\ \ln \pi_{22} \end{bmatrix} \quad (i)$$

yielding

$$\pi_{11} = \sqrt{\pi_{21}^{-1} \cdot \pi_{22}^{3}}; \quad \pi_{12} = \sqrt{\pi_{21}^{13} \cdot \pi_{22}^{-23}} \quad (j)$$

The reader can easily verify—and is urged to do so—that (g) and (j) are indeed compatible for all values of their ingredients.

⇑

Let us now revisit the case where in a Dimensional Set matrices **A** and **B** are interchanged. In Theorem 9-2 (Art. 9.1) we proved relations

$$\mathbf{T} = \mathbf{C}_2 \cdot \mathbf{D}_1^{-1} \quad \text{repeated (9-11a)}$$

$$\mathbf{T} = \mathbf{D}_2 \cdot \mathbf{C}_1^{-1} \quad \text{repeated (9-11b)}$$

by direct enumeration of the constituents of the Dimensional Sets. There the proof was reasonably straightforward, but a bit long. Now, based on Theorem 9-17 and relations (9-53), we offer a much more rapid proof.

An Alternative (and More Elegant) Proof of Theorem 9-2. If matrices **A** and **B** in a Dimensional Set are interchanged, then, by the construction of the Shift Matrix **S** (Fig. 9-2 and Fig. 9-3), the defined submatrices of **S** will be

$\mathbf{S}_1 = \mathbf{0}$ (null matrix); $\mathbf{S}_2 = \mathbf{I}$ (unit matrix); $\mathbf{S}_3 = \mathbf{I}$ (unit matrix); $\mathbf{S}_4 = \mathbf{0}$ (null matrix)

Hence the first and second parts of (9-53) yield, in order, $\mathbf{T} = \mathbf{D}_2 \cdot \mathbf{C}_1^{-1}$ and $\mathbf{T} = \mathbf{C}_2 \cdot \mathbf{D}_1^{-1}$—which are identical to (9-11a) and (9-11b), respectively, thereby proving Theorem 9-2. As observed, this proof is significantly more compact than what was presented previously.

We see from the last relation, that if $\mathbf{D}_1 = \mathbf{I}$, then the transformation matrix $\mathbf{T}$ is identical to $\mathbf{C}_2$. This means that to find $\mathbf{T}$, we only have to consider $\mathbf{C}_2$ as the transformation matrix. The following example highlights this very convenient property.

Example 9-28

Consider the following Dimensional Sets 1 and 2:

Dimensional Set 1

	V_1	V_2	V_3	V_4
d_1	1	-2	2	1
d_2	2	-5	1	0
π_{11}	1	0	-2	3
π_{12}	0	1	5	-8

Dimensional Set 2

	V_3	V_4	V_1	V_2
d_1	2	1	1	-2
d_2	1	0	2	-5
π_{21}	1	0	-8	-3
π_{22}	0	1	-5	-2

We note that $\mathbf{A}$ and $\mathbf{B}$ are interchanged between the two sets, i.e., $\mathbf{A}_2 = \mathbf{B}_1$ and $\mathbf{B}_2 = \mathbf{A}_1$. From the above sets

$$\left. \begin{array}{ll} \pi_{11} = V_1^1 \cdot V_2^0 \cdot V_3^{-2} \cdot V_4^3 ; & \pi_{21} = V_1^{-8} \cdot V_2^{-3} \cdot V_3^1 \cdot V_4^0 \\ \pi_{12} = V_1^0 \cdot V_2^1 \cdot V_3^5 \cdot V_4^{-8} ; & \pi_{22} = V_1^{-5} \cdot V_2^{-2} \cdot V_3^0 \cdot V_4^1 \end{array} \right\} \quad \text{(a)}$$

How are these two groups of dimensionless variables related? By (9-11a) and (9-11b), the transformation matrix and its inverse are

$$\mathbf{T} = \mathbf{C}_2 = \begin{bmatrix} -8 & -3 \\ -5 & -2 \end{bmatrix}; \quad \mathbf{T}^{-1} = \mathbf{C}_1 = \begin{bmatrix} -2 & 3 \\ 5 & -8 \end{bmatrix}$$

Hence, by (9-50) and (9-52),

$$\left. \begin{array}{ll} \pi_{21} = \pi_{11}^{-8} \cdot \pi_{12}^{-3} ; & \pi_{11} = \pi_{21}^{-2} \cdot \pi_{22}^{3} \\ \pi_{22} = \pi_{11}^{-5} \cdot \pi_{12}^{-2} ; & \pi_{12} = \pi_{21}^{5} \cdot \pi_{22}^{-8} \end{array} \right. \quad \text{(b)}$$

The reader can easily verify these relations by assigning any numerical values to $V_1, \ldots, V_4$, and then calculating $\pi_{11}, \pi_{12}, \pi_{21}, \pi_{22}$ by (a) and substituting them into (b).

⇑

Encouraged by the above development, we now inquire whether the proof of Theorem 9-16 can be "enhanced" (this theorem deals with Dimensional Sets differing only in their $\mathbf{D}$ matrices).

An Alternative (and More Elegant) Proof of Theorem 9-16. If Dimensional Sets 1 and 2 differ only in their $\mathbf{D}$ and $\mathbf{C}$ matrices—i.e., their respective dimensional matrices are identical—then by the construction of the Shift Matrix $\mathbf{S}$ (Fig. 9-2 and Fig. 9-3) the submatrices of $\mathbf{S}$ so defined will be:

$\mathbf{S}_1 = \mathbf{I}$ (unit matrix); $\mathbf{S}_2 = \mathbf{0}$ (null matrix); $\mathbf{S}_3 = \mathbf{0}$ (null matrix); $\mathbf{S}_4 = \mathbf{I}$ (unit matrix)

Hence *both* relations of (9-53) will yield identically

$$\mathbf{T} = \mathbf{D}_2 \cdot (\mathbf{D}_1 \cdot \mathbf{S}_1 + \mathbf{C}_1 \cdot \mathbf{S}_3)^{-1} = \mathbf{D}_2 \cdot \mathbf{D}_1^{-1}$$

$$\mathbf{T} = (\mathbf{D}_2 \cdot \mathbf{S}_1^T + \mathbf{C}_2 \cdot \mathbf{S}_2^T) \cdot \mathbf{D}_1^{-1} = \mathbf{D}_2 \cdot \mathbf{D}_1^{-1}$$

confirming (9-40) and thereby proving Theorem 9-16. Again, we observe—with some satisfaction—the brevity and simplicity of the proof.

We now state and prove a corollary of Theorem 9-17.

Corollary of Theorem 9-17. *If in a Dimensional Set (in which $\mathbf{D} = \mathbf{I}$) matrices $\mathbf{A}$ and $\mathbf{B}$ are interchanged, and then the columns (i.e., variables) of the new $\mathbf{A}$ and $\mathbf{B}$ matrices are rearranged in any manner whatever—but without intermixing the columns $\mathbf{A}$ and $\mathbf{B}$—then the transformation matrix is*

$$\mathbf{T} = \mathbf{C}_2 \cdot \mathbf{S}_2^T \qquad (9\text{-}54\text{a})$$

or, equivalently

$$\mathbf{T} = \mathbf{S}_3^T \cdot \mathbf{C}_1^{-1} \qquad (9\text{-}54\text{b})$$

where submatrices $\mathbf{S}_2$, $\mathbf{S}_3$, $\mathbf{C}_1$, and $\mathbf{C}_2$ are as defined in Theorem 9-17.

Proof. As a result of the indicated interchange of matrices $\mathbf{A}$ and $\mathbf{B}$ and their columns, and by the construction (definition) of the Shift Matrix $\mathbf{S}$ (Fig. 9-2 and Fig. 9-3), the submatrices of $\mathbf{S}$ will be:

$\mathbf{S}_1 = \mathbf{0}$ (null matrix),
$\mathbf{S}_2 =$ orthogonal matrix in which every row and every column has exactly one "1";
$\mathbf{S}_3 =$ orthogonal matrix in which every row and every column has exactly one "1";
$\mathbf{S}_4 = \mathbf{0}$ (null matrix).

Accordingly, the substitution of $\mathbf{S}_1 = \mathbf{0}$, $\mathbf{D}_1 = \mathbf{D}_2 = \mathbf{I}$ into the first of (9-53) will give

$$\mathbf{T} = (\mathbf{C}_1 \cdot \mathbf{S}_3)^{-1} = \mathbf{S}_3^{-1} \cdot \mathbf{C}_1^{-1} = \mathbf{S}_3^T \cdot \mathbf{C}_1^{-1} \qquad (a)$$

since $\mathbf{S}_3^{-1} = \mathbf{S}_3^T$ ($\mathbf{S}_3$ being orthogonal). Similar substitution into the second of (9-53) results in

$$\mathbf{T} = \mathbf{C}_2 \cdot \mathbf{S}_2^T \qquad (b)$$

But (a) and (b) are identical to (9-54b) and (9-54a), respectively, therefore the corollary is proved. The example below shows how this corollary is put to good use.

Example 9-29

In the Dimensional Sets 1 and 2 shown below, observe that matrices $\mathbf{A}$ and $\mathbf{B}$ are interchanged, and that the two columns (variables) of each matrix are then switched. Thus, in this situation, the Corollary of Theorem 9-17 [and hence relations (9-54a) and (9-54b)] apply.

220 APPLIED DIMENSIONAL ANALYSIS AND MODELING

Dimensional Set 1

	V_1	V_2	V_3	V_4
d_1	1	-2	2	1
d_2	2	-5	1	0
π_{11}	1	0	-2	3
π_{12}	0	1	5	-8

Dimensional Set 2

	V_4	V_3	V_2	V_1
d_1	1	2	-2	1
d_2	0	1	-5	2
π_{21}	1	0	-2	-5
π_{22}	0	1	-3	-8

To determine $\mathbf{S}_2$ and $\mathbf{S}_3$ we construct the shift matrix $\mathbf{S}$

		V_4	V_3	V_2	V_1	
$\mathbf{S}_1$ →	V_1	0	0	0	1	← $\mathbf{S}_2$
	V_2	0	0	1	0	
$\mathbf{S}_3$ →	V_3	0	1	0	0	← $\mathbf{S}_4$
	V_4	1	0	0	0	

We see that $\mathbf{S}_1$ and $\mathbf{S}_4$ are indeed null matrices, and that $\mathbf{S}_2$, $\mathbf{S}_3$ are orthogonal. By (9-54a) and the above data

$$\mathbf{T} = \mathbf{C}_2 \cdot \mathbf{S}_2^T = \begin{bmatrix} -2 & -5 \\ -3 & -8 \end{bmatrix} \cdot \begin{bmatrix} 0 & 1 \\ 1 & 0 \end{bmatrix}^T = \begin{bmatrix} -5 & -2 \\ -8 & -3 \end{bmatrix} \tag{a}$$

and by (9-54b),

$$\mathbf{T} = \mathbf{S}_3^T \cdot \mathbf{C}_1^{-1} = \begin{bmatrix} 0 & 1 \\ 1 & 0 \end{bmatrix} \cdot \begin{bmatrix} -2 & 3 \\ 5 & -8 \end{bmatrix}^{-1} = \begin{bmatrix} -5 & -2 \\ -8 & -3 \end{bmatrix} \tag{b}$$

We see that (a) and (b) are identical, as they should be. Hence we can write, by (9-50) and (9-52),

$$\begin{bmatrix} \ln \pi_{21} \\ \ln \pi_{22} \end{bmatrix} = \mathbf{T} \cdot \begin{bmatrix} \ln \pi_{11} \\ \ln \pi_{12} \end{bmatrix} \tag{c}$$

by which—considering (a) or (b)—the relations between the two sets of dimensionless variables are

$$\pi_{21} = \pi_{11}^{-5} \cdot \pi_{12}^{-2} ; \qquad \pi_{22} = \pi_{11}^{-8} \cdot \pi_{12}^{-3} \tag{d}$$

or, correspondingly, by (a) or (b),

$$\begin{bmatrix} \ln \pi_{11} \\ \ln \pi_{12} \end{bmatrix} = \mathbf{T}^{-1} \cdot \begin{bmatrix} \ln \pi_{21} \\ \ln \pi_{22} \end{bmatrix} = \begin{bmatrix} 3 & -2 \\ -8 & 5 \end{bmatrix} \cdot \begin{bmatrix} \ln \pi_{21} \\ \ln \pi_{22} \end{bmatrix}$$

yielding

$$\pi_{11} = \pi_{21}^{3} \cdot \pi_{22}^{-2} ; \qquad \pi_{12} = \pi_{21}^{-8} \cdot \pi_{22}^{5} \tag{e}$$

The persistent reader may wish to verify that (d) and (e) do confirm each other.

⇑

Theorem 9-18. *If in Dimensional Set 1, one column of the $\mathbf{C}_1$ matrix under variable V_A is all zeros, except for one nonzero element* q *in its* ith *row, and if variable*

TRANSFORMATIONS

V_A is exchanged with the ith variable V_B in the **B** matrix, then the transformation matrix (which is of size $N_P \times N_P$) is

$$\mathbf{T} = \begin{bmatrix} 1 & 0 & \cdots & & \cdots & 0 \\ 0 & 1 & \cdots & & \cdots & 0 \\ \vdots & & & & & \\ 0 & 0 & \cdots & \left(\dfrac{1}{q}\right) & \cdots & 0 \\ \vdots & & & & & \vdots \\ 0 & 0 & \cdots & 0 & \cdots & 1 \end{bmatrix} \quad \text{element in the } i\text{th row, } i\text{th column} \tag{9-55}$$

where N_P is the number of dimensionless variables.

It is evident by (9-55) that $\mathbf{T}$ is a diagonal matrix whose elements in the main diagonal are all "1," except in the ith row (and column) where it is $1/q$. Accordingly, we can write Dimensional Set 2, which now has the following interchange of variables:

$$\left. \begin{array}{rcl} \pi_{21} &=& \pi_{11} \\ &\vdots& \\ \pi_{2i} &=& \pi_{1i}^{1/q} \\ &\vdots& \\ \pi_{2N_P} &=& \pi_{1N_P} \end{array} \right\} \tag{9-56}$$

We see that really only one dimensionless group changed, and even this change is "cosmetic" rather than "essential"; for a dimensionless variable raised to any power, in effect, remains the same.

Proof. We consider $N_V = 6$ dimensionless variables and $N_d = 3$ dimensions for the proof, but, of course, the process outlined can involve any Dimensional Set of legitimate size and composition. Thus, in Set 1 we see that the second column in the $\mathbf{C}_1$ matrix is all zeros, except in its first row ($i = 1$), where the element is the nonzero q. We also notice that q appears under variable V_5. Note that in our notation $V_A = V_5$ and $V_B = V_1$.

Dimensional Set 1

	V_1	V_2	V_3	V_4	V_5	V_6
d_1						
d_2		$\mathbf{B}_1$			$\mathbf{A}_1$	
d_3						
π_{11}	1	0	0	c_{11}	q	c_{13}
π_{12}	0	1	0	c_{21}	0	c_{23}
π_{13}	0	0	1	c_{31}	0	c_{33}

Therefore, by Set 1 we have

$$\left. \begin{array}{l} \pi_{11} = V_1^1 \cdot V_4^{c_{11}} \cdot V_5^q \cdot V_6^{c_{13}} = V_B \cdot V_4^{c_{11}} \cdot V_A^q \cdot V_6^{c_{13}} \\ \pi_{12} = V_2 \cdot V_4^{c_{21}} \cdot V_6^{c_{23}} \\ \pi_{13} = V_3 \cdot V_4^{c_{31}} \cdot V_6^{c_{33}} \end{array} \right\} \tag{9-57}$$

Now let us interchange variables V_A and V_B. We immediately notice that this exchange affects only π_{11}, since neither π_{12} nor π_{13} has ingredients V_A or V_B. Hence we only concentrate on π_{11}. By the exchange and by (9-57), we write

$$\pi_{11} = V_A^q \cdot V_4^{c_{11}} \cdot V_B \cdot V_6^{c_{13}} \tag{9-58}$$

since the product of scalars is commutative (sequential order is unimportant). Next, we raise π_{11} to the $1/q$th power. This will yield

$$\pi_{11}^{1/q} = V_A \cdot V_4^{c_{11}/q} \cdot V_B^{1/q} \cdot V_6^{c_{13}/q} \tag{9-59}$$

Now we construct Set 2 in which the $V_A - V_B$ interchange (i.e., the $V_1 - V_5$ interchange) is effected.

Dimensional Set 2

	V_5	V_2	V_3	V_4	V_1	V_6
d_1						
d_2		$\mathbf{B}_2$			$\mathbf{A}_2$	
d_3						
π_{21}	1	0	0	$\dfrac{c_{11}}{q}$	$\dfrac{1}{q}$	$\dfrac{c_{13}}{q}$
π_{22}	0	1	0	c_{21}	0	c_{23}
π_{23}	0	0	1	c_{31}	0	c_{33}

from which

$$\left.\begin{array}{l} \pi_{21} = V_A \cdot V_4^{c_{11}/q} \cdot V_B^{1/q} \cdot V_6^{c_{13}/q} \\ \pi_{22} = V_2 \cdot V_4^{c_{21}} \cdot V_6^{c_{23}} \\ \pi_{23} = V_3 \cdot V_4^{c_{31}} \cdot V_6^{c_{33}} \end{array}\right\} \tag{9-60}$$

Comparing the first of these with (9-59), we obtain at once

$$\pi_{21} = \pi_{11}^{1/q} \tag{9-61}$$

Further comparison of the second and third in (9-60) with the corresponding expressions in (9-57) yields

$$\pi_{22} = \pi_{12} \;; \qquad \pi_{23} = \pi_{13} \tag{9-62}$$

Hence, by (9-61) and (9-62), we can write

$$\begin{bmatrix} \ln \pi_{21} \\ \ln \pi_{22} \\ \ln \pi_{23} \end{bmatrix} = \begin{bmatrix} \dfrac{1}{q} & 0 & 0 \\ 0 & 1 & 0 \\ 0 & 0 & 1 \end{bmatrix} \cdot \begin{bmatrix} \ln \pi_{11} \\ \ln \pi_{12} \\ \ln \pi_{13} \end{bmatrix} = \mathbf{T} \cdot \begin{bmatrix} \ln \pi_{11} \\ \ln \pi_{12} \\ \ln \pi_{13} \end{bmatrix} \tag{9-63}$$

As seen, with $i = 1$ as used in this proof, the transformation matrix $\mathbf{T}$ in (9-63) is identical to relation (9-55). This proves the theorem.

Example 9-30

Consider Dimensional Set 1

Dimensional Set 1

	V_1	V_2	V_3	V_4	V_5	V_6	V_7
d_1	8	3	−8	1	1	2	−2
d_2	−3	−1	−3	0	−1	2	1
d_3	−7	−3	−5	−2	1	2	1
π_{11}	1	0	0	0	2	0	5
π_{12}	0	1	0	0	1	0	2
π_{13}	0	0	1	0	1	2.5	−1
π_{14}	0	0	0	1	1	0	1

C_1 matrix

from which

$$\pi_{11} = V_1 \cdot V_5^2 \cdot V_7^5; \quad \pi_{12} = V_2 \cdot V_5 \cdot V_7^2; \quad \pi_{13} = V_3 \cdot V_5 \cdot V_6^{2.5} \cdot V_7^{-1}; \quad \pi_{14} = V_4 \cdot V_5 \cdot V_7 \quad \text{(a)}$$

We now notice that the second column in C_1 under variable V_6 is all zeros, except in the third row ($i = 3$), where the element is $q = 2.5$. Hence, if we exchange variables V_6 and V_3, we will obtain a new set of dimensionless variables $\pi_{21}, \pi_{22}, \pi_{23}, \pi_{24}$. By Theorem 9-18, the transformation matrix between these two sets is

$$\mathbf{T} = \begin{bmatrix} 1 & 0 & 0 & 0 \\ 0 & 1 & 0 & 0 \\ 0 & 0 & \frac{1}{2.5} & 0 \\ 0 & 0 & 0 & 1 \end{bmatrix} \quad \text{(b)}$$

since in the main diagonal all elements are 1, except in the third row (or column), where the element is $1/q = 1/2.5 = 0.4$. Hence we write, in view of (9-50),

$$\begin{bmatrix} \ln \pi_{21} \\ \ln \pi_{22} \\ \ln \pi_{23} \\ \ln \pi_{24} \end{bmatrix} = \begin{bmatrix} 1 & 0 & 0 & 0 \\ 0 & 1 & 0 & 0 \\ 0 & 0 & 0.4 & 0 \\ 0 & 0 & 0 & 1 \end{bmatrix} \cdot \begin{bmatrix} \ln \pi_{11} \\ \ln \pi_{12} \\ \ln \pi_{13} \\ \ln \pi_{14} \end{bmatrix} \quad \text{(c)}$$

which yields

$$\pi_{21} = \pi_{11}; \quad \pi_{22} = \pi_{12}; \quad \pi_{23} = \pi_{13}^{0.4}; \quad \pi_{24} = \pi_{14} \quad \text{(d)}$$

Now let us verify this by direct enumeration. Accordingly, we construct Dimensional Set 2 incorporating the above indicated $V_3 - V_6$ exchange of variables

Dimensional Set 2

	V_1	V_2	V_3	V_4	V_5	V_6	V_7
d_1	8	3	2	1	1	−8	−2
d_2	−3	−1	2	0	−1	−3	1
d_3	−7	−3	2	−2	1	−5	1
π_{21}	1	0	0	0	2	0	5
π_{22}	0	1	0	0	1	0	2
π_{23}	0	0	1	0	0.4	0.4	−0.4
π_{24}	0	0	0	1	1	0	1

from which

$$\pi_{21} = V_1 \cdot V_5^2 \cdot V_7^5 = \pi_{11} \; ; \quad \pi_{22} = V_2 \cdot V_5 \cdot V_7^2 = \pi_{12}$$
$$\pi_{23} = V_3^{0.4} \cdot V_5^{0.4} \cdot V_6 \cdot V_7^{-0.4} = (V_3 \cdot V_5 \cdot V_6^{2.5} \cdot V_7^{-1})^{0.4} = \pi_{13}^{0.4} \; ; \quad \pi_{24} = V_4 \cdot V_5 \cdot V_7 = \pi_{14}$$ (e)

Comparing this set with (a) and (d), we see that there is total agreement.

Finally, let us confirm transformation matrix **T** by the Shift Matrix method. From the construction defined in Fig. 9-2 and Fig. 9-3, as well as by the two Dimensional Sets in this example, the Shift Matrix and its submatrices are

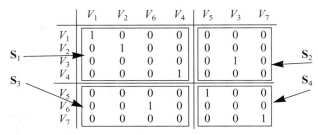

and therefore, by (9-53)

$$\mathbf{T} = (\mathbf{S}_1 + \mathbf{C}_1 \cdot \mathbf{S}_3)^{-1} = \begin{bmatrix} 1 & 0 & 0 & 0 \\ 0 & 1 & 0 & 0 \\ 0 & 0 & 0.4 & 0 \\ 0 & 0 & 0 & 1 \end{bmatrix}$$

which is identical to (b), as expected and required.

⇑

It is emphasized that by using the transformation matrix **T**, given in formulas (9-53), we can determine the transformed $\mathbf{C}_2$ matrix directly, without any reference to the dimensional composition of the involved variables. In this respect, the following theorem is applicable:

Theorem 9-19. *For Dimensional Sets 1, 2—in general—*

$$\mathbf{C}_2 = \mathbf{T} \cdot (\mathbf{D}_1 \cdot \mathbf{S}_2 + \mathbf{C}_1 \cdot \mathbf{S}_4)$$ (9-64)

or, if $\mathbf{D}_1 = \mathbf{D}_2 = \mathbf{I}$—*which is most often the case—*

$$\mathbf{C}_2 = \mathbf{T} \cdot (\mathbf{S}_2 + \mathbf{C}_1 \cdot \mathbf{S}_4) = (\mathbf{S}_1 + \mathbf{C}_1 \cdot \mathbf{S}_3)^{-1} \cdot (\mathbf{S}_2 + \mathbf{C}_1 \cdot \mathbf{S}_4)$$ (9-65)

In these expressions **T** is the transformation matrix defined in (9-53), and $\mathbf{S}_1$, $\mathbf{S}_2$, $\mathbf{S}_3$, and $\mathbf{S}_4$ are the submatrices of the *Shift Matrix* defined in Fig. 9-3. The proof of this theorem is in Appendix 7.

Example 9-31

Given particular **D** and **C** matrices and a sequence of variables for Set 1 as follows:

$$\mathbf{D}_1 = \begin{bmatrix} 3 & 5 \\ 2 & 8 \end{bmatrix} ; \quad \mathbf{C}_1 = \begin{bmatrix} -19 & 11 \\ -22 & 12 \end{bmatrix} ; \quad V_1, V_2, V_3, V_4$$ (a)

If the **D** matrix and the sequence are changed to

$$\mathbf{D}_2 = \begin{bmatrix} 5 & 1 \\ -3 & 4 \end{bmatrix}; \quad V_3, V_2, V_4, V_1 \tag{b}$$

what will the new $\mathbf{C}_2$ matrix be?

Shift Matrix **S** and its partitioning (providing the submatrices, see Fig. 9-3) are

$$\begin{array}{c} \mathbf{S}_1 \longrightarrow \begin{array}{c} V_1 \\ V_2 \end{array} \\ \mathbf{S}_3 \longrightarrow \begin{array}{c} V_3 \\ V_4 \end{array} \end{array} \begin{array}{|cc|cc|} \hline V_3 & V_2 & V_4 & V_1 \\ \hline 0 & 0 & 0 & 1 \\ 0 & 1 & 0 & 0 \\ \hline 1 & 0 & 0 & 0 \\ 0 & 0 & 1 & 0 \\ \hline \end{array} \begin{array}{c} \longleftarrow \mathbf{S}_2 \\ \\ \longleftarrow \mathbf{S}_4 \end{array} \tag{c}$$

Hence, by (9-53), (a), (b) and (c), the transformation matrix is

$$\mathbf{T} = \mathbf{D}_2 \cdot (\mathbf{D}_1 \cdot \mathbf{S}_1 + \mathbf{C}_1 \cdot \mathbf{S}_3)^{-1} = \frac{1}{42} \begin{bmatrix} -62 & 44 \\ -64 & 61 \end{bmatrix} \tag{d}$$

Thus, by (9-64), (c) and (d)

$$\mathbf{C}_2 = \mathbf{T} \cdot (\mathbf{D}_1 \cdot \mathbf{S}_2 + \mathbf{C}_1 \cdot \mathbf{S}_4) = \frac{1}{3} \begin{bmatrix} -11 & -7 \\ 2 & -5 \end{bmatrix} \tag{e}$$

It is noted that we were able to determine the transformed $\mathbf{C}_2$ matrix without knowing anything about the dimensional compositions of variables $V_1, \ldots, V_4$. Furthermore, we can even construct the new set of dimensionless variables. For, by (a)

$$\pi_{11} = V_1^3 \cdot V_2^5 \cdot V_3^{-19} \cdot V_4^{11}; \quad \pi_{12} = V_1^2 \cdot V_2^8 \cdot V_3^{-22} \cdot V_4^{12} \tag{f}$$

and by (b) and (e)

$$\pi_{21} = V_3^5 \cdot V_2^1 \cdot (V_4^{-11} \cdot V_1^{-7})^{1/3}; \quad \pi_{22} = V_3^{-3} \cdot V_2^4 \cdot (V_4^2 \cdot V_1^{-5})^{1/3} \tag{g}$$

⇑

9.4. INDEPENDENCE OF DIMENSIONLESS PRODUCTS OF THE DIMENSIONAL SYSTEM USED

In the foregoing three articles we explored the effects of various sequential arrangements of variables (in dimensional sets) on the composition of dimensionless variables. Now we are going to investigate what effects—if any—*dimensional systems* have on these variables. We state and prove the following important theorem:

Theorem 9-20. *The composition of dimensionless variables, their number, and their numerical values are independent of the dimensional system used.*

In other words, any variable that is dimensionless in a dimensional system will remain dimensionless—and its numerical value unchanged—if expressed in any other dimensional system *consistently* applied.

Proof. Suppose we have a dimensional matrix of $N_V = 5$ variables ($V_1, V_2, \ldots$), $N_d = 2$ dimensions (d_1, d_2), and a rank of $R_{DM} = 2$. Then the dimensional matrix is

$$\mathbf{B}_1 \text{ matrix} \longrightarrow \begin{array}{c|ccc|cc} & V_1 & V_2 & V_3 & V_4 & V_5 \\ \hline d_1 & b_{11} & b_{12} & b_{13} & a_{11} & a_{12} \\ d_2 & b_{21} & b_{22} & b_{23} & a_{21} & a_{22} \end{array} \longleftarrow \mathbf{A}_1 \text{ matrix}$$

Hence, by the above, the dimensions of the variables in $\mathbf{B}_1$ are

$$[V_1] = d_1^{b_{11}} \cdot d_2^{b_{21}} \; ; \qquad [V_2] = d_1^{b_{12}} \cdot d_2^{b_{22}} \; ; \qquad [V_3] = d_1^{b_{13}} \cdot d_2^{b_{23}}$$

This can be written in compact matrix form

$$\begin{bmatrix} \ln V_1 \\ \ln V_2 \\ \ln V_3 \end{bmatrix} = \begin{bmatrix} b_{11} & b_{21} \\ b_{12} & b_{22} \\ b_{13} & b_{23} \end{bmatrix} \cdot \begin{bmatrix} \ln d_1 \\ \ln d_2 \end{bmatrix} = \mathbf{B}_1^T \cdot \begin{bmatrix} \ln d_1 \\ \ln d_2 \end{bmatrix} \tag{9-66}$$

Now we have another dimensional system—call it number 2—whose dimensions Δ_1 and Δ_2 are connected to the dimensions d_1 and d_2 of the first system by

$$d_1 = \Delta_1^{e_{11}} \cdot \Delta_2^{e_{12}} \; ; \qquad d_2 = \Delta_1^{e_{21}} \cdot \Delta_2^{e_{22}}$$

This relation is put into compact matrix form

$$\begin{bmatrix} \ln d_1 \\ \ln d_2 \end{bmatrix} = \overbrace{\begin{bmatrix} e_{11} & e_{12} \\ e_{21} & e_{22} \end{bmatrix}}^{\mathbf{e}} \cdot \begin{bmatrix} \ln \Delta_1 \\ \ln \Delta_2 \end{bmatrix} = \mathbf{e} \cdot \begin{bmatrix} \ln \Delta_1 \\ \ln \Delta_2 \end{bmatrix} \tag{9-67}$$

By combining (9-66) and (9-67), we have

$$\begin{bmatrix} \ln V_1 \\ \ln V_2 \\ \ln V_3 \end{bmatrix} = (\mathbf{B}_1^T \cdot \mathbf{e}) \cdot \begin{bmatrix} \ln \Delta_1 \\ \ln \Delta_2 \end{bmatrix} \tag{9-68}$$

But this relation has the same form as (9-66), so for this second system we can write, in unison with (9-66),

$$\begin{bmatrix} \ln V_1 \\ \ln V_2 \\ \ln V_3 \end{bmatrix} = \mathbf{B}_2^T \cdot \begin{bmatrix} \ln \Delta_1 \\ \ln \Delta_2 \end{bmatrix} \tag{9-69}$$

where, by (9-68),

$$\mathbf{B}_2^T = \mathbf{B}_1^T \cdot \mathbf{e} \tag{9-70}$$

The same derivation can be done for the $\mathbf{A}$ matrix. Thus, similarly to (9-70),

$$\mathbf{A}_2^T = \mathbf{A}_1^T \cdot \mathbf{e} \tag{9-71}$$

and by (9-70) and (9-71)

$$\mathbf{B}_2 = \mathbf{e}^T \cdot \mathbf{B}_1 \; ; \qquad \mathbf{A}_2 = \mathbf{e}^T \cdot \mathbf{A}_1 \tag{9-72}$$

At this point we notice that *both* the $\mathbf{B}_1$ and $\mathbf{A}_1$ matrices are premultiplied by the *same* $\mathbf{e}^T$ matrix. Therefore, by Theorem 9-3, the respective $\mathbf{C}$ matrices of the two dimensional sets are identical. Hence $\mathbf{C}_1 = \mathbf{C}_2$, and since the dimensionless products are solely determined by the $\mathbf{C}$ matrix—if $\mathbf{D}$ is unaltered (which is the case here)—therefore the dimensionless products remain unchanged. This proves the theorem.

Example 9-32. Kinetic Energy of a Moving Ball

We have a non-rotating solid ball moving at a known speed. We wish to derive a formula for its kinetic energy.

The following variables are relevant in the SI and *force*-based Imperial system:

Variable	Symbol	Dimension	
		SI	in–lbf–s
kinetic energy of ball	E	$m^2 \cdot kg \cdot s^{-2}$	$in \cdot lbf$
density of ball	ρ	$m^{-3} \cdot kg$	$in^{-4} \cdot lbf \cdot s^2$
diameter of ball	D	m	in
speed of ball	v	$m \cdot s^{-1}$	$in \cdot s^{-1}$

from which the Dimensional Sets in the two systems are

SI

	E	ρ	D	v
m	2	–3	1	1
kg	1	1	0	0
s	–2	0	0	–1
π_{11}	1	–1	–3	–2	← $\mathbf{C}_1$ matrix

force-based Imperial

	E	ρ	D	v
in	1	–4	1	1
lbf	1	1	0	0
s	0	2	0	–1
π_{21}	1	–1	–3	–2	← $\mathbf{C}_2$ matrix

Thus, as seen, $\mathbf{C}_1 = \mathbf{C}_2$, and therefore

$$\pi_{11} = \frac{E}{\rho \cdot D^3 \cdot v^2} = \pi_{21}$$

i.e., the sole dimensionless product in the two systems are identical—as Theorem 9-20 prescribes.

⇑

CHAPTER 10
NUMBER OF SETS OF DIMENSIONLESS PRODUCTS OF VARIABLES

10.1. DISTINCT AND EQUIVALENT SETS

A set of dimensionless products for a *given* dimensional matrix (matrices **A** and **B**) is plainly defined by the **D** matrix—for any such a set is a function of matrices **A**, **B**, and **D** only. Now since an infinite number of *different* (nonsingular) **D** matrices is possible, the number of *distinct* sets of dimensionless products is also infinite.

As described in Art. 9.2. [equations (9-39) and (9-40)], the relation between any two Dimensional Sets of identical **A** and **B** matrices, but different **D** matrices, is

$$\begin{bmatrix} \ln \pi_{21} \\ \ln \pi_{22} \\ \vdots \\ \ln \pi_{2N_P} \end{bmatrix} = (\mathbf{D}_2 \cdot \mathbf{D}_1^{-1}) \cdot \begin{bmatrix} \ln \pi_{11} \\ \ln \pi_{12} \\ \vdots \\ \ln \pi_{1N_P} \end{bmatrix} \qquad (10\text{-}1)$$

where $\pi_{11}, \pi_{12}, \ldots$ and $\pi_{21}, \pi_{22}, \ldots$ are the dimensionless products in Set 1 and Set 2, respectively, N_P is the number of dimensionless products in each set, and $\mathbf{D}_1$ and $\mathbf{D}_2$ are the respective **D** matrices.

However, if **D** fixed, then the number of distinct sets of dimensionless variables becomes *finite*. Therefore the question naturally emerges: In case of fixed **D**, how many *distinct* sets are possible?

To answer, first we have to formulate the condition(s) for two sets to be called *distinct*. If two sets are not distinct, then they are called *equivalent* (for the purpose of this investigation). Therefore, any two sets are either *distinct* or *equivalent*, but not both.

Obviously, two sets are equivalent if their dimensionless products are the same. Suppose Set 1 consists of $\pi_{11}, \pi_{12}, \pi_{13}$, and Set 2 consists of $\pi_{21}, \pi_{22}, \pi_{23}$. Then if

$$\pi_{21} = \pi_{11}; \qquad \pi_{22} = \pi_{12}; \qquad \pi_{23} = \pi_{13} \qquad (10\text{-}2)$$

then the two sets are *equivalent*. We note that the transformation matrix corresponding to (10-2) is

$$\mathbf{T} = \begin{bmatrix} 1 & 0 & 0 \\ 0 & 1 & 0 \\ 0 & 0 & 1 \end{bmatrix} \quad (10\text{-}3)$$

Next, suppose we have

$$\pi_{21} = \pi_{13}; \qquad \pi_{22} = \pi_{11}; \qquad \pi_{23} = \pi_{12} \quad (10\text{-}4)$$

then we still judge the two sets to be equivalent, since the products are really the same, but for their sequential order, which, of course is immaterial. Therefore the transformation matrix for (10-4) is

$$\mathbf{T} = \begin{bmatrix} 0 & 0 & 1 \\ 1 & 0 & 0 \\ 0 & 1 & 0 \end{bmatrix} \quad (10\text{-}5)$$

Now we further complicate the issue. Consider the relations

$$\pi_{21} = \pi_{13}^2; \qquad \pi_{22} = \pi_{11}^3; \qquad \pi_{23} = \pi_{12}^4 \quad (10\text{-}6)$$

yielding the transformation matrix

$$\mathbf{T} = \begin{bmatrix} 0 & 0 & 2 \\ 3 & 0 & 0 \\ 0 & 4 & 0 \end{bmatrix} \quad (10\text{-}7)$$

Are the two sets connected by relations (10-6) equivalent? To answer, we reiterate that any power of a *single* dimensionless variable is still dimensionless and still *single*. With this in mind, the two sets in (10-6) must be judged to be *equivalent*. For example, if V, a, and L are speed, acceleration and length, then the two dimensionless variables

$$\pi_1 = \frac{v}{\sqrt{a \cdot L}}; \qquad \pi_2 = \frac{a \cdot L}{v^2}$$

are equivalent, since $\pi_2 = \pi_1^{-2}$, and on both sides of this expressions only a *single* dimensionless variable appears. In fact, frequently a dimensionless variable is raised to an (appropriate) power just to eliminate its fractional exponent.

Next, we consider

$$\pi_{21} = \pi_{11}^2 \cdot \pi_{12}; \qquad \pi_{22} = \pi_{11}; \qquad \pi_{23} = \pi_{13} \quad (10\text{-}8)$$

in which one of the dimensionless variables of Set 2 (π_{21}) is expressed as a function of *two* dimensionless variables of Set 1. Hence these sets are *distinct*. The transformation matrix corresponding to (10-8) is now

$$\mathbf{T} = \begin{bmatrix} 2 & 1 & 0 \\ 1 & 0 & 0 \\ 0 & 0 & 1 \end{bmatrix} \quad (10\text{-}9)$$

Considering the three transformation matrices of (10-3), (10-5), and (10-7)—all representing equivalent pairs of sets, and (10-9) which does not—we can easily draw the conclusion spelled out by the following definition:

Definition 10-1. *Two sets of dimensionless variables are equivalent if, and only if, the transformation matrix connecting them has exactly one nonzero element in each of its rows and columns.*

By this definition, then, the matrix in (10-9) has two nonzero elements in its first row, hence the two sets are *distinct,* as (10-8) shows. By contrast, the sets connected by transformation matrices (10-3), (10-5), and (10-7) are pair-wise *equivalent,* since these matrices fulfill the above definition, i.e., they have exactly one nonzero element in each of their rows and columns.

10.2. CHANGES IN A DIMENSIONAL SET NOT AFFECTING THE DIMENSIONLESS VARIABLES

There are certain changes in the sequence of variables in a Dimensional Set that leave the *nature* of the derived set of dimensionless variables intact. That is, by virtue of Definition 10-1, the transformed set is *equivalent* to the original set. In particular, we state and prove the following theorems (the *original* set of dimensionless variables is designated as Set 1, and the *transformed* one as Set 2):

Theorem 10-1. *If in Set 1 two variables in the **B** matrix are interchanged, then Set 2 will be equivalent to Set 1.*

We present 2 proofs.

Proof 1. By Theorem 9-6, the interchange of two *columns* (variables) in **B** causes the interchange of the corresponding *rows* in **C**. Say we interchange the first and second columns in **B**. Then the first and second rows in **C** will be interchanged. But by the construction of the unity **D** matrix, its corresponding two rows, i.e., rows 1 and 2, will also be interchanged. Hence in both **D** and **C** the *same* two rows will be interchanged, resulting in a transformation matrix **τ**, which can be derived from the unit matrix by interchanging the latter's two rows. Thus **τ** will have exactly one nonzero element in each of its rows and columns. Therefore, by Definition 10-1, the two sets are *equivalent.* This proves the theorem.

Example 10-1

Consider Dimensional Sets 1 and 2 as follows:

Dimensional Set 1

	V_1	V_2	V_3	V_4	V_5	V_6
d_1	−6	3	−1	1	−2	1
d_2	3	−9	−5	2	2	−1
d_3	−4	1	−2	1	−1	1
π_{11}	1	0	0	1	−2	1
π_{12}	0	1	0	2	2	−1
π_{13}	0	0	1	2	1	1

Dimensional Set 2

	V_3	V_2	V_1	V_4	V_5	V_6
d_1	−1	3	−6	1	−2	1
d_2	−5	−9	3	2	2	−1
d_3	−2	1	−4	1	−1	1
π_{21}	1	0	0	2	1	1
π_{22}	0	1	0	2	2	−1
π_{23}	0	0	1	1	−2	1

In Set 2, variables V_1, V_3 (*columns* 1 and 3 in matrix **B**) are interchanged with respect to Set 1. Consequently, in the **C** matrix of Set 2, *rows* 1 and 3 are also interchanged. Thus, by the above,

$$\left. \begin{aligned} \pi_{11} &= V_1 \cdot V_4 \cdot V_5^{-2} \cdot V_6 = \pi_{23} \\ \pi_{12} &= V_2 \cdot V_4^2 \cdot V_5^2 \cdot V_6^{-1} = \pi_{22} \\ \pi_{13} &= V_3 \cdot V_4^2 \cdot V_5 \cdot V_6 = \pi_{21} \end{aligned} \right\} \quad (a)$$

and we can write

$$\begin{bmatrix} \ln \pi_{21} \\ \ln \pi_{22} \\ \ln \pi_{23} \end{bmatrix} = \begin{bmatrix} 0 & 0 & 1 \\ 0 & 1 & 0 \\ 1 & 0 & 0 \end{bmatrix} \cdot \begin{bmatrix} \ln \pi_{11} \\ \ln \pi_{12} \\ \ln \pi_{13} \end{bmatrix} = \mathbf{T} \cdot \begin{bmatrix} \ln \pi_{11} \\ \ln \pi_{12} \\ \ln \pi_{13} \end{bmatrix} \quad (b)$$

The transformation matrix **T** now has exactly one nonzero element in each of its rows and columns, and hence the two sets of dimensionless variables are *equivalent*.
⇑

Proof 2. Let us again interchange the first and third columns in **B**. Then submatrices $\mathbf{S}_1$ and $\mathbf{S}_3$ of Shift Matrix **S** (Fig. 9-2, Fig. 9-3) will be (only the essential part of **S** is shown here)

	V_3	V_2	V_1	
V_1	0	0	1	
V_2	0	1	0	$\mathbf{S}_1$ submatrix
V_3	1	0	0	
V_4	0	0	0	
V_5	0	0	0	$\mathbf{S}_3$ submatrix
V_6	0	0	0	

Note that we did not bother to determine submatrices $\mathbf{S}_2$ and $\mathbf{S}_4$ since they are not needed for this proof. We now observe that $\mathbf{S}_3$ is a null matrix (this is *always* the case if there is no interchange between the columns of **A** and **B**). Therefore, by (9-53), since $\mathbf{S}_3 = \mathbf{0}$,

$$\mathbf{T} = (\mathbf{S}_1 + \mathbf{C}_1 \cdot \mathbf{S}_3)^{-1} = \mathbf{S}_1^{-1} \quad (10\text{-}10)$$

But $\mathbf{S}_1$ is a matrix which contains—by virtue of its construction—exactly one nonzero element in each of its rows and columns. Hence its inverse also has this property. Thus, in view of Definition 10-1 of the transformation matrix, the theorem is proven.

Example 10-2

Consider the two Dimensional Sets of Example 10-1. Accordingly, submatrices $\mathbf{S}_1$ and $\mathbf{S}_3$ of **S** are

NUMBER OF SETS OF DIMENSIONLESS PRODUCTS OF VARIABLES 233

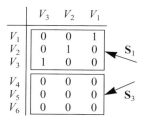

Consequently, by (10-10)

$$\mathbf{T} = \mathbf{S}_1^{-1} = \begin{bmatrix} 0 & 0 & 1 \\ 0 & 1 & 0 \\ 1 & 0 & 0 \end{bmatrix}^{-1} = \begin{bmatrix} 0 & 0 & 1 \\ 0 & 1 & 0 \\ 1 & 0 & 0 \end{bmatrix} \quad \text{(b)}$$

This is the same as in (b) in Example 10-1, as expected.

⇑

Theorem 10-2. *If, in a Dimensional Set, two variables in the* **A** *matrix are interchanged, then the transformed set of dimensionless variables will be equivalent to the original one.*

Proof. By the Corollary of Theorem 9-7, an interchange of two columns (variables) in the **A** matrix does not change the composition of the generated dimensionless variables. Hence the two sets of dimensionless variables must be identical. This proves the theorem.

Example 10-3

Consider the Dimensional Sets

Dimensional Set 1

	V_1	V_2	V_3	V_4	V_5	V_6
d_1	−6	3	−1	1	−2	1
d_2	3	−9	−5	2	2	−1
d_3	−4	1	−2	1	−1	1
π_{11}	1	0	0	1	−2	1
π_{12}	0	1	0	2	2	−1
π_{13}	0	0	1	2	1	1

Dimensional Set 2

	V_1	V_2	V_3	V_4	V_5	V_6
d_1	−6	3	−1	1	1	−2
d_2	3	−9	−5	2	−1	2
d_3	−4	1	−2	1	1	−1
π_{21}	1	0	0	1	1	−2
π_{22}	0	1	0	2	−1	2
π_{23}	0	0	1	2	1	1

In Set 2, variables V_5 and V_6 are interchanged with respect to Set 1. Consequently, in Set 2 the corresponding columns in the **C** matrix are also interchanged—as shown. Therefore we have

$$\left. \begin{array}{l} \pi_{11} = V_1 \cdot V_4 \cdot V_5^{-2} \cdot V_6 = \pi_{21} \\ \pi_{12} = V_2 \cdot V_4^2 \cdot V_5^2 \cdot V_6^{-1} = \pi_{22} \\ \pi_{13} = V_3 \cdot V_4^2 \cdot V_5 \cdot V_6 = \pi_{23} \end{array} \right\} \quad \text{(a)}$$

whence

$$\begin{bmatrix} \ln \pi_{21} \\ \ln \pi_{22} \\ \ln \pi_{23} \end{bmatrix} = \begin{bmatrix} 1 & 0 & 0 \\ 0 & 1 & 0 \\ 0 & 0 & 1 \end{bmatrix} \cdot \begin{bmatrix} \ln \pi_{11} \\ \ln \pi_{12} \\ \ln \pi_{13} \end{bmatrix} = \mathbf{T} \cdot \begin{bmatrix} \ln \pi_{11} \\ \ln \pi_{12} \\ \ln \pi_{13} \end{bmatrix} \quad \text{(b)}$$

As we see, transformation matrix **T** fulfills Definition 10-1, and hence these two sets are indeed *equivalent*.

⇑

Corollary of Theorem 10-2. *If, in a Dimensional Set, two variables in* **A** *are interchanged, then the transformation matrix represented by this interchange is a unit matrix.*

Proof. By the referred interchange, matrix **B** remains intact, and hence (by the construction of the S_1 and S_3 submatrices) $S_1 = I$ (unit matrix) and $S_3 = 0$ (null matrix). Consequently, by (9-53),

$$\mathbf{T} = (S_1 + C_1 \cdot S_3)^{-1} = S_1^{-1} = I^{-1} = I \quad (10\text{-}11)$$

which proves the corollary. The reader's attention is directed to relation (b) in Example 10-3, where it is shown that **T** is indeed a unit matrix.

Theorem 10-3. *If a column of the* **C** *matrix contains only one nonzero element in its ith row, and if the variable corresponding to this column is interchanged with the ith variable in the* **B** *matrix, then the generated new set of dimensionless variables will be equivalent to the original set of dimensionless variables.*

Proof. This case is the same as was dealt with to illustrate Theorem 9-18. There it was shown that the transformation matrix was diagonal. But in a diagonal matrix each row and column contains exactly one element, thus fulfilling Definition 10-1. Therefore the two sets of dimensional variables must be *equivalent*. This proves the theorem.

Example 10-4

Consider the two Dimensional Sets:

Dimensional Set 1

	V_1	V_2	V_3	V_4	V_5
d_1	−2	−11	−3	1	2
d_2	4	−3	6	−2	1
π_{11}	1	0	0	2	0
π_{12}	0	1	0	1	5
π_{13}	0	0	1	3	0

Dimensional Set 2

	V_1	V_5	V_3	V_4	V_2
d_1	−2	2	−3	1	−11
d_2	4	1	6	−2	−3
π_{21}	1	0	0	2	0
π_{22}	0	1	0	0.2	0.2
π_{23}	0	0	1	3	0

We see in Set 1 that the second column of **C** (under V_5) has only one nonzero element, which is in its second row ($i = 2$). Therefore, if we exchange variable V_5 with the second

NUMBER OF SETS OF DIMENSIONLESS PRODUCTS OF VARIABLES   **235**

variable of **B**, which is V_2, then the new set of dimensionless variables will be *equivalent* to the original one. Indeed, by the above sets

$$\left.\begin{array}{l} \pi_{21} = V_1 \cdot V_4^2 = \pi_{11} \\ \pi_{22} = V_5 \cdot (V_4 \cdot V_2)^{0.2} = (V_5^5 \cdot V_4 \cdot V_2)^{0.2} = \pi_{12}^{0.2} \\ \pi_{23} = V_3 \cdot V_4^3 = \pi_{13} \end{array}\right\} \quad (a)$$

which can be written

$$\begin{bmatrix} \ln \pi_{21} \\ \ln \pi_{22} \\ \ln \pi_{23} \end{bmatrix} = \overbrace{\begin{bmatrix} 1 & 0 & 0 \\ 0 & \dfrac{1}{5} & 0 \\ 0 & 0 & 1 \end{bmatrix}}^{\mathbf{T}} \cdot \begin{bmatrix} \ln \pi_{11} \\ \ln \pi_{12} \\ \ln \pi_{13} \end{bmatrix} \quad (b)$$

and we see that the transformation matrix **T** is a diagonal matrix, fulfilling Definition 10-1. Hence the two sets of dimensionless variables are *equivalent*.
⇑

Now we determine the number U_c of such possible exchanges of variables yielding equivalent sets.

Let us assume that the first row of the **C** matrix contains n_1 elements which *only* appear in columns whose all *other* elements are zeros. Similarly, the second row contains n_2 such elements, and so on. For example, if the **C** matrix is

$$\mathbf{C} = \begin{bmatrix} 0 & 0 & 0 & 5 & 0 & 0 \\ 0 & 4 & 8 & 0 & 0 & 0 \\ 9 & 0 & 0 & 0 & 3 & 2 \end{bmatrix}$$

then $n_1 = 1$, because the first row has one nonzero element (5), and it appears in a column where all other elements are zeros;

$n_2 = 2$, because the second row has two nonzero elements (4, 8), and they appear in columns where all other elements are zeros;

$n_3 = 3$, because the third row has three nonzero elements (9, 3, 2), and they appear in columns where all other elements are zeros.

Similarly, if

$$\mathbf{C} = \begin{bmatrix} 5 & 8 & 0 \\ 2 & 3 & 1 \\ 4 & 5 & 0 \end{bmatrix}$$

then $n_1 = 0$, $n_2 = 1$, $n_3 = 0$.

Next, assume a **C** matrix of $N_R = 3$ rows, where the respective numbers are n_1, n_2, and n_3, and where these numbers can have any value 0, 1, ... up to the number of columns in **C**. Now, if we take *one* element, then obviously we have $S_1 = n_1 + n_2 + n_3$ choices. If we take *two* elements, one from each row, then we have $S_2 = n_1 \cdot n_2 + n_1 \cdot n_3 + n_2 \cdot n_3$ choices. Finally, if we take three elements, one from each row, then we have $S_3 = n_1 \cdot n_2 \cdot n_3$ choices. In these expressions we notice that

$$S_1 \text{ comprises } \binom{N_R}{1} = \binom{3}{1} = 3 \text{ members}$$

$$S_2 \text{ comprises } \binom{N_R}{2} = \binom{3}{2} = 3 \text{ members}$$

$$S_3 \text{ comprises } \binom{N_R}{3} = \binom{3}{3} = 1 \text{ member}$$

where, in general, $\binom{n}{m}$ denotes the number of *combinations* of n objects taken m at a time. The above relations indicates the general formula

$$S = S_1 + S_2 + \ldots + S_{N_R} = \sum_{j=1}^{N_R} \binom{N_R}{j} \qquad (10\text{-}12)$$

where N_R is the number of *rows* in the **C** matrix. The table in Fig. 10-1 presents the values of S for $N_R = 1, 2, \ldots, 10$.

N_R	1	2	3	4	5	6	7	8	9	10
S	1	3	7	15	31	63	127	255	511	1023

Figure 10-1
Value of S in expression (10-12)
N_R = number of rows in the C matrix

It can be proven that (10-12) can also be written

$$S = 2^{N_R} - 1 \qquad (10\text{-}13)$$

The reader can easily verify this by the above table. For example, if $N_R = 9$, then $S = 2^9 - 1 = 512 - 1 = 511$.

It should be noted that some, or perhaps all, of the members in series (10-12) may be zero. Consider, for example, the **C** matrix

$$\mathbf{C} = \begin{bmatrix} 2 & 0 \\ 1 & 5 \\ 3 & 0 \end{bmatrix}$$

Here $N_R = 3$ and $n_1 = 0$, $n_2 = 1$, $n_3 = 0$. Hence, by (10-12), or (10-13), or by Fig. 10-1, we have $S = 7$. So, U_c has seven members as follows:

$$U_c = n_1 + n_2 + n_3 + n_1 \cdot n_2 + n_1 \cdot n_3 + n_2 \cdot n_3 + n_1 \cdot n_2 \cdot n_3 = 0 + 1 + 0 + 0 + 0 + 0 + 0 = 1$$

The following example further illustrates.

Example 10-5

Given the **C** matrix of a dimensional set as

$$\mathbf{C} = \begin{bmatrix} 5 & 6 & 0 & 8 & 0 \\ 0 & 6 & 7 & 0 & 9 \end{bmatrix} \quad \text{(a)}$$

How many interchanges of variables between **C** and **D** (i.e., between matrices **B** and **A**) are possible to yield *equivalent* sets of dimensionless variables? We have $N_R = 2$ (**C** has two rows) and by (a), $n_1 = 2$, $n_2 = 2$. So, by Fig. 10-1, U_c has $S = 3$ members:

$$U_c = n_1 + n_2 + n_1 \cdot n_2 = 2 + 2 + 4 = 8 \quad \text{(b)}$$

Thus, there are eight different exchanges possible, each resulting in a set of dimensionless variables *equivalent* to the original set. Specifically, we have

D matrix →

	V_1	V_2	V_3	V_4	V_5	V_6	V_7
π_{11}	1	0	5	6	0	8	0
π_{12}	0	1	0	6	7	0	9

← $\mathbf{C}_1$ matrix (c)

Therefore the eight possible exchanges yielding *equivalent* sets are:

variable V_5	in **C** is exchanged with variable	V_2	in **D**
variable V_7	in **C** is exchanged with variable	V_2	in **D**
variable V_3	in **C** is exchanged with variable	V_1	in **D**
variable V_6	in **C** is exchanged with variable	V_1	in **D**
variables V_3, V_5	in **C** are exchanged with variables	V_1, V_2	in **D** in this order
variables V_3, V_7	in **C** are exchanged with variables	V_1, V_2	in **D** in this order
variables V_5, V_6	in **C** are exchanged with variables	V_2, V_1	in **D** in this order
variables V_6, V_7	in **C** are exchanged with variables	V_1, V_2	in **D** in this order

As an illustration, let us verify that the exchange indicated in the last line in the above table yields an *equivalent* set of dimensionless variables. By (c),

$$\pi_{11} = V_1 \cdot V_3^5 \cdot V_4^6 \cdot V_6^8; \qquad \pi_{12} = V_2 \cdot V_4^6 \cdot V_5^7 \cdot V_7^9 \quad \text{(d)}$$

and by the last line of the table

	V_6	V_7	V_3	V_4	V_5	V_1	V_2
π_{21}	1	0			?		
π_{22}	0	1					

← $\mathbf{C}_2$ matrix (e)

where, of course, we do not know the $\mathbf{C}_2$ matrix, yet. But we do know submatrices $\mathbf{S}_1$ and $\mathbf{S}_3$ of the Shift Matrix **S**

	V_6	V_7
V_1	0	0
V_2	0	0	← $\mathbf{S}_1$
V_3	0	0
V_4	0	0
V_5	0	0	← $\mathbf{S}_3$
V_6	1	0
V_7	0	1

(f)

and hence, by the first part of (9-54), (c) and (f),

$$\mathbf{T} = (\mathbf{S}_1 + \mathbf{C}_1 \cdot \mathbf{S}_3)^{-1} = \frac{1}{72}\begin{bmatrix} 9 & 0 \\ 0 & 8 \end{bmatrix} \quad (g)$$

and by this and (9-50)

$$\begin{bmatrix} \ln \pi_{21} \\ \ln \pi_{22} \end{bmatrix} = \frac{1}{72}\begin{bmatrix} 9 & 0 \\ 0 & 8 \end{bmatrix} \cdot \begin{bmatrix} \ln \pi_{11} \\ \ln \pi_{12} \end{bmatrix} \quad (h)$$

so that

$$\left.\begin{array}{l} \pi_{21} = \sqrt[8]{\pi_{11}} = V_1^{1/8} \cdot V_3^{5/8} \cdot V_4^{3/4} \cdot V_6 \\ \pi_{22} = \sqrt[9]{\pi_{12}} = V_2^{1/9} \cdot V_4^{2/3} \cdot V_5^{7/9} \cdot V_7 \end{array}\right\} \quad (i)$$

We see in (g) that $\mathbf{T}$ is a diagonal matrix, hence, by Definition 10-1, the new set of dimensionless variables π_{21}, π_{22} is *equivalent* to the original one π_{11}, π_{12}—as relation (i) also shows.

A useful property of (i) is that it allows the determination of $\mathbf{C}_2$ without having any information on the dimensional composition of any of the variables V_1, V_2, ... To construct $\mathbf{C}_2$ in this particular case, we merely have to fill in the spaces in (e) with the respective exponents appearing in (i). Accordingly,

	V_6	V_7	V_3	V_4	V_5	V_1	V_2
π_{21}	1	0	$\frac{5}{8}$	$\frac{3}{4}$	0	$\frac{1}{8}$	0
π_{22}	0	1	0	$\frac{2}{3}$	$\frac{7}{9}$	0	$\frac{1}{9}$

$\mathbf{C}_2$ matrix

Note that the same result for $\mathbf{C}_2$ can be obtained by relation (9-65).

10.3. PROHIBITED CHANGES IN A DIMENSIONAL SET

Changes in a Dimensional Set that result in a singular $\mathbf{A}$ matrix are not permitted. The reason for this restriction resides in the Fundamental Formula, which contains the *inverse* of $\mathbf{A}$; if $\mathbf{A}$ is singular, it has no inverse, and hence the stated prohibition.

Now we recall Theorem 9-8, which states that if an element of the $\mathbf{C}$ matrix in its ith row and jth column is zero, then the variables corresponding to the ith column of the $\mathbf{B}$ matrix and the jth column of the $\mathbf{A}$ matrix cannot be interchanged, for if such an interchange took place, then it would result in a singular $\mathbf{A}$.

Example 10-6

Consider Dimensional Sets 1 and 2 as follows:

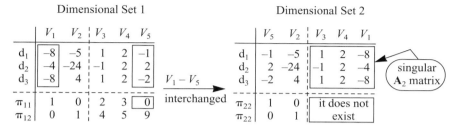

In Set 1 we have a zero in the first row and third column of the **C** matrix, thus $c_{13} = 0$. Hence, the first column in **B** (variable V_1) and the third column in **A** (variable V_5) cannot be interchanged, because such an interchange would result in a singular **A** matrix. Indeed, the determinant of matrix **A** in Set 2 is

$$|A_2| = \begin{vmatrix} 1 & 2 & -8 \\ -1 & 2 & -4 \\ 1 & 2 & -8 \end{vmatrix} = 0$$

⇑

Next, let us consider the case where there are N_Z zeros in a particular *row* of the **C** matrix. Exchanging any one of these zeros with the "1" in the **D** matrix would result in a singular **A** (Theorem 9-8). Now suppose we select j zeros ($j = 1, 2, \ldots, N_Z$) which can be done $\binom{N_Z}{j}$ ways, where the expression in parentheses is the number of *combinations* of N_Z objects taken j at a time. The numerical formula for this term is

$$\binom{N_Z}{j} = \frac{N_Z!}{j! \cdot (N_Z - j)!} \tag{10-14}$$

in which the exclamation mark "!" designates the *factorial* of a number; e.g., $4! = 1 \cdot 2 \cdot 3 \cdot 4 = 24$.

Now we must determine the number of different ways j number of zeros can be placed in a row of matrix **D**. Since **D** is square, it has as many rows as columns. Further, it has, by its construction, exactly N_P rows (N_P is the number of dimensionless variables in a set) and hence N_P columns. Therefore *every* row of **D** has exactly N_P elements of which $N_P - 1$ are zeros, and 1 is a "1."

Next, we consider j zeros placed in a row of **D** such that the sole "1" in the row will be replaced by a zero. How many ways can this be done? If 1 zero replaces the sole nonzero element "1," then obviously there are $j - 1$ remaining zeros which can be put in $N_P - 1$ remaining places. Hence, there are precisely $\binom{N_P - 1}{j - 1}$ ways to do this, where, as in (10-14),

$$\binom{N_P - 1}{j - 1} = \frac{(N_P - 1)!}{(j - 1)!\cdot(N_P - 1 - j + 1)!} = \frac{(N_P - 1)!}{(j - 1)!\cdot(N_P - j)!} \quad (10\text{-}15)$$

defines the number of combinations of $N_P - 1$ objects taken $j - 1$ at a time. Since, by (10-14), there are $\binom{N_Z}{j}$ ways to select j zeros, therefore ultimately we have

$$U_R = \sum_{j=1}^{N_Z} \binom{N_P - 1}{j - 1}\binom{N_Z}{j} \quad (10\text{-}16)$$

ways to select zeros in any *row* of matrix **C** which (the row) contains N_Z zeros.

If the above appears a bit confusing, the reader may be comforted by the author's admission that, *at first,* it also seemed "somewhat" complicated to him. But *only* at first! As the examples to follow show, the process is really quite simple.

Example 10-7

Suppose we have **D** and **C** matrices:

		V_1	V_2	V_3	V_4	V_5	V_6	V_7	
D matrix	π_1	1	0	0	2	0	0	0	
	π_2	0	1	0	5	4	3	2	**C** matrix
	π_3	0	0	1	1	4	3	3	

As seen $N_P = 3$, and in the first row of **C** we have $N_Z = 3$ zeros. Therefore

selecting one zero from the first row of **C** can be done in $\binom{N_Z}{1} = \binom{3}{1} = 3$ ways

selecting two zeros from the first row of **C** can be done in $\binom{N_Z}{2} = \binom{3}{2} = 3$ ways

selecting three zeros from the first row of **C** can be done in $\binom{N_Z}{3} = \binom{3}{3} = 1$ way

Therefore, by (10-16), we have

$$U_R = \binom{N_P - 1}{1 - 1}\binom{N_Z}{1} + \binom{N_P - 1}{2 - 1}\binom{N_Z}{2} + \binom{N_P - 1}{3 - 1}\binom{N_Z}{3} = (1)\cdot(3) + (2)\cdot(3) + (1)\cdot(1) = 10 \quad (a)$$

ways to place the three zeros in the first row of **D** such that the "1" in it would be replaced by a zero.

⇑

In order to facilitate the use of relation (10-16), a table showing the U_R values as a function of N_P and N_Z is presented in Fig 10-2.

To illustrate, in Example 10-7 we had $N_P = 3$ and $N_Z = 3$. Using the table in Fig. 10-2, $U_R = 10$, confirming relation (a) in that example.

When there is more than one row containing zeros in the **C** matrix, then the U_R values must be determined for each row separately and then these numbers added.

NUMBER OF SETS OF DIMENSIONLESS PRODUCTS OF VARIABLES 241

		N_P								
		1	2	3	4	5	6	7	8	9	10
N_Z	0	0	0	0	0	0	0	0	0	0	0
	1	1	1	1	1	1	1	1	1	1	1
	2	2	3	4	5	6	7	8	9	10	11
	3	3	6	10	15	21	28	36	45	55	66
	4	4	10	20	35	56	84	120	165	220	286
	5	5	15	35	70	126	210	330	495	715	1001
	6	6	21	56	126	252	462	792	1287	2002	3003
	7	7	28	84	210	462	924	1716	3003	5005	8008
	8	8	36	120	330	792	1716	3432	6435	11440	19448
	9	9	45	165	495	1287	3003	6435	12870	24310	43758
	10	10	55	220	715	2002	5005	11440	24310	48620	92378

Figure 10-2
Number of possible ways, U_R, of placing N_Z zeros in a row of matrix D such that the only nonzero element in that row would be replaced
N_P = number of dimensionless variables in the set

Example 10-8

Consider the **D** and **C** matrices as follows:

$$\begin{array}{c} \\ \text{D matrix} \longrightarrow \\ \\ \end{array} \begin{array}{c} \\ \pi_1 \\ \pi_2 \\ \pi_3 \end{array} \begin{array}{|ccc|} \hline V_1 & V_2 & V_3 \\ \hline 1 & 0 & 0 \\ 0 & 1 & 0 \\ 0 & 0 & 1 \\ \hline \end{array} \begin{array}{|cccc|} \hline V_4 & V_5 & V_6 & V_7 \\ \hline 2 & 0 & 0 & 0 \\ 0 & 0 & -3 & 2 \\ 0 & 5 & 0 & 0 \\ \hline \end{array} \longleftarrow \text{C matrix}$$

Here we have $N_P = 3$ and $N_Z = 3, 2, 3$ for the first, second, and third rows of **C**, respectively. From the table in Fig. 10-2,

$$U_R = \underbrace{10}_{\text{first}} + \underbrace{4}_{\text{second}} + \underbrace{10}_{\text{third rows in C}} = 24 \qquad (a)$$

It is interesting to observe that a typical element, $U_R\{N_Z, N_P\}$, in the table may be expressed as

$$U_R\{N_Z, N_P\} = U_R\{N_Z, N_P - 1\} + U_R\{N_Z - 1, N_P\} \qquad (10\text{-}17)$$

where braces { } designate the *argument* of U_R. For example, if $N_Z = 5$, $N_P = 3$, then

$$U_R\{N_Z, N_P\} = U_R\{5, 3\} = 35$$
$$U_R\{N_Z, N_P - 1\} = U_R\{5, 2\} = 15$$
$$U_R\{N_Z - 1, N_P\} = U_R\{4, 3\} = 20$$

and, by (10-17), 35 = 15 + 20.

By the table, another relation is also easily apparent:

$$U_R\{N_Z, N_P\} = \sum_{j=1}^{N_Z} U_R\{j, N_P - 1\} \qquad (10\text{-}18)$$

For example, if $N_Z = 5$, $N_P = 3$, then

$$U_R\{5, 3\} = \sum_{j=1}^{5} U_R\{j, 2\} = U_R\{1, 2\} + U_R\{2, 2\} + U_R\{3, 2\} + U_R\{4, 2\} + U_R\{5, 2\}$$
$$= 1 + 3 + 6 + 10 + 15 = 35$$

Finally, the mathematically motivated reader might enjoy *proving* the formulas found in the table of Fig. 10-3.

N_P	Formula for U_R
2	$\frac{1}{2}(N_Z^2 + N_Z)$
3	$\frac{1}{6}(N_Z^3 + 3 \cdot N_Z^2 + 2 \cdot N_Z)$
4	$\frac{1}{24}(N_Z^4 + 6 \cdot N_Z^3 + 11 \cdot N_Z^2 + 6 \cdot N_Z)$
5	$\frac{1}{120}(N_Z^5 + 10 \cdot N_Z^4 + 35 \cdot N_Z^3 + 50 \cdot N_Z^2 + 24 \cdot N_Z)$
6	$\frac{1}{720}(N_Z^6 + 15 \cdot N_Z^5 + 85 \cdot N_Z^4 + 225 \cdot N_Z^3 + 274 \cdot N_Z^2 + 120 \cdot N_Z)$
7	$\frac{1}{5040}(N_Z^7 + 21 \cdot N_Z^6 + 175 \cdot N_Z^5 + 735 \cdot N_Z^4 + 1624 \cdot N_Z^3 + 1764 \cdot N_Z^2 + 720 \cdot N_Z)$

Figure 10-3
U_R values as function of N_P and N_Z
For definition of symbols, see Fig. 10-2

10.3.1. Duplications

It may happen that combinations generated by interchanging variables between the **A** and **B** matrices, resulting in a singular **A**, occur in more than one way. This phenomenon is called *duplication* and its number is denoted by ϑ. Obviously, the number of duplications must be deducted from the U_R value (calculated in Art. 10.3.) since otherwise the *same* arrangement would be counted twice, or even more. The next example illustrates.

NUMBER OF SETS OF DIMENSIONLESS PRODUCTS OF VARIABLES   **243**

Example 10-9

Given the **D** and **C** matrices of a Dimensional Set as follows:

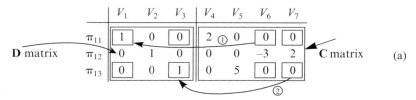

If we exchange V_6, V_7 with V_1, V_3, in this order, *considering the first row*, then we get a singular **A** (since a zero in **C** under V_6 replaces "1" under V_1 in **B**). By Theorem 9-8, however, this exchange is forbidden [arrow ① in (a)]. Similarly, in the third row, if we exchange the *same* two zeros (under V_6, V_7) with V_1, V_3, then again we obtain singular **A**. This exchange is also forbidden since a zero under V_7 replaces a "1" in **D** [arrow ② in (a)]. Therefore, in this example, the number of duplications is $\vartheta = 1$. Hence this number must be deducted from that previously determined for U_R.

⇑

Now the question naturally arises: is there a reasonably easy way to recognize duplications, so that their number is properly accounted for? The answer is *yes*. All we have to do is find duplications of two or more zeros occurring in the same group of *columns* in matrix **C**. For instance, in (a) of Example 10-9, we have the **C** matrix

$$\mathbf{C} = \begin{bmatrix} 2 & 0 & \boxed{0 \quad 0} \\ 0 & 0 & -3 & 2 \\ 0 & 5 & \boxed{0 \quad 0} \end{bmatrix} \text{duplications}$$

in which the two zeros in the boxes appear in *pairs* and in the same group of *columns*, viz., in the third and the fourth. Therefore, it is a duplication. We further see that the involved columns are of the variables V_6, V_7—as stipulated in Example 10-9.

Example 10-10

Given the **C** matrix

$$\mathbf{C} = \begin{bmatrix} 0 & 2 & 0 \\ 4 & 0 & 0 \\ 0 & 0 & 0 \\ 0 & 0 & 1 \end{bmatrix} \quad (a)$$

What is the number of duplications? The best way to find out is to graphically present the **C** matrix, and then box and match the pairs of appropriate entries.

$$\left(\begin{bmatrix} \boxed{0} & 2 & \boxed{0} \\ 4 & 0 & 0 \\ \boxed{0} & 0 & \boxed{0} \\ 0 & 0 & 1 \end{bmatrix} \right) \quad \begin{bmatrix} 0 & 2 & 0 \\ 4 & \boxed{0 \quad 0} \\ 0 & \boxed{0 \quad 0} \\ 0 & 0 & 1 \end{bmatrix} \quad \begin{bmatrix} 0 & 2 & 0 \\ 4 & 0 & 0 \\ \boxed{0 \quad 0} & 0 \\ \boxed{0 \quad 0} & 1 \end{bmatrix} \quad (b)$$

Thus, there are just three pair-wise matchings of zeros, hence the number of duplications is $\vartheta = 3$.

⇑

Example 10-11

Given the **C** matrix

$$C = \begin{bmatrix} 2 & 4 & 3 & 0 & 2 & 8 \\ 0 & 0 & 0 & 1 & 4 & 0 \\ 0 & 0 & 0 & 0 & 0 & 2 \end{bmatrix}$$

What is the number of duplications? Again we box and pair the involved groups of zeros.

$$\begin{bmatrix} 2 & 4 & 3 & 0 & 2 & 8 \\ 0 & \boxed{0 \ 0} & 1 & 4 & 0 \\ 0 & \boxed{0 \ 0} & 0 & 0 & 2 \end{bmatrix} \quad \begin{bmatrix} 2 & 4 & 3 & 0 & 2 & 8 \\ \boxed{0 \ 0} & 0 & 1 & 4 & 0 \\ \boxed{0 \ 0} & 0 & 0 & 0 & 2 \end{bmatrix} \quad \begin{bmatrix} 2 & 4 & 3 & 0 & 2 & 8 \\ \boxed{0 \ 0} & 0 & 1 & 4 & 0 \\ \boxed{0 \ 0} & 0 & 0 & 0 & 2 \end{bmatrix} \quad \begin{bmatrix} 2 & 4 & 3 & 0 & 2 & 8 \\ \boxed{0} & 0 & \boxed{0} & 1 & 4 & 0 \\ \boxed{0} & 0 & \boxed{0} & 0 & 0 & 2 \end{bmatrix}$$

There are three pairs and one trio. Hence the number of duplications is $\vartheta = 3 + 1 = 4$.

⇑

10.4. NUMBER OF DISTINCT SETS

We are now in a position to determine the number of *distinct* sets of dimensionless variables. Recognizing that interchange of variables *within* either the **A** or **B** matrices does not yield distinct sets (Theorem 10-1, Theorem 10-2), we see that we need only to consider *intermixing* the variables *between* these two matrices.

If we have N_V variables and N_d dimensions, then obviously the **B** matrix consists of $N_V - N_d$ columns. Therefore the maximum possible number of different arrangements, k, is the number of combinations k of N_V taken $N_V - N_d$ at a time. Thus

$$k = \binom{N_V}{N_V - N_d} = \binom{N_V}{N_d} \quad (10\text{-}19)$$

Of course, we assume that the rank of the dimensional matrix R_{DM} is equal to the number of dimensions N_d, which is tantamount to the existence of a nonsingular **A** matrix. The table in Fig 10-4 presents the k values in terms of N_V and N_d.

Example 10-12

Given the dimensional matrix of six variables and two dimensions:

	V_1	V_2	V_3	V_4	V_5	V_6
d_1	2	−1	0	4	3	2
d_2	5	2	6	2	1	0

What is the maximum possible number of arrangements of variables yielding *distinct* sets of dimensionless variables?

NUMBER OF SETS OF DIMENSIONLESS PRODUCTS OF VARIABLES

		N_V											
		2	3	4	5	6	7	8	9	10	11	12	13
N_d	1	2	3	4	5	6	7	8	9	10	11	12	13
	2		3	6	10	15	21	28	36	45	55	66	78
	3			4	10	20	35	56	84	120	165	220	286
	4				5	15	35	70	126	210	330	495	715
	5					6	21	56	126	252	462	792	1287
	6						7	28	84	210	462	924	1716
	7							8	36	120	330	792	1716
	8								9	45	165	495	1287
	9									10	55	220	715
	10										11	66	286

Figure 10-4
Maximum possible number of distinct sets of dimensionless variables
N_V = number of variables; N_d = number of dimensions

Here we have $N_V = 6$, $N_d = 2$, and **A** is not singular. From the table in Fig. 10-4, or by relation (10-19), there are $k = 15$ such combinations of variables. Note that by (10-19) and the table, the same number is obtained if the number of dimensions is 4.

⇑

The number of distinct sets of dimensionless variables is k, as defined in (10-19), provided the following two conditions are fulfilled:

(i) in matrix **C** the rows are independent;
(ii) matrix **C** contains no zeros.

The case where condition (i) is not fulfilled is discussed in Art.10.5.2. If condition (ii) is not fulfilled, then the number of *distinct* sets N_S is

$$N_S = k - (U_R - \vartheta) - U_c = k - U_R - U_c + \vartheta \tag{10-20}$$

where k is as defined in (10-19) and Fig. 10-4;
U_R is the number of *prohibited* interchanges of variables between matrices **A** and **B**. The value of U_R is given in (10-16) and by Fig. 10-2.
U_c is the number of *possible* interchanges of variables between matrices **A** and **B** yielding *equivalent* sets of dimensionless variables. The method for determining U_c is given in Art. 10.2.
ϑ is the number of *duplications* associated with the U_R value. The method for determining ϑ is found in Art. 10.3.1.

We now present several examples showing the use of relation (10-20).

Example 10-13

Given the Dimensional Set

	V_1	V_2	V_3	V_4	V_5	
d_1	−20	3	1	2	3	
d_2	3	1	4	−5	1	**A** matrix
d_3	−23	4	2	1	4	
π_{11}	1	0	2	3	4	**C** matrix
π_{12}	0	1	0	0	−1	

(a)

How many distinct sets of dimensionless variables can be formed, and what are they?

First, we check the singularity of **A**. The determinant of **A** is −7, which is different from zero, hence **A** is not singular and we may proceed.

Next, we observe that $N_V = 5$, $N_d = 3$, so from the table in Fig. 10-4

$$k = 10 \tag{b}$$

This is the number of *distinct* sets of dimensionless variables *if* there were no zeros in **C**. But there are zeros in **C**, therefore we must also determine U_R, U_c, and ϑ, and then use relation (10-20).

Accordingly, by Fig. 10-2, the number of prohibited interchanges is

$$U_R = 0 + 3 = 3 \tag{c}$$

where the number "0" represents the first row of **C**, while the following number "3" represents the second row. We see that there is no duplication of zeros in the corresponding columns of **C**, hence

$$\vartheta = 0 \tag{d}$$

Next, we calculate U_c; i.e., the number of interchanges yielding *equivalent* sets. In **C** the number of nonzero elements in the first *row* occupying a *column*, in which all the other elements are zero, is 2 (columns of variables V_3 and V_4). The number of similar elements in the second row is 0. Hence

$$n_1 = 2; \qquad n_2 = 0 \tag{e}$$

Thus

$$U_c = n_1 + n_2 + n_1 \cdot n_2 = 2 + 0 + 0 = 2 \tag{f}$$

Therefore, by (10-20), (b), (c), (d) and (f)

$$N_S = k - U_R - U_c + \vartheta = 10 - 3 - 2 + 0 = 5 \tag{g}$$

Thus, there are only five distinct sets of dimensionless variables. Since $k = 10$, we have the following 10 possible combinations of variables (excluding, of course, those which involve interchanging variables only within **A** or within **B**):

NUMBER OF SETS OF DIMENSIONLESS PRODUCTS OF VARIABLES

$$
\begin{array}{lll}
\#1 & V_1 - V_2 - V_3 - V_4 - V_5 & \text{possible, distinct} \\
\#2 & V_1 - V_3 - V_2 - V_4 - V_5 & \text{impossible} \\
\#3 & V_1 - V_4 - V_2 - V_3 - V_5 & \text{impossible} \\
\#4 & V_1 - V_5 - V_2 - V_3 - V_4 & \text{possible, distinct} \\
\#5 & V_2 - V_3 - V_1 - V_4 - V_5 & \text{possible, equivalent} \\
\#6 & V_2 - V_4 - V_1 - V_3 - V_5 & \text{possible, equivalent} \\
\#7 & V_2 - V_5 - V_1 - V_3 - V_4 & \text{possible, distinct} \\
\#8 & V_3 - V_4 - V_1 - V_2 - V_5 & \text{impossible} \\
\#9 & V_3 - V_5 - V_1 - V_2 - V_4 & \text{possible, distinct} \\
\#10 & V_4 - V_5 - V_1 - V_2 - V_3 & \text{possible, distinct}
\end{array}
\qquad (h)
$$

Thus, there are $U_R = 3$ *impossible* sequences (#2, #3, #8), $U_c = 2$ *equivalent* sequences (#5, #6), and $N_S = 5$ *distinct* sequences (#1, #4, #7, #9, #10). To illustrate, let us construct a distinct sequence (#7) and an equivalent sequence (#6) with respect to the "basic sequence" #1.

The basic sequence (#1) dimensional set is shown in (a):

$$\pi_{11} = V_1 \cdot V_3^2 \cdot V_4^3 \cdot V_5^4; \qquad \pi_{12} = V_2 \cdot V_5^{-1} \qquad (i)$$

The #6 and #7 Dimensional Sets are:

Dimensional Set 6

	V_2	V_4	V_1	V_3	V_5
d_1	3	2	-20	1	3
d_2	1	-5	3	4	1
d_3	4	1	-23	2	4
π_{61}	1	0	0	0	-1
π_{62}	0	1	$\tfrac{1}{3}$	$\tfrac{2}{3}$	$\tfrac{4}{3}$

Dimensional Set 7

	V_2	V_5	V_1	V_3	V_4
d_1	3	3	-20	1	2
d_2	1	1	3	4	-5
d_3	4	4	-23	2	1
π_{71}	1	0	$\tfrac{1}{4}$	$\tfrac{1}{2}$	$\tfrac{3}{4}$
π_{72}	0	1	$\tfrac{1}{4}$	$\tfrac{1}{2}$	$\tfrac{3}{4}$

(j)

Hence

$$
\begin{aligned}
\pi_{61} &= V_2 \cdot V_5^{-1}; & \pi_{71} &= V_1^{1/4} \cdot V_2 \cdot V_3^{1/2} \cdot V_4^{3/4} \\
\pi_{62} &= V_1^{1/3} \cdot V_3^{2/3} \cdot V_4 \cdot V_5^{4/3}; & \pi_{72} &= V_1^{1/4} \cdot V_3^{1/2} \cdot V_4^{3/4} \cdot V_5
\end{aligned}
\qquad (k)
$$

Thus, by (i) and (k)

$$\pi_{61} = \pi_{12}; \qquad \pi_{62} = \pi_{11}^{1/3} \qquad \text{therefore } \mathbf{T}_6 = \begin{bmatrix} 0 & 1 \\ \dfrac{1}{3} & 0 \end{bmatrix}$$

$$\pi_{71} = \pi_{11}^{1/4} \cdot \pi_{12}; \qquad \pi_{72} = \pi_{11}^{1/4} \qquad \text{therefore } \mathbf{T}_7 = \begin{bmatrix} \dfrac{1}{4} & 1 \\ \dfrac{1}{4} & 0 \end{bmatrix}$$

Observe that #6 is indeed *equivalent* to #1, but that #7 is *distinct* from it, since $\mathbf{T}_7$ does not fulfill Definition 10-1 for equivalence (there are *two*, instead of one, elements in one of its columns).

Example 10-14

How many distinct sets of dimensionless variables can be formed by the Dimensional Set given below?

	V_1	V_2	V_3	V_4	V_5	V_6	
d_1	−10	−8	−8	2	2	4	
d_2	−10	4	−4	−1	2	2	$\mathbf{A}_1$ matrix
d_3	10	−4	2	1	−2	−1	
π_{11}	1	0	0	0	5	0	
π_{12}	0	1	0	4	0	0	$\mathbf{C}_1$ matrix
π_{13}	0	0	1	0	0	2	

(a)

We have six variables and three dimensions, and $\mathbf{A}_1$ is not singular. Therefore, by Fig. 10-4

$$k = 20 \tag{b}$$

is the number of distinct sets *if* there were no zeros in the **C** matrix—which is not the case here. Hence we have to determine U_R, U_c, and ϑ. For U_R we use the table in Fig 10-2 by which

$$U_R = 4 + 4 + 4 = 12 \tag{c}$$

where the three middle members represent the first, second, and third rows of **C**, respectively. Since there is no *duplication*,

$$\vartheta = 0 \tag{d}$$

Thus, by (c), we have 12 impossible sequences (combinations) of the six variables.

Next, we see that the number of nonzero elements in the first row (occupying a *column* in which all the other elements are zero) is 1. Similarly for the second and third rows. Therefore

$$n_1 = n_2 = n_3 = 1 \tag{e}$$

so that, since the **C** matrix has $N_R = 3$ rows, therefore, by (10-13) and Fig. 10-1, U_c is the sum of a series of

$$S = 2^{N_R} - 1 = 2^3 - 1 = 7 \tag{f}$$

members. Now by (e)

$$U_c = n_1 + n_2 + n_3 + n_1 \cdot n_2 + n_1 \cdot n_3 + n_2 \cdot n_3 + n_1 \cdot n_2 \cdot n_3 = 7 \tag{g}$$

(it is only a coincidence that in this example $S = U_c = 7$). Now by (10-20) and the values of k in (b), U_R in (c), ϑ in (d), and U_c in (g),

$$N_S = k - U_R - U_c + \vartheta = 20 - 12 - 7 + 0 = 1 \tag{h}$$

Astonishingly, we have only *one* distinct set of dimensionless variables, because the seven other *possible* ones are actually equivalent to the one defined in (a). This singular set is

$$\pi_{11} = V_1 \cdot V_5^5; \qquad \pi_{12} = V_2 \cdot V_4^4; \qquad \pi_{13} = V_3 \cdot V_6^2 \tag{i}$$

NUMBER OF SETS OF DIMENSIONLESS PRODUCTS OF VARIABLES **249**

For example, let us see what happens if variables V_3, V_6 are interchanged. The relevant submatrices S_1 and S_3 are determined by construction of (part of) Shift Matrix **S**.

	V_1	V_2	V_6	
V_1	1	0	0	
V_2	0	1	0	S_1
V_3	0	0	0	
V_4	0	0	0	
V_5	0	0	0	S_3
V_6	0	0	1	

(j)

By (9-54), (a), and (j), the transformation matrix **T** is

$$\mathbf{T} = (\mathbf{S}_1 + \mathbf{C}_1 \cdot \mathbf{S}_3)^{-1} = \begin{bmatrix} 1 & 0 & 0 \\ 0 & 1 & 0 \\ 0 & 0 & \frac{1}{2} \end{bmatrix} \quad (k)$$

We see that this matrix is diagonal, and hence the two sets are indeed equivalent. The connecting relations therefore are $\pi_{21} = \pi_{11}$; $\pi_{22} = \pi_{12}$; $\pi_{23} = \sqrt{\pi_{13}}$.

⇑

Example 10-15

In Example 10-11 we presented a **C** matrix which, for convenience, is repeated here.

$$\mathbf{C} = \begin{bmatrix} 2 & 4 & 3 & 0 & 2 & 8 \\ 0 & 0 & 0 & 1 & 4 & 0 \\ 0 & 0 & 0 & 0 & 0 & 2 \end{bmatrix} \quad (a)$$

The number of *duplications* in the above matrix (as determined in Example 10-11) is

$$\vartheta = 4 \quad (b)$$

We now also wish to determine the number of *equivalent* sets of dimensionless variables.

Since matrix **C** has 3 rows, therefore the **D** matrix must also have three rows, and thus the number of dimensionless variables is $N_P = 3$. But **D** is square, therefore it must have three columns. The number of variables N_V is the sum of the columns of matrices **C** and **D** (see Fig. 8-2 and Fig. 8-3). Therefore, $N_V = 6 + 3 = 9$. Moreover, since the number of dimensions is always equivalent to the number of columns of **C** (which in the present case is $N_d = 6$), by (10-19) and Fig. 10-4, the theoretical maximum number of distinct sequences of variables is

$$k = \binom{N_V}{N_d} = \binom{9}{6} = 84 \quad (c)$$

To determine the number of impossible combinations of variables U_R, we consult the table in Fig. 10-2. Accordingly, since we have $N_Z = 1, 4, 5$ zeros in rows 1, 2, 3, respectively, and since $N_P = 3$, therefore

$$U_R = 1 + 20 + 35 = 56 \quad (d)$$

250 APPLIED DIMENSIONAL ANALYSIS AND MODELING

Next, we count the number of columns of **C** where there is only one *nonzero* element. We see that there are four such columns, viz., columns 1, 2, 3, and 4 and that this distinguished element occurs three times in row 1, once in row 2, and zero times in row 3. Hence,

$$n_1 = 3; \qquad n_2 = 1; \qquad n_3 = 0 \qquad (e)$$

Thus, the number of equivalent sets is

$$U_c = n_1 + n_2 + n_3 + n_1 \cdot n_2 + n_1 \cdot n_3 + n_2 \cdot n_3 + n_1 \cdot n_2 \cdot n_3 = 7 \qquad (f)$$

We now have all the ingredients required to calculate the number of distinct sets of dimensionless variables. By (10-20), (c), (d), (f), and (b), then

$$N_S = k - U_R - U_c + \vartheta = 84 - 56 - 7 + 4 = 25 \qquad (g)$$

Thus, of the theoretical maximum of 84, only 25 are *distinct,* and the remaining 59 are either *impossible,* or are *equivalent* to one another.

⇑

10.5. EXCEPTIONS

There are two types of situations for which the preceding rules must be modified somewhat. The first involves the presence of a *dimensionally irrelevant* variable, and the second where there is a variable whose dimension is a constant times of that of another. These two exceptions are now explained separately.

10.5.1. Dimensionally Irrelevant Variable

The relevancy of variables, in general, will be discussed in detail in Art. 11. Here it is only mentioned briefly that a variable is *dimensionally irrelevant* if, solely because of its dimensional composition, it cannot be matched with the other physical variables present. For this reason, the designation *"dimensionally impossible"* would be more appropriate. But, to be linguistically consistent with the term *physical irrelevancy* (also discussed in Art. 11), we shall henceforth (albeit reluctantly) use the term *"dimensionally irrelevant."* Thus, a variable which, by its dimensional characteristics *alone,* has no place in a formula connecting other dimensionally compatible physical variables, is a *dimensionally irrelevant* variable. The next example illustrates.

Example 10-16.   Oscillating Period of a Simple Pendulum

We wish to express the oscillating period of a simple pendulum. The assumed variables involved, their symbols and dimensions are as follows:

Variable	Symbol	Dimension
length	L	m
mass	M	kg
gravitational acceleration	g	m·s^{-2}
oscillating period	t	s

So we may write

$$t = k \cdot L^{\epsilon_1} \cdot M^{\epsilon_2} \cdot g^{\epsilon_3} \qquad (a)$$

where k, ϵ_1, ϵ_2, and ϵ_3 are dimensionless constants to be found. According to the Law of Homogeneity (Art. 6.1), the dimensions of both sides of any equation must be identical. Hence, by (a),

$$m^0 \cdot kg^0 \cdot s^1 = m^{\epsilon_1} \cdot kg^{\epsilon_2} \cdot m^{\epsilon_3} \cdot s^{-2\epsilon_3} \qquad (b)$$

It is now obvious that $\epsilon_2 = 0$, since on the left-hand side the exponent of "kg" is 0, so on the right-hand side it must also be 0. But the *only* variable which has "kg" as dimension is the mass M. Hence mass is a *dimensionally irrelevant* variable in any formula expressing the period of oscillation of a simple pendulum.

⇑

Two questions now emerge: How do we recognize the existence of a dimensionally irrelevant variable in a Dimensional Set, and if there is such a variable in the set, what do we do with it?

To answer the first question, we state and prove the following simple theorem.

Theorem 10-4. *A variable is dimensionally irrelevant, if—and only if—the column of the* **C** *matrix (of the Dimensional Set) under that variable is all zeros.*

Note the *bidirectionality* of this theorem. That is, if the variable is dimensionally irrelevant, then the **C** matrix *does* have a zero column under that variable, *and* if there is such a zero column in **C**, then the variable of that column *is* dimensionally irrelevant.

Proof. If a column of **C** is all zeros, then obviously the variable corresponding to that column *cannot* be part of any dimensionless variable, because if it were, then somewhere in the relevant column of the **C** matrix, at least one element would have to be a nonzero. But, by assumption, there is no nonzero element there. Similarly and for the same reason, if a variable is dimensionally irrelevant, then the column of **C** under the variable cannot contain any nonzero elements. This proves the theorem.

At this point the inquisitive reader might ask: What happens if the dimensionally irrelevant variable is not in the **A** matrix (which is exactly above **C**), but in the **B** matrix, which is above **D**? However, as Theorem 10-5 (see below) states, this cannot happen.

Theorem 10-5. *If a variable is dimensionally irrelevant, then it cannot appear in the* **B** *matrix, i.e., it must appear in the* **A** *matrix, if it appears at all.*

Proof. Suppose the *j*th column of **C** corresponds to the irrelevant variable. Then, by Theorem 10-4, the *j*th column of **C** is all zeros. Therefore, if c_{ij} is an element of **C** in its *i*th row and *j*th column, then

$$c_{ij} = 0 \qquad i = 1, 2, \ldots, N_P$$

where N_P is the number of dimensionless variables which, by the construction of **C** and **D**, is identical to the number of rows in these two matrices. Now, by Theorem

9-8, if $c_{ij} = 0$, then the ith column of **B** cannot be exchanged with the jth column of **A**. But, by construction of the Dimensional Set, the jth column of **C** is the jth column of **B**, so none of the variables in **B** can be exchanged with the jth column of **C**, and hence the irrelevant variable (which, by assumption, is the jth variable in **A**) cannot be in the **B** matrix. This proves the theorem.

In summation, if, and only if, a column in the **C** matrix is *all* zeros, then the corresponding variable is dimensionally irrelevant. Consequently, if a column in **C** is not all zeros, then the variable is *not* dimensionally irrelevant.

Example 10-17

Consider Dimensional Sets 1 and 2

Dimensional Set 1

		V_1	V_2	V_3	V_4	V_5	V_6	
B$_1$	d_1	4	8	10	2	−2	4	**A**$_1$
	d_2	2	4	5	1	−1	3	
π_{11}		1	0	0	0	2	0	
π_{12}		0	1	0	0	4	0	**C**$_1$
π_{13}		0	0	1	0	5	0	
π_{14}		0	0	0	1	1	0	

Dimensional Set 2

		V_1	V_6	V_3	V_4	V_5	V_2	
B$_2$	d_1	4	4	10	2	−2	8	singular **A**$_2$
	d_2	2	3	5	1	−1	4	
π_{21}		1	0	0	0			
π_{22}		0	1	0	0	this does		
π_{23}		0	0	1	0	not exist		**C**$_2$
π_{24}		0	0	0	1			

We see that in **C**$_1$ the column under V_6 is all zeros. Therefore, by Theorem 10-4, V_6 is a dimensionally irrelevant variable. Now if we attempted to place V_6 in the **B** matrix, then we would invariably get a singular **A**. For example, let us put V_6 in place of V_2 (which is in the **B**$_2$ matrix) as in Set 2. We see that **A**$_2$ is singular and hence **C**$_2$ does not exist. There are similar results if V_6 is interchanged with *any* of the variables in **B**$_1$, as the skeptical reader should verify.

⇑

Next, we deal with the *second* question posed earlier. What do we do if there is a dimensionally irrelevant variable in a Dimensional Set? To answer, we state and prove the following theorem.

Theorem 10-6. *Removal of n dimensionally irrelevant variables from a dimensional matrix diminishes its rank by n.*

Proof. By (7-26), the number of dimensionless variables N_P is

$$N_P = N_V - R_{DM} \qquad \text{repeated (7-26)}$$

where N_V is the number of variables (whether relevant or irrelevant) and R_{DM} is the rank of the dimensional matrix.

Obviously N_P is not changed by the removal of a dimensionally irrelevant variable, because the latter does not appear in any of the dimensionless products (variables). Therefore, by (7-26), the difference between N_V and R_{DM} must remain constant. Consequently, if N_V is diminished by n, then R_{DM} must also be diminished by n. This proves the theorem. The following example illustrates.

Example 10-18

Given the Dimensional Set in which V_4 is a dimensionally irrelevant variable, since the column of **C** under V_4 is all zeros:

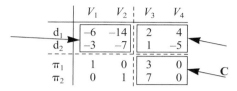

The dimensional matrix is

$$\begin{bmatrix} -6 & -14 & 2 & 4 \\ -3 & -7 & 1 & -5 \end{bmatrix} \leftarrow \text{dimensional matrix 1}$$
$$\mathbf{B}\mathbf{A}$$

whose rank is obviously 2, since **A** is nonsingular. Now we remove V_4 (the dimensionally irrelevant variable) from the set. The new dimensional matrix thus becomes

$$\begin{bmatrix} -6 & -14 & 2 \\ -3 & -7 & 1 \end{bmatrix} \leftarrow \text{dimensional matrix 2}$$

whose rank cannot be 2 since the first row is twice the second row, therefore the rank is 1. Thus, the rank of dimensional matrix 2 is one less than that of dimensional matrix 1, and we note that the number of dimensionally irrelevant variables removed from the set is also 1.

⇑

Therefore, if we have a dimensionally irrelevant variable, we should follow the following procedure:

Step 1. Remove the dimensionally irrelevant variable from the Dimensional Set.
Step 2. Remove one dimension from the Dimensional Set. This selected dimension should be such that the rank of the new dimensional matrix is 1 less than that of the original one.
Step 3. Construct the new Dimensional Set and see if there is still a dimensionally irrelevant variable in it. If "yes," go to Step 1; if "no," go to Step 4.
Step 4. Proceed normally, as described in Art. 10.4.

The following example illustrates this process.

Example 10-19

Given the Dimensional Set 1 as shown.

(i) How many dimensionally irrelevant variables are there?
(ii) How many distinct sets of dimensionless variables are there?

Dimensional Set 1

	V_1	V_2	V_3	V_4	V_5	V_6
$\mathbf{B_1}$ d_1	−14	7	−6	1	2	4
d_2	2	−6	−2	2	1	−2
d_3	−5	1	−3	1	−2	1
π_{11}	1	0	0	2	0	3
π_{12}	0	1	0	1	0	−2
π_{13}	0	0	1	2	0	1

with $\mathbf{A_1}$ and $\mathbf{C_1}$ labeled on the right. (a)

We see that V_5 is dimensionally irrelevant because the column in $\mathbf{C_1}$ under V_5 is all zeros. Moreover, V_5 is the *only* variable whose column in $\mathbf{C}$ is all zeros. Therefore, we have one dimensionally irrelevant variable in the set. We also note that the rank of the dimensional matrix is $(R_{DM})_1 = 3$.

The first *three* steps in the process are then:

Step 1. Remove variable V_5
Step 2. Remove dimension d_3
Step 3. Construct (new) Dimensional Set 2

By these steps we obtain

Dimensional Set 2

	V_1	V_2	V_3	V_4	V_6
$\mathbf{B_2}$ d_1	−14	7	−6	1	4
d_2	2	−6	−2	2	−2
π_{21}	1	0	0	2	3
π_{22}	0	1	0	1	−2
π_{23}	0	0	1	2	1

with $\mathbf{A_2}$ and $\mathbf{C_2}$ labeled on the right. (b)

We note that by (a) and (b)

$$\left. \begin{array}{l} \pi_{21} = \pi_{11} = V_1 \cdot V_4^2 \cdot V_6^3 \\ \pi_{22} = \pi_{12} = V_2 \cdot V_4 \cdot V_6^{-2} \\ \pi_{23} = \pi_{13} = V_3 \cdot V_4^2 \cdot V_6 \end{array} \right\} \quad (c)$$

Thus, as expected, the sets of dimensionless variables did not change. We also note that the rank of dimensional matrix 2 is $(R_{DM})_2 = 2$. Therefore,

$$(R_{DM})_1 - 1 = (R_{DM})_2$$

as prescribed by Theorem 10-6.

Step 4. We now have a set in which every variable is dimensionally relevant—so we can proceed. There are five variables and two dimensions, i.e., $N_V = 5$, $N_d = 2$, hence the maximum number of combinations of variables in Set 2 is, by Fig. 10-4 and (10-19)

$$k = 10 \quad (d)$$

The number of *prohibited* interchanges is, by Fig. 10-2

$$U_R = 0 \quad (e)$$

NUMBER OF SETS OF DIMENSIONLESS PRODUCTS OF VARIABLES **255**

Since there is no zero in $\mathbf{C}_1$, the number of *duplications*, as described in Art. 10.3.1, is

$$\vartheta = 0 \tag{f}$$

The number of sequences of variables yielding *equivalent* sets of dimensionless variables is

$$U_c = 0 \tag{g}$$

since there are no columns in $\mathbf{C}_2$ consisting of only one nonzero element. Consequently, by (10-20),

$$N_S = k - U_R - U_c + \vartheta = 10 - 0 - 0 + 0 = 10 \tag{h}$$

i.e., there are ten *distinct* sets of dimensionless variables. Note that since $N_S = k$, all possible exchanges of variables between matrices **A** and **B** yield distinct sets.

⇑

10.5.2. In Matrix C, One Row is a Multiple of Another Row

Sometimes in a Dimensional Set, in **C**, the elements of one row are a constant multiple of that of another row. As will be shown presently, this set can be always transformed into another one in which a row in **C** consists of only *one* nonzero element. Consequently, the original set cannot be viewed in a "regular" way, even though its two involved rows contain no zeros.

To deal with this type of case we state and prove a theorem, followed by two examples.

Theorem 10-7. *If in matrix **C** (of a Dimensional Set) the ith row is q times the jth row, and if the ith column of matrix **B** is interchanged with the kth column of matrix **A**, then the jth row of **C** will become all zeros, except for its kth element, which will be $(-1/q)$.*

This rather forbidding text can be easily tamed by the following simple "clarification." If, say, the second *row* of **C** is 4 times its third *row*, then the third column of **B** is four times its second column. We then simply interchange the second column in **B** to any (kth) column in **A**. This will result in all zeros in the third row of **C**, except the kth element, which will be $-\frac{1}{4}$. The examples following the proof will greatly mitigate any remaining confusion.

Proof. The proof is in two parts.

Part 1. We prove that if in $\mathbf{C}_2$ the *i*th row is q times the *j*th row, then in $\mathbf{B}_2$ the *i*th *column* is q times the *j*th *column*.

By Theorem 9-5, if

$$\mathbf{C}_2 = \mathbf{Z} \cdot \mathbf{C}_1 \tag{10-21}$$

then

$$\mathbf{B}_2 = \mathbf{B}_1 \cdot \mathbf{Z}^T \tag{10-22}$$

for whatever compatible matrices. If in (10-21) $\mathbf{Z}$ produces $\mathbf{C}_2$ in which the ith row is q times the jth row, then, by (10-22), in $\mathbf{B}_2$ the ith *column* is q times the jth column. This proves Part 1 of the theorem.

Example 10-20

Consider, in agreement with (10-21),

$$\underbrace{\begin{bmatrix} 5 & 6 & 8 \\ 6 & 4 & 2 \\ 3 & 2 & 1 \end{bmatrix}}_{\mathbf{C}_2} = \underbrace{\begin{bmatrix} 1 & 0 & 0 \\ 0 & 0 & 2 \\ 0 & 0 & 1 \end{bmatrix}}_{\mathbf{Z}} \cdot \underbrace{\begin{bmatrix} 5 & 6 & 8 \\ 29 & 37 & 77 \\ 3 & 2 & 1 \end{bmatrix}}_{\mathbf{C}_1}$$

We see that in $\mathbf{C}_2$ the second row is twice ($q = 2$) the third row. Therefore, by (10-22),

$$\underbrace{\begin{bmatrix} 6 & 56 & 28 \\ 3 & -12 & -6 \\ 13 & 18 & 9 \end{bmatrix}}_{\mathbf{B}_2} = \underbrace{\begin{bmatrix} 6 & 17 & 28 \\ 3 & 4 & -6 \\ 13 & 8 & 9 \end{bmatrix}}_{\mathbf{B}_1} \cdot \underbrace{\begin{bmatrix} 1 & 0 & 0 \\ 0 & 0 & 0 \\ 0 & 2 & 1 \end{bmatrix}}_{\mathbf{Z}^\mathrm{T}}$$

in which the second column of $\mathbf{B}_2$ is indeed twice ($q = 2$) the third column. ⇑

Part 2. We prove that if in $\mathbf{B}$ the jth column is q times the ith column, and if the jth column in $\mathbf{B}$ is interchanged with the kth column in $\mathbf{A}$, then the ith row in $\mathbf{C}$ will be all zeros, except for the kth element, which will be $(-1/q)$.

The reader will easily see that this point is really only different wording of Theorem 9-11, which was already proved. Therefore Part 2 is proved.

In summary, it has been shown that if in $\mathbf{C}$ one of the *rows* is a multiple of another *row*, then a transformation exists yielding a (new) $\mathbf{C}$ matrix in which there is a row composed of only one nonzero element. The following example illustrates:

Example 10-21

Consider the following two Dimensional Sets:

Dimensional Set 1

	V_1	V_2	V_3	V_4	V_5	V_6
d_1	2	0	4	1	2	3
d_2	3	4	6	4	-2	-1
d_3	1	6	2	1	2	4
π_{11}	1	0	0	$\dfrac{-7}{5}$	$\dfrac{-9}{5}$	1
π_{12}	0	1	0	$\dfrac{8}{5}$	$\dfrac{41}{5}$	-6
π_{13}	0	0	1	$\dfrac{-14}{5}$	$\dfrac{-18}{5}$	2

Dimensional Set 2

	V_1	V_2	V_3	V_4	V_5	V_6
d_1	2	0	2	1	4	3
d_2	3	4	-2	4	6	-1
d_3	1	6	2	1	2	4
π_{21}	1	0	0	0	$\dfrac{-1}{2}$	0
π_{22}	0	1	0	$\dfrac{-43}{9}$	$\dfrac{41}{18}$	$\dfrac{-13}{9}$
π_{23}	0	0	1	$\dfrac{7}{9}$	$\dfrac{-5}{18}$	$\dfrac{-5}{9}$

NUMBER OF SETS OF DIMENSIONLESS PRODUCTS OF VARIABLES   **257**

We see that in C_1 the third *row* is two times ($q = 2$) the first *row*. Therefore, in B_1 the third *column* is two times the first *column*. Now, if the third column of B_1 (variable V_3) is exchanged with the second ($k = 2$) column in A_1 (variable V_5), then the first row of C_2 will have only one nonzero element and this will occur in its second ($k = 2$) *column*, under the newly placed variable (V_3). The value of this element is $-(1/q) = -(1/2)$, as expected.

⇑

Therefore, the procedure is as follows: if in **C**, a *row* is a multiple of another *row*, then

Step 1. In **C** identify one of these *rows* as the *i*th, the other as the *j*th
Step 2. In **B** identify the *i*th and *j*th *columns*
Step 3. Exchange the *j*th column of **B** with the *k*th column of **A** (*k* is arbitrary)
Step 4. Determine the new **C** matrix, in which the *i*th row will have only one nonzero element occurring in the *k*th column
Step 5. Proceed as described in Art. 10.4

The following example demonstrates this process:

Example 10-22

Given the Dimensional Set 1

Dimensional Set 1

	V_1	V_2	V_3	V_4	V_5	V_6	V_7
d_1	13	−7	39	1	2	−2	−1
d_2	−2	−6	−6	0	2	4	1
d_3	−12	3	−36	2	−2	−1	2
d_4	−2	−3	−6	2	0	−1	1
π_{11}	1	0	0	1	−4	2	2
π_{12}	0	1	0	3	2	1	−2
π_{13}	0	0	1	3	−12	6	6

(a)

How many distinct sets of dimensionless variables are there?
We notice in C_1 that the third row is three times ($q = 3$) the first row. Thus we follow the five-step procedure outlined above:

Step 1. Identify the first row in C_1 as the *i*th row, and the third row as the *j*th row. Therefore $i = 1$ and $j = 3$
Step 2. Identify the first and third columns in **B**. These will be variables V_1 and V_3
Step 3. Exchange the third variable of **B** (V_3) with, say, the fourth variable of **A** (V_7), i.e., $k = 4$. Note that k can be any integer between 1 and 4 (since there are four columns in **A**)
Step 4. Determine the "new" **C** matrix (C_2) by constructing Dimensional Set 2, which incorporates the indicated exchanges (Step 3)

Dimensional Set 2

	V_1	V_2	V_7	V_4	V_5	V_6	V_3
d_1	13	−7	−1	1	2	−2	39
d_2	−2	−6	1	0	2	4	−6
d_3	−12	3	2	2	−2	−1	−36
d_4	−2	−3	1	2	0	−1	−6
π_{21}	1	0	0	0	0	0	$-\frac{1}{3}$
π_{22}	0	1	0	4	−2	3	$\frac{1}{3}$
π_{23}	0	0	1	$\frac{1}{2}$	−2	1	$\frac{1}{6}$

(b)

We notice—with satisfaction—that in the first ($i = 1$) row of $\mathbf{C}_2$, there is only one nonzero element, and this is found in the fourth ($k = 4$) column under variable V_3, which is the variable transported into matrix $\mathbf{A}$. The value of the only nonzero element (in the first row of $\mathbf{C}_2$) is $-(1/q) = -(1/3)$, as expected.

Step 5. Now we can determine the number of distinct sets of dimensionless variables by the method described in Art.10.4. Accordingly, by (10-20)

$$N_S = k - U_R - U_c + \vartheta \qquad \text{repeated (10-20)}$$

where the symbols are as defined in Art.10.4. In this specific case $N_V = 7$, $N_d = 4$ (i.e., seven variables and four dimensions), so, by Fig. 10-4,

$$k = 35 \qquad (c)$$

Next, we note that in (b)—i.e., Set 2—the numbers of zeros in $\mathbf{C}$ are

$$\left. \begin{array}{ll} \text{in the first row} & N_Z = 3 \\ \text{in the second row} & N_Z = 0 \\ \text{in the third row} & N_Z = 0 \end{array} \right\} \qquad (d)$$

Hence, by Fig 10-2, since $N_P = 3$

$$U_R = 10 + 0 + 0 = 10 \qquad (e)$$

and since there is no duplication,

$$\vartheta = 0 \qquad (f)$$

Finally, in $\mathbf{C}_2$ there is no column with only one nonzero element. Therefore

$$U_c = 0 \qquad (g)$$

Thus, by (10-20), (c), (e), (f), and (g),

$$N_S = k - U_R - U_c + \vartheta = 35 - 10 - 0 + 0 = 25 \qquad (h)$$

which is the number of distinct sets of dimensionless variables.

⇑

10.6. PROBLEMS

Solutions are in Appendix 6.

10/1 Prove that if **A** = **B** in a Dimensional Set, then—regardless of the number of variables—there is but one distinct set of dimensionless variables. Demonstrate this fact for a case of six variables.

10/2 With respect to the set of dimensionless variables

$$\pi_1 = \left(\frac{V_1^{24} \cdot V_4^{12} \cdot V_6^{18}}{V_5^{27}}\right)^{1/24}; \quad \pi_2 = \left(\frac{V_2^{24} \cdot V_5^{21} \cdot V_6^{10}}{V_4^{44}}\right)^{1/24}; \quad \pi_3 = \left(\frac{V_3^8 \cdot V_4^4}{V_5^5 \cdot V_6^2}\right)^{1/8}$$

which sets presented below are *distinct*, and which are *equivalent*?

(a) $\pi_1 = \dfrac{V_1 \cdot V_3^3 \cdot V_4^2}{V_5^3}; \quad \pi_2 = V_1^2 \cdot V_2 \cdot \left(\dfrac{V_6^{46}}{V_4^{20} \cdot V_3^{33}}\right)^{1/24}; \quad \pi_3 = V_1^2 \cdot V_2 \cdot V_3^3 \cdot \left(\dfrac{V_4^8 \cdot V_6^{14}}{V_5^{39}}\right)^{1/12}$

(b) $\pi_1 = V_1 \cdot \left(\dfrac{V_4^{12}}{V_5^{27} \cdot V_2^{27}}\right)^{1/24}; \quad \pi_2 = \dfrac{1}{V_3} \cdot \left(\dfrac{V_5^5 \cdot V_6^2}{V_4^4}\right)^{1/8}; \quad \pi_3 = \left(\dfrac{V_4^{44}}{V_2^{24} \cdot V_5^{21} \cdot V_6^{10}}\right)^{1/72}$

(c) $\pi_1 = \left(\dfrac{V_1^8 \cdot V_4^4 \cdot V_6^6}{V_5^9}\right)^{1/40}; \quad \pi_2 = \left(\dfrac{V_5^{55} \cdot V_6^{22}}{V_3^{88} \cdot V_4^{44}}\right)^{1/144}; \quad \pi_3 = \left(\dfrac{V_2^{24} \cdot V_5^{21}}{V_4^{44}}\right)^{1/8}$

(d) $\pi_1 = V_3 \cdot \left(\dfrac{V_1^8 \cdot V_4^{12} \cdot V_6^2}{V_5^{19}}\right)^{1/16}; \quad \pi_2 = \left(\dfrac{V_2^{24} \cdot V_5^{21} \cdot V_6^{10}}{V_4^{44}}\right)^{1/24}; \quad \pi_3 = \dfrac{1}{V_1^2 \cdot V_4} \left(\dfrac{V_5^9}{V_6^6}\right)^{1/4}$

(e) $\pi_1 = \left(\dfrac{V_5^9}{V_1^8 \cdot V_4^4 \cdot V_6^6}\right)^{1/24}; \quad \pi_2 = \left(\dfrac{V_4^{44}}{V_2^{24} \cdot V_5^{21} \cdot V_6^{10}}\right)^{1/288}; \quad \pi_3 = \left(\dfrac{V_3^8 \cdot V_4^4}{V_5^5 \cdot V_6^2}\right)^{1/216}$

10/3 Given the $\mathbf{C}_1$ matrix of Dimensional Set 1 comprised of five variables

	V_1	V_2	V_3	V_4	V_5
π_1	1	0	2	4	−3
π_2	0	1	4	8	−6

← $\mathbf{C}_1$

Based on the same five variables whose dimensional compositions are not given, construct matrices $\mathbf{C}_2$ and $\mathbf{C}_3$ such that (a) the first row of $\mathbf{C}_2$ will be [x 0 0]; (b) the second row of $\mathbf{C}_3$ will be [0 y 0]. Determine the numerical values of x and y. Hint: use Theorem 10-7 and relations (9-53), (9-64).

10/4 Given the **C** matrix of a Dimensional Set of six variables whose dimensional compositions are not known.

	V_1	V_2	V_3	V_4	V_5	V_6
π_1	1	0	0	4	5	8
π_2	0	1	0	0	6	10
π_3	0	0	1	0	7	−14

← **C**

- **(a)** What is the number of *impossible* combinations of variables?
- **(b)** With respect to the above given Set, how many *equivalent* sets of dimensional variables exist?
- **(c)** How many *distinct* sets of dimensionless variables exist?

10/5 Given the **C** matrix of a Dimensional Set

$$\mathbf{C} = \begin{bmatrix} 2 & 0 & 0 & 0 & 0 \\ 0 & 1 & 5 & 6 & 0 \\ 0 & 0 & 0 & 0 & 5 \end{bmatrix}$$

- **(a)** How many variables and dimensions are in the Dimensional Set?
- **(b)** How many dimensionless variables are in the Dimensional Set?
- **(c)** What is the number of *impossible* combinations of variables?
- **(d)** How many *duplications* are there?
- **(e)** How many *equivalent* sets of dimensionless variables exist?
- **(f)** How many *distinct* sets exist?
- **(g)** What is the *maximum* number of *distinct* sets of dimensionless variables if there are eight (physical) variables and five dimensions?

10/6 Given the following two sets of dimensionless variables:

$$\pi_{11} = \frac{V_1 \cdot V_4^2 \cdot V_7^2}{V_5}; \quad \pi_{12} = \frac{V_2 \cdot V_4^3 \cdot V_6 \cdot V_7^4}{V_5^2}; \quad \pi_{13} = \frac{V_3 \cdot V_4^2 \cdot V_6^2}{V_5}$$

$$\pi_{21} = \frac{V_1^2 \cdot V_4}{V_2 \cdot V_6}; \quad \pi_{22} = \frac{V_2 \cdot V_6 \cdot V_7}{\sqrt{V_1^3 \cdot V_5}}; \quad \pi_{23} = \frac{V_2^2 \cdot V_3 \cdot V_6^4}{V_1^4 \cdot V_5}$$

What is the relation between these two sets, i.e., what is the transformation matrix **T**? Hint: Use the method outlined in the solution of Problem 10/2.

10/7 Given the dimensional matrix of seven variables and three dimensions

	V_1	V_2	V_3	V_4	V_5	V_6	V_7
d_1	2	4	-4	8	1	-2	1
d_2	-2	-4	4	-8	3	2	-1
d_3	4	8	-8	16	1	-4	3

Which—if any—variable(s) is (are) *dimensionally irrelevant*?

10/8 Given the **C** matrix of a dimensional set

$$\mathbf{C} = \begin{bmatrix} 5 & 6 & 0 & 8 & 0 \\ 0 & 6 & 7 & 0 & 9 \end{bmatrix}$$

How many *distinct* sets of dimensionless variables are there?

10/9 Given the **C** matrix of a Dimensional Set

$$\mathbf{C} = \begin{bmatrix} 5 & -1 & 3 \\ 2 & 0 & 4 \\ 2 & 0 & -1 \end{bmatrix}$$

How many *distinct* sets of dimensionless variables are there?

10/10 Given the dimensional matrix of six variables and three dimensions

	V_1	V_2	V_3	V_4	V_5	V_6
d_1	−5	0	5	1	2	3
d_2	−2	8	−14	2	−1	−2
d_3	4	1	2	1	−2	2

How many *distinct* sets of dimensionless variables are there?

CHAPTER 11
RELEVANCY OF VARIABLES

In Art. 10.5.1 we briefly discussed the occurrence of dimensionally irrelevant variables in a Dimensional Set. We pointed out and proved that if a column of the **C** matrix in a Dimensional Set is all zeros, then the variable corresponding to that column is dimensionally irrelevant, i.e., the variable—because of its dimensional incompatibility—cannot occur in any relation connecting the remaining variables. In this chapter, we will discuss the general and particular relevancies of variables and other pertinent matters in greater detail.

11.1. DIMENSIONAL IRRELEVANCY

11.1.1. Condition

As mentioned, a variable is dimensionally irrelevant if *solely* because of its dimension, it *cannot* be part of any relation among the variables. By Theorem 10-4 if a variable is dimensionally irrelevant, then the column under that variable in the **C** matrix is all zeros—and conversely. Now we state and prove some other related and useful theorems, and then present some illustrative examples.

Theorem 11-1. *If a variable has a fundamental dimension which no other variable has, then this variable is dimensionally irrelevant.*

Here, of course, the distinction must be made between *fundamental dimension,* and *dimension*. The former, by definition, is a single entity, while the latter may comprise several fundamental dimensions joined by (maybe) repeated multiplications and divisions. For example, if v is a symbol for linear speed, then the dimension of v is $[v] = $ m/s, where "m" and "s" are fundamental dimensions, but m/s is the dimension of speed.

Proof. The proof involves four variables V_1, V_2, V_3, and V_4 and three fundamental dimensions d_1, d_2, and d_3 (which can be extended at will to any number of these ingredients). Suppose V_4 is expressed by the (arbitrary) power of dimensions $d_1^{\epsilon_1} \cdot d_2^{\epsilon_2} \cdot d_3^{\epsilon_3}$, and the rest of the variables only by dimensions d_1, d_2. Therefore, V_4 is the only variable which has the dimension d_3. Generally, for any physical relation among four variables we can write

$$\Psi\{V_1, V_2, V_3, V_4\} = c = \text{const} \qquad (11\text{-}1)$$

where Ψ is the symbol for function, and c is a dimensionless constant (i.e., it has the dimension of "1"). By the Law of Dimensional Homogeneity, both sides of (11-1) must have the same dimension. Since the right side has the dimension "1," then the left side must also have it. Hence

$$[V_1]^{\epsilon_1} \cdot [V_2]^{\epsilon_2} \cdot [V_3]^{\epsilon_3} \cdot [V_4]^{\epsilon_4} = d_1^0 \cdot d_2^0 \cdot d_3^0 = 1 \qquad (11\text{-}2)$$

where $\epsilon_1, \ldots, \epsilon_4$ are some exponents, and the designation $[x]$ for any x means "the dimension of x." In general, if $[V_4] = d_1^\alpha \cdot d_2^\beta \cdot d_3^\gamma$, then (11-2) can be written

$$[V_1]^{\epsilon_1} \cdot [V_2]^{\epsilon_2} \cdot [V_3]^{\epsilon_3} \cdot (d_1^\alpha \cdot d_2^\beta \cdot d_3^\gamma)^{\epsilon_4} = d_1^0 \cdot d_2^0 \cdot d_3^0 \qquad (11\text{-}3)$$

Now, by assumption, V_1, V_2, and V_3 do not contain d_3, only V_4 does. Therefore, by (11-3), $\gamma \cdot \epsilon_4 = 0$. But $\gamma \neq 0$, therefore $\epsilon_4 = 0$. Thus the exponent of V_4 is zero, and hence it cannot be part of any relation. This proves the theorem.

Example 11-1. **Free Vibration of a Massive Body on a Weightless Spring (I)**

A body, suspended on a weightless coil spring, vibrates freely in a vertical direction (Fig. 11-1). We wish to determine the frequency of this vibration. The assumed relevant variables are:

Variable	Symbol	Dimension
frequency of vibration	ν	s^{-1}
mass of body	M	kg
spring constant	k	$kg \cdot s^{-2}$
gravitational acceleration	g	$m \cdot s^{-2}$

Figure 11-1
Body suspended on a helical spring vibrates in a vertical direction

The spring constant k is the *force* necessary to deflect the spring in unit *distance*. Therefore, the dimension of k is

$$[k] = \frac{[\text{force}]}{[\text{distance}]} = \frac{\text{m·kg/s}^2}{\text{m}} = \text{kg·s}^{-2}$$

Gravitational acceleration is included, because it was *felt* that since the mass moves *vertically*, gravitation influences the motion.

Based on the above list of variables, the dimensional matrix will be

	v	M	k	g
m	0	0	0	1
kg	0	1	1	0
s	-1	0	-2	-2
π_1	1	$\frac{1}{2}$	$-\frac{1}{2}$	0

← C

We see that g is the only variable in which the (fundamental) dimension "m" appears; therefore, by Theorem 11-1, g is a *dimensionally irrelevant* variable. This also can be verified by the all-zero column under "g" in the **C** matrix. The fact that **C** consists of only one row is, of course, immaterial.

By the above Dimensional Set, the sole dimensionless "variable" (a constant in this case) is

$$\pi_1 = v \cdot \sqrt{\frac{M}{k}} = c = \text{constant} \qquad (a)$$

in which, of course, g does not appear. By (a),

$$v = c \cdot \sqrt{\frac{k}{M}} \qquad (b)$$

where the constant c happens to be $1/(2 \cdot \pi)$.

Relation (b) purports that—contrary to any of our *feelings*—the frequency of vibration is actually independent of the effect of gravitation; the mass vibrates with the same frequency on Earth as it does on the Moon.

Since g is *dimensionally irrelevant*, by Theorem 10-6, it can be removed. But g is the only variable in which the dimension "m" appears. Hence if there is no g, then there is no "m" in the Dimensional Set either. Thus, we have

	v	M	k
kg	0	1	1
s	-1	0	-2
π_1	1	$\frac{1}{2}$	$-\frac{1}{2}$

resulting in the dimensionless "variable"

$$\pi_1 = v \cdot \sqrt{\frac{M}{k}} = c = \text{constant} \qquad (c)$$

which, of course, is identical to (a), as expected and required.

Sometimes it is possible to recognize the presence of *and* identify a dimensionally irrelevant variable without constructing a Dimensional Set (and hence the **C** matrix), and even if the variable is not the "odd dimension" type (discussed in Example 11-1). The following example illustrates.

Example 11-2

Consider the dimensional matrix of six variables and three dimensions

	V_1	V_2	V_3	V_4	V_5	V_6
d_1	1	2	3	4	5	1
d_2	2	4	6	8	10	−3
d_3	3	2	−1	4	3	2

The rank of this matrix is clearly $R_{DM} = 3$, since the rightmost third-order determinant

$$\begin{vmatrix} 4 & 5 & 1 \\ 8 & 10 & -3 \\ 4 & 3 & 2 \end{vmatrix} = -40 \neq 0$$

However, if we remove variable V_6, then the rank becomes $R_{DM} = 2$ since, in this case, the second row becomes twice the first row. Therefore the removal of V_6 reduces the rank of the dimensional matrix and hence, by Theorem 10-6, V_6 is a *dimensionally irrelevant* variable. To verify this assertion, we construct the Dimensional Set

	V_1	V_2	V_3	V_4	V_5	V_6
d_1	1	2	3	4	5	1
d_2	2	4	6	8	10	−3
d_3	3	2	−1	4	3	2
π_1	1	0	0	$-\frac{3}{2}$	1	0
π_2	0	1	0	$-\frac{1}{2}$	0	0
π_3	0	0	1	$\frac{7}{4}$	−2	0

We see that the column of **C** under V_6 is all zeros, and hence the variable is indeed dimensionally irrelevant (see also Theorem 10-4).

⇑

A notable property of a dimensionally irrelevant variable is stated in the following theorem:

Theorem 11-2. *If a variable in a dimensional matrix is dimensionally irrelevant, then it can have any dimension and still remain dimensionally irrelevant.*

Proof. If a variable is dimensionally irrelevant, then it does not (cannot) appear in any dimensionless variable, nor in any relation connecting the rest of the variables.

If, however, the variable is not present, then obviously its dimension is immaterial. This proves the theorem. The following example illustrates:

Example 11-3

Consider the dimensional matrix of six variables and three dimensions

	V_1	V_2	V_3	V_4	V_5	V_6
d_1	1	2	3	4	5	11
d_2	2	4	6	8	10	7
d_3	3	2	−1	4	3	−19

This matrix is the same as presented in Example 11-2, except for the dimension of variable V_6, which was $d_1 \cdot d_2^{-3} \cdot d_3^2$ instead of the above $d_1^{11} \cdot d_2^7 \cdot d_3^{-19}$. But, as we saw in Example 11-2, V_6 was a dimensionally irrelevant variable, hence its dimension should not matter. Therefore in this set, V_6 should still be a dimensionally irrelevant variable. To verify this, we construct the corresponding Dimensional Set to see if the column in **C** under V_6 is still all zeros.

	V_1	V_2	V_3	V_4	V_5	V_6
d_1	1	2	3	4	5	11
d_2	2	4	6	8	10	7
d_3	3	2	−1	4	3	−19
π_1	1	0	0	−1.5	1	0
π_2	0	1	0	−0.5	0	0
π_3	0	0	1	1.75	−2	0

The column in **C** under V_6 is indeed all zeros, verifying the dimensional irrelevancy of that variable.

⇑

11.1.2. Adding a Dimensionally Irrelevant Variable to a Set of Relevant Variables

How do we *add* a dimensionally irrelevant variable to an existing set, the members of which are all relevant? The process is the exact reverse of what was described in Art. 10.5.1, where the technique of *removing* a dimensionally irrelevant variable from the Dimensional Set was given. Accordingly, the simple two-step procedure to add a dimensionally irrelevant variable to an existing set is:

Step 1. Add one dimension to the dimensional matrix such that its rank would *not* increase.

Step 2. Add one variable to the **A** matrix such that the generated new **A** would *not* be singular. Otherwise, the dimensional composition of this new variable can be arbitrary. This added variable is a *dimensionally irrelevant* one.

The following example demonstrates this process.

Example 11-4

Given the Dimensional Set of five variables and two dimensions

$$\begin{array}{c|ccccc} & V_1 & V_2 & V_3 & V_4 & V_5 \\ \hline d_1 & 1 & 2 & 3 & 1 & -3 \\ d_2 & 3 & 4 & 2 & 2 & 2 \end{array}$$ ← dimensional matrix rank = 2   (a)

The rank of this matrix is obviously 2, since the rightmost determinant is not zero. We now wish to add a dimensionally irrelevant variable, say V_6, to the set. To do this we follow the above two-step procedure.

Step 1. We add one dimension, say d_3, to the dimensional matrix such that its rank would not increase. The best way to do this (which leaves the dimensional composition of the original variables intact) is to assign all zeros to this new variable's dimension. Thus we have the new dimensional matrix

$$\begin{array}{c|ccccc} & V_1 & V_2 & V_3 & V_4 & V_5 \\ \hline d_1 & 1 & 2 & 3 & 1 & -3 \\ d_2 & 3 & 4 & 2 & 2 & 2 \\ d_3 & 0 & 0 & 0 & 0 & 0 \end{array}$$ ← dimensional matrix rank = 2 (still)

Note that the rank of this matrix is still 2.

Step 2. We add one variable, say V_6, to the **A** matrix such that the new **A** would not be singular. Suppose we add V_6 whose dimension is $d_1^4 \cdot d_2^9 \cdot d_3^{-2}$. Thus, we have the Dimensional Set

$$\begin{array}{c|ccc|ccc} & V_1 & V_2 & V_3 & V_4 & V_5 & V_6 \\ \hline d_1 & 1 & 2 & 3 & 1 & -3 & 4 \\ d_2 & 3 & 4 & 2 & 2 & 2 & 9 \\ d_3 & 0 & 0 & 0 & 0 & 0 & -2 \\ \hline \pi_1 & 1 & 0 & 0 & \frac{-11}{8} & \frac{-1}{8} & 0 \\ \pi_2 & 0 & 1 & 0 & -2 & 0 & 0 \\ \pi_3 & 0 & 0 & 1 & \frac{-3}{2} & \frac{1}{2} & 0 \end{array}$$

A (nonsingular) — points to V_4, V_5, V_6 in dimensional rows

all zero column — points to V_6 column in π rows

C — points to the π block

where the **A** matrix is *nonsingular*. We see that the column in **C** under V_6 is all zeros—verifying the dimensional irrelevancy of that variable.

⇑

11.1.3. The Cascading Effect

Suppose we have the dimensional matrix of five variables and four dimensions as follows:

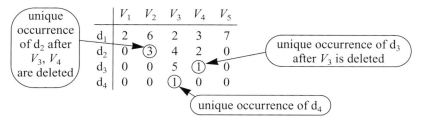

What can we see in this matrix, as far as dimensional irrelevance is concerned? We immediately see that dimension d_4 *only* appears in variable V_3. Therefore, by Theorem 11-1, V_3 is dimensionally irrelevant. So we remove V_3. But if V_3 is absent, then dimension d_3 becomes unique since it *only* appears in variable V_4—therefore V_4 is irrelevant, too. So we remove this variable, as well. We now have a dimensional matrix in which d_2 appears *only* in V_2—hence V_2 must also go. Finally only two variables, V_1 and V_5, and one dimension, d_1, remain. The resultant Dimensional Set is

	V_1	V_5
d_1	2	7
π_1	1	$-\dfrac{2}{7}$

yielding the sole dimensionless variable

$$\pi_1 = \frac{V_1}{\sqrt[7]{V_5^2}} = \text{const}$$

Let us see what the *direct* construction of the Dimensional Set based on the *original* set would provide.

	V_1	V_2	V_3	V_4	V_5
d_1	2	6	2	3	7
d_2	0	3	4	2	0
d_3	0	0	5	1	0
d_4	0	0	1	0	0
π_1	1	0	0	0	$-\dfrac{2}{7}$

C matrix

We observe that the columns of **C** under variables V_2, V_3, and V_4 are all zeros, and hence, by Theorem 10-4, these variables are all dimensionally irrelevant.

The above process, whereby the elimination of one variable causes the need to eliminate another variable which, in turn, necessitates the deletion of yet another variable, and so on, is *cascading,* and the resultant effect is called the *Cascading Effect.*

The following two examples further demonstrate how the Cascading Effect works.

Example 11-5. Critical Sliding Friction Coefficient of a Sphere Rolling Down an Incline

A sphere is rolling down an incline, as shown in Fig. 11-2. If the friction coefficient between the contacting bodies is *sufficiently* large, then the sphere will roll *without sliding;* else it will roll as well as slide.

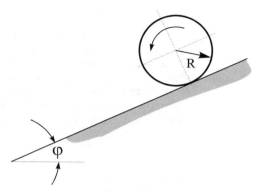

Figure 11-2
Sphere rolling down an incline

Question: for a given inclination φ, what must the *minimum* value of the friction coefficient be to prevent sliding? This value is then—by definition—the *critical* friction coefficient μ_c.

We assume the following relevant variables:

Variable	Symbol	Dimension
critical friction coefficient	μ_c	1
inclination angle	φ	1
radius of sphere	R	m
gravitational acceleration	g	m·s^{-2}
density of sphere	ρ	m^{-3}·kg

The dimensional matrix is therefore

	μ_c	φ	R	g	ρ
m	0	0	1	1	−3
kg	0	0	0	0	1
s	0	0	0	−2	0

We see that g is the only variable in which the dimension "s" appears, and "kg" is present only in ρ. So, by Theorem 11-1, g and ρ are *dimensionally irrelevant* variables. Therefore both g and ρ are deleted. But if there are no g and ρ, then R is the only variable in which "m" appears. Therefore, R must go, too. Ultimately, variables R, g, and ρ are all dimensionally irrelevant. This is a typical manifestation of the Cascading Effect; one variable's dimensional irrelevancy causes other variable(s)' similar "disqualification."

What does the Dimensional Set tell us? Since both μ_c and φ are dimensionless, they must appear in the **B** matrix, else **A** is singular. Thus we have

	μ_c	φ	R	g	ρ
m	0	0	1	1	-3
kg	0	0	0	0	1
s	0	0	0	-2	0
π_1	1	0	0	0	0
π_2	0	1	0	0	0

C

In the **C** matrix the columns under variables R, g, and ρ are all zeros, hence these variables are indeed dimensionally irrelevant, as expected.

From the above set, the two dimensionless variables are

$$\pi_1 = \mu_c; \qquad \pi_2 = \varphi \tag{a}$$

and therefore we can write $\pi_1 = \Psi\{\pi_2\}$, or

$$\mu_c = \Psi\{\varphi\} \tag{b}$$

where Ψ is an as-yet undefined function. We see from relation (b) that μ_c *only* depends on the inclination angle φ. Indeed, according to Timoshenko (Ref. 141, p. 435), for a sphere relation (b) has the form

$$\mu_c = \frac{2}{7}\cdot\tan\varphi \qquad \text{(for sphere)} \tag{c}$$

Similar relations can be easily obtained for a solid disk, and for a thin rim:

$$\mu_c = \frac{1}{3}\cdot\tan\varphi \qquad \text{(for disk)} \tag{d}$$

$$\mu_c = \frac{1}{2}\cdot\tan\varphi \qquad \text{(for rim)} \tag{e}$$

It is admittedly counter-intuitive that in all cases μ_c is independent not only of gravitational acceleration, but of the radius and density of the sphere, as well.

⇑

One benefit of the Cascading Effect is that sometimes it greatly facilitates the discovery that a *relevant* variable has been omitted from consideration by mistake. This may happen because when several variables are present, the intuitive "feel" of having one left out is relatively masked. But, by cascading, the number of variables is greatly reduced, and the falsehood of the result (often grotesque) is more easily detectable. The following example demonstrates this phenomenon.

Example 11-6. Elongation by Its Own Weight of a Suspended Bar

Consider a long prismatic bar of uniform cross-section. The bar is suspended at its upper end as illustrated in Fig. 11-3. We wish to find the bar's elongation by its own weight.

272 APPLIED DIMENSIONAL ANALYSIS AND MODELING

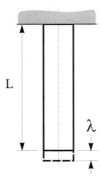

Figure 11-3
Elongation by its own weight of a suspended bar

The following variables are assumed to be relevant in this problem:

Variable	Symbol	Dimension
elongation	λ	m
density	ρ	$m^{-3} \cdot kg$
Young's modulus	E	$m^{-1} \cdot kg \cdot s^{-2}$
length	L	m

Therefore the dimensional matrix is

	λ	ρ	E	L
m	1	−3	−1	1
kg	0	1	1	0
s	0	0	−2	0

We see at once that E is the *only* variable which contains the dimension "s." Therefore, E is a dimensionally irrelevant variable and thus must go. But if there is no E, then ρ is the only variable which contains "kg," therefore ρ must also be discarded. The result is that the elongation is purported to be independent of density and Young's modulus, and is *only* influenced by the length. This is, of course, patent nonsense! The fault lies in the omission of the *relevant* gravitational acceleration g, without which—obviously—the pulling force generated by the weight of the bar cannot develop. To rectify the situation, we now include g and construct the correct Dimensional Set.

	λ	ρ	E	L	g
m	1	−3	−1	1	1
kg	0	1	1	0	0
s	0	0	−2	0	−2
π_1	1	0	0	−1	0
π_2	0	1	−1	1	1

C matrix

As none of the columns of **C** is all zeros, therefore none of the variables is dimensionally irrelevant—as none should be.

RELEVANCY OF VARIABLES

Occasionally, the Cascading Effect may be combined with other characteristics of the dimensional matrix to identify and eliminate dimensionally irrelevant variables. The following example demonstrates this technique.

Example 11-7

Consider the dimensional matrix of six variables and three dimensions

	V_1	V_2	V_3	V_4	V_5	V_6
d_1	-6	2	-3	2	1	-3
d_2	-12	4	-6	4	2	-1
d_3	0	0	0	1	0	2

(a)

Are there any dimensionally irrelevant variable here? The rank of this matrix is obviously 3, since the rightmost 3×3 determinant is $5 \neq 0$. But now a little scrutiny easily reveals that if V_6 is *disregarded,* then the second row becomes twice the first row. Hence, by Theorem 10-6, variable V_6 is dimensionally irrelevant. But if V_6 is no longer present, then V_4 is the only variable in which d_3 appears. Hence, by Theorem 11-1, V_4 is also dimensionally irrelevant. If we now remove V_4 (as we must), then the rank is further reduced to 1. Ultimately, therefore, we end up with four variables and one dimension. The resulting dimensional matrix is then

	V_1	V_2	V_3	V_5
d_1	-6	2	-3	1

(b)

in which all variables are now dimensionally relevant. The corresponding Dimensional Set is

	V_1	V_2	V_3	V_5
d_1	-6	2	-3	1
π_1	1	0	0	6
π_2	0	1	0	-2
π_3	0	0	1	3

(c)

The Dimensional Set constructed from the original dimensional matrix (a) is

	V_1	V_2	V_3	V_4	V_5	V_6
d_1	-6	2	-3	2	1	-3
d_2	-12	4	-6	4	2	-1
d_3	0	0	0	1	0	2
π_1	1	0	0	0	6	0
π_2	0	1	0	0	-2	0
π_3	0	0	1	0	3	0

(d)

As the reader can easily verify, the dimensionless variables generated by the two *very* different dimensional matrices (a) and (b)—which gave rise to the respective Dimensional Sets (c) and (d)—are identical, as required and expected.

11.2. PHYSICAL IRRELEVANCY

In general, establishing a variable's *physical irrelevancy* is a more difficult and intellectually challenging task than the corresponding undertaking for *dimensional irrelevancy*. Still, there are techniques which make the task manageable, at least most of the time. In this segment we describe these techniques in some detail.

First, we must establish what physical irrelevancy *is* for a variable.

Definition 11-1. *In a relation, an independent variable is physically irrelevant if its influence on the dependent variable is below a certain—usually low—threshold.*

At this point, two explanatory comments must be made: First, the definition mentions *dependent* variable. This is because the purpose of the very existence of a physical relation is to determine the *dependent* variable. Consequently, if any other variable has no, or negligible, effect on the dependent variable, then obviously this "other" variable must be judged *physically irrelevant.*

Second, the definition also uses the word *threshold.* This is because in theory there are very few—if any—variables which have *exactly* zero effect on any other variable. Does the mass of the planet Jupiter affect the weight of a loaf of bread on Earth? Yes, it does, but this influence is so small that it can be ignored. In other words, the effect is below a certain threshold, which, in this instance, is not defined—except that it is "low." However, we are satisfied that Jupiter's influence on the baker's scale is so small that we do not need to pursue this matter further.

The essence of the notion of "threshold," then, is that it does not have to be zero, but just sufficiently small, so that if the effect of any independent variable on the dependent variable falls below this limit, then this effect and the variable causing it can be ignored.

11.2.1. Condition

First we state a theorem establishing a *sufficient* condition for the physical irrelevancy of a variable, then we state and prove—by way of an example—another theorem showing that this condition is *not* a *necessary* one.

Theorem 11-3. *A sufficient condition for a variable to be physically irrelevant is that it be dimensionally irrelevant.*

Proof. Assume there is a variable which is physically relevant. Then it must appear in a relation expressing the dependent variable. Hence it cannot be dimensionally irrelevant, because the latter is absent from such a relation. This proves the theorem.

Theorem 11-4. *Dimensional irrelevancy is not a necessary condition for a variable to be physically irrelevant.*

Proof. The proof consists of showing an instance where a variable, although physically irrelevant, is not dimensionally so. This will prove the theorem.

Example 11-8. Gravitational Pull by a Solid Sphere on an External Material Point (I)

We have a massive homogeneous sphere and a material point outside of it at some distance from the sphere's center. Fig 11-4 shows the arrangement. The variables we consider are as follows:

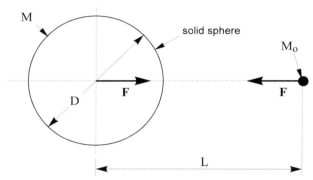

Figure 11-4
Solid sphere attracts an external material point

Variable	Symbol	Dimension	Remark
gravitational force	F	$m \cdot kg \cdot s^{-2}$	
mass of sphere	M	kg	homogeneous sphere
diameter of sphere	D	m	
distance of material point	L	m	from center of sphere
mass of material point	M_0	kg	
universal gravitational constant	k	$m^3 \cdot kg^{-1} \cdot s^{-2}$	

Hence the Dimensional Set will be

$$
\begin{array}{c|cccccc}
 & F & M & L & D & M_0 & k \\
\hline
m & 1 & 0 & 1 & 1 & 0 & 3 \\
kg & 1 & 1 & 0 & 0 & 1 & -1 \\
s & -2 & 0 & 0 & 0 & 0 & -2 \\
\hline
\pi_1 & 1 & 0 & 0 & 2 & -2 & -1 \\
\pi_2 & 0 & 1 & 0 & 0 & -1 & 0 \\
\pi_3 & 0 & 0 & 1 & -1 & 0 & 0 \\
\end{array} \quad \text{(a)}
$$

C matrix — not all zeros

We see that the column in matrix **C** under variable D is not all zeros, therefore, by Theorem 10-4, D is not dimensionally irrelevant. However, it is physically irrelevant, since in the famous formula for F, first established by Newton,

$$F = k \frac{M_0 \cdot M}{L^2} \quad \text{(b)}$$

D does not appear (Ref. 146, p. 161).

Thus it was demonstrated that there is at least one instance where a variable is not dimensionally irrelevant, although physically it is. Therefore dimensional irrelevancy is not a necessary condition for physical irrelevancy. This proves Theorem 11-4.

Understanding of different characteristics of dimensional and physical irrelevancies of a variable is facilitated by the Venn diagram shown in Fig 11-5.

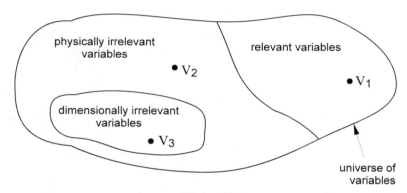

Figure 11-5
**A Venn diagram depicting relations among variables
of different types of irrelevancies**

In the diagram V_1 is a relevant variable; V_2 is a physically, but not dimensionally, irrelevant variable; V_3 is a variable which is both dimensionally and physically irrelevant.

We can see in the diagram that, although every dimensionally irrelevant variable must also be physically irrelevant, e.g., V_3 (Theorem 11-3), the reverse is not the case; i.e., not every physically irrelevant variable is also dimensionally irrelevant, e.g., V_2 (Theorem 11-4).

11.2.2. Techniques to Identify a Physically Irrelevant Variable

There are five ways to identify the physical irrelevancy of a variable:

- common sense
- existence of dimensional irrelevancy
- heuristic reasoning
- tests combined with deft interpretation of results
- analysis

Here we will discuss the first four; the fifth shall not be dealt with because if a relation for the dependent variable is available by analysis, then there is no need to scrutinize the encountered variables, since all of them are relevant, for otherwise they would not appear in a *derived* formula.

RELEVANCY OF VARIABLES

Common Sense. This is the first and maybe the most important stratagem to detect the existence of a physically irrelevant variable. The question the investigator must always ask first is: Do all the (assumed) independent variables influence the dependent variable to a *sufficiently* high degree? Very often the mere posing of such a question identifies the variable's physical irrelevancy.

For example, if an experimenter tries to collect the variables influencing the terminal speed of a 0.1 m diameter solid steel ball dropped from a height of 2 m, then common sense dictates that the *age* of the experimenter should be judged a physically irrelevant variable. A bit less clear in this example is the status of air density. Here again, common sense tells us that if the height from which the ball is dropped is rather small—as it is in this case—then air density is a negligible factor, and hence is physically irrelevant. This now brings up an important fact: a variable can be physically irrelevant in one setting, and relevant in another. For example, air density would not be irrelevant for a Ping-Pong ball dropped from 1848 m. Broadly speaking, this phenomenon is called a *scale effect* and will be further discussed in the section dealing with heuristic reasoning.

Existence of Dimensional Irrelevancy. Very often a physically irrelevant variable is also dimensionally irrelevant. If a variable is found to be dimensionally irrelevant, then, by Theorem 11-3, it *must* also be physically irrelevant. Therefore, the next thing an investigator should do (after the selected variables have passed the "common sense" scrutiny) is construct the Dimensional Set. If in this set the **C** matrix has a column of all zeros, then the variable at the top of this column is dimensionally, and hence physically, irrelevant. The following two examples illustrate.

Example 11-9. Hydroplaning of Tires on Flooded Surfaces

When an automobile drives over a well-wetted (i.e., flooded) road surface, beyond a certain speed the tires lose contact with the ground and the car becomes uncontrollable. This is a dangerous condition that should be avoided. The *minimum* speed at which hydroplaning occurs is called the *critical speed*. At critical speed, the *product* of hydrostatic pressure p under the tire and the contact area A, equals that part of the weight G of the car that is borne by the tire (we ignore aerodynamic vertical forces here). Thus

$$G = p \cdot A \tag{a}$$

Fig 11-6 shows the arrangement.

We will now determine the critical speed. The following variables are assumed to be relevant:

Variable	Symbol	Dimension
critical speed	v_c	$m \cdot s^{-1}$
tire pressure	p	$m^{-1} \cdot kg \cdot s^{-2}$
water density	ρ	$m^{-3} \cdot kg$
gravitational acceleration	g	$m \cdot s^{-2}$

It is assumed that the pressure inside the tire is identical to the contact pressure, in other words, the tire is completely flexible in bending.

We now construct the Dimensional Set based on the above table.

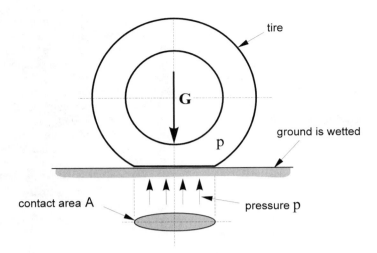

Figure 11-6
Hydroplaning of a tire (exaggerated view)

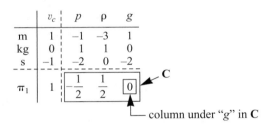

column under "g" in **C**

We see that the column in **C** under "g" is all zeros, therefore, by Theorem 10-4, g is a *dimensionally irrelevant* variable. For this reason, by Theorem 11-3, g is also a *physically irrelevant* variable; the critical speed is independent of g. A car on the Moon would hydroplane at the same speed as it does on Earth. How is this possible? Does not a larger g mean a *heavier* car and hence a higher critical speed? Fortunately, this riddle is easily solved. If the downward force (weight) is larger, the same pressure requires a larger surface area A—see relation (a)—which, in turn, requires a *lower* speed necessary to carry the weight. Thus the two effects cancel out nicely.

By the Dimensional Set the sole dimensionless variable (a constant in this case) is

$$\pi_1 = v_c \cdot \sqrt{\frac{\rho}{p}} = c_1 = \text{const} \tag{b}$$

which yields

$$v_c = c_1 \cdot \sqrt{\frac{p}{\rho}} = \frac{c_1}{\sqrt{\rho}} \cdot \sqrt{p} = c_2 \cdot \sqrt{p} \tag{c}$$

where c_2 is a dimensional constant whose value, according to Wong (Ref. 3, p. 41), is about

$$c_2 = 0.0557 \text{ m}^{3/2} \cdot \text{kg}^{-1/2} \tag{d}$$

Relation (c) is instructive because it purports that the critical speed is proportional to the square root of the tire pressure and is *independent* of the weight of the car and the size of the tire. These two variables therefore are dimensionally and hence physically irrelevant. For example, by (c) and (d), if $p = 207000$ Pa ($\cong 30$ psi), then the critical speed is

$$v_c = c_2 \cdot \sqrt{p} = 0.0557\sqrt{207000} = 25.34 \text{ m/s} = 91.2 \text{ km/h}$$

However, if the pressure drops to $p = 138000$ Pa ($\cong 20$ psi), then the critical speed reduces to 20.7 m/s = 74.5 km/h. This justifies the good advice: when driving on a wet surface, use the highest permitted tire pressure!

⇑

The next example is a little more complex, but equally instructive.

Example 11-10. Diameter of a Crater Caused by an Underground Explosion

Detonations by underground charges generate craters whose diameters are functions of energy released by the charge, gravitational parameters, soil characteristics, and the depth at which the charge is placed. Specifically, the following variables are assumed to be relevant:

Variable	Symbol	Dimension	Remark
crater diameter	D	m	
soil compressive strength	σ	$m^{-1}\cdot kg\cdot s^{-2}$	
soil pressure gradient	k	$m^{-2}\cdot kg\cdot s^{-2}$	with respect to depth
soil density	ρ	$m^{-3}\cdot kg$	
energy released	Q	$m^2\cdot kg\cdot s^{-2}$	
depth of charge	h	m	

Here the pressure gradient k is defined by

$$k = \frac{dp}{dh} \tag{a}$$

It is obvious that k incorporates gravitational effects, and hence the absence of g in the table. The Dimensional Set then is

	D	σ	k	ρ	Q	h
m	1	−1	−2	−3	2	1
kg	0	1	1	1	1	0
s	0	−2	−2	0	−2	0
π_1	1	0	0	0	0	−1
π_2	0	1	0	0	−1	3
π_3	0	0	1	0	−1	4

↑ all zeros — **C**

The column in **C** under ρ is all zeros, therefore, by Theorem 10-4, the density of soil is a *dimensionally irrelevant* variable, and so, by Theorem 11-3, it is also a *physically irrelevant* variable.

At first, this finding seems to defy experience and even common sense. Why does density—the "massiveness"—of soil not influence the size of the crater that a *given* explosive charge creates? The explanation is that the pressure gradient k already incorpo-

280   APPLIED DIMENSIONAL ANALYSIS AND MODELING

rates density since—obviously—the larger the density, the larger the pressure gradient (for liquids $k = \rho \cdot g$).

The set of dimensionless variables therefore are:

$$\pi_1 = \frac{D}{h}; \quad \pi_2 = \frac{\sigma \cdot h^3}{Q}; \quad \pi_3 = \frac{k \cdot h^4}{Q} \tag{b}$$

in which, of course, ρ does not appear, indicating its dimensional and hence physical irrelevancy (as long as the pressure gradient k is considered a *relevant* variable).

⇑

Heuristic reasoning. Heuristic reasoning combines common sense, technical knowledge, imagination, serendipity, an unquenchable thirst for the unusual, for the not-yet-tried, for the atypical, and for the unorthodox. It is where a good investigator can display mastery and use resourcefulness. So, it is indeed quite compelling and rewarding that, with just a few ingenious, but otherwise very simple schemes, so many significant results can be easily obtained. The several illustrative examples which follow demonstrate the use of these powerful stratagems.

Example 11-11.   Gravitational Pull by a Flat Infinite Plate

We have a flat thin plate of infinite size. A material point external to it is attracted by the mass of the plate. Fig 11-7 shows the arrangement. Note in the figure that the F force is perpendicular to the plate. This is because of symmetry; the horizontal components of F to the right and left are always equal, hence they cancel out—thus the resultant force is *normal* to the plate.

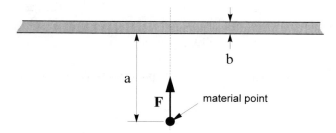

Figure 11-7
Infinite plate gravitationally attracts an external material point

The question is now: What is gravitational force F that the plate imparts to the point? To answer, we list the variables assumed to be relevant to define F.

Variable	Symbol	Dimension	Remark
gravitation force	F	m·kg·s^{-2}	normal to plate
distance of point to plate	a	m	normal to plate
mass of point	M	kg	
universal gravitational constant	k	m^3·kg^{-1}·s^{-2}	
thickness of plate	b	m	
density of plate	ρ	m^{-3}·kg	

RELEVANCY OF VARIABLES

We have six variables and three dimensions. Therefore, there are 6 – 3 = 3 dimensionless variables supplied by the Dimensional Set.

	F	a	M	k	b	ρ
m	1	1	0	3	1	-3
kg	1	0	1	-1	0	1
s	-2	0	0	-2	0	0
π_1	1	0	0	-1	-4	-2
π_2	0	1	0	0	-1	0
π_3	0	0	1	0	-3	-1

which yields

$$\pi_1 = \frac{F}{k \cdot b^4 \cdot \rho^2}; \quad \pi_2 = \frac{a}{b}; \quad \pi_3 = \frac{M}{b^3 \cdot \rho} \tag{a}$$

We can write, using monomial form

$$F = c \cdot k \cdot b^4 \cdot \rho^2 \cdot \left(\frac{a}{b}\right)^{\epsilon_2} \cdot \left(\frac{M}{b^3 \cdot \rho}\right)^{\epsilon_3} \tag{b}$$

in which c, ϵ_2, and ϵ_3 are dimensionless constants. Now we carry out some heuristic magic. Obviously, the gravitational force must be proportional to M, since a point twice as massive must generate twice the force. Hence it is necessary that $\epsilon_3 = 1$. Thus, by (b), we have

$$F = c \cdot k \cdot b^4 \cdot \rho^2 \cdot \left(\frac{a}{b}\right)^{\epsilon_2} \cdot \left(\frac{M}{b^3 \cdot \rho}\right)^{\epsilon_3} = c \cdot k \cdot b \cdot \rho \cdot M \cdot \left(\frac{a}{b}\right)^{\epsilon_2} \tag{c}$$

Furthermore, it is obvious that the gravitational force must be proportional to the mass of the plate which, in turn, is proportional to its thickness b. Therefore, F must be proportional to b. This condition, however, mandates that $\epsilon_2 = 0$ in (c). Thus, by this latest result and (c),

$$F = c \cdot k \cdot b \cdot \rho \cdot M \tag{d}$$

in which "a" does not appear; i.e., F is independent of "a," which is therefore a *physically irrelevant* variable. This is an astonishing result, since one would have expected that the *closer* the point is to the plate, the *larger* the gravitational force *towards the plate* should be. But there is another factor present here! If the point is close to the plate, most of the matter of the plate pulls the point at a *shallow* angle, whose normal component is small; and conversely, when the point is far away, the gravitational pull's normal component is large. It happens that these two effects exactly cancel out each other, and hence the independence of F upon "a."

The constant c in (d) is $2 \cdot \pi$ (Ref. 9, Vol. 1, p. 13–8). Finally, therefore, we have

$$F = 2 \cdot \pi \cdot k \cdot M \cdot b \cdot \rho \tag{e}$$

where $\pi = 3.14159\ldots$ is a dimensionless constant (it does *not* have a subscript!).

Now consider a ball made of the same material as the plate (i.e., density = ρ), and situated at a *unit distance* from point-mass M (Fig. 11-8). What diameter D_e must this ball have in order to attract point-mass M with the force F defined in (e)?

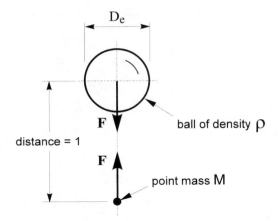

Figure 11-8
"Equivalent" ball attracts point-mass through unit distance

The diameter we seek is called the *equivalent diameter;* its value is provided by the relation

$$F = \frac{k \cdot M_e \cdot M}{1^2} = \underbrace{2 \cdot \pi \cdot k \cdot M \cdot b \cdot \rho}_{\text{by (e)}} \tag{f}$$

where M_e is the *equivalent mass* of the ball

$$M_e = D_e^3 \cdot \pi \cdot \frac{\rho}{6}$$

which, by (f), yields

$$D_e = \sqrt[3]{12 \cdot b} \tag{g}$$

i.e., the equivalent diameter is independent of the gravitational constant, the density and the mass of the material point; it is only influenced by the thickness of the plate. A truly astonishing fact! For example, a $b = 0.05$ m thick infinite plate generates a pull on a point-mass equivalent to that of a ball of $D_e = \sqrt[3]{(12) \cdot (0.05)} = 0.843$ m diameter.

In this example, by purely dimensional considerations, augmented by some modest heuristic reasoning, we managed to arrive at a very important result: the gravitational attraction of an infinite plate on an external point-mass is independent of the normal distance of that point-mass from the plate.

⇑

Example 11-12. Gravitational Pull by a Solid Sphere on an External Material Point (II)

We return to the problem in Example 11-8, where a sketch of the arrangement is shown in Fig. 11-4. The reader is advised to review this simple drawing before continuing.

It was noted in Example 11-8 that Newton analytically derived the important fact that a solid homogenous sphere attracts an external material point as if the entire mass of the sphere were concentrated at its center.

Here, by adroit heuristic reasoning, we will derive the same law, but without resorting to analytical derivation. This will demonstrate the great facility and prowess of the dimensional method.

At the outset we note that we should *not* assume *a priori* the validity of the classic

$$F = k \cdot \frac{M \cdot M_0}{L^2} \tag{a}$$

formula for the gravitational force between the center of a sphere of mass M and that of the material point-mass M_0, which are L distance apart. For some portion of the sphere (less than half) is closer than L to the point, and the rest is farther away. So it is conceivable that the diameter would influence the generated force at L distance—as illustrated in Fig. 11-4.

The Dimensional Set given in (a) of Example 11-8 gave rise to the following set of three dimensionless variables:

$$\pi_1 = \frac{F \cdot D^2}{M_0^2 \cdot k}; \qquad \pi_2 = \frac{M}{M_0}; \qquad \pi_3 = \frac{L}{D} \tag{b}$$

Now we notice that in (b), D occurs twice (in π_1 and π_3), but L occurs only once (in π_3). However, since it is D whose effect we wish to find out, it would be advisable to modify the Dimensional Set based on (b) such that D would occur in only *one* dimensionless variable. To do this, we invoke Theorem 9-10 which states in essence that if the dimensions of any two variables are identical, then they can change places in the set of dimensionless variables. We now utilize this useful theorem by noting that D and L have the same dimensions, hence they can be interchanged. Thus the modified dimensionless variables will be

$$\pi_1 = \frac{F \cdot L^2}{M_0^2 \cdot k}; \qquad \pi_2 = \frac{M}{M_0}; \qquad \pi_3 = \frac{D}{L} \tag{c}$$

by which, considering the monomial form,

$$\pi_1 = c \cdot \pi_2^{\epsilon_2} \cdot \pi_3^{\epsilon_3} \tag{d}$$

Thus, by (c),

$$F = c \cdot \frac{M_0^2 \cdot k}{L^2} \cdot \left(\frac{M}{M_0}\right)^{\epsilon_2} \left(\frac{D}{L}\right)^{\epsilon_3} \tag{e}$$

where c, ϵ_2, and ϵ_3 are constants yet to be determined.

Obviously F must be proportional to the mass of the sphere M. So, by (e),

$$\epsilon_2 = 1 \tag{f}$$

Next, assume that ϵ_3 is positive. In this case, if D is reduced to zero—but everything else is kept constant—then F would also be reduced to zero. This is tantamount to saying that the gravitational force between two massive points a finite distance apart is zero. This is nonsense. Therefore, ϵ_3 cannot be positive. Now assume that ϵ_3 is negative. Then, if D approaches zero, then F would increase without limit. This means that the gravitational force between two material points of finite masses and a nonzero distance apart is infinite. This is equally absurd. Hence, ϵ_3 cannot be either positive or negative; therefore it must be zero, i.e.

$$\epsilon_3 = 0 \tag{g}$$

By (f) and (g) now, our relation for F in (e) becomes

$$F = c \cdot \frac{M_0 \cdot M}{L^2} \cdot k \tag{h}$$

which says that the net effect of the distributed mass of a homogeneous sphere upon an external material point is the same as if the mass of the sphere were concentrated at its center.

Note that we arrived at this remarkable result [first proved by Newton (Ref. 146, p. 161) with $c = 1$] strictly on the basis of Newton's law of gravitation, which he composed to predict the gravitational force between two material *points*.

⇑

When a variable is found, or suspected, to be physically irrelevant, then the dimensionless variable in which it appears can, and should, be ignored. There is one important restriction, however: the physically irrelevant variable must appear in only *one* dimensionless variable. If it does not, then the Dimensional Set should be modified accordingly. The best and most straightforward way to do this is to place the physically irrelevant variables in the **B** matrix because, by the construction of the Dimensional Set, variables constituting the **B** matrix appear in only *one* dimensionless variable (provided the **D** matrix is an identity matrix, which is normally the case).

Example 11-13. Deformation by Wind of an Air-Supported Radome

The subject radome is essentially a protection device for dish antennas and other large-size structures. The shape of the radome is a truncated sphere (or very close to it). The level of truncation is characterized by the truncation angle φ. Fig. 11-9 shows the arrangement.

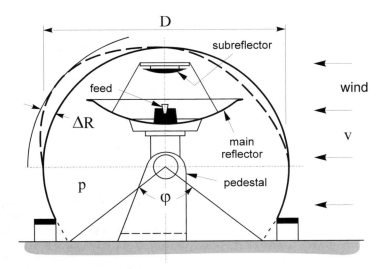

Figure 11-9
Radome supported by internal pressurized air is deformed by horizontal wind.
Dashed line indicates the deformed shape.

RELEVANCY OF VARIABLES

A horizontal wind is blowing from the right. The deformation of the radome is characterized by the difference in radii between the nondeformed shape and the circumscribed sphere of the deformed one. In the figure this characteristic deformation is denoted by ΔR. It is assumed that the centers of the two spheres as defined above are coincidental.

We now wish to derive a formula for ΔR. The first step is to list the variables presumed relevant.

Variable	Symbol	Dimension	Remark
deformation	ΔR	m	characteristic value
wind speed	v	m·s^{-1}	horizontal and uniform
gravitational acceleration	g	m·s^{-2}	
air density	ρ	kg·m^{-3}	
diameter	D	m	characteristic value
internal pressure	p	m^{-1}·kg·s^{-2}	above outside pressure

We assume geometrically similar structures, so that the truncation angle φ remains constant. Under this condition, a single linear dimension is sufficient to define size. This linear measure is then the *characteristic* linear dimension. Similarly, the deformation is characterized by the distance ΔR, as defined in Fig. 11-9, which is then the *characteristic* value of the deformation. Gravitational acceleration g is included because it is suspected that it affects the deformation, since movement of the radome takes place in a gravitational field. Finally, it is assumed that the radome's material is completely flexible, but inextensible and, further, that the deformation is small compared with the size of the radome.

In the above table we have six variables and three dimensions. Hence, there are $6 - 3 = 3$ dimensionless variables furnished by the dimensional set:

	ΔR	v	g	ρ	D	p
m	1	1	1	-3	1	-1
kg	0	0	0	1	0	1
s	0	-1	-2	0	0	-2
π_1	1	0	0	0	-1	0
π_2	0	1	0	$\frac{1}{2}$	0	$-\frac{1}{2}$
π_3	0	0	1	1	1	-1

(a)

from which

$$\pi_1 = \frac{\Delta R}{D}; \quad \pi_2 = v \cdot \sqrt{\frac{\rho}{p}}; \quad \pi_3 = \frac{g \cdot \rho \cdot D}{p} \qquad (b)$$

and we can write

$$\pi_1 = \Psi\{\pi_2, \pi_3\} \qquad (c)$$

where Ψ is a yet undefined function. Let us now assume a monomial form for Ψ, which is plausible. Accordingly, by (b),

$$\Delta R = c \cdot D \cdot \left(v \cdot \sqrt{\frac{\rho}{p}}\right)^{\epsilon_2} \cdot \left(\frac{g \cdot \rho \cdot D}{p}\right)^{\epsilon_3} \qquad (d)$$

where c, ϵ_2, and ϵ_3 are constants.

286   APPLIED DIMENSIONAL ANALYSIS AND MODELING

Now we carry out some simple heuristic reasoning. First, it is evident that if the deformation ΔR is small, then it must be proportional to the deforming force which is generated by wind. For Reynolds numbers larger than 1000 (which is the case here), the wind force is proportional to the square of air flow velocity. Hence, by (d), $\epsilon_2 = 2$. Moreover, the wind force is proportional to the air density, and the deformation is proportional to the wind force. Therefore the deformation is proportional to the air density. Consequently, in relation (d) we must have

$$\frac{1}{2} \cdot \epsilon_2 + \epsilon_3 = 1 \tag{e}$$

But we found above that $\epsilon_2 = 2$, therefore (e) provides $\epsilon_3 = 0$. Thus, by these findings and (d)

$$\Delta R = c \cdot \frac{D \cdot v^2 \cdot \rho}{p} \tag{f}$$

Since g does not appear in this relation, g is a *physically irrelevant* variable, although not a *dimensionally irrelevant* one. We also note that the deformation is inversely proportional to the internal pressure; i.e., the larger the pressure, the smaller the deformation. It is also evident that for a given assembly (constants c, D, and ρ) the deformation is proportional to the ratio v^2/p. This means that to maintain a given (i.e., allowable) deformation, the pressure must increase with the square of the wind speed. This is the governing factor in the design of air supply systems for radomes of large steerable antennas.

Since g is physically irrelevant, and since it appears in only *one* dimensionless variable, π_3, therefore π_3 can be ignored in (c) and (d). Thus we can write $\pi_1 = \Psi\{\pi_2\}$ or, since $\epsilon_2 = 2$, as was found above,

$$\Delta R = c \cdot D \cdot \left(v \cdot \sqrt{\frac{\rho}{p}} \right)^2 = c \cdot D \cdot v^2 \cdot \frac{\rho}{p} \tag{g}$$

which is identical to (f)—as expected and required.

But now suppose we compose the Dimensional Set differently from (a):

	ΔR	v	ρ	g	D	p
m	1	1	−3	1	1	−1
kg	0	0	1	0	0	1
s	0	−1	0	−2	0	−2
π_1	1	0	0	0	−1	0
π_2	0	1	0	$-\frac{1}{2}$	$-\frac{1}{2}$	0
π_3	0	0	1	1	1	−1

(h)

in which it is noted that g appears in the **A** matrix, and not in the **B** matrix as in (a). The dimensionless variables therefore, by (h), are

$$\pi_1 = \Delta R/D; \qquad \pi_2 = \frac{v}{\sqrt{g \cdot D}}; \qquad \pi_3 = \frac{\rho \cdot g \cdot D}{p} \tag{i}$$

Here we observe that g appears in *two* dimensionless variables, viz., π_2 and π_3. Now if, because of the physical irrelevancy of g, we discarded both π_2 and π_3, then only π_1 would remain, meaning $\pi_1 = $ const, or $\Delta R = $ const$\cdot D$. That is, the wind-caused deformation is independent of wind speed, internal pressure, and air density; it is a function only of the diameter of the radome. But this is absurd. We emphasize: the error occurred because we allowed the appearance of the physically irrelevant variable in *more than one* dimensionless variable. Therefore the Golden Rule is: The physically irrelevant variable must appear in only *one* dimensionless variable—which can then safely be ignored.

Example 11-14. Axial Reaction Force Generated by Thermal Load on a Straight Bar

A prismatic bar of uniform cross-section is placed between two *unyielding* surfaces, as shown in Fig. 11-10. At installation, the bar is of ambient temperature, so that the reaction force F is zero. Now the bar's temperature is raised uniformly along its length. What will the reaction force F be? The relevant variables are as listed below.

Variable	Symbol	Dimension	Remark
reaction force	F	m·kg·s^{-2}	axial
cross-section's area	A	m^2	uniform along length
temperature rise	Δt	°C	uniform along length
thermal expansion coefficient	α	°C^{-1}	
Young's modulus	E	m^{-1}·kg·s^{-2}	
length	L	m	

Thus the dimensional matrix is

	F	A	Δt	α	E	L
m	1	2	0	0	−1	1
kg	1	0	0	0	1	0
s	−2	0	0	0	−2	0
°C	0	0	1	−1	0	0

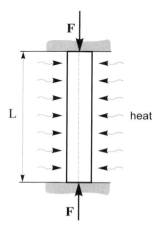

Figure 11-10
Reaction force by thermal load on a straight bar

The rank of this matrix is 3, whereas the number of dimensions is 4. Therefore we have to ignore one dimension (Art. 7.6). Let this unfortunate dimension be "s." Hence the Dimensional Set will be

	F	A	Δt	α	E	L
m	1	2	0	0	−1	1
kg	1	0	0	0	1	0
°C	0	0	1	−1	0	0
π_1	1	0	0	0	−1	−2
π_2	0	1	0	0	0	−2
π_3	0	0	1	1	0	0

(a)

which yields three dimensionless variables

$$\pi_1 = \frac{F}{E \cdot L^2}; \qquad \pi_2 = \frac{A}{L^2}; \qquad \pi_3 = \alpha \cdot \Delta t \tag{b}$$

from which we can write—by assuming the monomial form

$$F = c \cdot E \cdot L^2 \cdot \left(\frac{A}{L^2}\right)^{\epsilon_2} \cdot (\alpha \cdot \Delta t)^{\epsilon_3} \tag{c}$$

where c, ϵ_2, and ϵ_3 are constants.

Now we perform some heuristic deliberations. The force generated by F must vary linearly with Δt, since the thermal dilatation varies linearly with Δt, and in turn the dilatation is proportional to the force (within the elastic limit). Therefore, the force is proportional to Δt. This fact mandates that in (c)

$$\epsilon_3 = 1 \tag{d}$$

Further, F must also be proportional to the cross-section A, since a bar of twice the area of cross-section requires twice as much force to produce the *same* compression. Hence, by (c),

$$\epsilon_2 = 1 \tag{e}$$

Substituting (d) and (e) into (c), we obtain

$$F = c \cdot E \cdot L^2 \cdot \left(\frac{A}{L^2}\right) \cdot (\alpha \cdot \Delta t) = c \cdot \alpha \cdot (\Delta t) \cdot A \cdot E \tag{f}$$

in which on the far right L does not appear. Hence L is a *physically irrelevant* variable. This might astonish some people, since it could be argued that a longer bar, which dilates more, requires a larger force to compel it to remain at its original length. However, a longer bar is more "flexible" inasmuch as it absorbs a given contraction "easier" than a shorter bar does. So the two effects (the longer dilatation and the increased flexibility) conveniently cancel each other out, and the result is a relation for the reaction force from which the length is absent.

Since L is physically irrelevant, the dimensionless variable (singular!) in which it appears should be ignored. But in (b) we have two dimensionless variables (π_1 and π_2) in which L participates. This is because L is in the **A** matrix of the Dimensional Set (a), instead of in the **B** matrix. We did not follow our own good advice!

To remedy the situation we recast the sequence of variables in the Dimensional Set such that L appears in **B**. This will yield

RELEVANCY OF VARIABLES

	F	L	Δt	α	E	A
m	1	1	0	0	−1	2
kg	1	0	0	0	1	0
°C	0	0	1	−1	0	0
π_1	1	0	0	0	−1	−1
π_2	0	1	0	0	0	$-\dfrac{1}{2}$
π_3	0	0	1	1	0	0

(g)

by which we have the new set

$$\pi_1 = \frac{F}{E \cdot A}; \qquad \pi_2 = \frac{L}{\sqrt{A}}; \qquad \pi_3 = \alpha \cdot \Delta t \qquad (h)$$

Since L is physically irrelevant, we can ignore π_2, for it is the *only* dimensionless variable in which L appears. Hence, by (h), we now have

$$F = c \cdot E \cdot A \cdot (\alpha \cdot \Delta t)^{\epsilon_2} \qquad (i)$$

Again, since, by relation (e), $\epsilon_2 = 1$

$$F = c \cdot E \cdot A \cdot \alpha \cdot \Delta t \qquad (j)$$

which is identical to (f)—as expected and required. Note, however, that to arrive at the defining relation, this time we only had to use one, instead of two, "heuristic conditions." This subtle, but powerful, improvement was due to our *a priori* knowledge that the thermally generated reaction force is independent of the length of the bar.

⇑

The next two examples take us into the alluring field of biomechanics.

Example 11-15. Maximum Running Speed of Animals

To determine the maximum running speed of animals (including man) is a classic problem of comparative physiology; it is discussed in many texts, notably by Smith (Ref. 122, p. 11) and Schepartz (Ref. 43, p. 155). We now deal with this subject using the dimensional method which, as far as the present author knows, has not been done before. In any case, the conclusion is of course the same as that derived by others using different techniques.

To begin, two assumptions must be made. First, the energy stored in muscles per unit body mass is the same for all animals, and second, all animals are geometrically similar. Hence, by the second assumption, one linear measure called *characteristic length* defines the size of the animal. Once these assumptions are accepted, we proceed by listing the variables assumed to be relevant:

Variable	Symbol	Dimension	Remark
maximum running speed	v	m·s^{-1}	
energy available	$\overline{Q}$	m^2·s^{-2}	per unit body mass
characteristic length	L	m	e.g., height

Here we have three variables and two dimensions. Hence there is $3 - 2 = 1$ dimensionless variable, a constant. It is provided by the Dimensional Set

290 APPLIED DIMENSIONAL ANALYSIS AND MODELING

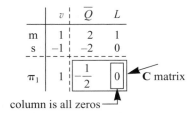

The column under L in the **C** matrix is all zeros (there is only one row, hence only one element under L). Therefore, by Theorem 10-4, L is a *dimensionally irrelevant* variable, and so, by Theorem 11-3, it is also a *physically irrelevant* variable. Consequently, the astonishing conclusion is that the maximum running speed of an animal is independent of its size, i.e., it is the same for *all* animals (including man). Of course, this conclusion is not exact, but considering the approximately identical running speeds of rabbits and horses, the counterintuitive truth in the result is inescapable.

⇑

Example 11-16. Jumping Heights of Animals (II)

Let us now investigate the height to which animals (including man) can jump. This problem was already discussed in Chapter 6, Example 6-19. But now we utilize the much more rapid technique accorded by the recognition of a dimensionally irrelevant variable in the dimensional matrix. The following assumptions are made:

- All animals are geometrically similar
- The energy stored in unit mass of animal muscle does not vary among species
- The ratio of muscle mass to body mass does not vary among species
- The density of an animal's body does not vary among species
- Air resistance is negligible

The variables considered relevant are as follows:

Variable	Symbol	Dimension
jumping height	h	m
mass of animal	M	kg
energy per mass of muscle	$\overline{Q}$	m²·s⁻²
gravitational acceleration	g	m·s⁻²

Therefore the Dimensional Matrix is

	h	$\overline{Q}$	M	g
m	1	2	0	1
kg	0	0	1	0
s	0	−2	0	−2

We see that M is the only variable in which the dimension "kg" appears. Thus, by Theorem 11-1, M is a *dimensionally irrelevant* variable and therefore, by Theorem 11-3, it is also a *physically irrelevant* variable. So, an animal's mass does not influence its jumping height and since (by assumption) the mass is the *only* quantity which varies among species, therefore all animals (including man) jump to the same height (which is about 2.5 m). This conclusion is well documented (see for example Ref. 139, p. 185 and Ref. 140, p. 952).

This startling result was reached by using only the most elementary dimensional reasoning, once more demonstrating the ease and practicality of this technique.

⇑

It often happens that a variable is physically irrelevant for particular ranges of variables or circumstances, but relevant under some other conditions. In these cases, especially keen critical thinking and technical judgement are called for.

For example, consider the terminal speed of a falling object. This speed is determined by the *distance* the object falls and by the gravitational acceleration. Consequently, the height at which the vertical journey starts is unimportant; i.e., it is *physically irrelevant*. This means that, if a body falls from altitude 2 m to 1 m, its terminal speed will be the same as if it were falling from 3 m to 2 m. But if we perform this simple experiment at a height of 1000 km, then we find that the terminal speed of the body completing its 1 m journey is only 86.4 % of the corresponding value obtained previously (in both cases air resistance is ignored). Therefore, when we practice heuristic cogitation to screen out physically irrelevant variables, we must always be careful to consider the different physical circumstances prevailing at different *ranges* of the independent variables present. This is one manifestation of the *scale effects* to which the investigator must always pay close attention. The following example gives a practical illustration of this effect.

Example 11-17. Velocity of Surface Waves

It is assumed that the depth of water is infinite, so that depth does not influence speed. The following variables are considered relevant:

Variable	Symbol	Dimension
speed of wave	v	m·s^{-1}
surface tension	σ	kg·s^{-2}
wavelength	λ	m
gravitational acceleration	g	m·s^{-2}
water density	ρ	m^{-3}·kg

Hence the Dimensional Set will be

	v	σ	λ	g	ρ
m	1	0	1	1	-3
kg	0	1	0	0	1
s	-1	-2	0	-2	0
π_1	1	0	$-\frac{1}{2}$	$-\frac{1}{2}$	0
π_2	0	1	-2	-1	-1

(a)

Thus we have two dimensionless variables

$$\pi_1 = \frac{v}{\sqrt{\lambda \cdot g}}; \quad \pi_2 = \frac{\sigma}{\lambda^2 \cdot g \cdot \rho} \qquad (b)$$

This will produce — assuming monomial form —

$$v = \text{const} \cdot \sqrt{g \cdot \lambda} \cdot \left(\frac{\sigma}{\lambda^2 \cdot g \cdot \rho} \right)^{\epsilon_2} \tag{c}$$

Now we perform some "double barrelled" heuristic cerebrations. First, we only consider *long* wavelengths in which capillary effects are negligible. In this case surface tension may be ignored, since it does not influence the velocity of waves. Hence in (c) we have

$$\epsilon_2 = 0 \tag{d}$$

So (c) can be written

$$v_{\text{long}} = \text{const} \cdot \sqrt{g \cdot \lambda} \tag{e}$$

where the constant is $1/\sqrt{2 \cdot \pi}$ (Ref. 9, Vol. 1, p. 51-7). We immediately notice that in (e) density ρ does not appear, either. Therefore we have established the first pretty nice conclusion: if the wavelengths are *long*, then both surface tension *and* density are *physically irrelevant* variables.

The physical irrelevancy of density seems astonishing at first. However, consider that mass influences both the wave-inducing force and the inertia (the restoring force) equally, but in opposite directions — so the two effects cancel out. This phenomenon is entirely similar to free fall, where the moving force (gravitation) and the resisting force (inertia) are both proportional to mass — and hence the latter has no effect on the movement. The end result is that all bodies, regardless of their masses, fall at the same rate of speed (ignoring air resistance).

Next, we consider the *short* wavelength case in which gravity effects may be neglected. In order to isolate g in (c), we recompose the formula as follows:

$$v = \text{const} \cdot g^{0.5 - \epsilon_2} \cdot \sqrt{\lambda} \cdot \left(\frac{\sigma}{\lambda^2 \cdot \rho} \right)^{\epsilon_2} \tag{f}$$

Now if v is independent of g, then it is necessary that $0.5 - \epsilon_2 = 0$ in (f). Thus

$$\epsilon_2 = 0.5 \tag{g}$$

We now insert this value into (c) or (f) to obtain

$$v_{\text{short}} = \text{const} \cdot \sqrt{\frac{\sigma}{\lambda \cdot \rho}} \tag{h}$$

where the constant happens to be $\sqrt{2 \cdot \pi}$ (Ref. 9. vol. 1, p. 51-8).

Therefore we see in (e) and (h) that wavelength influences the speed of wave propagation quite differently when the wavelengths are long, or when they are short. The *critical wavelength* λ_c separating the two regions can be obtained by equating (e) with (h) with the appropriate constants being used. Accordingly,

$$\sqrt{\frac{g \cdot \lambda_c}{2 \cdot \pi}} = \sqrt{2 \cdot \pi \cdot \frac{\sigma}{\lambda_c \cdot \rho}}$$

from which the critical wavelength is

$$\lambda_c = 2 \cdot \pi \cdot \sqrt{\frac{\sigma}{g \cdot \rho}} \tag{i}$$

For water at 20 °C, $\sigma = 0.0728$ kg·s^{-2} (Ref. 4, p. 209), $\rho = 998$ m^{-3}·kg (Ref. 124, p. 313), and $g = 9.81$ m·s^{-2}. Therefore the critical wavelength for water at 20 °C is

$$\lambda_c = 0.0171 \text{ m} \tag{j}$$

Thus if $\lambda < \lambda_c$, we say the wavelength is *short,* otherwise the wavelength is *long.* The speed corresponding to the critical wavelength is the *crossover* speed; it is supplied by either (e) or (h). In this way we derive the critical speed v_{cr} which, using the above supplied data, is

$$v_{cr} = \sqrt[4]{\frac{g \cdot \sigma}{\rho}} = 0.164 \text{ m/s} \tag{k}$$

It is emphasized that the formulas (e) and (h) are only approximations which progressively improve the more we deviate in either direction from the critical wavelength of 0.0171 m.

It can be shown (Ref. 9, Vol. 1, p. 51-7) that the exact relation encompassing both gravitational and capillary effects is

$$v = \sqrt{\frac{2 \cdot \pi \cdot \sigma}{\lambda \cdot \rho} + \frac{g \cdot \lambda}{2 \cdot \pi}} \tag{ℓ}$$

which, if λ_c from (i) is substituted to it, provides

$$(v_{cr})_{exact} = \sqrt{2} \cdot \sqrt[4]{\frac{g \cdot \sigma}{\rho}} = 0.231 \text{ m/s} \tag{m}$$

Fig. 11-11 shows the short and long wavelength approximations of (e) and (h), as well as the exact relation (ℓ), where the two effects (the capillary and the gravitational) are com-

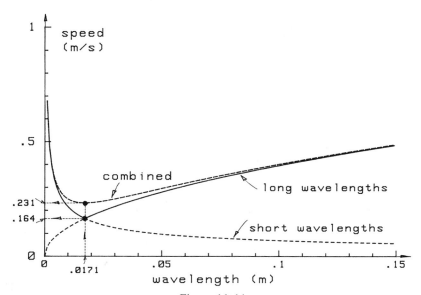

Figure 11-11
Wave velocity versus wavelength for water at 20 °C

bined. The figure also shows the critical wavelength $\lambda_c = 0.0171$ m value, as well as the corresponding v_{cr} and $(v_{cr})_{exact}$ speed data as given in (k) and (m).

Figure 11-11 shows that the approximations of (e) and (h) for long and short wavelengths are very good if the wavelengths are shorter than about 0.008 m, or longer than about 0.06 m. The largest error occurs at $\lambda_c = 0.0171$ m, where it is 29%.

The $v_{cr-exact}$ value—given in relation (m)—means that if an object moves slower than 0.231 m/s in water, it does not create waves. This fact is well known to submariners to the effect that a periscope will not create any tell-tale V-shaped wave as long as the speed of the submerged vessel does not exceed 0.231 m/s ~ 0.83 km/h ~ 0.45 knot.

Returning to the original object of this example, we have demonstrated that, depending on the *magnitude* of a wavelength, either gravitation, or surface tension *and* density, can be classified as *physically irrelevant*—thereby showing the significant influence of scale effect on the result.

⇑

Example 11-18. Frequency of Vibration of a Sphere of Liquid
(adapted from Ref. 71, p. 66).

Given a sphere of liquid which is not under the influence of any gravitational forces originating exteriorly. The sphere is in equilibrium and is levitating in empty space. Due to some external cause now, the sphere is set into vibration. What will be the period of this vibration?

To answer this question, we first consider the relevant variables:

Variable	Symbol	Dimension
period of vibration	T	s
sphere diameter	D	m
density	ρ	$m^{-3} \cdot kg$
gravitational constant	k	$m^3 \cdot kg^{-1} \cdot s^{-2}$
surface tension	σ	$kg \cdot s^{-2}$

We have five variables and three dimensions, therefore there are 5 – 3 = 2 dimensionless variables supplied by the Dimensional Set

	k	σ	T	D	ρ
m	3	0	0	1	–3
kg	–1	1	0	0	1
s	–2	–2	1	0	0
π_1	1	0	2	0	1
π_2	0	1	2	–3	–1

Note in the above that the dependent variable T is not in the **B** matrix. Why? Because we want to see the effects of k and σ on T *separately*. Since any variable in the **B** matrix appears only *once* in the generated dimensionless variables, by putting k and σ in the **B** matrix, their effects on T can be readily isolated.

By the above, then

$$\pi_1 = k \cdot T^2 \cdot \rho; \qquad \pi_2 = \frac{\sigma \cdot T^2}{D^3 \cdot \rho} \qquad \text{(a)}$$

RELEVANCY OF VARIABLES

Now two cases can be easily distinguished:

(i) The sphere is *small*. In this case, gravitational effects may be neglected, therefore the vibration is controlled by surface tension. Hence, k is a *physically irrelevant* variable, thus π_1, the *only* variable in which k appears, can be ignored. We are left with a single dimensionless variable π_2, which, therefore, becomes a constant. Thus,

$$\pi_2 = \frac{\sigma \cdot T^2}{D^3 \cdot \rho} = \text{const} \tag{b}$$

or

$$T = \text{const} \cdot \sqrt{\frac{D^3 \cdot \rho}{\sigma}} \tag{c}$$

The vibration period varies with the $\frac{3}{2}$ power of the diameter and with the square root of the density. A small ball that is *twice* as big as an even smaller one vibrates with a period that is $\sqrt{2^3} = \sqrt{8} = 2.828$ times longer.

(ii) The sphere is *large*. In this case the surface effects may be safely ignored, and hence internal gravitation controls the vibration. Thus σ becomes a *physically irrelevant* variable, and as a consequence π_2, the *only* dimensionless variable in which σ appears, may be ignored. But if π_2 is ignored, then the diameter D must also be ignored, since D appears only in π_2. Hence D also becomes a *physically irrelevant* variable.

So we again have a *single* dimensionless variable—this time it is π_1—and therefore it must be a constant. Now we write, by (a)

$$\pi_1 = k \cdot T^2 \cdot \rho = \text{const} \tag{d}$$

from which

$$T = \frac{\text{const}}{\sqrt{k \cdot \rho}} \tag{e}$$

Since k is constant for every possible occurrence in the universe, the vibratory period for a large ball of liquid varies with the inverse square root of the density and is independent of the diameter. For example, for mercury $\rho = 13530$ kg·m^{-3}, and for water $\rho = 1000$ kg·m^{-3}. Their ratio is 13.53 and hence a large mercury ball vibrates with a period $1/\sqrt{13.53} = 0.272$ times that of a large water ball. In other words, a mercury ball vibrates $1/0.272 = 3.678$ times faster than a water ball—both balls being large.

The interplay between the inertial and gravitational forces defined in relation (e) manifests itself nicely in a certain class of giant stars, the *Cepheid variables*. The diameters of these stars seem to contract and expand about $\pm 10\%$ with great regularity, with respect to their average values. For example, the δ Cepheids stars have a period of pulsation of 464,007.46 s. By (e) now

$$T^2 \cdot \rho = \text{const} \tag{f}$$

since $k = \text{const}$. Hence, the product of the density and the square of the period of pulsation is the same for all Cepheids, irrespective of their diameters. According to Schuring (Ref. 31, p. 197), the data of the 14 groups of Cepheid variables surveyed confirm formula (f) almost exactly. The results found in (c) and (e) were first given by Lord Rayleigh in 1915 (Ref. 71), but only verbatim and without derivation.

Finally, by (a), we can write

$$\pi_1 = \Psi\{\pi_2\} \tag{g}$$

where Ψ is an as-yet undefined function. Let us assume a monomial form for Ψ—which is likely to be valid. Thus

$$\pi_1 = c \cdot \pi_2^\epsilon \tag{h}$$

where c and ϵ are constants. By (a) and (h)

$$k \cdot T^2 \cdot \rho = c \cdot \left(\frac{\sigma \cdot T^2}{D^3 \cdot \rho}\right)^\epsilon \tag{i}$$

from which

$$T = \left(\frac{1}{k \cdot \rho}\right)^{\frac{1}{2 \cdot (1-\epsilon)}} \cdot \left(\frac{\sigma}{D^3 \cdot \rho}\right)^{\frac{\epsilon}{2 \cdot (1-\epsilon)}} \tag{j}$$

Now if D is small, then k has no effect on T; it is physically irrelevant! Hence, in this case, by (j), $1/[2 \cdot (1 - \epsilon)] = 0$, and thus $\epsilon = -\infty$. On the other hand, if D is large, then it is σ which is physically irrelevant since it has no effect on T. Hence, by (j) $\epsilon/[2 \cdot (1 - \epsilon)] = 0$, and thus $\epsilon = 0$. These facts suggest to the author that perhaps a relation in the form of

$$\epsilon = \frac{-b}{D} \tag{k}$$

may be valid. Here b is a liquid-specific positive dimensional constant of dimension "m." The value of b could be determined by experimentation or by observation.

Tests Combined with Deft Interpretation of Results. Often no matter how cleverly heuristic reasoning is carried out, the detection of physically irrelevant variables still remains elusive. In these cases, experimental determination of relation(s) among the involved dimensionless variables may be used to identify a physical irrelevancy. In particular, if we find that varying one dimensionless variable does not affect another one, then from this fact—by employing the theorem given below—the physically irrelevant variable can be identified.

Theorem 11-5. *If, in the relation $\pi_1 = \Psi\{\pi_2\}$, a change of π_2 leaves π_1 unchanged (i.e., constant), then any variable which is present in π_2, but not present in π_1, is physically irrelevant with respect to π_1.*

Proof. The proof is presented for a four-variable case, but of course it could be applied to any number of variables.
Suppose we have

$$\pi_1 = V_1^a \cdot V_2^b \cdot V_3^c = k = \text{const}$$

$$\pi_2 = V_2^\beta \cdot V_3^\gamma \cdot V_4^\delta$$

where $V_1, \ldots, V_4$ are variables and $a, b, c, \ldots, \beta, \gamma, \delta$ are constant exponents. Note that V_4 appears in π_2, but *not* in π_1. Next, we select any values for V_2 and V_3 (common to both π_1 and π_2). This will enable us to determine V_1 by

$$V_1 = \left(\frac{k}{V_2^b \cdot V_3^c}\right)^{1/a}$$

Now with fixed V_2 and V_3, any value of π_2 can be generated by simply substituting the appropriate V_4 into the expression for π_2 given above. This of course will not affect π_1 because V_4 is not part of π_1. Hence, the value of V_4 is immaterial for setting the value of π_1 and consequently it must be classed as *physically irrelevant*. This proves the theorem.

Note the important distinction: *only* the variable that is present in π_2, but absent from π_1, is physically irrelevant, but not the other way around. For example, V_1 is present in π_1, but not in π_2, yet V_1 is *not* physically irrelevant.

The following example illustrates the practical application of this theorem.

Example 11-19.   Minimum Pressure in an Inflated Radome (I)*

A radome is essentially a device to protect microwave dish antennas or other tall structures against atmospheric effects, viz., rain, dust, wind, etc. A radome is made of a flexible and pressure-tight material and is supported entirely by pressurized internal air. For this reason, no mechanical support structure is needed to keep the radome in position. Fig. 11-12 shows the arrangement.

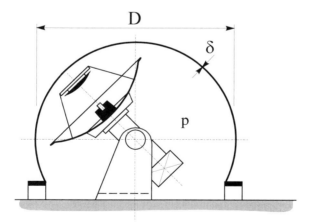

Figure 11-12
**A radome supported by pressurized air is protecting
an antenna from atmospheric effects**

---

*The author is grateful to Mr. G. Schmideg, P.Eng., of George Schmideg Engineering Inc., Toronto, Ontario, for providing him with some of the background material for this example and for Example 11-22.

298 APPLIED DIMENSIONAL ANALYSIS AND MODELING

The question naturally emerges: how much pressure will keep the radome in its desired configuration? Or more precisely, what is the *minimum* value of such a pressure? For obviously we would not want to over-stress the radome's material, nor do we want the radome to collapse for the lack of pressure. We shall now deal with this problem by using dimensional considerations.

The variables assumed to be relevant are as follows:

Variable	Symbol	Dimension	Remark
inside air pressure	p	$m^{-1} \cdot kg \cdot s^{-2}$	above outside pressure
radome diameter	D	m	
density of material	ρ	$m^{-3} \cdot kg$	
gravitational acceleration	g	$m \cdot s^{-2}$	
thickness of material	δ	m	

Here g is included, since pressure must act against gravitational forces to maintain the radome's shape. Also, diameter D is included because it is felt that the size of the structure influences the required minimum pressure. We have five variables and three dimensions, hence there are $5 - 3 = 2$ dimensionless variables supplied by the Dimensional Set

	p	D	ρ	g	δ
m	−1	1	−3	1	1
kg	1	0	1	0	0
s	−2	0	0	−2	0
π_1	1	0	−1	−1	−1
π_2	0	1	0	0	−1

Hence

$$\pi_1 = \frac{p}{\rho \cdot g \cdot \delta}; \qquad \pi_2 = \frac{D}{\delta} \tag{a}$$

or

$$p = c \cdot \rho \cdot g \cdot \delta \cdot \Psi\left\{\frac{D}{\delta}\right\} \tag{b}$$

where c is a constant and Ψ is a function yet to be determined. Let us assume a *monomial* form for this function—a very likely construct because of the physical arrangement. Thus (b) can be written

$$p = c \cdot \rho \cdot g \cdot \delta \cdot \left(\frac{D}{\delta}\right)^{\epsilon_2} \tag{c}$$

where ϵ_2 is an as-yet unknown exponent. Now let us carry out some elementary heuristic reasoning. If the thickness of the radome's material δ increases by a factor k, then the *weight* of the radome, against which the pressure must act, also increases by a factor k, since δ is much smaller than D. Hence p should be proportional to δ, mandating $\epsilon_2 = 0$ in (c). Therefore, by (c)

$$p = c \cdot \rho \cdot g \cdot \delta \tag{d}$$

Thus p is independent of the diameter D, i.e., D is a *physically irrelevant* variable. Note that it is *not* a dimensionally irrelevant variable since D is in the **B** matrix, and in **B** no dimensionally irrelevant variable may occur (Theorem 10-5, Art. 10.5.1).

With a little more heuristic deliberation it can be easily deduced that p must, indeed, be independent of D. The argument goes like this: If D increases k-fold, then the weight increases k^2-fold, since weight is proportional to the surface area, which, in turn, is proportional to D^2. The upward force F_u which supports the down-pointing weight is proportional to D^2 since $F_u = (D^2 \cdot \pi /4) \cdot p$. Therefore, the weight and the counteracting upward force *both* vary as the square of the diameter, so the two effects cancel each other out. Hence the physical irrelevancy of D.

Suppose we plotted the π_1 versus π_2 relation as a result of a series of experiments, but without having (of course) any knowledge of the nature of the Ψ function in (b). We would then get a single-curve plot, something like the one shown in Fig. 11-13 (the plot shows only the measured points marked by crosses). We see on this plot that $\pi_1 \cong$ const for whatever π_2. We also observe that, by (a), D is a variable present in π_2, but *absent* from π_1. Therefore, by Theorem 11-5, D is a *physically irrelevant* variable although, of course, not a dimensionally irrelevant one.

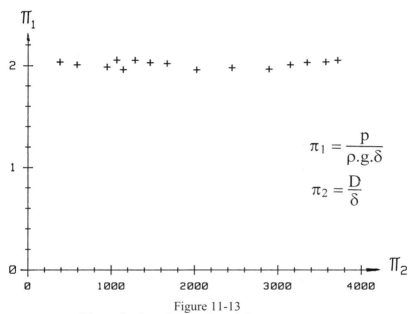

Figure 11-13
Dimensionless plot for the minimum necessary internal pressure of an inflatable radome
See table in this example for definitions of physical variables

Example 11-20. Linear Speed of a Sphere Rolling Down an Incline

We investigate a solid sphere rolling down an inclined plane under the influence of constant gravity. It is assumed that there is no *rolling* resistance, and that the sphere rolls without slipping. This latter condition makes it mandatory that the static friction coefficient between the sphere and the incline is not less than $\frac{2}{7} \cdot \tan \varphi$ (see Example 11-5 in Art. 11.1.3.). Fig. 11-14 shows the arrangement.

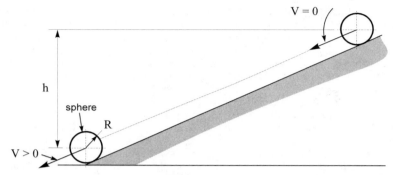

Figure 11-14
Sphere rolls down an incline without slipping

The following variables are assumed to be relevant:

Variable	Symbol	Dimension	Remark
terminal speed	v	$m \cdot s^{-1}$	linear
radius of sphere	R	m	
gravitational acceleration	g	$m \cdot s^{-2}$	= 9.80665
vertical travel	h	m	

We have four variables and two dimensions, hence there are 4 − 2 = 2 dimensionless variables supplied by the Dimensional Set

	v	R	g	h
m	1	1	1	1
s	−1	0	−2	0
π_1	1	0	$-\dfrac{1}{2}$	$-\dfrac{1}{2}$
π_2	0	1	0	−1

from which

$$\pi_1 = \frac{v}{\sqrt{g \cdot h}}; \qquad \pi_2 = \frac{R}{h} \tag{a}$$

therefore

$$\pi_1 = \Psi\{\pi_2\} \tag{b}$$

where Ψ is an as-yet undefined function. We now assume a *monomial* form for Ψ, hence, by (b),

$$\frac{v}{\sqrt{g \cdot h}} = c \cdot \left(\frac{R}{h}\right)^{\epsilon_2}$$

or

$$v = c \cdot \sqrt{g \cdot h} \cdot \left(\frac{R}{h}\right)^{\epsilon_2} \tag{c}$$

where c and ϵ_2 are constants to be found. Now suppose that ϵ_2 is positive. Then v increases with R *without limit*, which makes no sense. Hence ϵ_2 cannot be positive. If ϵ_2 is negative, then v increases without limit with decreasing R, and so we could achieve any terminal speed—however large—by merely selecting a small enough radius for the sphere. This is equally absurd.

Therefore, we are forced to conclude that $\epsilon_2 = 0$, since it cannot be either positive or negative. Thus, by (c), v is independent of R, i.e., R is a *physically irrelevant* variable. This is indeed so, as the simple derivation—not performed here—can show that

$$v = \sqrt{\frac{10}{7}} \cdot \sqrt{g \cdot h} \qquad (d)$$

This result is rather counterintuitive since one might feel that the radius of the sphere must somehow influence the linear speed. But a little reflection shows that the physical irrelevancy of R is actually very plausible. Although a larger radius means a slower rotation, but it is precisely because of the increased radius why the tangential speed (i.e., the linear speed of the center of the sphere) increases; the two effects of the change of radius offset each other, so that the linear speed remains constant.

Let us now assume that we do not have the exact analytical formula (d), but that we only know (c), which we obtained by purely dimensional means. The latter equation contains two constants: c and ϵ_2. We now determine these constants by conducting two *measurements*, in which not *both* R and h are the same. Thus let us vary h (easier) and keep R constant. The two tests, their inputs and results, are tabulated below.

	selected variables		measured variable
	R	h	v
Test 1	0.05 m	1.200 m	4.1 m/s
Test 2	0.05 m	2.005 m	5.3 m/s

By these data and (c)

$$4.1 = c \cdot (3.43045) \cdot \left(\frac{0.05}{1.2}\right)^{\epsilon_2} \quad \text{and} \quad 5.3 = c \cdot (4.43422) \cdot \left(\frac{0.05}{2.005}\right)^{\epsilon_2}$$

Dividing the first of these relations by the second yields $0.99994 = 1.67083^{\epsilon_2}$, from which the exponent is

$$\epsilon_2 = \frac{\ln 0.99994}{\ln 1.67083} = -0.00012 \cong 0$$

Therefore, by (c), v is independent of R, i.e., R is a *physically irrelevant* variable. Now we determine the constant c. By the first of the test result formulas, with $\epsilon_2 = 0$

$$c = \frac{4.1}{3.43045} = 1.19518$$

which is within 0.004% to the analytically determined exact value of $\sqrt{10/7} = 1.19523$.

It is remarkable that, by performing only two experiments, we not only determined the two constants in (c), but also proved the a nonsliding rolling ball's terminal speed is independent of its radius.

Finally, it is noted that the terminal speed of a *nonrolling* object sliding downward without friction is $v_0 = \sqrt{2 \cdot g \cdot h}$. Therefore, compared with free-falling terminal speed, nonsliding rolling reduces the terminal speed by a factor of $\sqrt{5/7} = 0.845$.

⇑

Example 11-21. Velocity of Fluid Flowing Through an Orifice

We now investigate the velocity of an inviscid (nonviscous) fluid flowing through an orifice. Fig. 11-15 shows the arrangement. The following variables are considered:

Variable	Symbol	Dimension	Remark
fluid velocity	v	m·s^{-1}	
fluid bulk modulus	β	m^{-1}·kg·s^{-2}	
pressure difference	Δp	m^{-1}·kg·s^{-2}	$\Delta p = p_2 - p_1$
fluid density	ρ	m^{-3}·kg	

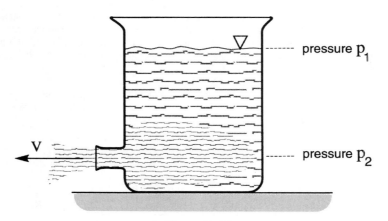

Figure 11-15
Velocity of fluid flowing through an orifice

Here the bulk modulus is included because it is felt that the *compressibility* of liquid might influence the discharge speed. The bulk modulus is defined by

$$\beta = \frac{p}{\left(\dfrac{\Delta\rho}{\rho}\right)} \quad \text{(a)}$$

where p is the gauge pressure which causes the density change $\Delta\rho$ of a liquid whose density at $p = 0$ is ρ. Consequently, the dimension of β is the same as that of pressure. For water $\beta = 2.06 \times 10^9$ Pa, i.e., water is about 100 times more compressible than steel.

The dimensional matrix is

	v	β	Δp	ρ
m	1	−1	−1	−3
kg	0	1	1	1
s	−1	−2	−2	0 ← (deleted row)

RELEVANCY OF VARIABLES

whose rank is 2, as the reader can easily verify. Therefore we have to delete one dimension, say we delete "s." The Dimensional Set therefore will be

	v	β	Δp	ρ
m	1	−1	−1	−3
kg	0	1	1	1
π_1	1	0	$-\dfrac{1}{2}$	$\dfrac{1}{2}$
π_2	0	1	−1	0

yielding two dimensionless variables

$$\pi_1 = v \cdot \sqrt{\frac{\rho}{\Delta p}}; \qquad \pi_2 = \frac{\beta}{\Delta p} \tag{b}$$

Therefore we can write

$$\pi_1 = \Psi\{\pi_2\} \tag{c}$$

where Ψ is a function to be determined.

If we now make measurements and plot the single-curve relationship (c), as Fig. 11-16 shows, we find that $\pi_1 \cong$ const. In other words, π_1 is independent of π_2. We see in (b) that β is a variable that occurs in π_2, but is absent from π_1. Hence, by Theorem 11-5, the bulk modulus β is a physically irrelevant variable, and, as a consequence, π_2, the *only* dimensionless variable in which it occurs, can be ignored. Thus, only one dimensionless variable π_1 remains, which therefore, must be a constant. We can write, then

$$\pi_1 = v \cdot \sqrt{\frac{\rho}{\Delta p}} = \text{const} \tag{d}$$

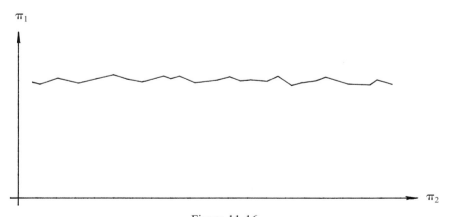

Figure 11-16
The π_1 versus π_2 relation showing the constancy of π_1
See formula (b) and table in this example for the definitions of symbols

304 APPLIED DIMENSIONAL ANALYSIS AND MODELING

from which

$$v = \text{const} \cdot \sqrt{\frac{\Delta p}{\rho}} \tag{e}$$

where the constant, of course, can be read off the curve presented in Fig. 11-16.

⇑

It frequently happens that although in the $\pi_1 = \Psi\{\pi_2\}$ relation π_1 is not constant, it can be nevertheless expressed with good accuracy by a simple analytic function, usually a polynomial or some other rational relation. In these cases, very valuable information can be obtained by equating π_1 with the experimentally found and analytically expressed function. The following examples demonstrate this feature.

Example 11-22. Minimum Pressure in an Inflatable Radome (II)

This problem was already dealt with in Example 11-19. This time we treat it a bit differently. Recall that our aim is to find the minimum pressure necessary inside an inflatable radome to prevent the completely flexible skin from collapsing. For convenience we again list the involved variables.

Variable	Symbol	Dimension	Remark
air pressure in radome	p	$m^{-1} \cdot kg \cdot s^{-2}$	over outside pressure
material thickness	δ	m	
material density	ρ	$m^{-3} \cdot kg$	
gravitational acceleration	g	$m \cdot s^{-2}$	
radome diameter	D	m	

This yields the Dimensional Set

	p	δ	ρ	g	D
m	−1	1	−3	1	1
kg	1	0	1	0	0
s	−2	0	0	−2	0
π_1	1	0	−1	−1	−1
π_2	0	1	0	0	−1

whence

$$\pi_1 = \frac{p}{\rho \cdot g \cdot D}; \qquad \pi_2 = \frac{\delta}{D} \tag{a}$$

The π_1 versus π_2 curve is shown in Fig. 11-17. Here we observe that π_1 is a linear function of π_2, viz.

$$\pi_1 = 2 \cdot \pi_2 \tag{b}$$

By (a) and (b)

$$\frac{p}{\rho \cdot g \cdot D} = 2 \cdot \frac{\delta}{D} \tag{c}$$

in which D cancels out. Thus D is a *physically irrelevant* variable—the same conclusion which was also reached in Example 11-19.

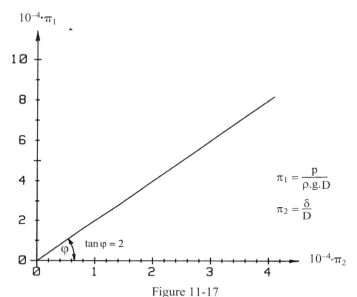

**Figure 11-17
Dimensionless plot for the minimum necessary
internal pressure of an inflatable radome**
See table in this example for definitions of physical variables

Example 11-23. Period of a Conical Pendulum

We investigate the period of a conical pendulum secured to its frictionless, spherical hinge by a weightless rod. The bob circles the vertical axis at a constant angular speed. Fig. 11-18 shows the arrangement.

The following variables are assumed to be relevant:

Variable	Symbol	Dimension
period	T	s
length of rod	L	m
height of cone	h	m
gravitational acceleration	g	m·s^{-2}

We have four variables and two dimensions, therefore there are $4 - 2 = 2$ dimensionless variables determined by the Dimensional Set

	T	h	L	g
m	0	1	1	1
s	1	0	0	−2
π_1	1	0	$-\dfrac{1}{2}$	$\dfrac{1}{2}$
π_2	0	1	−1	0

306 APPLIED DIMENSIONAL ANALYSIS AND MODELING

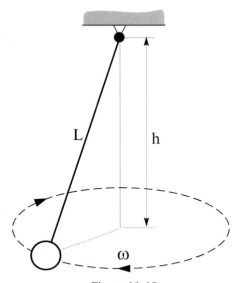

Figure 11-18
A conical pendulum

from which

$$\pi_1 = T \cdot \sqrt{\frac{g}{L}}; \qquad \pi_2 = \frac{h}{L} \qquad (a)$$

Therefore we can write

$$\pi_1 = \Psi\{\pi_2\} \qquad (b)$$

where Ψ is an as-yet unspecified function.

Now in order to determine Ψ we perform a number of simple measurements on a test piece of length $L = 1$ m. We swing (rotate) the bob at five different rotational speeds, characterized by the periods T, and then measure the conical heights h in each case. These tests result in the set of data presented below. By these data, the π_1 versus π_2 curve is plotted (see Fig. 11-19).

Given		Set	Measured	Calculated by (a)	
L	g	h	T	π_1	π_2
1	9.81	0.1	0.634	1.987	0.1
1	9.81	0.2	0.897	2.810	0.2
1	9.81	0.3	1.099	3.441	0.3
1	9.81	0.4	1.269	3.974	0.4
1	9.81	0.5	1.419	4.443	0.5

Looking at the plot, we notice its shape, which suggests the simple analytic form

$$\pi_1 = k \cdot \pi_2^n \qquad (c)$$

RELEVANCY OF VARIABLES

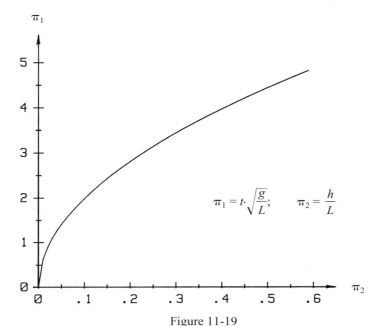

Figure 11-19
Dimensionless plot to determine the rotational period of a conical pendulum
See table in this example for definitions of physical variables

where k and n are constants to be determined. We have two unknowns so we consider two *lines* in the given test result table. Let us select $h = 0.1$ m and $h = 0.5$ m. Accordingly,

$$1.987 = k \cdot (0.1)^n; \qquad 4.443 = k \cdot (0.5)^n \qquad (d)$$

Division of the first by the second parts of the above yields $0.44722 = 0.2^n$, from which

$$n = \frac{\ln 0.44722}{\ln 0.2} = 0.4999 \cong 0.5 \qquad (e)$$

and by this result, the first part of (d) provides

$$k = \frac{1.987}{0.1^{0.5}} = 6.2834 \cong 2 \cdot \pi \qquad (f)$$

Therefore (c) can be written

$$\pi_1 = 2 \cdot \pi \cdot \sqrt{\pi_2} \qquad (g)$$

Let us check if this relation holds for the other data-points, as well. For example, for $\pi_2 = 0.2$, we have $\pi_1 = 2.81$ by the test, and 2.80993 by (g). The reader can be easily convinced that the remaining two data-points also satisfy (g).

308 APPLIED DIMENSIONAL ANALYSIS AND MODELING

Relation (g) has a significant consequence: if it is compared with (a), we get

$$T \cdot \sqrt{\frac{g}{L}} = 2 \cdot \pi \cdot \sqrt{\frac{h}{L}} \tag{h}$$

whence

$$T = 2 \cdot \pi \cdot \sqrt{\frac{h}{g}} \tag{i}$$

in which L does not appear. Therefore, L is a *physically irrelevant* variable.

Now we try another way to arrive at the same conclusion. By Theorem 9-10 we know that if two variables—one in matrix **A**, the other in **B**—have identical dimensions, then they may be *interchanged* in the set of dimensionless variables. We now apply this theorem by observing that L and h have identical dimensions, so they may be, and are now, interchanged. Hence, the new set of dimensionless variables will be, by (a),

$$\tilde{\pi}_1 = T \cdot \sqrt{\frac{g}{h}}; \qquad \tilde{\pi}_2 = \frac{L}{h} \tag{j}$$

We emphasize that this set is the same as was established previously in (a), except for the interchange of L and h. Our new set of data-points is

Given		Set	Measured	Calculated by (a)	
L	g	h	T	$\tilde{\pi}_1$	$\tilde{\pi}_2$
1	9.81	0.1	0.634	6.27948	10
1	9.81	0.2	0.897	6.28220	5
1	9.81	0.3	1.099	6.28451	3.3333
1	9.81	0.4	1.269	6.28443	2.5
1	9.81	0.5	1.419	6.28539	2

We observe (with delight) that $\tilde{\pi}_1 \cong 2 \cdot \pi$ = constant, so we now look for a *constituent* variable in $\tilde{\pi}_2$ which is *not* present in $\tilde{\pi}_1$. There *is* one such variable and it is L. Therefore, by Theorem 11-5, L is a *physically irrelevant* variable. Note that L is not dimensionally irrelevant, since in the dimensional set under L the elements are not all zeros (in this case, none of them is a zero).

Finally, we repeat the useful rule: if a variable is suspected of being *physically irrelevant*, then it should be put in the **B** matrix (not in the **A** matrix) because variables in the **B** matrix appear in only *one* dimensionless variable, and therefore their effects on the dependent variable can be easily *isolated*.

⇑

Example 11-24. Kepler's Third Law

Kepler's third law establishes a relation between the orbital period of a planet and its *average* distance from the Sun. It states that, for all planets, the *ratio* of the square of the orbital period and the cube of the average distance to the Sun is constant. Since the orbit is an ellipse (Kepler's first law), therefore the average distance equals half the major axis of the orbit (see Appendix 7).

Around 1620, Johann Kepler (German astronomer, 1572–1630) discovered his third law *experimentally* after 2 decades of tenacious work analysing the observational data of Tycho Brahe (Danish astronomer, 1546–1601). The law of course is easily *derivable* from Newton's law of universal gravitation, but that was only published 57 years after Kepler's death! Now we will use dimensional techniques to establish this relation. *Assume* the following relevant variables:

Variable	Symbol	Dimension
orbital period	T	s
mass of planet	M_p	kg
semi-major axis of orbit	a	m
mass of Sun	M_s	kg
universal gravitational constant	k	$m^3 \cdot kg^{-1} \cdot s^{-2}$

We have five variables and three dimensions. Therefore, there are 5 – 3 = 2 dimensionless variables describing the system. To obtain these variables, we construct the Dimensional Set.

	T	M_p	a	M_s	k
m	0	0	1	0	3
kg	0	1	0	1	–1
s	1	0	0	0	–2
π_1	1	0	$-\frac{3}{2}$	$\frac{1}{2}$	$\frac{1}{2}$
π_2	0	1	0	–1	0

from which

$$\pi_1 = T \cdot \sqrt{\frac{M_s \cdot k}{a^3}}; \qquad \pi_2 = \frac{M_p}{M_s} \qquad (a)$$

Note that since M_p and M_s have identical dimensions, both cannot be in the **A** matrix—else **A** is singular. We now calculate the above-defined π_1 and π_2 values for the nine planets using Ref. 17, p. 89 for "a," Ref. 17, p. 93 for M_p and M_s, and Ref. 20, p. 422 for T. The data so obtained are presented in the table of Fig. 11-20 (see next page).

We immediately observe from these data that π_1 is very nearly *constant* (its largest deviation from the mean is 0.9 %), while π_2 varies widely (it changes by a factor of more than 6000). Therefore π_1 may be considered independent of π_2. To find the *reason* for this independence, we merely search for a variable in π_2 which does *not* appear in π_1. By (a), such a variable is M_p (mass of the planet) which, by Theorem 11-5, is therefore a *physically irrelevant* variable. Consequently, π_2 should be ignored. Thus we are left with only *one* dimensionless variable π_1 which, owing to its singleness, must be a constant. Thus, by (a),

$$\pi_1 = T \cdot \sqrt{\frac{M_s}{k \cdot a^3}} = \text{const} \qquad (b)$$

yielding

$$\frac{T^2}{a^3} = \text{const} \cdot \frac{k}{M_s} = \text{const} \qquad (c)$$

since for all planets k and M_s are constant. Therefore we can say, on the strength of (c), that for all planets the ratio of the *square* of the orbital period to the *cube* of the semi-major axis of the orbit (i.e., the average distance from the Sun) is the same. This statement is Kepler's third law.

310 APPLIED DIMENSIONAL ANALYSIS AND MODELING

Planet	T	a	M_P	π_1	π_2
↓ ↓	1E6 s	1E10 m	1E24 kg	1	1E-7
Mercury	7.6	5.791	0.326	6.278	1.641
Venus	19.414	10.821	4.881	6.278	24.57
Earth	31.558	14.96	5.975	6.278	30.077
Mars	59.355	22.794	0.643	6.278	3.237
Jupiter	374.336	77.848	1986.7	6.273	10000.503
Saturn	929.595	143.321	567.6	6.237	2857.143
Uranus	2651.202	286.211	87.13	6.303	438.589
Neptune	5200.375	451.418	101.88	6.241	512.836
Pluto	7839.693	587.827	5.68	6.332	28.592

T = planet's orbital period
a = length of semi-major axis of planet's orbit
M_P = mass of planet
M_S = mass of Sun = 1.9866E30 kg
k = universal gravitational constant = 6.67E-11 m³/(s²·kg)

Figure 11-20
Dimensionless variables π_1, π_2 as derived from the orbital and physical properties of the nine planets
π_1 and π_2 are defined in relation (a) of this example

⇑

Example 11-25. Frequency of Transversal Vibration of a Stretched Wire

We wish to determine the frequency of lateral vibration of a stretched, completely flexible wire. The following relevant variables are assumed:

Variable	Symbol	Dimension
frequency	ν	s^{-1}
tension	F	$m \cdot kg \cdot s^{-2}$
length	L	m
mass per unit length	q	$m^{-1} \cdot kg$
Young's modulus	E	$m^{-1} \cdot kg \cdot s^{-2}$

We have five variables and three dimensions, therefore there are 5 – 3 = 2 dimensionless variables supplied by the Dimensional Set.

	ν	F	L	q	E
m	0	1	1	−1	−1
kg	0	1	0	1	1
s	−1	−2	0	0	−2
π_1	1	0	0	$\frac{1}{2}$	$-\frac{1}{2}$
π_2	0	1	−2	0	−1

from which
$$\pi_1 = v \cdot \sqrt{\frac{q}{E}}; \quad \pi_2 = \frac{F}{L^2 \cdot E} \quad (a)$$

Hence we write
$$\pi_1 = \Psi\{\pi_2\} \quad (b)$$

Let us assume the monomial form Ψ. Then we have
$$\pi_1 = c \cdot \pi_2^\epsilon \quad (c)$$

where c and ϵ are constants. In order to determine these two constants, we must perform (at least) two experiments. To this end, we select a 16-gauge steel wire of diameter $D = 0.001588$ m, unit-length mass $q = 0.01554$ kg/m, and Young's modulus $E = 2.08 \times 10^{11}$ Pa. This wire is now stretched by a constant $F = 2000$ N force. We select two lengths $L_1 = 0.20$ m and $L_2 = 0.44$ m, and in both cases we measure the lowest (natural) frequency. Assume we find $v_1 = 896.87$ Hz, and $v_2 = 407.67$ Hz. By these data, the following table can be prepared:

experiment #	L m	v Hz	$\pi_1(10^{-7})$ 1	$\pi_2(10^{-10})$ 1
1	0.2	896.87	2451.4	2403.8
2	0.44	407.67	1114.3	496.7

By (c), we now have the following two equations:
$$2451.4 \times 10^{-7} = c \cdot (2403.8 \times 10^{-10})^\epsilon; \quad 1114.3 \times 10^{-7} = c \cdot (496.7 \times 10^{-10})^\epsilon \quad (d)$$

Dividing the first of these by the second, we get
$$2.19995 = 4.83954^\epsilon \quad (e)$$

therefore
$$\epsilon = \frac{\ln 2.19995}{\ln 4.83954} = 0.50001 \cong 0.5 \quad (f)$$

By this and the first part of (d)
$$c = \frac{2451.4 \times 10^{-7}}{\sqrt{2403.8 \times 10^{-10}}} = 0.49999 \cong 0.5 \quad (g)$$

Hence, by (f) and (g), (c) can be written
$$\pi_1 = \frac{1}{2} \cdot \sqrt{\pi_2} \quad (h)$$

which, by appropriate substitution from (a), becomes
$$v \cdot \sqrt{\frac{q}{E}} = \frac{1}{2} \cdot \sqrt{\frac{F}{L^2 \cdot E}} \quad (i)$$

from where the frequency is
$$v = \frac{1}{2 \cdot L} \cdot \sqrt{\frac{F}{q}} \quad (j)$$

312 APPLIED DIMENSIONAL ANALYSIS AND MODELING

The remarkable thing about this formula is the absence of Young's modulus E. Therefore the frequency ν is independent of E, which is therefore a *physically irrelevant* variable.

Thus, having performed only two experiments, we have not only determined the two hitherto unknown constants in (c), but have also *proved* the *physical irrelevancy* of Young's modulus for the lateral vibration of a stretched wire.

There is one more thing we can do to boost our *confidence* in the *monomial* form of Ψ as it appears in (b) and (c). The easiest way to accomplish this is to select a few more lengths and *measure* the ensuing frequencies. For example let us select $L_3 = 0.32$ m, and measure the frequency $\nu_3 = 560.54$ Hz. With this pair of values $\pi_1 = 1532.1 \times 10^{-7}$ and $\pi_2 = 939.0 \times 10^{-10}$, fitting nicely into (h). This of course does not *prove* (h), it only makes it much less likely that it is false. This process can — in theory — continue *ad infinitum*, thereby approaching the *complete* verification of (c) with greater and greater certainty, without, however, ever fully achieving it.

⇑

11.3. PROBLEMS

Solutions are in Appendix 6.

11/1 **Period of oscillation of a fluid in a U tube.** Prove that the density of a fluid is a *physically irrelevant* variable in the determination of the oscillation of that fluid in a U tube.

11/2 **Velocity of disturbance along a stretched wire.** Prove that the lateral amplitude of a disturbance along a stretched wire does not influence the longitudinal speed of that disturbance.

11/3 **Geometry of a catenary.** It is conceivable that the sag U of a catenary is a function of its span λ, length L, density of its material ρ, and the gravitational acceleration g. Fig. 11-21 shows the arrangement. By purely dimensional

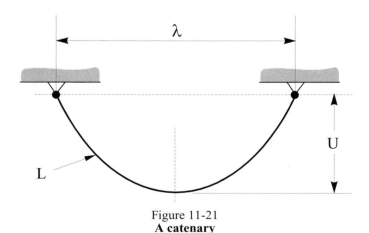

Figure 11-21
A catenary

considerations, decide which — if any — of the listed independent variables is (are) *physically irrelevant.*

11/4 **Stress in glass windows by wind.** Prove that wind-generated stresses in geometrically similar glass windows are independent of the size of those windows.

11/5 **Energy of a laterally vibrating stretched wire.** Assuming that the energy of a laterally vibrating stretched wire is a function of its length, linear density, amplitude of vibration, and tension force, which — if any — of these independent variables is (are) *physically irrelevant*?

11/6 **Sloshing frequency of fuel during take-off in missile tanks** (adapted from Ref. 31, p. 240). Fuel sloshing in fuel tanks of missiles could be rather dangerous during take-off, if the sloshing frequency is close to the missile's natural frequency. Assume that the sloshing frequency of fuel is influenced by the characteristic size of the fuel tank, density of fuel, and the *resultant* acceleration. Also assume that all missiles (and hence fuel tanks) are geometrically similar. The resultant acceleration is the vectorial sum of the gravitational acceleration (local g) and the actual linear acceleration of the missile.

Answer the following questions:

(a) Which, if any, of the listed independent variables is (are) *physically irrelevant*?
(b) How is the sloshing frequency influenced by the resultant acceleration of the missile, and by the size of the tank?

11/7 **Time scale of the universe.** Staciu (Ref. 33, p. 106) quotes an interesting problem due to Labocetta, who posed it in 1832. The same problem is also mentioned by Palacios (Ref. 37, p. 61), who calls it Labocetta's Problem.

How would all the time-related events of the universe be affected, if *all* lengths in the universe underwent a multiplication by a constant, and *all* densities remained unchanged? In other words, we wish to find out whether the pace of events in the universe would change, e.g., orbital periods of planets, oscillating periods of all pendulums, etc., and if so, by how much and in what direction (plus or minus).

(a) Solve this problem as stated.
(b) Extend the problem by assuming that all densities also changed by a factor S_ρ (Density Scale Factor).

Hint: Assume that density, length, and the universal gravitational constant are the only potential independent variables.

11/8 **Linear momentum of a quantum.** In considering the linear momentum of a quantum, it is assumed that the following variables are relevant:

Variable	Symbol	Dimension
linear momentum of quantum	q	$m \cdot kg \cdot s^{-1}$
frequency	ν	s^{-1}
speed of light	c	$m \cdot s^{-1}$
Planck's constant	h	$m^2 \cdot kg \cdot s^{-1}$
universal gravitational constant	k	$m^3 \cdot kg^{-1} \cdot s^{-2}$

There are five variables and three dimensions yielding 5 − 3 = 2 dimensionless variables, which are determined by the (here omitted) Dimensional Set

$$\pi_1 = q \cdot \sqrt{\frac{k}{h \cdot c^3}}; \qquad \pi_2 = v \cdot \sqrt{\frac{k \cdot h}{c^5}} \qquad (a)$$

When, upon careful measurements, the π_1 versus π_2 relation was established, it was found that

$$\pi_1 = u \cdot \pi_2 \qquad (b)$$

where u is a constant. Based on (a) and (b), answer the following three questions:

(i) Are any of the five listed physical variables *dimensionally irrelevant*?
(ii) Are any of the five listed physical variables *physically irrelevant*?
(iii) If frequency is not physically irrelevant, how does the linear momentum depend on it?

11/9 A general physical system. The behavior of a physical system is *assumedly* described by two dimensionless variables

$$\pi_1 = \frac{\omega}{g} \cdot \sqrt{\frac{\sigma}{\rho}}; \qquad \pi_2 = \frac{L \cdot g \cdot \rho}{\sigma} \qquad (a)$$

where the physical variables are as presented below:

Variable	Symbol	Dimension
angular speed	ω	s^{-1}
gravitational acceleration	g	$m \cdot s^{-2}$
stress (strength)	σ	$m^{-1} \cdot kg \cdot s^{-2}$
density	ρ	$m^{-3} \cdot kg$
size	L	m

When, by testing, the π_1 versus π_2 relationship is established, it is found that the graph of the relation is a rectangular hyperbola whose asymptotes are the coordinate axes. From the information given, answer the following questions:

(i) Which—if any—of the five named variables is (are) physically *irrelevant*?
(ii) How does stress vary with size?
(iii) How does angular speed vary with stress?
(iv) How does density vary with angular speed?

11/10 Stresses generated by the collision of two steel balls. Consider the stresses generated in two identical steel balls experiencing direct collision. The velocities of the balls are identical in magnitude and opposite in direction.

Assume the following relevant variables: normal stress, density of material, diameter of balls, speed of balls at impact, and Young's modulus of material.

- **(i)** How many independent dimensionless variables can be formed?
- **(ii)** Construct the above-determined number of dimensionless variables.
- **(iii)** Are any of the assumed variables (listed above) physically irrelevant?
- **(iv)** Is the total mass of the balls physically relevant?

CHAPTER 12
ECONOMY OF GRAPHICAL PRESENTATION

12.1. NUMBER OF CURVES AND CHARTS

One of the most significant benefits of using dimensionless versus physical variables is the very substantial saving of space and effort in the logistic in presenting the relations graphically. To illustrate, suppose we wish to find from a chart the dependent variable for k distinct values of the independent variable and the p number of *parameters,* each of which can also assume k distinct values. To present such a function graphically, how many curves do we need? The answer is found by the following simple argument:

if there are zero parameters, then we need $k^0 = 1$ curve
if there is 1 parameter, then we need $k^1 = k$ curves
if there are 2 parameters, then we need k^2 curves

Therefore, generally

if there are p parameters, then we need k^p curves

We can write, then

$$N_{\text{curv}} = k^p \qquad (12\text{-}1)$$

where N_{curv} = number of curves needed to present the given relationship;
k = number of distinct values for the dependent variable and each parameter;
p = number of parameters.

If N_V is the number of variables, then the number of parameters is generally

$$p = N_V - 2 \qquad (12\text{-}2)$$

since we consider *one* variable to be independent and *one* dependent. Thus, by (12-1) and (12-2),

$$N_{\text{curv}} = k^{N_V - 2} \qquad (12\text{-}3)$$

A chart can have one *family* of curves characterized by a *single* parameter. Thus each curve has a *particular* value of this parameter assigned to it, and therefore there are k curves in a chart. It follows that the number of charts required to present a relation of N_V variables is

$$\left. \begin{array}{ll} N_{\text{chart}} = k^{N_V - 3} & \text{if } N_v > 2 \\ N_{\text{chart}} = 1 & \text{if } N_V = 2 \end{array} \right\} \quad (12\text{-}4)$$

where k is the number of distinct values of each parameter. If $N_V = 2$, then of course there is only one curve and hence only one chart (a special case is dealt with in Example 12-5, where there are *unequal* numbers of distinct values for the parameters).

Example 12-1

How many curves and charts are necessary to plot a relationship among $N_V = 6$ variables if the number of distinct values of each parameter is $k = 5$?

The number of curves, by (12-3), is

$$N_{\text{curv}} = 5^{6-2} = 5^4 = 625$$

and the number of charts (each having $k = 5$ curves) is, by (12-4),

$$N_{\text{chart}} = 5^{6-3} = 5^3 = 125$$

⇑

Example 12-2. Volume of a Right Circular Cone

Consider volume V of a right circular cone of base diameter D and altitude h (Fig. 12-1)

$$V = \frac{\pi}{12} \cdot D^2 \cdot h \qquad (a)$$

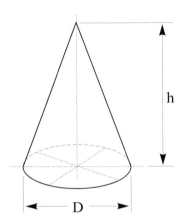

Figure 12-1
A right circular cone

By (a), we have $N_V = 3$ variables. We can consider h to be the independent variable and V the dependent variable. Therefore the parameter is D.

Now we select $k = 8$ distinct values for D. Therefore, by (12-3), the number of curves is

$$N_{\text{curv}} = k^{N_V - 2} = 8^{3-2} = 8 \qquad (b)$$

and the number of charts, by (12-4)

$$N_{\text{chart}} = k^{N_V - 3} = 8^{3-3} = 8^0 = 1 \qquad (c)$$

i.e., we need just one chart containing $N_{\text{curv}} = 8$ curves (see Fig. 12-2).

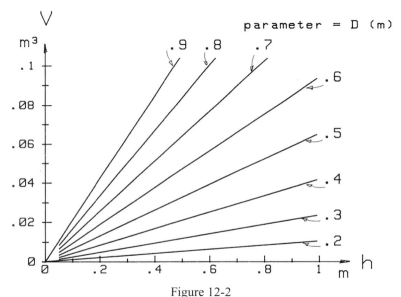

Figure 12-2
Volume of a right circular cone of base diameter D and altitude h

The table presented in Fig. 12-3 gives the number of curves and charts necessary to present a function graphically. The entries in the table were calculated by relations (12-3) and (12-4). For example, if we wished to plot a six-variable function ($N_v = 6$) with $k = 6$ distinct values for each of the $p = N_V - 2 = 6 - 2 = 4$ parameters, then we would need 216 charts, each composed of 6 curves, altogether amounting to 1296 curves.

One can see in this table that the number of curves and charts to be plotted grows rather vehemently with the number of variables—so much so that the situation soon becomes unmanageable. Referring to the above illustration, who could prepare, then deal with 216 charts? And this set deals with only six variables, which is not at all uncommon! Is there a way out of this predicament? Yes, there is, use *dimensionless* variables!

Number of variables	Number of distinct values of each variable				
	4	6	8	10	
2	1	1	1	1	number of curves
	1	1	1	1	number of charts
3	4	6	8	10	number of curves
	1	1	1	1	number of charts
4	16	36	64	100	number of curves
	4	6	8	10	number of charts
5	64	216	512	1000	number of curves
	16	36	64	100	number of charts
6	256	1296	4096	10,000	number of curves
	64	216	512	1000	number of charts
7	1024	7776	32,768	100,000	number of curves
	256	1296	4096	10,000	number of charts
8	4096	46,656	262,144	1,000,000	number of curves
	1024	7776	32,768	100,000	number of charts

Figure 12-3
Number of curves and charts necessary to plot a function for a given number of variables

As we saw in Art. 7.4

$$N_P = N_V - N_d \qquad \text{repeated (7-18)}$$

where N_P is the number of dimensionless variables which can be formed from (and by) N_v number of physical variables, and N_d is the number of dimensions.

Here we assumed that the rank of the dimensional matrix is not less than the number of dimensions N_d. If this condition is not fulfilled, then we must delete one or more dimensions to gain the equality of N_d with the rank of dimensional matrix (see Art.7.6).

So, if we have N_d dimensions in a relation, then the number of *dimensionless* variables that can be formed is always *less* than the number of variables in that relation. This results in a *very* significant improvement in the logistics of graphical presentation. If we substitute N_P from (7-18) for N_V in (12-3) and (12-4), we obtain

$$\left. \begin{array}{l} N'_{\text{curve}} = k^{N_P-2} = k^{N_V-N_d-2} \\ N'_{\text{chart}} = k^{N_P-3} = k^{N_V-N_d-3} \end{array} \right\} \qquad (12\text{-}5)$$

hence, by (12-5), (12-3), and (12-4)

$$z = \frac{N_{curv}}{N'_{curv}} = \frac{N_{chart}}{N'_{chart}} = k^{N_d} \quad (12\text{-}6)$$

It is noted that this ratio is independent of the number of variables, and is dependent only upon the number of dimensions N_d and the number of distinct values k for each parameter. For example, if we have a relation of three dimensions and six distinct values for each parameter (variable), then by *not* employing *dimensionless* variables, we would need $z = k^{N_d} = 6^3 = 216$ *times* (!) as many curves and charts as we would if we used only these variables.

The following two examples demonstrate this very substantial advantage.

Example 12-3. Gravitational Acceleration on a Celestial Body as a Function of Altitude (Positive or Negative)

To find the altitude-dependent gravitational acceleration on a celestial body, we deal with the following variables:

Variable	Symbol	Dimension	Remark
gravitational acceleration	g	m/s²	
distance from center	R	m	of celestial body
universal gravitational const.	k	m³/(kg·s²)	
mass	M	kg	of celestial body
radius	R_0	m	of celestial body

We have five variables and hence, by the table in Fig. 12-3, if the number of distinct values of variables is $k = 6$, then to represent the relation

$$g = \Psi\{R, k, M, R_0\} \quad (a)$$

graphically, we need 216 curves drawn on 36 charts. What do we have if we use dimensionless variables? There are five variables and three dimensions, therefore we have $5 - 3 = 2$ dimensionless variables supplied by the Dimensional Set

	g	R	k	M	R_0
m	1	1	3	0	1
kg	0	0	−1	1	0
s	−2	0	−2	0	0
π_1	1	0	−1	−1	2
π_2	0	1	0	0	−1

from which (b)

$$\pi_1 = \frac{g \cdot R_0^2}{k \cdot M}; \quad \pi_2 = \frac{R}{R_0}$$

So now we have only *two* (dimensionless) variables, and therefore (by using the same table) the relation can be plotted by a *single* curve—a 216-fold improvement!

To continue now, we can write

$$\pi_1 = \Psi\{\pi_2\} \quad (c)$$

which, by assuming monomial form, can be written

$$\pi_1 = c \cdot \pi_2^n \tag{d}$$

where c and n are absolute numbers still to be determined. As a simple analytical derivation (not presented here) can show, $c = 1$ and

$$\left. \begin{array}{ll} n = -2 & \text{if } R > R_0 \\ n = 1 & \text{if } R < R_0 \end{array} \right\} \tag{e}$$

Fig. 12-4 presents this "one-curve" plot. Note the interesting feature that n is not constant but varies—rather abruptly—with the *sign* of the altitude. The plot shows that g is maximum at "sea level" on any homogeneous celestial body.

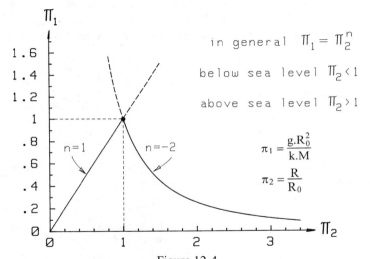

Figure 12-4
Dimensionless plot for the altitude-dependent gravitational acceleration on a celestial body
For definition of symbols, see dimensional table in this example

Example 12-4. Radial Deflection of a Semicircular Ring by a Concentrated Radial Force

Given a constant cross-section semicircular ring as illustrated in Fig. 12-5. The following variables are involved:

Variable	Symbol	Dimension	Remark
deflection	λ	m	radial
force (load)	F	N	radial
second moment of area of cross-section	I	m^4	constant
Young's modulus	E	N/m^2	
radius	R	m	

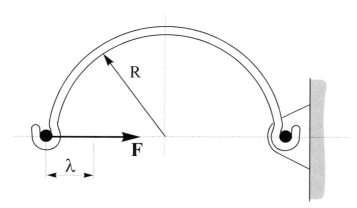

Figure 12-5
Semicircular ring loaded by concentrated radial force

We have five variables, thus if there are $k = 8$ distinct values for each parameter (variable), then we face the task of constructing 512 curves drawn on 64 charts (Fig. 12-3). To lighten this Herculean burden, we employ *dimensionless* variables supplied by the Dimensional Set

	λ	F	I	E	R
m	1	0	4	-2	1
N	0	1	0	1	0
π_1	1	0	0	0	-1
π_2	0	1	0	-1	-2
π_3	0	0	1	0	-4

As observed, there are two dimensions and five variables, therefore we have $5 - 2 = 3$ dimensionless variables. If we again consider $k = 8$ distinct values for each variable, then by Fig 12-3, to represent the relation connecting the given variables, we only need to draw eight curves which, in fact are straight lines all passing through the origin of the π_1 versus π_2 coordinate system.

To see this, we reason as follows: the three dimensionless variables by the Dimensional Set are

$$\pi_1 = \frac{\lambda}{R}; \qquad \pi_2 = \frac{F}{E \cdot R^2}; \qquad \pi_3 = \frac{I}{R^4} \qquad \text{(a)}$$

Assuming now a *monomial* form—which is plausible—we write

$$\pi_1 = c \cdot \pi_2^{n_2} \cdot \pi_3^{n_3} \qquad \text{(b)}$$

where c, n_2, and n_3 are as-yet undefined constants. By (a) and (b)

$$\frac{\lambda}{R} = c \cdot \left(\frac{F}{E \cdot R^2}\right)^{n_2} \left(\frac{I}{R^4}\right)^{n_3} \qquad \text{(c)}$$

324   APPLIED DIMENSIONAL ANALYSIS AND MODELING

It is obvious that for *small* deflection λ the load F must be proportional to the deflection. Hence in (c) we must have $n_2 = 1$. Moreover, λ must be inversely proportional to I, hence $n_3 = -1$. Thus, by (b),

$$\pi_1 = c \cdot \frac{\pi_2}{\pi_3} \tag{d}$$

If now we consider c/π_3 as the parameter, then we can write

$$\pi_1 = \left(\frac{c}{\pi_3}\right) \cdot \pi_2 \tag{e}$$

i.e., the plot is a straight line with a slope of $\varphi = \arctan(c/\pi_3)$ in the π_1 versus π_2 coordinate system. Finally, substitution of (a) into (e) yields

$$\lambda = c \cdot \frac{F \cdot R^3}{E \cdot I}$$

where c happens to be $\pi/2$. The interesting thing in this relation is that the deflection—quite unexpectedly—is proportional to the cube of the radius (i.e., size) of the spring.

⇑

Example 12-5. Blackbody Radiation Law

The emissive power radiated by a blackbody is proportional to the fourth power of its absolute temperature. This fact was experimentally established by Josef Stefan (Austrian physicist, 1835–1893) in 1879, and proved theoretically by Ludwig Boltzmann (Austrian physicist, 1844–1906). Hence the designation Stefan–Boltzmann law.

It is of some interest to find out how much of this emissive power lies between two *given* wavelengths; in particular, between λ and $\lambda + \Delta\lambda$, where λ is the wavelength and $\Delta\lambda$ is its infinitesimal increment. Let $W_\lambda \cdot d\lambda$ be the infinitesimal *monochromatic emissive power*, which is defined as the power radiated from a unit area of a blackbody in the wavelength interval specified above. Then, according to Planck's law (Max Planck, German physicist, 1858–1947) [Ref. 53, p. 59],

$$W_\lambda = 2 \cdot \pi \cdot \frac{h \cdot c^2}{\lambda^5 \cdot \left(e^{\frac{c \cdot h}{k_B \cdot \lambda \cdot T}} - 1\right)} \tag{a}$$

where:

Constant or variable	Symbol	Magnitude (if constant)	Dimension
monochromatic emissive power per unit area	W_λ	variable	kg/(m·s^3)
Boltzmann's constant	k_B	1.381E-23	m^2·kg/(s^2·K)
Planck's constant	h	6.626E-34	m^2·kg/s
speed of light	c	2.9979E8	m/s
wavelength	λ	variable	m
absolute temperature	T	variable	K

The first question is: how many curves and charts are necessary to plot relation (a)? Here we must consider that three of the constituent quantities in (a) are *constants*, i.e., they can each assume only *one* value. Hence we will have only three variables, one of which is the parameter of, say, $k = 8$ distinct values. It follows that the number of curves necessary to plot (a) is $N_{\text{curve}} = 8$, and the number of charts is $N_{\text{chart}} = 1$.

Since altogether we have six variables/constants and four dimensions, if the rank of the dimensional matrix is 4, then the number of dimensionless variables is $N_P = 6 - 4 = 2$. To verify this, we construct the dimensional matrix

	W_λ	k_B	h	c	λ	T
m	−1	2	2	1	1	0
kg	1	1	1	0	0	0
s	−3	−2	−1	−1	0	0
K	0	−1	0	0	0	1

and we see that the value of the rightmost 4 × 4 determinant is $-1 \neq 0$, hence the rank is 4, and thus we do indeed have two dimensionless variables, which we will now determine.

Ordinarily, we construct the Dimensional Set to obtain the dimensionless variables. However, in our present case, we can easily find these variables by merely inspecting relation (a). By Rule 4 of Dimensional Homogeneity (Art.6.1), exponents must be dimensionless. Thus $[(h \cdot c)/(k_B \cdot \lambda \cdot T)] = 1$. But in this case, again considering relation (a), $[(W_\lambda \cdot \lambda^5)/(h \cdot c^2)] = 1$, i.e., *this* bracketed expression must also be dimensionless. Hence we have our two dimensionless variables

$$\pi_1 = \frac{W_\lambda \cdot \lambda^5}{h \cdot c^2}; \qquad \pi_2 = \frac{k_B \cdot \lambda \cdot T}{h \cdot c} \qquad (b)$$

Considering (b) now, (a) can be written

$$\pi_1 = \frac{2 \cdot \pi}{e^{1/\pi_2} - 1} \qquad (c)$$

How many curves do we need to plot this relation? Since we have two variables, therefore, by the table in Fig.12-3, we only need *one* curve (presented in Fig.12-6). Compare this to the eight curves needed to plot relation (a)!

As seen, the curve has an asymptote of slope $\varphi = \arctan 2 \cdot \pi$ and it intercepts the abscissa at exactly $\pi_2 = 0.5$. The proofs of these enchanting characteristics are left as a rewarding exercise for the resourceful and deserving reader.

To illustrate the use of the plot, suppose we have a black body of temperature $T = 1223.5$ K (950.35°C). We wish to find the radiated power emitted by an $A = 2.3$ cm² surface area of this body between wavelengths $\lambda_1 = 9 \times 10^{-6}$ m and $\lambda_2 = 1.1 \times 10^{-5}$ m. The wavelength increment is $\Delta\lambda = \lambda_2 - \lambda_1 = 2 \times 10^{-6}$ m, and the average wavelength is $\lambda = (\lambda_1 + \lambda_2)/2 = 10^{-5}$ m. With these values and using the constants given earlier, the independent dimensionless variable π_2, by (b), is

$$\pi_2 = \frac{k_B \cdot \lambda \cdot T}{h \cdot c} = 0.85$$

For this, the plot gives $\pi_1 = 2.8$, as illustrated by dashed lines in Fig. 12-6. Therefore, by (b), the emitted power per unit area and per unit wavelength is

$$W_\lambda = \pi_1 \cdot \frac{h \cdot c^2}{\lambda^5} = 1.67 \times 10^9 \text{ W/m}^3$$

Hence, at the given temperature of T = 1223.5 K, the emitted power by the given surface area of $A = 0.00023$ m² and in the given wavelength *interval* of $\Delta\lambda = 2 \times 10^{-6}$ m is

$$W = W_\lambda \cdot A \cdot \Delta\lambda = 0.768 \text{ watts}$$

326 APPLIED DIMENSIONAL ANALYSIS AND MODELING

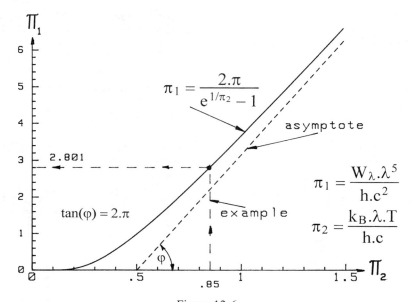

Figure 12-6
Dimensionless plot to determine the monochromatic power radiated by unit surface area of a blackbody
See table in this example for definition of symbols

Example 12-6. Deformation of an Elastic Foundation Under a Dropped Mass

The elastic foundation is represented by a coil spring of spring-constant k (= force/deflection). The spring rests on immovable ground. The maximum deflection of the spring (upon the mass dropped) is less than the spring's "iron-on-iron" compression. Only constant gravity governs the motion of the mass before it reaches the spring. Fig. 12-7 shows the arrangement.

To plot the deformation λ of the spring, how many curves do we need if we consider $k = 8$ distinct values for each variable? First we list the relevant variables (including g):

Variable	Symbol	Dimension	Remark
deflection of spring	λ	m	maximum (dynamic)
height	h	m	
spring-constant	k	kg/s^2	considered variable in this case
gravitational acceleration	g	m/s^2	constant
mass dropped	M	kg	

We see that there are four variables and one constant. Therefore, by Fig. 12-3, to plot the

$$\lambda = \Psi\{h, k, g, M\} \tag{a}$$

relation with 8 distinct values for each *variable,* we need 64 curves on 8 charts. Next question: If we use dimensionless variables, how many curves must we plot, and on how many charts? First we construct the Dimensional Set

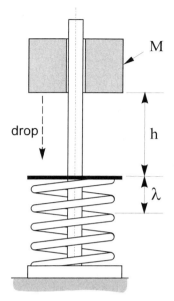

Figure 12-7
Mass dropped from height produces deflection in the spring

	λ	h	k	M	g
m	1	1	0	0	1
kg	0	0	1	1	0
s	0	0	-2	0	-2
π_1	1	0	1	-1	-1
π_2	0	1	1	-1	-1

The number of dimensions is three, the number of variables is five (g is considered a variable here), and matrix **A** is not singular (its determinant is 2), therefore there are $5 - 3 = 2$ dimensionless variables supplied by the above set. Accordingly, we have

$$\pi_1 = \frac{\lambda \cdot k}{M \cdot g}; \qquad \pi_2 = \frac{h \cdot k}{M \cdot g} \qquad (b)$$

and therefore we can write

$$\pi_1 = \Psi\{\pi_2\} \qquad (c)$$

To plot this relation, we need, by Fig. 12-3, only *one* curve. Thus by introducing dimensionless variables we managed to reduce the number of curves from 64 to 1!

At this point it is tempting to assume—for the sake of simplicity—the *monomial* form for formula (c). Unfortunately, as the following rather primitive heuristic reasoning proves, the monomial form is impossible in this case (a detailed treatment of the general problem of the *reconstruction* of relations of dimensionless variables will be given in Chapter 13).

328　　　APPLIED DIMENSIONAL ANALYSIS AND MODELING

Let us assume that (c) can be written in the monomial form

$$\pi_1 = c \cdot \pi_2^n \tag{d}$$

where c and n are constants. Thus by (d) and (b)

$$\lambda = c \cdot \frac{M \cdot g}{k} \cdot \left(\frac{h \cdot k}{M \cdot g}\right)^n \tag{e}$$

Obviously λ must increase with increasing h, therefore $n > 0$ is mandatory. But in this case if $h = 0$, then $\lambda = 0$ which cannot happen since the spring is compressed (somewhat) even if the mass is "dropped" onto it from zero height, i.e., the mass is put on the spring "gently" with zero speed. Therefore we have a contradiction, and consequently the monomial forms (d) and hence (e) are untenable. Indeed, as a simple analysis can attest (which is left as a mildly stimulating exercise for the curious reader), the deflection of the spring is

$$\lambda = \frac{1}{k} \cdot \left(M \cdot g + \sqrt{M \cdot g \cdot (M \cdot g + 2 \cdot k \cdot h)}\right) \tag{f}$$

which shows that if $h = 0$, then $\lambda = (2 \cdot M \cdot g)/k > 0$, as expected. By (b) now, (f) can be written simply

$$\pi_1 = 1 + \sqrt{1 + 2 \cdot \pi_2} \tag{g}$$

the graph of which is presented in Fig. 12-8.

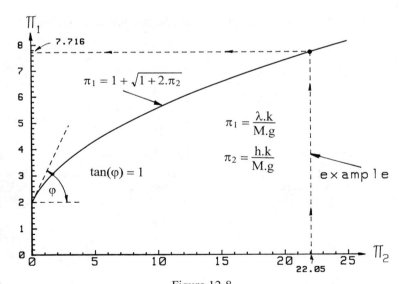

Figure 12-8
Dimensionless plot for the determination of the dynamic deformation of an elastic foundation by a dropped mass
For definition of symbols, see table in this example

To illustrate the use of this graph, assume an $M = 13.2$ kg mass dropped from $h = 0.143$ m height onto a foundation (spring) whose stiffness is $k = 19965$ N/m. Find the maximum (i.e., dynamic) deformation of the foundation. By the given data, with $g = 9.80665$ m/s²,

$$\pi_2 = \frac{h \cdot k}{M \cdot g} = \frac{(0.143) \cdot (19965)}{(13.2) \cdot (9.80665)} = 22.05 \quad \text{(h)}$$

By (g) and (h)

$$\pi_1 = 1 + \sqrt{1 + 2 \cdot \pi_2} = 1 + \sqrt{1 + (2) \cdot (22.05)} = 7.716 \quad \text{(i)}$$

Thus, by (b) and (i)

$$\lambda = \pi_1 \cdot \frac{M \cdot g}{k} = (7.716) \cdot \frac{(13.2) \cdot (9.80665)}{19965} = 0.05 \text{ m} \quad \text{(j)}$$

i.e., the foundation (spring) is dynamically deformed by 0.05 m (= 1.97 in). This checks with the explicit formula (f). Now the *static* deflection λ_S of the spring upon the weight of the mass M is

$$\lambda_S = \frac{M \cdot g}{k} \quad \text{(k)}$$

and the *dynamic* deflection is λ_D of course $\lambda_D = \lambda$. Therefore their *ratio*, by (k) and (b), is

$$\frac{\lambda_D}{\lambda_S} = \frac{\lambda}{\left(\dfrac{M \cdot g}{k}\right)} = \frac{\lambda \cdot k}{M \cdot g} = \pi_1 \quad \text{(l)}$$

In other words—and *very* interestingly—the ratio of the dynamic to the static deflection is simply the numerical value of the dimensionless variable π_1. Therefore, in the example just worked out, the *dynamic deflection* of the foundation (spring) exceeds the *static deflection* by a factor of exactly $\pi_1 = 7.716$.

From (g) it is also evident that for a large π_2 (large h), the maximum deflection varies as $\sqrt{\pi_2}$ (which means $\sqrt{h}$ since k, M, and g are constants), a fact which is certainly counterintuitive, since one would expect that a mass dropped from twice the height (and hence possessing twice the energy) would create twice the deflection on a *linear* absorber. But this is not the case.

⇧

12.2. PROBLEMS

Solutions are in Appendix 6.

12/1 In Example 12-2 we stated that the volume of a right circular cone of base diameter D and altitude h is $V = (\pi/12) \cdot D^2 \cdot h$.

(a) If dimensionless variables are used, how many curves and charts are necessary to plot the above relation?
(b) What are the dimensionless variables?
(c) Express the above relation using the variables obtained in (b), and plot the result in dimensionless form.

12/2 Given the dimensional matrix of a system

	V_1	V_2	V_3	V_4	V_5	V_6
d_1	0	1	3	-2	1	4
d_2	2	1	0	3	-2	-2
d_3	-2	2	9	-9	5	14

(a) How many curves and charts are necessary to plot the (here unspecified) relation among the given variables, if the number of distinct values for each variable is $k = 6$?

(b) How many curves and charts are necessary to plot the corresponding dimensionless relation? Use $k = 6$.

12/3 Power of a dynamo. Macagno (Ref. 61, p. 397) quotes Carvallo, who in 1891 showed that the power of a dynamo is

$$W^2 = E^2 \cdot I^2 - 4 \cdot \pi^2 \cdot \frac{L^2 \cdot I^4}{T^2} \tag{\#}$$

where the variables and their dimensions are as follows:

Variable	Symbol	Dimension
power	W	$m^2 \cdot kg/s^3$
electromotive force	E	$kg \cdot m^2/(A \cdot s^3)$
current	I	A
self-inductance	L	$m^2 \cdot kg/(s^2 \cdot A^2)$
period of alternating current	T	s

(a) Using the physical variables specified, how many curves and charts are necessary to plot relation (#)? Assume $k = 8$ distinct values for each variable.

(b) How many N_P independent dimensionless variables can be formed by the physical variables specified?

(c) Construct a complete set of dimensionless variables.

(d) Using the result of (c), convert relation (#) to dimensionless form.

(e) Using the result of (d), how many curves and charts are necessary to plot the relation?

(f) Construct the number of curves determined in (e).

12/4 Critical axial load on columns. Given a long, slender, and weightless column of uniform cross-section, clamped at its lower end and free at its upper end. The column is loaded with a centrally positioned axial force F at the free end, as illustrated in Fig. 12-9. By gradually increasing F, a particular value will be reached which causes the column to buckle. This is the *critical load* F_c.

(a) List the variables and their dimensions that influence the critical load.

(b) Compose the dimensional matrix.

(c) How many curves and charts are necessary to plot the relation determining F_c?

ECONOMY OF GRAPHICAL PRESENTATION **331**

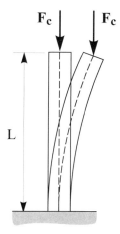

Figure 12-9
Long slender column is loaded by a concentrated axial force

- **(d)** Determine a complete set of dimensionless variables.
- **(e)** How many curves and charts are necessary to plot the relation among the dimensionless variables determined in (d)?
- **(f)** By heuristic reasoning (based on your general knowledge of the behavior of bending beams), derive a formula among the dimensionless variables determined in (d), and plot the appropriate curve(s).

12/5 **Relativistic mass.** We know from Einstein's (Albert Einstein, German–Swiss–American physicist, 1879–1955) relativity theory that the mass of a moving body is a function of its speed measured in a coordinate system which is at rest with respect to the observer. Einstein stated that mass is the smallest M_0 when it is at rest (resting mass), and infinite when its speed reaches that of light in vacuum. The following variables/constant are involved:

Variable/constant	Symbol	Dimension	Remark
relativistic mass	M	kg	
speed of mass	v	m/s	with respect to observer
mass at rest	M_0	kg	with respect to observer
speed of light	c	m/s	in vacuum

- **(a)** How many curves and charts are necessary to graphically present Einstein's relation

$$M = \frac{M_0 \cdot c}{\sqrt{c^2 - v^2}} \qquad (\#)$$

Assume $k = 8$ distinct values for the variables.

- **(b)** How many independent dimensionless variables can be formed by the variables and the constant listed? List one such set.
- **(c)** How many curves and charts are needed to plot the relation among the dimensionless variables obtained in (b) above? Assume $k = 8$ distinct values for each variable.
- **(d)** Prove that no monomial form of relation among the dimensionless variables can exist.
- **(e)** Plot the Einstein relation (#) in dimensionless form. Use the dimensionless variables obtained in (b) above. Verify that the number of curves is the same as determined in (c) above.

CHAPTER 13
FORMS OF DIMENSIONLESS RELATIONS

13.1. GENERAL CLASSIFICATION

Relations composed of dimensionless variables can assume one of two distinct forms: *monomial* or *nonmonomial*. In general, the monomial form is

$$\pi_1 = c \cdot \pi_2^{\epsilon_2} \cdot \pi_3^{\epsilon_3} \ldots \pi_{N_P}^{\epsilon_{N_P}} \tag{13-1}$$

where $\pi_1, \pi_2, \ldots, \pi_{N_P}$ are the dimensionless variables, N_P is the *number* of dimensionless variables, $\epsilon_2, \ldots, \epsilon_{N_P}$ are numeric exponents, and c is a numeric constant. Note that the number of exponents is $N_P - 1$, which is why the subscripts of these exponents start with 2.

An important characteristic of (13-1) is that there is no "+" or "−" sign in it; in fact it is the *absence* of these signs which is the *condition* for such a form to qualify as a monomial. Correspondingly, if there is a "+" or "−" sign in a dimensionless expression, then it is a *nonmonomial*. For example

$$\pi_1 = (1 + \pi_2)^2 \tag{13-2}$$

is a nonmonomial because it contains a "+" sign. The following three definitions are now formulated:

Definition 13-1. *If a relation, in its algebraically simplest form, contains no "+" (plus) or "−" (minus) sign, then it is a monomial.*

Definition 13-2. *If a relation, in its algebraically simplest form, contains at least one "+" (plus) or "−" (minus) sign, then it is a nonmonomial.*

The qualification "algebraically simplest form" in both of these above definitions is necessary, for otherwise $\pi_1 = \pi_2^2 + 5 - 5$ would not be a monomial, although of course it is because in its algebraically simplest form $\pi_1 = \pi_2^2$ there is no "+" or "−" sign.

Definition 13-3. *If a relation contains a transcendental function, then this relation is a nonmonomial.*

In essence this definition is an extension of Definition 13-2, since any transcendental function, if expressed by its Taylor series, is an infinite number of terms connected by "+" and "−" signs.

A very important and useful characteristic is stated in the following theorem.

Theorem 13-1. *If a relation among N_P dimensionless variables is a monomial, then it contains the least number of constants, and this number is N_P. If a relation is a nonmonomial, then the number of constants in it exceeds N_P.*

Proof. If a relation among N_P dimensionless variables is a monomial, then it has the form of (13-1). In this form the number of constant *exponents* is $N_P - 1$, and the number of constant *coefficients* is 1. Hence the number of constants is the *sum* of these two numbers, which is therefore $N_P - 1 + 1 = N_P$. If the relation is a nonmonomial, then it must contain at least one "+" or "−" sign. Hence the number of coefficients must be at least two, although the number of exponents does not diminish. Hence the sum of these numbers will exceed N_P. This proves the theorem.

Example 13-1

What is (are) the unknown constant(s) to be determined in a monomial of $N_p = 3$ dimensionless variables?

The cited monomial must have the form $\pi_1 = c \cdot \pi_2^a \cdot \pi_3^b$, and hence the unknown constants are c, a, and b, altogether three in number, which, in accordance with Theorem 13-1, is N_P.

⇑

Example 13-2

The constants in the *nonmonomial* relation of $N_P = 3$ dimensionless variables

$$\pi_1 = c_1 \cdot \pi_2^p + c_2 \cdot \pi_2^q + c_3 + c_4 \cdot \pi_3^r - c_5 \cdot e^{\pi_3}$$

are $c_1, c_2, c_3, c_4, c_5, p, q,$ and r—altogether eight in number, which is larger than 3.

⇑

Example 13-3

Is the relation $\pi_1 = e^{\pi_2}$ a monomial? Since the relation contains a transcendental function, therefore, by Definition 13-3, it is a nonmonomial. Indeed, in its algebraically simplest form it can be written as $\pi_1 = 1 + \pi_2 + \pi_2^2/2! + \pi_2^3/3! + \ldots$ and hence contains an infinite number of terms.

⇑

In engineering and the physical sciences it is very important to know whether an as-yet unknown relation among N_P dimensionless variables a monomial or nonmonomial. This is because if the relation is a *monomial*, then we only have to perform N_P experiments (measurements) to determine the valid relation among the dimensionless variables. If, however, the relation is a *nonmonomial*, then the task of finding out the relation experimentally is much more difficult. In the latter case, generally, not only do we not know the nature of the relation (is it trigonometric, logarithmic, exponential, or does it contain multiple exponents for the same variable, i.e., a polynomial, etc.?), but we do not even know the number of members in the relation.

Therefore, it is very important to find out the *nature* of the relation (yet to be formulated). If it is a *monomial*, then we must determine the N_P number of constants by either heuristic reasoning, by N_P (number of) measurements, or some combination of these two. If, however, the relation is a *nonmonomial*, then our choices are rather limited (although not entirely hopeless), as some of the following illustrative examples demonstrate.

13.2. MONOMIAL IS MANDATORY

In some situations, by purely speculative means, without measurement or analysis, it is possible to conclude that the sought-after relation *must* be a monomial. The following two examples illustrate:

Example 13-4. Flow of Fluid Over a Spillway

We wish to determine the quantity of fluid flowing over a spillway shown in Fig. 13-1. Assume that the reservoir out of which the fluid is flowing is sufficiently large, so that height h remains constant. Thus the cross-section of the flow is a rectangle of constant width and height, as given in the figure. Accordingly, we have the following relevant variables:

Variable	Symbol	Dimension	Remark
mass flow rate	Q	kg/s	
width of opening	a	m	
height of flow	h	m	constant
gravitational acceleration	g	m/s²	
density of fluid	ρ	kg/m³	

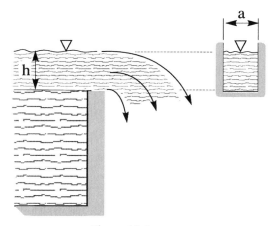

Figure 13-1
Fluid flowing over a spillway of constant height

336 APPLIED DIMENSIONAL ANALYSIS AND MODELING

We have five variables and three dimensions, therefore there are 5 – 3 = 2 dimensionless variables supplied by the Dimensional Set.

	Q	a	h	g	ρ
m	0	1	1	1	–3
kg	1	0	0	0	1
s	–1	0	0	–2	0
π_1	1	0	$-\frac{5}{2}$	$-\frac{1}{2}$	–1
π_2	0	1	–1	0	0

from which

$$\pi_1 = \frac{Q}{\rho \cdot \sqrt{h^5 \cdot g}}; \qquad \pi_2 = \frac{a}{h} \tag{a}$$

Therefore we can write

$$\pi_1 = \Psi\{\pi_2\} \tag{b}$$

where Ψ is an as-yet undefined function. By (a) and (b)

$$Q = k \cdot \rho \cdot \sqrt{h^5 \cdot g} \cdot \Psi\left(\frac{a}{h}\right) \tag{c}$$

where k is constant. Now we recognize that Q is proportional to "a" since, obviously, twice as *wide* a channel provides twice as much outflow of liquid per unit time. This condition is fulfilled *only* if Ψ in relation (c) is a *monomial* with an exponent of unity. Hence, from this finding and (c)

$$Q = k \cdot \rho \cdot \sqrt{h^5 \cdot g} \cdot \frac{a}{h} = k \cdot \rho \cdot a \cdot \sqrt{g \cdot h^3} \tag{d}$$

or

$$\frac{Q}{a} = \overline{Q} = k \cdot \rho \cdot \sqrt{g \cdot h^3} \tag{e}$$

where $\overline{Q}$ is the flow of mass per unit width of the opening per unit time. This example shows a case where the monomial form of the function Ψ in (b) is obligatory. We arrived at this very favourable conclusion by the simplest heuristic reasoning.

⇑

Example 13-5. Drag on a Moving Body Immersed in a Fluid

We determine the drag acting on a moving body fully immersed in viscous liquid (Fig 13-2). The following variables may be considered relevant:

Variable	Symbol	Dimension	Remark
drag force	F	m·kg/s²	
fluid viscosity	μ	kg/(m·s)	dynamic
speed of body	v	m/s	
density of fluid	ρ	kg/m³	uniform
length of body	L	m	characteristic value

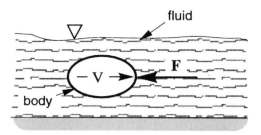

Figure 13-2
Drag force on a slowly moving body immersed in fluid

The number of variables is five and we have three dimensions, hence there are 5 − 3 = 2 dimensionless variables supplied by the Dimensional Set

	F	μ	v	ρ	L
m	1	−1	1	−3	1
kg	1	1	0	1	0
s	−2	−1	−1	0	0
π_1	1	0	−2	−1	−2
π_2	0	1	−1	−1	−1

from which

$$\pi_1 = \frac{F}{v^2 \cdot \rho \cdot L^2}; \qquad \pi_2 = \frac{\mu}{v \cdot \rho \cdot L} \qquad (a)$$

Thus, we write

$$\pi_1 = \Psi\{\pi_2\} \qquad (b)$$

or, by (a),

$$F = k \cdot v^2 \cdot L^2 \cdot \rho \cdot \Psi\left\{\frac{\mu}{v \cdot \rho \cdot L}\right\} \qquad (c)$$

where k is a constant and Ψ is an as-yet unknown function. If the speed is low, then inertial effects, represented by the fluid density ρ, can be disregarded. The only way ρ can disappear from (c) is if Ψ is a monomial with exponent 1. Accordingly,

$$F = k \cdot v^2 \cdot L^2 \cdot \rho \cdot \left(\frac{\mu}{v \cdot \rho \cdot L}\right) = k \cdot v \cdot L \cdot \mu \qquad (d)$$

in which ρ does not appear. If the speed is higher, then it is the viscosity μ that should be disregarded, since at high velocities the inertial effects dominate. The only way relation (c) can accommodate this condition is if Ψ is (again) a monomial with exponent 0. Accordingly,

$$F = k \cdot v^2 \cdot L^2 \cdot \rho \qquad (e)$$

in which, of course, μ does not appear. This is the relation used to determine the drag of a typical automobile, and the speed of a falling stone. In the case of a stone falling at constant (terminal) speed, drag is identical to weight which, in turn, is proportional to the cube of the dimension L. Hence, by (e), $L^3 = \text{const} \cdot v^2 \cdot L^2 \cdot \rho$, from which

$$v = \text{const} \cdot \sqrt{L} \tag{f}$$

However, if the stone is *small* (i.e., the speed is low), then (d) is applicable. Hence, in this case $L^3 = \text{const} \cdot v \cdot L \cdot \mu$, from which

$$v = \text{const} \cdot L^2 \tag{g}$$

Relations (f) and (g) can be plotted as in Fig. 13-3, which of course shows a qualitative relation only, since the constants—defined by the shape and density of the stone—must also be considered in each case separately.

In this example, the monomial form was mandated *differently* according to the physical conditions in which the independent variable L assumed different *ranges* of values.

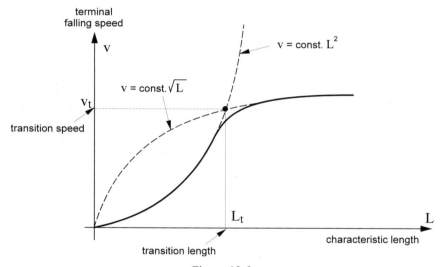

Figure 13-3
Terminal speed of small and medium sized falling stones
The transitional length L_t and associated speed v_t are
determined by the shape and density of the stone

13.3. MONOMIAL IS IMPOSSIBLE – PROVEN

Often the relation among dimensionless variables *cannot* be a monomial, and the sooner this is determined, the better, so one is not led to an erroneous relation for the sake of simplicity or easy logistic. The procedure is to scrutinize the assumed monomial form and try to prove that it leads to a contradiction—in other words, an impossibility. This process is called *reductio ad absurdum* in logic and it is proba-

FORMS OF DIMENSIONLESS RELATIONS   **339**

bly the most frequently used tool to prove or disprove an assertion. In our problem at hand, if our attempt is successful—i.e., the result shows an impossibility—then we will be certain that the assumed monomial form is *untenable*. However, if we cannot prove that the monomial is impossible, then this fact does *not* prove the *existence* of the monomial in question, i.e., the monomial could still be impossible. That is, a contradiction obtained is a *sufficient* but *not a necessary* condition for the nonexistence of the monomial form. Cases illustrating this situation are presented in Art. 13.4.

For the present, however, we now show by several examples, how the *nonexistence* of the monomial form can be *proven* by simple heuristic reasoning.

Example 13-6. Impact Velocity of a Meteorite*

The following variables are involved in determining the impact velocity of a meteorite at sea level:

Variable	Symbol	Dimension	Remark
speed of meteorite	v	m/s	at sea level
diameter of meteorite	D	m	
density of meteorite	ρ_m	kg/m^3	
drag coefficient	C_d	1	
entry angle to atmosphere	φ	1	to local horizon
density of air	ρ_a	kg/m^3	at sea level
scale height	h_0	m	see explanation below
interplanetary speed	v_0	m/s	maximum 56 km/s

Scale height h_0 is, by definition, that *particular altitude* at which the density of air is $1/e = 0.36788$ times its density at sea level. The *entry angle* is measured *locally*, i.e., at the entry point to the atmosphere; it is always larger than the *impact angle* at ground level.

We have eight variables and three dimensions. Therefore $8 - 3 = 5$ independent dimensionless variables can be formed; they are supplied by the Dimensional Set shown below. Note that the two dimensionless *physical* variables C_d and φ must be in the **B** matrix, for otherwise the **A** matrix would be singular (Art.14.1). Also, observe that variables of identical dimensions, such as D and h_0, v and v_0, ρ and ρ_0, do not appear together in **A**, for these, too, would make **A** singular.

	v	D	ρ_m	C_d	φ	ρ_a	h_0	v_0
m	1	1	-3	0	0	-3	1	1
kg	0	0	1	0	0	1	0	0
s	-1	0	0	0	0	0	0	-1
π_1	1	0	0	0	0	0	0	-1
π_2	0	1	0	0	0	0	-1	0
π_3	0	0	1	0	0	-1	0	0
π_4	0	0	0	1	0	0	0	0
π_5	0	0	0	0	1	0	0	0

---

*When an interplanetary small celestial object, a *meteoroid,* enters Earth's atmosphere, a part of it burns up and the rest reaches the surface of Earth. This latter part is a *meteorite.*

This will yield the dimensionless variables

$$\pi_1 = \frac{v}{v_0}; \quad \pi_2 = \frac{D}{h_0}; \quad \pi_3 = \frac{\rho_m}{\rho_a}; \quad \pi_4 = C_d; \quad \pi_5 = \varphi \tag{a}$$

Hence we can write

$$\pi_1 = \Psi\{\pi_2, \pi_3, \pi_4, \pi_5\} \tag{b}$$

Now we try the *monomial* form

$$\pi_1 = k \cdot \pi_2^{\epsilon_2} \cdot \pi_3^{\epsilon_3} \cdot \pi_4^{\epsilon_4} \cdot \pi_5^{\epsilon_5} \tag{c}$$

which, by (a), can be written

$$v = k \cdot v_0 \cdot \left(\frac{D}{h_0}\right)^{\epsilon_2} \cdot \left(\frac{\rho_m}{\rho_a}\right)^{\epsilon_3} \cdot C_d^{\epsilon_4} \cdot \varphi^{\epsilon_5} \tag{d}$$

If $\epsilon_2 > 0$, then with increasing meteoroid size D, the impact velocity would increase *without limit*—which is impossible. Therefore ϵ_2 cannot be positive. If $\epsilon_2 < 0$, then with increasing D the impact velocity would decrease, which is equally absurd. Therefore ϵ_2 cannot be negative. If now $\epsilon_2 = 0$, then the impact speed would be independent of both the diameter of the meteorite and the scale height of the atmosphere—a profound impossibility. Therefore ϵ_2 cannot be zero. Hence ϵ_2 cannot be any number, including zero. The conclusion we *must* draw is that the dependence of v upon D/h_0 is *not* of the monomial power form.

Next, we examine ϵ_3. If it is positive, then with increasing density of the meteoroid the impact speed would grow without limit, which cannot happen. Hence ϵ_3 cannot be positive. If ϵ_3 is negative, then with increasing atmospheric density the impact speed would increase. This is absurd. Therefore ϵ_3 cannot be negative. If $\epsilon_3 = 0$, then the impact speed would be independent of the density of the atmosphere, which is again impossible. Thus ϵ_3 cannot be any number, either. Therefore we again forced to conclude that the monomial power form of v upon ρ_m/ρ_a cannot exist. Similar reasoning can be performed for ϵ_4 and ϵ_5.

Thus we must conclude that the monomial formula for the dimensionless variables π_1, π_2, π_3, and π_4 is untenable. Indeed, as Schultz states (Ref. 50, p. 16.184), the valid relation—in terms of the physical variables given—is

$$v = v_0 \cdot e^{-\frac{3}{8} \cdot \frac{C_d \cdot \rho_a \cdot h_0}{\rho_m \cdot D \cdot \sin\varphi}} \tag{e}$$

which, if expressed by the dimensionless variables defined in (a), becomes

$$\pi_1 = e^{-\frac{3}{8} \cdot \frac{\pi_4}{\pi_2 \cdot \pi_3 \cdot \sin\pi_5}} \tag{f}$$

Now let us see how formula (e) and hence (f) stand up against the same heuristic reasoning. If ρ_a or h_0 or both increase, then the algebraic value of the exponent in relation (e) decreases, yielding a lower impact speed. This makes sense. Moreover, if ρ_a increases without limit, then v tends to zero—as it should. If the density of meteoroid ρ_m increases, then, by (e), the speed v increases, and in the limiting case where $\rho_m \to \infty$, the impact speed v will be the same as the interplanetary speed v_0, as it should be. Similarly for the size D of the meteoroid: if $D \gg 0$, then $v \approx v_0$, as it should be. Finally, if φ increases (i.e., the entry angle is becoming more vertical), then by (e) v increases, as it should. Thus, at least to the extent of this reasoning, relations (e) and hence (f) seem to be correct!

In this example then, by heuristic deliberation it was conclusively *proved* that the sea-level impact speed of a meteorite cannot be a monomial function of its determining variables.

Example 13-7. Area of a Triangle Whose Side-Lengths Form a Geometric Progression

Given a triangle by its sides a, b, and c. If these sides form a geometric progression, then

$$a = a; \qquad b = a \cdot k; \qquad c = a \cdot k^2 \tag{a}$$

where k is a constant factor larger than 1, and "a" is the shortest side.

We wish to express the area of the triangle T as a function of k and a. The relevant variables and their dimensions are:

Variable	Symbol	Dimension	Remark
common ratio of side-lengths	k	1	as defined in relation (a)
area of triangle	T	m²	
length of one side	a	m	designated (shortest) side

We have three variables and one dimension, therefore there are $3 - 1 = 2$ dimensionless variables, furnished by the Dimensional Set

	k	T	a
m	0	2	1
π_1	1	0	0
π_2	0	1	−2

Thus,

$$\pi_1 = k; \qquad \pi_2 = \frac{T}{a^2} \tag{b}$$

from which

$$\pi_2 = \Psi\{\pi_1\} \tag{c}$$

where Ψ represents some function yet to be found. In what follows we prove that Ψ cannot be a monomial, viz.

$$\pi_2 = c \cdot \pi_1^n \tag{d}$$

is impossible, no matter what constants c and n might be. By (d) and (b)

$$T = c \cdot a^2 \cdot k^n \tag{e}$$

The proof is in two steps. In Step 1 we prove that for any triangle, k must be within a certain well-defined range, i.e., k has lower and upper limits. Moreover—still in Step 1—we also determine these limits. In Step 2 we prove that if k is within the limits established in Step 1—as it must be—then no exponent n satisfies (e).

Step 1. We invoke an obvious—but often overlooked—elementary characteristic of any triangle, namely that the sum of the lengths of any two sides cannot be less than the length of the third side (Euclid's Proposition 20, Ref. 129, p. 12). That is, if a, b, and c are the lengths of the sides of a triangle, then

$$a + b \geq c; \qquad b + c \geq a \tag{f}$$

Now we have the side-lengths from (a), so by (a) and the first part of (f),

$$a + a \cdot k \geq a \cdot k^2$$

or, by simplification,

$$k^2 - k - 1 \leq 0$$

the solution of which for k is

$$k \leq \frac{\sqrt{5} + 1}{2} = 1.61803\ldots \tag{g}$$

Hence

$$k_{\max} = 1.61803\ldots \tag{h}$$

On the other hand, by (a) and the second relation of (f),

$$a \cdot k + a \cdot k^2 \geq a$$

or

$$k^2 + k - 1 \geq 0$$

the solution of which for k is

$$k \geq \frac{\sqrt{5} - 1}{2} = 0.61803\ldots \tag{i}$$

hence

$$k_{\min} = 0.61803\ldots \tag{j}$$

As the reader can easily verify, both of the limiting values of k, defined in (h) and (j), result in a straight line which may be considered a *degenerated* triangle of zero area. Now we are ready for Step 2 of our proof.

Step 2. If in (e) we have $n = 0$, then area T is independent of k, which is impossible, since k determines the lengths of sides b and c, which do influence the area. If in (e) we have $n > 0$, then T increases without limit with positive k. But this monotonic increase is impossible, because at $k = k_{\max}$ area T is zero. Hence, if k is slightly *less* than its maximum, then T must *increase* with *decreasing* k, a fact that contradicts formula (e) with positive k.

Now if in (e) we have $n < 0$, then T must monotonically *decrease* with *increasing* k at all values of k. However, this is impossible, since at $k = k_{\min}$ area T is zero and hence in the vicinity of minimum k area T must *increase* with *increasing* k. Consequently, formula (e) does not hold in the case of negative n.

We must conclude therefore that formula (e) is untenable, and hence the area of a triangle whose side-lengths form a geometric progression cannot be expressed by a monomial as in (d). Indeed, as the tenacious reader might wish to confirm, the formula fulfilling form (c), and employing the derived dimensionless variables π_1 and π_2 in (b), is

$$\pi_2 = \frac{1}{4} \cdot \sqrt{4 \cdot \pi_1^6 - (1 - \pi_1^2 - \pi_1^4)^2} \tag{k}$$

the plot of which is presented in Fig. 13-4. If $k = 1$, i.e., $\pi_1 = 1$, then $a = b = c$, and the triangle is equilateral. Area T of an equilateral triangle whose side-length is "a" is

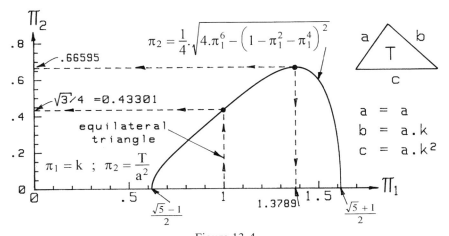

Figure 13-4
Dimensionless plot to determine the area of a triangle whose side-lengths form a geometric progression of common ratio k

$T = a^2 \cdot \sqrt{3}/4$, and hence $\pi_2 = T/a^2 = \sqrt{3}/4 = 0.43301$. This value is confirmed by relation (k), and shown in the plot by dashed lines. The plot also clearly indicates (as the profoundly indefatigable reader may also want to verify) that the curve *has* a maximum which occurs at

$$\pi_1 = \frac{1}{\sqrt{2}} \cdot \sqrt{1 + \left(4 + \sqrt{\frac{307}{27}}\right)^{1/3} + \left(4 - \sqrt{\frac{307}{27}}\right)^{1/3}} = 1.37887 \qquad (\ell)$$

yielding the maximum π_2, by (k),

$$(\pi_2)_{max} = 0.66595 \qquad (m)$$

This means that if a triangle's side-lengths form a geometric progression of common ratio k, as in (a), then this triangle has a maximum area if $k = 1.37887\ldots$, yielding $T_{max} = a^2 \cdot (0.66594\ldots)$, where "$a$" is the length of the *shortest* side of the triangle.

Finally, it is pointed out that the lower limit of k, defined in relation (j), is the famous *golden ratio*, or *golden cut*, known since antiquity. It is the only number equalling its reciprocal diminished by unity. Another noteworthy characteristic of this number is that if a rectangle has the *golden ratio* (width/length), then it is generally to be the most visually pleasing. A famous and well-quoted example of this is the Parthenon built 2500 years ago on the Acropolis at Athens. This classic monument with its triangular pediment fits almost exactly into a golden rectangle. There are numerous appearances of this sublime ratio in nature. Reference 130 is a highly recommended scholarly work on this subject.

⇑

Example 13-8. Boy Meets Girl

A boy and a girl have a rendezvous at the Nelson Monument in London, and then go to a dance. They agree to arrive at the Monument at random *points of time*, but within an agreed time *interval T* (say between 5 and 6 PM). Further, the boy will wait for the girl (if she is not yet there) only Δt_b minutes, and the girl will wait for the boy (if he is not yet there) only Δt_g minutes. What is the likelihood (probability) p that they will meet—and hence go on to the dance? What are the relevant variables and their dimensions in this problem? How many independent dimensionless variables can be formed, and what are they? Is the *monomial* form encompassing these dimensionless variables tenable? If not, why?

The relevant variables and their dimensions are listed below:

Variable	Symbol	Dimension	Remark
time interval	T	minute	
boy's waiting time for girl	Δt_b	minute	$\Delta t_b \leq T$
girl's waiting time for boy	Δt_g	minute	$\Delta t_g \leq T$
probability of boy meets girl	p	1	

The restrictions for waiting times in the "Remark" column merely indicate the obvious fact that neither partner will wait longer than the agreed time interval for the arrival of the other.

We have four variables and one dimension, therefore the number of dimensionless variables is $4 - 1 = 3$; they are supplied by the Dimensional Set

	p	Δt_b	Δt_g	T
minute	0	1	1	1
π_1	1	0	0	0
π_2	0	1	0	-1
π_3	0	0	1	-1

from which

$$\pi_1 = p; \qquad \pi_2 = \frac{\Delta t_b}{T}; \qquad \pi_3 = \frac{\Delta t_g}{T} \qquad (a)$$

where, in view of the restrictions imposed by the waiting times, the values of all three dimensionless variables must be between 0 and 1. By (a) we have

$$\pi_1 = \Psi\{\pi_2, \pi_3\} \qquad (b)$$

where Ψ is some function. Let us now see if the monomial form for this function is possible. Thus, we assume the existence of

$$\pi_1 = c \cdot \pi_2^{n_2} \cdot \pi_3^{n_3} \qquad (c)$$

where c and the exponents are yet to be determined numeric constants. By (a) and (c),

$$p = c \cdot \left(\frac{\Delta t_b}{T}\right)^{n_2} \cdot \left(\frac{\Delta t_g}{T}\right)^{n_3} \qquad (d)$$

Because of symmetry $n_2 = n_3 = n$ at all times, hence we have by (d),

$$p = c \cdot \left(\frac{\Delta t_b \cdot \Delta t_g}{T^2}\right)^n \qquad (e)$$

Now if either Δt_b or Δt_g increases, and the other does not, then obviously p must *increase*. Hence *n must* be positive. But if $n > 0$, and if either (but not both) Δt_b or Δt_g is zero, then

(e) yields zero probability which, however, is contrary to what happens in real life. For imagine that $\Delta t_b > 0$ and $\Delta t_g = 0$, i.e., the boy waits, but the girl does not. In this case the meeting might still take place, because when the girl arrives the boy might already be there waiting for her since *his* waiting time is larger than zero. Thus, in fact $p > 0$, even though $\Delta t_g = 0$. This scenario contradicts (e) with positive n. Hence n in (e) cannot be positive. But it must be positive, as already shown. Thus we have a contradiction, and therefore the *monomial* form (e) is untenable.

Indeed, as the inquisitive reader can derive (hint: use geometric interpretation of probability; in case of emergency see Appendix 7),

$$\pi_1 = \pi_2 + \pi_3 - \frac{1}{2} \cdot (\pi_2^2 + \pi_3^2) \tag{f}$$

which of course is not a monomial. To illustrate the use of this relation, suppose the time interval during which the boy and the girl have agreed to arrive is $T = 60$ minutes, the boy waits for the girl $\Delta t_b = 15$ minutes, and the girl waits for her date $\Delta t_g = 12$ minutes. Then, by (a),

$$\pi_2 = \frac{\Delta t_b}{T} = \frac{15}{60} = 0.25; \qquad \pi_3 = \frac{\Delta t_g}{T} = \frac{12}{60} = 0.2$$

and by (f)

$$\pi_1 = p = \frac{1}{4} + \frac{1}{5} - \frac{1}{2} \cdot \left(\frac{1}{16} + \frac{1}{25} \right) = \frac{319}{800} = 0.39875$$

i.e., there is a 39.875% chance that they will go dancing.*

The chart in Fig 13-5 presents the π_1 versus π_2 relation with π_3 as parameter. The numerical example worked out above is indicated by dashed lines. In the special case where the waiting times are equal, then $\Delta t_b = \Delta t_g$ and therefore $\pi_2 = \pi_3$. In this case (f) reduces to

$$\pi_1 = p = \pi_2 \cdot (2 - \pi_2) \tag{g}$$

from which

$$\pi_2 = 1 - \sqrt{1 - \pi_1} = 1 - \sqrt{1 - p} \tag{h}$$

What is the chance that the two will meet if each waits for the other for 30 minutes? By (a), $\pi_2 = 30/60 = 0.5$ and, by (g), $p = (0.5) \cdot (2 - 0.5) = 0.75$, i.e., they have a 75% chance of meeting. Similarly, how long would each partner have to wait for the other, if they wish to have an even (50%) chance to meet within a 60 minute time interval? With $p = 0.5$ relation (h) provides

$$\pi_2 = \frac{\Delta t_b}{T} = \frac{\Delta t_g}{T} = 1 - \sqrt{1 - 0.5} = 0.29289$$

by which $\Delta t_b = \Delta t_g = T \cdot \pi_2 = (60) \cdot (0.29289) = 17.57359$ minutes. In other words, each must agree to wait for the other for a little over 17.5 minutes.

---

*The author carried out a Monte Carlo simulation to verify this result. In 250,000 simulated rendezvous, the boy met the girl (and then took her to the dance) 99665 times, i.e., in 39.866% of the cases. This is within 0.0226% of the theoretical exact value.

346 APPLIED DIMENSIONAL ANALYSIS AND MODELING

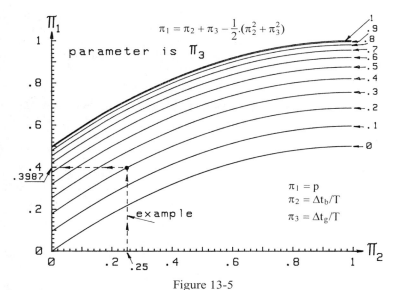

Figure 13-5
Dimensionless plot to determine the probability p that a boy and a girl will meet within an agreed time interval and at an agreed place
They arrive at random times within an interval of T minutes, and the boy and girl wait for each other Δt_b, Δt_g minutes, respectively

⇑

Example 13-9. Speed of a Vertically Ejected Particle (I)

We now investigate the speed v of a vertically ejected particle upon travelling a distance h. The motion of the particle is governed entirely by constant gravity, i.e., air resistance is ignored. The list of relevant variables is as follows:

Variable	Symbol	Dimension	Remark
speed	v	m/s	at altitude h
altitude	h	m	
ejection speed	v_0	m/s	at altitude zero
gravitational acceleration	g	m/s²	constant

Thus we have four variables and two dimensions yielding $4 - 2 = 2$ dimensionless variables; these are obtained from the Dimensional Set

	v	h	v_0	g
m	1	1	1	1
s	−1	0	−1	−2
π_1	1	0	−1	0
π_2	0	1	−2	1

from which the two dimensionless variables are

$$\pi_1 = \frac{v}{v_0}; \qquad \pi_2 = \frac{h \cdot g}{v_0^2} \tag{a}$$

FORMS OF DIMENSIONLESS RELATIONS

Hence, we can write

$$\pi_1 = \Psi\{\pi_2\} \tag{b}$$

where Ψ is a function yet to be determined. Let us try the *monomial* form for Ψ and see whether it is tenable. Accordingly,

$$\pi_1 = c \cdot \pi_2^n \tag{c}$$

where c and n are constants. By (a) and (c),

$$v = c \cdot v_0 \cdot \left(\frac{h \cdot g}{v_0^2}\right)^n \tag{d}$$

We can see that with increasing h, speed v decreases. Hence n must be negative. But if n is negative, then at $h = 0$ the speed is infinite, which cannot be since it is—by assumption—v_0. Consequently, the *monomial* form (c)—or (d)—is untenable. Indeed, as the reader can easily derive

$$v^2 = v_0^2 - 2 \cdot g \cdot h \tag{e}$$

which, with the aid of the dimensionless variables in (a), can be expressed as

$$\pi_1 = \sqrt{1 - 2 \cdot \pi_2} \tag{f}$$

This relation is not a monomial. Fig 13-6 presents the graph of (f). To illustrate the use of this plot, consider a projectile ejected vertically upward at speed $v_0 = 15.3$ m/s. What will the speed of this projectile be at altitude $h = 8.3$ m?

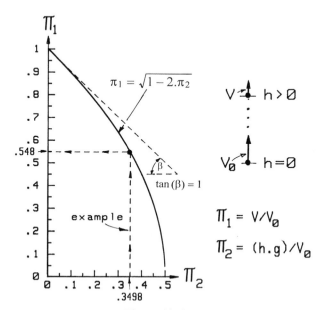

Figure 13-6
Dimensionless plot to determine the altitude-dependent speed of a projectile ejected vertically upward from the ground
$g = 9.80665$ m/s² constant gravitational acceleration

348 APPLIED DIMENSIONAL ANALYSIS AND MODELING

By the second relation of (a)

$$\pi_2 = \frac{h \cdot g}{v_0^2} = \frac{(8.35) \cdot (9.80665)}{15.3^2} = 0.3498$$

for which the plot supplies $\pi_1 = 0.548$. Thus, by this value and relation (a)

$$v = \pi_1 \cdot v_0 = (0.548) \cdot (15.3) = 8.384 \text{ m/s}$$

This example is marked by dashed lines in Fig. 13-6.

⇑

13.4. MONOMIAL IS IMPOSSIBLE – NOT PROVEN

Sometimes a monomial form among dimensionless variables does not exist, but proving this by heuristic reasoning cannot be done. As mentioned before, the fact that one is not able to prove the impossibility of a monomial form does not thereby imply that the monomial exists. The following examples demonstrate this fact.

Example 13-10. Capstan Drive

In a capstan drive, a strong, completely flexible but inextensible, rope is slung over a drum (the capstan), which is slowly rotated by an electric motor, or some other drive. Fig. 13-7 shows an arrangement in which the drum is rotated counterclockwise.

Because of the friction between the rope and the drum, the force (tension) in the two ends of the rope is different. On the left side the force is T, while on the right side it is

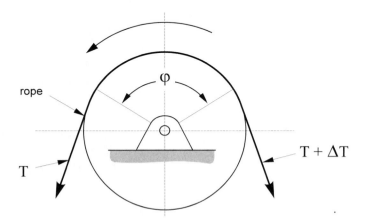

Figure 13-7
A capstan drive
The drum is turned counterclockwise by a motor

$T + \Delta T$. Thus the difference is ΔT. In determining this differential force, we consider the following variables:

Variable	Symbol	Dimension	Remark
contact angle	φ	1	can exceed $2\cdot\pi$
friction coefficient	μ	1	between rope and drum
differential rope force	ΔT	N	direction is opposite to rotation
rope force	T	N	

We have four variables and one dimension, therefore there are $4 - 1 = 3$ dimensionless variables; they are defined by the Dimensional Set as follows:

	φ	μ	ΔT	T
N	0	0	1	1
π_1	1	0	0	0
π_2	0	1	0	0
π_3	0	0	1	-1

from which

$$\pi_1 = \varphi; \qquad \pi_2 = \mu; \qquad \pi_3 = \frac{\Delta T}{T} \qquad (a)$$

Thus we can write

$$\pi_3 = \Psi\{\pi_1, \pi_2\} \qquad (b)$$

where Ψ is some function yet to be defined. First, let us assume the *monomial* form for Ψ, followed by some heuristic reasoning to test the validity of this assumption. Thus we write

$$\pi_3 = c\cdot\pi_1^{n_1}\cdot\pi_2^{n_2} \qquad (c)$$

where c, n_1, and n_2 are constants to be determined. By (c) and (a)

$$\Delta T = c\cdot T\cdot\varphi^{n_1}\cdot\mu^{n_2} \qquad (d)$$

Now if $n_1 > 0$, $n_2 > 0$, then with increasing φ and μ, ΔT increases, which makes sense. Further, if either φ or μ, or both, is zero, then ΔT is also zero, as expected. For obviously, if there is no friction ($\mu = 0$), there can be no difference between the forces acting on each end of the rope. Similarly, if there is no wrap-around ($\varphi = 0$), then the forces on both sides are co-linear and hence they must be numerically equal, so their difference again is zero.

The conclusion therefore might be that the monomial form (d), with some positive constants n_1, n_2, and c, is sound, and hence we only have to do three experiments to determine these three constants, and *voila!*—we have our valid formula. Unfortunately, this is not the case! For the correct formula is (see, for example, Ref. 131, p. 100)

$$\Delta T = T\cdot(e^{\mu\cdot\varphi} - 1) \qquad (e)$$

or, using the dimensionless variables derived in (a)

$$\pi_3 = e^{\pi_1 \pi_2} - 1 \qquad (f)$$

which are quite different from (d) and (c), respectively, although they yield the same results if $\pi_1 = \varphi = 0$ or $\pi_2 = \mu = 0$ are substituted into them. The *moral:* the fact that heuristic reasoning does not disprove a *monomial* relation does *not* prove that the said form exists, although the same reasoning can prove that such form is impossible and hence does not exist. In short, an affirmation offered by heuristic reasoning is a necessary, but not sufficient, condition for the existence of a monomial form among dimensionless variables.

⇑

Example 13-11. Time Available to Clear Out from the Path of a Falling Brick upon Hearing a Warning

Imagine the following scenario: A brick dislodges from the top of a building. A man, John, who is in the vicinity of the place from where the brick is about to fall, sees not only this, but also that another man, Harry, is standing on the ground directly in the path of the falling brick. To warn Harry, John loudly yells a warning at him to clear out. Will Harry have time to jump clear? More precisely: does the time difference Δt between the arrivals of the warning sound and of the brick exceed Harry's reaction time to move? We deal with this problem first by the dimensional method. The following simplifying assumptions are made:

- John is near the place from where the brick is falling off.
- John yells his warning simultaneously with the displacement of the brick.
- Air resistance is negligible.
- Gravitational acceleration g and speed of sound are constants.
- Only negligible movement by Harry is necessary to avoid being hit by the brick.

The relevant variables are as follows:

Variable	Symbol	Dimension	Remark
time difference	Δt	s	see Note 1
height brick falling from	h	m	see Note 2
speed of sound	c	m/s	constant
gravitational acceleration	g	m/s²	constant

Note 1: Between arrivals of warning yell and brick at Harry's head.
Note 2: Vertical distance between Harry's head and the brick's dislodgement point.

We have four variables/constants and two dimensions, hence there are 4 – 2 = 2 dimensionless variables; they are supplied by the following Dimensional Set

	Δt	h	c	g
m	0	1	1	1
s	1	0	–1	–2
π_1	1	0	–1	1
π_2	0	1	–2	1

from which

$$\pi_1 = \frac{\Delta t \cdot g}{c}; \qquad \pi_2 = \frac{h \cdot g}{c^2} \tag{a}$$

Therefore we can write

$$\pi_1 = \Psi\{\pi_2\} \tag{b}$$

where Ψ is an as-yet undefined function. Let us try the *monomial* form for this function and see whether it is tenable. Accordingly, by (a) and (b),

$$\Delta t = k \cdot \frac{c}{g} \cdot \left(\frac{h \cdot g}{c^2}\right)^n \tag{c}$$

where k and n are constants. Now if $h = 0$, then $\Delta t = 0$. Hence n must be positive. Next, we consider that if c increases, then Δt must also increase, since sound arrives faster and hence the time difference between the arrivals of sound and brick grows. Thus, by (c), $(1 - 2 \cdot n) > 0$, from which $n < 0.5$. If now g were to increase, then Δt would surely decrease since the object would fall faster. Therefore, $-1 + n < 0$ in (c), from which $n < 1$. To sum up

$$0 < n < \frac{1}{2} \tag{d}$$

seems to satisfy our requirements. Yet it does not! The time it takes the brick to reach Harry's head is

$$t_g = \sqrt{\frac{2 \cdot h}{g}} \tag{e}$$

and the time for the sound to reach Harry's ear is

$$t_s = \frac{h}{c} \tag{f}$$

and hence

$$\Delta t \equiv t_g - t_s = \sqrt{\frac{2 \cdot h}{g}} - \frac{h}{c} \tag{g}$$

which by (a) can be written

$$\pi_1 = \sqrt{2 \cdot \pi_2} - \pi_2 \tag{h}$$

and as we see, this relation is definitely not a monomial! The graph of (h) is shown in Fig. 13-8.

If the warning sound arrives before the falling brick lands, then $t_g > t_s$, and hence $\Delta t > 0$, thus also $\pi_1 > 0$. Therefore, if $\pi_1 < 0$, i.e., the curve is below the π_2 axis and the brick arrives first. According to the plot this happens if $\pi_2 > 2$.

If $g = 9.81$ m/s^2 and $c = 331.3$ m/s, then by (a), $\pi_2 = (h \cdot g)/c^2 \geq 2$, from which

$$h \geq \frac{2 \cdot c^2}{g} = \frac{2 \cdot (331.3)^2}{9.81} = 22{,}377.103 \text{ m}$$

Thus, above this height the falling brick arrives first, i.e., before the warning yell does. Let us conservatively assume a reaction time of 2.5 s (i.e., this is how long it takes for Harry to look up, realize the danger, and move away). Then $\Delta t \geq 2.5$ s or, by (a), $\pi_1 \geq 0.0740266$. For this, (h), being a quadratic, supplies two π_2 values:

$$\pi_2 = 1 - \pi_1 \pm \sqrt{1 - 2 \cdot \pi_1} \tag{i}$$

352 APPLIED DIMENSIONAL ANALYSIS AND MODELING

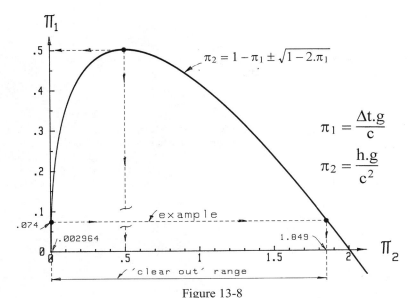

Figure 13-8
**Dimensionless plot to determine the time available to clear out
from under a falling brick upon hearing a warning**
Within the "clear-out" range—shown here for human reaction time 2.5 s—one can avoid being hit by the brick. For definition of symbols, see table in this example.

by which $(\pi_2)_1 = 0.0029635$ and $(\pi_2)_2 = 1.84898$. Therefore, within this escape range of $0.0029635 \leq \pi_2 \leq 1.84898$ it is possible to avoid being hit by the brick This range is shown in the plot by dashed lines. Now, by (a), the limiting values of π_2 can be translated into heights. Thus,

$$h_{min} = \frac{(\pi_2)_1 \cdot c^2}{g} = \frac{(0.0029635)\cdot(331.3)^2}{9.81} = 33.157 \text{ m}$$

$$h_{max} = \frac{(\pi_2)_2 \cdot c^2}{g} = \frac{(1.84898)\cdot(331.3)^2}{9.81} = 20{,}687.4 \text{ m}$$

that is, if the height h is between 33.157 m and 20,687.4 m, then being hit by the brick can be avoided. The plot also provides the *maximum* time difference and corresponding height. They are

$$(\pi_1)_{max} = 0.5; \qquad (\pi_2)_0 = 0.5 \qquad \qquad \text{(j)}$$

which correspond to

$$(\Delta t)_{max} = \frac{(\pi_1)_{max} \cdot c}{g} = \frac{(0.5)\cdot(331.3)}{9.81} = 16.886 \text{ s} \quad \text{and} \quad h_0 = \frac{(\pi_2)_0 \cdot c^2}{g} = 5594.276 \text{ m}$$

So, if $h = h_0 = 5594.276$ m, then the time to move away from under the falling brick is 16.886 s, which is the *maximum* (possible), i.e., the *safest* building to be standing beside is 5.6 km tall. *Moral:* If you *must* stand beside a building, be sure it is about 5.6 km high!

⇑

13.5. RECONSTRUCTIONS

13.5.1. Determination of Exponents of Monomials

As mentioned in Art.13.1, the monomial power form for dimensionless variables is the most efficient in terms of the number of *constants* to be determined. In particular, it was proved in Theorem 13-1, that this number—which is the *minimum* achievable—is equivalent to the number of dimensionless variables appearing in the relation.

The question now naturally arises: how do we determine these (minimum number of) constants? There are three ways to do this:

- by measurement,
- by analysis,
- by heuristic reasoning.

The Measurement Method. In the measurement method we carry out as many independent measurements as there are constants to be determined. The following example illustrates:

Example 13-12

Consider the relation

$$\pi_1 = c \cdot \pi_2^{\epsilon_2} \cdot \pi_3^{\epsilon_3} \tag{a}$$

in which the number of dimensionless variables is $N_P = 3$ and hence, by Theorem 13-1, there are three constants, viz., c, ϵ_2, and ϵ_3. To determine these constants we carry out three measurements with the following input data (independent variables) and results (dependent variable).

Measurement #	Independent variable		Dependent variable
	π_2	π_3	π_1
1	2	1	0.5
2	1	2	0.9
3	1	1	1.6

By (a) now

$$\ln \pi_1 = \ln c + \epsilon_2 \cdot \ln \pi_2 + \epsilon_3 \cdot \ln \pi_3 \tag{b}$$

therefore the data given in the above table can be written as

$$\left. \begin{array}{l} \ln 0.5 = \ln c + \epsilon_2 \cdot \ln 2 + \epsilon_3 \cdot \ln 1 \\ \ln 0.9 = \ln c + \epsilon_2 \cdot \ln 1 + \epsilon_3 \cdot \ln 2 \\ \ln 1.6 = \ln c + \epsilon_2 \cdot \ln 1 + \epsilon_3 \cdot \ln 1 \end{array} \right\} \tag{c}$$

or in matrix form

$$\begin{bmatrix} \ln 0.5 \\ \ln 0.9 \\ \ln 1.6 \end{bmatrix} = \underbrace{\begin{bmatrix} 1 & \ln 2 & \ln 1 \\ 1 & \ln 1 & \ln 2 \\ 1 & \ln 1 & \ln 1 \end{bmatrix}}_{\text{coefficient matrix}} \cdot \begin{bmatrix} \ln c \\ \epsilon_2 \\ \epsilon_3 \end{bmatrix} \tag{d}$$

from which

$$\begin{bmatrix} \ln c \\ \epsilon_2 \\ \epsilon_3 \end{bmatrix} = \begin{bmatrix} 1 & \ln 2 & \ln 1 \\ 1 & \ln 1 & \ln 2 \\ 1 & \ln 1 & \ln 1 \end{bmatrix}^{-1} \cdot \begin{bmatrix} \ln 0.5 \\ \ln 0.9 \\ \ln 1.6 \end{bmatrix} = \begin{bmatrix} 0.47000 \\ -1.67807 \\ -.83007 \end{bmatrix}$$

Therefore

$$c = e^{0.47} = 1.6; \qquad \epsilon_2 = -1.67807; \qquad \epsilon_3 = -0.83007 \tag{e}$$

Thus, by (e), relation (a) becomes

$$\pi_1 = \frac{1.6}{\pi_2^{1.67807} \cdot \pi_3^{0.83007}} \tag{f}$$

which is the result we wanted.

⇑

The measurement method always works, provided N_P different sets of *independent* input dimensionless variables are available for the tests. The adjective "independent" means that the *coefficient matrix* appearing in relation (d) must not be singular.

The Analytic Method. In the analytic method, we *derive* the formula in question—if we can—thereby solving *all* our problems! The key phrase here of course is "*if we can*." But what if we cannot? Then we fall back on either the *measurement method* discussed above, or the *heuristic reasoning method* discussed below.

The Heuristic Reasoning Method. Very often the constant *exponents*—and rarely even the constant *coefficient* (Example 13-16)—of a monomial can be found merely by heuristic reasoning. This way at the most we only have to determine the *single* constant coefficient, an activity which requires only *one* measurement, regardless of the number of dimensionless variables in the relation. It must be mentioned, though, that this reasoning does sometimes call for rather high levels of sophistication, imagination, an inquisitive if not iconoclastic mind, and perhaps some knowledge of the relevant physical laws.

The following examples demonstrate this process and its beneficial and powerful results.

Example 13-13. Deflection of a Simply Supported Uniform Beam Loaded by its Own Weight

We wish to determine the maximum deflection of a simply supported beam of uniform cross-section, loaded only by its own weight. Fig. 13-9 shows the arrangement.

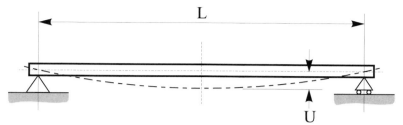

Figure 13-9
Simply supported uniform beam is loaded by its own weight
Centerline of the deformed shape is indicated by dashed line

The relevant variables are as follows:

Variable	Symbol	Dimension	Remark
deflection	U	m	maximum
length	L	m	
Young's modulus	E	N/m²	
cross-section's second moment of area	I	m⁴	constant
load per unit length	q	N/m	constant

The number of variables is five, the number of dimensions is two, therefore we have $5 - 2 = 3$ dimensionless variables supplied by the Dimensional Set

	U	L	E	I	q
m	1	1	−2	4	−1
N	0	0	1	0	1
π_1	1	0	0	$-\frac{1}{4}$	0
π_2	0	1	0	$-\frac{1}{4}$	0
π_3	0	0	1	$\frac{1}{4}$	−1

from which

$$\pi_1 = \frac{U}{\sqrt[4]{I}}; \quad \pi_2 = \frac{L}{\sqrt[4]{I}}; \quad \pi_3 = \frac{E}{q} \cdot \sqrt[4]{I} \tag{a}$$

Thus, we can write

$$\pi_1 = \Psi\{\pi_2, \pi_3\} \tag{b}$$

where Ψ is an as-yet undefined function. If a monomial is now assumed—which is plausible because of the nature of the problem—then, by (b),

$$\pi_1 = c \cdot \pi_2^{\epsilon_2} \cdot \pi_3^{\epsilon_3} \tag{c}$$

356 APPLIED DIMENSIONAL ANALYSIS AND MODELING

or, by (a) and (c),

$$U = c \cdot \sqrt[4]{I} \cdot \left(\frac{L}{\sqrt[4]{I}}\right)^{\epsilon_2} \cdot \left(\frac{E}{q} \cdot \sqrt[4]{I}\right)^{\epsilon_3} \tag{d}$$

where c, ϵ_2, and ϵ_3 are constants.

It stands to reason that if everything else is kept constant, a beam which is twice as heavy deflects twice as much; i.e., U is proportional to q. From this condition $\epsilon_3 = -1$. Further, it is also anticipated that U is inversely proportional to I. By applying this condition in (d)

$$\frac{1}{4} - \frac{\epsilon_2}{4} + \frac{\epsilon_3}{4} = -1 \tag{e}$$

from which, since $\epsilon_3 = -1$,

$$\epsilon_2 = 5 + \epsilon_3 = 5 - 1 = 4 \tag{f}$$

Hence (d) becomes

$$U = c \cdot \sqrt[4]{I} \left(\frac{L}{\sqrt[4]{I}}\right)^4 \cdot \left(\frac{E}{q} \cdot \sqrt[4]{I}\right)^{-1} = c \cdot \frac{L^4 \cdot q}{I \cdot E} \tag{g}$$

where the constant c, if determined by a *single* measurement, will be found to be $c = 5/384$.

In this illustration, through a *single* experiment, and by employing some elementary heuristic reasoning, we managed to derive a numeric relation among five (!) physical variables; a remarkable achievement demonstrating the prowess of the method.

⇑

Example 13-14. **The Coffee Warmer (I)**

A coffee mug with a lid is put on an electric hot plate transmitting Q_1 heat per unit time to the coffee through the bottom of the mug (Fig. 13-10). The heat transmitted away from the coffee through the *side* of the mug per unit time is Q_2. The evaporative heat loss Q_3 through the upper surface of the coffee is prevented by the lid, considered a perfect insulator, i.e., $Q_3 = 0$. Thus, by the equivalence of heat input and output: $Q_1 = Q_2 = Q$.

The heat loss per unit time across the side of the mug is influenced by heat-transfer coefficient h, surface area A, and temperature difference Δt between the coffee's t and that of the ambient t_1. Since it is assumed that the coffee is warmer than the ambient air, therefore $t > t_1$, and hence $\Delta t = t - t_1 > 0$. Also, the surface area through which the heat is transmitted from the mug is obviously a function of the mug's diameter D and height b of the coffee in the mug.

Altogether, therefore, we have the following relevant variables and their dimensions:

Variable	Symbol	Dimension	Remark
temperature difference	Δt	°C	between coffee and surroundings
height of coffee in the mug	b	m	
heat transmitted	Q	m²·kg/s³	by heater per unit time
heat transfer coefficient	h	kg/(s³·°C)	across side of mug
diameter of the mug	D	m	

FORMS OF DIMENSIONLESS RELATIONS

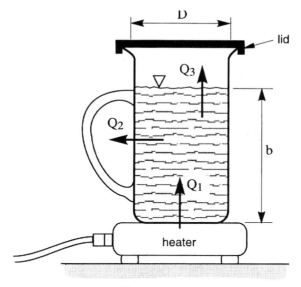

Figure 13-10
The coffee warmer

The dimensional matrix is

	Δt	b	Q	h	D
m	0	1	2	0	1
kg	0	0	1	1	0
s	0	0	-3	-3	0
°C	1	0	0	-1	0

Here we see that the rank of this matrix is three, although the number of dimensions is four. Therefore we must delete one dimension, say "s." This will result in five variables and three dimensions, and hence we will have $5 - 3 = 2$ dimensionless variables which are supplied by the Dimensional Set

	Δt	b	Q	h	D
m	0	1	2	0	1
kg	0	0	1	1	0
°C	1	0	0	-1	0
π_1	1	0	-1	1	2
π_2	0	1	0	0	-1

yielding

$$\pi_1 = \frac{\Delta t \cdot h \cdot D^2}{Q}; \qquad \pi_2 = \frac{b}{D} \qquad (a)$$

Thus in general we can write,

$$\pi_1 = \Psi\{\pi_2\} \tag{b}$$

where Ψ is a function to be defined. Let us assume a *monomial* form for this function. Hence, by (a) and (b)

$$Q = c \cdot \Delta t \cdot h \cdot D^2 \cdot \left(\frac{b}{D}\right)^\epsilon \tag{c}$$

where c and ϵ are constants.

We now embark upon some heuristic reasoning to determine the exponent ϵ. Assuming the coffee temperature is uniform, then it is evident that the transmitted heat Q will be proportional to surface area A through which the heat is removed from the coffee. This area is

$$A = D \cdot b \cdot \pi \tag{d}$$

Hence Q must be proportional to both D and b. Consequently, in (c), the exponent of both D and b must be 1. Accordingly,

$$2 - \epsilon = 1; \quad \epsilon = 1 \tag{e}$$

Both of these equations yield $\epsilon = 1$. Thus, by (c) and (e),

$$Q = c \cdot \Delta t \cdot h \cdot D^2 \cdot \left(\frac{b}{D}\right) = c \cdot \Delta t \cdot h \cdot D \cdot b \tag{f}$$

from which

$$\Delta t = \frac{Q}{c \cdot h \cdot D \cdot b} \tag{g}$$

Now for a given mug and hot plate, h and D are constants, and so of course is c. Therefore we can say that the temperature difference between the coffee and the ambient air is inversely proportional to the height of the coffee in the mug b. The lower the level of coffee in the mug, the hotter the coffee is. Further, since b is proportional to the *quantity* of coffee, the temperature difference is inversely proportional to the quantity of coffee in the mug. This finding confirms the author's *practical* experience: the more coffee he drank from his mug, the hotter the coffee became, i.e., the less the coffee, the hotter it is. The product of constants c and h in (g) can be determined by a single experiment.

⇑

Example 13-15. Relativistic Red Shift

According to Einstein's relativity theory, light rays emitted by a star are retarded by gravitation and hence lose energy. This phenomenon manifests itself in the rays' reduced frequency and—since the speed of light remains constant—increased wavelength. The latter causes the *color* of the light to shift toward red; hence the name of this phenomenon—"*red shift*." Assume that the following variables and constants are involved in a relativistic red shift.

Variable	Symbol	Dimension	Remark
change of wavelength	$\Delta\lambda$	m	always positive
wavelength	λ	m	reference value
universal gravitational constant	k	m³/(kg·s²)	
radius of star emitting radiation	R	m	
mass of star emitting radiation	M	kg	
speed of light	c	m/s	in vacuum

We have six variables and three dimensions, therefore there are 6 – 3 = 3 dimensionless variables supplied by the Dimensional Set

	$\Delta\lambda$	$\Delta\lambda$	k	R	M	c
m	1	1	3	1	0	1
kg	0	0	−1	0	1	0
s	0	0	−2	0	0	−1
π_1	1	0	0	−1	0	0
π_2	0	1	0	−1	0	0
π_3	0	0	1	−1	1	−2

from which the three dimensionless variables are

$$\pi_1 = \frac{\Delta\lambda}{R}; \quad \pi_2 = \frac{\lambda}{R}; \quad \pi_3 = \frac{k \cdot M}{R \cdot c^2} \tag{a}$$

Thus, we can write

$$\pi_1 = \Psi\{\pi_2, \pi_3\} \tag{b}$$

where Ψ defines an as-yet undefined function. Assuming *monomial* form for Ψ, which is plausible, (b) can be written

$$\pi_1 = b \cdot \pi_2^{\epsilon_2} \cdot \pi_3^{\epsilon_3} \tag{c}$$

where b, ϵ_2, and ϵ_3 are constants to be determined. By (a) this relation becomes

$$\Delta\lambda = b \cdot R \cdot \left(\frac{\lambda}{R}\right)^{\epsilon_2} \cdot \left(\frac{k \cdot M}{R \cdot c^2}\right)^{\epsilon_3} \tag{d}$$

Now we undertake some heuristic activity. First, it is reasonable to assume that the change of wavelength, $\Delta\lambda$, is proportional to the reference value λ. Hence the exponent of λ in (d) must be $\epsilon_2 = 1$. Next, we also consider that $\Delta\lambda$ should be proportional to the gravitational constant k, since it is, after all, gravitation which *retards* the emitted light. This consideration mandates that $\epsilon_3 = 1$. Thus (d) becomes

$$\Delta\lambda = b \cdot R \cdot \left(\frac{\lambda}{R}\right) \cdot \left(\frac{k \cdot M}{R \cdot c^2}\right) = b \cdot \frac{\lambda \cdot k \cdot M}{R \cdot c^2}$$

where the constant b happens to be 1 (Ref. 16, p. 562). Hence the *relative* relativistic red shift is

$$\frac{\Delta\lambda}{\lambda} = \frac{k \cdot M}{R \cdot c^2} \tag{e}$$

This relative shift is extremely small. We have $k = 6.67 \times 10^{-11}$ m^3/(s^2·kg) and $c = 3 \times 10^8$ m/s, resulting in the following typical values:

	Earth	Sun	Sirius B
relative red shift	6.96E-10	2.11E-6	5.9E-5

The physical measurement (observation) of this value for Earth is out of the question. For the Sun it is marginal. However, the American astronomer Walter Adams (1876–1956) measured this shift in 1925 for the "companion" of the star Sirius—named *Sirius B*—which is a *white dwarf*. This star is as small as the Earth (small R), but 40,000 (!) times as massive (large M). Therefore, the relative red shift is about 28 times as large for Sirius B as it is for the Sun. Incidentally, local gravitational acceleration g on Sirius B is 236,000 m/s^2, which is 24,000 times as much as on Earth. Adams' measured values for Sirius B were in close agreement with that predicted by relation (e).

⇑

Example 13-16. Kepler's Second Law

Kepler's second law states that for any planet the radius-vector pointing to the planet (from the Sun) *sweeps* equal areas in unit time. We shall now derive this law by dimensional considerations.

By Kepler's first law, the orbit of any planet is an ellipse, and the Sun is located at one of the two foci (Fig. 13-11). In general, the tangential (i.e., orbital) speed of the planet is v and, in particular, at *perihelion* (point of orbit nearest to the Sun) it is v_p. At perihelion the orbital speed is perpendicular to the radius-vector ρ_p. At this point let the time be $t = 0$, so that at a general point, characterized by the radius-vector ρ, the time will be $t > 0$. Quantities v_p and ρ_p—together with the solar mass M_S and gravitational constant k—define not only the ellipse (i.e., orbit), but the (tangential) speed of the planet at any time t, as well.

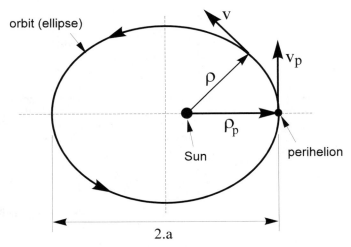

Figure 13-11
The orbit of a planet around the Sun

FORMS OF DIMENSIONLESS RELATIONS

We now introduce variable q, which is the area *swept* by radius-vector ρ in unit time. Our aim is to find the relation between q and t. The following quantities are considered:

Variable or constant	Symbol	Dimension	Remark
time	t	s	reckoned from perihelion
area swept by radius-vector	q	m²/s	per unit time
speed of planet at perihelion	v_p	m/s	tangential
length of radius-vector	ρ_p	m	at perihelion
mass of Sun	M_S	kg	
universal gravitational constant	k	m³/(kg·s²)	

We have six variables and constants and three dimensions, therefore there are $6 - 3 = 3$ dimensionless variables characterizing the system; they are supplied by the Dimensional Set

	t	q	v_p	ρ_p	M_S	k
m	0	2	1	1	0	3
kg	0	0	0	0	1	-1
s	1	-1	-1	0	0	-2
π_1	1	0	0	$-\dfrac{3}{2}$	$\dfrac{1}{2}$	$\dfrac{1}{2}$
π_2	0	1	0	$-\dfrac{1}{2}$	$-\dfrac{1}{2}$	$-\dfrac{1}{2}$
π_3	0	0	1	$\dfrac{1}{2}$	$-\dfrac{1}{2}$	$-\dfrac{1}{2}$

from which

$$\pi_1 = t \cdot \sqrt{\frac{M_S \cdot k}{\rho_p^3}}; \qquad \pi_2 = \frac{q}{\sqrt{\rho_p \cdot M_S \cdot k}}; \qquad \pi_3 = v_p \cdot \sqrt{\frac{\rho_p}{M_S \cdot k}} \qquad \text{(a)}$$

Thus, we have

$$\pi_2 = \Psi\{\pi_1, \pi_3\}$$

where Ψ is some function yet to be determined. Let us assume the monomial form for Ψ. Hence

$$\pi_2 = \gamma \cdot \pi_1^{\epsilon_1} \cdot \pi_3^{\epsilon_3} \qquad \text{(b)}$$

where γ, ϵ_1, and ϵ_3 are constants, as yet unknown. By (a) and (b),

$$q = \gamma \cdot \sqrt{\rho_p \cdot M_S \cdot k} \cdot \left(t \cdot \sqrt{\frac{M_S \cdot k}{\rho_p^3}} \right)^{\epsilon_1} \cdot \left(v_p \cdot \sqrt{\frac{\rho_p}{M_S \cdot k}} \right)^{\epsilon_3} \qquad \text{(c)}$$

This frightening expression can be mellowed down by some heuristic activity. It is obvious that ϵ_1 cannot be positive since in this case at perihelion ($t = 0$), q would be zero, i.e., the planet would be stationary, which is impossible. If ϵ_1 is negative, then at perihelion the speed would be infinite—which is equally absurd. Therefore, since ϵ_1 can be neither positive, nor negative, it must be zero. Thus, relation (c) becomes

$$q = \gamma \cdot \sqrt{\rho_p \cdot M_S \cdot k} \cdot \left(v_p \cdot \sqrt{\frac{\rho_p}{M_S \cdot k}} \right)^{\epsilon_3} \qquad \text{(d)}$$

in which t does *not* appear; i.e., q is *independent* of t. In other words, for any *particular planet*, q = constant—since, for any selected planet, all quantities on the right-side of (d) are constants. But this is clearly Kepler's second law, which posits the constancy of q for any particular planet. Thus we managed to derive this important law using only the dimensional method.

Although we could deduce Kepler's second law by purely dimensional considerations—which was our principal objective—we are still ignorant about exponent ϵ_3 and constant γ in (d). Can we also squeeze these values out of our still forbidding looking relation (d)?

It is evident that ϵ_3 cannot be zero, because if it were, then q would be independent of v_p, which would be impossible. However, we can reason as follows: If q is constant, then it must have the *same* value at $t = 0$ (perihelion) as anywhere else. But at perihelion, v_p is *perpendicular* to the radius-vector ρ_p, and hence the area of the elementary triangle swept by radius-vector ρ_p is

$$dA = \frac{1}{2} \cdot \rho_p \cdot dU \tag{e}$$

where dU is the infinitesimal motion (travel) of the planet in infinitesimal dt time (Fig. 13-12). Thus, we can write

$$dU = v_p \cdot dt \tag{f}$$

since dt is small and therefore dU can be considered a straight line. By (e) and (f)

$$dA = \frac{1}{2} \cdot \rho_p \cdot v_p \cdot dt \tag{g}$$

But obviously $q = dA/dt$, therefore, by (g),

$$q = \frac{1}{2} \cdot \rho_p \cdot v_p \tag{h}$$

Thus, in (d) we have $\gamma = \frac{1}{2}$ and $\epsilon_3 = 1$. For example, for Earth

semi-major axis of orbit (ellipse): $a = 1.49674 \times 10^{11}$ m
eccentricity of orbit: $e = 0.016728$

therefore at perihelion the radius-vector's length is (Ref. 20, p. 99)

$$\rho_p = a \cdot (1 - e) = 1.4717 \times 10^{11} \text{ m} \tag{i}$$

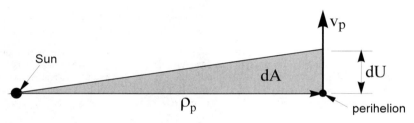

Figure 13-12
Geometry at a planet's perihelion position

FORMS OF DIMENSIONLESS RELATIONS 363

which is then the Sun–Earth shortest distance. The Earth's speed at perihelion is (Ref 20, p. 109)

$$v_p = \sqrt{\frac{k \cdot M_S}{a} \cdot \frac{1+e}{1-e}} = 30261 \text{ m/s} \quad \text{(j)}$$

where

$k = 6.67259 \times 10^{-11}$ m³/(s²·kg);   the universal gravitational constant, and
$M_S = 1.9866 \times 10^{30}$ kg;   the mass of Sun.

By (i), (j), and (h) now

$$q = \frac{1}{2} \cdot \rho_p \cdot v_p = \frac{1}{2} \cdot (1.4717 \times 10^{11}) \cdot (30261) = 2.22681 \times 10^{15} \text{ m}^2/\text{s}$$

which is then the swept area of the radius-vector in 1 second, regardless of the position of Earth in its orbit.

⇑

From physics and engineering we now venture into the alluring field of *biomechanics*. There we will find the dimensional method no less rewarding and equally useful.

Example 13-17. Relative Load-Carrying Capabilities of Animals

The relative load-carrying capability of an animal—including man—is characterized by the *number* of *identical* animals it can carry on its back. We shall determine this number by purely dimensional considerations. To this end we first list the relevant variables:

Variable	Symbol	Dimension	Remark
number of animals carried	n	1	identical to host animal
density of animal	ρ	kg/m³	
size of animal	L	m	characteristic value
max stress in bones and tendons	σ	kg/(m·s²)	characteristic value
gravitational acceleration	g	m/s²	

We have five variables and three dimensions, therefore there are $5 - 3 = 2$ dimensionless variables describing the system. Note that one of the physical variables n is already dimensionless. Nevertheless, for the sake of consistency it is included in the list. The two dimensionless variables are provided by the Dimensional Set as follows:

	n	ρ	L	σ	g
m	0	−3	1	−1	1
kg	0	1	0	1	0
s	0	0	0	−2	−2
π_1	1	0	0	0	0
π_2	0	1	1	−1	1

which yields

$$\pi_1 = n; \qquad \pi_2 = \frac{\rho \cdot L \cdot g}{\sigma} \quad \text{(a)}$$

and therefore in general we can write

$$\pi_1 = \Psi\{\pi_2\} \tag{b}$$

where Ψ is an as-yet undefined function. Let us now assume a *monomial* form for this function—which is plausible. Accordingly,

$$n = c \cdot \left(\frac{\rho \cdot L \cdot g}{\sigma}\right)^\epsilon \tag{c}$$

where c and ϵ are constants to be determined. Obviously, we are mainly interested in the exponent ϵ. To find ϵ we now undertake some heuristic speculation. If everything else remains unchanged, doubling g will halve n. For it is plain that if the strength of the carrying animal remains the same, then it can carry only half as many double-heavy similar animals. Therefore n must be inversely proportional to g, thus $\epsilon = -1$ in (c). It follows that

$$n = c \cdot \frac{\sigma}{\rho \cdot g \cdot L} \tag{d}$$

Further, since for all animals c, σ, ρ, and g can be considered constants, relation (d) can be written

$$n = \frac{\text{const}}{L} \tag{e}$$

This interesting relation tells us that the number of identical (to itself) animals an animal can carry is inversely proportional to its linear size. Thus, *relative* strength diminishes with increasing size—a fact which is perhaps contrary to intuition, for strength is commonly associated with size.* A 1.8 m man can carry another similar man on his back ($n = 1$), and a 3.5 m horse can carry only about half a horse ($n = 0.5$), but a 0.8 m dog can carry at least two other dogs ($n \geq 2$)—not to mention an ant which is famous for its ability to carry maybe 50 times its own weight ($n = 50$). If a man were as strong as an ant—relatively speaking—he could carry six grand pianos on his back!

⇑

Example 13-18. Injuries to Animals Falling to the Ground

The extent of injuries to an animal—including man—from falling to the ground is influenced by many factors, viz., the resiliency of the ground (is it soft or hard?), the height from which the body falls, air resistance, etc. But the most important characteristic is the energy-absorbing ability of the muscles, tendons, and cartilage of the body, for these anatomical parts act as shock absorbers or buffers. Therefore the critical factor for injuries—or more appropriately, for the avoidance of injuries—is whether the impact energy absorbed per unit mass of the body does or does not exceed a *critical value,* beyond which injuries result. We shall investigate this problem by using the dimensional method. It is assumed that the impact energy absorbed by the ground is negligible. As usual, we list the relevant variables.

---

*It may be remarked, however, that *absolute* strength varies as the *square* of linear size, since the absolute load-carrying ability is determined by the cross-sectional *areas* of bones and muscles.

FORMS OF DIMENSIONLESS RELATIONS

Variable	Symbol	Dimension	Remark
absorbed impact energy	q	m²/s²	per unit body mass
density of animal	ρ_b	kg/m³	
height body falls from	h	m	
animal's characteristic size	L	m	animals are geometrically similar
gravitational acceleration	g	m/s²	constant
density of air	ρ_a	kg/m³	constant during the fall

We have six variables and three dimensions. Hence there are $6 - 3 = 3$ dimensionless variables, supplied by the Dimensional Set

	q	ρ_b	h	L	g	ρ_a
m	2	-3	1	1	1	-3
kg	0	1	0	0	0	1
s	-2	0	0	0	-2	0
π_1	1	0	0	-1	-1	0
π_2	0	1	0	0	0	-1
π_3	0	0	1	-1	0	0

from which

$$\pi_1 = \frac{q}{L \cdot g}; \quad \pi_2 = \frac{\rho_b}{\rho_a}; \quad \pi_3 = \frac{h}{L} \qquad (a)$$

If monomial form is assumed, then we can write

$$\pi_1 = c, \ \pi_2^{\epsilon_2} \cdot \pi_3^{\epsilon_3} \qquad (b)$$

where c, ϵ_2, and ϵ_3 are numeric constants yet to be determined. Now two cases are distinguished:

- air resistance is not significant,
- air resistance is significant.

Air resistance is not significant. In the case of a relatively large animal, e.g., horse, man, dog, etc., air resistance is insignificant if the fall is from low or even medium heights. Hence the density of air ρ_a is a physically *irrelevant* variable. We now observe that π_2 is the *only* dimensionless variable in which ρ_a appears, hence π_2 can be ignored (see Art.11.2 and in particular Example 11-13). Thus (b) is written

$$\pi_1 = c \cdot \pi_3^{\epsilon_3} \qquad (c)$$

where c and ϵ_3 are numeric constants. By (a) and (c)

$$q = c \cdot L \cdot g \cdot \left(\frac{h}{L}\right)^{\epsilon_3} \qquad (d)$$

Now it is obvious that if there is no air resistance, then the absorbed energy on impact must be equal to the *potential* energy at height h (energy converted to heat is negligible). Therefore this energy must be *proportional* to h. Hence $\epsilon_3 = 1$ in (d). From this it follows that

$$q = c \cdot L \cdot g \cdot \left(\frac{h}{L}\right) = c \cdot g \cdot h \qquad (e)$$

The remarkable thing in this relation is the absence of L (since it cancels out), which therefore becomes a *physically irrelevant* variable. Thus the absorbed energy per unit mass of an animal is independent of its size. A mouse falling *in vacuum* onto hard ground would injure itself to the same extent as a horse would.

Air resistance is significant. In the case of a relatively small animal (e.g., a mouse), or if the fall is from a great height, then air resistance cannot be ignored, for the falling animal will attain a terminal speed at which the drag pointing upward will be numerically equal to the weight pointing downward. Once this terminal speed is reached, height is no longer relevant, and thus it becomes *physically irrelevant*. Consequently, it is now π_3 which can be ignored in (a). By (b), then

$$\pi_1 = c \cdot \pi_2^{\epsilon_2} \tag{f}$$

where c and ϵ_2 are numeric constants. By (a) and (f)

$$q = c \cdot L \cdot g \cdot \left(\frac{\rho_b}{\rho_a}\right)^{\epsilon_2} \tag{g}$$

It is evident that the denser the air, the smaller the absorbed energy at impact, since denser air means reduced terminal speed. Hence $\epsilon_2 > 0$; in fact, a quite simple analytical derivation (not presented here) can show that $\epsilon_2 = 1$. The important thing in (g) is that q is *proportional* to size L. Thus, the required amount of absorbed energy *per unit body mass* grows *linearly* with size. It follows that where a small animal might survive a fall from a given height, a large animal will not. As J. B. S. Haldane (English geneticist, 1892–1964) remarked (Ref. 140, vol. 2, p. 952):

> You can drop a mouse down a thousand-yard shaft; and on arriving at the bottom, it gets a slight shock and walks away. A rat would probably be killed, though it can fall safely from the eleventh story of a building; a man is killed, a horse splashes.

This example demonstrated a typical case in which, although the *monomial* form was valid, the respective exponents underwent drastic change according to the *ranges* of physical variables. This phenomenon, called "scale effect," will be discussed in more detail in Art. 17.5. For our present example, the table below summarizes the obtained results:

Exponents in relation (b)	Air resistance	
	no	yes
ϵ_2	0	1
ϵ_3	1	0

13.5.2. Determination of Some Nonmonomials

As explained in Art. 13.5.1, if the form of the dimensionless relation is a *monomial*, then we can determine the exponents and the constant coefficient by N_P measurements, where N_P is the number of dimensionless variables in the relation. However, if the form is not a *nonmonomial*, then generally we have a much more difficult task to determine (reconstruct) the relationship. In these cases the stratagem to follow is to *intelligently* guess an appropriate *polynomial* of a sufficiently large degree, and then determine the coefficients of this polynomial by measurements. Generally, and unfortunately, this method does not work if the relation sought involves fractional or negative exponents, or transcendental functions. Nevertheless, in some notable instances, the sought-after relation is a polynomial and hence, based on some sensible assumptions as to its degree, it can be reconstructed quite easily by taking measurements.

The number of measurements one must take to define a polynomial of degree n is $n + 1$, since an nth degree polynomial has $n + 1$ coefficients. It is a good policy to be conservative and assume (if in doubt) a larger, rather than a smaller n. For although by using this approach the number of measurements may increase slightly, it is a small price to pay to avoid getting an erroneous relation. If the assumed degree of the polynomial is larger than the *de facto* requirement, then the determined coefficient(s) of the term, which should be absent, will be several *orders of magnitude smaller* in absolute value than the largest absolute value of the rest. For example if in

$$\pi_1 = c_0 + c_1 \cdot \pi_2 + c_2 \cdot \pi_2^2 + c_3 \cdot \pi_2^3$$

$|c_3|$ is *much* smaller than the largest absolute value of the other coefficients, then the above relation can be written

$$\pi_1 = c_0 + c_1 \cdot \pi_2 + c_2 \cdot \pi_2^2$$

The following example demonstrates this technique.

Example 13-19. The Power Required to Lift a Given Mass to a Given Height within a Given Time

A mass resting on the ground is lifted by a *constant* force which is *larger* than the weight of the mass (the force must exceed the weight, else the mass would remain on ground). The force F is just large enough to lift mass M to a given height h *within* given time t (Fig. 13-13). What is the power required for such an undertaking?

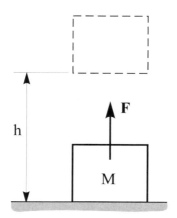

Figure 13-13
A given mass is lifted to a given height within a given time

368   APPLIED DIMENSIONAL ANALYSIS AND MODELING

The relevant variables and their dimensions are listed in the table below:

Variable	Symbol	Dimension	Remark
power applied	P	$m^2 \cdot kg/s^3$	constant
gravitational acceleration	g	m/s^2	constant
mass	M	kg	
given height	h	m	
given time	t	s	

Note the conspicuous absence of the uplifting force F in the above table. This is because F is a dependent variable upon the independent variables M, h, t, and g, and hence its presence would just complicate matters by needlessly increasing the number of dimensionless variables. If the question had been to *find F*, on the condition that the mass is lifted to a prescribed height h within the prescribed time t, then F would be a relevant (the dependent) variable, but P would not be. As it stands, we have five variables and three dimensions, therefore there are $5 - 3 = 2$ dimensionless variables determined by the Dimensional Set

	P	g	M	h	t
m	2	1	0	1	0
kg	1	0	1	0	0
s	-3	-2	0	0	1
π_1	1	0	-1	-2	3
π_2	0	1	0	-1	2

from which

$$\pi_1 = \frac{P \cdot t^3}{M \cdot h^2}; \quad \pi_2 = \frac{g \cdot t^2}{h} \tag{a}$$

Hence we can write

$$\pi_1 = \Psi\{\pi_2\} \tag{b}$$

where Ψ is an as-yet undetermined function. Let us try the *monomial* form for Ψ. Thus

$$\pi_1 = c \cdot \pi_2^\epsilon \tag{c}$$

where c and ϵ are constants to be determined. By (a), this last relation can be written

$$P = c \cdot \frac{M \cdot h^2}{t^3} \cdot \left(\frac{g \cdot t^2}{h}\right)^\epsilon \tag{d}$$

Now if t decreases, then P must increase, since the shorter the time, the larger the power must be to lift the mass to the same height. From this condition, for the exponents of t in (d), we have $-3 + 2 \cdot \epsilon < 0$, from which

$$\epsilon < \frac{3}{2} \tag{e}$$

Similarly, the greater the height, the larger the P. Hence, by (d), again

$$\epsilon < 2 \tag{f}$$

Finally, and equally obviously, the greater the g, the larger the P, hence

$$\epsilon > 1 \tag{g}$$

Therefore, by (e), (f) and (g), it seems that the monomial form of Ψ as it appears in (c) is tenable, as long as the exponent ϵ is between 1 and 1.5. To verify this assertion, we conduct three experiments. Here of course the *minimum* number of experiments must be three, since if we perform only two, then the magnitudes of c and ϵ could be determined with certainty *regardless* of the validity of relation (c). However, if c and ϵ are found to *not* satisfy the results of three (or more) experiments, then this fact would *prove* the falsehood of the *monomial* form (c). The results of the three measurements conducted are as follows:

Measurement #	π_1	π_2
1	19.658	17.658
2	30.304	28.304
3	46.211	44.211

Considering the results of #1 and #2, we write

$$\left. \begin{array}{l} 19.658 = c \cdot (17.658)^\epsilon \\ 30.304 = c \cdot (28.304)^\epsilon \end{array} \right\} \tag{h}$$

from which, by simple arithmetic

$$\left. \begin{array}{l} c = 1.41163 \\ \epsilon = 0.91730 \end{array} \right\} \tag{i}$$

On the other hand, if we consider measurements #2 and #3, then by similar process

$$\left. \begin{array}{l} 30.304 = c \cdot (28.304)^\epsilon \\ 46.211 = c \cdot (44.211)^\epsilon \end{array} \right\} \tag{j}$$

from which we get

$$\left. \begin{array}{l} c = 1.28200 \\ \epsilon = 0.94611 \end{array} \right\} \tag{k}$$

It is seen that the corresponding data in (i) and (k) are significantly different, hence the monomial form of Ψ defined in (b)—and as it appears in (c)—is untenable.

This example illustrates the fact that the lack of contradiction offered by even the most elaborate and ingenious heuristic reasoning, is not a proof of the existence of a monomial. In other words, the lack of contradiction—brought about by however inventive deliberation—is a necessary, but not sufficient condition for the existence of a monomial form.

Now let us try to fit the experimentally obtained data into some analytical form. A glance at the numbers in the preceding table reveals that within a three-decimal accuracy

$$\pi_1 = 2 + \pi_2 \tag{l}$$

Therefore, by definitions of π_1 and π_2 in (a),

$$\frac{P \cdot t^3}{M \cdot h^2} = 2 + \frac{g \cdot t^2}{h}$$

from which, the power is

$$P = \frac{M \cdot h}{t} \cdot \left(\frac{2 \cdot h}{t^2} + g \right) \qquad (m)$$

The reader should note that from the results of only three experiments, by dimensional techniques the derivation of a rather complex formula (*m*), involving five physical variables and one constant, could be accomplished.

⇑

Example 13-20. Deflection of a Simply Supported Beam Loaded by Two Symmetrically Placed Identical Forces

Given a slender, simply supported beam loaded by two forces, which are concentrated, identical, symmetrically placed, and lateral—as illustrated in Fig. 13-14.

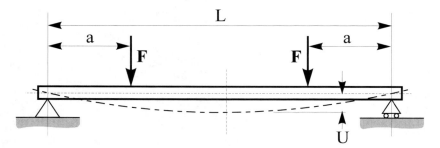

Figure 13-14
A simply supported beam is loaded laterally by two symmetrically placed identical forces

We wish to determine the maximum lateral deflection of this beam. The relevant variables are:

Variable	Symbol	Dimension	Remark
deflection	U	m	at middle of the beam
lateral forces	F	N	symmetrically placed
position of forces	a	m	from respective ends of beam
length of beam	L	m	
rigidity of beam	Z	m²·N	uniform, see definition below

Here "*rigidity*" (sometimes called "*flexural stiffness*") is defined as $Z = I \cdot E$, where E is the Young's modulus of material, and I is the cross-section's second moment of area. Accordingly, the dimension of Z is

$$[Z] = [I \cdot E] = \mathrm{m}^4 \cdot \frac{\mathrm{N}}{\mathrm{m}^2} = \mathrm{m}^2 \cdot \mathrm{N}$$

The determination of the dependence of U upon four independent variables would be a manifestly Herculean task. However, if we use dimensionless variables, this task can be carried out with relative ease, as will be now demonstrated:

Since we have five variables and two dimensions, therefore there are $5 - 2 = 3$ dimensionless variables supplied by the Dimensional Set

	U	F	a	L	Z
m	1	0	1	1	2
N	0	1	0	0	1
π_1	1	0	0	-1	0
π_2	0	1	0	2	-1
π_3	0	0	1	-1	0

from which the three dimensionless variables are

$$\pi_1 = \frac{U}{L}; \quad \pi_2 = \frac{F \cdot L^2}{Z}; \quad \pi_3 = \frac{a}{L} \quad \text{(a)}$$

Now we carry out some tests on a $L = 3.4$ m long steel beam whose cross-section is a rectangle of 0.1 m width and 0.06 m thickness. Hence $I = 1.8 \times 10^{-6}$ m^4. The Young's modulus for steel is $E = 2.07 \times 10^{11}$ N/m^2. Thus, $Z = I \cdot E = (1.8 \times 10^{-6}) \cdot (2.07 \times 10^{11}) = 372{,}600$ m$^2 \cdot$N.

It is now noted that if $\pi_3 = $ const, then, for *small* deformations, π_1 must be proportional to π_2. Therefore

$$\pi_1 = k \cdot \pi_2 \quad \text{(b)}$$

where k is constant and whose value depends only on π_3. Let us select four values for π_3, and a constant force $F = 27100$ N. Then the "a" positions of the two symmetrical forces on the beam are also defined, since $\pi_3 = a/L$ and L is given. In the actual tests, we measure the deflections at exactly mid-length under the beam. These measurements can be executed with great precision and assumed uncertainty of less than about 0.01 mm ($\approx$ 0.0004 in). From the measurements, the following table can be prepared:

Quantity	Symbol	Dimension	Source	Measurement # 1	2	3	4
beam length	L	m	chosen	3.4			
rigidity	Z	m$^2 \cdot$N	chosen	372,600			
force	F	N	chosen	27,100			
dimensionless force position	π_3	1	chosen	0.1	0.2	0.3	0.4
	a	m	calculated by (a)	0.34	0.68	1.02	1.36
deflection	U	m	measured	0.03526	0.06766	0.09434	0.11244
dimensionless	π_1	1	calculated by (a)	0.0103706	0.0199	0.0277471	0.0330706
dimensionless	π_2	1	calculated by (a)	0.8407837			
a constant	k	1	calculated by (b)	0.0123344	0.0236684	0.0330015	0.0393331

As mentioned, deflection U is measured to five decimal accuracy (in meters), while the calculated values are given to seven decimals, because by giving more decimals, the precision of end-results is enhanced.

Now the function of k [defined in relation (b)] upon π_3 is to be determined. Let us assume a polynomial relation, which seems justified in this case. To be conservative, we consider terms up to fourth degree. Hence we write

$$k = b_0 + b_1 \cdot \pi_3 + b_2 \cdot \pi_3^2 + b_3 \cdot \pi_3^3 + b_4 \cdot \pi_3^4 \qquad \text{(c)}$$

where the b's are as-yet unknown coefficients. We immediately note that $b_0 = 0$, since otherwise in the case of $\pi_3 = 0$ (i.e., $a = 0$), the deflection caused by F would *not* be zero, which cannot happen because then the load (forces) would be directly over the supports. Consequently, from the above table we can have the following set of equations:

$$\left.\begin{aligned}
0.0123344 &= b_1 \cdot (0.1) + b_2 \cdot (0.1)^2 + b_3 \cdot (0.1)^3 + b_4 \cdot (0.1)^4 \\
0.0236684 &= b_1 \cdot (0.2) + b_2 \cdot (0.2)^2 + b_3 \cdot (0.2)^3 + b_4 \cdot (0.2)^4 \\
0.0330015 &= b_1 \cdot (0.3) + b_2 \cdot (0.3)^2 + b_3 \cdot (0.3)^3 + b_4 \cdot (0.3)^4 \\
0.0393331 &= b_1 \cdot (0.4) + b_2 \cdot (0.4)^2 + b_3 \cdot (0.4)^3 + b_4 \cdot (0.4)^4
\end{aligned}\right\} \qquad \text{(d)}$$

the solution of which is facilitated by its matrix form

$$\underbrace{\begin{bmatrix} 0.1 & 0.1^2 & 0.1^3 & 0.1^4 \\ 0.2 & 0.2^2 & 0.2^3 & 0.2^4 \\ 0.3 & 0.3^2 & 0.3^3 & 0.3^4 \\ 0.4 & 0.4^2 & 0.4^3 & 0.4^4 \end{bmatrix}}_{\mathbf{P} \text{ matrix}} \cdot \begin{bmatrix} b_1 \\ b_2 \\ b_3 \\ b_4 \end{bmatrix} = \underbrace{\begin{bmatrix} 0.0123344 \\ 0.0236684 \\ 0.0330015 \\ 0.0393331 \end{bmatrix}}_{\mathbf{Q} \text{ matrix}}$$

from which

$$\begin{bmatrix} b_1 \\ b_2 \\ b_3 \\ b_4 \end{bmatrix} = \mathbf{P}^{-1} \cdot \mathbf{Q} = \begin{bmatrix} 0.1250113 \\ 0.0000004 \\ -0.1667250 \\ -0.0000417 \end{bmatrix} \qquad \text{(e)}$$

So evidently,

$$b_1 = \frac{1}{8}; \qquad b_2 = 0; \qquad b_3 = -\frac{1}{6}; \qquad b_4 = 0 \qquad \text{(f)}$$

since the absolute values of b_2 and b_4 are several orders of magnitude less than the absolute values of the other coefficients. By (f) and (c) now

$$k = \frac{1}{8} \cdot \pi_3 - \frac{1}{6} \cdot \pi_3^3 \qquad \text{(g)}$$

We now substitute k from (g) into (b). This yields

$$\pi_1 = \left(\frac{1}{8} \cdot \pi_3 - \frac{1}{6} \cdot \pi_3^3 \right) \cdot \pi_2 \qquad \text{(h)}$$

or, by (a),

$$\frac{U}{L} = \left(\frac{1}{8} \cdot \frac{a}{L} - \frac{1}{6} \cdot \frac{a^3}{L^3} \right) \cdot \frac{F \cdot L^2}{Z} \qquad \text{(i)}$$

and therefore, finally

$$U = \frac{F \cdot a}{24 \cdot Z} \cdot (3 \cdot L^2 - 4 \cdot a^2) \tag{j}$$

which is the explicit formula for the lateral deformation of the beam as shown in Fig. 13-14. The ramifications of this result should be fully understood at this point: relation (j) was derived by the adroit use of dimensionless variables, a fact which is even more remarkable if one considers that, although the formula obtained contains five variables and three numeric constants, it was determined on the strength of only four measurements in which only *one* variable was changed! This fact forcefully demonstrates the power and potential of the dimensional method.

⇑

13.6. PROBLEMS

Solutions are in Appendix 6.

13/1 Deflection of a simply supported beam loaded by a lateral force. Given an $L = 3.4$ m long simply supported beam. The beam is loaded by a single lateral concentrated force $F = 40{,}000$ N, as illustrated in Fig. 13-15. By taking precise measurements, deflections U, under the given load F, are determined at four different "a" values (i.e., F is shifted along the beam). The input and results data are as follows (all values are in meters).

Test #	a	U
1	0.2	0.0040157
2	0.4	0.0141176
3	0.7	0.0350206
4	1	0.0564706

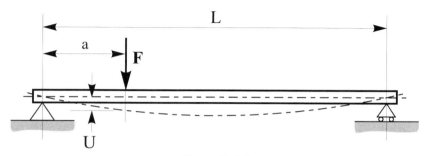

Figure 13-15
Simply supported beam is loaded by a single lateral force

Based on these test results, on the other given input data and on the assumed "rigidity" $Z = I \cdot E = 4 \times 10^5$ m$^2 \cdot$N, derive the general and explicit formula for deflection U; i.e., determine function Ψ in the relation $U = \Psi\{F, a, L, Z\}$.

13/2 Fundamental lateral frequencies of geometrically similar cantilevers. How do natural vibratory lateral frequencies of geometrically similar cantilevers vary with their sizes (Fig. 13-16)? Assume all cantilevers have uniform cross-sections and identical material. Also assume that the frequency is determined by characteristic size (length), Young's modulus, and density of material.

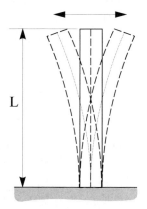

Figure 13-16
Cantilever is vibrating laterally

13/3 Area of an elliptic segment. Determine the shaded area of the segment of an ellipse as illustrated in Fig. 13-17.

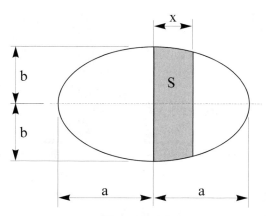

Figure 13-17
Area of an elliptic segment

FORMS OF DIMENSIONLESS RELATIONS **375**

(a) List the relevant variables and their dimensions.
(b) Determine a complete set of dimensionless variables.
(c) Prove that the *monomial* form connecting the dimensionless variables [derived in (b)] is untenable.
(d) Area S, derived analytically, is $S = \frac{b}{a} \cdot (x \cdot \sqrt{a^2 - x^2} + a^2 \cdot \arcsin \frac{x}{a})$. Express this formula by the dimensionless variable(s) derived in (b) above. Verify your dimensionless formula for the values $x = 7$ m, $a = 12$ m, $b = 6$ m.
(e) Draw a dimensionless plot to determine the area of an arbitrary segment. Check the accuracy of the plot using the values given in (d) above.

13/4 Torus volume—normal, degenerating, and degenerated. A torus is a surface of revolution generated by (usually) a circle (radius R) revolving about a line in its plane. We use the symbol h for the distance between this line and the center of the generating circle (see Fig. 13-18). Then, by definition, the torus is

normal,	if $h \geq R$;	see Fig. 13-18 (a) and (b)
degenerating,	if $0 < h < R$;	see Fig. 13-18 (c)
degenerated,	if $h = 0$;	see Fig. 13-18 (d)

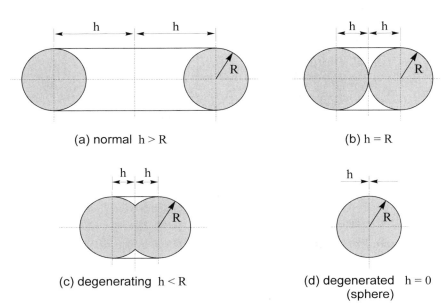

(a) normal h > R

(b) h = R

(c) degenerating h < R

(d) degenerated h = 0
(sphere)

Figure 13-18
Torus geometry in four characteristic configurations

(a) List the relevant variables and their dimensions for the determination of the volume of the torus in the four of the configurations depicted in the figure.
(b) Determine a complete set of dimensionless variables.
(c) If the torus is *normal* (i.e., $h \geq R$), then the volume is $V = 2 \cdot \pi^2 \cdot h \cdot R^2$. Convert this relation to dimensionless form using the dimensionless variables derived in (b) above. Is the obtained formula a monomial?
(d) If the torus is degenerating or degenerated (i.e., $0 \leq h < R$), then prove that the relation connecting the dimensionless variables cannot be a monomial.
(e) The volume of a degenerating or a degenerated torus (derivable analytically with a modicum of effort) is

$$V = 2 \cdot \pi^2 \cdot h \cdot R^2 + 2 \cdot \pi \cdot R^3 \cdot \left[\left(\frac{h}{R}\right)^2 \cdot \sqrt{1 - \left(\frac{h}{R}\right)^2} - \frac{h}{R} \cdot \arccos\frac{h}{R} + \frac{2}{3} \cdot \left(1 - \frac{h^2}{R^2}\right)^{1.5} \right]$$

Express this formula using the dimensionless variables derived in (b) above. Check your formula for numerical accuracy in the degenerated case (a sphere).
(f) Construct a dimensionless plot to determine the volume of a torus covering the range $0 \leq h/R \leq 2.5$.

13/5 Bolometric luminosity of a star. Bolometric luminosity L of a star is the energy flux emitted by that star in the form of electromagnetic radiation—especially in the microwave and infrared ranges. It follows that L is *power*. The following variables and physical constants are relevant in this case.

Variable	Symbol	Dimension
bolometric luminosity	L	$m^2 \cdot kg/s^3$
radius of star	R	m
Boltzmann's constant	k	$m^2 kg/(s^2 \cdot K)$
star's surface temperature	T	K
speed of light	c	m/s
Planck's constant	h	$m^2 \cdot kg/s$

(a) Compose the dimensional set, and from this determine a complete set of dimensionless variables.
(b) Assume the *monomial* form among the variables determined in (a) above. Then, by considering the plausible fact that radiated power is proportional to the area of the radiating surface, determine how the emitted power varies with (i) temperature; (ii) speed of light.

13/6 Volume of a conical wedge. Given a wedge cut from a right circular cone, as illustrated in Fig. 13-19.

(a) Name all the physical variables relevant in determining the volume of the wedge.
(b) Construct the Dimensional Set and, on the basis of this set, determine a complete set of dimensionless variables.

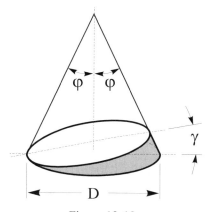

Figure 13-19
A conical wedge

- **(c)** Prove that the *monomial* form among these dimensionless variables is untenable.
- **(d)** The volume of the wedge is

$$V = \frac{\pi \cdot D^3}{24 \cdot \tan \varphi} \cdot \left[1 - \left(\frac{1 - \tan \varphi \cdot \tan \gamma}{1 + \tan \varphi \cdot \tan \gamma} \right)^{1.5} \right]$$

Express this relation in dimensionless form. Use the dimensionless variables obtained in (b) above. Check the result using a numerical example.
- **(e)** Draw the dimensionless plot of your result in (d). Confirm the correctness of this graph by the numerical example used in (d).

13/7 Volume of the frustum of a right circular cone. Given the frustum of a right circular cone, as illustrated in Fig. 13-20.

- **(a)** List the physical variables relevant in the determination of the volume of the frustum.
- **(b)** Define an independent set of dimensionless variables.
- **(c)** Prove that the *monomial* form connecting the dimensionless variables derived in (b) is untenable.
- **(d)** The volume of the illustrated frustum is

$$V = \frac{\pi}{12} \cdot h \cdot (d_2^2 + d_1 \cdot d_2 + d_1^2) \qquad (\#)$$

Using the dimensionless variables derived in (b), express formula (#) in dimensionless form.
- **(e)** From the results of (d) and by employing the established dimensionless variables [as in (b) above], express the volume of an *intact* cone

378   APPLIED DIMENSIONAL ANALYSIS AND MODELING

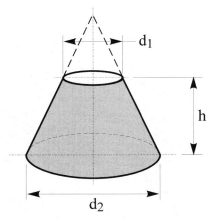

Figure 13-20
The frustum of a right circular cone

and a *cylinder*. Check your results against the well-known formulas for the volume of a cone and of a cylinder.

13/8 Discharge of a capacitor. Consider an electric circuit consisting of ohmic resistance R, capacitance C, direct current supply, and a switch (Fig. 13-21). In the illustrated position of the switch, a potential difference V_0 exists across capacitance C. If the switch is now set to position a-b, then an electric current i will flow in the closed circuit. This current is obviously time dependent.

(a) Establish the relevant physical variables to define current i as a function of time t.

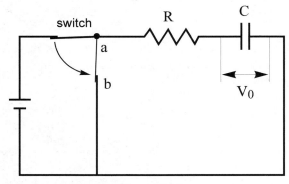

Figure 13-21
A direct current RC circuit

(b) Establish a complete set of dimensionless variables concordant with the physical variables defined in (a).
(c) Prove that the *monomial* form connecting the dimensionless variables derived in (b) is untenable.
(d) Assume relation $i = \frac{V_0}{R} \cdot e^{-\frac{t}{RC}}$, then convert it to dimensionless form.
(e) Construct a dimensionless plot based on the result obtained in (d).

13/9 Area of a triangle whose side-lengths form an arithmetic progression.
Given a triangle whose side-lengths a, b, and c form an *arithmetic* progression, i.e.,

$$a = a; \qquad b = a + d; \qquad c = a + 2 \cdot d$$

where $d \geq 0$ is the *common difference* (see Fig. 13-22).

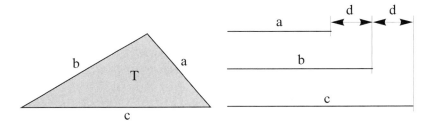

Figure 13-22
A triangle whose side-lengths form an arithmetic progression

(a) What are the relevant variables in an expression for the area of such a triangle?
(b) Construct a complete set of dimensionless variables based on the set obtained in (a).
(c) Prove that the relationship among the dimensionless variables established in (b) cannot be a *monomial* one.

CHAPTER 14
SEQUENCE OF VARIABLES IN THE DIMENSIONAL SET

It was mentioned briefly in Art. 7.2 that the sequence of physical variables in a dimensional matrix—and hence in a Dimensional Set—should be such that the dependent variable is in the leftmost position. Of course, this statement assumes that there is only *one dependent* variable in the set. But what happens if there are two such variables? This short chapter will give some useful rules and guidelines for sequencing the physical variables in a dimensional set.

14.1. DIMENSIONLESS PHYSICAL VARIABLE IS PRESENT

It often happens that among the relevant *physical* variables in a problem, there is one (or more) that is (are) dimensionless. Such physical variables are, for example, strain, angle, friction coefficient, Poisson's ratio, efficiency, refractive index, emissivity, etc. Do we have to incorporate this type of variable into the Dimensional Set, and if so, how do we do it?

Since these variables are dimensionless to begin with, strictly speaking they do not have to be part of the Dimensional Set. We could just consider all the other dimensional variables as usual and, at the end, when the other dimensionless variables have been determined, just add the former (dimensionless physical) ones to the obtained set. The following example illustrates this simple process:

Example 14-1. Mass of Electrochemically Deposited Material

When a current is sent through an electrolytic solution, material will migrate between the electrodes. What is the quantity of this transferred mass? To find the answer, we first define the relevant physical variables.

Variable	Symbol	Dimension
transferred mass between electrodes	M	kg
quantity of electricity	q	A·s
atomic weight of element transferred	w	kg/mol
Faraday's constant	F	A·s/mol
valence	v	1

382   APPLIED DIMENSIONAL ANALYSIS AND MODELING

The dimensional matrix is

	M	q	w	F	v
kg	1	0	1	0	0
s	0	1	0	1	0
A	0	1	0	1	0
mol	0	0	−1	−1	0

We observe that the variable *valence* is dimensionless, so immediately we can write

$$\pi_1 = v \qquad (a)$$

and remove v from the dimensional matrix, which then becomes a square 4 × 4 matrix. Thus, by Theorem 7-5 (Art. 7.11), this dimensional matrix must be singular, else there is no solution. If we remove v from the above matrix, we get the array

$$\begin{vmatrix} 1 & 0 & 1 & 0 \\ 0 & 1 & 0 & 1 \\ 0 & 1 & 0 & 1 \\ 0 & 0 & -1 & -1 \end{vmatrix} = 0$$

whose determinant—as seen—is zero. Therefore there *is* a solution. Proceeding, we remove one dimension—say A (ampere)—from the dimensional matrix (Art. 7.11), and then construct the Dimensional Set

	M	q	w	F
kg	1	0	1	0
s	0	1	0	1
mol	0	0	−1	−1
π_2	1	−1	−1	1

Thus, by (a) and the above dimensional set, the two dimensionless variables are

$$\pi_1 = v; \qquad \pi_2 = \frac{M \cdot F}{q \cdot w} \qquad (b)$$

from which we can write

$$\pi_2 = \Psi\{\pi_1\}$$

where Ψ is a function. If Ψ is a *monomial,* which is plausible, then by (b)

$$M = c \cdot \frac{q \cdot w}{F} \cdot (v)^\epsilon \qquad (c)$$

where c and ϵ are numeric constants to be determined by analysis or experimentation. Easy analysis (Ref. 4, pp. 401–402) in fact results in $c = 1$ and $\epsilon = -1$. Therefore (c) becomes

$$M = \frac{q \cdot w}{F \cdot v} \qquad (d)$$

which is, then, the relation we were looking for.

But there is another way to handle a dimensionless physical variable; it can be incorporated in the Dimensional Set by applying the following theorem:

Theorem 14-1. *If there is a dimensionless physical variable in the dimensional set, then this variable must be the* **B** *matrix.*

Proof. By definition, the column of a dimensionless variable in the dimensional matrix is all zeros. Therefore, if it were in the **A** matrix, then **A** would be singular, which is forbidden (Art. 7.6). A variable which is in the dimensional matrix must be either in the **A** matrix or in the **B** matrix. Since it cannot be in the **A** matrix, it must be in the **B** matrix. This proves the theorem.

Therefore a dimensionless physical variable, if it is in a Dimensional Set (which is made up by the dimensional matrix and **D** and **C** matrices) at all, must be in the **B** matrix, in which case the following theorem applies:

Theorem 14-2. *If there is a dimensionless physical variable in the* ith *column of the* **B** *matrix of a Dimensional Set, then the* ith *row of the* **C** *matrix of this set is all zeros.*

Proof. If the ith *column* of matrix **B** is all zeros, then, obviously, **B** = **B·Z**, where **Z** is a compatible diagonal matrix in which the ith *column* (and hence the ith row, as well), is all zeros and all other elements in the main diagonal is 1. In this case, by Theorem 9-5 (Art. 9.1), the **C** matrix can be written as **C** = **Z^T·X**, where **X** is some matrix. Thus, no matter what **X** is, the ith *row* of **C** will be all zeros. This proves the theorem.

The next example deals with an application whereby a physical dimensionless variable is incorporated onto the Dimensional Set.

Example 14-2. Maximum Thickness of Wet Paint on an Inclined Surface

The thickness of liquid paint on a surface of inclination $\varphi > 0$ (the surface is not a *floor*) cannot exceed a critical (maximum) value, for otherwise the paint would run off the surface (Fig. 14-1).

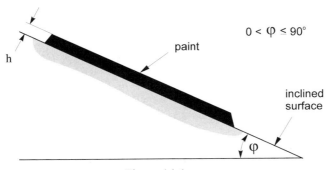

Figure 14-1
Layer of liquid paint on an inclined surface

Using the dimensional method we now derive a relation among the following physical variables:

Variable	Symbol	Dimension	Remark
thickness of wet paint	h	m	maximum value
inclination of surface	φ	1	to horizontal
shear strength of paint	τ	kg/(m·s²)	liquid
density of paint	ρ	kg/m³	liquid
gravitational acceleration	g	m/s²	

We have five variables and three dimensions, therefore there are $5 - 3 = 2$ dimensionless variables supplied by the Dimensional Set

	h	φ	τ	ρ	g
m	1	0	−1	−3	1
kg	0	0	1	1	0
s	0	0	−2	0	−2
π_1	1	0	−1	1	1
π_2	0	1	0	0	0

from which

$$\pi_1 = \frac{h \cdot \rho \cdot g}{\tau}; \qquad \pi_2 = \varphi \tag{a}$$

Note that φ is a dimensionless variable, therefore (by Theorem 14-1) it must be part of the **B** matrix. Since φ is in the second *column* of **B**, the second *row* of the **C** matrix is all zeros (Theorem 14-2), as seen above. By (a), now

$$h = c \cdot \frac{\tau}{\rho \cdot g} \cdot \Psi\{\varphi\} \tag{b}$$

where c is a numeric constant and Ψ is a function to be determined by other means. By simple analytic derivation, we have

$$c = 1; \qquad \Psi\{\varphi\} = \frac{1}{\sin \varphi} \tag{c}$$

With this and (b)

$$h = \frac{\tau}{\rho \cdot g \cdot \sin \varphi} \tag{d}$$

or by (a) and (c)

$$\pi_1 = \frac{1}{\sin \pi_2} \tag{e}$$

If, for example, $\pi_2 = \pi/2 = 1.5708 \ldots$ ($= 90°$), then $\pi_1 = 1$ and by (d), $h = \tau/(\rho \cdot g)$. If, however $\pi_2 = \pi/4 = 0.7854 \ldots$ ($= 45°$), then $\pi_1 = \sqrt{2} = 1.4142 \ldots$ and by (d), $h = (1.4142) \cdot \tau/(\rho \cdot g)$, i.e., the maximum thickness of paint on a 45° surface is 41.4% more than on a vertical surface (i.e., "regular" wall). Note that $0 < \varphi \leq 90°$; i.e., "hanging" paints are not included in these considerations.

14.2. PHYSICAL VARIABLES OF IDENTICAL DIMENSIONS ARE PRESENT

It often happens that two or more physical variables have identical dimensions. In these cases the following theorem must be considered.

Theorem 14-3. *In a Dimensional Set, no two or more variables in matrix* **A** *may have identical dimensions.*

In other words, if two or more physical variables have identical dimensions, then *at the most* only *one* of them can be in matrix **A**.

Proof. If two or more variables in the **A** matrix have identical dimensions, then at least two columns of **A** are identical, and hence **A** is singular—which cannot be allowed to happen (Art. 7.6). This proves the theorem.

It should be noted that there is no restriction against placing any number of physical variables with identical dimensions in the **B** matrix. The only effect of identical *columns* in **B** is the appearance of the same number of identical *rows* in the **C** matrix. In this respect, the following theorem applies:

Theorem 14-4. *If in a Dimensional Set, the ith and the jth columns of matrix* **B** *are identical, then the ith and jth rows of the matrix* **C** *are also identical.*

That is, if two variables in the **B** matrix have identical dimensions, i.e., the respective *columns* in **B** are the same, then the respective *rows* in matrix **C** are also the same.

Proof. Recall from the definition of matrix multiplication that for any *compatible* matrices **P·Q** = **R**, if two columns of **Q** are identical, then the same two columns of **R** will also be identical. For example, in

$$\underbrace{\begin{bmatrix} 4 & 8 & 3 \\ 2 & 6 & 7 \\ 2 & 8 & 3 \end{bmatrix}}_{\mathbf{P}} \cdot \underbrace{\begin{bmatrix} 2 & 6 & 8 & 6 \\ 6 & 3 & 7 & 3 \\ 4 & 2 & 1 & 2 \end{bmatrix}}_{\mathbf{Q}} = \underbrace{\begin{bmatrix} 68 & 54 & 91 & 54 \\ 68 & 44 & 65 & 44 \\ 64 & 42 & 75 & 42 \end{bmatrix}}_{\mathbf{R}}$$

the second and fourth columns of **Q** are identical. Therefore the second and fourth columns of **R** are also identical.

Next, we consider the *Fundamental Formula* $\mathbf{C} = -(\mathbf{A}^{-1}\cdot\mathbf{B})^T$. If the *i*th and *j*th *columns* of **B** are identical, then the same two *columns* of $\mathbf{A}^{-1}\cdot\mathbf{B}$ are also identical. Therefore the *i*th and *j*th *rows* of $(\mathbf{A}^{-1}\cdot\mathbf{B})^T$ are identical, since matrices $(\mathbf{A}^{-1}\cdot\mathbf{B})$ and $(\mathbf{A}^{-1}\cdot\mathbf{B})^T$ differ only in the interchange of their rows with columns. Thus, in matrix **C** the *i*th and *j*th rows are also identical since **C** and $(\mathbf{A}^{-1}\cdot\mathbf{B})^T$ differ only in the signs of their elements. This proves the theorem.

The following two examples demonstrate this characteristic of the Dimensional Set.

Example 14-3

Consider the Dimensional Set of six variables $V_1, V_2, \ldots, V_6$ and two dimensions d_1, d_2.

	V_1	V_2	V_3	V_4	V_5	V_6
d_1	2	−3	2	0	2	1
d_2	−1	4	−1	3	−1	2
π_1	1	0	0	0	−1	0
π_2	0	1	0	0	2	−1
π_3	0	0	1	0	−1	0
π_4	0	0	0	1	0.6	−1.2

As we see, variables V_1 and V_3 have identical dimensions, and therefore the first and third *columns* of matrix **B** are identical. Hence, by Theorem 14-4, the first and third *rows* of matrix **C** are also identical. Note that *both* V_1 and V_3 must not be in matrix **A**, for this would violate Theorem 14-3.

⇑

Example 14-4. The Coffee Warmer (II)

In Example 13-14 (Art. 13.5.1) we investigated the coffee warmer from the point of view of establishing the constants of the *monomial* relation existing between the relevant dimensionless variables. Now we revisit this problem to show the effect of identical physical variables on the **C** matrix of the Dimensional Set. The arrangement is shown in Fig. 14-2.

The warmer is essentially a constant-temperature hot-plate. The task is to determine the equilibrium temperature of the coffee in the mug. We investigate this problem using dimensional considerations. But first we must assess the conditions of thermal equilibrium. Heat Q_1 enters the coffee from the hot-plate through the bottom of the mug. The heat then dissipates through the cylindrical wall of the container Q_2, and by evaporation Q_3

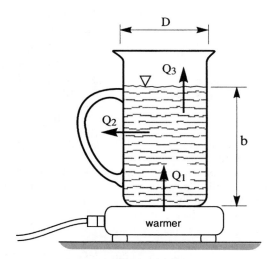

Figure 14-2
A mug containing coffee on a hot-plate

through the exposed top surface of the coffee. More heat is dissipated by the nonwetted surface of the mug, but this can be neglected because of the high heat conducting resistance of the mug. Therefore, we can say that at thermal equilibrium

$$Q_1 = Q_2 + Q_3 \tag{a}$$

The relevant variables must be those which define the three quantities of heat transmitted per unit time, and so we have the following list of variables:

Variable	Symbol	Dimension
coffee temperature	t	K
air temperature	t_a	K
hot-plate temperature	t_p	K
heat transfer coefficient of hot-plate	h_p	kg/(s³·K)
heat transfer coefficient of mug	h_g	kg/(s³·K)
heat transfer coefficient of evaporation	h_e	kg/(s³·K)
diameter of mug	D	m
level of coffee in mug	b	m

Here h_p is the heat transfer coefficient for heat Q_1, h_g for Q_2 and h_e for Q_3. For these coefficients it is assumed that the heat transmitted is proportional to the respective temperature differences. It is also assumed that *steady state* conditions exist, i.e., there is thermal equilibrium between the bodies in contact.

Based on the above variables and their dimensions, the following dimensional matrix can be constructed:

	t	t_a	t_p	h_g	h_p	h_e	D	b
m	0	0	0	0	0	0	1	1
kg	0	0	0	1	1	1	0	0
s	0	0	0	−3	−3	−3	0	0
K	1	1	1	−1	−1	−1	0	0

Obviously, the rank of this matrix is less than 4, since there are only three variables of differing dimensions (temperature, heat transfer coefficient and length). Therefore one dimension must be abandoned. This dimension cannot be either "m" or "K," because in the former case both D and b are eliminated, while in latter case only these variables are retained. Thus, we delete "s."

But now we note that if the sequence of variables were left intact, then in matrix **A**—which is composed of the *rightmost* 3 × 3 determinant of the dimensional matrix—two variables (D and b) would have the *same* dimension which, by Theorem 14-3, cannot happen. Thus, we rearrange the Set to eliminate this duplication in **A**. The result is the following Dimensional Set:

	t	t_a	h_g	h_e	b	t_p	h_p	D
m	0	0	0	0	1	0	0	1
kg	0	0	1	1	0	0	1	0
K	1	1	−1	−1	0	1	−1	0
π_1	1	0	0	0	0	−1	0	0
π_2	0	1	0	0	0	−1	0	0
π_3	0	0	1	0	0	0	−1	0
π_4	0	0	0	1	0	0	−1	0
π_5	0	0	0	0	1	0	0	−1

We see that in the **A** matrix no two variables have identical dimensions (obeying Theorem 14-3), and, by Theorem 14-4, since

- the first and second *columns* of **B** are identical, therefore the first and second *rows* of **C** are the same;
- the third and fourth *columns* of **B** are identical, therefore the third and fourth *rows* of **C** are the same.

By the Dimensional Set, it is also seen that we have eight variables and three dimensions, therefore there are $8 - 3 = 5$ dimensionless variables provided by the set:

$$\pi_1 = \frac{t}{t_p}; \quad \pi_2 = \frac{t_a}{t_p}; \quad \pi_3 = \frac{h_g}{h_p}; \quad \pi_4 = \frac{h_e}{h_p}; \quad \pi_5 = \frac{b}{D} \quad \text{(b)}$$

Now if one drinks some of the coffee from the mug, level b decreases and thus π_5 is the independent dimensionless variable. By similar reasoning, the dependent variable is the temperature of coffee t, and hence the dependent dimensionless variable is π_1. Therefore, for a given physical setup (hot-plate plus mug), temperature, pressure, and moisture content of the air—π_2, π_3 and π_4 are constants.* Hence, we can write

$$\pi_1 = \Psi\{\pi_2, \pi_3, \pi_4, \pi_5\} = \Psi\{\text{constants}, \pi_5\} \quad \text{(c)}$$

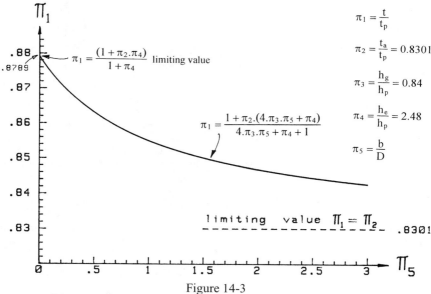

Figure 14-3
Dimensionless plot for the steady state temperature of coffee warmed in a mug on a thermally controlled hot-plate
Parameters π_2, π_3, and π_4 are fixed values—as shown—and are selected for illustration only. Physical variables are as defined in the table in this example

---

*Strictly speaking, the coefficient of evaporation h_e is not constant, for it varies with temperature.

where Ψ is a function. Thus, for a given setup and atmospheric conditions, (c) can be represented by a *single* curve in the π_5 versus π_1 coordinate system.

The following simple relations exist for the energy transfers

$$Q_1 = (t_p - t) \cdot h_p \cdot \frac{D^2 \cdot \pi}{4}; \qquad Q_2 = (t - t_a) \cdot h_g \cdot D \cdot b \cdot \pi; \qquad Q_3 = (t - t_a) \cdot h_e \cdot \frac{D^2 \cdot \pi}{4} \qquad (d)$$

Then, by (a), t can be explicitly expressed as

$$t = \frac{t_p \cdot h_p \cdot D + 4 \cdot t_a \cdot D \cdot b \cdot h_g + t_a \cdot h_e \cdot D}{4 \cdot b \cdot h_g + h_e \cdot D + h_p \cdot D} \qquad (e)$$

which by (b) can be put into the dimensionless form of (c)

$$\pi_1 = \frac{1 + \pi_2 \cdot (4 \cdot \pi_3 \cdot \pi_5 + \pi_4)}{4 \cdot \pi_3 \cdot \pi_5 + \pi_4 + 1} \qquad (f)$$

The single-curve graph of this relation, with assumed constants π_2, π_3, and π_4, is shown in Fig. 14-3.

14.3. INDEPENDENT AND DEPENDENT VARIABLES

If there is more than one dimensionless variable, then the sequential order of physical variables placed in the dimensional set is strongly influenced by the dependent or independent nature of the physical variables. However, if there is only one dimensionless variable, then the sequence is inconsequential. This because a single dimensionless variable is a constant (Theorem 7-4, Art. 7.10), and therefore must be a monomial function of the physical variables. In a monomial—being a scalar product—the resultant value is impervious to the sequential order.

Example 14-5. Bursting Speed of a Flywheel

If the rotational speed of even a perfectly balanced flywheel is increased beyond a critical value, the flywheel will disintegrate, i.e., it will *burst*. This is called the *bursting speed*. We will now examine how this speed is influenced by the relevant physical variables listed below.

Variable	Symbol	Dimension	Remark
bursting speed	ω	1/s	
diameter	D	m	
strength	σ	kg/(m·s^2)	constant
density	ρ	kg/m^3	constant

We have four variables and three dimensions, therefore there is $4 - 3 = 1$ dimensionless "variable"; it is supplied by Dimensional Set 1.

Dimensional Set 1

	ω	D	σ	ρ
m	0	1	−1	−3
kg	0	0	1	1
s	−1	0	−2	0
π_{11}	1	1	$-\dfrac{1}{2}$	$\dfrac{1}{2}$

(a)

from which

$$\pi_{11} = \omega \cdot D \cdot \sqrt{\dfrac{\rho}{\sigma}} = \text{const} \qquad (b)$$

Now we use a *different* sequence to form Dimensional Set 2.

Dimensional Set 2

	D	σ	ρ	ω
m	1	−1	−3	0
kg	0	1	1	0
s	0	−2	0	−1
π_{21}	1	$-\dfrac{1}{2}$	$\dfrac{1}{2}$	1

(c)

from which

$$\pi_{21} = D \cdot \omega \cdot \sqrt{\dfrac{\rho}{\sigma}} = \text{const} \qquad (d)$$

Thus, by (b) and (d)

$$\pi_{11} = \pi_{21} \qquad (e)$$

as expected and required. Of course, the same result should be obtained if *transformation matrix* **T** is used (Art. 9.3), and just for the exercise we shall verify this now. *Shift Matrix* **S** is

	D	σ	ρ	ω
ω	0	0	0	1
D	1	0	0	0
σ	0	1	0	0
ρ	0	0	1	0

and hence, by Fig. 9-3, submatrices $\mathbf{S}_1$ and $\mathbf{S}_3$ are

$$\mathbf{S}_1 = [0]; \qquad \mathbf{S}_3 = \begin{bmatrix} 1 \\ 0 \\ 0 \end{bmatrix} \qquad (f)$$

Thus, by (9-54), the transformation matrix is

$$\mathbf{T} = (\mathbf{S}_1 + \mathbf{C}_1 \cdot \mathbf{S}_3)^{-1} = \left([0] + \begin{bmatrix} 1 & -\dfrac{1}{2} & \dfrac{1}{2} \end{bmatrix} \cdot \begin{bmatrix} 1 \\ 0 \\ 0 \end{bmatrix} \right)^{-1} = [1]^{-1} = [1]$$

and therefore

$$\pi_{21} = \mathbf{T} \cdot \pi_{11} = [1] \cdot \pi_{11} = \pi_{11} \qquad (g)$$

which is the same as (e) above, as expected and required.

It is interesting to observe that, by either (b) or (d), the bursting speed of a flywheel varies with the inverse of its diameter, with the square root of its strength and with the inverse square root of its density. Since the ratios of strength to density for steel and aluminium are about the same, therefore, provided their diameters are identical, a steel and an aluminium flywheel will fly apart at the same rotational speed.

⇑

Next, we deal with a more involved case where *more than one* dimensionless variable is obtained from the Dimensional Set. Generally, but with exceptions (discussed presently), the *dependent* variable (singular!) should be placed in the **B** matrix, because variables in **B** appear in *only* one dimensionless variable, and hence they (the variables) can be conveniently *isolated*. If there is more than one dependent variable present, then special attention should be paid. This latter event will be discussed separately. For now, we will only deal with the *one dependent variable* case. Consider the following example:

Example 14-6. Longitudinal Deformation of an Axially Loaded Bar

We have a bar of uniform cross-section, as illustrated in Fig. 14-4. We wish to determine its longitudinal deformation upon a concentrated axial load applied at its free end. The table lists the relevant variables:

Variable	Symbol	Dimension	Remark
length of bar	L	m	
axial load	F	N	centrally applied
area of cross-section	A	m²	constant
Young's modulus of material	E	N/m²	
axial deformation	λ	m	

We have five variables and two dimensions, hence there are $5 - 2 = 3$ dimensionless variables supplied by Dimensional Set 1.

Dimensional Set 1

	L	F	A	E	λ	
m	1	0	2	−2	1	
N	0	1	0	1	0	
π_{11}	1	0	0	0	−1	(a)
π_{12}	0	1	0	−1	−2	
π_{13}	0	0	1	0	−2	

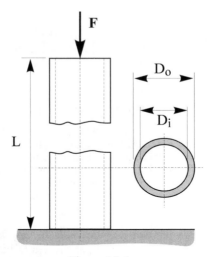

Figure 14-4
An axially loaded straight bar of uniform cross-section

which yields

$$\pi_{11} = \frac{L}{\lambda}; \qquad \pi_{12} = \frac{F}{E \cdot \lambda^2}; \qquad \pi_{13} = \frac{A}{\lambda^2} \qquad (b)$$

and hence we can write $\pi_{11} = \Psi\{\pi_{12}, \pi_{13}\}$, or by (b),

$$\frac{L}{\lambda} = \Psi\left\{\frac{F}{E \cdot \lambda^2}, \frac{A}{\lambda^2}\right\} \qquad (c)$$

We note that this formula is rather inconvenient, because the *dependent* variable λ appears in more than one dimensionless variable—here in all three of them. To isolate λ, even if we consider the *monomial* form for (c), is rather cumbersome, to say the least. The reason for this unsatisfactory arrangement is of course the placement of the dependent variable in the **A** matrix, instead of in the **B** matrix. If λ *were* in **B**, then by necessity it would appear in only one dimensional variable and hence it could be conveniently *isolated*. To demonstrate, we recast the sequence of variables in Dimensional Set 2 thus

Dimensional Set 2

	λ	F	A	E	L
m	1	0	2	−2	1
N	0	1	0	1	0
π_{21}	1	0	0	0	−1
π_{22}	0	1	0	−1	−2
π_{23}	0	0	1	0	−2

(d)

in which λ appears in the **B** matrix. Hence we have

$$\pi_{21} = \frac{\lambda}{L} \; ; \quad \pi_{22} = \frac{F}{E \cdot L^2} \; ; \quad \pi_{23} = \frac{A}{L^2} \tag{e}$$

and we write

$$\frac{\lambda}{L} = \Psi\left\{\frac{F}{E \cdot L^2}, \frac{A}{L^2}\right\} \tag{f}$$

where Ψ is a function. We now adopt the *monomial* form for Ψ, which is plausible. Then

$$\lambda = k \cdot L \cdot \left(\frac{F}{E \cdot L^2}\right)^{\epsilon_2} \cdot \left(\frac{A}{L^2}\right)^{\epsilon_3} \tag{g}$$

where k, ϵ_2, and ϵ_3 are numerical constants.

We see in (e) that λ appears in only *one* dimensionless variable and therefore it is conveniently explicit in (g), as it should be.

This example showed the significant advantage of placing the dependent variable in the **B** matrix.

⇑

A case is special when there is more than one dependent variable. In these cases great care must be exercised not to generate a formula in which more than one *dependent* variable appear. For, if such a formula is used, then the result obtained, by assigning at will values to the *dependent* variables, as if they were *independent* ones, will be erroneous in the great majority of cases.

We have already presented a relatively simple example (Example 6-10 in Art. 6.1) of the erroneous results obtained where more than one dependent *physical* variable is used in a relationship. The following is a slightly more complex case.

Example 14-7. Volume and Surface of the Frustum of a Cone

This topic appeared as a problem in Chapter 13. We approach it now from a different viewpoint. The volume and surface of the frustum of a right circular cone (as in Fig. 13-20) are

$$V = \frac{\pi}{12} \cdot h \cdot (d_1^2 + d_1 \cdot d_2 + d_2^2) \quad \text{[volume]} \tag{a}$$

$$A = \frac{\pi}{4} \cdot (d_1 + d_2) \cdot \sqrt{(d_2 - d_1)^2 + 4 \cdot h^2} \quad \text{[surface]} \tag{b}$$

Thus the following physical variables are involved:

Variable	Symbol	Dimension
volume of frustum	V	m³
surface of frustum	A	m²
base diameter of frustum	d_2	m
top diameter of frustum	d_1	m
height of frustum	h	m

We have five variables and one dimension, hence there are $5 - 1 = 4$ dimensionless variables supplied by the Dimensional Set:

394   APPLIED DIMENSIONAL ANALYSIS AND MODELING

		dependent			
		↓	↓		
	V	A	d_1	d_2	h
m	3	2	1	1	1
π_1	1	0	0	0	−3
π_2	0	1	0	0	−2
π_3	0	0	1	0	−1
π_4	0	0	0	1	−1

(c)

yielding

$$\pi_1 = \frac{V}{h^3}; \quad \pi_2 = \frac{A}{h^2}; \quad \pi_3 = \frac{d_1}{h}; \quad \pi_4 = \frac{d_2}{h} \quad\quad (d)$$

With these dimensionless variables, (a) and (b) can now be written

$$\pi_1 = \frac{\pi}{12}\cdot(\pi_3^2 + \pi_3\cdot\pi_4 + \pi_4^2) \quad\quad (e)$$

$$\pi_2 = \frac{\pi}{4}\cdot(\pi_3 + \pi_4)\cdot\sqrt{(\pi_4 - \pi_3)^2 + 4} \quad\quad (f)$$

Both of these formulas are correct as they each contain only *one* dependent dimensionless variable, namely π_1 in (e) and π_2 in (f). However, we can now write (e) and (f), respectively, as

$$\pi = \frac{12\cdot\pi_1}{\pi_3^2 + \pi_3\cdot\pi_4 + \pi_4^2}; \quad \pi = \frac{4\cdot\pi_2}{(\pi_3 + \pi_4)\cdot\sqrt{(\pi_4 - \pi_3)^2 + 4}}$$

from which, by equating the right sides and some rearranging, we obtain

$$\pi_1 = \frac{\pi_2\cdot(\pi_3^2 + \pi_3\cdot\pi_4 + \pi_4^2)}{3\cdot(\pi_3 + \pi_4)\cdot\sqrt{(\pi_4 - \pi_3)^2 + 4}} \quad\quad (g)$$

In this relation we have *two* independent dimensionless variables, namely π_1 and π_2. Thus, this relation would give, in general, a false result for π_1 if arbitrary values were assigned for π_2, π_3, and π_4 [on the right side of (g)]. To illustrate this anomaly, we input the following values into (g):

$$\pi_2 = 2; \quad \pi_3 = 3; \quad \pi_4 = 4 \quad\quad (h)$$

Then, $\pi_1 = 1.5759$ by (g), but $\pi_1 = 9.68658$ by (e). The fault of course is caused by our considering π_2 in (g) as an *independent* dimensionless variable, although it is a *dependent* one upon π_3 and π_4, as relation (e) clearly shows.

To reiterate: one *must not* employ a *formula* which contains more than one dependent dimensionless variable.

⇑

In some special circumstances several dependent variables can be usefully incorporated in a single Dimensional Set. This occurs when *every* dependent variable is in the **B** matrix, and *every* independent variable is in the **A** matrix. In such cases *each* dependent variable can be considered *separately* from the others and therefore no more than *one* dependent variable will be present in each of the obtained dimensionless variable and, further, every dimensionless variable will contain exactly one dependent variable.

The following example clearly demonstrates this important fact.

Example 14-8. Free Fall

We now investigate the characteristics of free fall of a mass through distance L (Fig. 14-5).

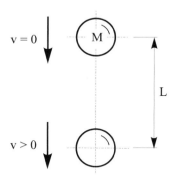

Figure 14-5
Free fall of mass M through distance L

We assume constant gravitational acceleration, initial zero speed and negligible air resistance. The following physical variables are considered relevant:

Variable	Symbol	Dimension	Remark
time of fall	t	s	dependent variable
terminal speed	v	m/s	dependent variable
kinetic energy	E_k	m²·kg/s²	dependent variable
distance of fall	L	m	independent variable
gravitational acceleration	g	m/s²	independent variable
mass	M	kg	independent variable

Note that variables L, g, and M are *independent*, and t, v, and E_k are *dependent*. The Dimensional Set is

	dependent ↓ ↓ ↓			independent ↓ ↓ ↓		
	t	v	E_k	L	g	M
m	0	1	2	1	1	0
kg	0	0	1	0	0	1
s	1	−1	−2	0	−2	0
π_1	1	0	0	$-\dfrac{1}{2}$	$\dfrac{1}{2}$	0
π_2	0	1	0	$-\dfrac{1}{2}$	$-\dfrac{1}{2}$	0
π_3	0	0	1	−1	−1	−1

We have six variables and three dimensions, therefore there are 6 − 3 = 3 dimensionless variables obtained from the above set.

$$\pi_1 = t \cdot \sqrt{\frac{g}{L}}; \quad \pi_2 = \frac{v}{\sqrt{g \cdot L}}; \quad \pi_3 = \frac{E_k}{L \cdot g \cdot M} \tag{a}$$

As seen in the Dimensional Set, all the dependent variables appear in the **B** matrix, and therefore are nicely isolated. This allows us to consider them *separately*. Thus, by (a), we may write

$$\left.\begin{array}{l} t \cdot \sqrt{\dfrac{g}{L}} = \text{const, and therefore } t = \text{const} \cdot \sqrt{\dfrac{L}{g}} \\[1em] \dfrac{v}{\sqrt{g \cdot L}} = \text{const, and therefore } v = \text{const} \cdot \sqrt{g \cdot L} \\[1em] \dfrac{E_k}{L \cdot g \cdot M} = \text{const, and therefore } E_k = \text{const} \cdot L \cdot g \cdot M \end{array}\right\} \tag{b}$$

Each of the relations of (b) is valid *separately*, since each contains only *one* dependent variable. For example, the second of (b) is

$$v = \text{const} \cdot \sqrt{g \cdot L} = \sqrt{2 \cdot g \cdot L} \tag{c}$$

i.e., the constant is $\sqrt{2}$.

Of course, each of the three relations in (b) could have been obtained separately by considering the Dimensional Set as being composed of only the three independent variables (L, g, and M) and *one* of the dependent variables. For example, if we are only interested in time t, then we can have

	t	L	g	M
m	0	1	1	0
kg	0	0	0	1
s	1	0	−2	0
π_1	1	$-\frac{1}{2}$	$\frac{1}{2}$	0

yielding $\pi_1 = t \cdot \sqrt{g/L}$, which is identical to the previously obtained relation in (a).

Incidentally, note that the column under M in the **C** matrix of the above Dimensional Set is all zeros. Therefore, by Theorem 10-4 (Art. 10.5), mass M is a *dimensionally irrelevant* variable, and because of this, by Theorem 11-3 (Art. 11.2), it is also a *physically irrelevant* variable. In other words, mass does not influence the time it takes to fall a given distance. This epoch-making conclusion was first reached by Galilei around 1586 (Galileo Galilei, Italian astronomer and physicist, 1564–1642).

⇑.

Occasionally it is advantageous to place the independent variable in the **B** matrix, and the *sole* dependent variable in the **A** matrix. This arrangement is warranted when *selective considerations* of independent variables are necessary. In this way one or more dimensionless variables may be ignored, an act which does not affect the remaining independent variable (singular!) since each of these physical variables appears in only *one* dimensionless variable. The following example demonstrates this very important feature.

Example 14-9. Stresses in a Structure Loaded by its Own Weight, External Forces, and Moments

Given a structure of general configuration carrying its own weight, some external forces and moments, as Fig. 14-6 depicts. What are the stresses?

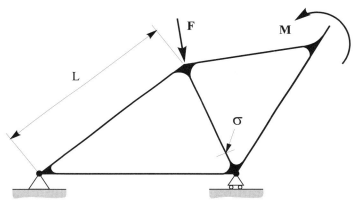

Figure 14-6
Heavy structure of characteristic length L is carrying a characteristic force F and characteristic moment M
Characteristic stress σ may be at location shown

It is assumed that the loads are small and that the deformations obey Hooke's law, i.e., they are within the elastic limits. The relevant variables are as follows:

Variable	Symbol	Dimension	Remark
force	F	m·kg/s^2	characteristic value
moment	M	m^2·kg/s^2	characteristic value
density of material	ρ	kg/m^3	
gravitational acceleration	g	m/s^2	
length	L	m	characteristic value
stress	σ	kg/(m·s^2)	characteristic value

It is noted that the variables stress, force, moment, and length have the attribute "characteristic." This means that these physical quantities may occur in several places on the structure, but that their magnitudes are *relatively* fixed numbers. For example, if there is j number of external forces $F_1, F_2, \ldots, F_j$ and F_c is the characteristic force, then

$$F_1 = q_1 \cdot F_c; \qquad F_2 = q_2 \cdot F_c; \ldots; \qquad F_j = q_j \cdot F_c \tag{a}$$

where factors $q_1, q_2, \ldots, q_j$ are constant numerics as we go from one structure to another geometrically similar one. Usually F_c is selected to be one of the forces—say F_1—in which case $F_c = F_1 = F$, and $q_1 = 1$.

In the above table we have six variables and three dimensions. Therefore there are 6 − 3 = 3 dimensionless variables supplied by the Dimensional Set.

	F	M	ρ	g	L	σ
m	1	2	−3	1	1	−1
kg	1	1	1	0	0	1
s	−2	−2	0	−2	0	−2
π_1	1	0	0	0	−2	−1
π_2	0	1	0	0	−3	−1
π_3	0	0	1	1	1	−1

(The first four columns headed F, M, ρ, g are independent; L is independent; σ is dependent.)

from which

$$\pi_1 = \frac{F}{L^2 \cdot \sigma}; \qquad \pi_2 = \frac{M}{L^3 \cdot \sigma}; \qquad \pi_3 = \frac{\rho \cdot g \cdot L}{\sigma} \qquad (b)$$

Note that we have only *one* dependent variable, σ, which appears (contrary to our often-repeated advice) not in the **B** matrix, but in the **A** matrix. Why? Because we wish to *isolate* the three "*causing agents*" (F, M, and ρ) responsible for the stress. If they are put in the **B** matrix, then, by necessity, they will appear only in *one* dimensionless variable—and this is exactly what we want. Indeed as we see in (b)

F appears only in π_1

M appears only in π_2

ρ appears only in π_3

We can proceed:

- In a structure if the effects of external forces and moments are negligible, i.e., the stresses are generated mainly by the structure's *weight* (e.g., a large bridge), then π_1 and π_2 may be disregarded and only π_3 should be considered as relevant. Hence π_3 = const, and by the third part of (b),

$$\sigma = \text{const} \cdot \rho \cdot g \cdot L \qquad (c)$$

i.e., stress is proportional to the linear size. In a structure which is twice as large as another geometrically similar one (and of the same material), the stress caused by its own weight will also be twice as much.

- In a structure if the effects of its own weight and external forces are negligible, i.e., stresses are generated by external *moments* only, then by (b), π_1 and π_3 may be ignored, and only π_2 remains relevant. Hence π_2 = const, or by the second part of (b),

$$\sigma = \text{const} \cdot \frac{M}{L^3} \qquad (d)$$

i.e., the stresses are inversely proportional to the *cube* of linear size. In a structure A which is three times the size of a geometrically similar smaller one B, the stresses generated in A by the *same* external moment will be less by a factor of 27.

- In a structure if the effects of its own weight and external moments are negligible, i.e., the stresses are generated by external *forces* only, then in (b) π_2 and π_3 may be ignored, and therefore only π_1 remains relevant. Hence π_1 = const, thus, by the first part of (b),

$$\sigma = \text{const} \cdot \frac{F}{L^2} \qquad (e)$$

i.e., the stresses are inversely proportional to the *square* of linear size. In a structure A which is twice the size of a geometrically similar smaller one B, the stresses generated in A by *identical* (i.e., unchanging) external force(s) are less by a factor of 4.

Note that these extremely useful conclusions could be derived very easily *because* the *dependent* variables were *separated* by virtue of being in the **B** matrix.

⇑

14.4. PROBLEMS

Solutions are in Appendix 6.

14/1 Terminal speed of a mass sliding down a nonfrictionless inclined surface. A given mass M is sliding down a nonfrictionless surface. The surface is inclined to horizontal φ radian. Fig. 14-7 shows the arrangement.

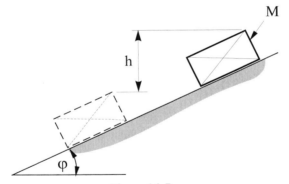

Figure 14-7
Mass sliding down a nonfrictionless inclined surface

The terminal speed upon traveling h *vertical* distance is defined by constant gravitational acceleration g, friction coefficient μ, and inclination φ. On the other hand, frictional energy E_f is a function of all of the above, plus mass M.

(a) Can either μ or φ, or both, be part of the **A** matrix of the Dimensional Set containing all the above physical variables?
(b) Of the cited physical variables, which is (are) dependent, and which is (are) independent?
(c) Construct a Dimensional Set comprising all of the cited physical variables.

- **(d)** Define a set of independent dimensionless variables on the basis of the Dimensional Set determined in (c).
- **(e)** Define the possible valid relation(s) among the dimensionless variables obtained in (d).

14/2 The **C** matrix of a Dimensional Set is

$$\mathbf{C} = \begin{bmatrix} 2 & 1 & 3 & 2 \\ 4 & -3 & 2 & 0 \\ 0 & 0 & 0 & 0 \end{bmatrix}$$

- **(a)** How many physical variables are there?
- **(b)** How many dimensions are there?
- **(c)** How many dimensionless physical variables are there?
- **(d)** If there is (are) dimensionless physical variable(s), what is (are) its (their) position(s) in the **B** matrix?
- **(e)** Are there two or more physical variables in the **A** matrix whose dimensions are identical?
- **(f)** Are there two or more physical variables whose dimensions are identical?

14/3 Surface area, volume, and weight of a right circular cylinder. Given a solid, right circular cylinder.

- **(a)** Establish relations among the *independent* variables of diameter, height, density and gravitational acceleration, and the *dependent* variables of surface area, volume and weight.
- **(b)** Construct the Dimensional Set incorporating all the physical variables, and then identify the dependent and independent physical variables.
- **(c)** Construct a set of independent dimensionless variables based upon the Dimensional Set established in (b).
- **(d)** Construct independent dimensionless relations based upon the results of (a) and (c).
- **(e)** Construct a single relation in which *all* the dimensionless variable established in (d) participate.
- **(f)** Prove, by way of a numeric example, that the relation obtained in (e) can lead to wrong results, and then give reason for it.

CHAPTER 15
ALTERNATE DIMENSIONS

There are problems in which dimensions other than the "conventional" mass, force, length, time, etc. occur. Can these problems be treated by methods so far expounded in these pages, even though the dimensional systems hitherto introduced do not have the "special" units needed to treat these problems? The answer is "yes." Nothing prevents us from introducing and using whatever dimension we wish, provided it fulfills the following three conditions:

Condition 1. The dimension has a physical meaning, i.e., the affected variable is *quantifiable*.
Condition 2. The dimension is used *consistently* in all relevant variables.
Condition 3. The dimension occurs in at least *two* variables.

Condition 1 is justified because if there is no physical meaning, i.e., the variable in which the dimension appears is not quantifiable, then, obviously, no *numerical* results can be obtained. For example, "fear" is not a quantifiable variable and therefore any dimension to describe it is meaningless. We cannot say *"Mary's fear of snakes is 2.45 times as much as John's."* For if we could, then we would first have to agree on the use of a *dimension* of fear, which—as far as the author knows—does not yet exist.

Condition 2 stems from the obvious requirement (elaborated on in Chapter 2) that for *all* variables the *same* applicable dimensions are used. Thus if length is expressed in feet, then speed cannot be expressed in inches per second.

Condition 3 emanates from Theorem 11-1 (Art. 11.1) which states in essence that if a variable has a fundamental dimension that no other variable has, then this variable cannot be part of a relation encompassing the other variables. That is, if this condition is not fulfilled, then the variable in question is *dimensionally irrelevant* and hence it must be left out. A further consequence of this is that the dimension that occurs *only* in this discarded variable cannot appear in the dimensional matrix, either.

We emphatically note, however, that it is *possible* to use alternate dimension(s) *together* with other "conventional" dimensional systems, most notably with SI.

The following examples introduce and illustrate the use of some alternate dimensions.

Example 15-1. Charge of Free Electrons in a Wire

Given a metallic conductor wire. We wish to determine the total charge of free electrons per unit length. The following variables and their dimensions are considered:

Variable	Symbol	Dimension	Remark
total charge of electrons	q	A·s/m	per unit length
area of cross-section	b	m²	
free electrons	n	c/m³	per unit volume
charge of electron	e	A·s/c	

Note that dimension n includes "c" which stands for "count." Thus the dimension of n is c per unit volume, or formally $[n]$ = c/m³. Here we introduced an *alternate* dimension—"count"—which in this case refers to electrons. Accordingly, the dimensional matrix is

	q	b	n	e
c	0	0	1	−1
m	−1	2	−3	0
s	1	0	0	1
A	1	0	0	1

and we see that the rank of this matrix cannot be 4 since two rows are identical. The fact that the rank must be 3 can be verified at once by considering the top-right 3 × 3 determinant whose value is −2. We therefore have for rank R_{DM} = 3, and for dimensions N_d = 4. Thus, the quantity Δ, defined in relation (7-22), Art. 7.6, is

$$\Delta = N_d - R_{DM} = 4 - 3 = 1$$

and therefore one dimension must be deleted. Obviously this dimension cannot be either "c," or "m," because in this case two rows would remain identical. We choose "A" for deletion, and hence the Dimensional Set is

	q	b	n	e
c	0	0	1	−1
m	−1	2	−3	0
s	1	0	0	1
π_1	1	−1	−1	−1

from which the sole dimensionless variable—a constant in this case—is

$$\pi_1 = \frac{q}{b \cdot n \cdot e} = \text{const}$$

or

$$q = \text{const} \cdot b \cdot n \cdot e$$

In other words, the charge that free electrons represent is a linear function of the *product* of the cross-sectional area, the number of free electrons per unit volume, and the elementary charge.

Example 15-2. Compound Interest

We wish to find out the amount of money accumulated in a bank savings account over a given length of time. The bank pays *interest* on the *existing* amount on account q times per year. This is called *compounding,* and q is the *compounding frequency.* Thus, if interest is paid yearly, every 6 months, every 4 months, every 3 months, or monthly, then q is 1, 2, 3, 4, or 12, respectively. The relevant variables and their dimensions are as follows:

Variable	Symbol	Dimension	Remark
amount available	A	$	
elapsed time	n	year	
compounding frequency	q	1/year	
interest rate	i	1/year	
initial deposit	A_0	$	also called "principal"

We now elaborate upon these dimensions:

- *Elapsed time n* is measured in *years*. Note that n does not have to be an integer. For example, for 3 months we have $n = 0.25$.
- Compounding frequency q is the number of *times* per *year* interest is added to the money on account. Therefore the dimension of q is $[q] = 1/\text{year}$.
- Interest rate i is the amount of money paid per year for each dollar on account. Therefore the dimension of i is $[i] = \$/(\$\cdot\text{year}) = 1/\text{year}$. For example, if there is 22,000 \$ on account, and the interest paid is 1870 \$/year, then the interest rate is $i = 1870/22{,}000 = 0.085$ 1/year. Note that i is not "percent" p, although it can also be expressed as such. In particular

$$p = 100 \cdot i \qquad (a)$$

At this point, the reader might be interested to learn a bit of completely useless jargon: interest rate i is also called "perunage" by at least one professional economist (Ref. 36, p. 79).

- Initial deposit A_0—called "principal"—is the amount of money given to the bank to earn interest; hence its dimension is \$.

From the above, we have five variables and two dimensions—none of which, incidentally, is part of SI. Thus, there are $5 - 2 = 3$ dimensionless variables supplied by the Dimensional Set.

	A	n	q	i	A_0
$	1	0	0	0	1
year	0	1	−1	−1	0
π_1	1	0	0	0	−1
π_2	0	1	0	1	0
π_3	0	0	1	−1	0

which yields

$$\pi_1 = \frac{A}{A_0}; \qquad \pi_2 = n \cdot i; \qquad \pi_3 = \frac{q}{i} \qquad (b)$$

Thus we can write

$$\pi_1 = \Psi\{\pi_2, \pi_3\} \qquad (c)$$

where Ψ is an as-yet undefined function. By the well-known formula (e.g., Ref. 36, p. 80)

$$A = A_0 \cdot \left(1 + \frac{i}{q}\right)^{n \cdot q} \qquad (d)$$

which by (b), can be written

$$\pi_1 = \left(1 + \frac{1}{\pi_3}\right)^{\pi_2 \cdot \pi_3} \qquad (e)$$

Note that whereas (d) contains five variables, (e) contains only three. Therefore the latter can be plotted on a single chart, as shown in Fig. 15-1.

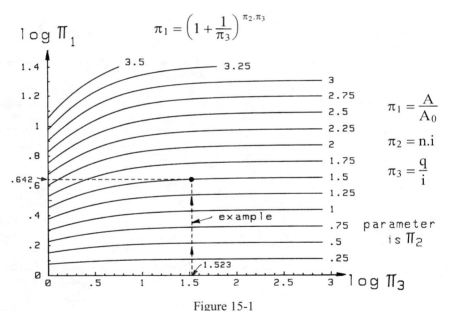

Figure 15-1
Dimensionless plot for the relative growth of money earning compound interest
For definitions of symbols, see table in this example; the logarithmic scales are of base 10

The use of this chart is illustrated by an example: Suppose Mr. Higgs deposits $A_0 = 24{,}500$ \$ at fixed interest rate $i = 0.12$ ($p = 12\%$) for $n = 12.5$ years. The bank compounds his money every 3 months, i.e., $q = 4$ 1/year. How much has Mr. Higgs on account at the end of his 12.5 year term? By (b),

$$\pi_2 = n \cdot i = (12.5) \cdot (0.12) = 1.5$$

$$\pi_3 = \frac{q}{i} = \frac{4}{0.12} = 33.33333$$

Hence log π_3 = 1.52288. This is the abscissa of the plot. The parameter is π_2 = 1.5. With these data, the plot provides log π_1 = 0.642. Hence π_1 = $10^{0.642}$ = 4.38531. Since $\pi_1 = A/A_0$, therefore Mr. Higgs will receive $A = \pi_1 \cdot A_0$ = (4.38531)·(24,500) = 107,440 \$ [actually, the exact formula (d) provides 107,405.70 \$, which is within 0.03% of the value provided by the plot]. This numeric example is shown on the plot by dashed lines.

Finally, observe that each π_2 = const line in the chart apparently reaches an asymptotic maximum for π_1. This π_2-specific maximum is

$$(\pi_1)_{max} = \lim_{\pi_3 \to \infty} \pi_1 = e^{\pi_2} \tag{f}$$

where e = 2.71828 ... is the base of natural logarithm. Relation (f) follows from one of the definitions of e. Obviously, $(\pi_1)_{max}$ represents the *theoretical maximum* earnings of the money in the bank. This would happen if the money were compounded *infinitely* often.

We now define k as the *ratio* of the actual to the theoretical maximum π_1 values. We call this ratio the *Principal's Utilization Factor* (note the apostrophe). If k = 1, the principal is utilized to the maximum extent. If $k < 1$, then the bank does not pay as much as it could (or should) for the use of the client's money. By (e) and (f), we can write

$$k = \frac{\pi_1}{(\pi_1)_{max}} = \frac{1}{e^{\pi_2}} \cdot \left(1 + \frac{1}{\pi_3}\right)^{\pi_2 \cdot \pi_3} \tag{g}$$

where the definitions of dimensionless variables are as given in (b). The chart in Fig. 15-2 presents the plot of relation (g), with parameter π_2.

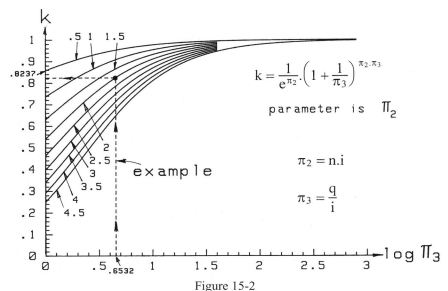

Figure 15-2
**Dimensionless plot for Principal's Utilisation Factor k
of money earning compound interest**
See table in this example for definition of symbols; the logarithmic scale is of base 10

To illustrate, consider the following example: a man buys a $A_0 = 3600$ \$ bond locked in for 18 years. The bank pays $i = \frac{1}{9}$ ($= 11.111\%$) interest compounded biennially (every second year), so that $q = 0.5$ 1/year. What is the *Principal's Utilization Factor k*? How much will our man get at the end of the term, and how much *would* he have got had k been 1? By (b), and the given data

$$\pi_2 = n \cdot i = 18 \cdot \left(\frac{1}{9}\right) = 2; \qquad \pi_3 = \frac{q}{i} = \frac{0.5}{\left(\frac{1}{9}\right)} = 4.5$$

With these values (g) provides

$$k = (0.13534) \cdot \left(1 + \frac{1}{4.5}\right)^9 = 0.82369$$

That is, our man gets only 82.369 % of the money he would have got, had the bank compounded his money more frequently. As it stands, by (e), $\pi_1 = 6.08628$ and therefore he gets

$$A = A_0 \cdot \pi_1 = (3600) \cdot (6.08628) = 21{,}910.59 \text{ \$}$$

However, if his deposit of $A_0 = 3600$ \$ had been compounded continuously (i.e., much more frequently), he would have received, by (f),

$$A = A_0 \cdot e^{\pi_2} = (3600) \cdot e^2 = 26{,}600.60 \text{ \$}$$

Thus, he gets 4690.01 \$ less than he could have. This case is shown in Fig. 15-2 by dashed lines.

⇑

Example 15-3. Optimal Density and Location Pattern for Retail Shops

In an area where the population is evenly distributed, the question naturally arises: what is the density of retail shops (i.e., retail shops per unit area) that *minimizes* the *maximum* distance customers must travel to reach the *nearest* store. The "obvious" answer that immediately comes to mind is to have a very large shop density resulting in shops situated very close to each other. In this scheme people would only have to travel very short distances to reach a store. But unfortunately this simplistic pattern would not work, for beyond a particular number of stores per unit area—the *critical density*—the population's buying power would not sustain all the shops. Therefore the question really should be: what is the maximum shop density (the critical density) that a given population (having a set of relevant characteristics) could support? Moreover, having found this density, what is (are) the location pattern(s) of the stores, and what would be the *maximum* and *average* distances customers would have to travel to reach the nearest store?

Again we will use the dimensional method. Thus first, as always, we list the relevant variables and their dimensions:

Variable	Symbol	Dimension	Remark
shop density	ρ_s	n_s/m^2	
population density	ρ_p	n_p/m^2	
threshold of good	q	$\$/(n_s \cdot \text{day})$	
spending rate	σ	$\$/(n_p \cdot \text{day})$	time invariant

The reader will notice some unusual dimensions here. Specifically:

- n_s is the dimension of number of shops;
- n_p is the dimension of number of persons (potential customers);
- \$ is the dimension of money (dollars);
- "day" is the dimension of time.

Moreover, the variables also require some explanation:

- ρ_s (shop density) is the number of shops per unit area;
- ρ_p (population density) is the number of persons (potential customers) per unit area;
- σ (spending rate) is the average amount of money each customer spends per unit time;
- q (threshold of good) is the minimum amount of gross sales (in dollars per store and per unit time) sufficient to keep the store in business;

From the above we construct the dimensional matrix

	ρ_s	q	ρ_p	σ
n_p	0	0	1	−1
n_s	1	−1	0	0
\$	0	1	0	1
m	−2	0	−2	0
day	0	−1	0	−1

We can see that the number of dimensions is $N_d = 5$, and the rank of the dimensional matrix is $R_{DM} = 3$. Therefore, quantity Δ defined by (7-22) in Art. 7.6 is $\Delta = N_d - R_{DM} = 5 - 3 = 2$ and consequently, we have to jettison two dimensions, say "m" and "day." We now have four variables and three dimensions, therefore there is but one dimensionless variable—a constant in this case—supplied by the Dimensional Set:

	ρ_s	q	ρ_p	σ
n_p	0	0	1	−1
n_s	1	−1	0	0
\$	0	1	0	1
π_1	1	1	−1	−1

from which

$$\pi_1 = \frac{\rho_s \cdot q}{\rho_p \cdot \sigma} = \text{const} \qquad (a)$$

or

$$\rho_s = \text{const} \cdot \frac{\rho_p \cdot \sigma}{q} \qquad (b)$$

where the constant, by definition, is 1. This relation tells us—as expected—that the critical shop density is proportional to the product of population density and average buying power of the public (spending rate), and inversely proportional to the "threshold of good." This latter conclusion simply means that the greater the "overhead," the fewer stores a given population can support.

Next, we consider the patterns of shop locations. If the area is large enough, then ρ_s = constant, i.e., the distribution of shops is *isotropic*. This condition is satisfied only if the shops are located at the vertices of adjoined *equilateral triangles,* or *squares,* or *regular hexagons* (Fig. 15-3). Reason: by regular polygons an infinite plane can be tessellated *only* if the polygon is any of the above plane figures (the reader may find the proof of this assertion an "interesting" task).

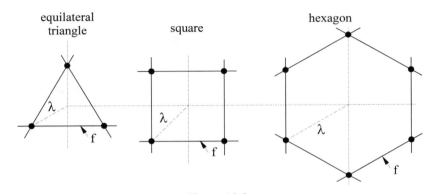

Figure 15-3
The three possible isotropic patterns for shop locations
λ = maximum distance customer must travel to reach nearest shop;
f = distance between adjacent shops; • indicates shop location

It is now instructive to find the *maximum* distance λ that a customer has to travel to reach the *nearest* shop. For example, in the case of triangular locations of shops, this distance is obviously 2/3 of the altitude of the triangle; for a square, it is half the diagonal, and for a hexagon it is the radius of the circumscribed circle. If these distances are now expressed as a function of separation f of adjacent shops, then we get the relations found in the third row of the table in Fig. 15-4. For example, for a triangular pattern, λ is 57.7 % of shop separation f. This number is 70.7% for a square pattern, and 100% for hexagonal arrangement. It follows that from the customers' point of view, the triangular pattern is the best.

In the last three rows of the table in Fig. 15-4, the maximum distance λ a customer must travel to reach the nearest shop is given in terms of the physical variables introduced. For example, in the case of the triangular shop allocation,

$$\lambda = \frac{\sqrt{2}}{\sqrt[4]{27}} \cdot \sqrt{\frac{q}{\rho_p \cdot \sigma}} = (0.62) \cdot \sqrt{\frac{q}{\rho_p \cdot \sigma}} \qquad (c)$$

This means—rather interestingly—that λ varies as a square root of q (threshold of good), i.e., if by some miracle q decreases two-fold (for example by increasing productivity), it will translate for the customer into reducing his maximum length of travel by 41.4%. The table also shows that the triangular arrangement is the most advantageous for the customer; everything else being identical, the maximum distance travelled by a customer in a triangular shop allocation pattern is only 70.7% of that in a hexagonal pattern.

So far we have only considered the *maximum* distance customers must travel to reach the nearest shop. But it might also be interesting to know the *average* of such value. This value, of course, must be based on a large number of customers, as opposed to a maxi-

ALTERNATE DIMENSIONS 409

	Equilateral triangle	Square	Regular hexagon
A	$\frac{\sqrt{3}}{4} \cdot f^2 = (0.433) \cdot f^2$	f^2	$\frac{\sqrt{27}}{2} \cdot f^2 = (2.598) \cdot f^2$
ρ_s	$\frac{1}{2 \cdot A} = \frac{2}{\sqrt{3}} \cdot \frac{1}{f^2} = \frac{1.155}{f^2}$	$\frac{1}{A} = \frac{1}{f^2}$	$\frac{2}{A} = \frac{4}{\sqrt{27}} \cdot \frac{1}{f^2} = \frac{0.770}{f^2}$
λ	$\frac{1}{\sqrt{3}} \cdot f = (0.577) \cdot f$	$\frac{1}{\sqrt{2}} \cdot f = (0.707) \cdot f$	f
λ	$\left(\frac{4}{27}\right)^{0.25} \cdot \frac{1}{\sqrt{\rho_s}} = \frac{0.620}{\sqrt{\rho_s}}$	$\frac{1}{\sqrt{2}} \cdot \frac{1}{\sqrt{\rho_s}} = \frac{0.707}{\sqrt{\rho_s}}$	$\left(\frac{16}{27}\right)^{0.25} \cdot \frac{1}{\sqrt{\rho_s}} = \frac{0.877}{\sqrt{\rho_s}}$
λ	$\left(\frac{4}{27}\right)^{0.25} \cdot \sqrt{\frac{q}{\rho_p \cdot \sigma}} = \frac{0.62 \cdot \sqrt{q}}{\sqrt{\rho_p \cdot \sigma}}$	$\frac{1}{\sqrt{2}} \cdot \sqrt{\frac{q}{\rho_p \cdot \sigma}} = (0.707) \cdot \sqrt{\frac{q}{\rho_p \cdot \sigma}}$	$\left(\frac{16}{27}\right)^{0.25} \cdot \sqrt{\frac{q}{\rho_p \cdot \sigma}} = \frac{(0.877) \cdot \sqrt{q}}{\sqrt{\rho_p \cdot \sigma}}$

Figure 15-4
Some geometric characteristics of shop location patterns seen in Fig. 15-3
A = area of respective pattern seen in Fig. 15-3; f = distance between adjacent shops;
λ = maximum distance customer to travel to nearest shop;
for definitions of other symbols, see table in this example

mum value, which, by its nature, may be based on only a very few. In Fig. 15-5 the respective average values are given in terms of distance between adjacent shops (derivations are left as a *very* rewarding exercise for the tireless and imaginative reader having a free afternoon, or several).

The formulas given in Fig. 15-5 for the average distances λ_{av} are rather complex and quite unexpected. For example, how does logarithm get involved in defining a *distance in a square*? Confirming these relations was therefore deemed advisable by the author, so he

Shop allocation pattern	Average travel for customers to reach nearest shop
equilateral triangle	$\lambda_{av} = \dfrac{4 + 3 \cdot \ln 3}{12 \cdot \sqrt{3}} \cdot f = (0.35102) \cdot f$
square	$\lambda_{av} = \dfrac{\sqrt{2} + \ln(1 + \sqrt{2})}{6} \cdot f = (0.3826) \cdot f$
regular hexagon	$\lambda_{av} = \dfrac{2 \cdot \sqrt{3} + \ln(2 + \sqrt{3})}{6 \cdot \sqrt{3}} \cdot f = (0.46006) \cdot f$

Figure 15-5
Average travel for customers to reach nearest shop for the three possible isotropic shop allocation patterns
f = distance between adjacent shops

ran 100,000 Monte Carlo simulations. He found that for the triangular, square, and hexagonal patterns, the simulated results were within 0.048%, 0.034% and 0.009%, respectively, to the theoretically predicted values. Therefore it may be said that the formulas—strange as they come—are correct.

It is interesting that the results tabulated are somewhat counterintuitive. For one might argue that the average distance should be longer—not shorter—than half of the distance between adjacent shops, since the great majority of customers do not live directly on the straight line connecting neighboring shop locations.

⇑

Finally we present an example which demonstrates that great care is required to include *all* the relevant variables when using the dimensional technique. Full attention is especially important if we are dealing with *unconventional* dimensions, since in these cases—due to lack of experience—our judgment is less acute in identifying all the influencing factors.

Example 15-4. Land Price

We wish to derive a relation for the price of land using the dimensional method. For this task, we consider the following relevant variables:

Variable	Symbol	Dimension
land price	P	$/m^2$
population density	ρ	n/m^2
average income	i	$/(n \cdot year)$
inflation rate	R	$1/year$

where

n is the dimension of population (1 n = 1 individual);

$ is the dimension of money (dollar).

The dimensional matrix therefore is

	P	ρ	i	R
n	0	1	−1	0
$	1	0	1	0
m	−2	−2	0	0
year	0	0	−1	−1

We have four dimensions and four variables, therefore this matrix must be singular, otherwise there is no solution. The determinant of the above matrix is zero, so there is a solution. Thus, we can eliminate one dimension, say "n." The resultant Dimensional Set is then

	P	ρ	i	R
$	1	0	1	0
m	−2	−2	0	0
year	0	0	−1	−1
π_1	1	−1	−1	1

from which

$$\pi_1 = \frac{P \cdot R}{\rho \cdot i} = \text{const} \tag{a}$$

or

$$P = \text{const} \cdot \frac{\rho \cdot i}{R} \tag{b}$$

This is a very nice and confidence-generating relation, simple and smooth. Yet, it is nonsense! Why? Because it was derived by leaving out several relevant variables, e.g., available land area, general economic conditions, political climate, mortgage rates, etc.

This example demonstrates that a relation derived by the dimensional method may appear quite legitimate—especially if some unconventional dimensions or variables are involved—and yet be patently false. For example, in the present case, our impressive-looking formula (b) states that increasing inflation R decreases the price of land—which is contrary to both common sense and experience.

CHAPTER 16
METHODS OF REDUCING THE NUMBER OF DIMENSIONLESS VARIABLES

It was mentioned in Art. 7.9 that the smaller the number N_P of independent dimensionless variables describing the behavior of a physical system, the better. The reason is simple: fewer variables make the expression easier to handle both analytically and graphically. *Very* significant labor savings can be achieved when plotting a relationship which has fewer variables. For example, if we have eight values for each variable, then to plot a three-variable function we need eight curves presented on one chart, but a five-variable function would need 512 curves plotted on 64 (!) charts (this subject was discussed in detail in Chapter 12; in particular see Fig. 12-3). Also, and maybe even more importantly, there is a corresponding labor saving in the experimentation needed to determine the dependent variable. Therefore, it is of *paramount* importance to reduce the number of variables as much as possible.

The *first* step to achieve this is to introduce dimensionless variables—which is *the* reason these variables are used at all. As we saw in relation (7-26), Art. 7.7, the number of dimensionless variables N_P is

$$N_P = N_V - R_{DM} \qquad \text{repeated (7-26)}$$

where N_V is the number of physical variables, and R_{DM} is the rank of the dimensional matrix. Under ordinary circumstances R_{DM} equals the number of dimensions N_d, thus

$$N_P = N_V - N_d \qquad (16\text{-}1)$$

Therefore the number of dimensionless variables is the *difference* between the number of physical variables and the dimensions. The best possible situation is when N_P is minimum, i.e., $N_P = 1$ [see relation (7-32)]. This brings us to the *second* step. How do we reduce N_P? Relation (16-1) offers two ways:

- reduce the number of physical variables N_V;
- increase the number of dimensions N_d;

414   APPLIED DIMENSIONAL ANALYSIS AND MODELING

and to this, we may add a third way, which on occasion can be used effectively to improve the logistics of graphical presentation:

- fuse dimensionless variables.

In what follows we describe these methods in some detail and demonstrate their use by several illustrative examples. The reader is strongly urged to follow these illustrations closely.

16.1. REDUCTION OF THE NUMBER OF PHYSICAL VARIABLES

The method of reducing the number of physical variables is known as *variable fusion*. It is essentially based on *a priori* knowledge of how two (or rarely, more) variables appear *as a group* in the formula connecting all variables. This group of variables can then be assigned a *name,* which then becomes a *new* variable replacing all the original ones. In this way, the number of variables is reduced by (at least) one. The following four examples demonstrate the efficacy of this process.

Example 16-1.   Terminal Velocity of a Raindrop

In order to determine the terminal velocity of a raindrop, we assume

- the drop is a sphere of constant diameter;
- gravitational acceleration is constant;
- air density is constant.

The following are the relevant variables:

Variable	Symbol	Dimension
terminal speed	v	m/s
dynamic viscosity of air	μ	kg/(m·s)
radius of drop	R	m
density of water	ρ	kg/m³
gravitational acceleration	g	m/s²

We have five variables and three dimensions, therefore there are $5 - 3 = 2$ dimensionless variables furnished by the Dimensional Set.

	v	μ	R	ρ	g
m	1	−1	1	−3	1
kg	0	1	0	1	0
s	−1	−1	0	0	−2
π_1	1	0	$-\dfrac{1}{2}$	0	$-\dfrac{1}{2}$
π_2	0	1	$-\dfrac{3}{2}$	−1	$-\dfrac{1}{2}$

from which

$$\pi_1 = \frac{v}{\sqrt{R \cdot g}}; \quad \pi_2 = \frac{\mu}{\rho \cdot \sqrt{R^3 \cdot g}} \qquad (a)$$

Thus, we can write

$$v = \sqrt{R \cdot g} \cdot \Psi\left\{\frac{\mu}{\rho \cdot \sqrt{R^3 \cdot g}}\right\} \qquad (b)$$

where Ψ is an unknown function.

This formula does not tell us much. However, we observe that the resistance F influencing the terminal speed of the raindrop is defined by only v, μ, and R, and at equilibrium, of course, the resistance must be equal to the *weight* G of the drop. Thus

$$F = G = \frac{4}{3} \cdot \pi \cdot R^3 \cdot \rho \cdot g \qquad (c)$$

It follows that ρ and g in the Dimensional Set can be replaced by G, i.e., two variables can be *fused* into *one*, namely G. Accordingly, we now have the Dimensional Set

	v	μ	R	G
m	1	−1	1	1
kg	0	1	0	1
s	−1	−1	0	−2
π_1	1	1	1	−1

in which we have only *one* dimensionless variable—a constant. We write

$$\pi_1 = \frac{v \cdot \mu \cdot R}{G} = \text{const}$$

from which

$$v = \text{const} \cdot \frac{G}{R \cdot \mu} \qquad (d)$$

We now again consider G [in relation (c)], by which (d) can be written

$$v = \frac{4 \cdot \pi}{3} \cdot \frac{R^2 \cdot \rho \cdot g}{\mu} \qquad (e)$$

Note that (e) is a vastly more *informative* formula than the "original" relation (b), for (e) tells us what (b) did not: *how* the terminal speed varies with *all* the determining variables. In particular, we now know that the speed is *proportional* to the *square* of the radius of the raindrop. This fact is rather startling, since one would have expected that the larger the surface (i.e., R^2), the larger the resistance and hence the lower the terminal speed. But, as (e) attests, this is not so: a drop twice as large falls four times as fast.

⇑

Example 16-2. Mass of Fluid Flowing Through a Tube (I)

We want to determine the quantity (mass) of an incompressible liquid flowing through a tube in unit time. The tube is of uniform circular cross-section. The following physical variables are involved:

Variable	Symbol	Dimension	Remark
mass flow	Q	kg/s	per unit time
pressure difference	Δp	kg/(m²·s²)	per unit length
density of liquid	ρ	kg/m³	constant
viscosity of liquid	μ	kg/(m·s)	dynamic
diameter of tube	D	m	inner

We have five variables and three dimensions, therefore there are 5 – 3 = 2 dimensionless variables furnished by the Dimensional Set

	Q	Δp	ρ	μ	D
m	0	–2	–3	–1	1
kg	1	1	1	1	0
s	–1	–2	0	–1	0
π_1	1	0	0	–1	–1
π_2	0	1	1	–2	3

from which

$$\pi_1 = \frac{Q}{\mu \cdot D}; \qquad \pi_2 = \frac{\Delta p \cdot \rho \cdot D^3}{\mu^2} \qquad (a)$$

Thus, we may write

$$Q = \mu \cdot D \cdot \Psi\left\{\frac{\Delta p \cdot \rho \cdot D^3}{\mu^2}\right\} \qquad (b)$$

where Ψ is an as-yet unknown function. But this relation is useless without knowing Ψ, and to learn this, extensive experimental or analytical work would be required. Can something be done to improve the situation? Yes; we can consider the *volume* instead of the *mass* of the fluid. Thus we write

$$V = \frac{Q}{\rho} \qquad (c)$$

where V is the required *new* variable (dimension: m³/s) incorporating both Q and ρ, which are now *fused* into the (new) single variable V. Accordingly, we have the modified Dimensional Set

	V	Δp	μ	D
m	3	–2	–1	1
kg	0	1	1	0
s	–1	–2	–1	0
π_1	1	–1	1	–4

yielding a *single* dimensionless variable—a constant

$$\pi_1 = \frac{V \cdot \mu}{\Delta p \cdot D^4} = \text{const} \qquad (d)$$

from which

$$V = \text{const} \cdot \frac{\Delta p \cdot D^4}{\mu}$$

and by (c)

$$Q = \text{const} \cdot \frac{\rho \cdot \Delta p \cdot D^4}{\mu} \qquad (e)$$

This is of course a *significantly* more informative formula, than the "original" (b), for it gives us the important fact that the mass flow of a fluid through a tube upon a pressure gradient Δp is proportional to the fourth power of the inner diameter of that tube [formula (e) is known as the *Poiseuille* equation, named after the French physicist Jean Louis Marie Poiseuille, 1799–1869].

⇑

Example 16-3.   Moment of Inertia of a Thin Lamina

A *thin* lamina, shown in Fig. 16-1, is rotated about its longitudinal axis of symmetry *k–k*. What is its polar moment of inertia about this axis?

We consider the following relevant variables:

Variable	Symbol	Dimension	Remark
moment of inertia	I	m²·kg	about axis *k–k*
length	L	m	along axis *k–k*
mass	q	kg/m²	per unit area
width	b	m	

We have four variables and two dimensions, thus there are two dimensionless variables supplied by the Dimensional Set

	I	L	q	b
m	2	1	−2	1
kg	1	0	1	0
π_1	1	0	−1	−4
π_2	0	1	0	−1

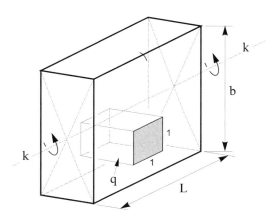

Figure 16-1
Thin lamina is rotated about its longitudinal axis of symmetry

yielding

$$\pi_1 = \frac{I}{q \cdot b^4}; \quad \pi_2 = \frac{L}{b} \tag{a}$$

Thus

$$I = q \cdot b^4 \cdot \Psi\left\{\frac{L}{b}\right\} \tag{b}$$

where Ψ is an as-yet unknown function. This formula is not very useful; it only tells us that *if* L/b is constant, then the inertia varies linearly with q, which is hardly a surprise. Can we do better?

Suppose we modify the question *What is the inertia?* to *What is the inertia per unit length?*, since a lamina which is twice as *long* represents twice as large inertia. Thus we *fuse* variables I and L into a *new* variable $\bar{I}$ such that

$$\bar{I} = \frac{I}{L} \tag{c}$$

thereby reducing the number of variables by one. The new Dimensional Set is now

	$\bar{I}$	q	b
m	1	-2	1
kg	1	1	0
π_1	1	-1	-3

yielding the *single* dimensionless variable $\pi_1 = \bar{I}/(q \cdot b^3) = \text{const}$, whereupon $\bar{I} = \text{const} \cdot q \cdot b^3$. By (c) now

$$I = \text{const} \cdot q \cdot L \cdot b^3 \tag{d}$$

We see that this relation is *much* more informative than (b), for (d) tells us what (b) does not, that the lamina's moment of inertia is *proportional* to the length which is *parallel* to the axis of rotation, and to the *cube* of the length, which is *perpendicular* to it.

⇑

Example 16-4. Path of an Electron in a Magnetic Field

Given magnetic field H perpendicular to plane P, as shown in Fig. 16-2. An electron with speed v—perpendicular to H—enters this field. By magnetic forces generated, the path of this electron will be a circle of radius R. Find R using the dimensional method.

We consider the following variables:

Variable	Symbol	Dimension	Remark
radius of orbit (path)	R	m	perpendicular to H
tangential speed	V	m/s	perpendicular to H
magnetic field strength	H	A/m	
charge of electron	e	A·s	
mass of electron	M	kg	
magnetic permeability	μ	m·kg/(A^2·s^2)	

We have six variables and four dimensions, therefore there are two dimensionless variables determined by the Dimensional Set

METHODS OF REDUCING THE NUMBER OF DIMENSIONLESS VARIABLES

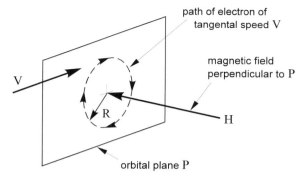

Figure 16-2
Path of an electron in a magnetic field

	R	V	H	e	M	μ
m	1	1	−1	0	0	1
kg	0	0	0	0	1	1
s	0	−1	0	1	0	−2
A	0	0	1	1	0	−2
π_1	1	0	0	−2	1	−1
π_2	0	1	−1	−3	2	−2

by which

$$\pi_1 = \frac{R \cdot M}{e^2 \cdot \mu}; \quad \pi_2 = \frac{V \cdot M^2}{H \cdot e^3 \cdot \mu^2} \tag{a}$$

By assuming *monomial* form—which is plausible—

$$R = c \cdot \frac{e^2 \cdot \mu}{M} \cdot \left(\frac{V \cdot M^2}{H \cdot e^3 \cdot \mu^2} \right)^n \tag{b}$$

where c and n are numeric constants that, in theory, can have any positive value. Why only positive? Obviously c cannot be either negative or zero, and hence it must be positive. For n, if it were zero, then R would be independent of V and H, which is impossible; if it were negative, then R would increase with magnetic intensity, and decrease with tangential speed. Both of these consequences are absurd. Therefore n must be positive.

Apart from these qualitative inferences, (b), as it stands, is rather useless. To improve the situation we now use *variable fusion*. It will reduce number of variables from six to five and, since the number of dimensions remains four, the number of dimensionless variables will be one—the ideal number!

Accordingly, we note that the first two variables in the dimensional set can be connected by

$$V = R \cdot \omega \tag{c}$$

where ω is the angular velocity of the orbiting electron. Therefore, instead of entering two variables R and V into the dimensional set, we will enter only one, namely ω. Thus, the new Dimensional Set is

	ω	H	e	M	μ
m	0	−1	0	0	1
kg	0	0	0	1	1
s	−1	0	1	0	−2
A	0	1	1	0	−2
π_1	1	−1	−1	1	−1

yielding

$$\pi_1 = \frac{\omega \cdot M}{H \cdot e \cdot \mu} = \text{const} \tag{d}$$

from which

$$\omega = \text{const} \cdot \frac{H \cdot e \cdot \mu}{M}$$

or, by (c) and simple rearrangement,

$$R = \text{const} \cdot \frac{M}{H \cdot e \cdot \mu} \cdot V \tag{e}$$

This formula is *significantly* more informative than (b)—a fact showing the great power of the method of *variable fusion*. To determine the constant, it is enough to execute *one* measurement.

⇑

There are times when the fusion of variables, although possible, is not recommended. This can happen when the number of dimensionless variables is already the *minimum*, namely $N_P = 1$. Therefore, reducing the number of physical variables would not improve the situation since, by relation (7-32), the number of dimensionless variables must remain 1.

But what would happen if we fuse two variables anyway, thus reduce the number of physical variables by one? Since $N_P = 1$, the number of variables must exceed the number of dimensions by exactly one. Hence, if the number of variables is reduced by one, then the number of variables and dimensions would become identical, and the dimensional matrix a square and singular. In this case the only thing to do would be to delete one dimension from the dimensional matrix. Thus both the number of variables and dimensions would be reduced by one and therefore the number of dimensionless variables (being the difference of these two magnitudes) would remain unaltered, i.e., one. We gained nothing! The following example demonstrates this futile process.

Example 16-5. Density and Size of the Universe

By the expansion of the universe from the *Big Bang*—the hypothesis proposed by the American astronomer Edwin Powell Hubble (1889–1953) in 1929—it is conceivable that the mass and size of the universe is somehow related to its age and the speed of light. Let us see what dimensional considerations alone can tell us, without reference to the related physics. To start, we list the relevant variables

METHODS OF REDUCING THE NUMBER OF DIMENSIONLESS VARIABLES   421

Variable	Symbol	Dimension	Remark
mass of universe	M	kg	
density of universe	ρ	kg/m^3	average value
speed of light	c	m/s	in vacuum
age of universe	t	s	reckoned from the Big Bang

We have four variables and three dimensions, thus there is 4 – 3 = 1 dimensionless variable—a constant; it is by the Dimensional Set

$$\begin{array}{c|cc|cc} & M & \rho & c & t \\ \hline m & 0 & -3 & 1 & 0 \\ kg & 1 & 1 & 0 & 0 \\ s & 0 & 0 & -1 & 1 \\ \hline \pi_1 & 1 & -1 & -3 & -3 \end{array}$$

yielding the only dimensionless variable

$$\pi_1 = \frac{M}{\rho \cdot c^3 \cdot t^3} = \text{const} \tag{a}$$

or, by assuming constant = 1 (since we are only interested in the order of magnitude),

$$M = \rho \cdot c^3 \cdot t^3 \tag{b}$$

This equation was first proposed by Sir Arthur Stanley Eddington, English astronomer and physicist (1882–1944), who derived it analytically.

Now let us employ *variable fusion,* and see where it leads us. We notice that volume V of the universe can be expressed as

$$V = \frac{M}{\rho} \tag{c}$$

Thus, variables M and ρ can be *fused* into one variable V. The modified dimensional matrix therefore will be

$$\begin{array}{c|ccc} & V & c & t \\ \hline m & 3 & 1 & 0 \\ kg & 0 & 0 & 0 \\ s & 0 & -1 & 1 \end{array}$$

which is *singular* (the second row is all zeros). From this we can infer that the system *has* a solution. To find this solution we must delete a dimension, which is obviously "kg." Hence, the Dimensional Set will be

$$\begin{array}{c|c|cc} & V & c & t \\ \hline m & 3 & 1 & 0 \\ s & 0 & -1 & 1 \\ \hline \pi_1 & 1 & -3 & -3 \end{array}$$

yielding *one* (still only one!) dimensionless variable

$$\pi_1 = \frac{V}{c^3 \cdot t^3} = \text{const} \tag{d}$$

or, by (c) and (d), $M/(\rho \cdot c^3 \cdot t^3) = \text{const}$, and hence

$$M = \text{const} \cdot \rho \cdot c^3 \cdot t^3 \tag{e}$$

which is identical—with const = 1—to Eddington's formula (b). Thus, in this case we did not gain anything by the method of variable fusion. Why? Because we had only one dimensionless variable originally—the minimum in *all* circumstances—and so this number could not be reduced further.

To develop the subject just a little further and show the usefulness of formula (b), consider the age of the universe between 10 and 20 billion years, say 15 billion. So, $t = 4.73 \times 10^{17}$ s. The speed of light is $c = 3 \times 10^8$ m/s. Therefore the mass of the universe, by (b), is

$$M = \rho \cdot (2.86 \times 10^{78}) \text{ kg} \tag{f}$$

or, since the volume is the ratio of mass to density

$$V = \frac{M}{\rho} = 2.86 \times 10^{78} \text{ m}^3 \tag{g}$$

If now we assume a *spherical* universe of diameter D, then $D = (6 \cdot V/\pi)^{1/3}$ and hence, by V calculated in (g),

$$D = \left(\frac{6 \cdot (2.86 \times 10^{78})}{\pi} \right)^{1/3} = 1.76 \times 10^{26} \text{ m} \tag{h}$$

which is then the "end-to-end" size of the universe, equalling about 18.6 billion light-years.

⇑

The foregoing examples might create an impression in the reader's mind that if $N_P > 1$ and a combination of variables is possible, then this combination will necessarily reduce the number of generated dimensionless variables. This impression is false. For a combination (fusion) to be useful (in this respect) it must appear in *exactly* the same format as in the analytically correct expression—regardless of whether this expression is known or available. In other words, the act of the fusion dictates that the fused variables themselves must *disappear* from the expression, whether this latter is available (i.e., known). This restriction means that if the chosen fusion (combination) does not appear exactly as defined in the analytical expression (reduced to the simplest form), then this fusion is worthless as it will not reduce the number of variables contained in the ensuing relation. The following example illustrates this sad fact.

Example 16-6. Deflection of a Cantilever upon a Concentrated Lateral Load (VII)

Consider the arrangement shown in Fig. 16-3. The beam is of constant cross-section, and the load F is at its free end and strictly lateral. The aim is to find the deflection U. The following variables are relevant:

METHODS OF REDUCING THE NUMBER OF DIMENSIONLESS VARIABLES

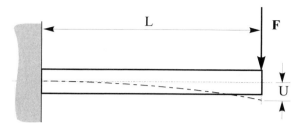

Figure 16-3
Cantilever is loaded laterally by a concentrated force

Variable	Symbol	Dimension	Remark
deflection	U	m	lateral
second moment of area of cross-section	I	m⁴	constant
concentrated load	F	N	lateral
length	L	m	
Young's modulus	E	N/m²	

We have five variables and two dimensions, hence there are 5 – 2 = 3 dimensionless variables, supplied by Dimensional Set 1.

Dimensional Set 1

	U	I	F	L	E
m	1	4	0	1	-2
N	0	0	1	0	1
π_1	1	0	0	-1	0
π_2	0	1	0	-4	0
π_3	0	0	1	-2	-1

which yields

$$\pi_1 = \frac{U}{L}; \quad \pi_2 = \frac{I}{L^4}; \quad \pi_3 = \frac{F}{L^2 \cdot E} \quad (a)$$

and we can write

$$\pi_1 = \Psi\{\pi_2, \pi_3\} \quad (b)$$

where Ψ is an as-yet unknown function. We know the simple analytically derived relation

$$U = \frac{1}{3} \cdot \frac{F \cdot L^3}{I \cdot E} \quad (c)$$

Thus (b) can be represented by the defined [in (a)] dimensionless variables as

$$\pi_1 = \frac{\pi_3}{3 \cdot \pi_2} \quad (d)$$

comprising three dimensionless variables which can be plotted on one chart having a number of curves.

424 APPLIED DIMENSIONAL ANALYSIS AND MODELING

We now want to reduce the number of dimensionless variables. How do we do that? We see that in (c) *I* and *E* appear as a product. This suggests the *fusion* of *I* and *E* into *one* variable *Z*, usually called "rigidity" (or "flexural stiffness")

$$Z = I \cdot E \quad \text{(e)}$$

whose dimension is m²·N. If we substitute *Z* into (c) we obtain

$$U = \frac{1}{3} \cdot \frac{F \cdot L^3}{Z}$$

in which neither *I* nor *E* appears. Therefore the fusion of *I* and *E* is feasible since it reduces the number of variables by one and, as a consequence, the number of *dimensionless* variables by one, also.

With this new variable now, Dimensional Set 2 will be

Dimensional Set 2

	U	F	L	Z
m	1	0	1	2
N	0	1	0	1
π_1	1	0	−1	0
π_2	0	1	2	−1

yielding *two* dimensionless variables (one less than by Dimensional Set 1):

$$\pi_1 = \frac{U}{L}; \quad \pi_2 = \frac{F \cdot L^2}{Z} \quad \text{(f)}$$

We fit these into our analytical formula (c), which yields

$$\pi_1 = \frac{\pi_2}{3} \quad \text{(g)}$$

As this relation contains only two dimensionless variables, it can be plotted by a *single* curve.

But now we try to *fuse* variables *F* and *L*, by introducing the *new* variable (dimension: N·m)

$$q = F \cdot L \quad \text{(h)}$$

Thus our third Dimensional Set will be

Dimensional Set 3

	U	I	E	q
m	1	4	−2	1
N	0	0	1	1
π_1	1	0	$\frac{1}{3}$	$-\frac{1}{3}$
π_2	0	1	$\frac{4}{3}$	$-\frac{4}{3}$

yielding two dimensionless variables

$$\pi_1 = U \cdot \sqrt[3]{\frac{E}{q}}; \qquad \pi_2 = I \cdot \sqrt[3]{\frac{E^4}{q^4}} \tag{i}$$

If we try to express formula (c) by the dimensionless variables defined in (i), we will fail; the best we can have is

$$\pi_1 = \frac{1}{3 \cdot \pi_2} \cdot L^3 \cdot \left(\frac{E}{q}\right)^{3/5} \tag{j}$$

Thus, the analytical relation (c) *cannot* be expressed solely by the defined dimensionless variables, as given in (j). Reason: The defined new variable q in (h), although it fuses F and L, does not eliminate *both* from the analytical formula. Indeed, by (h) and (c), we have $U = \frac{1}{3} \cdot (q \cdot L^2)/(I \cdot E)$, in which L still appears. Thus the "fusion" defined in (h) is useless.

The general conclusion can be drawn from this example: A fusion of two variables is useful *only* if it eliminates both variables from the analytical formula—regardless of whether the latter is known.

⇑

In the examples presented so far, the fused variables always appeared as a product or a ratio. However, this may not always be the case, as the variables to be combined may also appear as a sum or difference. The following example is a typical of such a case.

Example 16-7. Terminal Velocity of a Sinking Ball in a Viscous Liquid (I)

In Example 16-1 we determined the terminal velocity of a falling raindrop. In that investigation we ignored the buoyancy of the drop, since the density difference between the materials of the drop (water) and the surrounding medium (air) was so large that the upward-pointing force (due to buoyancy) was negligible. However, if this force is not negligible, then the densities of both the ball and the surrounding medium must be considered.

In this example we wish to find the terminal speed of a ball sinking in a viscous liquid whose density is comparable to that of the ball. Fig. 16-4 shows the arrangement.

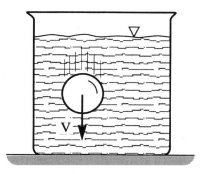

Figure 16-4
Ball is descending at constant speed in a viscous liquid

When the ball is first released, its speed increases to the point where the sum of the viscous forces and buoyancy equals the gravitational force (weight) acting on it. This is the terminal and constant speed of the ball. We assume that the ball's movement is slow, so the inertial effects opposing its motion may be ignored.

The following variables are involved:

Variable	Symbol	Dimension	Remark
diameter of ball	D	m	
density of ball	ρ_b	kg/m^3	
terminal speed of ball	v	m/s	constant
density of liquid	ρ_{liq}	kg/m^3	constant
gravitational acceleration	g	m/s^2	constant
viscosity of liquid	μ	kg/(m·s)	dynamic

This will give rise to the Dimensional Set

	D	ρ_b	v	ρ_{liq}	g	μ
m	1	−3	1	−3	1	−1
kg	0	1	0	1	0	1
s	0	0	−1	0	−2	−1
π_1	1	0	0	$\frac{2}{3}$	$\frac{1}{3}$	$-\frac{2}{3}$
π_2	0	1	0	−1	0	0
π_3	0	0	1	$\frac{1}{3}$	$-\frac{1}{3}$	$-\frac{1}{3}$

yielding three dimensionless variables

$$\pi_1 = D \cdot \sqrt[3]{\frac{\rho_{liq}^2 \cdot g}{\mu^2}}; \quad \pi_2 = \frac{\rho_b}{\rho_{liq}}; \quad \pi_3 = v \cdot \sqrt[3]{\frac{\rho_{liq}}{g \cdot \mu}} \quad \text{(a)}$$

Hence we can write

$$\pi_3 = \Psi\{\pi_1, \pi_2\} \quad \text{(b)}$$

where Ψ is an as-yet undefined function. This last relation is not a useful one because we have no notion about the nature of Ψ. Even if we assumed a *monomial* form for it, we would hardly be in a better position! For we would have, by (a) and (b),

$$v = c \cdot \sqrt[3]{\frac{g \cdot \mu}{\rho_{liq}}} \cdot \left(\frac{\rho_b}{\rho_{liq}}\right)^{\epsilon_2} \cdot \left(D \cdot \sqrt[3]{\frac{\rho_{liq}^2 \cdot g}{\mu^2}}\right)^{\epsilon_3} \quad \text{(c)}$$

where ϵ_2 and ϵ_3 are constant exponents. Let us assume that the densities of the ball and the liquid are identical. In this case the ball is in static equilibrium and hence the speed is zero. However, this contradicts (c) which does *not* yield zero speed if $\rho_b/\rho_{liq} = 1$. Therefore (c) is untenable, and hence Ψ is not a monomial.

We should now do a minuscule of heuristic meditation. Obviously the downward-pointing resultant force on the ball is proportional to the *difference* of densities $\Delta\rho = \rho_b - \rho_{liq}$. Thus we can *fuse* the two variables ρ_b and ρ_{liq} into one $\Delta\rho$. The modified Dimensional Set will then be

	v	$\Delta\rho$	D	g	μ
m	1	−3	1	1	−1
kg	0	1	0	0	1
s	−1	0	0	−2	−1
π_1	1	0	$-\frac{1}{2}$	$-\frac{1}{2}$	0
π_2	0	1	$\frac{3}{2}$	$\frac{1}{2}$	−1

from which

$$\pi_1 = \frac{v}{\sqrt{D \cdot g}}; \qquad \pi_2 = \frac{\Delta\rho}{\mu} \cdot \sqrt{D^3 \cdot g} \qquad (d)$$

Thus we may write

$$\pi_1 = \Psi\{\pi_2\} \qquad (e)$$

where Ψ is a function.

We see that, whereas (b) contained three dimensionless variables, (e) contains only two—a substantial improvement. This was achieved by *fusing* the two variables ρ_b and ρ_{liq} into one variable $\Delta\rho$, which then appears as a *difference*.

⇑

One word of warning, though. In some circumstances—especially when *transcendental* functions are involved—the number of dimensionless variables cannot be reduced below $N_P = 2$. Consider, for instance, the relation among dimensional variables V_1, V_2, V_3, and V_4

$$V_1 = V_2 \cdot \Psi\{V_3, V_4\}$$

where Ψ is a transcendental function. Therefore some combination of V_3 and V_4 must be dimensionless because these two variables make up the argument of Ψ, and the argument of a transcendental function *must* be dimensionless (Rule 4 of Dimensional Homogeneity, Art. 6.1). Consequently, by the above formula, V_1/V_2 must also be dimensionless. Thus the minimum number of such variables is (in this instance) $N_P = 2$. Problem 16/4 (in Art. 16.4) deals with this type of case, with the solution given in Appendix 6.

16.2. FUSION OF DIMENSIONLESS VARIABLES

It sometimes happens that if there are more than two dimensionless variables in a relation, then some of them *only* appear as part of a group. In these cases it is possible to represent this group by a newly defined dimensionless variable, denoted (usually) by the symbol π_c—where subscript "c" means "combined." Thus, in this case we *fuse* not the physical, but the dimensionless variables. The advantage gained by this fusion is especially evident in graphical representations of relations

428 APPLIED DIMENSIONAL ANALYSIS AND MODELING

where—as discussed in Chapter 12—a reduction of number of variables *very* significantly lessens the number of required curves.

The following two examples demonstrate.

Example 16-8. Jumping Heights of Animals (III) (air resistance is not ignored)

In Examples 6-19 and 11-16 (Art. 6.2 and Art. 11.2) we dealt with the problem of the heights to which animals—including man—can jump. We concluded that, regardless of size, all animals can jump to (very nearly) the same height. We arrived at this rather astonishing result by purely dimensional considerations and by assuming a number of simplifying conditions, one of which was negligible air resistance. This particular assumption—as we shall see presently—is fully justified in the case of large animals and even for smaller ones (e.g., frogs), but for *very* small creatures (e.g., fleas) air resistance should be considered (incidentally, this is a typical manifestation of the *Scale Effect*—discussed in more detail in Art. 17.5).

We now revisit the problem, but this time air resistance is no longer ignored. The following variables are considered relevant:

Variable	Symbol	Dimension
jumping height	h	m
mass	M	kg
projected surface area	A	m²
drag coefficient	C_d	1
upward speed at take-off	v	m/s
air density	ρ_a	kg/m³
gravitational acceleration	g	m/s²

We have seven variables and three dimensions, therefore there are $7 - 3 = 4$ dimensionless variables, supplied by the Dimensional Set

	h	M	A	C_d	v	ρ_a	g
m	1	0	2	0	1	−3	1
kg	0	1	0	0	0	1	0
s	0	0	0	0	−1	0	−2
π_1	1	0	0	0	−2	0	1
π_2	0	1	0	0	−6	−1	3
π_3	0	0	1	0	−4	0	2
π_4	0	0	0	1	0	0	0

from which

$$\pi_1 = \frac{h \cdot g}{v^2}; \quad \pi_2 = \frac{M \cdot g^3}{v^6 \cdot \rho_a}; \quad \pi_3 = \frac{A \cdot g^2}{v^4}; \quad \pi_4 = C_d \quad \text{(a)}$$

Note that C_d itself is a dimensionless variable; we only assigned π_4 to it for the sake of consistency (see also Art. 14.1).

The actual jumping height h is (Ref. 139, p. 200)

$$h = \frac{M}{\rho_a \cdot A \cdot C_d} \cdot \ln\left(\frac{\rho_a \cdot C_d \cdot A \cdot v^2}{2 \cdot M \cdot g} + 1\right) \quad \text{(b)}$$

METHODS OF REDUCING THE NUMBER OF DIMENSIONLESS VARIABLES   **429**

which, by (a), can be written

$$\pi_1 = \frac{\pi_2}{\pi_3 \cdot \pi_4} \cdot \ln\left(\frac{\pi_3 \cdot \pi_4}{2 \cdot \pi_2} + 1\right) \quad (c)$$

This relation is composed of four dimensionless variables, therefore to present them graphically—assuming eight distinct values for each—would require 64 curves appearing in eight charts (see Fig. 12-3 in Art. 12.1)—a logistic task of considerable magnitude. What can be done?

We see that π_2, π_3, and π_4 only appear in a *single* format of $(\pi_3 \cdot \pi_4)/\pi_2$. Therefore this group can be represented by a *single* dimensionless variable

$$\pi_c \equiv \frac{\pi_3 \cdot \pi_4}{\pi_2} \quad (d)$$

where the subscript "c" designates the adjective "combined." Thus, the three dimensionless variables π_2, π_3, and π_4 are *fused* into the one dimensionless variable π_c. By (c) and (d) now

$$\pi_1 = \frac{1}{\pi_c} \cdot \ln\left(\frac{\pi_c}{2} + 1\right) \quad (e)$$

This relation now has only two variables and hence can be represented graphically by a *single* curve—a *very* substantial improvement over its predecessor.

If there were no air resistance, then the theoretical height h_t of a jump would be

$$h_t = \frac{v^2}{2 \cdot g} \quad (f)$$

With air resistance we have h ($< h_t$), which is, by (a),

$$h = \pi_1 \cdot \frac{v^2}{g} \quad (g)$$

Therefore

$$\frac{h}{h_t} = \frac{\pi_1 \cdot \dfrac{v^2}{g}}{\dfrac{v^2}{2 \cdot g}} = 2 \cdot \pi_1$$

or by (e),

$$\frac{h}{h_t} = \frac{2}{\pi_c} \cdot \ln\left(\frac{\pi_c}{2} + 1\right) \quad (h)$$

If we consider h/h_t (< 1) as a *variable,* then relation (h) can also be represented by a *single* curve—as shown in Fig. 16-5.

When the frontal area to mass ratio A/M for an animal is very small, then $\pi_c \cong 0$ and $h/h_t \cong 1$. For example, for leopard $A/M \cong 0.0015$ m^2/kg, $C_d \cong 1$, $\rho_a \cong 1.2$ kg/m^3, $v \cong 7$ m/s, $g = 9.81$ m/s^2 and hence, with these values and by (d) and (a),

$$\pi_c = \frac{\pi_3 \pi_4}{\pi_2} = \frac{A}{M} \cdot \frac{C_d \cdot \rho_a \cdot v^2}{g} = 0.00899 \quad (i)$$

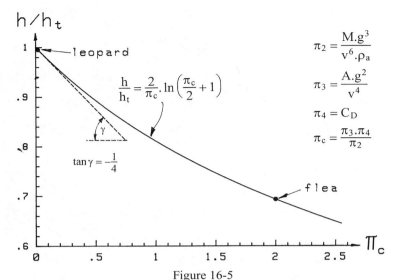

Figure 16-5
**Dimensionless plot for the jumping heights of animals
in the presence of air resistance**
h = jumping height with air resistance; h_t = jumping heights without air resistance.
See table in this example for definitions of other symbols

Therefore, by (h)

$$\left(\frac{h}{h_t}\right)_{leop} = \frac{2}{0.00899} \cdot \ln\left(\frac{0.00899}{2} + 1\right) = 0.99776$$

So, for a leopard, air resistance is not much of an impediment. But for a flea, the situation is vastly different. For a flea $A/M \cong 0.35$ m²/kg. Thus, for the flea, having identical other variables as for the leopard, $\pi_c \cong 2.098$ yielding, by (h),

$$\left(\frac{h}{h_t}\right)_{flea} = 0.684$$

We conclude, therefore, that a flea jumps only about 68% of the height it could clear in vacuum.

Incidentally, the h/h_t versus π_c curve in Fig. 16-5 has a slope of tangent exactly -0.25 at $\pi_c = 0$. The proof of this intriguing fact is left as a reward for the reader.

This example showed that by the adroit *fusing* of dimensionless variables, a seven-variable problem could be reduced to a simple two-variable one.

⇑

Example 16-9. Transient Velocity of a Sinking Ball in a Viscous Liquid

In Example 16-7 we dealt with the problem of determining the terminal speed of a sinking ball in a viscous liquid. The terminal speed is attained if a sufficiently long period of time has elapsed, so that the resistive forces against the motion of the ball are exactly counterbalanced by the negative buoyancy. In the present illustrative example we investigate the

METHODS OF REDUCING THE NUMBER OF DIMENSIONLESS VARIABLES

transient portion of the ball's movement, i.e., the (initial) time period which precedes the onset of terminal speed. We shall mainly use the dimensional method.

The differential equation describing the motion of the ball is (Ref. 33, p. 209)

$$\frac{\pi}{6} \cdot D^3 \cdot g \cdot \Delta\rho - 3 \cdot \pi \cdot D \cdot \mu \cdot v = M \cdot \frac{dv}{dt} \quad (a)$$

where the variables are:

Variable	Symbol	Dimension
diameter of ball	D	m
differential density	$\Delta\rho$	kg/m^3
dynamic viscosity of fluid	μ	kg/(m·s)
mass of ball	M	kg
velocity of ball	v	m/s
time	t	s
gravitational acceleration	g	m/s^2

The differential density is defined by

$$\Delta\rho = \rho_b - \rho_{liq} \quad (b)$$

where ρ_b and ρ_{liq} are the respective densities of the ball and the liquid. In (a) the first member on the left is the accelerating force, while the second member is the (opposing) force due to viscosity.

The differential equation of (a) is separable, therefore the solution is rather straightforward, and is

$$v = \frac{D^2 \cdot g \cdot \Delta\rho}{18 \cdot \mu} \cdot \left(1 - e^{\frac{-3\cdot\pi\cdot D\cdot\mu\cdot t}{M}}\right) \quad (c)$$

This relation consists of seven physical variables, therefore its graphical representation is totally hopeless. How could the dimensional method help us here? First, we construct the Dimensional Set

	v	g	$\Delta\rho$	t	D	μ	M
m	1	1	−3	0	1	−1	0
kg	0	0	1	0	0	1	1
s	−1	−2	0	1	0	−1	0
π_1	1	0	0	0	−2	−1	1
π_2	0	1	0	0	−3	−2	2
π_3	0	0	1	0	3	0	−1
π_4	0	0	0	1	1	1	−1

from which we obtain the required 7 − 3 = 4 dimensionless variables

$$\pi_1 = \frac{v \cdot M}{D^2 \cdot \mu}; \quad \pi_2 = \frac{g \cdot M^2}{D^3 \cdot \mu^2}; \quad \pi_3 = \frac{\Delta\rho \cdot D^3}{M}; \quad \pi_4 = \frac{t \cdot D \cdot \mu}{M} \quad (d)$$

By these dimensionless variables, relation (c) can be written in a neat form:

$$\pi_1 = \frac{\pi_2 \cdot \pi_3}{18} \cdot (1 - e^{-3\cdot\pi\cdot\pi_4}) \quad (e)$$

432 APPLIED DIMENSIONAL ANALYSIS AND MODELING

Note that in the exponent of e, the first π—*with no subscript*—is not a dimensionless *variable*, but the dimensionless constant 3.1415. . . .

Relation (e) is now quite compact, but it contains four variables and therefore still cannot be plotted on a single chart. What can be done? We notice that π_2 and π_3 occur only as a product, i.e., they do not appear separately in the expression. This suggests that we can *fuse* them into the one variable

$$\pi_c = \pi_2 \cdot \pi_3 \qquad (f)$$

where subscript "c" stands for "combined." By (d) and (e) now

$$\pi_c = \frac{g \cdot M^2}{D^3 \cdot \mu^2} \cdot \frac{\Delta \rho \cdot D^3}{M} = \frac{g \cdot M \cdot \Delta \rho}{\mu^2} \qquad (g)$$

Next, we combine (e) and (f)

$$\pi_1 = \frac{\pi_c}{18} \cdot (1 - e^{-3 \cdot \pi \cdot \pi_4}) \qquad (h)$$

This formula now consists of only three variables, and therefore it can be neatly displayed with parameter π_c in a chart, as Fig. 16-6 shows.

If t is very large, then π_4 is also very large and the value of the expression in parentheses on the right side of (h) is unity. Hence, the obtained π_1 value represents the terminal speed of the ball for a given π_c, i.e., for a given set of g, M, $\Delta\rho$, and μ values [see relation (g)].

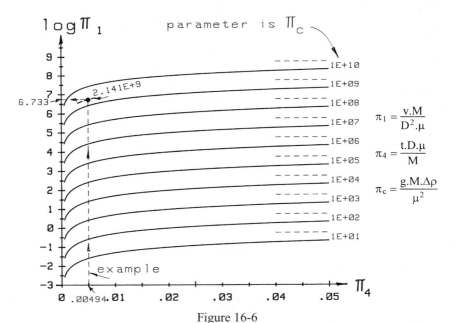

Figure 16-6
Dimensionless plot to determine the transient velocity of the descent of a ball in a viscous liquid
The short horizontal dashed lines represent terminal velocities for the parameters indicated. See table in this example for definitions of symbols of physical variables

For example, if for a ball, the diameter is $D = 0.02$ m and density $\rho_b = 7800$ kg/m^3, then its mass is $M = 0.03267$ kg. The ball is then immersed in water whose density is $\rho_{liq} = 1000$ kg/m^3, and viscosity $\mu = 0.001009$ kg/(m·s). We assume $g = 9.81$ m/s^2. Questions: at 8 s, what will be the ball's speed, and what will be the ball's *terminal* speed?

We consider relations (g), (b), (d), and the above data, by which $\pi_c = 2.1406 \times 10^9$, $\pi_4 = 0.00494$. Thus, by (h), and the above details, $\pi_1 = 5.41 \times 10^6$. The speed v of the ball is now obtained from the definition of π_1 in (d). Accordingly, $v = 66.83$ m/s, which is the answer to the first question. On the left side of the plot in Fig. 16-6, this result is indicated by dashed lines.

To answer the second question, we consider relation (h) with $\pi_4 \to \infty$ and $\pi_c = 2.1406 \times 10^9$ as calculated above. Accordingly,

$$(\pi_1)_{\text{terminal}} = \frac{\pi_c}{18} = 1.189 \times 10^8$$

and hence, by (d) again,

$$v_{\text{terminal}} = 1468.9 \text{ m/s}$$

By contrast, a free-falling object without air resistance reaches this speed in about 149.7 s.

This example thus showed the great logistic advantage offered by the fusion of dimensionless variables.

⇑

16.3. INCREASING THE NUMBER OF DIMENSIONS

We recall the relation

$$N_P = N_V - N_d \qquad \text{repeated (16-1)}$$

where N_P = number of dimensionless variables,
N_V = number of physical variables,
N_d = number of dimensions.

Our perennial goal is to reduce N_P. By the above formula, we can do this in two ways: by decreasing N_V, or by increasing N_d. The former process was discussed in Art. 16.2; now we deal with the latter method.

Increasing the number of dimensions can be done in three ways:

- Dimension splitting;
- Importing a new dimension;
- Intermixing dimensional systems.

We now explain these three techniques in some detail, and demonstrate their applications in several examples.

16.3.1. Dimension Splitting

The dimensions (in a Dimensional Set) that can be split into components are *length* and *mass*. Of these two, the decomposition of length is the more common, and perhaps the little less risky. For splitting dimensions is *always* a perilous undertaking; it must be done with *extreme* care, and the results checked, *and rechecked,* for possible errors. The investigator must also ask the questions: Does the relation obtained make sense? Does the result satisfy the *boundary* conditions where the values are known *a priori*?

Splitting the dimension of *length* is done on the basis of *direction*. To this end it is always necessary to attach a coordinate system to the structure, assembly, mechanism, etc.—provided such association makes sense. If two directions are equally applicable, then both, individually, must have the power of ½; in the case of three directions, the respective powers should be ⅓.

If this sounds a bit too abstract, the following examples should greatly increase the reader's understanding and appreciation of both the underlying idea and the practicality of the method.

Example 16-10. Rise of a Fluid in a Capillary Tube (adapted from Ref. 2, p. 87)

Given a glass tube with liquid in it (Fig. 16-7). We wish to determine the height of the *meniscus* due to surface tension.

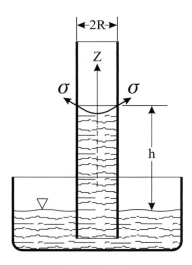

Figure 16-7
Capillary tube containing a liquid (exaggerated presentation)

We have the following list of relevant variables:

Variable	Symbol	Dimension
height of meniscus	h	m
density of liquid	ρ	kg/m^3
inner radius of tube	R	m
gravitational acceleration	g	m/s^2
surface tension of liquid	σ	kg/s^2

There are five variables and three dimensions, hence we have 5 − 3 = 2 dimensionless variables to deal with. They are determined by the Dimensional Set

	h	ρ	R	g	σ
m	1	−3	1	1	0
kg	0	1	0	0	1
s	0	0	0	−2	−2
π_1	1	0	−1	0	0
π_2	0	1	2	1	−1

from which

$$\pi_1 = \frac{h}{R}; \qquad \pi_2 = \frac{\rho \cdot R^2 \cdot g}{\sigma} \qquad (a)$$

and therefore

$$\pi_1 = \Psi\{\pi_2\} \qquad (b)$$

where Ψ is an as-yet undefined function. By (a) and (b) now

$$h = R \cdot \Psi\left\{\frac{\rho \cdot R^2 \cdot g}{\sigma}\right\} \qquad (c)$$

Without knowing Ψ, this relation is useless, for we have absolutely no idea how, for example, h varies with R or σ.

To improve our standing, let us use *directional* linear dimensions. Fig. 16-7 shows two specified directions: *axial* and *radial*. Accordingly, we now have a modified set of variables and dimensions:

Variable	Symbol	Dimension
height of meniscus	h	m_z
density of liquid	ρ	$kg/(m_r^2 \cdot m_z)$
inner radius of tube	R	m_r
gravitational acceleration	g	m_z/s^2
surface tension of liquid	σ	$m_z \cdot kg/(m_r \cdot s^2)$

Regarding the directional assignments of dimensions, a few explanatory comments are in order:

- Subscripts "r" and "z" designate "radial" and "axial" (z-directional) directions, respectively.
- For h and g, the dimensions are "axial."
- For density ρ, the axial and radial dimensions have the exponents 1 and 2, respectively. Reason: volume equals the *product* of z direction and cross-sectional area, the latter being the radial dimension squared.
- For surface tension σ, we are interested in its the z-directional component only, thus

$$\sigma = \frac{\text{z-directional force}}{\text{length of circumference of meniscus}}$$

436 APPLIED DIMENSIONAL ANALYSIS AND MODELING

The dimension of the z-directional force is obviously $(m_z \cdot kg)/s^2$, while that of the length of the meniscus" perimeter is m_r, since the latter exists in the cross-section of the liquid. Hence, for the dimension of σ we have

$$[\sigma] = \frac{m_z \cdot kg}{m_r \cdot s^2}$$

Based on the above, we have the modified Dimensional Set

	h	ρ	R	g	σ
m_r	0	−2	1	0	−1
m_z	1	−1	0	1	1
kg	0	1	0	0	1
s	0	0	0	−2	−2
π_1	1	1	1	1	−1

We observe that, while the number of variables is still $N_V = 5$, the number of dimensions increased from $N_d = 3$ to $N_d = 4$. This was achieved by *splitting* the single linear dimension "m" into "m_r" and "m_z." Therefore the number of dimensionless variables is now $N_p = N_v - N_d = 5 - 4 = 1$, the ideal number! By the above Dimensional Set, then, this sole dimensionless variable is

$$\pi_1 = \frac{h \cdot \rho \cdot R \cdot g}{\sigma} = \text{const}$$

from which

$$h = \text{const} \cdot \frac{\sigma}{\rho \cdot R \cdot g} \tag{d}$$

This formula is vastly more informative than the corresponding (c), which was derived on the basis of nondirectional (i.e., scalar) linear dimensions. For example, by (d), we now know that the meniscus' height h is inversely proportional to the inner radius of tube R.

⇑

Example 16-11. Range of a Horizontally Ejected Bullet (adapted from Ref. 2, p. 76)
A bullet is ejected at altitude h with horizontal speed U (Fig. 16-8). If we neglect air resistance and assume constant gravitational field, what will the horizontal range (distance travelled) be?
First, we list the relevant variables:

Variable	Symbol	Dimension	Remark
range	b	m	horizontal
ejection speed	U	m/s	horizontal
gravitational acceleration	g	m/s²	constant
ejection altitude	h	m	

We have four variables and two dimensions, hence there are $4 - 2 = 2$ dimensionless variables supplied by the Dimensional Set

METHODS OF REDUCING THE NUMBER OF DIMENSIONLESS VARIABLES

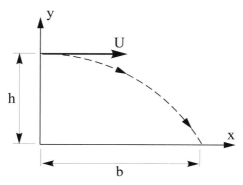

Figure 16-8
A bullet is horizontally ejected at an altitude

	b	U	g	h
m	1	1	1	1
s	0	−1	−2	0
π_1	1	0	0	−1
π_2	0	1	$-\dfrac{1}{2}$	$-\dfrac{1}{2}$

yielding

$$\pi_1 = \frac{b}{h}; \qquad \pi_2 = \frac{U}{\sqrt{g \cdot h}} \tag{a}$$

Thus we have

$$b = h \cdot \Psi\left\{ \frac{U}{\sqrt{g \cdot h}} \right\} \tag{b}$$

where Ψ is an unknown function. Therefore this relation is useless. To improve the situation, let us try to use directional dimensions for length. We then have the variables and dimensions as follows:

Variable	Symbol	Dimension	Remark
range	b	m_x	horizontal
ejection speed	U	m_x/s	horizontal
gravitational acceleration	g	m_y/s^2	constant
ejection altitude	h	m_y	

where subscripts "x" and "y" designate the respective directions as defined in the coordinate system shown in Fig. 16-8.

In this table we see that we now have three dimensions instead of the former two, although the number of variables is still four. Therefore there are 4 − 3 = 1 dimensionless variable—one less than before and the ideal number! This sole variable is furnished by the (modified) Dimensional Set as follows:

438   APPLIED DIMENSIONAL ANALYSIS AND MODELING

	b	U	g	h
m_x	1	1	0	0
m_y	0	0	1	1
s	0	−1	−2	0
π_1	1	−1	$\frac{1}{2}$	$-\frac{1}{2}$

yielding the solitary dimensionless "variable"—a constant in this case

$$\pi_1 = \frac{b}{U} \cdot \sqrt{\frac{g}{h}} = \text{const} \qquad (c)$$

from which

$$b = \text{const} \cdot U \cdot \sqrt{\frac{h}{g}} \qquad (d)$$

Note that this formula contains *much* more information than relation (b) does. For example, (d) tells us that range b grows *linearly* with ejection speed U, and with the *square root* of altitude h; no such information can be extracted from formula (b). This signal improvement was achieved by employing directional linear dimensions.   ⇑

Example 16-12. Tangential Speed of a Conical Pendulum (adapted from Ref. 2, p. 88)

Given a conical pendulum, as illustrated in Fig. 16-9. We wish to determine the bob's tangential speed as a function of its orbital radius, the pendulum's vertical length and gravitational acceleration. Accordingly, we have the following list of relevant variables:

Variable	Symbol	Dimension	Remark
speed	v	m/s	tangential
orbital radius	R	m	
height of pendulum	h	m	vertical
gravitational acceleration	g	m/s²	

We have four variables and two dimensions, therefore there are 4 − 2 = 2 dimensionless variables obtained from the Dimensional Set

	v	R	h	g
m	1	1	1	1
s	−1	0	0	−2
π_1	1	0	$-\frac{1}{2}$	$-\frac{1}{2}$
π_2	0	1	−1	0

supplying

$$\pi_1 = \frac{v}{\sqrt{h \cdot g}}; \qquad \pi_2 = \frac{R}{h} \qquad (a)$$

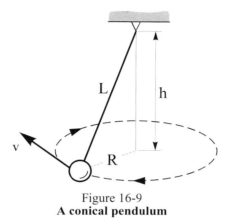

Figure 16-9
A conical pendulum

from which

$$\pi_1 = \Psi\{\pi_2\}$$

where Ψ is an as-yet unknown function. By (a) this relation can be written

$$v = c \cdot \sqrt{h \cdot g} \cdot \Psi\left\{\frac{R}{h}\right\} \qquad (b)$$

where c is a numeric constant.

We can see that (b) is not very useful since both R and h are inside the unknown function Ψ. Except for the fact that the tangential speed varies as a square root of gravitational acceleration, no other information can be extracted from (b). To improve our status, we now try to use directional linear dimensions. We note that v and R are *horizontal*, whereas g and h are *vertical*. These facts suggest the following list of dimensions for our variables:

Variable	Symbol	Dimension	Remark
speed	v	m_H/s	tangential
orbital radius	R	m_H	
height of pendulum	h	m_V	vertical
gravitational acceleration	g	m_V/s^2	

where subscripts "H" and "V" designate *horizontal* and *vertical* directions, respectively. There are now three dimensions instead of the previous two, and we still have four variables. Hence now there is only $4 - 3 = 1$ dimensionless variable (which is a constant), obtained by the Dimensional Set:

	v	R	h	g
m_H	1	1	0	0
m_V	0	0	1	1
s	−1	0	0	−2
π_1	1	−1	$\frac{1}{2}$	$-\frac{1}{2}$

yielding

$$\pi_1 = \frac{v}{R} \cdot \sqrt{\frac{h}{g}} = \text{const}$$

from which

$$v = \text{const} \cdot R \cdot \sqrt{\frac{g}{h}} \qquad (c)$$

As the reader will note, this relation is *much* more informative than the corresponding formula (b). For we now know how *each* independent variable influences the dependent variable. This improvement was achieved by the adroit use of directional linear dimensions.

⇑

Sometimes directions of length can be assigned to dimensionless physical variables, that is to physical variables whose dimensions are 1 (e.g., friction coefficient). This way, the number of dimensions can be increased, with the attendant benefits. The following example demonstrates this feature.

Example 16-13. An Elegant Method to Determine the Kinetic Friction Coefficient

The kinetic friction coefficient μ_k is the ratio of the *friction* force to the *normal* force experienced by a body *moving* on a dry, non-smooth surface.

Thus, if F_f and F_n are the respective frictional and normal forces, then by definition,

$$\mu_k \equiv \frac{F_f}{F_n} \qquad (a)$$

The determination of μ_k can be done quite accurately by a rather ingenious mechanism described by Timoshenko (Ref. 141, p. 286). A slab of uniform density is placed on two rotating rollers. The rollers turn at the same speed but in *opposite* directions (Fig. 16-10). The axes of the rollers are parallel, horizontal and "a" distance apart.

Obviously, if the slab is symmetrically positioned with respect to the rotating rollers

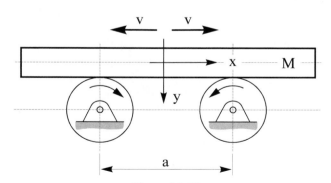

Figure 16-10
An apparatus to determine the kinetic friction coefficient between slab and rollers
Rollers are rotated at equal speed, but in opposite directions

METHODS OF REDUCING THE NUMBER OF DIMENSIONLESS VARIABLES

(as shown in the figure) then the slab is in dynamic equilibrium, since the *y*-directional *normal* forces between the slab and the rollers are identical and therefore the *tangential* (*x*-directional) forces are also identical in magnitude, but opposite in direction (the *x, y* coordinate system is defined in the figure).

However, if the symmetry is disturbed—for any reason—then this dynamic balance will no longer keep the slab stationary. Assume the slab has moved slightly to the right. Then the friction force on the right (pointing left) will be larger than the friction force on the left (pointing right). Thus, the net force will point to the opposite direction than the displacement. Consequently, if the slab is moved to the right, then the restoring force points left, and vice versa. It follows that the resultant horizontal force always tends to restore the equilibrium of the slab, and hence once the slab is displaced from its central position, it will execute a simple harmonic motion about its equilibrium position. It is not difficult to determine the period of this motion analytically (see for example Ref. 141, p 286), but here let us arrive at some useful conclusions by purely dimensional considerations. First we list the *assumed* relevant variables.

Variable	Symbol	Dimension	Remark
period of oscillation of slab	T	s	
kinetic friction coefficient	μ_k	1	independent of relative speed
distance between rollers	a	m	between centers of rollers
mass of slab	M	kg	
gravitational acceleration	g	m/s^2	

We see at once that M is the only variable which has the dimension "kg." Therefore, by Theorem 11-1 (Art. 11.1), M is a *dimensionally irrelevant* variable, and so, by Theorem 11-3 (Art. 11.2.1), it is also a *physically irrelevant* variable. Thus, we have four variables and two dimensions, and therefore there are $4-2=2$ dimensionless variables supplied by the Dimensional Set

	T	μ_k	a	g
m	0	0	1	1
s	1	0	0	-2
π_1	1	0	$-\frac{1}{2}$	$\frac{1}{2}$
π_2	0	1	0	0

from which

$$\pi_1 = T \cdot \sqrt{\frac{g}{a}}; \qquad \pi_2 = \mu_k \tag{b}$$

and hence

$$\pi_1 = \Psi\{\pi_2\} \tag{c}$$

which, in view of (b), can be written

$$T = c \cdot \sqrt{\frac{a}{g}} \cdot \Psi\{\mu_k\} \tag{d}$$

where c is a numeric constant and Ψ is an as-yet unknown function. Although (d) does not enable us to determine T from μ_k, or vice versa, but by it we can conclude that, provided

442 APPLIED DIMENSIONAL ANALYSIS AND MODELING

μ_k is constant, period T is proportional to the square root of distance "a" between the rollers, and inversely proportional to the square root of g. Thus, by (d), and everything else being the same, the period of the slab's oscillation on the Moon is

$$\sqrt{\frac{g_{\text{Earth}}}{g_{\text{Moon}}}} = \sqrt{\frac{9.81}{1.60}} = 2.476$$

times longer than on Earth.

We will now refine our approach. We notice that, by (a), the friction coefficient is really the quotient of the *frictional* and *gravitational* forces. We also note that the former is x-directional, while the latter is y-directional. This suggests that, instead of treating μ_k as a dimensionless physical variable, we could in fact assign direction-dependent dimensions to it. Accordingly,

$$[\mu_k] = \frac{[x\text{-directional force}]}{[y\text{-directional force}]} = \frac{(m_x \cdot \text{kg})/\text{s}^2}{(m_y \cdot \text{kg})/\text{s}^2} = \frac{m_x}{m_y} \quad (e)$$

where m_x and m_y are the respective x and y directional linear dimensions and—as usual—the brackets designate dimensions. Based on (e), there is the following list of variables:

Variable	Symbol	Dimension	Remark
period of oscillation of slab	T	s	
kinetic friction coefficient	μ_k	m_x/m_y	independent of relative speed between centers
distance between rollers	a	m_x	
gravitational acceleration	g	m_y/s^2	

We have now four variables and three dimensions (before we had only two) and hence there is $4 - 3 = 1$ dimensionless "variable"—a constant. It is supplied by the modified Dimensional Set

	T	μ_k	a	g
m_x	0	1	1	0
m_y	0	−1	0	1
s	1	0	0	−2
π_1	1	$\frac{1}{2}$	$-\frac{1}{2}$	$\frac{1}{2}$

from which

$$\pi_1 = T \cdot \sqrt{\frac{\mu_k \cdot g}{a}} = c = \text{const} \quad (f)$$

where c is a constant. Thus, by (f)

$$\mu_k = c^2 \cdot \frac{a}{g \cdot T^2} \quad (g)$$

It is emphatically pointed out that this relation is *much* more informative than (d). For now we know that the kinetic friction coefficient is inversely proportional to the square of period T. This significant improvement was achieved by using directional-dependent linear dimensions.

METHODS OF REDUCING THE NUMBER OF DIMENSIONLESS VARIABLES 443

Constant c in (f) can be determined (either experimentally or analytically) to be $2\cdot\pi$ (see Ref. 141, p. 286). Also, note that relation (g) makes the *accurate* determination of the kinetic friction coefficient quite easy, since both "*a*" and "*T*" can be measured with great precision. This elegant and efficient method is also discussed by Kármán in Ref. 143, p. 159.

⇑

Splitting the dimension of *mass* is less common, but on occasion and with *great care*, it can yield useful results. Here we consider two distinct functions of mass: its role associated with *inertial* effects, and its role associated with *quantity* of matter. If these two manifestations can be separated in a problem, then the dimension of mass "kg" can be split into *inertial* mass "kg_i" and *quantity* mass "kg_q." The following example shows a practical application of this technique.

Example 16-14. Mass of Fluid Flowing Through a Tube (II)

We already dealt with this problem in Example 16-2 (Art. 16.1) where it was solved by the method of *fusion of variables*. Now we solve the problem by the method of *splitting a dimension*. For the reader's convenience, we again list the variables and the Dimensional Set.

Variable	Symbol	Dimension	Remark
mass of fluid through tube	Q	kg/s	in unit time
differential pressure	Δp	kg/(m²·s²)	in unit length
density of fluid	ρ	kg/m³	constant
viscosity of fluid	μ	kg/(m·s)	dynamic
diameter of tube	D	m	inner

	Q	Δp	ρ	μ	D
m	0	−2	−3	−1	1
kg	1	1	1	1	0
s	−1	−2	0	−1	0
π_1	1	0	0	−1	−1
π_2	0	1	1	−2	3

from which

$$\pi_1 = \frac{Q}{\mu\cdot D}; \quad \pi_2 = \frac{\Delta p\cdot\rho\cdot D^3}{\mu^2} \quad (a)$$

Therefore we can write

$$Q = c\cdot\mu\cdot D\cdot\Psi\left\{\frac{\Delta p\cdot\rho\cdot D^3}{\mu^2}\right\} \quad (b)$$

where c is a constant and Ψ is a function not yet known. As it stands, this formula is useless, for it gives no information about how the mass flow varies with pressure difference, density and viscosity. But now we consider that Q and ρ are related to the *quantity* of matter, whereas Δp and μ involve the *inertial* characteristics of mass. These facts suggest that we split the dimension of mass "kg" into "kg_q" (quantity mass) and "kg_i" (inertial mass). Thus, the modified dimensional matrix will be

444   APPLIED DIMENSIONAL ANALYSIS AND MODELING

		Q	Δp	ρ	μ	D
	m	0	−2	−3	−1	1
inertial mass →	kg_i	0	1	0	1	0
"quantity" mass →	kg_q	1	0	1	0	0
	s	−1	−2	0	−1	0
	π_1	1	−1	−1	1	−4

and we see that the number of dimensionless variables is reduced to 1—the ideal number. We now have

$$\pi_1 = \frac{Q \cdot \mu}{\Delta p \cdot \rho \cdot D^4} = \text{const}$$

from which

$$Q = \text{const} \cdot \frac{\Delta p \cdot \rho \cdot D^4}{\mu} \tag{c}$$

This relation, of course, is identical to relation (e) of Example 16-2—known as Poiseuille's equation with constant = $\pi/128$ (Ref. 124, p. 118).

Again we point out the *very* significant difference between (b) and (c); for (b) is practically useless, whereas (c) gives us *all* the information we could ask for concerning the effects of each of the four independent variables upon the dependent variable. And as for finding the constant, it can be done by a *single* experiment.

⇑

It was pointed out earlier that splitting dimensions must be done with great care, as the method is rather perilous. It was also stated that extra scrutiny is always advisable before accepting and *using* the results. The following example demonstrates the serious dangers inherent in this technique.

Example 16-15. Torsion of a Prismatic Bar

We have a bar of a constant, nonhollow circular cross-section, clamped at one end, free at the other. Then a torsion moment is applied at the free end. We wish to find the angle of twist the torque produces at the free end. Fig. 16-11 shows the arrangement. The relevant variables are as follows:

Variable	Symbol	Dimension	Remark
angle of twist	φ	1	at free end
length	L	m	
twisting torque	T	m·N	at free end
cross-section's polar second moment of area	I_p	m⁴	about axis
shear modulus	G	N/m²	

We have five variables and two dimensions, therefore there are $5 - 2 = 3$ dimensionless variables, determined by the Dimensional Set

METHODS OF REDUCING THE NUMBER OF DIMENSIONLESS VARIABLES

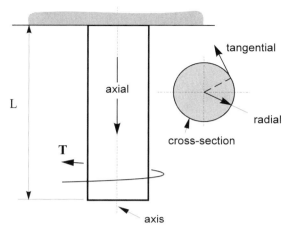

Figure 16-11
Prismatic bar loaded by torsional moment

	φ	L	T	I_p	G
m	0	1	1	4	-2
N	0	0	1	0	1
π_1	1	0	0	0	0
π_2	0	1	0	$-\frac{1}{4}$	0
π_3	0	0	1	$-\frac{3}{4}$	-1

yielding

$$\pi_1 = \varphi; \quad \pi_2 = \frac{L}{(I_p)^{1/4}}; \quad \pi_3 = \frac{T}{G \cdot (I_p)^{3/4}} \quad \text{(a)}$$

Hence we have

$$\pi_1 = \Psi\{\pi_2, \pi_3\} \quad \text{(b)}$$

where Ψ is a function. By (a) and (b),

$$\varphi = \Psi\left\{\frac{L}{(I_p)^{1/4}}, \frac{T}{G \cdot (I_p)^{3/4}}\right\} \quad \text{(c)}$$

As it stands, this equation is totally useless; it does not tell us anything, except the obvious that φ is a function of the independent variables L, I_p, T, and G.

To improve the situation, we apply directional dimensions. But we must first define a suitable coordinate system. To this end, we designate directions as *axial*, *radial*, and *tangential*, as shown in Fig. 16-11. Then we *carefully* assign oriented dimensions to the variables as follows (subscripts "a," "r," and "t" are used, respectively, for the cited directions):

446 APPLIED DIMENSIONAL ANALYSIS AND MODELING

Torsional angle φ is the ratio of tangential to radial dimensions. Therefore

$$[\varphi] = \frac{m_t}{m_r} \tag{d}$$

Length L occurs in the axial direction. Thus

$$[L] = m_a \tag{e}$$

Torsional moment T is the product of radial length and tangential force. But [radial length] = m_r and [tangential force] = N, therefore

$$[T] = m_r \cdot N \tag{f}$$

Polar second moment of area I_p is defined with the aid of Fig. 16-12. By definition

$$I_p = \int_0^R r^2 \cdot dA \tag{g}$$

where $dA = 2 \cdot r \cdot \pi \cdot dr$ is the elementary area. Therefore $[dA] = m_r^2$ and $[r^2] = m_r^2$. So, by (g),

$$[I_p] = m_r^4 \tag{h}$$

Modulus of shear G is defined by

$$G = \frac{\text{shear stress}}{\text{shear strain}} \tag{i}$$

Furthermore, shear stress = shear force/shear area and shear strain = tangential move/axial length. We now list the dimensions:

[shear force] = N

[shear area] = m_r^2 (since the radial dimension is two-directional)

[tangential move] = m_t

[axial length] = m_a

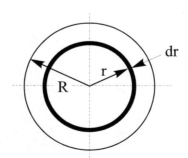

Figure 16-12
Explanatory sketch for the determination of polar second moment of area of a solid circle

If these are now substituted into (i), we obtain the dimension of G

$$[G] = \frac{m_a \cdot N}{m_t \cdot m_r^2} \tag{j}$$

Thus, by (d), (e), (f), (h), and (j), we can now construct the Dimensional Set

	φ	L	T	I_p	G
m_t	1	0	0	0	−1
m_r	−1	0	1	4	−2
m_a	0	1	0	0	1
N	0	0	1	0	1
π_1	1	−1	−1	1	1

resulting in the solitary dimensionless "variable"

$$\pi_1 = \frac{\varphi \cdot I_p \cdot G}{L \cdot T} = \text{const}$$

by which the sought-after twist angle is

$$\varphi = \text{const} \cdot \frac{L \cdot T}{I_p \cdot G} \tag{k}$$

This is the very formula we were seeking because, with const = 1, it gives angular deformation φ at the free end of the bar upon imposed torsional moment T. We arrived at this *valid* result by the *careful* selection of the direction-dependent length dimensions. Please note: the emphasis is on the adjective "careful," since any wrongly assigned dimension could—and most probably would—result in a formula which, although dimensionally correct, would be factually wrong.

To demonstrate the error in the result from directionally faulty dimensioning of linear variables, let us assume that we erroneously use the dimension $(m_a \cdot N)/(m_t^2 \cdot m_r)$ for shear modulus G. In this case we would have, instead of the last Dimensional Set, the arrangement:

	φ	L	T	I_p	G
m_t	1	0	0	0	−2
m_r	−1	0	1	4	−1
m_a	0	1	0	0	1
N	0	0	1	0	1
π_1	1	$-\frac{1}{2}$	$-\frac{1}{2}$	$\frac{1}{2}$	$\frac{1}{2}$

so that

$$\pi_1 = \varphi \cdot \sqrt{\frac{I_p \cdot G}{L \cdot T}} = \text{const}$$

from which

$$\varphi = \text{const} \cdot \sqrt{\frac{L \cdot T}{I_p \cdot G}} \tag{l}$$

This equation is dimensionally correct, but factually wrong. For it says that the twist angle φ is no longer proportional to either length L, or twisting torque T. These results are obviously at variance with experience.

448 APPLIED DIMENSIONAL ANALYSIS AND MODELING

Next, let us see what happens if we err in dimensioning the polar second moment of area I_p. Again we use definition (g), but now we erroneously define the elementary area as $\Delta A = \Delta r \cdot \Delta t$, where Δr is the radial dimension and Δt is the tangential dimension, as in Fig. 16-13.

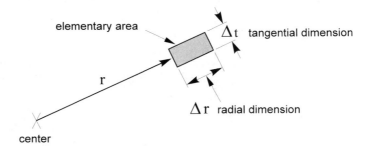

Figure 16-13
An erroneous definition of the elementary area for the determination of the polar second moment of area of a solid circle
(the correct definition is seen in Fig. 16-12)

Therefore the dimension of I_p according to this definition is

$$[I_p] = m_r^3 \cdot m_t \qquad (m)$$

which results in the dimensional matrix

	φ	L	T	I_p	G
m_t	1	0	0	1	−1
m_r	−1	0	1	3	−2
m_a	0	1	0	0	1
N	0	0	1	0	1

A matrix

We see that the **A** matrix here is *singular*, although the rank of the dimensional matrix is 4. Therefore it is possible to make **A** *nonsingular* by interchanging column φ with the column of any of the other variables (see Art. 7.6). Say we interchange φ and G. The new Dimensional Set is then

	G	L	T	I_p	φ
m_t	−1	0	0	1	1
m_r	−2	0	1	3	−1
m_a	1	1	0	0	0
N	1	0	1	0	0
π_1	1	−1	−1	1	0

C matrix

Here we see that the column of φ in matrix **C** is all zeros. Therefore, by Theorem 10-4 (Art. 10.5.1), φ is a dimensionally irrelevant variable, and further, by Theorem 11-3 (Art. 11.2.1), it is a physically irrelevant variable. But how can this be? How can a *dependent* variable be physically irrelevant? Of course, it cannot! The fault lies in the erroneous directional dimensioning of the polar second moment of area, as given in relation (m).

The above example illustrates the potential danger of faulty dimensioning. Note that this type of error does not occur if we use "nondirectional," i.e., "nonsplit" dimensions. But when we do split a dimension, then the selection of correct entries for the dimensional set demands not only extra precaution, but also more detailed knowledge of the theoretical background of the problem at hand.

16.3.2. Importation of New Dimensions

In some cases, reducing the number of dimensionless variables N_P can be accomplished by the adroit introduction (importation) of a new dimension. By this means the number of dimensions N_d is increased and N_P is decreased (in accordance with equation $N_P = N_V - N_d$). Here again, some knowledge of the underlying physical principles and increased caution are necessary. The following examples demonstrate this method and its benefits.

Example 16-16. Number of Free Electrons in a Metal Conductor (adapted from Ref. 4, p. 394)

This problem was already dealt with in Example 15-1; now we treat it from a different viewpoint. We again have a metal wire (conductor) and we wish to determine the number of free electrons in unit volume of this conductor. The following variables are considered:

Variable	Symbol	Dimension	Remark
number of free electrons	n	1/m³	in unit volume of conductor
number of free electrons per atom	f	1	
number of molecules per mol	N_a	1/mol	Avogadro number
atom mass	W	kg/mol	
density of conductor	ρ	kg/m³	

We have five variables and three dimensions, therefore there are $5 - 3 = 2$ dimensionless variables determined by the Dimensional Set

	n	f	N_a	W	ρ
m	−3	0	0	0	−3
kg	0	0	0	1	1
mol	0	0	−1	−1	0
π_1	1	0	−1	1	−1
π_2	0	1	0	0	0

yielding

$$\pi_1 = \frac{n \cdot W}{N_a \cdot \rho}; \qquad \pi_2 = f \qquad (a)$$

Thus, we can write

$$n = \frac{N_a \cdot \rho}{W} \cdot \Psi\{f\} \qquad (b)$$

where Ψ is an as-yet unknown function.

450   APPLIED DIMENSIONAL ANALYSIS AND MODELING

At this point we could assume a *monomial* form for Ψ—which in fact is the case here—and then derive the valid formula by employing rather simple heuristic reasoning, but instead we follow a different route: we will introduce a new dimension, the *electron number* n_e. Note that this is not the same as the *variable* of the number of electrons, n. Here n_e is a *dimension*. More specifically, n_e is the dimension of n. To emphasize and clarify the difference, two analogies may be worth mentioning here: "population" is a variable and its dimension is "person"; "amount of money" is a variable and its dimension is "dollar." Thus "person" is not equivalent to "population," and "dollar" is not equivalent to "amount of money." With this in mind, the modified dimensional set is

	n	f	N_a	W	ρ
m	−3	0	0	0	−3
kg	0	0	0	1	1
mol	0	0	−1	−1	0
n_e	1	1	0	0	0
π_1	1	−1	−1	1	−1

We have five variables, and now four dimensions (instead of the previous three). Therefore there is but one dimensionless variable,

$$\pi_1 = \frac{n \cdot W}{f \cdot N_a \cdot \rho} = \text{const} \tag{c}$$

by which

$$n = \text{const} \cdot \frac{f \cdot N_a \cdot \rho}{W} \tag{d}$$

As we see, (d) is much more informative than the corresponding relation (b). If const = 1 and $f = 1\ n_e$, then for copper, for example, we have

Avogadro number: $N_a = 6.022 \times 10^{23}$ 1/mol
atom weight: $W = 0.0635$ kg/mol
density: $\rho = 8960$ kg/m^3

Hence, by (d), a copper conductor contains

$$n = \frac{(1) \cdot (6.022 \times 10^{23}) \cdot (8960)}{0.0635} = 8.497 \times 10^{28}\ n_e/\text{m}^3\ (= \text{free electrons/m}^3)$$

⇑

Example 16-17. Heat Energy Input by Pouring Hot Liquid into a Tank

We pour hot liquid into a tank at a constant rate. Question: what is the amount of thermal energy entering this container? We can solve this simple problem in two ways:

Solution 1. Standard dimensions are used. The variables are:

Variable	Symbol	Dimension	Remark
quantity of heat added	q	m^2·kg/s^2	
time duration	t	s	
density of liquid	ρ	kg/m^3	
specific energy of liquid	U	m^2/s^2	energy per unit mass
flow volume rate	$\dot{V}$	m^3/s	

METHODS OF REDUCING THE NUMBER OF DIMENSIONLESS VARIABLES **451**

Note that the *specific energy* is also called *calorific value* (see Fig. 3-13a), although use of the latter is not recommended. By the above table we have five variables as three dimensions, therefore there are 5 – 3 = 2 dimensionless variables supplied by the Dimensional Set

	q	t	ρ	U	$\dot{V}$
m	2	0	–3	2	3
kg	1	0	1	0	0
s	–2	1	0	–2	–1
π_1	1	0	–1	$-\frac{1}{4}$	$-\frac{3}{2}$
π_2	0	1	0	$\frac{3}{4}$	$-\frac{1}{2}$

yielding

$$\pi_1 = \frac{q}{\rho} \cdot \frac{1}{\sqrt[4]{U} \cdot \dot{V}^{3/2}}; \qquad \pi_2 = \frac{t \cdot U^{3/4}}{\sqrt{\dot{V}}} \qquad (a)$$

Therefore, generally

$$\pi_1 = \Psi\{\pi_2\} \qquad (b)$$

where Ψ is an as-yet unknown function. By (a) and (b), now

$$q = c \cdot \rho \cdot \sqrt[4]{U} \cdot \dot{V}^{3/2} \cdot \Psi\left\{\frac{t \cdot U^{3/4}}{\sqrt{\dot{V}}}\right\} \qquad (c)$$

where c is a numeric constant.

This rather frightening expression can be tamed easily by either resorting to some heuristic reasoning, or by the importation of a new variable. In this particular case, obviously, we select the latter method, which is, then, Solution 2.

Solution 2. We note that *energy* here is in the form of *heat energy*, which for our present purposes, can be considered independent from the *mechanical energy* expressed by the dimension m^2·kg/s^2. Accordingly, we can have the following modified list of variables and dimensions:

Variable	Symbol	Dimension	Remark
quantity of heat added	q	H	
time duration	t	s	
density of liquid	ρ	kg/m^3	
specific energy of liquid	U	H/kg	energy per unit mass
flow volume rate	$\dot{V}$	m^3/s	

Note that we use H for the dimension of heat energy and, accordingly, the dimension of q is H, and of U is H/kg.

We still have five variables, but now the number of dimensions has increased from three to four. Thus, the number of dimensionless variables is 5 – 4 = 1, which is a constant. It is obtained from the Dimensional Set as follows:

452 APPLIED DIMENSIONAL ANALYSIS AND MODELING

	q	t	ρ	U	$\dot{V}$
m	0	0	−3	0	3
kg	0	0	1	−1	0
s	0	1	0	0	−1
H	1	0	0	1	0
π_1	1	−1	−1	−1	−1

yielding

$$\pi_1 = \frac{q}{t \cdot \rho \cdot U \cdot \dot{V}} = \text{const}$$

and hence we can write

$$q = \text{const} \cdot t \cdot \rho \cdot U \cdot \dot{V} \qquad (d)$$

Observe that this relation is not only much simpler, but also *much* more informative than the corresponding formula (c), which was obtained by using only the original set of "conventional" dimensions.

⇑

Example 16-18. Free Vibration of a Massive Body on a Weightless Spring (II)
(adapted from Ref. 44, p. 59)

We dealt with this problem in Example 11-1 (Art. 11.1). Now we treat the subject a little differently. Given a weightless container, of volume V, filled with liquid. The container is suspended on a helical weightless spring and is set to vertical motion. Fig. 16-14 shows the arrangement.

Figure 16-14
Container with liquid, suspended on a spring, oscillates vertically

METHODS OF REDUCING THE NUMBER OF DIMENSIONLESS VARIABLES   **453**

Question: What will the oscillation period be?
The relevant variables and their dimensions are as follows:

Variable	Symbol	Dimension
oscillation period	T	s
spring constant	k	kg/s^2
volume of container	V	m^3
density of liquid	ρ	kg/m^3
gravitational acceleration	g	m/s^2

We have five variables and three dimensions. Hence there are $5 - 3 = 2$ dimensionless variables determined by the Dimensional Set

	T	k	V	ρ	g
m	0	0	3	−3	1
kg	0	1	0	1	0
s	1	−2	0	0	−2
π_1	1	0	$-\frac{1}{6}$	0	$\frac{1}{2}$
π_2	0	1	$-\frac{2}{3}$	−1	−1

from which

$$\pi_1 = T \cdot \frac{\sqrt{g}}{\sqrt[6]{V}}; \qquad \pi_2 = \frac{k}{\rho \cdot g \cdot V^{2/3}} \qquad (a)$$

and hence we can write

$$\pi_1 = \Psi\{\pi_2\} \qquad (b)$$

where Ψ is an as-yet undefined function. By (a) and (b) now

$$T = c \cdot \frac{V^{1/6}}{\sqrt{g}} \cdot \Psi\left\{\frac{k}{\rho \cdot g \cdot V^{2/3}}\right\} \qquad (c)$$

where c is a numeric constant. This formula is not very useful. In fact it is not useful at all, since the unknown function Ψ encompasses *all* the independent variables. What can be done?
We increase the number of dimensions by introducing a new *dimension:* volume, symbol v. Thus, the Dimensional Set becomes

	T	k	V	ρ	g
m	0	0	0	0	1
v	0	0	1	−1	0
kg	0	1	0	1	0
s	1	−2	0	0	−2
π_1	1	$\frac{1}{2}$	$\frac{1}{2}$	$-\frac{1}{2}$	0

Carefully note three important characteristics of the above set:

454   APPLIED DIMENSIONAL ANALYSIS AND MODELING

First, we used symbol V for the *variable,* and v (lowercase!) for the *dimension* of the volume.

Second, the dimension of density ρ has changed from kg/m³ to kg/v. Why? Because if we had used kg/m³, then dimension v would have appeared in *only one* variable V, whereupon, by Theorem 11-1, Art. 11.1, V would have been rendered *dimensionally irrelevant,* which of course is nonsense. Therefore we had to have at least two variables in which the new dimension v appears. One was obviously volume V, and the other, by necessity, density ρ.

Third, the column in matrix **C** under variable g is all zeros. (there is only one row, thus there is only one zero). According to Theorem 10-4 (Art. 10.5), this means that g is a *dimensionally irrelevant* variable and therefore, by Theorem 11-3 (Art. 11.2.1), it is also a *physically irrelevant* variable. In other words, gravity does not influence the container's period of oscillation (this conclusion was also reached in Example 11-1).

By the above set, then, we have five variables and four dimensions. Therefore there is only one dimensionless variable—a constant.

$$\pi_1 = T \cdot \sqrt{\frac{k}{V \cdot \rho}} = \text{const} \tag{d}$$

from which

$$T = \text{const} \cdot \sqrt{\frac{V \cdot \rho}{k}} \tag{e}$$

Note that this formula is *much* more informative than the corresponding relation (c). The improvement was achieved—in this case—by the appropriate introduction (importation) of a *new dimension*.

⇑

16.3.3. Using Both Mass and Force Dimensions

Another *possible* method for increasing the number of dimensions is to use, for *selected* variables, *both* the dimension of mass "kg" and the dimension of force "N," e.g., for variable V_1 we use m, kg, s, and for variable V_2 we *replace* any of these three dimensions with another *trio* of dimensions in which N appears. In this way both "kg" and "N" will be in the dimensional matrix, thus the number of dimensions is increased from three to four.

The basis of these replacements is Newton's equation

$$F = M \cdot \frac{dv}{dt} \tag{16-2}$$

which dimensionally homogeneous of course. Therefore the dimensions on both sides of this formula must be identical. Thus, we can write for the dimensions

$$N = \text{kg} \cdot \frac{m}{s^2} \tag{16-3}$$

which can be transformed into any of the following forms:

$$m = N \cdot s^2 \cdot kg^{-1} \tag{16-4}$$

METHODS OF REDUCING THE NUMBER OF DIMENSIONLESS VARIABLES

$$kg = N \cdot s^2 \cdot m^{-1} \quad (16\text{-}5)$$

$$s = m^{1/2} \cdot kg^{1/2} \cdot N^{-1/2} \quad (16\text{-}6)$$

Next, suppose we have a variable V_1 whose dimension is $m^a \cdot kg^b \cdot s^c$, where the exponents are known numbers, not all zeros. There can be three cases now: we can replace either "m," or "kg," or "s."

Case 1. We replace the dimension length "m," Hence, we use (16-4)

$$m^a \cdot kg^b \cdot s^c = (N \cdot s^2 \cdot kg^{-1})^a \cdot kg^b \cdot s^c = N^x \cdot kg^y \cdot s^z \quad (16\text{-}7)$$

where x, y, and z are exponents to be determined. By this relation, since the respective exponents must be identical,

$$x = a; \quad y = -a + b; \quad z = 2 \cdot a + c \quad (16\text{-}8)$$

which can be conveniently written

$$\begin{bmatrix} x \\ y \\ z \end{bmatrix} = \mathbf{T}_m \cdot \begin{bmatrix} a \\ b \\ c \end{bmatrix} \quad (16\text{-}9)$$

where $\mathbf{T}_m$ is the *transformation matrix* with respect to *length*.

$$\mathbf{T}_m = \begin{bmatrix} 1 & 0 & 0 \\ -1 & 1 & 0 \\ 2 & 0 & 1 \end{bmatrix} \quad (16\text{-}10)$$

By (16-9) we can also have the reverse formula

$$\begin{bmatrix} a \\ b \\ c \end{bmatrix} = \mathbf{T}_m^{-1} \cdot \begin{bmatrix} x \\ y \\ z \end{bmatrix} = \begin{bmatrix} 1 & 0 & 0 \\ 1 & 1 & 0 \\ -2 & 0 & 1 \end{bmatrix} \cdot \begin{bmatrix} x \\ y \\ z \end{bmatrix} \quad (16\text{-}11)$$

Note that $\mathbf{T}_m$ is used to transform a *mass*-based system to a *force*-based one, while $\mathbf{T}_m^{-1}$ is used to perform the reverse transformation.

Example 16-19

What is 9.81 m/s² expressed in a system in which "m" is replaced by its proper combination of "N," "kg," and "s"? We have $a = 1$, $b = 0$, $c = -2$ and hence, by (16-9) and (16-10),

$$\begin{bmatrix} x \\ y \\ z \end{bmatrix} = \begin{bmatrix} 1 & 0 & 0 \\ -1 & 1 & 0 \\ 2 & 0 & 1 \end{bmatrix} \cdot \begin{bmatrix} 1 \\ 0 \\ -2 \end{bmatrix} = \begin{bmatrix} 1 \\ -1 \\ 0 \end{bmatrix}$$

Therefore $x = 1$, $y = -1$, $z = 0$, and thus, by (16-7),

$$9.81 \text{ m/s}^2 = 9.81 \text{ N}^1 \cdot \text{kg}^{-1} \cdot \text{s}^0 = 9.81 \text{ N/kg}$$

Note that *only* the dimension changes, the magnitude does not!

Case 2. We replace the dimension of mass "kg," thus (16-5) is applicable. We write

$$m^a \cdot kg^b \cdot s^c = m^a \cdot (N \cdot s^2 \cdot m^{-1})^b \cdot s^c = m^x \cdot N^y \cdot s^z \tag{16-12}$$

where x, y, and z are exponents to be determined. Since the respective exponents must be identical on both sides of the formula,

$$x = a - b; \quad y = b; \quad z = 2 \cdot b + c \tag{16-13}$$

which can be written

$$\begin{bmatrix} x \\ y \\ z \end{bmatrix} = \mathbf{T}_{kg} \cdot \begin{bmatrix} a \\ b \\ c \end{bmatrix} \tag{16-14}$$

where $\mathbf{T}_{kg}$ is the *transformation matrix* with respect to *mass*

$$\mathbf{T}_{kg} = \begin{bmatrix} 1 & -1 & 0 \\ 0 & 1 & 0 \\ 0 & 2 & 1 \end{bmatrix} \tag{16-15}$$

By (16-14) and (16-15), the *reverse* relation is

$$\begin{bmatrix} a \\ b \\ c \end{bmatrix} = \mathbf{T}_{kg}^{-1} \cdot \begin{bmatrix} x \\ y \\ z \end{bmatrix} = \begin{bmatrix} 1 & 1 & 0 \\ 0 & 1 & 0 \\ 0 & -2 & 1 \end{bmatrix} \cdot \begin{bmatrix} x \\ y \\ z \end{bmatrix} \tag{16-16}$$

Example 16-20

The dimension of the universal gravitational constant expressed in "m," "N," "s" is $[k] = m^4 \cdot N^{-1} \cdot s^{-4}$. What is the dimension of k using "m," "kg," and "s"? By (16-12), $x = 4$, $y = -1$, $z = -4$. Thus, by (16-16) and (16-15),

$$\begin{bmatrix} a \\ b \\ c \end{bmatrix} = \begin{bmatrix} 1 & 1 & 0 \\ 0 & 1 & 0 \\ 0 & -2 & 1 \end{bmatrix} \cdot \begin{bmatrix} 4 \\ -1 \\ -4 \end{bmatrix} = \begin{bmatrix} 3 \\ -1 \\ -2 \end{bmatrix}$$

from which $a = 3$, $b = -1$, $c = -2$ and so, by (16-12), the desired dimension of k is $[k] = m^3/(kg \cdot s^2)$.

⇑

Case 3. If the dimension of time "s" is replaced, then (16-6) is applicable. We can write

$$m^a \cdot kg^b \cdot s^c = m^a \cdot kg^b \cdot (m^{1/2} \cdot kg^{1/2} \cdot N^{-1/2})^c = m^x \cdot kg^y \cdot N^z \tag{16-17}$$

where x, y, and z are exponents to be determined. Since the respective exponents on both sides must be identical,

$$x = a + \frac{c}{2}; \quad y = b + \frac{c}{2}; \quad z = -\frac{c}{2} \tag{16-18}$$

METHODS OF REDUCING THE NUMBER OF DIMENSIONLESS VARIABLES **457**

This can be written

$$\begin{bmatrix} x \\ y \\ z \end{bmatrix} = \mathbf{T}_s \cdot \begin{bmatrix} a \\ b \\ c \end{bmatrix} \quad (16\text{-}19)$$

where $\mathbf{T}_s$ is the *transformation matrix* with respect to *time*

$$\mathbf{T}_s = \begin{bmatrix} 1 & 0 & \frac{1}{2} \\ 0 & 1 & \frac{1}{2} \\ 0 & 0 & -\frac{1}{2} \end{bmatrix} \quad (16\text{-}20)$$

By (16-19) and (16-20), the reverse relation is

$$\begin{bmatrix} a \\ b \\ c \end{bmatrix} = \mathbf{T}_s^{-1} \cdot \begin{bmatrix} x \\ y \\ z \end{bmatrix} = \begin{bmatrix} 1 & 0 & 1 \\ 0 & 1 & 1 \\ 0 & 0 & -2 \end{bmatrix} \cdot \begin{bmatrix} x \\ y \\ z \end{bmatrix} \quad (16\text{-}21)$$

Example 16-21

The speed of an object is $v = 13$ m/s. What is the dimension of this speed in a system which uses the dimensions "m," "kg," "N," i.e., does not include "s"? By (16-17), $a = 1$, $b = 0$, $c = -1$ and hence, by (16-19) and (16-20)

$$\begin{bmatrix} x \\ y \\ z \end{bmatrix} = \begin{bmatrix} 1 & 0 & \frac{1}{2} \\ 0 & 1 & \frac{1}{2} \\ 0 & 0 & -\frac{1}{2} \end{bmatrix} \cdot \begin{bmatrix} 1 \\ 0 \\ -1 \end{bmatrix} = \begin{bmatrix} \frac{1}{2} \\ -\frac{1}{2} \\ \frac{1}{2} \end{bmatrix}$$

Thus $x = \frac{1}{2}$, $y = -\frac{1}{2}$, $z = \frac{1}{2}$, and therefore, by (16-17)

$$v = 13 \ \frac{m}{s} = 13 \ \frac{\sqrt{m \cdot N}}{\sqrt{kg}}$$

in which the dimension "s" does not appear on the right side.

⇑

But now we must sound an emphatic warning. The method discussed in this heading—i.e., using *both* mass-based and force-based dimensions—is *extremely* risky. Any conclusion drawn from dimensional analysis which employed this type of "hybrid dimensioning" must be thoroughly scrutinized, and—*before* accepting the results—should be verified by *other* means as well. This being the case, why use the method at all if it is so vulnerable to error? The answer is: don't! But, for mainly academic interest, and because at least one reputable scientist used the technique, the present author feels that this method should be included in a book such as this which deals in depth with dimensional topics. Hence this brief chapter.

The following example illustrates the method—and its perils.

458   APPLIED DIMENSIONAL ANALYSIS AND MODELING

Example 16-22.   Terminal Velocity of a Sphere Slowly Descending in a Viscous Liquid (II) (adapted from Ref. 44, p. 66)

This problem was dealt with already in Example 16-7 (Art. 16.1). We shall now use the method of "hybrid dimensioning" to arrive at the same, and correct, result. The sphere in question descends *slowly* in the liquid, so inertial effect can be neglected. The following variables are considered relevant:

Variable	Symbol	Dimension	Remark
terminal speed	v	m/s	
diameter of sphere	D	m	
differential density	$\Delta\rho$	kg/m^3	between the sphere and liquid
viscosity of liquid	μ	kg/(m·s)	dynamic
gravitational acceleration	g	m/s^2	

We have five variables and three dimensions, therefore there are $5 - 3 = 2$ dimensionless variables supplied by the Dimensional Set

	v	D	$\Delta\rho$	μ	g
m	1	1	-3	-1	1
kg	0	0	1	1	0
s	-1	0	0	-1	-2
π_1	1	0	$\frac{1}{3}$	$-\frac{1}{3}$	$-\frac{1}{3}$
π_2	0	1	$\frac{2}{3}$	$-\frac{2}{3}$	$\frac{1}{3}$

from which

$$\pi_1 = v \cdot \sqrt[3]{\frac{\Delta\rho}{\mu \cdot g}}; \qquad \pi_2 = D \cdot \sqrt[3]{\frac{g \cdot (\Delta\rho)^2}{\mu^2}} \qquad (a)$$

Thus, we can write

$$\pi_1 = \Psi\{\pi_2\} \qquad (b)$$

or by (a),

$$v = c \cdot \left(\frac{\mu \cdot g}{\Delta\rho}\right)^{1/3} \cdot \Psi\left\{D \cdot \sqrt[3]{\frac{g \cdot (\Delta\rho)^2}{\mu^2}}\right\} \qquad (c)$$

where c is a numeric constant and Ψ is an as-yet unknown function. We see that this complex relation is not useful at all since all the independent variables are under the unknown function Ψ. To improve our status, we now increase the number of dimensions by using the extra dimension "N" (for force). At least two variables will be involved, because if only one variable were affected, then it would be the only *one* having N as a dimension, and hence, by Theorem 10-4 (Art. 10.5.1), this variable would be rendered *dimensionally irrelevant*.

So, we select v and g, and eliminate—say—the dimension of length "m." Case 1 therefore applies for v and hence its dimension will be $[v] = $ m^1·kg^0·s^{-1} ≡ m^a·kg^{b}s^c. Consequently, $a = 1$, $b = 0$, $c = -1$, thus, by (16-9) and (16-10),

METHODS OF REDUCING THE NUMBER OF DIMENSIONLESS VARIABLES   **459**

$$\begin{bmatrix} x \\ y \\ z \end{bmatrix} = \mathbf{T}_m \cdot \begin{bmatrix} 1 \\ 0 \\ -1 \end{bmatrix} = \begin{bmatrix} 1 & 0 & 0 \\ -1 & 1 & 0 \\ 2 & 0 & 1 \end{bmatrix} \cdot \begin{bmatrix} 1 \\ 0 \\ -1 \end{bmatrix} \tag{d}$$

from which $x = 1$, $y = -1$, $z = 1$. By (16-7)

$$[v] = N^x \cdot kg^y \cdot s^z = N^1 \cdot kg^{-1} \cdot s^1 = \frac{N \cdot s}{kg} \tag{e}$$

For g we have $[g] = m^1 \cdot kg^0 \cdot s^{-2}$ giving us $a = 1$, $b = 0$, $c = -2$. Thus, by (16-9) and (16-10),

$$\begin{bmatrix} x \\ y \\ z \end{bmatrix} = \mathbf{T}_m \cdot \begin{bmatrix} 1 \\ 0 \\ -2 \end{bmatrix} = \begin{bmatrix} 1 & 0 & 0 \\ -1 & 1 & 0 \\ 2 & 0 & 1 \end{bmatrix} \cdot \begin{bmatrix} 1 \\ 0 \\ -2 \end{bmatrix} = \begin{bmatrix} 1 \\ -1 \\ 0 \end{bmatrix} \tag{f}$$

and, by (16-7) its dimension is

$$[g] = N^x \cdot kg^y \cdot s^z = N^1 \cdot kg^{-1} \cdot s^0 = \frac{N}{kg} \tag{g}$$

Now we are ready to compose the Dimensional Set

	v	D	$\Delta\rho$	μ	g
m	0	1	-3	-1	0
kg	-1	0	1	1	-1
N	1	0	0	0	1
s	1	0	0	-1	0
π_1	1	-2	-1	1	-1

yielding

$$\overline{\pi}_1 = \frac{v \cdot \mu}{D^2 \cdot \Delta\rho \cdot g} = \text{const} \tag{h}$$

from which

$$v = \text{const} \cdot \frac{D^2 \cdot \Delta\rho \cdot g}{\mu} \tag{i}$$

This is a valid relation, and—as is readily seen—it is *much* more informative (and simpler) than the corresponding formula (c), which is useless.

At this point, the reader may ask if the process is really *this* simple? Do we merely have to select two variables, express them in a way which involves dimension of force N, and we have reached our goal? The answer is an emphatic NO! Here is a counterexample: Suppose we select D and $\Delta\rho$ from which the dimension "m" is eliminated, i.e., it is replaced by the suitable combination of "kg," "N," and "s." Then, by the procedure used for v and g, we obtain the dimensions of D and $\Delta\rho$ as follows: $[D] = N^1 \cdot kg^{-1} \cdot s^2$ and $[\Delta\rho] = N^{-3} \cdot kg^4 \cdot s^{-6}$. Therefore the Dimensional Set will be

	v	D	$\Delta\rho$	μ	g
m	1	0	0	-1	1
kg	0	-1	4	1	0
N	0	1	-3	0	0
s	-1	2	-6	-1	-2
$\overline{\overline{\pi}}_1$	1	-1	$-\dfrac{1}{3}$	$\dfrac{1}{3}$	$-\dfrac{2}{3}$

yielding

$$\bar{\bar{\pi}}_1 = \frac{v}{D} \cdot \sqrt[3]{\frac{\mu}{\Delta \rho \cdot g^2}} = \text{const} \qquad (j)$$

from which

$$v = \text{const} \cdot D \cdot \sqrt[3]{\frac{\Delta \rho \cdot g^2}{\mu}} \qquad (k)$$

This relation is very much different from (i), therefore both cannot be right. But (i) is right, therefore (k) is wrong! Note that *dimensionally* (k) is correct, but *numerically* it is false.

If we now consider the two independent dimensionless variables π_1 and π_2 derived in (a), we see that $\bar{\pi}_1$, defined in (h), can be expressed as

$$\bar{\pi}_1 = \pi_1 \cdot \pi_2^{-2} \qquad (l)$$

and $\bar{\bar{\pi}}_1$, defined in (j), as

$$\bar{\bar{\pi}}_1 = \pi_1 \cdot \pi_2^{-1} \qquad (m)$$

and we see again that (l) and (m) are different. But (l) is right, therefore (m) must be wrong, although dimensionally both are correct.

⇑

In summary, it is very tempting to employ the method of "hybrid dimensioning"—for it is neat, simple, and elegant—but, unfortunately, wrong (most of the time). *Occasionally,* it provides the right answer, but one cannot be sure whether this will happens in a particular application at hand. Therefore the considered advice of *this* author: Don't use it!

Despite the above elaborated pitfalls, some writers advocates its use. For example Bridgeman—one of the most distinguished authorities in the field—spends a major portion of a chapter in his book, *Dimensional Analysis,* extolling the attributes of the method, and totally ignoring its manifestly obvious dangers (Ref. 44, p. 66).

16.4. PROBLEMS

Solutions are in Appendix 6.

16/1 Terminal velocity of a sphere slowly descending in a viscous liquid (III). The subject sphere descends so slowly that inertial effects can be neglected. The arrangement and variables are identical to those in Example 16-22. By elementary heuristic reasoning prove that the (constant) *single* dimensionless "variable" (obtained by the elimination of the indicated dimension) is *false* in each of the following cases.

  (a) Dimension "m" is eliminated from variables μ and g;
  (b) Dimension "kg" is eliminated from variables D and μ;

(c) Dimension "s" is eliminated from variables v and μ;
(d) Dimension "s" is eliminated from variables $\Delta\rho$ and μ;
(e) Dimension "s" is eliminated from variables v and g.

16/2 Given the *monomial* relation among five variables $V_1, V_2, \ldots, V_5$

$$V_1 = \frac{V_2^3 \cdot V_4^2}{V_3^2 \cdot V_5^4}$$

and the fact that the rank of their dimensional matrix is 3. Decide whether the indicated *fusions* of variables reduce the number of dimensionless variables by 1.

(a) $V_c = V_1 \cdot V_3^2$
(b) $V_c = V_2 \cdot V_3$
(c) $V_c = V_4/V_3$
(d) $V_c = V_3 \cdot V_5^2$
(e) $V_c = V_4^3/V_5^2$
(f) $V_c = V_3^2 \cdot V_5$
(g) $V_c = V_2^3/V_1$
(h) $V_c = \sqrt{V_3^3 \cdot V_4}$
(i) $V_c = \sqrt{V_2^3 \cdot V_4}$
(j) $V_c = V_1^{2.803} \cdot V_5^{11.212}$
(k) $V_c = (V_1^{16}/V_2^{48})^{1/25}$
(l) $V_c = V_5^{1.304} \cdot V_1^{0.326}$

16/3 Given the relation among five variables $V_1, V_2, \ldots, V_5$

$$V_1 = \frac{V_2^2 \cdot V_3^2}{V_5^2} + \frac{V_3^2 \cdot V_5^2}{V_4}$$

and the fact that the rank of their dimensional matrix is 5. Decide in each of the cases below whether the indicated fusion results in a reduction in the number of dimensionless variables.

(a) $V_c = V_2 \cdot V_4$
(b) $V_c = V_3^2 \cdot V_5^2$
(c) $V_c = V_1 \cdot V_5^2$
(d) $V_c = V_1 \cdot V_4$
(e) $V_c = V_2/V_5$
(f) $V_c = V_3/V_4$
(g) $V_c = V_5^2/V_4$
(h) $V_c = V_4^2/V_5^2$

16/4 Given the formula

$$V_1 = \frac{V_2 \cdot V_3 \cdot V_5^2}{V_4 \cdot \cos(V_2 \cdot V_3)}$$

where $V_1, V_2, \ldots, V_5$ are *dimensional* physical variables. Prove that the number of *dimensionless* variables formed by $V_1, V_2, \ldots, V_5$ cannot be less than two.

16/5 Given the formula

$$V_1 = \frac{V_2 \cdot V_3^2}{V_4} + \frac{V_5 \cdot V_4^2}{V_3^4}$$

where $V_1, V_2, \ldots V_5$ are *dimensional* physical variables. How many different *fusions* of any two of the variables $V_1, V_2, \ldots V_5$ exist which reduces the number of *dimensionless* variables representing the above relation?

16/6 Curvature of a bimetallic thermometer (adapted from Ref. 22, p. 73). The operation of a bimetallic thermometer is based on the *difference* in linear thermal expansion coefficients between two strips of metal secured to each other. At a temperature change Δt, the differential length (caused by the different linear expansions of the strips) generates the shape of a circular arc with a subtended angle α (Fig. 16-15).

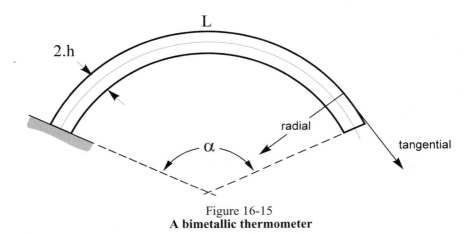

Figure 16-15
A bimetallic thermometer

(a) By using directional dimensions for lengths, derive a dimensionless relation incorporating physical variables h, α, L, Δt, and $\Delta\beta$, where Δt and $\Delta\beta$ are the temperature change and the differential linear expansion coefficient, respectively.

(b) Assume a *monomial* form among the dimensionless variables derived in (a), and the fact that α is proportional to Δt, then derive the functional relation for the cited physical variables.

CHAPTER 17
DIMENSIONAL MODELING

17.1. INTRODUCTORY REMARKS

The word modeling has many meanings; four frequently used ones are:

- mathematical modeling;
- mock-up modeling;
- garment and cosmetic modeling;
- dimensional modeling.

HERMAN © Jim Unger/Dist by Newspaper Enterprise Association, Inc.

Mathematical modeling is just a fancy expression for analytical derivation, and hence it falls outside the main topic of this book; it is treated extensively by a *very* large number of excellent publications.

Mock-up modeling is the construction of geometrically similar nonfunctional pieces to study mainly the aesthetic appearance or the ergonomic characteristics of a future product. For example, a clay model of an automobile may be made for executives to win their final and expert approvals for production of the vehicle. Supposedly, Ford's *Edsel* was such an automobile.

Garment and cosmetic modeling is certainly unrelated to our subject matter. It is nonetheless remarked that this activity is the only one of the four mentioned here that may present some *real* fascination for both the participant and the observer, both male and female—albeit, it is hoped, for different reasons. Regretfully and reluctantly, however, this type of modeling shall not be pursued further in these pages.

Dimensional modeling. The purpose for dimensional modeling is to be able to experiment on a scaled replica—called *model*—of the original construction—called *prototype,* and then to *project* the results obtained from the model to the prototype. According to Baker (Ref. 34, p. 1),

> A model is a device which is so related to a physical system [prototype] that observations on the model may be used to predict accurately the performance of a physical system in the desired respect.

We shall, of course, only be dealing with dimensional modeling. What follows, then, is a description of this very useful technique.

One may reasonably ask: why model? The use of models to facilitate the design and testing of engineering or physical systems is often *very* beneficial. Experimentation on models that are properly designed, constructed, and used significantly reduces the likelihood of committing costly mistakes on the prototype. As Baekeland stated: "Commit your blunders on a small scale, and make your profits on a large scale" (Ref. 41, p. 210). Moreover, there are cases when testing the full-scale product (prototype) is not only impractical, but impossible. For example, predicting the erosion rate of a river bank would not be possible without relying on model experimentation and studies of pertinent river characteristics.

The greatest benefits of using models occur in cases when the analytic expression of the sought-after characteristics or variables are not available or only inaccurately known. The above-mentioned example is such an instance.

In general, modeling is warranted when we wish to obtain

- Experimental data valid for the prototype (e.g., deflection of a beam loaded in a complex way)
- Behavior of a physical system (e.g., vibration of a drop of liquid constrained by surface tension)
- Functional relationship among variables, if the analytic form is too complex, inaccurate, or unknown (e.g., heat-transfer problems)

Therefore, in general, modeling is advisable when

- The prototype is too small or too large
- The prototype is not accessible
- The magnitudes of variables on the prototype are too small or too large to be measured
- Testing on the prototype would take too long or too short a time

It is a misconception that the model must be physically smaller than the prototype. Although *small-scale* models are the most common, they are by no means the universal choice; indeed, it is often neither possible nor even desirable to construct a model smaller than the prototype. For example, to determine the dynamic and flexural characteristics of miniature mechanisms, obviously, a *large-scale* model is appropriate.

Another false belief is that a model must be geometrically similar to the prototype. Again, although geometric similarity is often desirable, it is not always possible to achieve, and yet a useful model may still be easily constructed and experimented on (See Art. 17.4, Example 17-3). Indeed, sometimes the greatest advantage of a model is that it is *geometrically dissimilar* to the prototype. As this story unfolds, we will see many applications of geometrically dissimilar models.

Models are used in an extremely wide variety of engineering design and testing activities. Just a few of the most beneficial applications relate to:

- Nuclear reactor vessels and other reinforced and prestressed concrete containers
- Underwater structures (e.g., dams, platforms)
- Structures that, in general, are large or complex configurations (e.g., bridges)
- Events associated in general with the aerodynamic, hydrodynamic, and impact-related effects, such as blasts, wind, waves, etc.
- Problems involving heat propagation, in particular conduction and convection (both natural and forced)

Modeling has also found some very ingenious applications in areas as diverse as biomechanics, electromagnetism, physiology, mathematics, and geometry.

On the other hand, some subjects are generally unsuited to modeling experimentation. Among these we mention

- Crack propagation in structures
- Creep effects
- Shrinkage effects
- Adhesive and bond-splitting effects

There are also other limitations that restrict the use of models. In this respect the following points should be mentioned:

- To model, it is *necessary* to understand the basic physical structure of the phenomenon to be modeled, to the extent that *only* the relevant, but *all* the relevant, variables, parameters, and constants are considered. *Not necessary* is a *detailed* analytical knowledge of how the chosen variables, constants, and parameters influence the results. This is one of the most attractive features of the modeling method.
- The cost of experimentation on the model must be realistically estimated. These projections should include, in particular:
  1. Theoretical work to design the model, derivation of the Model Law (see Art.17.4.2);
  2. Building the model (material, labor, assembly fixtures and rigs, etc.);
  3. Building and acquisition of test set-up, instrumentation, and computing equipment;
  4. Execution of tests and experimentation;
  5. Data reduction, computation, conclusion;
  6. Preparation of documentation and reports.

  Of the above, probably (2) and (3) cost the most, while the cost of the rest could be quite modest.
- Is there a theory that could be applied to reliably predict the outcome of the experimentation on the model? If there is, modeling would be a waste of effort.
- Is modeling of phenomenon possible at all? Not every phenomenon in nature can be modeled.

To conclude this introduction, the sequential process required in modeling must be mentioned. To be effective, modeling should be carried out in distinct *phases* and in a well-established *chronological* order:

1. Establish the theoretical background, including—most importantly—the Model Law;
2. Design the model;
3. Build the model; this includes test rigs, instrumentation, and computing equipment;
4. Execute the experiments and tests;
5. Evaluate test data (data reduction);
6. Draw conclusions;
7. Prepare documentation and reports.

Very often the conclusions indicate the need to modify the model, in which case the above protocol must be repeated starting at (2) or (3). Indeed, the principal advantage of having a model at all is that it is very much simpler, faster, more economical, etc., to modify the model than the prototype itself.

Of the above items, (1), (2), and (5) are discussed in the following pages. The other items—for obvious reasons—fall outside the scope of this book.

17.2. HOMOLOGY

Broadly speaking homology means *correspondence*. If the prototype and its model are geometrically similar, then a point-to-point correspondence exists between these two bodies. The pairs of points so associated are said to be *homologous* points. Other variables, like stress, deformation, temperature, mass, time, speed, acceleration, heat transferred, etc. can also be made to correspond between the prototype and its model. Hence these characteristics can also be homologous. But here some caution is in order. For example, stresses experienced at homologous points in two bodies are homologous only if they occur at *homologous times*.

The concept of homologous times may be difficult to comprehend at first, since it does not ordinarily entail *simultaneity*. This fact may be astonishing to some, since we are all accustomed to a world in which for *all* events time passes at an *unvarying rate* from the infinite past to the infinite future. Why, then, could any instant not be considered to be homologous for two processes (bodies)? The answer can be found in the following definition:

Definition 17-1. *Two processes (bodies) assume homologous states (positions, configurations, etc.) at homologous times.*

This is a somewhat circular definition, since it defines homologous times in terms of homologous states, and homologous states in terms of homologous times. But, for the sake of expediency we shall conveniently overlook this philosophical blemish, since the essence of the general idea is clearly evident. The following example illuminates the concept of homology in the time-domain.

Example 17-1

Two geometrically similar straight beams A and B are vibrating laterally at 1 Hz and 2 Hz frequencies, respectively (Fig. 17-1).

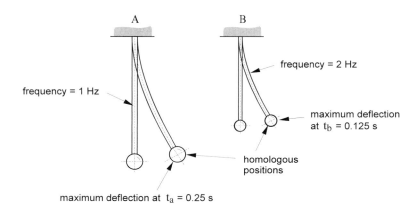

Figure 17-1
Two vibrating beams assume homologous positions at homologous times
(although times t_a and t_b are unequal, they are homologous)

Obviously, the straight positions of the beams are homologous, as are the maximum deflection positions. If the time is reckoned from the straight positions of the beams, then the maximum deflections of beams A and B will occur at $t_a = 0.25$ s and $t_b = 0.125$ s, respectively. Therefore t_a and t_b are *homologous times* (although they are unequal), since at these times the beams are at *homologous positions* (Definition 17-1).

Similarly, and by the same definition, the maximum deflection configurations are deemed to be homologous, because the beams reach these positions at homologous times. Here we clearly see the circularity of the reasoning, but this does not interfere with the process. There is one thing that should be noted though: It is usually easier to identify homologous positions, configurations, deformations, shapes, etc., than other physical attributes, e.g., times, temperatures, stress, etc. That is, the *space-domain* can usually be considered as a reference frame in which other physical variables can be expressed. In the above example, the positions of maximum deformations of the beams are easy to *identify* as homologous configurations and therefore can be conveniently used to define homologous times.

⇑

17.3. SPECIFIC SIMILARITIES

When are two bodies, systems, structures, etc. similar? Broadly speaking, when we hear somebody say "A is similar to B," we automatically interpret it as "A *looks* similar to B, but it is larger (or smaller)." Here of course geometric similarity is meant. But there are many other types of so-called "specific" similarities. There is a very large number of (possible) variables, and since if even *one* is brought to correspondence for two given systems, then we can say that the two systems are *similar* with respect to that variable. Therefore there exists, in theory, a very large (practically infinite) number of similarities between any two systems.

Since this loose definition does not serve our present restricted purpose (to furnish workable relationships between a prototype and its model), we must be a little more specific in defining types of similarities.

First, for general consumption, we define certain specific similarities which are frequently used.

17.3.1. Geometric Similarity

Two different-sized objects are geometrically similar if, with sufficient enlargement, the smaller can be brought to exact coincidence with the larger. For example, two circles are always similar, because even if they are of different sizes, with proper magnification the smaller can be enlarged so that the two circles can be brought into exact coincidence. Similarly, two spheres are always similar, as are two cubes (regular hexahedrons), and pairs of all the other regular solids. These are simple observations. However, we are a little less certain about parabolas. Are two parabolas *always* similar? To find the answer we proceed as follows:

Given an arbitrary plane curve defined in a polar coordinate system as shown in Fig. 17-2. Thus, if ρ is the radius-vector pointing from the *pole* to a running point of the curve, and φ is the subtended angle between ρ and a *reference direction,* then the curve is defined by

DIMENSIONAL MODELING

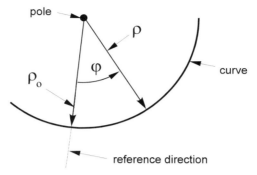

Figure 17-2
An arbitrary plane curve is defined in a polar coordinate system

$$\rho = \Psi\{\varphi\} \qquad (17\text{-}1)$$

where Ψ designates a function. Now if $\varphi = 0$, then $\rho = \rho_0 = \Psi\{0\}$, which is the *reference length* of the radius-vector. In view of this, we can write

$$\frac{\rho}{\rho_0} = \frac{\Psi\{\varphi\}}{\Psi\{0\}} = \Phi\{\varphi\} \qquad (17\text{-}2)$$

where Φ designates a function. Therefore two curves are *geometrically similar* if, and only if, their Φ functions are identical. Keeping this in mind, let us return to the question of the parabola. The polar equation of a parabola is

$$\frac{\rho}{F} = \frac{2 \cdot (1 - \cos \varphi)}{\sin^2 \varphi} \qquad (17\text{-}3)$$

where the reference direction is the axis, and the reference length the focal length F (Fig. 17-3).

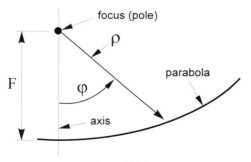

Figure 17-3
Parabola is defined in a polar coordinate system

470   APPLIED DIMENSIONAL ANALYSIS AND MODELING

In relation (17-3) we see that

$$\Phi\{\varphi\} = \frac{2 \cdot (1 - \cos\varphi)}{\sin^2\varphi} \qquad (17\text{-}4)$$

is independent of the focal length, therefore $\Phi(\varphi)$ is the same for all parabolas, and hence all parabolas are geometrically similar. In Fig. 17-4 there are two parabolas. Their sizes are much different, but their shapes are similar; if we magnify parabola 2 by 16, we get parabola 1, exactly.

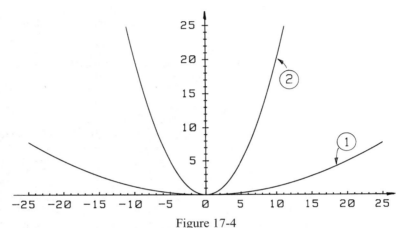

Figure 17-4
Two geometrically similar parabolas, differing only in size
Parabola 1 is exactly 16 times the enlargement of parabola 2

Of course not every curve has this self-similar type of characteristic. For example, an ellipse is defined by two independent parameters, which can be the lengths of the semimajor and semiminor axes "a" and "b." Thus the polar equation of an ellipse, if the pole is at the center, is

$$\frac{\rho}{a} = \left[\cos^2\varphi + \left(\frac{a}{b}\right)^2 \cdot \sin^2\varphi\right]^{-1/2} \qquad (17\text{-}5)$$

where "a" is the selected reference length. Therefore, by (17-2), the Φ function is

$$\Phi\{\varphi\} = \left[\cos^2\varphi + \left(\frac{a}{b}\right)^2 \cdot \sin^2\varphi\right]^{-1/2} \qquad (17\text{-}6)$$

and we see that Φ is independent of the defining parameters "a" and "b" *only* if the a/b ratio is constant. Consequently, not all ellipses are similar. For example, the two ellipses seen in Fig. 17-5 are dissimilar, since $a_1/b_1 = 20/16 = 1.25$, whereas $a_2/b_2 = 16/4 = 4$, and $1.25 \neq 4$.

Of course the same process can be used to prove that every paraboloid of surface of revolution is similar, and correspondingly, in general, two ellipsoids are dissimilar, unless the *ratios* of the lengths of their respective diameters are the same.

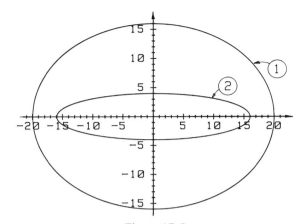

Figure 17-5
Two geometrically dissimilar ellipses
They are dissimilar because the *ratios* of the lengths
of their major and minor axes are different

17.3.2. Kinematic Similarity

Kinematics deals with motions. Therefore, the involved variables are length and time; forces play no part. It follows that two moving systems are kinematically similar if homologous points experience the same motion in homologous times. This criterion encompasses both linear and angular displacements, velocities, and accelerations. It is emphasised that simultaneity is usually not maintained; i.e., homologous times in general are not equivalent to chronological times (see Example 17-1). This topic will be detailed in Art. 17.5.

17.3.3. Dynamic Similarity

Dynamics deals with forces. Therefore, two bodies are dynamically similar if their homologous points experience the same forces in homologous times. Here "force" includes concentrated force, distributed force, and moments (originating from whatever source). Thus, we consider forces and moments by

- externally applied (bodies in contact)
- hydrostatic effects (pressure)
- inertial effects
- internal elastic effects
- gravitational effects
- surface tension (molecular attraction)
- buoyancy effects
- frictional and viscous effects

If ratios from combinations of these forces are identical for two systems, then we say that the two systems are dynamically similar with respect to the involved forces. Some of these ratios are called *numbers* and are associated with *names* of famous scientists who contributed significantly to the relevant discipline. The more important ones are listed in Fig. 17-6.

Ratio of forces	Name
inertial to viscous	Reynolds' number
inertial to elastic	Cauchy's number
inertial to gravitational	Froude's number
inertial to surface tension	Weber's number
pressure to inertial	Euler's number

Figure 17-6
Some named ratios of various manifestations of forces
A more extensive list, with descriptions and compositions, is in Appendix 3

17.3.4. Thermal (or Thermic) Similarity

Two bodies are thermally similar if at their homologous points, or surfaces, they have equal or homologous temperatures at homologous times. This is a rather loose definition, and unfortunately it is difficult to make it more specific. The main idea is this: if the heat-flow pattern in two bodies is similar, then these bodies can be considered thermally similar.

17.4. DIMENSIONAL SIMILARITY

Dimensional similarity is the cornerstone of dimensional modeling, and therefore we are going to discuss it in some detail. Moreover, the text will be augmented with several illustrative examples, which the reader is strongly urged to follow.

As we saw in the preceding chapters, the behavior of a physical system is defined by the complete set of *dimensionless* variables formed by the relevant *physical* variables. This fact suggests that if two systems have the *same* numerical values for all the defining dimensionless variables, then these two systems are *dimensionally similar*. Further, if there are two dimensionally similar systems, then their behavior can be closely correlated and hence the results of measurements on either one can be "projected" to the other. This is the *essence* of dimensional modeling. Therefore the first step in a model experiment is to construct the complete set of dimensionless variables relevant to the system. These dimensionless variables are then made to be equal for the model and the prototype.

We now have to assume that the list of relevant physical variables contains only *one* dependent variable. This is a very important condition, which should be emphasized. For although in theory multiple dependent-variable cases can also be dealt with, the numerous attendant complications are sometimes difficult to avoid. Therefore we formulate the following rules:

Rule 1 of Dimensional Modeling. *When forming a complete set of dimensionless variables for a system, be sure that only one dependent physical variable exists in the set.*

Rule 2 of Dimensional Modeling. *When forming a complete set of dimensionless variables for a system, be sure that the sole dependent variable (see Rule 1) appears in only one of the dimensionless variables.*

Rule 1 is simple and obvious; if by chance there is more than one dependent variable in the set, then to avoid complications this set should be broken up such that in each newly formed group there is only *one* dependent variable. However, with *proper* precautions, exceptions can be handled (see Art 14.3 and especially Example 14-9).

Rule 2 can be easily followed if the dependent physical variable is put in the **B** matrix of the Dimensional Set. As was detailed in Chapter 14, variables in the **B** matrix automatically appear in only one dimensionless variable.

From these 2 rules the following definition can be formulated:

Definition 17-2. *Two systems are dimensionally similar if all the dimensionless variables describing the systems are identical in construction and in magnitude.*

Moreover, if Rules 1 and 2 of dimensional modeling are obeyed, then we have the relevant theorem:

Theorem 17-1. *If in two systems (prototype and model) all corresponding pairs of dimensionless variables containing only independent variables are identical, then the dimensionless variables forming the remaining pair (containing the dependent variable) are also identical.*

That is, if the prototype and its model are both described by N_P dimensionless variables of pair-wise identical compositions, then if $N_P - 1$ pairs of them are pair-wise numerically identical (i.e., for the prototype and the model), then the members of the remaining one pair will also be numerically identical. Together, therefore, all N_P pairs will be numerically identical.

Proof. Behavior of a system can be described by (Buckingham's theorem)

$$\pi_1 = \Psi\{\pi_2, \pi_3, \ldots \pi_{N_P}\}$$

where N_P is the number of dimensionless variables, of which only π_1 contains the dependent physical variable. Therefore $N_P - 1$ dimensionless variables (i.e., $\pi_2, \ldots, \pi_{N_P}$) contain only independent physical variables. Consequently, if this latter group is identical in both systems, then π_1 must also be identical in both systems. This proves the theorem.

We now have a useful Corollary of Theorem 17-1

Corollary of Theorem 17-1. *If two systems have* $N_P - 1$ *dimensionless variables of pair-wise identical construction and magnitude (and containing only independent variables), then these two systems are dimensionally similar.*

Proof. By Theorem 17-1, if $N_P - 1$ dimensionless variables are pair-wise identical in two systems, then the same is true for N_P of such pairs. However, in this case, by Definition 17-2, the two systems are dimensionally similar. This proves the corollary.

Finally, we sharpen Definition 17-2 of dimensional similarity.

Definition 17-3. *If two systems (of* N_P *dimensionless variables of pair-wise identical constructions) have* n *pairs of dimensionless variables of pair-wise identical numerical values, then*

if $n = N_P - 1$,   then the two systems are *dimensionally similar;*

if $0 < n < N_P - 1$, then the two systems are *dimensionally partially similar;*

if $n = 0$,   then the two systems are *dimensionally dissimilar.*

The following example demonstrates the use of this definition (subscripts 1 and 2 designate prototype and model, respectively).

Example 17-2. Deflection of a Cantilever Upon a Concentrated Lateral Load (VIII)

Given a prototype steel ($E_1 = 2 \times 10^{11}$ N/m²) cantilever $L_1 = 2.4$ m long and of uniform rectangular cross-section of width $a_1 = 0.15$ m and thickness $b_1 = 0.05$ m, as seen in Fig. 17-7. The cantilever carries a load $F_1 = 6400$ N. By modeling experiment, we wish to determine its deflection U_1 under this load.

The first step is to determine the complete set of dimensionless variables. To this end, we list the relevant physical variables and then construct the Dimensional Set.

Variable	Symbol	Dimension	Remark
deflection	U	m	lateral
width of cross-section	a	m	uniform
length	L	m	
concentrated load	F	N	lateral
thickness of cross-section	b	m	uniform
Young's modulus	E	N/m²	

We have six variables and two dimensions, therefore there are $6 - 2 = 4$ dimensionless variables obtained by the Dimensional Set

	U	a	L	F	b	E
m	1	1	1	0	1	−2
N	0	0	0	1	0	1
π_1	1	0	0	0	−1	0
π_2	0	1	0	0	−1	0
π_3	0	0	1	0	−1	0
π_4	0	0	0	1	−2	−1

DIMENSIONAL MODELING

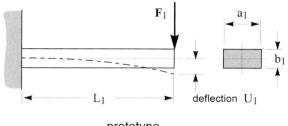

prototype

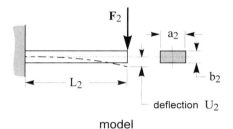

model

Figure 17-7
Modeling of a weightless laterally loaded cantilever

yielding

$$\pi_1 = \frac{U}{b}; \quad \pi_2 = \frac{a}{b}; \quad \pi_3 = \frac{L}{b}; \quad \pi_4 = \frac{F}{b^2 \cdot E} \quad \text{(a)}$$

From the given data for the *prototype*

$$\pi_2 = \frac{a_1}{b_1} = \frac{0.15}{0.05} = 3$$

We *select* for the model $E_2 = 6.65 \times 10^{10}$ N/m² (aluminium) and the width of its cross-section $a_2 = 0.03$ m. If π_2 is to be maintained

$$\pi_2 = \frac{a_2}{b_2} = 3$$

from which

$$b_2 = \frac{a_2}{\pi_2} = \frac{0.03}{3} = 0.01 \text{ m}$$

Next, if π_3 is maintained for the model, then we must have

$$\pi_3 = \frac{L_1}{b_1} = \frac{2.4}{0.05} = 48 = \frac{L_2}{b_2}$$

from which

$$L_2 = \pi_3 \cdot b_2 = 48 \cdot (0.01) = 0.48 \text{ m}$$

Finally, from the data for the prototype and the condition of constancy of π_4

$$\pi_4 = \frac{F_1}{b_1^2 \cdot E_1} = \frac{6400}{(0.05^2) \cdot (2 \times 10^{11})} = 1.28 \times 10^{-5} = \frac{F_2}{b_2^2 \cdot E_2}$$

from which

$$F_2 = \pi_4 \cdot b_2^2 \cdot E_2 = (1.28 \times 10^{-5}) \cdot (0.01^2) \cdot (6.65 \times 10^{10}) = 85.12 \text{ N}$$

Therefore if we now set $b_2 = 0.01$ m, $L_2 = 0.48$ m and $F_2 = 85.12$ N for the model, then we get the same numerical values for π_2, π_3, π_4. But the total number of dimensionless variables is $N_P = 4$. Therefore if $N_P - 1 = 3$ of them are identical—as they are here—then, by the Corollary of Theorem 17-1, our model is dimensionally similar to its prototype. Now π_2, π_3, π_4 contain only independent physical variables, therefore, by Theorem 17-1, the remaining dimensionless variable—π_1 in this case—will also be identical for both prototype and model. Hence we write

$$\pi_1 = \frac{U_1}{b_1} = \frac{U_2}{b_2}$$

from which

$$U_1 = U_2 \cdot \frac{b_1}{b_2}$$

Now we *measure* the deflection of the model—specified above and which is loaded with 85.12 N concentrated force. Say we find $U_2 = 0.0189$ m. With this datum, the above formula provides

$$U_1 = (0.0189) \cdot \frac{0.05}{0.01} = 0.0945 \text{ m}$$

If we had selected a different load for the model—not the $F_2 = 85.12$ N prescribed by complete similarity—then we could not have had the desired equivalence for three dimensionless variables, since the constancy of π_4 could not have been maintained. For example, if we had selected $F_2 = 140$ N for the model, then for the prototype we would have got $(\pi_4)_{\text{prot}} = 1.28 \times 10^{-5}$, and $(\pi_4)_{\text{mod}} = 2.10526 \times 10^{-5}$ for the model, so $(\pi_4)_{\text{prot}} \neq (\pi_4)_{\text{mod}}$; we could only have achieved *partial similarity*, and hence, the determination of the prototype's deformation (based on the measurement of the deformation of the model) would not have been possible. Note that even though *dimensional* similarity was *not* achieved in this alternative case, *geometric* similarity was.

⇑

At this point we emphasise an important fact: A model can be *dimensionally* similar to the prototype, without being *geometrically* similar to it, and vice versa. In other words, geometric similarity is neither a sufficient nor a necessary condition for dimensional similarity. The previous example demonstrated that geometric similarity was not a *sufficient* condition for dimensional similarity; the following example demonstrates that it is not a *necessary* condition, either.

Example 17-3. Thermally Induced Reaction Force in a Curved Beam

Given the *prototype* of a semicircular beam of uniform hollow cross-section. The beam is pivoted at both ends by frictionless but otherwise immovable hinges A and B—as shown in Fig. 17-8.

DIMENSIONAL MODELING

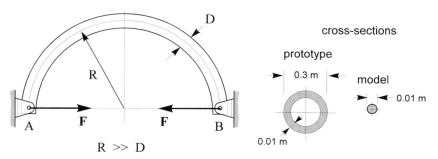

Figure 17-8
Temperature increase in a hinged semicircular beam produces radial reaction forces

Due to ambient change, the beam's temperature increases by $\Delta t = 75$ °C. It is required to construct a model to determine the radial reaction forces generated by this rise of temperature.

The cross-sections of the prototype and of the model are shown in Fig. 17-8. We see that the model is *geometrically dissimilar* to the prototype, yet—as demonstrated presently—*dimensional similarity* can be easily achieved. From this it follows that a valid model experiment can be carried out to determine the reaction forces on the prototype.

The relevant variables and their dimensions are as follows:

Variable	Symbol	Dimension	Remark
temperature increase	Δt	°C	uniform
reaction force	F	N	radial
second moment of area of cross-section	I	m⁴	uniform
Young's modulus	E	N/m²	
thermal expansion coefficient	α	1/°C	linear
centroidal radius	R	m	

We have six variables and three dimensions, therefore there are $6 - 3 = 3$ dimensionless variables characterising the system. The Dimensional Set is

	Δt	F	I	E	α	R
m	0	0	4	−2	0	1
N	0	1	0	1	0	0
°C	1	0	0	0	−1	0
π_1	1	0	0	0	1	0
π_2	0	1	0	−1	0	−2
π_3	0	0	1	0	0	−4

by which

$$\pi_1 = \Delta t \cdot \alpha; \qquad \pi_2 = \frac{F}{E \cdot R^2}; \qquad \pi_3 = \frac{I}{R^4} \qquad \text{(a)}$$

Note that the dependent variable F is in π_2. Therefore to have our model dimensionally similar to the prototype, we must have the numerical pair-wise equality of π_1 and π_3 in both of these structures.

The prototype is made of mild steel of $E_1 = 2.068 \times 10^{11}$ Pa, and $\alpha_1 = 0.00117$ 1/°C. For the model we choose aluminium of $E_2 = 6.895 \times 10^{10}$ Pa, and $\alpha_2 = 0.00231$ 1/°C. The centroidal radius of the prototype is $R_1 = 50$ m. For the given cross-sections (Fig. 17-8), we have for the prototype and model, respectively

$$I_1 = \frac{(0.3^4 - 0.28^4)\cdot\pi}{64} = 9.58893 \times 10^{-5} \text{ m}^4; \qquad I_2 = \frac{(0.01^4)\cdot\pi}{64} = 4.90874 \times 10^{-10} \text{ m}^4$$

Therefore, for the prototype, by the given data and (a)

$$\pi_3 = \frac{I_1}{R_1^4} = \frac{9.58893 \times 10^{-5}}{50^4} = 1.53423 \times 10^{-11}$$

For the model, if π_3 is maintained,

$$\pi_3 = \frac{I_2}{R_2^4} = 1.53423 \times 10^{-11}$$

from which the *obligatory* centroidal radius of the model is

$$R_2 = \sqrt[4]{\frac{I_2}{\pi_3}} = \sqrt[4]{\frac{4.90874 \times 10^{-10}}{1.53423 \times 10^{-11}}} = 2.37832 \text{ m}$$

Next, by (a) and the data given for the prototype

$$\pi_1 = (\Delta t_1)\cdot(\alpha_1) = 75\cdot(0.00117) = 0.08775$$

and if π_1 is to remain the same for the model

$$\pi_1 = (\Delta t_2)\cdot(\alpha_2)$$

from which the *obligatory* temperature rise for the model is

$$\Delta t_2 = \frac{\pi_1}{\alpha_2} = \frac{0.08775}{0.00231} = 37.987 \text{ °C}$$

Now we are ready to construct the model (since all of its geometric and material properties are known), install it into a jig, increase its temperature by 37.987 °C (above ambient), and measure the ensuing reaction force F_2. Say we measure $F_2 = 0.334$ N. By (a) then, for the model

$$\pi_2 = \frac{F_2}{E_2\cdot R_2^2} = \frac{0.334}{(6.895 \times 10^{10})\cdot(2.37832^2)} = 8.5639 \times 10^{-13}$$

and for the prototype—since π_2 must be maintained

$$\pi_2 = \frac{F_1}{E_1\cdot R_1^2} = 8.5639 \times 10^{-13}$$

from which the sought-after reaction force of the prototype is

$$F_1 = \pi_2 \cdot E_1 \cdot R_1^2 = (8.5639 \times 10^{-13})\cdot(2.068 \times 10^{11})\cdot(50^2) = 442.754 \text{ N}$$

We emphasize that although in this application there is no geometric similarity between the prototype and the model, dimensional similarity was nevertheless easily achieved and used to determine the prototype's reaction force by a modeling experiment.

⇑

17.4.1. Scale Factors

The use of scale factors greatly facilitates and enhances the modeling procedure; scale factors make the handling of data easier, the technique more comprehensible, and the process itself simpler.

A scale factor *always* refers to a particular physical variable. There is *no* such thing as a "scale factor of the model." Consequently, in any dimensional modeling experiment there are exactly as many scale factors as there are physical variables. The scale factor for a *particular* physical variable is always written with capitalized first letters, for example Density Scale Factor, or Young's Modulus Scale Factor. The capital letter *S* with appropriate (usually mnemonic) subscript is the symbol for scale factors. For example, if L is the symbol for length, then S_L is the symbol for the Length Scale Factor.

The definition of a scale factor is as follows:

Definition 17-4. *The scale factor with respect to a particular physical variable is the quotient of the magnitudes of that variable for the prototype and its model.*

For example, if the length of the prototype is $L_1 = 3.4$ m, and of its model $L_2 = 1.7$ m, then the Length Scale Factor is

$$S_L = \frac{L_2}{L_1} = \frac{1.7}{3.4} = 0.5$$

Note that in the quotient the model's value is *always* in the *numerator,* and bears the subscript 2. Correspondingly, the prototype's value is *always* in the *denominator,* and bears the subscript 1. By strictly following this convention, confusion is avoided, even in complex cases. Note also that a scale factor is *always* a *dimensionless number.* This follows from Definition 17-4.

17.4.2. Model Law

By definition, the Model Law is a relation, or a set of relations, among scale factors relevant to a particular modeling instance. Even though there may be more than one relation among scale factors, there is only one Model Law. Thus a Model Law may comprise several relations, each of which is part of the same Model Law. Note that we use capital first letters for the expression "Model Law." This is because of its uniqueness in characterizing a particular inter-relation between the prototype and its model.

The Model Law is the principal tool in the design of a modeling experiment, and therefore its determination *must precede* any physical preparation and execution of tests. In addition, a Model Law can often be used to compare the behavior and general performance of existing, or even hypothetical, systems. In these cases, of course, comparisons can be done without constructing a "model" or carrying out tests. In brief: Model Law *can* be useful without tests, but tests *cannot* be useful without Model Law. We will present a few examples to illuminate this fact.

The determination of a Model Law is *always* done on the basis of dimensionless variables which, in turn, are obtained from the Dimensional Set.

480 APPLIED DIMENSIONAL ANALYSIS AND MODELING

The best way to describe the technique of deriving the Model Law is by an example:

Example 17-4. Power Required to Tow a Barge

By modeling experiment we wish to determine the power necessary to tow a $G = 1.1768 \times 10^6$ N (= 120 metric tons) barge at a speed of $v = 3.858$ m/s (= 7.5 knots). We assume that all barges are geometrically similar, i.e., their total gross weights are uniquely defined by their characteristic linear sizes (Fig. 17-9).

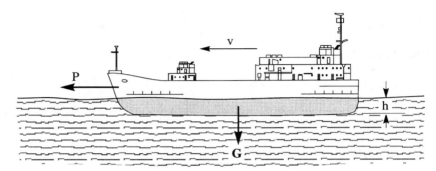

Figure 17-9
Power required to tow a barge of given weight

The relevant variables are as listed in the table below:

Variable	Symbol	Dimension	Remark
power to tow	P	m²·kg/s³	
weight of barge	G	m·kg/s²	gross
speed of barge	v	m/s	with respect to water
density of water	ρ	kg/m³	
gravitational acceleration	g	m/s²	

In the above, g is included because the barge generates waves which, by their vertical movement, are obviously affected by gravitation. Also, ρ is included because of its inertial effect as well as its influence on the "draft"—marked "h" in the figure—which, in turn, influences the towing force and hence the power needed.

We have five variables and three dimensions. Therefore there are $5 - 3 = 2$ dimensionless variables, determined by the Dimensional Set

	P	G	v	ρ	g
m	2	1	1	-3	1
kg	1	1	0	1	0
s	-3	-2	-1	0	-2
π_1	1	0	-7	-1	2
π_2	0	1	-6	-1	2

by which

$$\pi_1 = \frac{P \cdot g^2}{v^7 \cdot \rho}; \qquad \pi_2 = \frac{G \cdot g^2}{v^6 \cdot \rho} \qquad \text{(a)}$$

If we now wish to have a dimensionally similar model, the magnitude of π_2 must be *made* equal for both prototype and model. This is because $N_P = 2$, and by the Corollary of Theorem 17-1, $N_P - 1 = 1$ dimensionless variable must be identical. Moreover, this solo dimensionless variable must contain only *independent* physical variables. Thus, in our present case, this dimensionless variable is π_2. But, if π_2 is the same (for the prototype and model), then π_1 must also be the same (Theorem 17-1). Hence, if subscripts 1 and 2 are for the prototype and model, then by (a),

$$\pi_1 = \frac{P_1 \cdot g_1^2}{v_1^7 \cdot \rho_1} = \frac{P_2 \cdot g_2^2}{v_2^7 \cdot \rho_2} \tag{b}$$

$$\pi_2 = \frac{G_1 \cdot g_1^2}{v_1^6 \cdot \rho_1} = \frac{G_2 \cdot g_2^2}{v_2^6 \cdot \rho_2} \tag{c}$$

By (b)

$$\left(\frac{v_2}{v_1}\right)^7 \cdot \frac{\rho_2}{\rho_1} = \frac{P_2}{P_1} \cdot \left(\frac{g_2}{g_1}\right)^2 \tag{d}$$

and by (c)

$$\left(\frac{v_2}{v_1}\right)^6 \cdot \frac{\rho_2}{\rho_1} = \frac{G_2}{G_1} \cdot \left(\frac{g_2}{g_1}\right)^2 \tag{e}$$

If now

$S_v = \dfrac{v_2}{v_1}$ is the Velocity Scale Factor

$S_\rho = \dfrac{\rho_2}{\rho_1}$ is the Density Scale Factor

$S_P = \dfrac{P_2}{P_1}$ is the Power Scale Factor

$S_g = \dfrac{g_2}{g_1}$ is the Gravitational Acceleration Scale Factor

$S_G = \dfrac{G_2}{G_1}$ is the Weight Scale Factor

then relations (d) and (e) can be written

$$S_v^7 \cdot S_\rho = S_P \cdot S_g^2 \ ; \qquad S_v^6 \cdot S_\rho = S_G \cdot S_g^2 \tag{f}$$

Relations in (f), by definition, constitute the Model Law of the system. In this particular case, obviously $S_g = S_\rho = 1$; therefore (f) simplifies to

$$S_v^7 = S_P \ ; \qquad S_v^6 = S_G \tag{g}$$

We wish to use a model whose weight is $G_2 = 333.62$ N. Thus, the Weight Scale Factor is

$$S_G = \frac{G_2}{G_1} = \frac{333.62}{1.1768 \times 10^6} = 2.83498 \times 10^{-4} \tag{h}$$

Therefore, by the second relation of (g)

$$S_v = \frac{v_2}{v_1} = (S_G)^{1/6} = (2.83498 \times 10^{-4})^{1/6} = 0.25631 \tag{i}$$

by which the speed of the model *must* be

$$v_2 = S_v \cdot v_1 = (0.25631) \cdot (3.858) = 0.98884 \text{ m/s} \tag{j}$$

Now we can *build* the 333.62 N heavy model, tow it through water at 0.98884 m/s speed* and *measure* the (required) pulling force F_2. Say we measure $F_2 = 20.62$ N. Therefore

$$P_2 = F_2 \cdot v_2 = (20.62) \cdot (0.98884) = 20.39 \text{ W} \tag{k}$$

But by (i) and the first part of (g), the Power Scale Factor is

$$S_P = \frac{P_2}{P_1} = S_v^7 = (0.25631)^7 = 7.26706 \times 10^{-5} \tag{ℓ}$$

and hence the power to tow the prototype barge is

$$P_1 = \frac{P_2}{S_P} = \frac{20.39}{7.26706 \times 10^{-5}} = 280\,581 \text{ W} \; (= 376.3 \text{ HP}) \tag{m}$$

Since the model is geometrically similar to the prototype, its characteristic linear size (length) will obey relation

$$S_L^3 = S_G \tag{n}$$

and by (n) and (h)

$$S_L = \frac{L_2}{L_1} = \sqrt[3]{S_G} = \sqrt[3]{2.83498 \times 10^{-4}} = 0.06569$$

i.e., the model's length is only 6.6% of that of the prototype.

Finally, as a check, we calculate the π_1 and π_2 values for the prototype and model. We should find them pair-wise identical. Accordingly, for the prototype, by the above data,

$$\pi_1 = \frac{P_1 \cdot g_1^2}{v_1^7 \cdot \rho_1} = \frac{(280\,581) \cdot (9.81^2)}{(3.858^7) \cdot (1000)} = 2.12256 \tag{o}$$

$$\pi_2 = \frac{G_1 \cdot g_1^2}{v_1^6 \cdot \rho_1} = \frac{(1.1768 \times 10^6) \cdot (9.81^2)}{(3.858^6) \cdot (1000)} = 34.3453 \tag{p}$$

and for the model

$$\pi_1 = \frac{P_2 \cdot g_2^2}{v_2^7 \cdot \rho_2} = \frac{(20.39) \cdot (9.81^2)}{(0.98884^7) \cdot (1000)} = 2.12262 \tag{q}$$

$$\pi_2 = \frac{G_2 \cdot g_2^2}{v_2^6 \cdot \rho_2} = \frac{(333.62) \cdot (9.81^2)}{(0.98884^6) \cdot (1000)} = 34.34266 \tag{r}$$

---

*The practical way to do this is to hold the model *stationary* in water running at constant speed 0.98884 m/s, and then *measure* the drag on the model.

As seen, (o) and (q) differ by less than 0.003%, and (p) and (r) by less than 0.01%. Thus, within rounding-errors both dimensionless variables are indeed identical, as expected and required.

⇑

The use of scale factors is further demonstrated by the following examples:

Example 17-5.   Deflection of a curved bar upon a concentrated force

We now design and execute a model experiment to determine the vertical deflection of a curved bar clamped at one end and free to move at the other, as shown in Fig. 17-10.

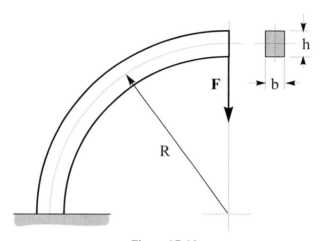

Figure 17-10
Curved bar is clamped at one end, loaded at the other by a concentrated force

The "prototype" steel bar is loaded at the free end where the vertical deflection is to be determined. The material's Young's and shear moduli are $E_1 = 2.05 \times 10^{11}$ Pa and $G_1 = 7.88 \times 10^{10}$ Pa. Furthermore, the force is $F_1 = 65000$ N, the centroidal radius is $R_1 = 12$ m, and the cross-section is a rectangle of $b_1 = 0.2$ m width and $h_1 = 0.8$ m height. The relevant variables are as follows:

Variable	Symbol	Dimension	Remark
vertical deflection	U	m	at free end
load	F	N	at free end, vertical
radius	R	m	to center of cross-section
Young's modulus	E	N/m²	
width of cross-section	b	m	uniform
height of cross-section	h	m	uniform
shear modulus	G	N/m²	

We have seven variables and two dimensions, therefore there are $7 - 2 = 5$ dimensionless variables determined by the Dimensional Set

	U	F	R	E	b	h	G
m	1	0	1	−2	1	1	−2
N	0	1	0	1	0	0	1
π_1	1	0	0	0	0	−1	0
π_2	0	1	0	0	0	−2	−1
π_3	0	0	1	0	0	−1	0
π_4	0	0	0	1	0	0	−1
π_5	0	0	0	0	1	−1	0

by which

$$\pi_1 = \frac{U}{h}; \quad \pi_2 = \frac{F}{h^2 \cdot G}; \quad \pi_3 = \frac{R}{h}; \quad \pi_4 = \frac{E}{G}; \quad \pi_5 = \frac{b}{h} \quad \text{(a)}$$

Therefore we have the Model Law

$$S_U = S_h; \quad S_F = S_h^2 \cdot S_G; \quad S_R = S_h; \quad S_E = S_G; \quad S_b = S_h \quad \text{(b)}$$

where S_U is the Deflection Scale Factor
S_h is the Height of Cross-section Scale Factor
S_F is the Load Scale Factor
S_G is the Shear Modulus Scale Factor
S_E is the Young's Modulus Scale Factor
S_R is the Centroidal Radius Scale Factor
S_b is the Width of Cross-section Scale Factor

Now we *design* the model. We *select* aluminium for its material (to increase deflection) with moduli of $E_2 = 6.83 \times 10^{10}$ Pa and $G_2 = 2.63 \times 10^{10}$ Pa. This results in

$$S_E = \frac{E_2}{E_1} = \frac{6.83 \times 10^{10}}{2.05 \times 10^{11}} = \frac{1}{3} \quad \text{(c)}$$

and

$$S_G = \frac{G_2}{G_1} = \frac{2.63 \times 10^{10}}{7.88 \times 10^{10}} = \frac{1}{3} \quad \text{(d)}$$

Thus, the fourth relation of the Model Law (b) is satisfied. Next we select another characteristic of the model, the height of its cross-section $h_2 = 0.05$ m. Hence

$$S_h = \frac{h_2}{h_1} = \frac{0.05}{0.80} = \frac{1}{16} \quad \text{(e)}$$

Thus, by the fifth relation of Model Law (b), $S_b = b_2/b_1 = S_h$. By this and (e),

$$b_2 = b_1 \cdot S_h = (0.2) \cdot \frac{1}{16} = \frac{1}{80} = 0.0125 \text{ m} \quad \text{(f)}$$

Next, by the second relation of the Model Law (b) and (e), $S_F = F_2/F_1 = S_h^2 \cdot S_G$ from which, with appropriate substitution, the force to be applied on the model is obtained

$$F_2 = F_1 \cdot S_F = F_1 \cdot S_h^2 \cdot S_G = (65\ 000) \cdot \left(\frac{1}{16}\right)^2 \cdot \frac{1}{3} = 84.6354 \text{ N} \quad \text{(g)}$$

The model's centroidal radius R_2 is derived by (e) and the third relation of Model Law (b). Accordingly,

$$R_2 = R_1 \cdot S_h = (12) \cdot \frac{1}{16} = 0.75 \text{ m} \tag{h}$$

We are now ready to construct the model, since we know all its geometric and material characteristics. Thus, we prepare our model, load it with the prescribed force [relation (g)] of $F_2 = 84.6354$ N, and *measure* the vertical deformation U_2 under this force. Say we find

$$U_2 = 0.00315 \text{ m} \tag{i}$$

By the first relation of the Model Law $S_U = U_2/U_1 = S_h$, from which and by (e) the sought-after deformation of the prototype is

$$U_1 = \frac{U_2}{S_h} = \frac{0.00315}{1/16} = 0.05040 \text{ m} \tag{j}$$

Finally, as a verification, we now determine the values of all the dimensionless variables of the prototype and model. If our calculations were correct, then these variables will be pair-wise identical. By substituting the appropriate values into formulas given, we obtain the following values for the prototype and model, respectively,

$\pi_1 = 0.06306$; $\quad \pi_2 = 1.28886 \times 10^{-6}$; $\quad \pi_3 = 15$; $\quad \pi_4 = 2.60152$; $\quad \pi_5 = 0.25$

$\pi_1 = 0.06300$; $\quad \pi_2 = 1.28723 \times 10^{-6}$; $\quad \pi_3 = 15$; $\quad \pi_4 = 2.59696$; $\quad \pi_5 = 0.25$

As seen, the values agree within 0.2%, verifying the accuracy of the executed modeling process.

⇑

Very often the Model Law can be used to compare the *performance* (behavior) of two *existing* systems. In this case, of course there is no need to construct a model; all we have to do is *call* one of these systems the "prototype" and the other the "model," and then proceed. The following example is a good illustration of this efficient and useful application of Model Law.

Example 17-6. Frequency of Respiration of Warm-blooded Animals

The temperature of a *homoithermal* (i.e., warm-blooded) animal, including man, is mainly constant and is therefore independent of its surroundings, as opposed to a *poikilothermal* (i.e., cold-blooded) creature whose body temperature varies in unison with its environment.

The frequency of the resting respiration of a homoitherm is obviously set by its metabolic rate since oxygen consumption is directly related to the utilization of energy. The resting energy *output* of an animal's body must be equal to its heat loss through its external (skin) and internal (lungs) surface areas. Therefore, the energy output in unit time P (which is power) can be expressed as

$$P = k_1 \cdot L^2 \tag{a}$$

where L is the characteristic linear size of the animal and k_1 is a dimensional constant whose dimension is kg/s³ [this follows directly from (a)].

On the other hand, the oxygen (i.e., energy) *input* in one respiratory cycle—one inhalation and one exhalation—must be proportional to the volume of the body, therefore

$$E = k_2 \cdot L^3 \tag{b}$$

where k_2 is a dimensional constant whose dimension is kg/(s²·m) [this follows from (b)]. With this information we can list the relevant variables and dimensional constants.

Variable	Symbol	Dimension	Remark
respiratory frequency	n	1/s	complete in-out cycles
dimensional constant	k_1	kg/s³	defined in relation (a)
dimensional constant	k_2	kg/(m·s²)	defined in relation (b)
characteristic size of animal	L	m	linear measure

We have four variables (including dimensional constants) and three dimensions. Therefore there is $4 - 3 = 1$ dimensionless variable—a constant. This is obtained from the Dimensional Set

	n	k_1	k_2	L
m	0	0	−1	1
kg	0	1	1	0
s	−1	−3	−2	0
π_1	1	−1	1	1

from which

$$\pi_1 = \frac{n \cdot k_2 \cdot L}{k_1} = c = \text{const} \tag{c}$$

Therefore the Model Law is

$$S_n \cdot S_{k_2} \cdot S_L = S_{k_1} \tag{d}$$

where S_n is the Respiratory Frequency Scale Factor
S_L is the Size Scale Factor
S_{k_1} is the Dimensional Constant k_1 Scale Factor
S_{k_2} is the Dimensional Constant k_2 Scale Factor

But k_1 and k_2 are constants for all animals, therefore

$$S_{k_1} = S_{k_2} = 1 \tag{e}$$

and therefore (d) can be simplified to

$$S_n \cdot S_L = 1 \tag{f}$$

or

$$S_n = \frac{1}{S_L} \tag{g}$$

which means that the Respiratory Frequency Scale Factor is inversely proportional to the Size Scale Factor. A large animal, having twice the length of a smaller one, breathes half as frequently.

Some may find this to be counterintuitive, since it could be reasoned that a large animal requires more air and therefore must breathe *more* frequently.

Let us now consider the *masses* of *geometrically similar* animals. If S_M is the Mass Scale Factor, then obviously

$$S_M = S_L^3 \tag{h}$$

Therefore the Model Law (f) can be written

$$S_n \cdot \sqrt[3]{S_M} = 1 \qquad (i)$$

Let us test the veracity of this formula for two animals, one small (rabbit), and one large (horse). We tag the rabbit "prototype," and the horse "model." By observation we know that an $M_1 = 3$ kg rabbit breathes approximately once per a second, therefore $n_1 = 1$ 1/s. Also, an $M_2 = 600$ kg horse breathes approximately 10 times per minute, i.e., $n_2 = \frac{1}{6}$ 1/s. Therefore $S_n = n_2/n_1 = \frac{1}{6}/1 = \frac{1}{6}$ and $S_M = M_2/M_1 = 600/3 = 200$. By (i) then $S_n \cdot \sqrt[3]{S_M} = (1/6) \cdot \sqrt[3]{200} = 0.975$, which is close enough to 1 predicted by relation (i). This is rather remarkable, since the horse's mass is 200 times of that of a rabbit!

⇑

Let us now consider the modeling of a system in which time is a relevant variable.

Example 17-7. Torsional Pendulum of Double Suspension
(adapted from Ref. 132, p. 38)

On the Moon a bar of uniform cross-section of length L is suspended by two parallel strings h long. The bar is now set into oscillatory motion about *vertical* axis z (see Fig. 17-11). By a modeling experiment on Earth we wish to determine the oscillatory period T on the Moon. The following are the relevant variables:

Variable	Symbol	Dimension	Remark
period of oscillation	T	s	about z axis
distance between suspensions	a	m	
length of bar	L	m	uniform cross-section and density
length of suspension	h	m	
gravitational acceleration	g	m/s^2	

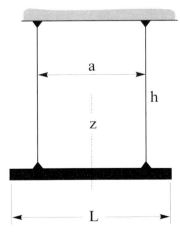

Figure 17-11
A doubly suspended torsional pendulum

488 APPLIED DIMENSIONAL ANALYSIS AND MODELING

We have five variables and two dimensions, therefore there are $5 - 2 = 3$ dimensionless variables, determined by the Dimensional Set

	T	a	L	h	g
m	0	1	1	1	1
s	1	0	0	0	-2
π_1	1	0	0	$-\dfrac{1}{2}$	$\dfrac{1}{2}$
π_2	0	1	0	-1	0
π_3	0	0	1	-1	0

from which

$$\pi_1 = T \cdot \sqrt{\dfrac{g}{h}}; \qquad \pi_2 = \dfrac{a}{h}; \qquad \pi_3 = \dfrac{L}{h} \tag{a}$$

Therefore the Model Law is

$$S_T = \sqrt{\dfrac{S_h}{S_g}}; \qquad S_a = S_h; \qquad S_L = S_h \tag{b}$$

where S_T is the Oscillatory Period Scale Factor
 S_h is the Height of Pendulum Scale Factor
 S_a is the Separation of Suspension Scale Factor
 S_L is the Bar's Length Scale Factor
 S_g is the Gravitational Acceleration Scale Factor

The prototype is on the Moon ($g_1 = 1.6$ m/s²) and has the following geometric data

$$a_1 = 0.2 \text{ m}; \qquad L_1 = 0.6 \text{ m}; \qquad h_1 = 0.8 \text{ m} \tag{c}$$

For the model (on Earth) we select bar length $L_2 = 2.4$ m, and hence

$$S_L = \dfrac{L_2}{L_1} = \dfrac{2.4}{0.6} = 4 \tag{d}$$

and by (d) and the third relation of Model Law (b)

$$S_L = S_h = \dfrac{h_2}{h_1} = 4 \tag{e}$$

from which, using (d) and (c), $h_2 = h_1 \cdot S_L = (0.8) \cdot (4) = 3.2$ m.
 Also, by the second relation of Model Law (b), $S_a = a_2/a_1 = S_h$, from which

$$a_2 = S_h \cdot a_1 = (4) \cdot (0.2) = 0.8 \text{ m} \tag{f}$$

We are now ready to build the model and measure its oscillatory period T_2. We measure, say,

$$T_2 = 6.216 \text{ s} \tag{g}$$

From the given gravitational data

$$S_g = \dfrac{g_2}{g_1} = \dfrac{9.81}{1.6} = 6.13125 \tag{h}$$

And by this and (b), the Oscillation Period Scale Factor is

$$S_T = \sqrt{\frac{S_h}{S_g}} = \sqrt{\frac{4}{6.13125}} = 0.80771 \tag{i}$$

But $S_T = T_2/T_1$ and hence, by (i) and (g), the oscillation period of the prototype on the Moon is

$$T_1 = \frac{T_2}{S_T} = \frac{6.216}{0.80771} = 7.696 \text{ s} \tag{j}$$

Therefore the prototype moves *slower* than the model. This is always the case if $S_T < 1$. On the other hand, if $S_T > 1$, then the prototype moves *faster*. Finally, if $S_T = 1$, then the prototype and the model move at the *same* speed. In view of this classification, we say that, with respect to the prototype,

- if $S_T < 1$, then the model moves *faster* than the prototype; we have *time contraction*;
- if $S_T = 1$, then the model and the prototype move at the *same* speed; we have *simultaneity*;
- if $S_T > 1$, then the model moves *slower* than the prototype; we have *time dilatation*.

In the case at hand, the model (on Earth) moves faster, therefore we have *time contraction*. To show an instance when time dilatation occurs, we place our prototype on Earth, where of course $\bar{g}_1 = g_2 = 9.81$ m/s². Therefore, in this alternate situation

$$\bar{S}_g = \frac{g_2}{\bar{g}_1} = \frac{9.81}{9.81} = 1 \tag{k}$$

while the other scale factors of course remain unaltered. Thus, by the first part of (b), and (k) and (e),

$$\bar{S}_T = \sqrt{\frac{S_h}{S_g}} = \sqrt{\frac{4}{1}} = 2 \tag{ℓ}$$

and we see that in this configuration $S_T > 1$, and hence the model moves *slower* (by a factor of 2) than the prototype; there is *time dilatation*.

This feature can be used to investigate fast processes by making the Time Scale Factor larger than 1; on the other hand, slow processes can be speeded up on the model by making $S_T < 1$.

⇑

17.4.3. Categories and Relations

Categories. In the preceding examples, the attentive reader probably observed that in modeling experiments, variables fall into three categories: some of them are *given,* some are *determined* by the Model Law (derived for this very purpose), and some are found by *measurement* on the model. It is also evident that we do not measure the dependent variable on the prototype. There are two reasons for this. First, the prototype is usually inaccessible, too expensive to instrument, too cumbersome to handle, or impossible to induce in it the desired physical condi-

490 APPLIED DIMENSIONAL ANALYSIS AND MODELING

tions the effect of which we want to determine. The second reason is much more mundane: if we can measure the dependent variable on the prototype, there is no reason to model!

In view of the above, we *define* the following possible categories for all variables (or physical constants) occurring in the modeling process:

Category 1. If a variable can be freely chosen, or *a priori* given, or is determinable by some means unrelated to modeling, then it is in Category 1. In short, the numerical value of a Category 1 variable is known, or can be found, at the outset—*before* the design of the modeling experiment commences. For example, mass is determinable by volume and density. Therefore if the latter two are known, then the mass in question is determinable outside of the modeling experiment, and hence it is a Category 1 variable.

Category 2. If the numerical value of a variable is determined by the application of the Model Law, then it is in Category 2. Note that for a variable to be in this category, it is not necessary that the variable's magnitude be determined *solely* by the Model Law. If the Model Law is utilized in any way for the determination, then the subject variable belongs to Category 2.

Category 3. If a variable is determined by *measurement* on the *model,* then it is of Category 3. It follows that only the model can have a Category 3 variable. If the numerical value of a variable is determined by measurement on the *prototype,* then this variable is in Category 1, since it is determined by means independent of the modeling process. For example, the length of a prototype cantilever can be determined by measuring it, but this has nothing to do with modeling, and hence the variable "length of prototype" is in Category 1.

The following example illustrates the allocation of categories among the variables, as defined above.

Example 17-8. Deflection of a Cantilever Upon a Concentrated Lateral Load (IX)

Consider the prototype cantilever seen in Fig. 17-12. We wish to determine the deflection of this prototype by a modeling experiment.

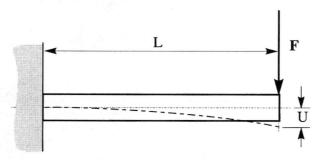

Figure 17-12
Cantilever loaded by a concentrated force

The following variables are relevant:

Variable	Symbol	Dimension	Remark
lateral deflection	U	m	under the load
load	F	N	lateral
length of beam	L	m	
Young's modulus	E	N/m^2	
second moment of area of cross-section	I	m^4	uniform

We have five variables and two dimensions, therefore there are $5 - 2 = 3$ dimensionless variables, supplied by the Dimensional Set

	U	F	L	E	I
m	1	0	1	-2	4
N	0	1	0	1	0
π_1	1	0	0	0	$-\frac{1}{4}$
π_2	0	1	0	-1	$-\frac{1}{2}$
π_3	0	0	1	0	$-\frac{1}{4}$

yielding

$$\pi_1 = \frac{U}{\sqrt[4]{I}}; \qquad \pi_2 = \frac{F}{E \cdot \sqrt{I}}; \qquad \pi_3 = \frac{L}{\sqrt[4]{I}} \qquad (a)$$

Thus the Model Law is

$$S_U = \sqrt[4]{S_I}; \qquad S_F = S_E \cdot \sqrt{S_I}; \qquad S_L = \sqrt[4]{S_I} \qquad (b)$$

where S_U is the Deflection Scale Factor
S_I is the Second Moment of Area Scale Factor
S_F is the Load Scale Factor
S_E is the Young's Modulus Scale Factor
S_L is the Length Scale Factor

The prototype has the following *given* characteristics:

$$F_1 = 50000 \text{ N } (1)\,; L_1 = 3.1 \text{ m } (1)\,; I_1 = 5.5 \times 10^{-5} \text{ m}^4 \, (1)\,; E_1 = 2 \times 10^{11} \text{ N/m}^2 \, (1) \qquad (c)$$

Since all these variables are *given,* they are all in Category 1. To indicate this allocation, we attached the symbol "(1)" to each of these variables.

Next, we *choose* the material and cross-section for the model. Therefore we have—say—

$$E_2 = 6.5 \times 10^{10} \text{ N/m}^2 \, (1)\,; \qquad I_2 = 3 \times 10^{-7} \text{ m}^4 \, (1) \qquad (d)$$

Note that we also attached "(1)" to both of these variables because they were *selected,* hence are in Category 1. By the data in (c) and (d) now

$$S_E = \frac{E_2}{E_1} = \frac{6.5 \times 10^{10}}{2 \times 10^{11}} = 0.325 \tag{e}$$

$$S_I = \frac{I_2}{I_1} = \frac{3 \times 10^{-7}}{5.5 \times 10^{-5}} = 5.45455 \times 10^{-3} \tag{f}$$

and, by (f) and the third relation of Model Law (b),

$$S_L = \sqrt[4]{S_I} = \sqrt[4]{5.45455 \times 10^{-3}} = 0.27176 \tag{g}$$

But $S_L = L_2/L_1$, therefore by (g) and (c),

$$L_2 = L_1 \cdot S_L = (3.1) \cdot (0.27176) = 0.84246 \text{ m } (2) \tag{h}$$

and we attached "(2)" to this variable, because it was determined by the Model Law. Using the second relation of the Model Law now, and by (e) and (f), we obtain

$$S_F = S_E \cdot \sqrt{S_I} = (0.325) \cdot \sqrt{5.45455 \times 10^{-3}} = 0.024 \tag{i}$$

But by definition $S_F = F_2/F_1$, therefore the load on the model by (c) and (i) is

$$F_2 = F_1 \cdot S_F = (50000) \cdot (0.024) = 1200 \text{ N } (2) \tag{j}$$

Again, we affixed "(2)" to this variable indicating that it is of Category 2. Now we can prepare our model, load it with 1200 N and measure its deflection U_2. Say we find

$$U_2 = 0.0123 \text{ m } (3) \tag{k}$$

Since U_2 is *measured* on the *model*, this variable is in Category 3, as indicated by "(3)." Finally, we consider (f) and the first relation of the Model Law (b).

$$S_U = \sqrt[4]{S_I} = \sqrt[4]{5.45455 \times 10^{-3}} = 0.27176 \tag{l}$$

But, by definition, $S_U = U_2/U_1$, therefore the required deflection of the prototype is

$$U_1 = \frac{U_2}{S_U} = \frac{0.0123}{0.27176} = 0.0453 \text{ m } (2) \tag{m}$$

This variable was determined by the application of the Model Law and hence it is in Category 2, which is indicated by the symbol "(2)."

⇑

Relations. We saw in the previous example that the of total of 10 values of variables appearing in the modeling experiment (for both the prototype and the model), 6 were in Category 1, 3 were in Category 2, and 1 was in Category 3. This prompts the question: Are there some relations defining these "membership-counts" for the categories? For if there were such relations, then we could create a stratagem to *reduce* the number of Category 2 variables, i.e., those fixed by the Model Law. This in turn would enable us to *increase* our *freedom* to select the characteristics of the *model*. We will now establish these relations in which we shall use the following symbols:

DIMENSIONAL MODELING

N_P = number of dimensionless variables
N_S = number of scale factors
N_V = number of variables
N_d = number of dimensions
N_e = number of equations in the Model Law
$(N_V)_1$ = total number of Category 1 variables in the prototype and model
$(N_V)_2$ = total number of Category 2 variables in the prototype and model
$(N_V)_3$ = total number of Category 3 variables in the prototype and model

Now it is obvious that

$$N_S = N_V \qquad (17\text{-}7)$$

since every variable generates its own scale factor. Moreover,

$$N_e = N_P \qquad (17\text{-}8)$$

since every dimensionless variable generates a relation (equation) of the Model Law (Art. 17.4.2). Also, by Buckingham's theorem (Art. 7.7),

$$N_P = N_V - N_d \qquad (17\text{-}9)$$

where we assumed that the rank of dimensional matrix equals the number of dimensions—a condition which is always satisfied in a valid Dimensional Set.

Next, it is evident that if there are N_e equations in the Model Law and altogether $2 \cdot N_V$ variables characterizing the prototype and the model, then there must be exactly $2 \cdot N_V - N_e$ variables that are *not* prescribed by the Model Law. Thus, by the notation defined above,

$$(N_V)_1 + (N_V)_3 = 2 \cdot N_V - N_e \qquad (17\text{-}10)$$

which, by (17-8) and (17-9), can be written

$$(N_V)_1 + (N_V)_3 = 2 \cdot N_V - (N_V - N_d) = N_V + N_d \qquad (17\text{-}11)$$

The number of *measured* variables is 1 (the measurement is done on the model). Therefore

$$(N_V)_3 = 1 \qquad (17\text{-}12)$$

and hence (17-11) becomes

$$(N_V)_1 = N_V + N_d - 1 \qquad (17\text{-}13)$$

On the other hand, obviously,

$$(N_V)_1 + (N_V)_2 + (N_V)_3 = 2 \cdot N_V$$

so $(N_V)_2 = 2 \cdot N_V - (N_V)_1 - (N_V)_3$ which, by (17-12), (17-13) and (17-9), can be written

$$(N_V)_2 = N_V - N_d = N_P \qquad (17\text{-}14)$$

i.e., the total number of variables—in both prototype and model—*imposed* by the Model Law, equals the number of dimensionless variables.

To illustrate the use of the above-derived category-related relations, we revisit Example 17-8. In that example we have $N_V = 5$ (number of variables), $N_d = 2$ (number of dimensions) and $N_P = 3$ (number of dimensionless variables). Therefore the total number of variables for both prototype and model is $2 \cdot N_V = 2 \times 5 = 10$, of which

$$(N_V)_1 = N_V + N_d - 1 = 5 + 2 - 1 = 6 \quad \text{[by relation (17-13)]}$$

are *selectable* (Category 1),

$$(N_V)_2 = N_P = 3 \quad \text{[by relation (17-14)]}$$

are *prescribed* by the Model Law (Category 2), and

$$(N_V)_3 = 1 \quad \text{[by relation (17-12)]}$$

is determined by *measurement* on the model (Category 3).

If two variables V_1 and V_2 are fused into a new one—say Z—then the number of variables N_V is reduced by 1, and therefore, by (17-13), the *total* number of Category 1 variables is also reduced by 1. Assuming that for the *prototype* all fused variables are Category 1, there can be three cases:

Case 1. For the *model,* both V_1 and V_2 are of Category 1. In this case the new Z variable will also be Category 1. By the deletion of V_1 and V_2, N_V is *reduced* by 4 (since both prototype and model are affected), and by the addition of Z, N_V is *increased* by 2. Therefore the net reduction of N_V is 2. But by (17-13) this reduction must be only 1. Therefore (17-13) can only be satisfied if one variable in the model changes from Category 2 to Category 1. The consequence of this is very beneficial, for this means that one variable in the model is "freed," and hence becomes selectable; thus, we have more freedom to shape and construct the model.

Case 2. For the *model,* one of the fused variables V_1 and V_2 is of Category 1, the other (initially) is of Category 2. In this case the new Z variable will be in Category 1 since one of its constituents—say V_1—is freely selectable. It follows that V_2 also becomes selectable since it is part of a selectable composition. Therefore a variable which was in Category 2 is now in Category 1; it is "freed" and hence becomes selectable. Again, this is very beneficial from the point of view of constructing the model.

Case 3. For the *model,* both fused variables V_1 and V_2 are (initially) of Category 2. In this case the new Z variable for the model will also be of Category 2. But if Z is fixed and it is composed of V_1, V_2, then one of these two—say V_1—can be freely chosen (Category 1), and the other adjusted such that Z remains intact. Thus, again, one variable (V_1 in this case) changes from Category 2 to Category 1. This results in—similarly to Cases 1 and 2—more freedom in selecting the characteristics of the model.

In summation, we saw in each case that the fusion of two variables freed a hitherto predetermined (by Model Law) variable, thus making it selectable.

In the next example (Example 17-9 in Art. 17.4.4) *fusions* of variables as in Case 2 and Case 3 will be demonstrated.

17.4.4. Modeling Data Table

It is very convenient and enhancing for the organization and execution of a model experiment, if all pertinent data for physical and dimensionless variables (values, categories, symbols, dimensions, and scale factors) are displayed in a single table, called the *Modeling Data Table*. It is therefore wise to prepare this table for every modeling experiment and, if necessary, compare it with others (obtained from differing model designs), thereby greatly facilitating the evaluation process. A specimen *blank* Modeling Data Table is given in Appendix 8, which the reader is free to copy.

The ready use of the Modeling Data Table is illustrated in the following example, which also demonstrates the advantage and the "cost" of *fusing* variables.

Example 17-9. Deflection of a Cantilever Upon a Concentrated Lateral Load (X)

The subject of this example is again our perennial cantilever. But now we will investigate the effects of variable fusions in several modeling versions of the same set-up shown in Fig. 17-13.

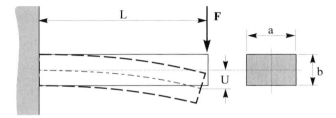

Figure 17-13
Cantilever loaded by a concentrated force

Our goal is to determine the deflection generated by the applied lateral concentrated force. The table below lists the relevant variables, symbols and dimensions.

Variable	Symbol	Dimension	Remark
lateral load	F	N	concentrated
width of cross-section	a	m	uniform
thickness of cross-section	b	m	uniform
lateral deflection	U	m	under the load
length	L	m	
Young's modulus	E	N/m²	

We have six variables and two dimensions, therefore there are 6 − 2 = 4 dimensionless variables supplied by the Dimensional Set

	F	a	b	U	L	E
m	0	1	1	1	1	−2
N	1	0	0	0	0	1
π_1	1	0	0	0	−2	−1
π_2	0	1	0	0	−1	0
π_3	0	0	1	0	−1	0
π_4	0	0	0	1	−1	0

496 APPLIED DIMENSIONAL ANALYSIS AND MODELING

Note that the dependent variable U is not the first in the line-up, but this does not matter because U is still in the **B** matrix (see Fig. 8-1 and Fig. 8-3 in Art. 8.1), thus it appears in only *one* dimensionless variable, as a well behaving *dependent* variable should. By the above set

$$\pi_1 = \frac{F}{L^2 \cdot E} \; ; \qquad \pi_2 = \frac{a}{L} \; ; \qquad \pi_3 = \frac{b}{L} \; ; \qquad \pi_4 = \frac{U}{L} \tag{a}$$

and the Model Law is

$$S_F = S_L^2 \cdot S_E \; ; \qquad S_a = S_L \; ; \qquad S_b = S_L \; ; \qquad S_U = S_L \tag{b}$$

where S_F, S_L, S_E, S_a, S_b, and S_U are the scale factors defined by the respective subscripts.

In (b) we see that $S_a = S_b = S_L$, i.e., the model *must* be *geometrically similar* to the prototype. In other words, if we want to determine the deflection of the prototype by measuring the deflection of the model, then the model must be an *exactly* scaled replica of the prototype. To illustrate, suppose we have a prototype with the following characteristics:

$$\left.\begin{array}{l} L_1 = 8 \text{ m} \\ a_1 = 0.2032 \text{ m} (= 8 \text{ in}) \\ b_1 = 0.254 \text{ m} (= 10 \text{ in}) \\ E_1 = 2 \times 10^{11} \text{ N/m}^2 \text{ (for steel)} \\ F_1 = 94600 \text{ N} \end{array}\right\} \tag{c}$$

We have $N_V = 6$ variables and $N_d = 2$ dimensions, therefore, by (17-13), altogether there can be $(N_V)_1 = N_V + N_d - 1 = 6 + 2 - 1 = 7$ *selectable* (Category 1) variables for both prototype and model. The prototype has 5 *given* variables (Category 1); 1 variable (deflection) will be determined by Model Law (Category 2) upon a model *measurement* (Category 3). Thus, the model has $(N_V)_1 - 5 = 2$ selectable (Category 1) variables. We *select* these 2 variables to be length $L_2 = 1.2$ m, and material (aluminium) $E_2 = 6.7 \times 10^{10}$ N/m². Since one variable (deflection) is *measured* on the model, therefore the remaining $N_V - 2 - 1 = 6 - 2 - 1 = 3$ variables must be Category 2—i.e., imposed by the Model Law (b). Accordingly, by (c) and the data given

$$S_L = \frac{L_2}{L_1} = \frac{1.2}{8} = 0.15 \tag{d}$$

and

$$S_E = \frac{E_2}{E_1} = \frac{6.7 \times 10^{10}}{2 \times 10^{11}} = 0.335 \tag{e}$$

With these two scale factors, the first relation of the Model Law (b) provides the Force Scale Factor

$$S_F = S_L^2 \cdot S_E = (0.15^2) \cdot (0.335) = 7.5375 \times 10^{-3} \tag{f}$$

Since $S_F = F_2/F_1$, therefore, by (f) and (c), the load on the model *must* be

$$F_2 = F_1 \cdot S_F = (94600) \cdot (7.5375 \times 10^{-3}) = 713.048 \text{ N} \tag{g}$$

Next, by (b) and (d), $S_a = a_2/a_1 = S_L = 0.15$, from which the cross-section's width can be obtained

$$a_2 = a_1 \cdot S_L = (0.2032) \cdot (0.15) = 0.03048 \text{ m} \tag{h}$$

DIMENSIONAL MODELING

Similarly, for the model cross-section's thickness

$$b_2 = b_1 \cdot S_L = (0.254) \cdot (0.15) = 0.0381 \text{ m} \quad \text{(i)}$$

where a_1 and b_1 are as given in relation (c).

We have now determined all the *geometric* characteristics of the model, and are ready to construct it, apply the mandatory load 713.048 N, and then *measure* the deflection U_2, which this load produces. Say we measure $U_2 = 0.0436$ m. Then, by the fourth relation of the Model Law (b) and (d), $S_U = U_2/U_1 = S_L$ from which, by the measured U_2 and (d), the prototype's deflection is

$$U_1 = \frac{U_2}{S_L} = \frac{0.0436}{0.15} = 0.29067 \text{ m (2)} \quad \text{(j)}$$

where the symbol "(2)" indicates Category 2 variable [since U_1 was obtained by the application of the Model Law (b)].

All of the above results are displayed in the Modeling Data Table (Fig. 17-14). Moreover, as a check, we listed the numerical values of all four dimensionless variables defined in (a). We see in the table that all four of these values are identical pair-wise (within

Variable					Scale factor S	Category	
name	symbol	dimension	prototype	model	model/prototype	prototype	model
length	L	m	8	1.2	0.15	1	1
width of cross-section	a	m	0.2032	0.03048	0.15	1	2
thickness of cross-section	b	m	0.254	0.0381	0.15	1	2
Young's modulus	E	N/m²	2.E11	6.7 E10	0.335	1	1
lateral force	F	N	94,600	713.048	7.5375E-3	1	2
lateral deflection	U	m	0.29067	0.0436	0.15	2	3
dimensionless	π_1	1	7.39E-9	7.39E-9	1	—	—
dimensionless	π_2	1	0.0254	0.0254	1	—	—
dimensionless	π_3	1	0.03175	0.03175	1	—	—
dimensionless	π_4	1	0.036334	0.036333	0.99997	—	—
categories of variables		1	freely chosen, *a priori* given, or determined independently				
		2	determined by application of Model Law				
		3	determined by measurement on the model				

Figure 17-14
Modeling Data Table for the cantilever experiment (Version 1)

rounding errors), as expected and required. The reader should also substitute the respective scale factor values into the Model Law (b) and verify that all four relations are numerically satisfied, as they must be.

We will now "improve" our model by fusing two of its variables. The well-known analytical formula for deflection U is (Ref. 133, Vol. 1, p. 150)

$$U = \frac{4 \cdot F \cdot L^3}{a \cdot b^3 \cdot E} \tag{k}$$

in which we note that variables "a" and "b" appear only in the form of "$a \cdot b^3$." To utilize this "information" as *Version 2* of our modeling experiment, we construct a Dimensional Set in which variables "a" and "b" are fused into a new auxiliary variable α

$$\alpha = a \cdot b^3 \tag{\ell}$$

whose dimension is m⁴. Accordingly, we have the Version 2 Dimensional Set

	F	α	U	L	E
m	0	4	1	1	−2
N	1	0	0	0	1
π_1	1	0	0	−2	−1
π_2	0	1	0	−4	0
π_3	0	0	1	−1	0

Note that both "a" and "b" are Category 2 variables, therefore we have here Case 3 (Art. 17.4.3). Consequently, as explained in the text in the referenced Article, one of these variables is "freed," i.e., becomes *selectable* (Category 1). We will use this feature as our story develops.

The number of dimensionless variables N_P is the difference between N_V and N_d and thus we have $5 - 2 = 3$ dimensionless variables, obtained from the Dimensional Set

$$\pi_1 = \frac{F}{L^2 \cdot E}; \qquad \pi_2 = \frac{\alpha}{L^4}; \qquad \pi_3 = \frac{U}{L} \tag{m}$$

We now we have only three dimensionless variables, instead of the four in Version 1. The Model Law is then, by (m)

$$S_F = S_L^2 \cdot S_E; \qquad S_\alpha = S_L^4; \qquad S_U = S_L \tag{n}$$

where S_α is the Auxiliary Variable Scale Factor and the other scale factors are the same as before. This set of equations is now the *improved* Model Law of the system. Why improved? Because (n) comprises one fewer conditions (equations) than (b) does. In particular, we no longer have the "imposition" of $S_a = S_b$ occurring in (b). This means, *importantly*, that geometric similarity, although allowed, is no longer *required*.

To illustrate the benefits of this feature, suppose we have the same beam prototype as before and a model whose length and material are also the same. Thus $L_2 = 1.2$ m, $E_2 = 6.7 \times 10^{10}$ N/m² and therefore the scale factors $S_L = 0.15$, $S_E = 0.335$ and $S_F = 7.5375 \times 10^{-3}$ remain unaltered. Therefore the load on the model also remains the same $F_2 = 713.048$ N.

But now, by the second relation of the Model Law (n),

$$S_\alpha = S_L^4 = 0.15^4 = 5.0625 \times 10^{-4} = \frac{\alpha_2}{\alpha_1} \tag{o}$$

The given data for the prototype and (ℓ) furnishes

$$\alpha_1 = a_1 \cdot b_1^3 = (0.2032) \cdot (0.254^3) = 3.32985 \times 10^{-3} \text{ m}^4 \qquad (p)$$

and hence, by (o) and (p),

$$\alpha_2 = \alpha_1 \cdot S_\alpha = (3.32985 \times 10^{-3}) \cdot (5.0625 \times 10^{-4}) = 1.68574 \times 10^{-6} \text{ m}^4 \qquad (q)$$

Observe now that we can select *any* cross-section for the model as long as its α_2 value is as above determined. This fact gives us a freedom that we did not have in Version 1. For example, we can now *select* an $a_2 = 0.04445$ m (= 1.75 in) wide aluminium strip for the model. What should the thickness of b_2 be now? By (q) and (l), $\alpha_2 = 1.68574 \times 10^{-6} = a_2 \cdot b_2^3$, thus

$$b_2 = \sqrt[3]{\frac{\alpha_2}{a_2}} = \sqrt[3]{\frac{1.68574 \times 10^{-6}}{0.04445}} = 0.03360 \text{ m } (= 1.323 \text{ in}) \qquad (r)$$

Next, we build the model, load it with 713.048 N force, and *measure* its deflection U_2 under the load. Say we find $U_2 = 0.04364$ m (= 1.718 in). Then, by the third relation of Model Law (n), $S_U = U_2/U_1 = S_L = 0.15$, from which the deflection of the prototype is

$$U_1 = \frac{U_2}{S_U} = \frac{0.04364}{0.15} = 0.29093 \text{ m}$$

identical, of course, to the previously obtained value (in Version 1). The Modeling Data Table in Fig. 17-15 summarizes all of the above results.

From this table the following observations can be made:

- Variables "*a*" and "*b*" are no longer "active" since they are *fused* to (replaced by) auxiliary variable α [see (ℓ)]. Therefore "*a*" and "*b*" appear for "reference only."
- For the model we have two freely *selected* (Category 1) variables, two *Model Law* determined (Category 2) variables, and one *measured* on the model (Category 3) variable.
- The numerical values of all three dimensionless variables π_1, π_2, and π_3 are identical (within rounding errors) confirming the *dimensional similarity* of the prototype and the model.
- The width/thickness ratio (*a/b*) of the cross-section is 0.8 for the prototype and 1.32292 for the model. Therefore the prototype and the model are *geometrically dissimilar*.

In view of the findings in this example, we can now formulate three important conclusions:

The *first* conclusion is that *dimensional similarity* can be achieved without *geometric similarity*. For to gain dimensional similarity it is only necessary to satisfy the Model Law, thus apart from this requirement, the experimenter has a free hand to shape the model.

The *second* conclusion is that it is best to have as few physical variables as possible. The number of Model Law-imposed (Category 2) variables is $(N_V)_2 = N_P = N_V - N_d$ [relations (17-9) and (17-14)], therefore fewer N_V means fewer $(N_V)_2$.

The *third* conclusion is this: the more *information* we inject into the process of formulating the relevant variables (including auxiliary ones), the more *freedom* we have for selecting the physical characteristics of the model. To illustrate, in Version 1 we did not consider any *a priori* information offered by the exact analytical formula (k) for deformation. Therefore we had four constraining equations in the Model Law (b). As a consequence, in this version we had the least freedom, which meant that, of the six physical variables for

Variable					Scale factor S	Category	
name	symbol	dimension	prototype	model	model/prototype	prototype	model
length	L	m	8	1.2	0.15	1	1
Young's modulus	E	N/m^2	2E11	6.7E10	0.335	1	1
lateral force	F	N	94,600	713.048	7.5375E-3	1	2
lateral deflection	U	m	0.29067	0.0436	0.15	2	3
auxiliary	α	m^4	3.32985E-3	1.68574E-6	5.0625E-4	1	2
cross-section width	a	m	0.2032	0.04445	for reference only		
cross-section thickness	b	m	0.254	0.0336	for reference only		
dimensionless	π_1	1	7.39063E-9	7.39063E-9	1	—	—
dimensionless	π_2	1	8.12952E-7	8.12953E-7	1	—	—
dimensionless	π_3	1	3.63338E-2	3.63333E-2	0.99999	—	—
categories of variables	1	freely chosen, *a priori* given, or determined independently					
	2	determined by application of Model Law					
	3	determined by measurement on the model					

Figure 17-15
Modeling Data Table for the cantilever experiment (Version 2)

the model, we had the freedom to select only two, viz., E and L; the other four were prescribed for us (Category 2). In particular, the model had to be *geometrically similar* to the prototype. In contrast, in Version 2 we used our *knowledge* that in the exact formula (k) expressing deflection, the term $a \cdot b^3$ occurs, and neither "a" nor "b" appears separately. So, we called this term an *auxiliary variable,* and used this variable in the Dimensional Set to construct a new Model Law (Version 2) that was less restrictive than in Version 1. In particular, it accorded us the luxury of using a model that was *dimensionally dissimilar* to the prototype.

From the above discussion the question naturally arises: can this process continue to attain the "maximum freedom" in sizing the model? The answer is "yes," it can, but there is a price to pay. The more freedom we have for our modeling routine, the more information we must enter into the formulation of the Model Law. Also, and importantly, the more *a priori* information is used, the less feasible the modeling itself becomes. In the extreme, if we use the maximum, i.e., *all* the possible information, we would not need to model at all, since all the information was available *a priori*—although in this case, of course, we would have *complete* freedom to shape the utterly unnecessary model. Diagram of Fig. 17-16 *qualitatively* illustrates the main idea behind this reasoning.

DIMENSIONAL MODELING

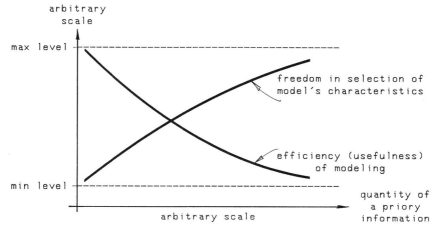

Figure 17-16
Interdependence of general modeling characteristics
(qualitative only illustration)

To further illustrate the effect of this "liberalization" process, let us have a *Version 3* in which we introduce yet another auxiliary variable into equation (k) expressing deflection

$$U = \frac{4 \cdot F \cdot L^3}{a \cdot b^3 \cdot E} \tag{k}$$

Let this second auxiliary variable be

$$\beta = F \cdot L^3 \tag{s}$$

with dimension N·m³. Therefore (k) can be written

$$U = \frac{4 \cdot \beta}{\alpha \cdot E} \tag{t}$$

where α is as defined in (ℓ). As we saw in Fig. 17-15, L is a Category 1 variable, and F is a Category 2 variable. Therefore we have a Case 2 situation as defined in Art. 17.4.3 and, as pointed out in there, one Category 2 variable (F in this instance) now becomes "free"— i.e., *selectable*. The Dimensional Set is then

	U	β	α	E
m	1	3	4	−2
N	0	1	0	1
π_1	1	0	$-\frac{1}{4}$	0
π_2	0	1	$-\frac{5}{4}$	−1

from which

$$\pi_1 = \frac{U}{\sqrt[4]{\alpha}}; \qquad \pi_2 = \frac{\beta}{E\cdot\sqrt[4]{\alpha^5}} \qquad (u)$$

yielding the Model Law (Version 3)

$$S_U = (S_\alpha)^{1/4}; \qquad S_\beta = S_E \cdot (S_\alpha)^{5/4} \qquad (v)$$

where S_U, S_α, S_β, and S_E are the respective scale factors. The second relation of this can be written

$$S_\alpha = \left(\frac{S_\beta}{S_E}\right)^{4/5} \qquad (w)$$

For the prototype suppose we have everything as before, and for the model we also keep E_2 and L_2 (as in Versions 1 and 2), but we now *select* the load F_2 for the model—instead of what previously was imposed by the Model Law. So, we select $F_2 = 600$ N. With this datum,

for the prototype: $\beta_1 = F_1 \cdot L_1^3 = (94{,}600)\cdot(8^3) = 48{,}435{,}200$ N·m³
for the model: $\beta_2 = F_2 \cdot L_2^3 = (600)\cdot(1.2^3) = 1036.8$ N·m³

Therefore

$$S_\beta = \frac{\beta_2}{\beta_1} = \frac{1036.8}{48{,}435{,}200} = 2.14059 \times 10^{-5}$$

and from the Modeling Data Table for Version 2 (Fig. 17-15), $S_E = 0.335$, since the materials of prototype and model remained the same. By (w),

$$S_\alpha = \left(\frac{2.14059 \times 10^{-5}}{0.335}\right)^{4/5} = 4.40952 \times 10^{-4}$$

We now have the width of the cross-section of the model as before $a_2 = 0.04445$ m and wish to determine the corresponding thickness b_2. Hence, by (ℓ), $S_\alpha = \alpha_2/\alpha_1 = (a_2 \cdot b_2^3)/\alpha_1$, from which

$$b_2 = \sqrt[3]{\frac{S_\alpha \cdot \alpha_1}{a_2}}$$

yielding, by the substitution of α_1 (Fig. 17-15), $b_2 = 0.03209$ m. Therefore by (ℓ)

$$\alpha_2 = a_2 \cdot b_2^3 = (0.04445)\cdot(0.03209^3) = 1.4683 \times 10^{-6} \text{ m}^4$$

By the first formula of Model Law (v) now

$$S_U = \frac{U_2}{U_1} = \sqrt[4]{S_\alpha} = \sqrt[4]{4.40952 \times 10^{-4}} = 0.14491 \qquad (x)$$

We now *measure* deflection $U_2 = 0.04216$ m of the model upon $F_2 = 600$ N concentrated load. Therefore the sought-after deflection of the prototype, by (x), is

$$U_1 = \frac{U_2}{S_U} = \frac{0.4216}{0.14491} = 0.29094 \text{ m}$$

the *same* as for Versions 2 and 3, as required and expected.

Variable					Scale factor S	Category	
name	symbol	dimension	prototype	model	model/prototype	prototype	model
deflection	U	m	0.29094	0.04216	0.14491	2	3
Young's modulus	E	N/m²	2E11	6.7E10	0.335	1	1
auxiliary variable	α	m⁴	3.32985E-3	1.4683E-6	4.40952E-4	1	2
auxiliary variable	β	m³·N	4.84352E7	1036.8	2.14059E-5	1	1
cross-section thickness	b	m	0.254	0.03209	for reference only		2
cross-section width	a	m	0.2032	0.04445	for reference only		1
lateral force	F	N	94,600	600	for reference only		1
length	L	m	8	1.2	for reference only		1
dimensionless	π_1	1	1.21115	1.21115	1	—	—
dimensionless	π_2	1	0.30276	0.30276	1	—	—
categories of variables		1	freely chosen, *a priori* given, or determined independently				
		2	determined by application of Model Law				
		3	determined by measurement on the model				

Figure 17-17
Modeling Data Table for the cantilever experiment (Version 3)

The Modeling Data Table in Fig. 17-17 conveniently collects all the above results. Note that again the ratio *a/b* for the prototype is 0.8, whereas for the model it is 1.38534. Therefore the prototype and the model are dissimilar *geometrically,* although of course similar *dimensionally,* since the two dimensionless variables π_1 and π_2 are pair-wise identical—as Fig. 17-17 shows. Also note that in this version, *force* on the model was a *selectable* variable (Category 1), in contrast with Versions 1 and 2 where it was *imposed* (Category 2) by the respective Model Laws.

In Fig. 17-18, the most important characteristics of the three versions of this modeling experiment are summarized. The main conclusions are:

Version 1. In Version 1 we assumed nothing; we merely listed and dealt with all the relevant variables. They were length *L,* width "*a*" and thickness "*b*" of cross-section, Young's modulus *E,* lateral force *F,* and deflection *U.* The Model Law consisted of four relations (conditions), and hence altogether four variables—three for the model (*a, b, F*) and one for the prototype (*U*)—were determined by the Model Law.

Version 2. In Version 2 we used the *information* (knowledge) that the term $a \cdot b^3$ appears in the exact formula for deflection *U.* Thus, we could introduce the auxiliary vari-

Variable		Version 1	Version 2	Version 3
length	L	Category 1		
cross-section width	a	Category 2	Category 1	
cross-section thickness	b	Category 2		
Young's modulus	E	Category 1		
lateral load	F	Category 2		Category 1
lateral deflection	U	Category 3		
number of auxiliary variables		0	1	2
number of selectable variables		2	3	4
categories of variables	1	freely chosen, *a priori* given, or determined independently		
	2	determined by application of Model Law		
	3	determined by measurement on the model		

Figure 17-18
Three versions of variable-related characteristics of the model of a cantilever loaded with a concentrated force

able $\alpha = a \cdot b^3$. Hence the number of variables was reduced by one, the number of constraining relations of the Model law became three, and the number of imposed variables was reduced to two (b, F)—the variable "a" being "freed."

Version 3. In Version 3 we used another property of the exact formula for deflection U, namely that in it the term $F \cdot L^3$ appears. This enabled us to introduce yet another auxiliary variable $\beta = F \cdot L^3$. Thus, the number of imposed characteristics on the model was further reduced by one, inasmuch as the only remaining imposed (Category 2) variable was the thickness of cross-section b—the lateral force being "freed."

So, as we injected more and more information into the process of formulating the variables, our freedom to shape the model gradually increased, while the actual usefulness of the modeling decreased (since we professed more and more knowledge about the very physical relation we wished to establish by modeling).

Another point is that in formulating and introducing auxiliary variables (to improve our freedom in designing model experiments) one must not compose a variable which does not occur—or probably does not occur—in the expression, known or not. For example, in (k), the combination of variables F/L does not occur (with the condition that neither F nor L would be "left behind"). Therefore the auxiliary variable $\gamma = F/L$, as a replacement for F and L, is not allowed. This topic was discussed in more detail in Art. 16.1.

⇑

The next example not only demonstrates the use of the Modeling Data Table, but also presents an application where the dependent variable of the prototype (i.e., the variable whose value is to be determined by the modeling experiment) is *not* the one that is *measured* on the model.

Example 17-10. Size and Impact Velocity of a Meteorite

In Example 13-6 (Art. 13.3) we dealt with the problem of determining the impact speed of a meteorite. Now we outline a way by which not only the impact speed but also the size of the impactor (meteorite) can be found experimentally. Since "full scale" experimentation would be "time-consuming" in the extreme, we will do a modeling experiment.

We construct the model by a process in which a small pellet is shot at a given angle into a box of crater material. Then we measure the crater's size, and from this information—and by the Model Law to be established—we determine both the impact speed and the size of the meteorite.

First, we list the relevant variables:

Variable	Symbol	Dimension	Remark
impact angle	φ	1 (rad)	to local horizon
mean diameter of crater	D	m	Note 1 below
speed of meteorite	v	m/s	at impact
density of meteorite	ρ_m	kg/m^3	
diameter of meteorite	d	m	Note 2 below
gravitational acceleration	g	m/s^2	at sea level
density of crater material	ρ_c	kg/m^3	

Note 1: If the crater is an ellipse, then D is the *geometric* mean of the major and minor axes.

Note 2: If the meteorite is not a sphere, then d is the diameter of a sphere of equal volume.

We have seven variables and three dimensions, therefore there are $7 - 3 = 4$ dimensionless variables, obtained from the Dimensional Set

	φ	D	v	ρ_m	d	g	ρ_c
m	0	1	1	−3	1	1	−3
kg	0	0	0	1	0	0	1
s	0	0	−1	0	0	−2	0
π_1	1	0	0	0	0	0	0
π_2	0	1	0	0	−1	0	0
π_3	0	0	1	0	$-\frac{1}{2}$	$-\frac{1}{2}$	0
π_4	0	0	0	1	0	0	−1

yielding

$$\pi_1 = \varphi \ ; \quad \pi_2 = \frac{D}{d} \ ; \quad \pi_3 = \frac{v}{\sqrt{d \cdot g}} \ ; \quad \pi_4 = \frac{\rho_m}{\rho_c} \qquad (a)$$

The Model Law is

$$S_\varphi = 1 \ ; \quad S_D = S_d \ ; \quad S_v = \sqrt{S_d S_g} \ ; \quad S_{\rho m} = S_{\rho c} \qquad (b)$$

where S_φ is the Impact Angle Scale Factor
       S_D is the Crater Diameter Scale Factor
       S_d is the Meteorite Diameter Scale Factor
       S_v is the Impact Speed Scale Factor

S_g is the Gravitational Acceleration Scale Factor
$S_{\rho m}$ is the Meteorite Density Scale Factor
$S_{\rho c}$ is the Crater Density Scale Factor

For the prototype we consider the famous Arizona meteorite crater, whose known characteristics are approximately as follows:

$$\left. \begin{array}{ll} \text{impact angle:} & \varphi_1 = 1.169 \text{ rad} (\cong 67 \text{ deg}) \\ \text{crater diameter:} & D_1 = 1200 \text{ m} \\ \text{meteorite density:} & (\rho_m)_1 = 7860 \text{ kg/m}^3 \text{ (iron)} \\ \text{crater density:} & (\rho_c)_1 = 2800 \text{ kg/m}^3 \\ \text{gravitational acceleration:} & g_1 = 9.81 \text{ m/s}^2 \end{array} \right\} \quad (c)$$

These variables are all Category 1. For the model, we *must* choose the same impact angle $\varphi_2 = 67°$ [first relation of Model Law (b)]. Moreover, we *select* the same material for the pellet as that of the meteorite (iron). Thus, $(\rho_m)_2 = 7860$ kg/m³, and hence

$$S_{\rho m} = \frac{(\rho_m)_2}{(\rho_m)_1} = \frac{7860}{7860} = 1 \quad (d)$$

By (c), (d) and the fourth relation of the Model Law (b)

$$S_{\rho c} = \frac{(\rho_c)_2}{(\rho_c)_1} = S_{\rho m} = 1 \quad (e)$$

from which

$$(\rho_c)_2 = (\rho_c)_1 = 2800 \text{ kg/m}^3 \quad (f)$$

Therefore both φ_2 and $(\rho_c)_2$ are Category 2 variables. We select a $d_2 = 0.004$ m diameter pellet and at $\varphi_2 = 67°$ shoot it into a box of crater material of the density defined in (f). The pellet's speed is $v_2 = 211.6$ m/s, as determined by ballistic test. Thus, d_2 and v_2 are Category 1 variables. Next, we *measure* the crater which the pellet created in the box. Say we find that the diameter of the crater is $D_2 = 0.166$ m. At this point we have all the information required to find impact speed v_1 and diameter d_1 of the prototype (i.e., the meteorite). By the given data

$$S_D = \frac{D_2}{D_1} = \frac{0.166}{1200} = 0.0001383 \quad (g)$$

From second relation of the Model Law (b)

$$S_d = \frac{d_2}{d_1} = S_D = 0.0001383 \quad (h)$$

Thus, the meteorite's diameter is

$$d_1 = \frac{d_2}{S_D} = \frac{0.004}{0.0001383} = 28.93 \text{ m} \quad (i)$$

By the third relation of the Model Law (b) now

$$S_v = \sqrt{S_d \cdot S_g} = \sqrt{(0.0001383) \cdot (1)} = 0.01176 \quad (j)$$

since $S_g = 1$, of course. Further, $S_v = v_2/v_1$, therefore by (j)

$$v_1 = \frac{v_2}{S_v} = \frac{211.6}{0.01176} = 17993 \text{ m/s} \cong 18 \text{ km/s} \tag{k}$$

The Modeling Data Table (Fig. 17-19) conveniently summarizes the above input and results data.

Variable					Scale factor S	Category	
name	symbol	dimension	prototype	model	model/prototype	prototype	model
meteorite diameter	d	m	28.93	0.004	0.0001383	2	1
crater diameter	D	m	1200	0.166	0.0001383	1	3
impact angle	φ	1 (rad)	1.169	1.169	1	1	2
impact speed	v	m/s	17993	211.6	0.01176	2	1
meteorite material density	ρ_m	kg/m³	7860	7860	1	1	1
crater material density	ρ_c	kg/m³	2800	2800	1	1	2
gravitational acceleration	g	m/s²	9.81	9.81	1	1	1
dimensionless	π_1	1	1.169	1.169	1	—	—
dimensionless	π_2	1	41.47943	41.5	1.0005	—	—
dimensionless	π_3	1	1068.058	1068.197	1.00013	—	—
dimensionless	π_4	1	2.80714	2.80714	1	—	—
categories of variables		1	freely chosen, *a priori* given, or determined independently				
		2	determined by application of Model Law				
		3	determined by measurement on the model				

Figure 17-19
Modeling Data Table for the experimental determination of the size and impact speed of a meteorite

Note the identical values (within rounding) of all four dimensionless variables—demonstrating the model's *dimensional* similarity to the prototype. Also observe that the measured (Category 3) variable of the model is the crater diameter, but the *derived* variable (from the Model Law) is the meteorite diameter of the prototype. Finally, it is seen that there are $(N_v)_2 = N_P = 4$ variables on the model imposed by the Model Law (b). This agrees with relation (17-14).

The next example demonstrates the way one model experiment can predict the behavior of many (in theory infinite number of) prototypes.

508 APPLIED DIMENSIONAL ANALYSIS AND MODELING

Example 17-11. Roasting Time for Turkey

In general, the roasting time for turkey is a function of the size of the bird. Let us determine this function using purely dimensional considerations. First, the basic conditions and assumptions are stated:

- All turkeys are geometrically similar and their material properties are identical
- The oven has enough heat input to maintain the temperature at the set level
- At the start of roasting, the bird is at room temperature
- The oven is preheated to 204 °C = 400 °F (Ref. 147, p. A24)

Therefore the oven's required heat input per unit time (power) is not an influencing parameter—as long as this power is adequate—which is assumed (second condition above).

Based on these assumptions and conditions, the following variables can be considered relevant:

Variable	Symbol	Dimension	Remark
roasting time	t	s	see Note 1
density of turkey	ρ	kg/m^3	
mass of turkey	M	kg	
thermal conductivity	k	m·kg/(s^3·°C)	see Note 2
specific heat capacity	c	m^2/(s^2·°C)	see Note 3

Note 1: We use "s" for dimension of time, although in "culinary science" time is customarily measured in minutes. However, this difference will present no difficulty in the presentation of results.

Note 2: The variable *thermal conductivity* is included because it influences the rate of heat transfer from the exterior to the interior of the turkey.

Note 3: The variable *specific heat capacity* is included because it determines the temperature increase of the bird upon a given heat input, and hence upon a given time.

Linear size L of the bird is not listed because it is defined by the relation, true for all geometrically similar turkeys,

$$M = a \cdot \rho \cdot L^3 \qquad (a)$$

where "a" is a numeric constant. Alternatively, of course, we could have considered any two of the triplet M, L, ρ because, by (a), any two such variables determine the third. We selected the pair M, ρ, because ρ is constant and M is easily measurable and is usually known at the time of purchase of the turkey.

By the table we see that we have five variables and four dimensions, therefore there is $5 - 4 = 1$ dimensionless variable, which by Theorem 7-4 (Art. 7.10) is a constant. This sole dimensionless "variable" is obtained by the Dimensional Set

	t	ρ	M	k	c
m	0	−3	0	1	2
kg	0	1	1	1	0
s	1	0	0	−3	−2
°C	0	0	0	−1	−1
π_1	1	$-\dfrac{1}{3}$	$-\dfrac{2}{3}$	1	−1

yielding

$$\pi_1 = \frac{t \cdot k}{c \cdot \sqrt[3]{\rho \cdot M^2}} \qquad (b)$$

From this relation the Model Law is

$$S_t \cdot S_k = S_c \cdot S_\rho^{1/3} \cdot S_M^{2/3} \qquad (c)$$

where S_t is the Time Scale Factor
  S_k is the Thermal Conductivity Scale Factor
  S_c is the Specific Heat Capacity Scale Factor
  S_ρ is the Density Scale Factor
  S_M is the Mass Scale Factor

By the conditions and assumptions

$$S_k = S_\rho = S_c = 1 \qquad (d)$$

Therefore the Model Law (c) simplifies to

$$S_t = S_M^{2/3} \qquad (e)$$

This Model Law states that the roasting time of a turkey varies as the $\frac{2}{3}$ power of its mass (weight), and since $\frac{2}{3} < 1$, this result confirms the general advice given in cook books: *roasting time per pound decreases as the weight of the bird increases.*

Lucy Waverman, who is a contributing columnist in the Domestic Science department of Toronto's daily newspaper, *The Globe and Mail,* gives the following advice: in an oven preheated to 400 °F, roast the turkey for 15 minutes per pound for the first 10 pounds, and 7 minutes per pound for the rest (Ref. 147, p. A24). This instruction translates to

$$t = 80 + (15.43) \cdot M \qquad (f)$$

where t is the roasting time in minutes, and M is the weight (mass) of the turkey in *kilograms*. For example, a 14 lb (M = 6.35 kg) bird requires $t = 177.99$ minutes to roast. Let us call *this* turkey the "model." Thus, $t_2 = 177.99$ min and $M_2 = 6.35$ kg. Then the prototype can be *any* bird we desire. By the Model Law (e)

$$S_t = \frac{t_2}{t_1} = \left(\frac{M_2}{M_1}\right)^{2/3}$$

from which

$$t_1 = \frac{t_2}{\sqrt[3]{M_2^2}} \cdot \sqrt[3]{M_1^2}$$

or, by the prototype data given above

$$t_1 = \frac{177.99}{\sqrt[3]{6.35^2}} \cdot \sqrt[3]{M_1^2} = 51.905 \cdot \sqrt[3]{M_1^2} \qquad (g)$$

Thus, an $M_1 = 7.94$ kg (= 17.5 lb) turkey requires

$$t_1 = (51.905) \cdot (7.94^{2/3}) = 206.5 \text{ min}$$

roasting at 204 °C (= 400 °F).

To check if the dimensional similarity was indeed achieved in this case, we calculate the sole dimensionless variable of (b) for both prototype and model. Accordingly, assuming dummy values of 1 for k, ρ, and c (they cancel out anyway), we have for the *prototype*

$$\pi_1 = \frac{t_1 \cdot k_1}{c_1 \cdot \sqrt[3]{\rho_1 \cdot M_1^3}} = \frac{(206.5) \cdot (1)}{(1) \cdot \sqrt[3]{(1) \cdot (7.94^2)}} = 51.885$$

and for the *model*

$$\pi_1 = \frac{t_2 \cdot k_2}{c_2 \cdot \sqrt[3]{\rho_2 \cdot M_2^3}} = \frac{(177.99) \cdot (1)}{(1) \cdot \sqrt[3]{(1) \cdot (6.35^2)}} = 51.905$$

and we see that the two π_1 values agree nicely.

All the above inputs and results are now summarized in the Modeling Data Table of Fig. 17-20.

Variable					Scale factor S model/prototype	Category	
name	symbol	dimension	prototype	model		prototype	model
roasting time	t	min	206.5	177.99	0.86194	2	3
mass of turkey	M	kg	7.94	6.35	0.79975	1	1
heat conductivity	k	m·kg/(s^3·°C)	1 (dummy)	1 (dummy)	1	1	1
density	ρ	kg/m^3	1 (dummy)	1 (dummy)	1	1	1
specific heat capacity	c	m^2/(s^2·°C)	1 (dummy)	1 (dummy)	1	1	1
dimensionless	π_1	1	51.885	51.905	1.00039	–	–
categories of variables	1	freely chosen, *a priori* given, or determined independently					
	2	determined by application of Model Law					
	3	determined by measurement on the model					

Figure 17-20
Modeling Data Table for roasting time for turkey

In this example we have

$N_V = 5$ (number of variables)

$N_d = 4$ (number of dimensions)

$N_P = 1$ (number of dimensionless variables)

Thus, the number of values for variables for the prototype *and* the model is $2 \cdot N_V = 10$, of which

$(N_V)_1 = N_V + N_d - 1 = 5 + 4 - 1 = 8$ are *selectable,* i.e., Category 1 [relation (17-13)]

$(N_V)_2 = N_P = 1$ is determined by the *Model Law,* i.e., Category 2 [relation (17-14)]

$(N_V)_3 = 1$ is determined by *measurement* on the model, i.e., Category 3 [relation (17-12)]

The presented Modeling Data Table (Fig. 17-20) confirms these findings.

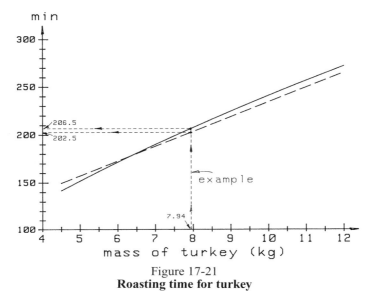

Figure 17-21
Roasting time for turkey
Solid line: relation (g), obtained by modeling experiment using the Model Law
[relation (e)]. Dashed line: relation (f), obtained by the author
from Lucy Waverman's recipe in Toronto's *The Globe and Mail*

Relation (g)—which is based on the Model Law and a single model measurement—allows us to determine the roasting time of *any* turkey. The *solid* line of Fig. 17-21 represents relation (g). The advice given by Lucy Waverman in *The Globe and Mail* (Ref. 147, p. A24) is the *dashed* line in the plot. As seen, the curves are remarkably close, proving both the high practical worth of the dimensional technique applied and the theoretical foundation of Lucy Waverman's recipe.

The numeric example, cited earlier, regarding the 7.94 kg (= 17.5 lb) turkey is illustrated in the graph.

Finally, it should be noted that some cook books advise different roasting temperatures and times for turkey. For example, the *Playboy Gourmet* (Ref. 126, p. 209), which is a rather acceptable source for food aesthetes, advocates a lower temperature and longer time than Lucy Waverman advises. However, the *Playboy* book calls for a higher temperature at 232 °C (= 450 °F) in the initial 30 minutes of roasting.

17.5. SCALE EFFECTS

In general, scale effects make our prediction of the prototype's behavior inaccurate, because some factors may influence a system differently when the system changes size or nature. A variable, which is entirely inconsequential in one setting, may be a dominating influence in another. It is therefore the major role of the investigator to critically evaluate the influence of every potentially relevant variable, and then ignore the ones whose effects on the system is marginal.

On occasion, the *direction* of the effect (dependent variable) *abruptly* changes as the magnitude of the independent variable *continuously* changes. To illustrate, in Example 12-3 (Art. 12.1) it was shown that local gravitational acceleration is

$$g = \frac{k \cdot M}{R_0^2} \cdot \left(\frac{R}{R_0}\right)^n \qquad (17\text{-}15)$$

where k is the universal gravitational constant;
M is the mass of the celestial body;
R_0 is the radius of the celestial body considered a perfect sphere;
R is the distance between the center of celestial body and the point at which g is determined;
n is a numeric constant.

Now if $R < R_0$, then $n = 1$; if $R = R_0$, then $n = 0$; and if $R > R_0$, then $n = -2$ (see Fig. 12-4). Thus, an investigator who is ignorant of the law of universal gravitation and who is below the surface of the celestial body (e.g., he is in a mine) will find—by experimentation—that

$$g = \frac{k \cdot M \cdot R}{R_0^3} \qquad (17\text{-}16)$$

On the other hand, another investigator who is above the surface (e.g., on the top of a tall tower) will find—also by valid experiments—that

$$g = \frac{k \cdot M}{R^2} \qquad (17\text{-}17)$$

The above two relations are different, since g in (17-16) is proportional to R, whereas in (17-17) it is inversely proportional to the square of it. Therefore if either investigator attempts to step over the surface barrier (i.e., $R = R_0$) in either direction, he will find that his formula is gravely wrong. This is an example of scale effect, when the validity of a formula obtained experimentally (or other means) is valid only within a specific range of the independent variable—beyond which it is false.

In other instances, the judicious elimination of variables in different ranges of other parameters is called for. An instance of this type of scale effect was presented in Example 11-17 (Art. 11.2) which dealt with the velocity of surface waves. There we saw that when the wave length was "short," gravity effects could be ignored, but surface tension and density could not; when the wave length was "long," then surface tension and density could be ignored, but gravity could not.

Consequently, the relation of velocity of wave propagation as a function of these variables is very different for the "short" and "long" regions of wavelengths. Thus, if by experiment one of these relations is derived for a particular wavelength, the same formula may be false for another wavelength. In this and all similar cases judicious consideration of all the pertinent parameters, and even the underlying physical laws, is called for.

For a prudent experimenter the best policy to follow is to always ask the question: do changes in independent variables influence the dependent variable in the same *direction* and in the same *way* as these changes increase (or decrease) in size?

Another subject of enquiry may be to find out whether any *abrupt* change occurs in the physical process of determining the dependent variable. For example, consider the torus geometry shown in Fig. 13-18 (Art. 13.6). When the hole diameter of the torus is *positive* (i.e., there *is* a hole), the relation defining the volume is entirely different from that when the torus is *degenerating* or *degenerated,* in which cases the hole is *negative* [see Fig. 13-18 (c) and (d)]. It is evident therefore that any formula determined by an experiment based on the hole being positive, will be wrong when the hole is negative.

Sometimes a change in physical structure is very gradual and subtle, so detecting it may require a very broad survey of the independent variables. The following example illustrates a typical case.

Example 17-12. Relative Mass of Mammalian Skeletons
(adapted from Ref. 24, p. 5.)

One might think that the mass of skeletons of *geometrically similar* land animals is proportional to the mass of their bodies. But here the scale effect plays an important part because the mass of a skeleton must grow faster with size than that of the total body. Consequently, proportionality does not exist.

Fig. 17-22 shows the skeleton's mass versus total body mass of land animals (including man). The "factual line" based on case measurements is the solid line. This line is described by the linear regression formula

$$M_s = (0.1) \cdot M_b^{1.13} \text{ kg} \quad \text{(a)}$$

where M_s is the mass of skeleton and M_b is the mass of body, including skeleton (kg). For example, an $M_b = 70$ kg man has a skeleton mass (according to the above formula) of approximately $M_s = 12.1$ kg. If perfect proportionality existed, then we would have

$$M_s = k \cdot M_b \quad \text{(b)}$$

where k is a numeric constant for all mammals (dashed line). The difference between skeleton masses M_s expressed by (a) and (b) is due to *scale effect,* which manifests itself with the progressively heavier bone structures of animals as their sizes increase. As shown in Fig. 17-22 an elephant's skeleton is about 2.5 times heavier than geometric similarity would predict.

To shed some light on *why* this scale effect exists, we look for what the exponent of M_b *would* be in (a), if the skeleton had constant stress generated by the weight of body. The capacity of a bone varies linearly with its *area* of cross section A, which, to maintain the same stress level, must be proportional to body mass. Thus

$$A = k_1 \cdot M_b \quad \text{(c)}$$

where k_1 is some dimensional constant. If L is the characteristic linear size of the animal, then

$$M_s = k_2 \cdot L \cdot A \quad \text{(d)}$$

and also of course

$$M_b = k_3 \cdot L^3 \quad \text{(e)}$$

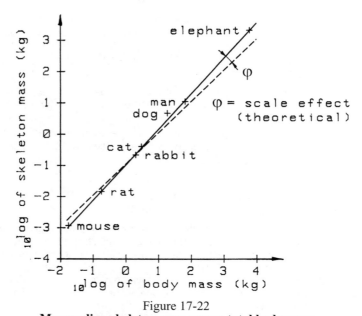

Figure 17-22
Mammalian skeleton mass versus total body mass
Solid line: best fit by measured data; dashed line: theoretical prediction
from geometric similarity (adapted from Ref. 24, p. 5)

where k_2 and k_3 are some dimensional constants. We now substitute A from (c) and L from (e) into (d). This operation will yield

$$M_s = k_2 \cdot L \cdot A = k_2 \cdot \left(\frac{M_b}{k_3}\right)^{1/3} \cdot k_1 \cdot M_b = k_4 \cdot M_b^{4/3} \tag{f}$$

where k_4 is a dimensional constant. This relation tells us that if the skeleton were maintained at the same *strength* to carry the body mass—regardless of its size—then the bone mass would have to grow as the 1.333 power of the body mass. But in fact—as the solid line in Fig. 17-22 shows—it only grows as the 1.13 power. Therefore the *factual scale effect* here is not that the exponent is *more* than it should be, but that it is *less* (only 1.13, instead of 1.333). What is the reason for this?

The most likely reason is that skeletons carry not only *static* weight (due to gravitation), but also *inertial* loads caused by the continual acceleration and deceleration of limbs and other body parts of the animal during jumping, running, etc. The smaller the animal, the larger the portion of all loads is carried by the skeleton due to inertial—as opposed to gravitational—effects. If an elephant jumped like a monkey, it would break most of its bones in one instalment. So it does not jump, and hence it can afford to have an "underdimensioned" skeleton structure.

⇑

The next example deals with a case in which scale effects modify the characteristics of a system so much that its performance becomes unacceptable.

DIMENSIONAL MODELING

Example 17-13. Ballet on the Moon

Some futuristic minds foresee man's colonization of the Moon. Naturally, this project would also involve cultural activities. As noted by a "space visionary"—who, for his own sake, shall remain nameless here—these events would include performances of Tchaikovsky's ballet *The Sleeping Beauty,* staged in a gigantic sealed dome. Is anything wrong with this scenario?

In general, the movements of ballet dancers are characterized by the following factors:

- time duration to execute their movements
- energy available in their bodies to execute movements
- their individual body mass
- local gravitational acceleration

Gravitation is included since vertical, and to a lesser extent horizontal, movements of body parts (leaps, jumps, etc., the basic ingredients of a ballet) must take place against gravitational forces. The list of variables, their symbols and dimensions is as follows:

Variable	Symbol	Dimension	Remark
dancer's energy	Q	m²·kg/s²	see Note 1
dancer's speed	v	m/s	see Note 2
distance of movement	h	m	see Note 3
time duration	t	s	see Note 4
gravitational acceleration	g	m/s²	
dancer's mass	M	kg	

Note 1: Q represents the dancer's *expended* energy, which may be less than his *available* energy.

Note 2: v represents the speed of movement that the expended energy Q produces.

Note 3: h represents the dancer's distance of "travel" in one typical movement, e.g., a "leap."

Note 4: t is the time duration needed to execute a typical movement, e.g., a "leap."

In the above table we have six variables and three dimensions, therefore there are $6 - 3 = 3$ dimensionless variables determined by the Dimensional Set

	Q	v	h	t	g	M
m	2	1	1	0	1	0
kg	1	0	0	0	0	1
s	−2	−1	0	1	−2	0
π_1	1	0	0	−2	−2	−1
π_2	0	1	0	−1	−1	0
π_3	0	0	1	−2	−1	0

from which

$$\pi_1 = \frac{Q}{t^2 \cdot g^2 \cdot M} \; ; \qquad \pi_2 = \frac{v}{t \cdot g} \; ; \qquad \pi_3 = \frac{h}{t^2 \cdot g} \qquad (a)$$

Thus the Model Law is

$$S_Q = S_t^2 \cdot S_g^2 \cdot S_M \; ; \qquad S_v = S_t \cdot S_g \; ; \qquad S_h = S_t^2 \cdot S_g \qquad (b)$$

where S_Q is the Dancer's Expended Energy Scale Factor;
S_t is the Time Scale Factor;
S_g is the Gravitational Acceleration Scale Factor;
S_M is the Dancer's Mass Scale Factor;
S_v is the Dancer's Speed Scale Factor;
S_h is the Dancer's Moving Distance Scale Factor.

Let us now designate *The Sleeping Beauty* performed on Earth as the *prototype,* and on the Moon as the *model.* We have the usual subscripts 1, 2 to designate the prototype and the model, respectively. Thus, g_1 = 9.81 m/s², g_2 = 1.6 m/s². Therefore $S_g = g_2/g_1$ = 1.6/9.81 = 0.1631. Moreover, S_M = 1, since the mass of the dancers does not change by being on the Moon.

We have six variables and hence six scale factors, but only three (constraining) relations in the Model Law (b). Therefore in theory we can select three scale factors, and the other three will be determined by the Modal Law. However, as S_M and S_g are already imposed upon us, only *one* of the remaining *four* scale factors can be freely chosen These four are: S_v, S_Q, S_t, and S_h. In theory then, we can keep constant either speed, energy, time, or distance—i.e., S_v = 1, or S_Q = 1, or S_t = 1, or S_h = 1, respectively. But if $S_t \cdot S_g$ from the second relation of the Model Law (b) is substituted into the first, then we have

$$S_Q = S_v^2 \cdot S_M \qquad (c)$$

Therefore S_v and S_Q are connected since S_M = 1 at all times. Thus, it is enough to restrict our selection to three scale factors, viz., S_Q, S_t, and S_h. Accordingly, *The Sleeping Beauty* on the Moon can be performed under any of the following three different conditions:

Condition 1: S_Q = 1. The *energy* expended by the dancer for a typical movement on the Moon is the same as on Earth.

Condition 2: S_t = 1. The *pace* of dancing on the Moon is the same as on Earth. For example, a "leap" takes the same time to execute on the Moon as on Earth.

Condition 3: S_h = 1. The *distances* travelled during individual ballet movements on the Moon and on Earth are the same.

Based on the Model Law (b), for these three basic conditions three sets of scale factors can be established (see Fig. 17-23). We now discuss these three conditions separately.

Condition 1: The dancer's expended energy remains constant. Therefore, by Fig. 17-23, the Time Scale Factor is S_t = 6.135. This means that corresponding movements on the Moon will take 6.135 times longer to perform. A ballerina—accustomed to exerting a given amount of energy for a "leap"—will stay 6.135 times longer in the air, and consequently (since S_h = 6.135) will land 6.135 times farther away horizontally. Similarly, a dancer will jump 6.135 times higher, and land that much later. Further, it would be necessary to enlarge the stage *area* by a factor of S_h^2 = 37.64 and raise the working *height* of the ceiling by a factor of S_h = 6.135.

But all these vicissitudes pale in comparison to what would happen to the *music.* For the music would *have to be* played 6.135 times *slower,* resulting in *The Sleeping Beauty* becoming *The Sleeping Audience,* since the performance now lasts a little over 15 hours! Then there is the problem of *pitch.* "Analogue" recorded music playing would be out of question of course, since this would cause the pitch to drop log S_t/log 2 = 2.617 octaves. But even if—by digital techniques, or a live orchestra in the pit—the pitch were maintained, how would Tchaikovsky's glorious music *sound* if played 6.135 time slower?

Therefore it is unthinkable that the ballet would be performed on the Moon under Condition 1 (constant energy), because neither the pace of the dancing nor the pitch of the music could be replicated. This calls for a look at Condition 2.

	Condition 1	**Condition 2**	**Condition 3**
by condition	$S_Q = 1$	$S_t = 1$	$S_h = 1$
given a priori		$S_M = 1$	
		$S_g = 0.163$	
determined by Model Law	$S_t = \dfrac{1}{S_g} \cdot \sqrt{\dfrac{S_Q}{S_M}} = \dfrac{1}{S_g} = 6.135$	$S_Q = S_t^2 \cdot S_g^2 \cdot S_M = S_g^2 = 0.0266$	$S_t = \sqrt{\dfrac{S_h}{S_g}} = \dfrac{1}{\sqrt{S_g}} = 2.477$
	$S_v = S_t \cdot S_g = \dfrac{1}{S_g} \cdot S_g = 1$	$S_v = S_t \cdot S_g = S_g = 0.163$	$S_v = S_t \cdot S_g = \dfrac{1}{\sqrt{S_g}} \cdot S_g = 0.404$
	$S_h = S_t^2 \cdot S_g = \dfrac{1}{S_g^2} \cdot S_g = \dfrac{1}{S_g} = 6.135$	$S_h = S_t^2 \cdot S_g = S_g = 0.163$	$S_Q = \dfrac{1}{S_g} \cdot S_g^2 \cdot S_M = S_g = 0.163$

Figure 17-23
Three sets of scale factors corresponding to the three basic conditions for performing ballet on the Moon
For definitions of scale factors, see list following relation (b)

Condition 2: We have the same pace of dancing and pitch of music. Let us examine the consequences of maintaining the same Time Scale Factor. i.e., $S_t = 1$. By the table of Fig. 17-23, the linear scale of all movements (vertical as well as horizontal) must be *reduced* to 16.3% of their respective values on Earth. In other words, the linear dimensions of all jumps, leaps, etc. on the Moon must be only 16.3% of those executed at the Lincoln Center in New York City; a 2-m (= 6.5 ft) leap will be reduced to a mere 0.32 m (= 13 in) hop, a 0.75-m (= 2.5 ft) jump to a minuscule skip of 0.12 m (= 4.8 in). Of course, the spectators will consider this ballet a rather stationary endeavor since the performers will hardly move at all, especially if viewed from a distance. Further, the *linear* dimension of the stage will be smaller by a factor of $1/S_h = 6.135$, and the *area* by a factor of 37.64! Moreover, the ceiling of the stage will be (to save costs) as close as 0.3 m (= 1 ft) to the tallest man's vertical standing reach.

Since the Speed Scale Factor is $S_v = 0.163$, the speed of the dancers' movements will also be only 16.3% of their "earthly" values. This fact would further exaggerate the grotesqueness of the performance, since the dancers would be moving in a very much slowed motion (by a factor of 6.135) on the rhythm of the music played normally.

The dancers will also have to adjust to the fact that the energy required to execute their drastically abbreviated movements will be reduced by a factor of $1/S_Q = 37.64$, i.e., to less than 2.7% (!) of their corresponding values on Earth (since $S_Q = 0.0266$). As a consequence, the training regimen for dancers performing on the Moon will be unnecessary since muscle exertion would be manifestly negligible. Competition to be a member of a Moon-based *corps de ballet* would be fierce, for *everybody* would qualify.

It is therefore safe to conclude that ballet on the Moon under Condition 2 (constant pace) would not be a tenable endeavour.

Condition 3: Finally, we consider the equal distance condition, whereby bodily movements on the Moon would have the same spatial dimensions as on Earth, i.e., $S_h = 1$. Since we have $S_t = 2.477$ and $S_v = 0.404$ (see table of Fig. 17-23), therefore the velocities of movements would be reduced to 40.4% of their original values and, correspondingly, time duration would increase by a factor of 2.477. Moreover, the expended energies to move would be reduced to 16.3% of their original values and hence, again, intensive retraining of all the dancers would be mandatory. Further, problems with the music as in Condition 1 would also occur here—although to a lesser extent. The pitch of analogue music would be reduced by log S_t/log 2 = 1.309 octaves and the music itself would have to be played slower by a factor of S_t = 2.477. For these reasons, ballet under Condition 3 could not be performed on the Moon.

The end result of course is that no matter which basic condition we choose, our new colony on the Moon—indeed on *any* celestial body whose surface gravitational acceleration is significantly different from 9.81 m/s²—would have to subsist—regretfully—without live ballet performances, including Tchaikovsky's masterpiece, *The Sleeping Beauty*.

⇑

Example 17-14. Most Comfortable Walking Speed (this material appeared in a greatly condensed form in Ref. 148, p. 37)

It is easy to observe that the most comfortable walking speeds of human beings depend on their sizes. In general, a man walks faster than a woman, who, in turn, walks faster than a child (when each walks alone). An obvious question now emerges: How does size affect the walking speeds of geometrically similar human beings? Although one can elect to walk slower or faster, there is a *particular* speed at which the energy expended per unit *distance* traveled is *minimum*. By definition, this is the *most comfortable walking speed*.

To determine this value, we consider the following variables:

Variable	Symbol	Dimension
most comfortable walking speed	v	m/s
characteristic linear size (e.g., height)	L	m
gravitational acceleration	g	m/s²

Note that we included gravitational acceleration because in walking the mass-center of the body also moves vertically, and so gravitation affects the expended energy. We have three variables and two dimensions, and hence there is only one dimensionless variable, which is therefore must be a constant. The Dimensional Set is

	v	L	g
m	1	1	1
s	−1	0	−2
π_1	1	$-\frac{1}{2}$	$-\frac{1}{2}$

from which

$$\pi_1 = \frac{v}{\sqrt{L \cdot g}} = k_1 \qquad (a)$$

where k_1 is a numeric constant. The *Model Law* is therefore

$$S_v = \sqrt{S_L \cdot S_g} \qquad (b)$$

where S_v is the Most Comfortable Walking Speed Scale Factor
S_L is the Characteristic Length Scale Factor
S_g is the Gravitational Acceleration Scale Factor

If we compare walking on the *same* celestial body (e.g., Earth), then $S_g = 1$, and the Model Law (b) simplifies to $S_v = \sqrt{S_L}$; i.e., the most comfortable walking speed is proportional to the square root of a person's characteristic length, say height. For example, a 2 m tall man walks about 1.41 times faster than a 1 m tall child, since $\sqrt{S_L} = \sqrt{2} = 1.41$ (Fig. 17-24).

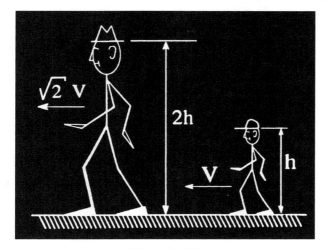

Figure 17-24
Modeling of the most comfortable walking speeds
The man is twice as tall, therefore he walks $\sqrt{2} = 1.41$ times faster than the boy
(on the same celestial body)

However, on the Moon, the *same* man ($S_L = 1$) walks only 0.404 times as fast as he does on Earth (i.e., 59.6% slower), since $g_{\text{Moon}} = 1.6$ m/s², hence $\sqrt{S_g} = \sqrt{\dfrac{g_{\text{Moon}}}{g_{\text{Earth}}}} = \sqrt{\dfrac{1.6}{9.81}} = 0.404$.

Now in addition to geometric similarity we assume that all human beings are equally dense. This is very nearly true, not only for man, but for all mammals as well. The value is approximately $\rho = 1000$ kg/m³, i.e., the density of water. Hence $S_\rho = 1$. In this case, then, the mass of an individual is proportional to the cube of his characteristic length L. Thus,

$$M = k_2 \cdot \rho \cdot L^3 \qquad (c)$$

where k_2 is a numeric constant. Thus, if S_M is the Mass Scale Factor, we can write

$$S_M = S_\rho \cdot S_L^3 \qquad (d)$$

or

$$S_L = \sqrt[3]{\dfrac{S_M}{S_\rho}} \qquad (e)$$

which, if substituted into (b) yields

$$S_v = \sqrt{S_L} \cdot \sqrt{S_g} = \sqrt[6]{\dfrac{S_M}{S_\rho}} \cdot \sqrt{S_g} \qquad (f)$$

Thus, the most comfortable walking speed is proportional to the sixth root of the individual's mass.

520   APPLIED DIMENSIONAL ANALYSIS AND MODELING

Now, by (f), for an individual

$$v = k_3 \cdot \sqrt[6]{\frac{M}{\rho}} \cdot \sqrt{g} \qquad (g)$$

Taylor quotes (Ref. 138, p. 129) Margaria et al., who measured the oxygen consumption in unit time of a $M = 70$ kg subject at speeds between 1 and 9 km/h (0.278 – 2.5 m/s). Since 1 liter of oxygen represents about 20100 J energy, what we have in this case is really the $P = \Psi\{v\}$ relation, where P is power, v is the speed, and Ψ is the symbol of a function. Further, if E is energy, t is time, and x is distance, then we can write

$$P = \Psi\{v\} = \frac{dE}{dt} = \frac{dE}{dt} \cdot \frac{dx}{dx} = \frac{dE}{dx} \cdot \frac{dx}{dt} = \frac{dE}{dx} \cdot v$$

Hence

$$\frac{dE}{dx} = \frac{\Psi\{v\}}{v} = \overline{E} \qquad (h)$$

Therefore the energy expended in unit distance $\overline{E}$ can be easily calculated from the published graph (ibid.) point-by-point. In Fig. 17-25 the tangent of slope φ of arbitrary point A is obviously, by (h)

$$\tan \varphi = \frac{P}{v} = \frac{\Psi\{v\}}{v} = \frac{dE}{dx}$$

It follows therefore that the *minimum* energy in unit distance travelled is defined by the position of point B, since at this point the slope—hence tan φ as well—is minimum. So

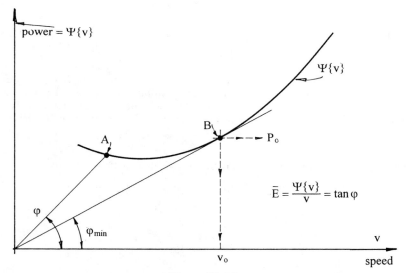

Figure 17-25
**Sketch to show how the $\overline{E}$ value is obtained
from the given (as above) $P = \Psi\{v\}$ plot**
$\overline{E}$ = energy expended in unit distance, P = power, v = speed; minimum $\overline{E}$ is at point B

$\overline{E}_{min} = \tan \varphi_{min}$. The construction of the $\overline{E} = \overline{E}\{v\}$ plot by the above method was done by the present author and the result is the graph shown in Fig. 17-26.

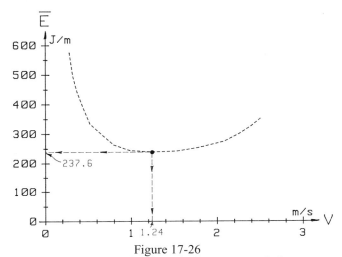

Figure 17-26
Energy expended in unit distance traveled versus walking speed by a person of 70 kg mass
Curve was constructed by the author from information obtained from Ref. 138

According to the plot, minimum energy expenditure is achieved at walking speed $v = 1.24$ m/s = 4.464 km/h; a slower or faster waking speed—especially slower (!)—increases the energy needed. We also see that the minimum amount of energy required by the subject was $\overline{E} = 237.6$ J/m. These data now enable us to determine the numeric constant k_3 in relation (g). Accordingly, this dimensionless constant is

$$k_3 = \frac{v}{\sqrt{g}} \cdot \sqrt[6]{\frac{\rho}{M}} = \frac{1.24}{\sqrt{9.81}} \cdot \sqrt[6]{\frac{1000}{70}} = 0.6167$$

therefore

$$v = 0.6167 \cdot \sqrt{g} \cdot \sqrt[6]{\frac{M}{\rho}} \quad \text{(i)}$$

Since for both man and mammals $\rho \cong 1000$ kg/m³, we can write

$$v = 0.19502 \cdot \sqrt{g} \cdot \sqrt[6]{M} \text{ m/s} \quad \text{(j)}$$

where, of course, 0.19502 is no longer dimensionless since $[0.19502] = \text{m}^{1/2} \cdot \text{kg}^{-1/6}$.

Let us now derive a relation for the walking energy expended in unit length covered at the most comfortable walking speed. We consider the following variables:

Variable	Symbol	Dimension	Remark
energy expended in unit distance	$\overline{E}$	m·kg/s²	at minimum energy point
mass of walker	M	kg	
gravitational acceleration	g	m/s²	

Note the absence of density (it is included in the mass). The dimensional matrix is then

	$\overline{E}$	M	g
m	1	0	1
kg	1	1	0
s	−2	0	−2

in which the third row is −2 time the first row. Therefore the matrix is singular and, as can be easily observed, its rank is 2. Thus, we must eliminate *one* dimension, which cannot be "kg," because by doing so the rank is reduced to 1. We therefore eliminate "s," and then have three variables and two dimensions. This arrangement will yield only 3 − 2 = 1 dimensionless variable—a constant. The Dimensional Set will be

	$\overline{E}$	M	g
m	1	0	1
kg	1	1	0
π_2	1	−1	−1

from which

$$\pi_2 = \frac{\overline{E}}{M \cdot g} = k_4 = \text{constant} \tag{k}$$

where we used subscript "2" for the dimensionless variable to distinguish it from π_1 defined in (a).

By the data in Fig. 17-26, $k_4 = \overline{E}/(M \cdot g) = 237.6/[(70) \cdot (9.81)] = 0.346$. Hence

$$\overline{E} = (0.346) \cdot M \cdot g \tag{ℓ}$$

Note that while the most comfortable walking speed varies with the square-root of g [relation (j)], the expended energy corresponding to this speed varies linearly with g [relation (ℓ)].

To illustrate the usefulness of relations (j) and (l), we will answer the following question: How long would it take, and how much energy would an $M = 85$ kg astronaut need to walk $u = 1850$ m on the Moon inside a radome? Answer: on the Moon $g = 1.6$ m/s², therefore by (j)

$$v = (0.19502) \cdot \sqrt{g} \cdot \sqrt[6]{M} = (0.19502) \cdot \sqrt{1.6} \cdot \sqrt[6]{85} = 0.5173 \text{ m/s} \; (= 1.86 \text{ km/h})$$

Hence the time necessary to walk $u = 1850$ m is $t = u/v = 1850/0.5173 = 3576.5$ s = 59 min 36.5 s. The energy expended in unit distance is, by (ℓ), $\overline{E} = (0.346) \cdot M \cdot g = (0.346) \cdot (85) \cdot (1.6) = 47.056$ J/m. Thus, the energy needed to walk 1850 m is $E = \overline{E} \cdot u = (47.056) \cdot (1850) = 87053.6$ J ($\cong 20.8$ kcal).

In comparison, on Earth the same astronaut would walk with 1.28 m/s speed (= 4.61 km/h) for 1444.4 s (= 24 min 4.4 s), and would consume 533747 J (= 127.5 kcal) total energy, i.e., he would walk 2.47 times faster and would expend 6.13 times more energy.

Finally, let us answer a truly challenging question: How fast would *Tyrannosaurus rex*—a dinosaur—have walked? Based on fossils, its height was about $L_2 = 6$ m (Fig. 17-27), and we know that its form of locomotion was bipedal—like that of *Homo sapiens*.

The mass of a "typical" $L_1 = 1.78$ m tall man (the author) is $M_1 = 88$ kg. Therefore, assuming a modicum of geometric—only geometric!—similarity between the dinosaur and

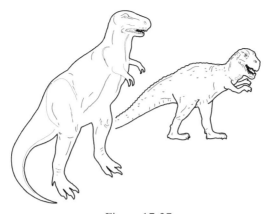

Figure 17-27
Dinosaur, 6 m tall
What was its "most comfortable" walking speed?

the author, the Length Scale Factor is $S_L = L_2/L_1 = 6/1.78 = 3.371$. Since the Density Scale Factor can be assumed to be unity, $S_\rho = 1$, hence, by (d), the Mass Scale Factor is

$$S_M = S_\rho \cdot S_L^3 = (1) \cdot (3.371^3) = 38.307 = \frac{M_2}{M_1}$$

Thus the mass of the dinosaur was

$$M_2 = S_M \cdot M_1 = (38.307) \cdot (88) = 3371 \text{ kg} \qquad \text{(m)}$$

With this datum and (i) then, its walking speed was

$$v_2 = (0.6167) \cdot \sqrt{g} \cdot \sqrt[6]{\frac{M_2}{\rho_2}} = (0.6167) \cdot \sqrt{9.81} \cdot \sqrt[6]{\frac{3371}{1000}} = 2.36 \text{ m/s} \qquad \text{(n)}$$

This value fits well within the range of 1 – 3.6 m/s derived by Schepartz (Ref. 43. p. 155) using footprint spacing. Moreover, the energy requirement for the dinosaur to walk a unit distance can now be easily determined. Thus, by (ℓ), and (m)

$$\overline{E} = (0.346) \cdot M_2 \cdot g = (0.346) \cdot (3371) \cdot (9.81) = 11442 \text{ J/m}$$

⇑

17.6. PROBLEMS

Solutions are in Appendix 6.

17/1 **To crack a window.** An $M_2 = 0.1$ kg stone is thrown in a perpendicular direction against a glass window, which cracks on impact. The speed of the stone is $v_2 = 12.6$ m/s and the characteristic size of the window is $L_2 = 1.4$ m.

(a) What will be the speed of an $M_1 = 0.2$ kg stone which cracks a geometrically similar window of characteristic size $L_1 = 0.70$ m? Assume the material of the window is unchanged, and the stone's direction of motion remains perpendicular to the plane of the window.

(b) Construct the Modeling Data Table and include in it the categories of all the variables for both the prototype and the model. Check that the numbers in these categories agree with those prescribed by relations (17-12), (17-13), and (17-14).

(c) Check whether the values of the dimensionless variables(s) for both prototype and model is (are) identical.

Assume the following variables are relevant: mass and speed of the impacting stone, cracking stress and Young's modulus of window's material, characteristic length of the window.

17/2 State the condition(s) for two rectangular hyperbolas to be geometrically similar.

17/3 A homogeneous sphere rolling down without sliding on a slope inclined to horizontal φ degrees. Set up a modeling experiment to determine the time to complete h vertical travel. Assume that the radius of the sphere is a physically irrelevant variable.

(a) How many physical variable are there and what are their dimensions?
(b) How many dependent and independent variables are there?
(c) How many variables are in Categories 1, 2, and 3?
(d) How many independent dimensionless variables can be formed, and what are they?
(e) Establish the Model Law.

17/4 Contact time of impacting balls. Given two identical steel balls (prototype) of radius $R_1 = 0.6$ m, density $\rho_1 = 7850$ kg/m^3, and Young's modulus $E_1 = 2 \times 10^{11}$ N/m^2. The balls travel toward each other with an individual speed of $v_1 = 2.8$ m/s. Set up a modeling experiment to determine the duration of contact (contact time) t_1. The model is two equal aluminium balls $R_2 = 0.05$ m, $\rho_2 = 2600$ kg/m^3, $E_2 = 6.7 \times 10^{10}$ N/m^2. Assume the same coefficient of restitution for both prototype and model.

(a) What are the relevant variables and their dimensions?
(b) How many Category 1, 2, 3 variables are there (total for prototype and model)?
(c) How many Category 1, 2, 3 variables are there for the prototype? For the model?
(d) How many dimensionless variables are there, and what are they?
(e) Establish the Model Law.
(f) What should the speed of the model balls be?
(g) If the measured contact time for the model, travelling at a speed determined in (f) above, is $t_2 = 4.412 \times 10^{-5}$ s, what is the contact time t_1 for the prototype?

(h) Calculate the values of the dimensionless variables, determined in (d) above, and check whether they are pair-wise identical for the prototype and the model (they should be).

(i) Construct the Modeling Data Table incorporating all the above input data and calculated results.

17/5 **Gravitational collapse of a star.** It is observed that a given star of diameter D_1, internal stress σ_1 and density ρ_1, collapsed under its own gravitational forces (became "unstable" is another expression for the same thing). If another star of diameter $D_2 = 20 \cdot D_1$ and density $\rho_2 = (0.08) \cdot \rho_1$ also collapsed, what was its internal stress σ_2 (in terms of σ_1) at the instant of collapse?

CHAPTER 18
FIFTY-TWO ADDITIONAL APPLICATIONS

This chapter presents 52 additional illustrations in which the usefulness and great facility of the dimensional method in solving problems in the physical sciences is further demonstrated. These examples show the versatility of the technique that was developed and discussed in the preceding chapters of this book.

The dimensional techniques applied in the examples are broadly classed into 26 "topics" and then each example is identified accordingly. The first table below gives *topic descriptions* and their symbols. For example, topic "R" is *heuristic reasoning to find exponents in monomial functions of dimensionless variables.* The second table presents the *allocations of topics* among the examples. Thus, Examples 18-6, 18-12, and 18-13 deal with topic "R."

The above classifications notwithstanding, the reader may wish to read the entire collection, thereby gaining invaluable experience and the skill to utilize the enormous potential offered by the dimensional method.

Symbols for categories of topics

Topic symbol	Topic description
A	Reduction of the number of variables by the dimensional method
B	There is only one dimensionless "variable," which is therefore a constant
C	Proof of a mathematical or geometric theorem
D	Use of directional dimensions (splitting dimensions)
E	Selective consideration of variables
F	Historical reference
G	Modeling, formulation of Model Law, Modeling Data Table
H	Monomial form of dimensional variables is proved untenable
I	Fusing variables, heuristic reasoning to reduce the number of variables
J	Dimensionally (and hence physically) irrelevant variable is present
K	Rank of dimensional matrix is less than the number of dimensions
L	Result is counterintuitive
M	Number of variables equals the number of dimensions
N	Usual sequence of variables in the Dimensional Set is changed
O	Experimental determination of constants, or infusion of other information
P	Fusion of dimensionless variables

(table is continued on next page)

(continued from preceding page)

Topic symbol	Topic description
Q	Geometric dissimilarity of prototype and model
R	Heuristic reasoning to find exponents of monomials
S	Splitting variables
T	Dimensionless physical variable is present
U	Omission of a relevant variable
V	General application; derivations of some named dimensionless variables
W	Mathematical derivation is presented
X	Fusing dimensions
Y	Numerical application is included
Z	**D** is not an identity matrix; forced composition of dimensionless variables

Topics of the 52 illustrative applications in this chapter. Definitions of topic *symbols* are in the preceding table.

Topic	Example number
A	7 19 29 41
B	2 3 4 5 9 10 11 14 16 17 24 27 28 33 34 38 46 48
C	1 52
D	6 28 44 47 48
E	35 38 42
F	4 16 17 19 49
G	18 23 27 30 31 32 36 37 39 40 43 45 50 51
H	15 25
I	49
J	2 20
K	4 7 8 9 11 26 28 33 34 46 49
L	20 34 37 39 41 42 52
M	28 33 34
N	38
O	39
P	41
Q	29 32
R	6 12 13 44
S	6
T	1 6 20 43
U	14
V	21 22
W	17 26 41 42 43 49
X	19
Y	17 18 23 24 26 27 29 30 31 32 33 34 36 37 40 41 42 43 45 49 50 51 52
Z	22

Example 18-1.   Proof of Pythagorean Theorem (adapted from Ref. 1, p. 47)

This theorem is named after Pythagoras of Samos (Greek philosopher, 582–497 B.C.); it states that the square of the length of the hypotenuse of a right triangle is equal to the sum of the squares of the lengths of its sides (Fig. 18-1). To determine the area of the triangle we use the following variables:

Variable	Symbol	Dimension	Remark
area	A	m²	
hypotenuse	c	m	
angle on hypotenuse	φ	1	see Fig. 18-1

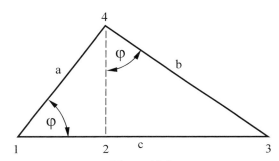

Figure 18-1
A right triangle

We have three variables and one dimension. Therefore among the variables listed there are $3 - 1 = 2$ dimensionless variables; they are determined by the Dimensional Set

	A	φ	c
m	2	0	1
π_1	1	0	-2
π_2	0	1	0

from which

$$\pi_1 = \frac{A}{c^2}; \qquad \pi_2 = \varphi \qquad (a)$$

Therefore

$$\pi_1 = \Psi\{\pi_2\}$$

where Ψ is an as-yet unknown function. By this relation and (a), we write

$$\frac{A}{c^2} = \Psi\{\varphi\}$$

or

$$A = c^2 \cdot \Psi\{\varphi\} \qquad (b)$$

Now we consider that triangles 1-3-4, 1-2-4, and 2-3-4 are *similar*. Therefore, by (b), we can write for the respective areas A_c, A_a, A_b

$$A_c = c^2 \cdot \Psi\{\varphi\}; \qquad A_a = a^2 \cdot \Psi\{\varphi\}; \qquad A_b = b^2 \cdot \Psi\{\varphi\} \qquad (c)$$

But obviously

$$A_c = A_a + A_b \qquad (d)$$

Hence, by (c) and (d)

$$c^2 = a^2 + b^2$$

which is the Pythagorean Theorem.

⇑

Example 18-2. Time to Fall a Given Distance

We assume constant gravitational acceleration and no air resistance. The variables are as follows:

Variable	Symbol	Dimension
time of fall	t	s
distance of fall	h	m
gravitational acceleration	g	m/s²
mass	M	kg

We have four variables and three dimensions, therefore there is but one dimensionless variable obtained by the Dimensional Set

	t	h	g	M
m	0	1	1	0
kg	0	0	0	1
s	1	0	−2	0
π_1	1	$-\dfrac{1}{2}$	$\dfrac{1}{2}$	0

from which the sole dimensionless "variable"—a constant in this case—is

$$\pi_1 = t \cdot \sqrt{\dfrac{g}{h}} = \text{const} \qquad (a)$$

or

$$t = \text{const} \cdot \sqrt{\dfrac{h}{g}} \qquad (b)$$

where the constant happens to be $\sqrt{2}$. Note that the column under M in the **C** matrix of the Dimensional Set is all zeros. Therefore, by Theorem 10-4, the mass of a falling object is a *dimensionally irrelevant* variable, and hence, by Theorem 11-3, it is also a *physically irrelevant* variable (this fact was first established by Galilei).

⇑

Example 18-3. Tension in a Wire Ring Rotated About its Central Axis Perpendicular to its Plane (adapted from Ref. 2, p. 50)

Given a thin inextensible wire ring which is rotated about its central axis perpendicular to its plane (see Fig. 18-2). We wish to find axial tension T generated by this rotation. The variables are:

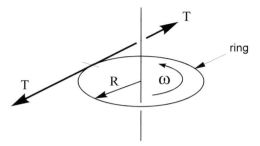

Figure 18-2
Axial tension in a rotating ring

Variable	Symbol	Dimension	Remark
tension force	T	m·kg/s²	
rotational speed	ω	rad/s	
linear mass	q	kg/m	per unit length
radius	R	m	

We have four variables and three dimensions, therefore there is only one dimensionless variable, in this case a constant. The Dimensional Set is

	T	ω	q	R
m	1	0	−1	1
kg	1	0	1	0
s	−2	−1	0	0
π_1	1	−2	−1	−2

which yields the sole dimensionless variable

$$\pi_1 = \frac{T}{\omega^2 \cdot q \cdot R^2} = \text{const}$$

Hence the tension force in the wire is

$$T = \text{const} \cdot q \cdot \omega^2 \cdot R^2$$

where the constant is 2.

⇑

Example 18-4. Velocity of Disturbance in a Liquid (adapted from Ref. 2, p. 65)

We consider the following relevant variables:

Variable	Symbol	Dimension
velocity of disturbance	v	m/s
density of fluid	ρ	kg/m³
bulk modulus of liquid	β	kg/(m·s²)

The bulk modulus of a fluid, by definition, is the ratio of pressure to the *relative* volume change this pressure produces. Therefore the dimension of the bulk modulus is the same as that of pressure. The dimensional matrix is

	v	ρ	β
m	1	−3	−1
kg	0	1	1
s	−1	0	−2

and we see that its determinant is zero. Therefore the dimensional matrix is *singular* and, by Theorem 7-5, the system has a solution. To this end, we determine the rank of this matrix, which is 2 (the bottom-right 2 × 2 determinant is not zero). Therefore one dimension must be deleted, say "kg." Thus, the Dimensional Set is

	v	ρ	β
m	1	−3	−1
s	−1	0	−2
π_1	1	$\frac{1}{2}$	$-\frac{1}{2}$

yielding $\pi_1 = v \cdot \sqrt{\rho/\beta}$ = const, or $v = \text{const} \cdot \sqrt{\beta/\rho}$, where the constant is 1. For water, since $\beta = 2.068 \times 10^9$ Pa (Ref. 124. p.5, i.e., water is about 100 times more compressible than steel) and $\rho = 1000$ kg/m³, the above relation supplies $v \cong 1438$ m/s.

⇑

Example 18-5. Reverberation Period in a Room (adapted from Ref. 2, p. 68)

We assume geometrically similar rooms and uniform atmospheric conditions. The variables are then

Variable	Symbol	Dimension
reverberation period	T	s
velocity of sound	v	m/s
volume of room	Q	m³

The Dimensional Set is

	T	v	Q
m	0	1	3
s	1	−1	0
π_1	1	1	$-\frac{1}{3}$

yielding

$$\pi_1 = \frac{T \cdot v}{\sqrt[3]{Q}} = \text{const} \qquad\qquad (a)$$

or

$$T = \text{const} \cdot \frac{\sqrt[3]{Q}}{v} \tag{b}$$

Thus, the reverberation period is proportional to the cube-root of volume, which, if the rooms are geometrically similar, means the *linear* measure of the room. Hence, (b) states that the reverberation period is proportional to the linear size of the room and inversely proportional to the speed of sound—hardly astonishing results!

⇑

Example 18-6. Flying Time of a Bullet Landing on an Inclined Plane (adapted from Ref. 2, p. 78)

What is the flight time of a bullet ejected at an angle and landing on an inclined plane (Fig. 18-3)?

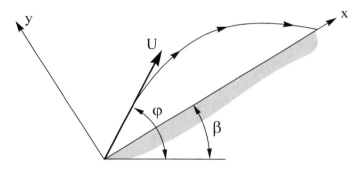

Figure 18-3
Bullet ejected at an angle lands on an inclined plane

We have the following list of variables:

Variable	Symbol	Dimension
ejection angle	φ	1
target plane inclination	β	1
flying time	t	s
ejection speed	U	m/s
gravitational acceleration	g	m/s²

The Dimensional Set therefore is

	φ	β	t	U	g
m	0	0	0	1	1
s	0	0	1	−1	−2
π_1	1	0	0	0	0
π_2	0	1	0	0	0
π_3	0	0	1	−1	1

yielding three dimensionless variables

$$\pi_1 = \varphi \; ; \qquad \pi_2 = \beta \; ; \qquad \pi_3 = \frac{t \cdot g}{U} \tag{a}$$

Hence, we can write

$$t = k \cdot \frac{U}{g} \cdot \Psi\{\varphi, \beta\} \tag{b}$$

where k is a numeric constant and Ψ is an as-yet undefined function.

We see that relation (b) is not very useful, for besides the almost obvious fact that t is proportional to ejection speed U and inversely proportional to g, it does not tell us anything. In particular, it does not provide information about how angles φ and β influence flight time t.

To improve our standing of the matter, we now employ *both directional variables* and *directional dimensions*. By this way, the number of dimensionless variables will decrease, and we will be able to extract much more information from the obtained result—as we shall see. Accordingly, using the coordinate system defined in Fig. 18-3, we resolve the variables U and g into directional components U_x, U_y, and g_x, g_y. The associated directed dimensions are m_x, m_y. Thus, we compose the modified Dimensional Set

	U_y	t	U_x	g_x	g_y
m_x	0	0	1	1	0
m_y	1	0	0	0	1
s	−1	1	−1	−2	−2
π_1	1	0	−1	1	−1
π_2	0	1	−1	1	0

yielding

$$\pi_1 = \frac{U_y \cdot g_x}{U_x \cdot g_y} \; ; \qquad \pi_2 = \frac{t \cdot g_x}{U_x} \tag{c}$$

If now a *monomial* form is assumed (which is plausible), then we can write

$$\pi_2 = k \cdot \pi_1^n \tag{d}$$

where k and n are numeric constants. By (c) and (d)

$$t = k \cdot \frac{U_x}{g_x} \cdot \left(\frac{U_y \cdot g_x}{U_x \cdot g_y} \right)^n \tag{e}$$

Next, we resolve speed U and gravitational acceleration g into their directional components

$$\left. \begin{array}{l} U_x = U \cdot \cos(\varphi - \beta) \\ U_y = U \cdot \sin(\varphi - \beta) \\ g_x = g \cdot \sin \beta \\ g_y = g \cdot \cos \beta \end{array} \right\} \tag{f}$$

With these details, (e) can be written

$$t = k \cdot \frac{U}{g} \cdot \frac{\cos(\varphi - \beta)}{\sin \beta} \cdot \left(\frac{\sin(\varphi - \beta) \cdot \sin \beta}{\cos(\varphi - \beta) \cdot \cos \beta} \right)^n \tag{g}$$

We now perform some heuristic meditation: if $\varphi > 0$ and $\beta \cong 0$, then we can write

$$t = k \cdot \frac{U}{g} \cdot \cos^{1-n}\varphi \cdot \sin^{n-1}\beta \cdot \sin^{n}\varphi \tag{h}$$

Let us first consider $n > 1$. Then $t \cong 0$, which makes no sense, since the flight time of a bullet ejected at an angle $\varphi > 0$ and reaching a target level with the ejector (gun) is certainly not zero. Next, consider $n < 1$. Then (h) can be written

$$t = \frac{k \cdot U}{g} \cdot \frac{\cos^{1-n}\varphi \cdot \sin^{n}\varphi}{\sin^{1-n}\beta} \tag{i}$$

where the numerator of the trigonometric term on the right is positive and the denominator tends to zero as β approaches zero. Therefore t tends to infinite, which is equally absurd. Consequently, n cannot be either less or more than 1, hence it must be 1. In this case (g) simplifies to

$$t = k \cdot \frac{U}{g} \cdot \frac{\sin(\varphi - \beta)}{\cos \beta} \tag{j}$$

Therefore we only have to determine constant k, which we can do with just a little more heuristic reasoning.

Since (j) must be true for all angles φ, therefore it must be true if $\varphi = 90°$, i.e., the bullet is shot *vertically* upward. In this case

$$t = k \cdot \frac{U}{g} \cdot \frac{\sin(90° - \beta)}{\cos \beta} = k \cdot \frac{U}{g} \cdot \frac{\cos \beta}{\cos \beta} = k \cdot \frac{U}{g} \tag{k}$$

But if $\varphi = 90°$, then $t = 2 \cdot (U/g)$, hence $k = 2$. With this then we can write, by (j)

$$t = 2 \cdot \frac{U}{g} \cdot \frac{\sin(\varphi - \beta)}{\cos \beta} \tag{ℓ}$$

which is the relation we were looking for. Thus, we succeeded to obtain the rather complex formula (ℓ) without any analytical derivation or even measurements, merely by some modest brain work and the adroit use of dimensions.

⇑

Example 18-7. Heat Transfer in a Calorifer (adapted from Ref. 2, p. 123)

A calorifer is a device in which heat is transferred from a liquid flowing in a tube to the outside through the wall and into another liquid in a tank. Fig. 18-4 shows the arrangement. It is assumed that the temperature is constant inside the tube and in the tank, and that the tube's conduction resistance is negligible (i.e., its thermal conductivity is very high).

The transmitted heat is governed by

$$Q = h \cdot A \cdot t \cdot \Delta\theta \tag{a}$$

where

Variable	Symbol	Dimension
heat transmitted	Q	$m^2 \cdot kg/s^2$
film heat transfer coefficient	h	$kg/(s^3 \cdot °C)$
area through which heat is transmitted	A	m^2
time	t	s
temperature difference	$\Delta\theta$	$°C$

536 APPLIED DIMENSIONAL ANALYSIS AND MODELING

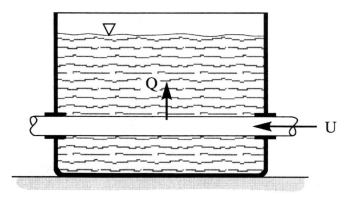

Figure 18-4
A calorifer

The problem is now to determine the *heat transfer coefficient h* for a given set of characteristics of the calorifer. The following variables are involved:

Variable	Symbol	Dimension
film heat transfer coefficient	h	kg/(s^3·°C)
fluid specific heat capacity	c	m^2/(s^2·°C)
fluid flow mass rate per unit area	U	kg/(m^2·s)
pipe diameter	D	m
fluid heat conductivity	k	m·kg/(s^3·°C)
fluid dynamic viscosity	μ	kg/(m·s)

The dimensional matrix is

	h	c	U	D	k	μ
m	0	2	−2	1	1	−1
kg	1	0	1	0	1	1
s	−3	−2	−1	0	−3	−1
°C	−1	−1	0	0	−1	0

whose rank is 3—as can be determined easily by the method described in Art. 1.3. Since the number of dimensions cannot be more than the rank, one dimension must be deleted. Let this unfortunate dimension be "°C." Thus we have six variables and three dimensions and therefore there are 6 − 3 = 3 dimensionless variables supplied by the Dimensional Set:

	h	c	U	D	k	μ
m	0	2	−2	1	1	−1
kg	1	0	1	0	1	1
s	−3	−2	−1	0	−3	−1
π_1	1	0	0	1	−1	0
π_2	0	1	0	0	−1	1
π_3	0	0	1	1	0	−1

from which

$$\pi_1 = \frac{h \cdot D}{k} \ ; \quad \pi_2 = \frac{c \cdot \mu}{k} \ ; \quad \pi_3 = \frac{U \cdot D}{\mu} \tag{b}$$

and hence we can write

$$\pi_1 = \Psi\{\pi_2, \pi_3\} \tag{c}$$

where Ψ is a function to be determined experimentally.

At this point, one should pause to reflect on the colossal saving of effort and resources in presenting and determining the *dimensionless* relation Ψ in (c) versus of the *dimensional* form

$$h = \Psi_1\{c, U, D, k, \mu\} \tag{d}$$

We see that (c) has three variables and (d) has six. If we assume eight distinct values for each variable, then (by the table in Fig. 12-3) to plot (d) we need 4096 curves appearing on 512 charts. In contrast, to plot (c) we need only eight curves on one chart!

Example 18-8. Electric Field Generated by a Dipole at a Point that Is on the Line Joining Them

Given an electric dipole of charge Q. What is the electric field strength generated by this dipole at point P? Fig. 18-5 shows the arrangement.

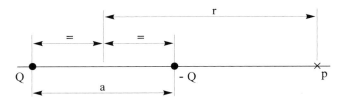

Figure 18-5
Electric field at point P generated by charges Q and $-Q$

The variables involved are:

Variable	Symbol	Dimension	Remark
electric field strength	E	m·kg/(s³·A)	at point P
distance between charges	a	m	
electric dipole charge	Q	A·s	
permittivity of vacuum	ϵ_0	A²·s⁴/(m³·kg)	
distance of point P	r	m	as defined in Fig. 18-5

Thus the dimensional matrix is

	E	a	Q	ϵ_0	r
m	1	1	0	−3	1
kg	1	0	0	−1	0
s	−3	0	1	4	0
A	−1	0	1	2	0

The rank of this matrix is 3—as the method described in Art. 1.3 can verify. Therefore one dimension—say A—must go. We are then left with five variables and three dimensions, and hence we have $5 - 3 = 2$ dimensionless variables supplied by the Dimensional Set

	E	a	Q	ϵ_0	r
m	1	1	0	-3	1
kg	1	0	0	-1	0
s	-3	0	1	4	0
π_1	1	0	-1	1	2
π_2	0	1	0	0	-1

from which

$$\pi_1 = \frac{E \cdot \epsilon_0 \cdot r^2}{Q} \; ; \qquad \pi_2 = \frac{a}{r} \qquad (a)$$

We now can write

$$\pi_1 = \Psi\{\pi_2\} \qquad (b)$$

where Ψ is some function. Let us assume a *monomial* form for this function. Thus, we can write

$$\pi_1 = k \cdot \pi_2^n \qquad (c)$$

where k and n are numerical constants. By (a) and (b)

$$E = k \cdot \frac{Q}{\epsilon_0 \cdot r^2} \cdot \left(\frac{a}{r}\right)^n \qquad (d)$$

With only two measurements (!) numeric constants k and n can now be determined and thus the relatively complex explicit formula joining five physical variables is established. (By easy analytical derivation we know that if $r \gg a$, then $k = 1/(2 \cdot \pi)$ and $n = 1$).

⇑

Example 18-9. Maximum Velocity of an Electron in a Vacuum Tube

We now consider a vacuum tube in which electrons leave the heated cathode 1 at negligible speed and arrive at the anode plate 2. The electrons move under the influence of potential difference U (Fig. 18-6). What are the electrons' impact speed on the anode?

First we list the relevant variables:

Variable	Symbol	Dimension	Remark
speed of electrons	v	m/s	at the anode
potential difference between electrodes	U	m²·kg/(s³·A)	
charge of an electron	e	A·s	
mass of an electron	M	kg	

and we have the dimensional matrix

	v	U	e	M
m	1	2	0	0
kg	0	1	0	1
s	-1	-3	1	0
A	0	-1	1	0

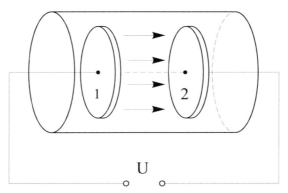

Figure 18-6
Electric potential U accelerates electrons between electrodes 1 and 2 in a vacuum tube

whose rank is 3. Therefore, one of the four dimensions must be sacrificed—say "A." The Dimensional Set is then

	v	U	e	M
m	1	2	0	0
kg	0	1	0	1
s	−1	−3	1	0
π_1	1	$-\dfrac{1}{2}$	$-\dfrac{1}{2}$	$\dfrac{1}{2}$

yielding the sole dimensionless variable—a constant in this case

$$\pi_1 = v \cdot \sqrt{\frac{M}{U \cdot e}} = \text{const} \qquad (a)$$

from which

$$v = \text{const} \cdot \sqrt{\frac{U \cdot e}{M}} \qquad (b)$$

The maximum speed (at impact on the anode) is proportional to the square root of the product $U \cdot e$, and inversely proportional to the square root of the mass of the electron. The constant can be determined by a single measurement or by analysis; it is 2 (Ref. 4, p. 377).

⇑

Example 18-10. Period of Oscillation of a Magnetic Torsional Dipole

A permanent magnet of pole strength p and length L is pivoted at its center. The magnet is placed in magnetic field H and then its $\varphi = 0$ equilibrium position is changed to $\varphi > 0$, as Fig. 18-7 shows. What will be the magnet's torsional oscillatory period?

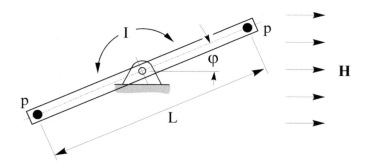

Figure 18-7
Permanent-magnet dipole oscillates in a magnetic field

The relevant variables are as follows:

Variable	Symbol	Dimension
period of oscillation	T	s
moment of inertia	I	$m^2 \cdot kg$
dipole length	L	m
magnetic pole strength	p	$m^2 \cdot kg/(A \cdot s^2)$
magnetic field intensity	H	A/m

We have five variables and four dimensions, therefore there is only $5 - 4 = 1$ dimensionless variable—a constant—which is supplied by the Dimensional Set

	T	I	L	p	H
m	0	2	1	2	-1
kg	0	1	0	1	0
s	1	0	0	-2	0
A	0	0	0	-1	1
π_1	1	$-\dfrac{1}{2}$	$\dfrac{1}{2}$	$\dfrac{1}{2}$	$\dfrac{1}{2}$

Therefore

$$\pi_1 = T \cdot \sqrt{\frac{L \cdot p \cdot H}{I}} = \text{const} \qquad (a)$$

from which

$$T = \text{const} \cdot \sqrt{\frac{I}{L \cdot p \cdot H}} \qquad (b)$$

where the constant is $2 \cdot \pi$ (Ref. 4, p. 425).

Example 18-11. Force on a Straight Wire Carrying Current and Placed in a Uniform Magnetic Field

We place a wire carrying a constant current in a uniform magnetic field whose direction is perpendicular to the wire. What will be the magnetically induced force per unit length of the wire? Fig. 18-8 shows the physical arrangement.

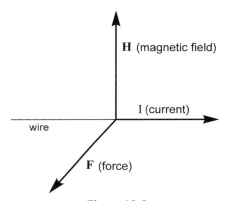

Figure 18-8
Straight wire carrying current in a magnetic field experiences a force normal to both the wire and the magnetic field

The relevant variables are

Variable	Symbol	Dimension	Remark
current	I	A	
force	F	kg/s²	per unit length of wire
permeability of vacuum	μ_0	m·kg/(A²·s²)	
magnetic field intensity	H	A/m	perpendicular to wire

We have the dimensional matrix

	I	F	μ_0	H
m	0	0	1	−1
kg	0	1	1	0
s	0	−2	−2	0
A	1	0	−2	1

whose rank is 3. There are four dimensions, therefore one dimension must go—say "kg." The Dimensional Set is then

	I	F	μ_0	H
m	0	0	1	−1
s	0	−2	−2	0
A	1	0	−2	1
π_1	1	−1	1	1

542 APPLIED DIMENSIONAL ANALYSIS AND MODELING

We have four variables and three dimensions, therefore there is but one dimensionless variable—a constant. By the above set it is

$$\pi_1 = \frac{I \cdot \mu_0 \cdot H}{F} = \text{const}$$

or since const = 1

$$F = I \cdot \mu_0 \cdot H$$

⇑

Example 18-12. Force Between Two Parallel Wires Carrying a Current

Given two parallel wires carrying a current in the same direction. Due to the generated magnetic field, the wires will be attracted toward each other. Question: what will this attractive force be?

First, we list the variables.

Variable	Symbol	Dimension	Remark
attractive force	F	m·kg/s²	perpendicular to wire
distance between wires	λ	m	
permeability of vacuum	μ_0	m·kg/(s²·A²)	
current	I	A	same direction
length of wire	L	m	each

Then we have the dimensional matrix

	F	λ	μ_0	I	L
m	1	1	1	0	1
kg	1	0	1	0	0
s	–2	0	–2	0	0
A	0	0	–2	1	0

whose rank is 3. Since the number of dimensions may not exceed the rank, we must eliminate one dimension, say "kg." This will leave three dimensions and five variables, resulting in two dimensionless variables. The Dimensional Set is then

	F	λ	μ_0	I	L
m	1	1	1	0	1
s	–2	0	–2	0	0
A	0	0	–2	1	0
π_1	1	0	–1	–2	0
π_2	0	1	0	0	–1

yielding

$$\pi_1 = \frac{F}{\mu_0 \cdot I^2}; \qquad \pi_2 = \frac{\lambda}{L} \tag{a}$$

and therefore we can write

$$\pi_1 = \Psi\{\pi_2\} \tag{b}$$

where Ψ is a function to be found.

Let us assume a monomial form for Ψ, which is plausible, so that (b) can be written

$$\pi_1 = k \cdot \pi_2^n \tag{c}$$

where k and n are numeric constants. By (a) and (c) now

$$F = k \cdot \mu_0 \cdot I^2 \cdot \left(\frac{\lambda}{L}\right)^n \tag{d}$$

We now consider the obvious fact that force F should be proportional to the length of wire. Therefore $n = -1$, and (d) becomes

$$F = k \cdot \frac{\mu_0 \cdot I^2 \cdot L}{\lambda} \tag{e}$$

where $k = 1/(4 \cdot \pi)$, which can be determined by either a *single* measurement or analysis.

⇑

Example 18-13. Buckling Load of a Hinged Column

Given a long, slender, hinged column of uniform cross-section. The column is loaded by axial force F. At some value of F, the column will buckle; we call this F the critical (buckling) load. Fig. 18-9 shows the arrangement.

We wish to determine this critical load. The first step is to list all the relevant variables:

Variable	Symbol	Dimension	Remark
critical load	F	N	concentric and axial
length	L	m	
Young's modulus	E	N/m²	uniform
cross-section's second moment of area	I	m⁴	uniform

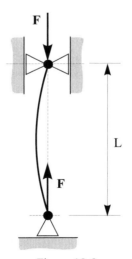

Figure 18-9
A long, slender, hinged column buckles under the critical axial load

Note that we use SI unit N (newton) for the *dimension* of force. The Dimensional Set is

	F	L	E	I
m	0	1	−2	4
N	1	0	1	0
π_1	1	0	−1	$-\dfrac{1}{2}$
π_2	0	1	0	$-\dfrac{1}{4}$

from which the two dimensionless variables are

$$\pi_1 = \frac{F}{E \cdot \sqrt{I}}; \qquad \pi_2 = \frac{L}{\sqrt[4]{I}} \qquad (a)$$

We write now

$$\pi_1 = \Psi\{\pi_2\} \qquad (b)$$

where Ψ is some function. Let us presume it is a *monomial*. Then (b) becomes

$$\pi_1 = k \cdot \pi_2^n \qquad (c)$$

where k and n are numeric constants. Thus, by (a) and (c)

$$F = k \cdot E \cdot \sqrt{I} \cdot \left(\frac{L}{\sqrt[4]{I}}\right)^n \qquad (d)$$

It is obvious that k must be positive and n cannot be zero. That k cannot be negative is evident since we assigned positive sign to the *compressive* force and of course *tension* (i.e., negative) force cannot produce buckling. As for n, if it were zero, then F would be independent of L, which is plainly impossible.

Apart from these rather primitive conclusions, in this instance dimensional considerations offer us nothing useful. But let us now consider the *very likely* fact that the critical load is *proportional* to the cross-section's second moment of area I. Accordingly, (d) provides $0.5 - (0.25) \cdot n = 1$, by which $n = -2$. Thus (d) becomes

$$F = k \cdot E \cdot \sqrt{I} \cdot \left(\frac{L}{\sqrt[4]{I}}\right)^{-2} = k \cdot \frac{E \cdot I}{L^2} \qquad (e)$$

The value of k is π^2 as can be determined by a *single* measurement or by analysis.

⇑

Example 18-14. Centrifugal Force Acting on a Point Mass

Given a point mass moving on a circle of radius R at tangential speed v (Fig. 18-10). What is the centrifugal force? The variables are:

Variable	Symbol	Dimension
centrifugal force	F	m·kg/s^2
mass	M	kg
tangential speed	v	m/s
radius	R	m

FIFTY-TWO ADDITIONAL APPLICATIONS **545**

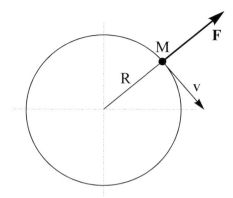

Figure 18-10
Centrifugal force acting on a point mass moving on a circle

We have four variables and three dimensions, hence there is only one dimensionless variable—a constant in this case. The Dimensional Set is

	F	M	v	R
m	1	0	1	1
kg	1	1	0	0
s	−2	0	−1	0
π_1	1	−1	−2	1

hence the sole dimensionless variable is

$$\pi_1 = \frac{F \cdot R}{M \cdot v^2} = \text{const} \tag{a}$$

thus we can write

$$F = \text{const} \cdot \frac{M \cdot v^2}{R} \tag{b}$$

where the constant is 1.

Now let us assume that we accidentally omitted the *relevant* variable R. We would then have the dimensional matrix

	F	M	v
m	1	0	1
kg	1	1	0
s	−2	0	−1

(c)

Here the number of dimensions equals the number of variables, therefore to have a solution at all it is necessary that this matrix be singular, i.e., its determinant be zero. But the determinant is not zero (it is 1), and hence the matrix is not singular. This means that there cannot be a relation among the variables considered in (c). In other words, there cannot be a relation $\Psi\{F, M, v\} = 0$, where Ψ is *any* function.

546   APPLIED DIMENSIONAL ANALYSIS AND MODELING

This example shows that when the rank of the dimensional matrix is not less than the number of variables, then there is an error somewhere—most likely the omission of one or more *relevant* variables.

⇑

Example 18-15. Work Done by a Piston Isothermally Compressing Ideal Gas

We wish to determine the energy necessary to compress a given volume V_1 of ideal gas isothermally from pressure p_1 to p_2 (Fig. 18-11). The following variables are relevant:

Variable	Symbol	Dimension
work done by piston	Q	m·N
initial pressure	p_1	N/m²
initial volume	V_1	m³
final pressure	p_2	N/m²

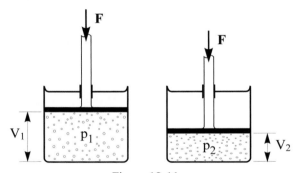

Figure 18-11
Piston isothermally compressing ideal gas

We have four variables and two dimensions, therefore there are 4 – 2 = 2 dimensionless variables, supplied by the Dimensional Set

	Q	p_1	V_1	p_2
m	1	–2	3	–2
N	1	1	0	1
π_1	1	0	–1	–1
π_2	0	1	0	–1

from which

$$\pi_1 = \frac{Q}{V_1 \cdot p_2}; \qquad \pi_2 = \frac{p_1}{p_2} \qquad \text{(a)}$$

and we can write

$$Q = V_1 \cdot p_2 \cdot \Psi\left\{\frac{p_1}{p_2}\right\} \qquad \text{(b)}$$

where Ψ is a function. Let us investigate if the *monomial* form for Ψ is possible. Thus, we assume the existence of

$$Q = k \cdot V_1 \cdot p_2 \cdot \left(\frac{p_1}{p_2}\right)^n \tag{c}$$

where k and n are nonzero numeric constants and $k > 0$. Imagine that we do *not* move the piston at all, i.e., $V_1 = V_2$ and $p_1 = p_2$. In this case (c) supplies $Q = k \cdot V_1 \cdot p_2 > 0$ (all quantities are positive). But this is impossible, since if we do not move the piston, then the amount of work performed by it is zero. Therefore the monomial form (c) is untenable.

Indeed, as straightforward analytic derivation shows, (b) has the form of

$$Q = V_1 \cdot p_2 \cdot \frac{p_1}{p_2} \cdot \ln \frac{p_2}{p_1} \tag{d}$$

Thus, if $p_1 = p_2$, then $p_2/p_1 = 1$ and hence Q is zero, as required.

⇑

Example 18-16. Energy Levels of Electrons in a Bohr Atom

As electrons circle the nucleus of an atom, they occupy distinct orbits that correspond to distinct energy values. This is called the *Bohr atom*—named after Niels Henrik David Bohr (Danish physicist, 1885–1962). The following variables are involved in finding the energy levels of an electron in a Bohr atom:

Variable	Symbol	Dimension
electron's energy level	Q	$m^2 \cdot kg/s^2$
mass of electron	M	kg
charge of electron	e	$A \cdot s$
permittivity of vacuum	ϵ_0	$A^2 \cdot s^4/(m^3 \cdot kg)$
Planck's constant	h	$m^2 \cdot kg/s$

We have five variables and four dimensions, therefore there is only one dimensionless variable—a constant in this case. The Dimensional Set is

	Q	M	e	ϵ_0	h
m	2	0	0	−3	2
kg	1	1	0	−1	1
s	−2	0	1	4	−1
A	0	0	1	2	0
π_1	1	−1	−4	2	2

yielding the sole dimensionless variable

$$\pi_1 = \frac{Q \cdot \epsilon_0^2 \cdot h^2}{M \cdot e^4} = \text{const} \tag{a}$$

from which

$$Q = \text{const} \cdot \frac{M \cdot e^4}{\epsilon_0^2 \cdot h^2} \tag{b}$$

where the constant equals $1/(8 \cdot n^2)$ with parameter $n = 1, 2, 3, \ldots$ for the successive energy levels (Ref. 4, p. 645). Equation (b) was first derived by Bohr by analysis.

⇑

548   APPLIED DIMENSIONAL ANALYSIS AND MODELING

Example 18-17. Existence Criteria for Black Holes

It is fascinating to speculate on the criteria for the existence of black holes. A black hole—as it is discussed in these pages—is a star so massive that its gravitational field prevents even light from leaving its surface. Using only dimensional considerations, let us try to formulate the criterion for the existence of such a star. We consider the following variables:

Variable	Symbol	Dimension	Remark
mass of star	M	kg	
speed of light	c	m/s	in vacuum
universal gravitational constant	k	m³/(s²·kg)	
radius of star	R	m	

We have four variables and three dimensions, therefore there is only one dimensionless variable—which is a constant. The Dimensional Set is

$$\begin{array}{c|cccc} & M & c & k & R \\ \hline m & 0 & 1 & 3 & 1 \\ kg & 1 & 0 & -1 & 0 \\ s & 0 & -1 & -2 & 0 \\ \hline \pi_1 & 1 & -2 & 1 & -1 \end{array}$$

yielding the sole dimensionless variable

$$\pi_1 = \frac{M \cdot k}{c^2 \cdot R} = \text{const} \qquad (a)$$

from which

$$\frac{M}{R} = \text{const} \cdot \frac{c^2}{k} \qquad (b)$$

Obviously if the radius is kept constant, then the *larger* the mass, the greater the gravitational effects. Conversely, if the mass is kept constant, then a *smaller* radius means that a point on the surface is closer to the center of mass, and hence there is a greater gravitational effect on the point. From this elementary argument and by formula (b), we can infer that the condition for the existence of a black hole is

$$\frac{M}{R} \geq \text{const} \cdot \frac{c^2}{k} \qquad (c)$$

Now we do a little heuristic cogitation. The mass of a star of density ρ and radius R is

$$M = \frac{4 \cdot \pi}{3} \cdot \rho \cdot R^3 \qquad (d)$$

Therefore

$$\frac{M}{R} = \frac{4 \cdot \pi \cdot \rho}{3} \cdot R^2 \qquad (e)$$

or, by (c) and (e),

$$\rho \cdot R^2 \geq \frac{3}{4 \cdot \pi} \cdot \frac{c^2}{k} \cdot \text{const} \qquad (f)$$

Since both *c* and *k* are constants, therefore (f) reduces to

$$\rho \cdot R^2 \geq \text{Const} \tag{g}$$

This means that for a star to qualify as a *black hole,* the product of its density and the square of its radius must exceed a certain critical value. Now treating a *photon* as a mass point (although it has no mass) and equating its escape velocity with that of light, makes the constant in (g)

$$\text{Const} = 1.6078 \times 10^{26} \text{ kg/m} \tag{h}$$

This datum yields the numerical criterion for the existence of a black hole. Thus, by (g),

$$\rho \cdot R^2 \geq 1.6078 \times 10^{26} \text{ kg/m} \tag{i}$$

To illustrate the severity of this condition, consider the following question: what density would a star the size of Earth have to possess to be a black hole? The Earth's radius is $R = 6.375 \times 10^6$ m, therefore, by (i), for the density we have

$$\rho \geq \frac{1.6078 \times 10^{26}}{(6.375 \times 10^6)^2} = 3.956 \times 10^{12} \text{ kg/m}^3$$

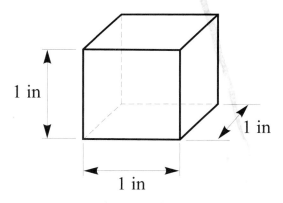

Figure 18-12
One cubic inch of material (shown here in full scale) of an Earth-size black hole must weigh about 65,000 tonnes = 143 million pounds

This means that 1 cubic inch of material would have to weigh about 64,827 metric tonnes = 143 million pounds (Fig. 18-12). The constant in (c) is in fact 0.5, which now allows us to write

$$R = R_S \leq \frac{2 \cdot k}{c^2} \cdot M \tag{j}$$

where R_S is the *Schwarzschild radius,* named after Karl Schwarzschild (German astronomer, 1873–1916). The Schwarzschild radius of a star of Earth's mass is 8.9 mm, i.e., if the Earth shrunk to an 1.8 cm diameter ball, it would become a black hole. This radius of the Sun is 2.95 km.

Example 18-18. Axial Thrust of a Screw Propeller

We wish to determine the axial thrust of a *prototype* propeller by an experiment executed on a geometrically similar *model*. The prototype is operating in deep water, thus surface waves generated are negligible and therefore gravitational effects may be ignored. The prototype's diameter is $D_1 = 1.15$ m and has a particular geometry (pitch, blade number, blade profile, etc.); it is rotated at $\omega_1 = 12.566$ rad/s (= 120 RPM) speed in water of temperature $t_1 = 20$ °C. The speed of the ship (axial speed of the prototype propeller) is $v_1 = 9.774$ m/s (= 19 knot).

The following variables are relevant:

Variable	Symbol	Dimension	Remark
propeller thrust	F	m·kg/s²	
propeller rotational speed	ω	rad/s	
water viscosity	μ	kg/(m·s)	dynamic
propeller diameter	D	m	
propeller axial speed	v	m/s	speed of ship
water density	ρ	kg/m³	

We have six variables and three dimensions, therefore there are $6 - 3 = 3$ dimensionless variables. The Dimensional Set is

	F	ω	μ	D	v	ρ
m	1	0	−1	1	1	−3
kg	1	0	1	0	0	1
s	−2	−1	−1	0	−1	0
π_1	1	0	0	−2	−2	−1
π_2	0	1	0	1	−1	0
π_3	0	0	1	−1	−1	−1

yielding

$$\pi_1 = \frac{F}{D^2 \cdot v^2 \cdot \rho}; \quad \pi_2 = \frac{\omega \cdot D}{v}; \quad \pi_3 = \frac{\mu}{D \cdot v \cdot \rho} \quad \text{(a)}$$

Therefore the Model Law is

$$S_F = S_D^2 \cdot S_v^2 \cdot S_\rho; \quad S_\omega \cdot S_D = S_v; \quad S_\mu = S_D \cdot S_v \cdot S_\rho \quad \text{(b)}$$

where S_F is the Thrust Scale Factor
  S_D is the Diameter Scale Factor
  S_v is the Axial Speed Scale Factor
  S_ρ is the Density Scale Factor
  S_ω is the Rotational Speed Scale Factor
  S_μ is the Dynamic Viscosity Scale Factor

Next, we *choose* the diameter of the model's propeller to be $D_2 = 0.45$ m and immerse this model into deep water of $t_2 = 60$ °C temperature (to decrease the water's viscosity). By the above data, the viscosity will be $\mu_1 = 0.001$ kg/(m·s) for the prototype, and $\mu_2 = 0.0005$ kg/(m·s) for the model (Ref. 11, p. 823). Also, the densities are $\rho_1 = 998.2$ kg/m³ and $\rho_2 = 983.2$ kg/m³. Therefore

$$S_\mu = \frac{\mu_2}{\mu_1} = \frac{0.0005}{0.001} = 0.5 \qquad \text{(c)}$$

$$S_\rho = \frac{\rho_2}{\rho_1} = \frac{983.2}{998.2} = 0.98497 \qquad \text{(d)}$$

$$S_D = \frac{D_2}{D_1} = \frac{0.45}{1.15} = 0.391304 \qquad \text{(e)}$$

By the third relation of Model Law (b), as well as by (c), (d), and (e),

$$S_v = \frac{S_\mu}{S_D \cdot S_\rho} = \frac{0.5}{(0.391304)\cdot(0.98497)} = 1.29728 \qquad \text{(f)}$$

But, by definition, $S_v = v_2/v_1$, therefore the speed v_2 of the model *must* be

$$v_2 = S_v \cdot v_1 = (1.29728)\cdot(9.774) = 12.67959 \text{ m/s } (= 24.65 \text{ knot}) \qquad \text{(g)}$$

By (f), (e), and the second relation of Model Law (b)

$$S_\omega = \frac{S_v}{S_D} = \frac{1.29728}{0.391304} = 3.31527 \qquad \text{(h)}$$

But $S_\omega = \omega_2/\omega_1$, therefore the rotational speed ω_2 of the model *must* be

$$\omega_2 = S_\omega \cdot \omega_1 = (3.31527)\cdot(12.566) = 41.65968 \text{ rad/s } (= 397.82 \text{ RPM})$$

Finally, by (e), (f), (d), and the first relation of Model Law (b)

$$S_F = S_D^2 \cdot S_v^2 \cdot S_\rho = (0.391304^2)\cdot(1.29728^2)\cdot(0.98497) = 0.25382 \qquad \text{(i)}$$

At *this* time we perform the experiment on the model; we immerse it into 60 °C water flowing at $v_2 = 12.6796$ m/s speed, rotate it at $\omega_2 = 41.65968$ rad/s speed and *measure* the generated axial force F_2. Suppose we measure $F_2 = 175.7$ N (= 39.5 lbf).

Since $S_F = F_2/F_1$, therefore by (i) the axial force F_1 on the prototype will be

$$F_1 = \frac{F_2}{S_F} = \frac{175.7}{0.25382} = 692.2337 \text{ N } (= 155.62 \text{ lbf})$$

To verify these results, we calculate the values of the dimensionless variables for both the prototype and the model; we find they are identical within rounding errors.

The *Modeling Data Table* in Fig. 18-13 summarizes the foregoing data. Note that there are three "*Category 2*" designations—in accordance with the fact that there are three relations (conditions) in Model Law (b). This problem was also discussed by Buckingham (Ref. 70, p. 352) and Isaacson (Ref. 22, p. 54).

⇑

Example 18-19. Boussinesq's Problem

We wish to determine the amount of heat transmitted by a constant temperature heat conductor immersed in an inviscid and incompressible fluid flowing at constant speed normal to that conductor (see Fig. 18-14). This problem is referred to by Lord Rayleigh as *Boussinesq's problem* (Ref. 71, p. 67). The following variables in this problem are relevant:

552   APPLIED DIMENSIONAL ANALYSIS AND MODELING

Variable					Scale factor S	Category	
name	symbol	dimension	prototype	model	model/ prototype	prototype	model
diameter of propeller	D	m	1.15	0.45	0.3913	1	1
axial speed	v	m/s	9.774	12.67959	1.29728	1	2
water density	ρ	kg/m³	998.2	983.2	0.98497	1	1
rotational speed	ω	rad/s	12.566	41.65968	3.31527	1	2
axial force (thrust)	F	m·kg/s²	692.2337	175.7	0.25382	2	3
water viscosity	μ	kg/(m·s)	0.001	0.0005	0.5	1	1
dimensionless	π_1	1	5.48902E-3	5.48902E-3	1		
dimensionless	π_2	1	1.4785	1.47851	1.00001		
dimensionless	π_3	1	8.91276E-8	8.91272E-8	1		
categories of variables	1	freely chosen, *a priori* given, or determined independently					
	2	determined by application of Model Law					
	3	determined by measurement on the model					

Figure 18-13
**Modeling Data Table for the determination
of the thrust of a deeply submerged propeller**

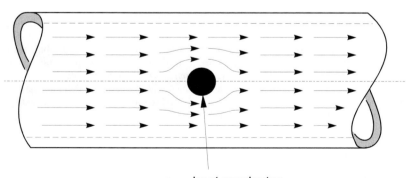

heat conductor

Figure 18-14
Heat is transmitted from conductor to liquid flowing in direction normal to it

Variable	Symbol	Dimension	Remark
heat transmitted	q	$m^2 \cdot kg/s^3$	in unit time
speed of fluid	v	m/s	normal to conductor
specific heat capacity of fluid	c	$kg/(s^2 \cdot K \cdot m)$	with respect to volume
heat conductivity of fluid	k	$m \cdot kg/(s^3 \cdot K)$	
characteristic size of conductor	L	m	linear
temperature differential (conductor versus fluid)	t	K	steady state

We have six variables and four dimensions, therefore there are 6 − 4 = 2 dimensionless variables supplied by the Dimensional Set

	q	v	c	k	L	t
m	2	1	−1	1	1	0
kg	1	0	1	1	0	0
s	−3	−1	−2	−3	0	0
K	0	0	−1	−1	0	1
π_1	1	0	0	−1	−1	−1
π_2	0	1	1	−1	1	0

from which

$$\pi_1 = \frac{q}{k \cdot L \cdot t} \; ; \quad \pi_2 = \frac{v \cdot c \cdot L}{k} \qquad (a)$$

Therefore we can write

$$\pi_1 = \Psi\{\pi_2\} \qquad (b)$$

where Ψ is a function to be determined experimentally.

By (a) and (b) now

$$q = k \cdot L \cdot t \cdot \Psi\left\{\frac{v \cdot c \cdot L}{k}\right\} \qquad (c)$$

Since in (b) there are only two variables, the plot for this relation is a single curve—despite the fact that there are six (!) physical variables involved.

Formula (c) was first derived—by using strictly dimensional means—by Lord Rayleigh (formerly John William Strutt, English physicist, 1842–1919), who was one of the principal developers of dimensional techniques (Ref. 71, p. 67).

Now let us see what happens if we consider temperature a manifestation of the mean kinetic *energy* of molecules. In this case the number of *dimensions* is reduced by one—since K no longer appears—although of course the number of *variables* remains unchanged. Therefore the number of dimensionless variables increases to three, so that the generated formula no longer representable by a *single* curve, but only by a *chart* comprising a *family* of curves.

We again list the variables and their dimensions:

Variable	Symbol	Dimension	Remark
heat transmitted	q	m²·kg/s³	in unit time
speed of fluid	v	m/s	normal to conductor
specific heat capacity of fluid	c	1/m³	with respect to volume
heat conductivity of fluid	k	1/(m·s)	
characteristic size of conductor	L	m	linear
temperature differential (conductor versus fluid)	t	m²·kg/s²	appears as energy

Note that the dimensions of c and k are now

$$[c] = \frac{[\text{heat energy}]}{[\text{volume}]\cdot[\text{temperature}]} = \frac{[\text{energy}]}{[\text{volume}]\cdot[\text{energy}]} = \frac{1}{[\text{volume}]} = \frac{1}{\text{m}^3}$$

$$[k] = \frac{[\text{heat energy}]}{[\text{time}]\cdot[\text{length}]\cdot[\text{temperature}]} = \frac{[\text{energy}]}{[\text{time}]\cdot[\text{length}]\cdot[\text{energy}]} = \frac{1}{[\text{time}]\cdot[\text{length}]} = \frac{1}{\text{s}\cdot\text{m}}$$

We have six variables and three dimensions, therefore 6 − 3 = 3 dimensionless variables characterize the system. To obtain these variables we construct the Dimensional Set

	q	v	c	k	L	t
m	2	1	−3	−1	1	2
kg	1	0	0	0	0	1
s	−3	−1	0	−1	0	−2
π_1	1	0	0	−1	−1	−1
π_2	0	1	0	−1	−2	0
π_3	0	0	1	0	3	0

yielding

$$\pi_1 = \frac{q}{k\cdot L\cdot t}\ ;\qquad \pi_2 = \frac{v}{k\cdot L^2}\ ;\qquad \pi_3 = c\cdot L^3 \tag{d}$$

and therefore we can write

$$\pi_1 = \Psi\{\pi_2, \pi_3\} \tag{e}$$

or by (d)

$$q = k\cdot L\cdot t\cdot\Psi\left\{\frac{v}{k\cdot L^2}, c\cdot L^2\right\} \tag{f}$$

Comparing (c) with (f) we immediately see that (f) makes π_1 dependent on 2 variables, whereas in (c) π_1 depends on only one. Therefore plotting (f) involves a *family* of curves, whereas to plot (c) only a *single* curve is needed. This shows the great advantage of making temperature a fundamental dimension.

There is an interesting historical event that might be appropriate to recount here. Shortly after Lord Rayleigh published his formula in the 1915 March issue of *Nature* (Ref. 71, p. 66), which, except for notation, was identical to (c), the Russian physicist and mathematician Riabouchinsky made himself immortal by writing following letter to the Editor of the journal (Ref. 103, p. 591):

The Principle of Similitude

In *NATURE* of March 18, Lord Rayleigh gives this formula $h = \kappa\cdot\alpha\cdot\theta\cdot F(\alpha\cdot v\cdot c/\kappa)$, considering heat, temperature, length, and time as four "independent" units. If we suppose that only three of these quantities are "really independent," we obtain a different result. For example, if the temperature is defined as the mean kinetic energy of the molecules, the principle of similitude allows us only to affirm that $h = \kappa \cdot \alpha \cdot \theta \cdot F(v/\kappa \cdot \alpha^2, c \cdot \alpha^3)$ [F here is the symbol for a function].

Here, of course, Dr. Riabouchinsky simply affirmed our statement (Chapter 16) that the fewer the number of dimensions used, the lesser is the "efficiency" of using dimensional considerations in general, and in creating dimensionless variables in particular.

Somewhat surprisingly, Lord Rayleigh's reply (Ref. 109, p. 644) did not refer to this elementary fact—as the only logical explanation. Instead, he skirted the issue with some rather vague language:

> The question raised by Dr. Riabouchinsky (*NATURE,* July 29, p. 591) belongs rather to the logic than to the use of the principle of similitude, which I was mainly concerned (*NATURE,* March 18, p. 66). It would be well worthy of further discussion. The conclusion that I gave follows on the basis of the usual Fourier equations for conduction of heat, in which heat and temperature are regarded "sui generis." It would indeed be a paradox if the further knowledge of the nature of heat afforded by the molecular theory put us in a worse position than before in dealing with a particular problem. The solution would seem to be that the Fourier equations embody something as to the nature of heat and temperature which is ignored in the alternative argument of Dr. Riabouchinsky.

Example 18-20. Rolling Resistance of Automobile Tires

The rolling resistance of an automobile tire is assumed to be characterized by the following variables:

Variable	Symbol	Dimension	Remark
Rolling Resistance Coefficient	μ_R	1	Note 1
slip angle	φ	1	Note 2
tire pressure	p	kg/(m·s²)	
load on tire	G	m·kg/s²	vertical
speed of automobile	v	m/s	
diameter of tire	D	m	

Note 1: Rolling Resistance Coefficient μ_R is defined by $\mu_R = F/G$, where F is the force necessary to pull the axle of a tire horizontally in the *direction of travel,* and G is the vertical load on the tire which is assumed to roll on a flat horizontal surface.

Note 2: For a rolling tire the slip angle is defined as the *angular distance* between the *heading* direction and *travel* direction (see Fig. 18-15).

We have six variables and three dimensions, therefore there are 6 – 3 = 3 dimensionless variables—supplied by the Dimensional Set

	μ_R	φ	p	G	v	D
m	0	0	–1	1	1	1
kg	0	0	1	1	0	0
s	0	0	–2	–2	–1	0
π_1	1	0	0	0	0	0
π_2	0	1	0	0	0	0
π_3	0	0	1	–1	0	2

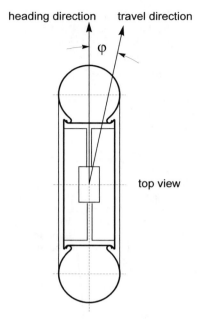

Figure 18-15
Definition of the slip angle φ of a rolling tire

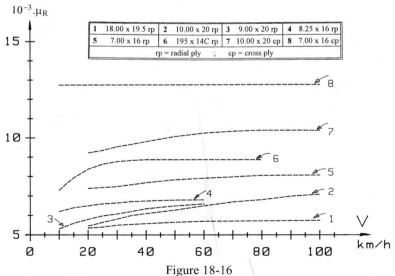

Figure 18-16
Rolling Resistance Coefficient μ_R versus speed v for various commercial tires
Note the practical independence of μ_R upon speed. Adapted from Ref. 48, Fig. 10

Hence

$$\pi_1 = \mu_R; \qquad \pi_2 = \varphi; \qquad \pi_3 = \frac{p \cdot D^2}{G} \qquad \text{(b)}$$

Note that the dimensionless physical variables μ_R and φ must be placed in the **B** matrix (Art. 14.1), or the **A** matrix would be singular.

Another important observation is that the column in the **C** matrix under variable v is all zeros. This means that—by Theorem 10-4 (Art. 10.5)—the speed of the automobile is a *dimensionally irrelevant* variable and, as such, by Theorem 11-3, it is also a *physically irrelevant* variable. This fact was confirmed experimentally by Ramshaw and Williams (Ref. 48, p. 10). Of course, the independence of the rolling resistance upon speed is not total, for it may be influenced by such factors as temperature, aerodynamic effects, etc. Fig. 18-16 shows the referenced data.

⇑

Example 18-21. General Fluid Flow Characteristics

In most fluid flow problems involving time-independent (i.e., ergodic) phenomena, there appear the following physical variables:

Variable	Symbol	Dimension
pressure difference	Δp	kg/(m·s²)
gravitational acceleration	g	m/s²
dynamic viscosity	μ	kg/(m·s)
surface tension	σ	kg/s²
bulk modulus (compressibility)	β	kg/(m·s²)
characteristic linear size	L	m
density	ρ	kg/m³
velocity	v	m/s

We have eight variables and three dimensions, therefore there are $8 - 3 = 5$ dimensionless variables supplied by the Dimensional Set

	Δp	g	μ	σ	β	L	ρ	v
m	−1	1	−1	0	−1	1	−3	1
kg	1	0	1	1	1	0	1	0
s	−2	−2	−1	−2	−2	0	0	−1
π_1	1	0	0	0	0	0	−1	−2
π_2	0	1	0	0	0	1	0	−2
π_3	0	0	1	0	0	−1	−1	−1
π_4	0	0	0	1	0	−1	−1	−2
π_5	0	0	0	0	1	0	−1	−2

Therefore the dimensionless variables are (see also Appendix 3):

$$\pi_1 = \frac{\Delta p}{\rho \cdot v^2} \qquad \text{is called the } \textit{Euler number, } \text{Eu}$$

$$\pi_2 = \frac{g \cdot L}{v^2} = \left(\frac{v}{\sqrt{g \cdot L}}\right)^{-2}, \quad \text{where the expression in parentheses is called the } \textit{Froude number, } \text{Fr}$$

$$\pi_3 = \frac{\mu}{L \cdot \rho \cdot v} = \left(\frac{L \cdot \rho \cdot v}{\mu}\right)^{-1}$$ where the expression in parentheses is called the *Reynolds number*, Re

$$\pi_4 = \frac{\sigma}{L \cdot \rho \cdot v^2} = \left(\frac{L \cdot \rho \cdot v^2}{\sigma}\right)^{-1},$$ where the expression in parentheses is called the *Weber number*, We

$$\pi_5 = \frac{\beta}{\rho \cdot v^2} = \left(\frac{\rho \cdot v^2}{\beta}\right)^{-1},$$ where the expression in parentheses is called the *Cauchy number*, Ca

Accordingly, we can write

$$\Psi\{\pi_1 \cdot \pi_2 \cdot \pi_3 \cdot \pi_4 \cdot \pi_5\} = 0 \qquad (a)$$

or

$$\Psi\{\text{Eu, Fr, Re, We, Ca}\} = 0 \qquad (b)$$

It is very rare that all five of these *named* dimensionless variables occur in a single problem. For example, in cases where the effects of surface tension σ and viscosity μ can be ignored, the Weber number and the Reynolds number are omitted, etc.

⇑

Example 18-22. Heat Transfer to a Fluid Flowing in a Pipe

We now examine heat transfer from a hot pipe to a cold incompressible fluid flowing in a pipe. Fig. 18-17 schematically illustrates the arrangement.

The heat transferred is

$$Q = h \cdot A \cdot \Delta t \cdot \Delta \theta \qquad (a)$$

where h is the convective heat transfer coefficient between the pipe and the moving liquid in the pipe;

A is the area through which the heat transfer takes place;

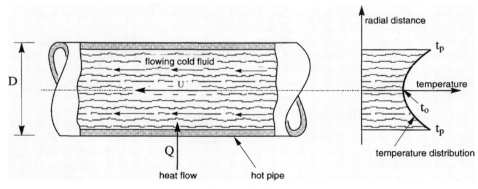

Figure 18-17
Heat transfer from a hot pipe to a cold liquid
The parabolic distribution of temperature is seen on the right

Δt is the temperature difference between the pipe's wall and the *average* temperature of the moving liquid. If we assume parabolic temperature distribution (see right side of Fig. 18-17), then the average fluid temperature in the pipe is

$$t_f = \frac{t_0 + t_p}{2} \quad \text{(b)}$$

where t_0 is the temperature of the fluid at the centerline of the pipe,
t_p is the temperature of the pipe (fluid's temperature very near to the pipe's wall),

$\Delta\theta$ is the time duration.

Since A, Δt, and $\Delta\theta$ are easily measurable, to determine Q we only have to find h. To this end, we consider the following variables:

Variable	Symbol	Dimension	Remark
film heat transfer coefficient	h	kg/(s^3·K)	
fluid density	ρ	kg/m^3	assumed constant
fluid specific heat capacity	c	m^2/(s^2·K)	
pipe temperature	t_p	K	
heat conductivity of fluid	k	m·kg/(s^3·K)	
fluid viscosity	μ	kg/(m·s)	dynamic
velocity of fluid flow	u	m/s	
fluid average temperature	t_f	K	by relation (b)
pipe diameter	D	m	inner

We have nine variables and four dimensions, therefore there are $9 - 4 = 5$ dimensionless variables describing the system.

But now, instead of writing down the Dimensional Set and determining the five dimensionless variables from it, we reverse the process. That is, we want to construct a dimensional set that will yield the *Nusselt number* (ratio of total heat transfer and conductive heat transfer), the *Reynolds number*, (ratio of inertia force and viscous force) and the *Prandtl number* (ratio of kinematic viscosity and thermometric conductivity $k/(\rho \cdot c)$). Other dimensionless variables we wish to include are the ratio of absolute temperatures t_p, t_f, and the ratio of kinetic energy of the moving fluid to its thermal energy $u^2/(c \cdot t_f)$. Altogether therefore in this system there are the following (desired) dimensionless variables:

$$\pi_1 = \frac{h \cdot D}{k}; \qquad \text{Nusselt number}$$

$$\pi_2 = \frac{\rho \cdot u \cdot D}{\mu}; \qquad \text{Reynolds number}$$

$$\pi_3 = \frac{c \cdot \mu}{k}; \qquad \text{Prandtl number} \quad \text{(c)}$$

$$\pi_4 = \frac{t_p}{t_f}$$

$$\pi_5 = \frac{u^2}{t_f \cdot c}$$

Two questions now naturally emerge: are these dimensionless variables
- independent?
- forming a complete set?

Since, as we know, the number of independent variables is five, and by (c) we have five variables, therefore if they are independent in (c), then they do form a complete set. So, if the answer to the first question is "yes," then it is also the answer to the second question. To answer the first question we form the Dimensional Set for relations (c).

		h	ρ	c	t_p	k	μ	u	t_f	D
B matrix	m	0	−3	2	0	1	−1	1	0	1
	kg	1	1	0	0	1	1	0	0	0
	s	−3	0	−2	0	−3	−1	−1	0	0
	K	−1	0	−1	1	−1	0	0	1	0
D matrix	π_1	1	0	0	0	−1	0	0	0	1
	π_2	0	1	0	0	0	−1	1	0	1
	π_3	0	0	1	0	−1	1	0	0	0
	π_4	0	0	0	1	0	0	0	−1	0
	π_5	0	0	−1	0	0	0	2	−1	0

A matrix on the right; C matrix on the right.

Note that **D** is no longer an identity matrix, but is constructed to fulfil the requirements of the predetermined dimensionless variables shown in (c). To decide whether the variables in (c) are independent, all we have to do is test matrix **D** for singularity. If **D** is not singular, (i.e., its determinant is not zero), then the $\pi_1, \pi_2, \ldots, \pi_5$ dimensionless variables in (c) are independent, otherwise they are dependent. In our case $|\mathbf{D}| = -1 \neq 0$, therefore the variables in (c) are independent, and hence form a complete set. We can then write

$$\pi_1 = \Psi\{\pi_2, \pi_3, \pi_4, \pi_5\} \tag{d}$$

or, by (c),

$$h = \frac{k}{D} \cdot \Psi\left\{\frac{\rho \cdot u \cdot D}{k}, \frac{c \cdot \mu}{k}, \frac{t_p}{t_f}, \frac{u^2}{t_f \cdot c}\right\} \tag{e}$$

where Ψ designates a function to be determined experimentally.

Now the last term in the right side of (e) is the ratio of kinetic energy (numerator) to its thermal energy (denominator), both of which are related to unit mass of the flowing fluid. However, the kinetic energy equivalent to Δt temperature rise corresponds to a speed increase (from zero) $v = \sqrt{2 \cdot c \cdot \Delta t}$, c being the specific heat capacity, which for water is $c_w = 4186.8$ m²/(s²·K). It follows therefore that for water a 1 K temperature raise corresponds to a speed increase (from zero) of 91.51 m/s = 329.4 km/h, which is manifestly too large a value to be encountered in practice. Consequently, for an incompressible liquid, this term can be safely ignored.

Moreover, it can be generally assumed that the $\pi_4 = t_p/t_f$ term has very little influence on the heat transfer coefficient h (Ref. 34, p. 276). Therefore, ultimately, relation (e) is simplified to

$$\text{Nu} = \Psi\{\text{Re}, \text{Pr}\} \tag{f}$$

which, if Pr = const, can be graphically represented by a single curve, as Fig. 18-18 shows.

FIFTY-TWO ADDITIONAL APPLICATIONS   **561**

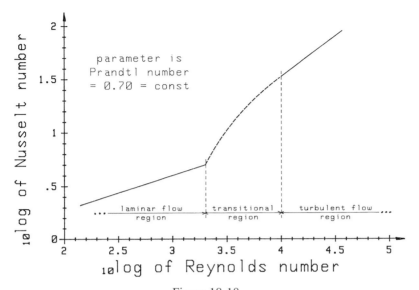

Figure 18-18
Nusselt number versus Reynolds number for a liquid flowing in a pipe
Parameter is Prandtl number = 0.70 = constant. Adapted from Ref. 34. p. 277

Example 18-23. Volume of a Torus of Arbitrary Cross-Section

By a modeling experiment, let us determine the volume of a *torus* of irregular, constant, but otherwise arbitrary cross-section (Fig. 18-19). The relevant variables are as follows:

Variable	Symbol	Dimension	Remark
volume	V	m^3	
generating diameter	D	m	characteristic value
linear size of cross-section	d	m	characteristic value

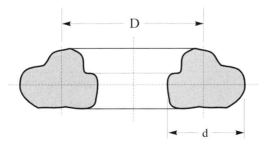

Figure 18-19
A torus of irregular, but constant, cross-section

We have three variables and one dimension, therefore there are 3 − 1 = 2 dimensionless variables, supplied by the Dimensional Set

	V	D	d
m	3	1	1
π_1	1	0	−3
π_2	0	1	−1

Hence

$$\pi_1 = \frac{V}{d^3}; \qquad \pi_2 = \frac{D}{d} \qquad \text{(a)}$$

The Model Law therefore is

$$S_V = S_d^3; \qquad S_D = S_d \qquad \text{(b)}$$

where S_V is the Volume Scale Factor
S_d is the Cross-section Size Scale Factor
S_D is the Generating Diameter Scale Factor

Now suppose we have to determine volume V_1 of a $D_1 = 33.2$ m diameter *prototype* torus whose characteristic cross-section size is $d_1 = 2.08$ m. Because of the extremely large size of the prototype, we build a model whose cross-section size is $d_2 = 0.0254$ m (= 1 inch). Therefore

$$S_d = \frac{d_2}{d_1} = \frac{0.0254}{2.08} = 0.0122115 \qquad \text{(c)}$$

Variable					Scale factor S	Category	
name	symbol	dimension	prototype	model	model/ prototype	prototype	model
volume	V	m³	340.47454	6.2E-4	1.82099E-6	2	3
generating diameter	D	m	33.2	0.40542	1.22114E-2	1	2
cross-section size	d	m	2.08	0.0254	1.22115E-2	1	1
dimensionless	π_1	1	37.83508	37.83472	0.99999	—	—
dimensionless	π_2	1	15.96154	15.96142	0.99999	—	—
categories of variables	1	freely chosen, *a priori* given, or determined independently					
	2	determined by application of Model Law					
	3	determined by measurement on the model					

Figure 18-20
Modeling Data Table for the determination of the volume of a torus of arbitrary cross-section by a modeling experiment

By this result and the second relation of Model Law (b)

$$S_D = \frac{D_2}{D_1} = \frac{D_2}{33.2} = S_d = 0.0122115 \qquad (d)$$

from which the generating diameter of the model *must* be

$$D_2 = D_1 \cdot S_d = (33.2) \cdot (0.0122115) = 0.40542 \text{ m} \qquad (e)$$

Now we can build the model and measure its volume by immersing it into a tank of water. Say we measure $V_2 = 0.62$ liter $= 0.00062$ m^3. Next, by the first relation of Model Law (b) $S_V = V_2/V_1 = S_d^3$ from which, by (c), the prototype's volume is

$$V_1 = \frac{V_2}{S_d^3} = \frac{0.00062}{0.0122115^3} = 340.47454 \text{ m}^3$$

The Modeling Data Table shown in Fig. 18-20 conveniently summarizes all the relevant data.

Example 18-24. Pitch of a Kettledrum

The kettledrum (*timpano*, plural *timpani*) is a musical instrument, and the only drum which can produce musical notes of definite frequency (pitch). The drum is constructed by stretching a skin over a steel rim which is mounted over a basin-shaped copper or brass shell (Fig. 18-21). The *radial* tension of the skin is controlled by peripheral screws—called "taps," or, in modern times, by pedals. In this way the pitch—i.e., the frequency of transverse vibration—of the drum can be sensitively adjusted. The drum is "played" by hitting the skin with the elaborately padded ends (heads) of cane drum-sticks.

Figure 18-21
A kettledrum

How does radial tension influence pitch? We answer this question using dimensional considerations. First we list the relevant variables:

Variable	Symbol	Dimension	Remark
frequency (pitch)	f	1/s	transversal
radial tension of skin	σ	kg/s^2	per unit length of perimeter
diameter of drum	D	m	
mass of skin	M	kg/m^2	per unit area

We have four variables and three dimensions, therefore there is only one dimensionless variable, a constant in this case. The Dimensional Set is

	f	σ	D	M
m	0	0	1	-2
kg	0	1	0	1
s	-1	-2	0	0
π_1	1	$-\dfrac{1}{2}$	1	$\dfrac{1}{2}$

Hence the sole dimensionless variable is

$$\pi_1 = f \cdot D \cdot \sqrt{\frac{M}{\sigma}} = \text{const} \tag{a}$$

from which the frequency

$$f = \text{const} \cdot \frac{1}{D} \cdot \sqrt{\frac{\sigma}{M}} \tag{b}$$

This relation tells us that the natural frequency of the drum varies
- with the inverse of its diameter;
- with the square root of the radial tension of the skin;
- with the inverse of the square root of the skin's mass per unit area.

Of these three possibilities, of course, only the second one can be used to change the pitch of an existing drum. For example, if one wants to raise the pitch by a *semitone*—say from D to D sharp—then the frequency must be increased in an *"equally tempered scale"* by the factor 1.059463, which is the 12th root of 2 (there are 12 semitones in an *octave* and an octave means exactly doubling the frequency). Therefore the tension of the membrane of the drum must be increased by the factor $1.059463^2 = 1.122462$, i.e., 12.246%. In other words, every 12.246% increase of the radial tension of the skin increases the pitch by exactly a semitone, no matter what the pitch is.

A regular size kettledrum in a symphony orchestra playing 18th century classical music has a range of 7 semitones (between B flat and F), and hence the tension of the membrane of the drum must be changed by peripheral screws (or pedal) by a factor of $1.122462^7 = 2.244924$, i.e., 124.492%. Specifically, if the orchestra is "tuned" to middle A's frequency of say $f_A = 445$ Hz, then the drum must be "tuneable" between $f_{Bflat} = 117.865$ Hz and $f_F = 176.598$ Hz, a ratio of 1.498307, which is exactly the seventh power of the semitone pitch ratio of 1.059463 mentioned above.

We saw that by relation (b) the pitch varies inversely with the diameter. Therefore, if we want to have a "larger" drum whose highest tone is C (i.e., $f_C = 132.299$ Hz), then the ratio of the highest frequencies of the "larger" and "smaller" drums is $176.5984/132.2992 = 1.33481$. Therefore, the diameter of the larger drum must be 33.48% larger than that of the smaller one, provided of course, that the stretching force and mass per unit area of the skin are the same.

⇑

Example 18-25. Volume of a Spherical Segment

Given a spherical segment characterized by its thickness, position and radius of the sphere—as illustrated in Fig. 18-22. Our task is to establish a complete set of dimension-

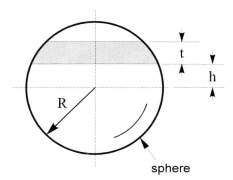

Figure 18-22
A spherical segment

less variables related to the *volume* of the segment and then prove that the *monomial* form connecting these variables is untenable. We have the following relevant variables:

Variable	Symbol	Dimension	Remark
volume of segment	V	m^3	
radius of sphere	R	m	
position of segment	h	m	as defined in Fig. 18-22
thickness of segment	t	m	

We have four variables and one dimension, therefore there are three dimensionless variables in this system. The Dimensional Set is

	V	R	h	t
m	3	1	1	1
π_1	1	0	0	-3
π_2	0	1	0	-1
π_3	0	0	1	-1

which yields

$$\pi_1 = \frac{V}{t^3} \ ; \qquad \pi_2 = \frac{R}{t} \ ; \qquad \pi_3 = \frac{h}{t} \qquad (a)$$

Hence we can write

$$\pi_1 = \Psi\{\pi_2, \pi_3\} \qquad (b)$$

where Ψ is a function to be determined. If this function is a monomial, then, by (a) and (b),

$$V = k \cdot t^3 \cdot \left(\frac{R}{t}\right)^{n_1} \cdot \left(\frac{h}{t}\right)^{n_2} \qquad (c)$$

where k, n_1, and n_2 are numeric constants.

Now we observe in Fig. 18-22 that if h increases and t and R remain constant, then V must decrease. This fact mandates a negative n_2 in (c). But if n_2 is negative and $h = 0$, then relation (c) yields infinite volume, which is impossible. Therefore the monomial form for Ψ in (b) is untenable, as we intended to demonstrate. Indeed, as the reader may verify, the correct formula for volume is

$$V = \frac{\pi}{3} \cdot t \cdot [3 \cdot t \cdot (R - h) + 3 \cdot h \cdot (2 \cdot R - h) - t^2] \tag{d}$$

which, with the aid of the dimensionless variables defined in (a), can be written

$$\pi_1 = \frac{\pi}{3} \cdot [3 \cdot (\pi_2 - \pi_3) + 3 \cdot \pi_3 \cdot (2 \cdot \pi_2 - \pi_3) - 1] \tag{e}$$

and, as seen, (e) is not a monomial.

⇑

Example 18-26. Force-Related Characteristics and Geometry of a Catenary

In Problem 11.3 (solution is in Appendix 6) we dealt with the *shape* of a catenary expressed in dimensionless form of two such variables

$$\pi_1 = \frac{\lambda}{L} \quad ; \quad \pi_2 = \frac{U}{L} \tag{a1}$$

where λ, L, and U are as defined in Fig. 18-23. In the solution of that problem (Appendix 6) we proved that the *shape* of a catenary is independent from gravitational effects, i.e., a heavy rope assumes the same configuration on the Moon as it does on Earth, provided of course that the rope's length L and the span of suspension λ are identical in both locations. Nevertheless, the fact that gravitation does not effect the shape does not mean that for the *determination* of the catenary's geometry the gravitation caused forces should not be considered.

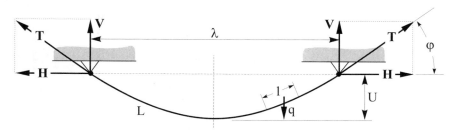

Figure 18-23
Force-related characteristics and geometry of a catenary

We pointed out in the solution of the problem that the shape of a catenary is completely described by the dimensionless relation

$$\pi_2 = \Psi\{\pi_1\} \tag{b1}$$

where Ψ is some function, and π_1 and π_2 are as defined in (a1). We also proved that Ψ cannot possible be a *monomial* function.

The question now is: What is Ψ? Unfortunately, there is no explicit formula which uses only "established" functions to fulfill (b1). What we have instead is a *parametric* system which can be—with some adroit manipulation—made into a convenient dimensionless form. The process is as follows.

We first introduce two force-related physical variables: weight of the cable per unit length q and the horizontal component H of the anchor force T (see Fig. 18-23). Thus we have the following set of variables:

Variable	Symbol	Dimension
span	λ	m
sag	U	m
cable weight per unit length	q	N/m
length	L	m
horizontal component of anchor force	H	N

We have five variables and two dimensions, therefore there are $5 - 2 = 3$ dimensionless variables supplied by the Dimensional Set

	λ	U	q	L	H
m	1	1	−1	1	0
N	0	0	1	0	1
π_1	1	0	0	−1	0
π_2	0	1	0	−1	0
π_3	0	0	1	1	−1

from which

$$\pi_1 = \frac{\lambda}{L} \; ; \quad \pi_2 = \frac{U}{L} \; ; \quad \pi_3 = \frac{q \cdot L}{H} \tag{c1}$$

Using a rather straightforward and well documented analysis (see for example Ref. 131, p. 63), the following relations among the relevant physical variables can be established:

$$L = \frac{2 \cdot H}{q} \cdot \sinh \frac{q \cdot \lambda}{2 \cdot H} \tag{d1}$$

$$U = \frac{H}{q} \cdot \left(\cosh \frac{q \cdot \lambda}{2 \cdot H} - 1 \right) \tag{e1}$$

Referring to (c1) now, (d1) can be written in the dimensionless form $\pi_3/2 = \sinh(\pi_1 \cdot \pi_3)/2$ from which

$$\pi_1 = \frac{2}{\pi_3} \cdot \ln \left(\frac{\pi_3}{2} + \frac{1}{2} \cdot \sqrt{4 + \pi_3^2} \right) \tag{f1}$$

Similarly, (e1) can be written—using (c1)

$$\pi_2 = \frac{1}{\pi_3} \cdot \left(\cosh \frac{\pi_1 \cdot \pi_3}{2} - 1 \right) \tag{g1}$$

Therefore, by (f1) and (g1) the Ψ function in (b1) can be expressed in parametric dimensionless forms, the parameter being π_3. Thus, the two-step process is as follows:

Step 1. Assign an arbitrary value to π_3 (parameter), and then calculate π_1 by (f1);
Step 2. By the now-known π_1 and π_3, calculate π_2 by (g1).

With this simple protocol, the corresponding π_1 and π_2, as well as π_1 and π_3 pairs can be easily determined to any desired degree of accuracy. To illustrate, say $\pi_3 = 15$. Hence, by (f1), we calculate $\pi_1 = 0.36166$ (Step 1), and then (g1) furnishes $\pi_2 = 0.43776$ (Step 2).

The table of Fig. 18-24 gives corresponding values for dimensionless variables π_1, π_2, and π_3 defined in (c1). For π_3 the given data are exact; for π_1 and π_2 they are rounded to three decimal places.

Tension T on the cable will be maximum at the anchor points. This maximum T can be easily determined by considering

$$T = \sqrt{V^2 + H^2} \tag{h1}$$

where V and H are the vertical and horizontal components of the reaction force (= tension) T at the anchors. But now, obviously

$$V = \frac{q \cdot L}{2} \tag{i1}$$

and by (c1)

$$H = \frac{q \cdot L}{\pi_3} \tag{j1}$$

These two values are now substituted into (h1) yielding

$$T = q \cdot L \cdot \sqrt{\frac{1}{4} + \frac{1}{\pi_3^2}}$$

which can be written

$$\pi_4 = \frac{q \cdot L}{T} = \frac{2 \cdot \pi_3}{\sqrt{4 + \pi_3^2}} \tag{k1}$$

or, by solving (k1) for π_3,

$$\pi_3 = \frac{2 \cdot \pi_4}{\sqrt{4 - \pi_4^2}} \tag{\ell1}$$

where π_4 is a new dimensionless variable introduced merely for convenience. Note that π_4 is not independent, for, as (k1) shows, it can be expressed by another dimensionless variable, viz., π_3.

The plot presented in Fig. 18-25 shows the π_2 versus π_1 and the π_3 versus π_1 relations. The easy and efficient use of these graphs and the table in Fig. 18-24 is demonstrated by the following illustrations.

Illustration A. Suppose we have an $L = 145.2$ m long chain weighing $q = 335.66$ N/m (= 23 lbf/ft). We intend to hang this chain between two points at the same elevation and $\lambda = 60$ m apart. What will be the values of sag U, horizontal anchor force H, vertical anchor force V and maximum tension force T?

By (c1), with three-decimal accuracy, $\pi_1 = \lambda/L = 60/145.2 = 0.413$ for which the plot in Fig. 18-25 furnishes $\pi_2 = 0.424$, and $\pi_3 = 12.1$, and $\pi_4 = 1.973$ (these values are indicated by dashed lines). By (c1) now, the sag is

$$U = \pi_2 \cdot L = (0.424) \cdot (145.2) = 61.57 \text{ m}$$

FIFTY-TWO ADDITIONAL APPLICATIONS

π_1	π_2	π_3	π_1	π_2	π_3	π_1	π_2	π_3	π_1	π_2	π_3
1.000	.000	0.00	.420	.422	11.80	.267	.460	23.80	.200	.473	35.80
.998	.025	.20	.415	.424	12.00	.265	.460	24.00	.199	.473	36.00
.993	.050	.40	.411	.425	12.20	.263	.460	24.20	.198	.473	36.20
.986	.073	.60	.407	.426	12.40	.262	.461	24.40	.198	.473	36.40
.975	.096	.80	.403	.427	12.60	.261	.461	24.60	.197	.473	36.60
.962	.118	1.00	.399	.428	12.80	.259	.461	24.80	.196	.474	36.80
.948	.138	1.20	.396	.429	13.00	.258	.462	25.00	.195	.474	37.00
.932	.158	1.40	.392	.430	13.20	.256	.462	25.20	.194	.474	37.20
.916	.175	1.60	.388	.431	13.40	.255	.462	25.40	.194	.474	37.40
.899	.192	1.80	.385	.432	13.60	.253	.462	25.60	.193	.474	37.60
.881	.207	2.00	.381	.433	13.80	.252	.463	25.80	.192	.474	37.80
.864	.221	2.20	.378	.434	14.00	.251	.463	26.00	.191	.474	38.00
.847	.234	2.40	.374	.435	14.20	.249	.463	26.20	.191	.475	38.20
.830	.246	2.60	.371	.435	14.40	.248	.464	26.40	.190	.475	38.40
.813	.257	2.80	.368	.436	14.60	.247	.464	26.60	.189	.475	38.60
.797	.268	3.00	.365	.437	14.80	.246	.464	26.80	.189	.475	38.80
.781	.277	3.20	.362	.438	15.00	.244	.464	27.00	.188	.475	39.00
.765	.286	3.40	.359	.439	15.20	.243	.465	27.20	.187	.475	39.20
.750	.294	3.60	.356	.439	15.40	.242	.465	27.40	.187	.475	39.40
.736	.302	3.80	.353	.440	15.60	.241	.465	27.60	.186	.475	39.60
.722	.309	4.00	.350	.441	15.80	.239	.465	27.80	.185	.476	39.80
.708	.316	4.20	.347	.441	16.00	.238	.466	28.00	.184	.476	40.00
.695	.322	4.40	.344	.442	16.20	.237	.466	28.20	.184	.476	40.20
.683	.328	4.60	.342	.443	16.40	.236	.466	28.40	.183	.476	40.40
.671	.333	4.80	.339	.443	16.60	.235	.466	28.60	.182	.476	40.60
.659	.339	5.00	.336	.444	16.80	.233	.466	28.80	.182	.476	40.80
.648	.343	5.20	.334	.445	17.00	.232	.467	29.00	.181	.476	41.00
.637	.348	5.40	.331	.445	17.20	.231	.467	29.20	.181	.476	41.20
.626	.352	5.60	.329	.446	17.40	.230	.467	29.40	.180	.476	41.40
.616	.356	5.80	.326	.446	17.60	.229	.467	29.60	.179	.477	41.60
.606	.360	6.00	.324	.447	17.80	.228	.468	29.80	.179	.477	41.80
.597	.364	6.20	.321	.448	18.00	.227	.468	30.00	.178	.477	42.00
.587	.368	6.40	.319	.448	18.20	.226	.468	30.20	.177	.477	42.20
.579	.371	6.60	.317	.449	18.40	.225	.468	30.40	.177	.477	42.40
.570	.374	6.80	.315	.449	18.60	.224	.468	30.60	.176	.477	42.60
.562	.377	7.00	.312	.450	18.80	.223	.469	30.80	.176	.477	42.80
.554	.380	7.20	.310	.450	19.00	.222	.469	31.00	.175	.477	43.00
.546	.383	7.40	.308	.451	19.20	.221	.469	31.20	.174	.477	43.20
.538	.385	7.60	.306	.451	19.40	.220	.469	31.40	.174	.477	43.40
.531	.388	7.80	.304	.452	19.60	.219	.469	31.60	.173	.478	43.60
.524	.390	8.00	.302	.452	19.80	.218	.470	31.80	.173	.478	43.80
.517	.393	8.20	.300	.452	20.00	.217	.470	32.00	.172	.478	44.00
.510	.395	8.40	.298	.453	20.20	.216	.470	32.20	.171	.478	44.20
.503	.397	8.60	.296	.453	20.40	.215	.470	32.40	.171	.478	44.40
.497	.399	8.80	.294	.454	20.60	.214	.470	32.60	.170	.478	44.60
.491	.401	9.00	.292	.454	20.80	.213	.470	32.80	.170	.478	44.80
.485	.403	9.20	.290	.455	21.00	.212	.471	33.00	.169	.478	45.00
.479	.405	9.40	.288	.455	21.20	.211	.471	33.20	.169	.478	45.20
.473	.407	9.60	.287	.455	21.40	.210	.471	33.40	.168	.478	45.40
.468	.408	9.80	.285	.456	21.60	.209	.471	33.60	.168	.479	45.60
.462	.410	10.00	.283	.456	21.80	.208	.471	33.80	.167	.479	45.80
.457	.411	10.20	.281	.457	22.00	.207	.471	34.00	.166	.479	46.00
.452	.413	10.40	.279	.457	22.20	.207	.472	34.20	.166	.479	46.20
.447	.414	10.60	.278	.457	22.40	.206	.472	34.40	.165	.479	46.40
.442	.416	10.80	.276	.458	22.60	.205	.472	34.60	.165	.479	46.60
.437	.417	11.00	.274	.458	22.80	.204	.472	34.80	.164	.479	46.80
.433	.419	11.20	.273	.458	23.00	.203	.472	35.00	.164	.479	47.00
.428	.420	11.40	.271	.459	23.20	.202	.472	35.20	.163	.479	47.20
.424	.421	11.60	.270	.459	23.40	.202	.473	35.40	.163	.479	47.40

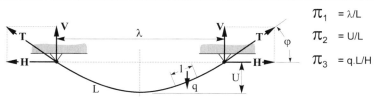

$\pi_1 = \lambda/L$
$\pi_2 = U/L$
$\pi_3 = q.L/H$

Figure 18-24
Corresponding values for dimensionless variables π_1, π_2, and π_3 characterizing the shape and some force-related properties of a catenary

570 APPLIED DIMENSIONAL ANALYSIS AND MODELING

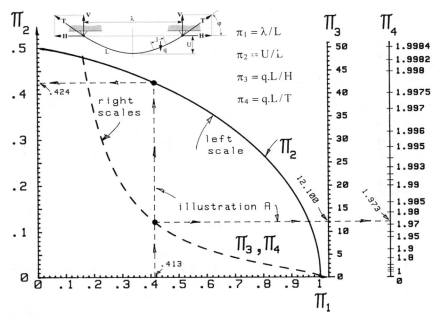

**Figure 18-25
Dimensionless plot for the determination of geometry
and force-related characteristics of a catenary**

the horizontal anchor force is by (j1)

$$H = \frac{q \cdot L}{\pi_3} = \frac{(335.66) \cdot (145.2)}{12.1} = 4028 \text{ N} (= 905.5 \text{ lbf})$$

and, by (k1), the maximum tension force in the chain is

$$T = \frac{q \cdot L}{\pi_4} = \frac{(335.66) \cdot (145.2)}{1.973} = 24{,}702.398 \text{ N} (= 5553.32 \text{ lbf})$$

Since the safe load in this chain is 149015 N, the safety factor is about 6.03.
The vertical anchor force V is by (i1)

$$V = \frac{q \cdot L}{2} = \frac{(335.66) \cdot (145.2)}{2} = 24{,}368.916 \text{ N} (= 5478.35 \text{ lbf})$$

The angle φ at the anchor can also be easily determined. For obviously $\tan \varphi = V/H$, which by (i1) and (j1) can be written

$$\tan \varphi = \frac{V}{H} = \frac{(q \cdot L)/2}{(q \cdot L)/\pi_3} = \frac{\pi_3}{2} \tag{m1}$$

and hence

$$\varphi = \arctan \frac{\pi_3}{2} \tag{n1}$$

In the numerical case mentioned above, since $\pi_3 = 12.1$, therefore $\varphi = \arctan 12.1/2 = 80.6$ deg. Similar derivation also yields

$$\sin \varphi = \frac{\pi_4}{2} \tag{o1}$$

By this and (m1), we get $\arcsin \pi_4/2 = \arctan \pi_3/2$. Thus

$$\pi_3 = 2 \cdot \tan\left(\arcsin \frac{\pi_4}{2}\right); \qquad \pi_4 = 2 \cdot \sin\left(\arctan \frac{\pi_3}{2}\right) \tag{p1}$$

which are convenient alternatives to (k1) and (ℓ1).

Illustration B. We wish to suspend a cable weighing $q = 300$ N/m between poles $\lambda = 80$ m apart. Moreover, we limit the *horizontal* anchor force to $H = 5500$ N. What will be the cable's *minimum* allowable length L, corresponding sag U, tension T, vertical anchor load V, and anchor angle φ?

Although none of the dimensionless variables in (c1) is known, we know by the given data that

$$\pi_1 \cdot \pi_3 = \frac{\lambda}{L} \cdot \frac{q \cdot L}{H} = \frac{\lambda \cdot q}{H} = \frac{(80) \cdot (300)}{5500} = 4.36364$$

and therefore, by (g1),

$$\pi_2 \cdot \pi_3 = \cosh \frac{\pi_1 \cdot \pi_3}{2} - 1 = 3.48763$$

thus, from the last two relations,

$$\frac{\pi_2 \cdot \pi_3}{\pi_1 \cdot \pi_3} = \frac{\pi_2}{\pi_1} = \frac{3.48763}{4.36364} = 0.79925$$

We now observe that in the π_2 versus π_1 coordinate system (Fig. 18-26) a $\pi_2/\pi_1 = $ const relation is represented by a straight line (through the origin) whose slope β to the π_1 axis is $\tan \beta = \pi_2/\pi_1$. So, through the origin we draw a straight line whose inclination to the π_1 axis is $\beta = \arctan \pi_2/\pi_1$. The coordinate of the *intersect* of this line with π_2 versus π_1 curve will provide the desired π_1, π_2 values. From the plot presented in Fig. 18-26, we have $\pi_1 = 0.499$ and $\pi_2 = .399$. Therefore, with appropriate rounding to three decimals

$$\pi_3 = \frac{\pi_2 \cdot \pi_3}{\pi_2} = \frac{3.48763}{0.399} = 8.741$$

and hence, by (p1) and (k1)

$$\pi_4 = 1.950$$

These values are indicated by dashed lines in the explanatory plot in Fig. 18-26.
These numerical results now enable us to determine all the sought-after properties of the hanging cable. Accordingly, the length of the cable cannot be less than

$$L = \frac{\lambda}{\pi_1} = \frac{80}{0.499} = 160.32 \text{ m}$$

for otherwise the imposed minimum H value would be exceeded. The associated sag U is

$$U = L \cdot \pi_2 = (160.32) \cdot (0.399) = 63.968 \text{ m}$$

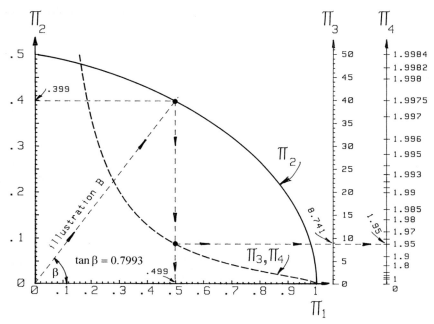

Figure 18-26
Dimensionless explanatory plot attached to Illustration B
(For definition of dimensionless variables, see Fig. 18-25)

Vertical anchor load V is by (i1)

$$V = \frac{q \cdot L}{2} = \frac{(300) \cdot (160.32)}{2} = 24{,}038 \text{ N}$$

and horizontal anchor force H is by (j1)

$$H = \frac{q \cdot L}{\pi_3} = \frac{(300) \cdot (160.32)}{8.741} = 5502.3 \text{ N}$$

which is within 0.042% of the imposed maximum of 5500 N. The maximum tension T of the cable is by (k1)

$$T = \frac{q \cdot L}{\pi_4} = \frac{(300) \cdot (160.32)}{1.950} = 24664 \text{ N}$$

Finally, anchor angle φ is by

$$\varphi = \arctan \frac{V}{H} = \arctan \frac{24{,}038}{5502.3} = \arctan 4.36872 = 77.1071°$$

Illustration C. Often one is faced with a problem of this nature: a suspended cable spans $\lambda = 90$ m and weighs $q = 60$ N/m. It is now required to determine its length L, sag U, the horizontal anchor force *H*, and vertical anchor force *V* with the condition that the tension in the cable does not exceed $T = 6625.8$ N.

From the given physical characteristics (λ, q, T), none of the hitherto established dimensionless variables can be determined and therefore it seems that the problem cannot be solved. However, when *direct* approach is not possible, maybe an *indirect* one is! To this end we introduce a *new* dimensionless variable

$$\pi_5 \equiv \pi_1 \cdot \pi_4 \tag{q1}$$

which, with the aid of (c1) and (k1), can be written

$$\pi_5 = \frac{\lambda}{L} \cdot \frac{q \cdot L}{T} = \frac{\lambda \cdot q}{T} \tag{r1}$$

in which the "ingredients" are all known. Note that, as with π_4, this new π_5 is not an independent dimensionless variable, for it can be expressed in terms of already established π_1, π_2, and π_3. Let us now express this dependence, by which we will be able to determine the desired physical characteristics of the cable.

We start with relation (f1) and (ℓ1), which are repeated here for convenience

$$\pi_1 = \frac{2}{\pi_3} \cdot \ln\left(\frac{\pi_3}{2} + \frac{1}{2} \cdot \sqrt{4 + \pi_3^2}\right) \qquad \text{repeated (f1)}$$

$$\pi_3 = \frac{2 \cdot \pi_4}{\sqrt{4 - \pi_4^2}} \qquad \text{repeated (ℓ1)}$$

Substitution of π_3 from (ℓ1) into (f1) yields—after some elementary simplification—

$$\pi_1 = \frac{\sqrt{4 - \pi_4^2}}{2 \cdot \pi_4} \cdot \ln \frac{2 + \pi_4}{2 - \pi_4} \tag{s1}$$

But by (q1) $\pi_4 = \pi_5/\pi_1$, which is now inserted into (s1). This results in

$$\pi_1 = \frac{\sqrt{4 \cdot \pi_1^2 - \pi_5^2}}{2 \cdot \pi_5} \cdot \ln \frac{2 \cdot \pi_1 + \pi_5}{2 \cdot \pi_1 - \pi_5} \tag{t1}$$

This important implicit relation can be solved numerically for π_5 (if π_1 is given), or for π_1 (if π_5 is given). However, as explained presently, while the former activity yields a unique result, the latter generally yields two. The table in Fig. 18-27 gives the corresponding π_1 and π_5 values appearing in (t1) for the range $0.04 \leq \pi_1 \leq 1.000$ in increments of 0.004 in π_1.

The plot showing the π_5 versus π_1 relation is given in Fig. 18-28. This plot, of course, is fully compatible with the numerical table presented in Fig. 18-27. In this present illustration, we have

$$\pi_5 = \frac{\lambda \cdot q}{T} = \frac{(90) \cdot (60)}{6625.8} = 0.815$$

for which the table in Fig. 18-27, with linear interpolation, supplies *two* π_1 values: $(\pi_1)_{\text{low}} = 0.413$ and $(\pi_1)_{\text{high}} = 0.9671$ (see dashed lines in Fig. 18-28).

Since, in general, $\pi_1 = \lambda/L$ and λ is constant, therefore $(\pi_1)_{\text{low}}$ and $(\pi_1)_{\text{high}}$ yield maximum $L_{\max}$ and minimum $L_{\min}$ lengths, respectively. Thus $(\pi_1)_{\text{low}} = \lambda/L_{\max}$, from which

$$L_{\max} = \frac{\lambda}{(\pi_1)_{\text{low}}} = \frac{90}{0.413} = 217.92 \text{ m}$$

574 APPLIED DIMENSIONAL ANALYSIS AND MODELING

π_1	π_5	π_1	π_5	π_1	π_5	π_1	π_5	π_1	π_5	π_1	π_5
.040	.0800	.280	.5577	.520	1.0097	.760	1.3168	.881	1.2473	.941	1.0210
.044	.0880	.284	.5656	.524	1.0166	.764	1.3186	.882	1.2451	.942	1.0150
.048	.0960	.288	.5735	.528	1.0235	.768	1.3202	.883	1.2429	.943	1.0087
.052	.1040	.292	.5813	.532	1.0304	.772	1.3216	.884	1.2407	.944	1.0023
.056	.1120	.296	.5892	.536	1.0372	.776	1.3228	.885	1.2383	.945	.9958
.060	.1200	.300	.5970	.540	1.0440	.780	1.3238	.886	1.2360	.946	.9893
.064	.1280	.304	.6049	.544	1.0508	.784	1.3246	.887	1.2337	.947	.9826
.068	.1360	.308	.6127	.548	1.0575	.788	1.3251	.888	1.2312	.948	.9758
.072	.1440	.312	.6205	.552	1.0641	.792	1.3254	.889	1.2288	.949	.9689
.076	.1520	.316	.6283	.556	1.0708	.796	1.3255	.890	1.2263	.950	.9618
.080	.1600	.320	.6361	.560	1.0774	.800	1.3253	.891	1.2237	.951	.9545
.084	.1680	.324	.6439	.564	1.0839	.804	1.3249	.892	1.2211	.952	.9472
.088	.1760	.328	.6517	.568	1.0904	.808	1.3241	.893	1.2185	.953	.9397
.092	.1840	.332	.6595	.572	1.0969	.812	1.3231	.894	1.2158	.954	.9320
.096	.1920	.336	.6673	.576	1.1033	.816	1.3218	.895	1.2131	.955	.9242
.100	.2000	.340	.6751	.580	1.1097	.820	1.3202	.896	1.2103	.956	.9162
.104	.2079	.344	.6828	.584	1.1160	.824	1.3183	.897	1.2075	.957	.9080
.108	.2159	.348	.6906	.588	1.1223	.828	1.3161	.898	1.2047	.958	.8998
.112	.2239	.352	.6983	.592	1.1285	.832	1.3135	.899	1.2017	.959	.8913
.116	.2319	.356	.7061	.596	1.1347	.836	1.3105	.900	1.1987	.960	.8825
.120	.2399	.360	.7138	.600	1.1408	.841	1.3063	.901	1.1957	.961	.8738
.124	.2479	.364	.7215	.604	1.1469	.842	1.3054	.902	1.1926	.962	.8647
.128	.2559	.368	.7292	.608	1.1529	.843	1.3044	.903	1.1895	.963	.8554
.132	.2639	.372	.7369	.612	1.1588	.844	1.3035	.904	1.1863	.964	.8460
.136	.2719	.376	.7446	.616	1.1647	.845	1.3025	.905	1.1831	.965	.8364
.140	.2798	.380	.7522	.620	1.1705	.846	1.3015	.906	1.1797	.966	.8265
.144	.2878	.384	.7599	.624	1.1763	.847	1.3005	.907	1.1764	.967	.8162
.148	.2958	.388	.7675	.628	1.1820	.848	1.2994	.908	1.1730	.968	.8059
.152	.3038	.392	.7751	.632	1.1876	.849	1.2983	.909	1.1695	.969	.7952
.156	.3118	.396	.7828	.636	1.1932	.850	1.2972	.910	1.1660	.970	.7844
.160	.3197	.400	.7904	.640	1.1986	.851	1.2960	.911	1.1624	.971	.7732
.164	.3277	.404	.7979	.644	1.2041	.852	1.2948	.912	1.1588	.972	.7618
.168	.3357	.408	.8055	.648	1.2094	.853	1.2936	.913	1.1551	.973	.7501
.172	.3436	.412	.8131	.652	1.2147	.854	1.2924	.914	1.1513	.974	.7377
.176	.3516	.416	.8206	.656	1.2198	.855	1.2911	.915	1.1475	.975	.7255
.180	.3596	.420	.8281	.660	1.2249	.856	1.2898	.916	1.1436	.976	.7127
.184	.3675	.424	.8357	.664	1.2299	.857	1.2885	.917	1.1396	.977	.6994
.188	.3755	.428	.8432	.668	1.2349	.858	1.2872	.918	1.1356	.978	.6859
.192	.3835	.432	.8506	.672	1.2397	.859	1.2858	.919	1.1315	.979	.6718
.196	.3914	.436	.8581	.676	1.2445	.860	1.2844	.920	1.1274	.980	.6572
.200	.3994	.440	.8655	.680	1.2491	.861	1.2830	.921	1.1231	.981	.6422
.204	.4073	.444	.8730	.684	1.2537	.862	1.2815	.922	1.1188	.982	.6269
.208	.4153	.448	.8804	.688	1.2581	.863	1.2800	.923	1.1144	.983	.6108
.212	.4232	.452	.8877	.692	1.2624	.864	1.2785	.924	1.1099	.984	.5942
.216	.4312	.456	.8951	.696	1.2667	.865	1.2769	.925	1.1054	.985	.5767
.220	.4391	.460	.9025	.700	1.2708	.866	1.2753	.926	1.1007	.986	.5587
.224	.4470	.464	.9098	.704	1.2748	.867	1.2737	.927	1.0961	.987	.5397
.228	.4550	.468	.9171	.708	1.2787	.868	1.2720	.928	1.0913	.988	.5201
.232	.4629	.472	.9244	.712	1.2825	.869	1.2703	.929	1.0865	.989	.4989
.236	.4708	.476	.9316	.716	1.2862	.870	1.2686	.930	1.0815	.990	.4771
.240	.4788	.480	.9388	.720	1.2897	.871	1.2668	.931	1.0765	.991	.4537
.244	.4867	.484	.9461	.724	1.2931	.872	1.2650	.932	1.0714	.992	.4292
.248	.4946	.488	.9532	.728	1.2963	.873	1.2632	.933	1.0662	.993	.4023
.252	.5025	.492	.9604	.732	1.2994	.874	1.2613	.934	1.0609	.994	.3735
.256	.5104	.496	.9675	.736	1.3024	.875	1.2595	.935	1.0555	.995	.3420
.260	.5183	.500	.9746	.740	1.3052	.876	1.2575	.936	1.0500	.996	.3064
.264	.5262	.504	.9817	.744	1.3079	.877	1.2556	.937	1.0444	.997	.2666
.268	.5341	.508	.9888	.748	1.3103	.878	1.2535	.938	1.0387	.998	.2175
.272	.5420	.512	.9958	.752	1.3127	.879	1.2515	.939	1.0328	.999	.1546
.276	.5499	.516	1.0028	.756	1.3148	.880	1.2494	.940	1.0270	1.00	.0000

Figure 18-27
Corresponding values of dimensionless variables
$\pi_1 = \lambda/L$ and $\pi_5 = (q \cdot L)/T$ **of a catenary**
Data were obtained by the numerical solution of relation (t1).
For definition of symbols, see Fig. 18-25

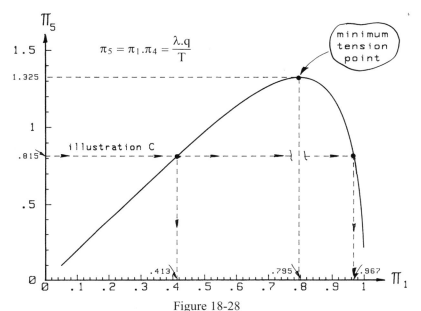

Figure 18-28
**Plot showing relation between dimensionless variables
$\pi_1 = \lambda/L$ and $\pi_5 = (\lambda \cdot q)/T$ of a catenary**
Data were obtained by the numerical solution of relation (t1).
For definition of symbols, see Fig. 18-25

and similarly

$$L_{min} = \frac{\lambda}{(\pi_1)_{high}} = \frac{90}{0.9671} = 93.06 \text{ m}$$

These results mean that with a *given* span $\lambda = 90$ m and unit weight $q = 60$ N/m, if the length of the cable is between 93.06 m and 217.92 m, then the tension will not exceed the imposed $T = 6625.8$ N limit. To test this statement, let us have $L = 150$ m—which is certainly between the specified limits. We therefore have $\pi_1 = \lambda/L = 90/150 = 0.6$ for which the table in Fig. 18-27 and the plot in Fig. 18-28 give $\pi_5 = 1.1408$. Hence, by (r1), $\pi_5 = 1.1408 = (\lambda \cdot q)/T$ from which

$$T = \frac{\lambda \cdot q}{\pi_5} = \frac{(90) \cdot (60)}{1.1408} = 4733.5 \text{ N}$$

and which is, indeed, less than the allowed 6625.8 N.

The two sag values corresponding to the *two* π_1 values is of course different and can be easily obtained from the plot in Fig. 18-25. For $(\pi_1)_{low} = 0.413$ the plot supplies (dashed line) $\pi_2 = 0.424$ and $\pi_3 = 12.1$. Hence, by (c1), the sag is

$$U = L_{max} \cdot \pi_2 = (217.92) \cdot (0.424) = 92.4 \text{ m}$$

The horizontal anchor force is by (j1)

$$H = \frac{q \cdot L_{max}}{\pi_3} = \frac{(60) \cdot (217.92)}{12.1} = 1080.6 \text{ N}$$

The vertical anchor force is by (i1)

$$V = \frac{q \cdot L_{max}}{2} = \frac{(60) \cdot (217.92)}{2} = 6537.6 \text{ N}$$

To verify, we determine tensile force T in the cable

$$T = \sqrt{V^2 + H^2} = \sqrt{6537.6^2 + 1080.6^2} = 6626.3 \text{ N}$$

which is the required 6625.8 N value (within 0.008%).

An identical procedure yields corresponding results for the $(\pi_1)_{high} = 0.9671$ value. Thus, by (f1) $\pi_3 = 0.92962$, and by (g1) $\pi_2 = 0.11052$. So, since $L = L_{min} = 93.06$ m, the sag is

$$U = L_{min} \cdot \pi_2 = (93.06) \cdot (0.11052) = 10.28 \text{ m}$$

The horizontal H and vertical V anchor forces are

$$H = \frac{q \cdot L_{min}}{\pi_3} = \frac{(60) \cdot (93.06)}{0.92962} = 6006.33 \text{ N and } V = \frac{q \cdot L_{min}}{2} = \frac{(60) \cdot (93.06)}{2} = 2791.8 \text{ N}$$

Again, we verify the cable tension

$$T = \sqrt{V^2 + H^2} = \sqrt{6006.33^2 + 2791.8^2} = 6623.4 \text{ N}$$

which checks nicely, since it is within 0.036% of the given 6625.8 N value.

It is obvious that if sag U is very small, then horizontal anchor force H must be very large, and hence cable tension T will also be very large. On the other hand, if U is very large, then the length of the cable and therefore the weight of the cable will be very large. Hence, again, cable tension T will be very large. This argument suggests that somewhere between the two extrema there must be a *particular* cable length L_0, which for a *given* span λ and unit weight q will yield *minimum* tension T_{min}.

In order to determine this optimum point we consider (d1)

$$L = \frac{2 \cdot H}{q} \sinh \frac{q \cdot \lambda}{2 \cdot H} \qquad \text{repeated (d1)}$$

and by (h1) and (i1)

$$T = \sqrt{\left(\frac{q \cdot L}{2}\right)^2 + H^2} \qquad \text{(u1)}$$

from which

$$L = \frac{2}{q} \cdot \sqrt{T^2 - H^2} \qquad \text{(v1)}$$

Now by (d1) and (v1)

$$\sqrt{T^2 - H^2} = H \cdot \sinh \frac{q \cdot \lambda}{2 \cdot H} \qquad \text{(w1)}$$

which can be easily rearranged to

$$T^2 = H^2 \cdot \left(\sinh^2 \frac{q \cdot \lambda}{2 \cdot H} + 1\right) \qquad \text{(x1)}$$

FIFTY-TWO ADDITIONAL APPLICATIONS

Now we consider the identity valid for all x

$$\cosh^2 x - \sinh^2 x = 1 \tag{y1}$$

Therefore, by (x1) and (y1)

$$T = H \cdot \cosh \frac{\lambda \cdot q}{2 \cdot H} \tag{z1}$$

In order to find the minimum T we differentiate this expression with respect to H and then equate the result to zero. Accordingly,

$$\frac{dT}{dH} = \cosh \frac{\lambda \cdot q}{2 \cdot H} + H \cdot \sinh \frac{\lambda \cdot q}{2 \cdot H} \cdot \left(-\frac{\lambda \cdot q}{2 \cdot H^2}\right) = 0$$

from which, by using definitions of the dimensionless variables π_1, π_2, and π_3 in (c1), we obtain

$$\frac{\pi_1 \cdot \pi_3}{2} \cdot \tanh \frac{\pi_1 \cdot \pi_3}{2} = 1 \tag{a2}$$

This implicit formula cannot be solved analytically, but by any of the standard numerical methods the root $\pi_1 \cdot \pi_3$ can be found rather easily. Thus the root of (a2) in terms of $\pi_1 \cdot \pi_3$ is

$$(\pi_1 \cdot \pi_3)_0 = 2.39935728052 \tag{b2}$$

which, by (g1) provides

$$(\pi_2 \cdot \pi_3)_0 = 0.8101705807 \tag{c2}$$

By (f1) now

$$(\pi_3)_0 = 3.01775912308 \tag{d2}$$

Hence

$$(\pi_2)_0 = \frac{(\pi_2 \cdot \pi_3)_0}{(\pi_3)_0} = 0.268467610455 \tag{e2}$$

$$(\pi_1)_0 = \frac{(\pi_1 \cdot \pi_3)_0}{(\pi_3)_0} = 0.795079124165 \tag{f2}$$

and by (k1) and (d2)

$$(\pi_4)_0 = 1.6671131192 \tag{g2}$$

Finally, by (q1), (f2), and (g2)

$$(\pi_5)_0 = 1.3254868387 \tag{h2}$$

The subscript "0" in all of the above data designates *minimum* cable force, i.e., $T = T_{\min}$. The corresponding minimum tension point on the π_5 versus π_1 curve in Fig. 18-28 is the *maximum* point, since T appears in the denominator of the expression of π_5.

At minimum tension, the anchor angle φ_0 of the cable is by (m1)

$$\tan \varphi_0 = \frac{(\pi_3)_0}{2} = 1.50887956154$$

from which

$$\varphi_0 = 56.465835128 \text{ deg} = 0.985514738 \text{ rad} \qquad (i2)$$

This last result means that if the cable's anchor angle is $\varphi_0 = 56.465835128$ deg, then for the prevailing span λ and unit weight q, the tension is the minimum possible. Specifically, by (h2) and (r1),

$$T = T_{\min} = \frac{q \cdot \lambda}{(\pi_5)_0} = (0.754439781) \cdot q \cdot \lambda$$

i.e., with very good approximation it is 75.4 % of the weight of the cable whose length equals span λ. In short, the minimum achievable tensile force in a cable spanning $\lambda = 100$ m is the weight of 75.444 m of *that* cable.

Sag U_0 corresponding to *minimum* tension is supplied by

$$\frac{U_0}{\lambda} = \frac{(\pi_2)_0}{(\pi_1)_0} = 0.33766150097 \qquad (j2)$$

i.e., the sag at minimum tension is 33.766 % of the span. The table in Fig. 18-29 summarizes the above results.

Illustration D. **The problem:** To span a distance of $\lambda = 90$ m with a cable of unit weight $q = 60$ N/m. If we minimize the tension of the cable T (anchor force), what will be cable length L, sag U, cable tension T, horizontal anchor force H, vertical anchor force V, and anchor angle φ?

The solution: From the table in Fig. 18-29

length: $\quad L = \dfrac{\lambda}{(\pi_1)_0} = \dfrac{90}{0.7950791} = 113.19628$ m

sag: $\quad U = L \cdot (\pi_2)_0 = (113.19628) \cdot (0.2684676) = 30.390$ m

minimum tension: $\quad T = \dfrac{\lambda \cdot q}{(\pi_5)_0} = \dfrac{(90) \cdot (60)}{1.3254868} = 4073.975$ N

horizontal anchor force: $\quad H = \dfrac{\lambda \cdot q}{(\pi_1)_0 \cdot (\pi_3)_0} = \dfrac{(90) \cdot (60)}{2.3993573} = 2250.603$ N

vertical anchor force: $\quad V = \dfrac{q \cdot L}{2} = \dfrac{(60) \cdot (113.196)}{2} = 3395.88$ N

anchor angle: $\quad \varphi_0 = 56.46583$ deg $= 0.9855147$ rad

Illustration E. We have an $L = 100$ m long chain whose unit weight is $q = 240$ N/m and maximum allowable tension force is $T = 12600$ N. Determine maximum achievable span λ, corresponding sag U, anchor angle φ, and horizontal anchor force H.

Dimensionless variable	Definition, expression	Value	Reference
$(\pi_1)_0$	λ/L	0.795079124	(f2)
$(\pi_2)_0$	U/L	0.26846761	(e2)
$(\pi_3)_0$	$q \cdot L/H$	3.017759123	(d2)
$(\pi_4)_0$	$q \cdot L/T$	1.667113119	(g2)
$(\pi_5)_0$	$\lambda \cdot q/T$	1.325486839	(h2)
$(\pi_1)_0 \cdot (\pi_3)_0$	$\lambda \cdot q/H$	2.399357281	(f2), (d2)
$(\pi_2)_0 \cdot (\pi_3)_0$	$U \cdot q/H$	0.810170581	(e2), (d2)
$(\pi_2)_0/(\pi_1)_0$	U/λ	0.337661501	(e2), (f2)
φ_0	$\arctan \dfrac{(\pi_3)_0}{2}$	56.465835128 deg 0.985514738 rad	(i2)

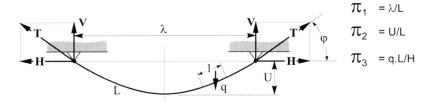

Figure 18-29
High-precision geometry and force-related dimensionless characteristics of a minimum tension catenary
Subscript "0" designates values specific for minimum tension configuration

From the given data

$$\pi_4 = \frac{q \cdot L}{T} = \frac{(240) \cdot (100)}{12600} = 1.90470$$

from which by (ℓ1) $\pi_3 = 6.24695$, and the anchor angle by (n1),

$$\varphi = \arctan \frac{\pi_3}{2} = 72.247 \text{ deg}$$

and by the definition of π_3 in (c1)

$$H = \frac{q \cdot L}{\pi_3} = \frac{(240) \cdot (100)}{6.24695} = 3841.87 \text{ N}$$

The plot in Fig. 18-30 supplies $\pi_1 = 0.594$ and $\pi_2 = 0.365$ (dashed lines). Hence, by (c1),

the span: $\lambda = \pi_1 \cdot L = (0.594) \cdot (100) = 59.4$ m
the sag: $U = \pi_2 \cdot L = (0.365) \cdot (100) = 36.5$ m

580 APPLIED DIMENSIONAL ANALYSIS AND MODELING

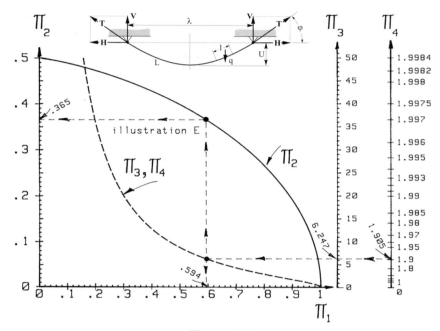

Figure 18-30
Explanatory dimensionless plot attached to Illustration *E*
For definition of dimensionless variables, see Fig. 18-25

The obtained $\lambda = 59.4$ m span is the maximum for the imposed conditions, for if λ were longer, then π_1 would also be larger and hence, by the π_4 versus π_1 curve in Fig. 18-30, π_4 would be smaller. But, since q and L are constants and T is in the denominator of π_4, a smaller π_4 would yield a larger T, therefore the imposed maximum allowable T value would be exceeded.

⇑

Example 18-27. Speed of a Vertically Ejected Projectile (II)

The problem: Design an experiment on Earth to determine the vertical distance achieved of a vertically ejected projectile on Mars. In particular, if on Earth ($g_2 = 9.81$ m/s²) a vertically ejected particle with initial speed of $v_2 = 7.2$ m/s is measured to reach an altitude of $h_2 = 2.642$ m, what would be the height achieved on Mars ($g_1 = 3.73$ m/s²), if the initial vertical speed there is $v_1 = 6.3$ m/s? In all cases ignore atmospheric resistance and assume constant gravitational effects.

The solution: First, we list the relevant variables.

Variable	Symbol	Dimension
altitude reached	h	m
initial vertical speed	v	m/s
gravitational acceleration	g	m/s²

We have three variables and two dimensions, therefore there is only one dimensionless variable—a constant in this case; it is supplied by the Dimensional Set

	h	v	g
m	1	1	1
s	0	−1	−2
π_1	1	−2	1

from which

$$\pi_1 = \frac{h \cdot g}{v^2} = \text{const} \tag{a}$$

The Model Law therefore is

$$S_h \cdot S_g = S_v^2 \tag{b}$$

where S_h is the Altitude Scale Factor;
S_g is the Gravitational Acceleration Scale Factor;
S_v is the Initial Speed Scale Factor.

By the given data

$$S_v = \frac{v_2}{v_1} = \frac{7.2}{6.3} = 1.14286 \tag{c}$$

$$S_g = \frac{g_2}{g_1} = \frac{9.81}{3.73} = 2.63003 \tag{d}$$

Therefore, by (b), (c) and (d)

$$S_h = \frac{S_v^2}{S_g} = \frac{1.14286^2}{2.63003} = 0.49662 \tag{e}$$

Variable					Scale factor S	Category	
name	symbol	dimension	prototype	model	model/prototype	prototype	model
initial speed	v	m/s	6.3	7.2	1.14286	1	1
altitude reached	h	m	5.32	2.642	0.49662	2	3
gravitational acceleration	g	m/s²	3.73	9.81	2.63003	1	1
dimensionless	π_1	1	0.49996	0.49996	1	—	—
categories of variables	1	freely chosen, *a priori* given, or determined independently					
	2	determined by application of Model Law					
	3	determined by measurement on the model					

Figure 18-31
Modeling Data Table for determining of the altitude achieved by a vertically ejected projectile on Mars, by a modeling experiment on Earth

582 APPLIED DIMENSIONAL ANALYSIS AND MODELING

But $S_h = h_2/h_1$, therefore by (e)

$$h_1 = \frac{h_2}{S_h} = \frac{2.642}{0.49662} = 5.31996 \text{ m}$$

which is the answer we were looking for. The Modeling Data Table (Fig. 18-31) summarizes.

Note the identical values for dimensionless variable π_1 for the prototype (Mars) and the model (Earth). Also note that there is no variable for the model which is set (i.e., defined) by Model Law (b). Here the Model Law is used only to determine the dependent variable (height) of the prototype (Mars) upon the *measured* and *a priori* given data for the model on Earth.

⇑

Example 18-28. Drag on a Flat Plate in a Parallel Fluid Flow

We investigate the drag on a smooth flat plate of negligible thickness immersed in a *slowly* flowing liquid. The flow is parallel to the plate. Because of the low speed of the moving fluid, the flow is laminar. The geometry and assigned directions are as illustrated in Fig. 18-32.

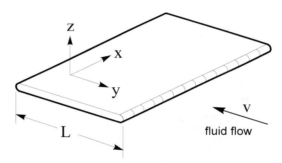

Figure 18-32
Flat thin plate is immersed in a fluid stream

In this example we demonstrate the use of directional dimensions. Accordingly, the following variables and their dimensions are considered.

Drag F per unit *x*-directional width of the plate. Hence

$$[F] = \frac{[\text{force}_y]}{[\text{width}_x]} = \frac{m_y \cdot kg}{s^2 \cdot m_x}$$

Velocity v of the streaming liquid. Hence

$$[v] = \frac{[\text{length}_y]}{[\text{time}]} = \frac{m_y}{s}$$

Density of fluid ρ. Because of symmetry, all three directions must take part equally. Hence

$$[\rho] = \frac{[\text{mass}]}{[\text{length}_x] \cdot [\text{length}_y] \cdot [\text{length}_z]} = \frac{kg}{m_x \cdot m_y \cdot m_z}$$

FIFTY-TWO ADDITIONAL APPLICATIONS

Shear stress τ, defined as

$$\tau = \frac{(\text{shear force})_y}{(\text{surface area})} = \frac{[\text{shear force}]_y}{[\text{length}_x]\cdot[\text{length}_y]}$$

Therefore

$$[\tau] = \frac{m_y \cdot kg}{s^2 \cdot m_x \cdot m_y} = \frac{kg}{s^2 \cdot m_x}$$

Dynamic viscosity μ, defined by Newton's law $\tau = \mu \cdot (dv/dz)$, from which $\mu = \tau \cdot (dz/dv)$. Using the dimensions derived above, we obtain

$$[\mu] = [\tau] \cdot \frac{[\text{length}_z]}{[\text{speed}_y]} = \frac{kg}{s^2 \cdot m_x} \cdot \frac{m_z}{(m_y/s)} = \frac{m_z \cdot kg}{s \cdot m_x \cdot m_y}$$

Length, y directional. Hence

$$[L] = [\text{length}_y] = m_y$$

The above is summarized in the table:

Variable	Symbol	Dimension
drag force	F	$m_y \cdot kg/(s^2 \cdot m_x)$
fluid velocity	v	m_y/s
fluid density	ρ	$kg/(m_x \cdot m_y \cdot m_z)$
fluid dynamic viscosity	μ	$m_z \cdot kg/(s \cdot m_x \cdot m_y)$
body's characteristic length	L	m_y

From these data, we can now compose the dimensional matrix

	F	v	ρ	μ	L
m_x	−1	0	−1	−1	0
m_y	1	1	−1	−1	1
m_z	0	0	−1	1	0
kg	1	0	1	1	0
s	−2	−1	0	−1	0

We see that the number of variables equals the number of dimensions, therefore to have any valid relations at all, this matrix must be singular, which it is. That the rank is 4 can be confirmed by eliminating the fourth row and the first column, and then calculating the value of the resulting determinant, which is 2 (i.e., not zero). Thus, we have to eliminate one dimension—say "kg." This results in five variables and four dimensions, and hence there is $5 - 4 = 1$ dimensionless variable—a constant in this case (Theorem 7-4). The Dimensional Set is then

	F	v	ρ	μ	L
m_x	−1	0	−1	−1	0
m_y	1	1	−1	−1	1
m_z	0	0	−1	1	0
s	−2	−1	0	−1	0
π_1	1	$-\dfrac{3}{2}$	$-\dfrac{1}{2}$	$-\dfrac{1}{2}$	$-\dfrac{1}{2}$

yielding the sole dimensionless variable

$$\pi_1 = \frac{F}{\sqrt{v^3 \cdot \rho \cdot \mu \cdot L}} = \text{const} \qquad (a)$$

from which the drag force per unit width is

$$F = \text{const} \cdot \sqrt{v^3 \cdot \rho \cdot \mu \cdot L} \qquad (b)$$

Then by performing a *single* experiment, we can determine the constant, resulting in an *explicit* formula of five physical variables derived without any analytical considerations! This example was adapted from Ref. 27, p. 29.

⇑

Example 18-29. Natural Frequency of a Transversally Vibrating Simply Supported Beam

We wish to find out the natural transversal frequency of a simply supported beam of uniform cross-section (Fig. 18-33). The relevant variables are as follows:

Variable	Symbol	Dimension	Remark
frequency	f	1/s	transversal
length	L	m	
Young's modulus	E	kg/(m·s^2)	
mass	M	kg	
cross-section's second moment of area	I	m^4	uniform

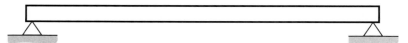

Figure 18-33
A simply supported beam of uniform cross-section

We have five variables and three dimensions, therefore there are 5 − 3 = 2 dimensionless variables determined by the Dimensional Set

	f	L	E	M	I
m	0	1	−1	0	4
kg	0	0	1	1	0
s	−1	0	−2	0	0
π_1	1	0	$-\frac{1}{2}$	$\frac{1}{2}$	$-\frac{1}{8}$
π_2	0	1	0	0	$-\frac{1}{4}$

yielding

$$\pi_1 = f \cdot \sqrt[8]{\frac{M^4}{E^4 \cdot I}}; \quad \pi_2 = \frac{L}{\sqrt[4]{I}} \qquad (a)$$

FIFTY-TWO ADDITIONAL APPLICATIONS

We can write

$$\pi_1 = \Psi\{\pi_2\} \tag{b}$$

where Ψ is an as yet unknown function. Let us assume that this function is a *monomial*, which is plausible. In this case we have

$$\pi_1 = c \cdot \pi_2^n \tag{c}$$

where c and n are numeric constants. By (a) and (c)

$$f = c \cdot \sqrt[8]{\frac{E^4 \cdot I}{M^4}} \cdot \left(\frac{L}{\sqrt[4]{I}}\right)^n \tag{d}$$

Constants c and n can be determined by either performing *two* measurements or by analysis, the latter being a slightly more demanding task in this particular case (see for example Ref. 132, p. 203). In either way, c and n can be determined to be

$$c = \frac{\pi}{2} = 1.57080; \qquad n = -\frac{3}{2} \tag{e}$$

Thus, by (d) and (e)

$$f = \frac{\pi}{2} \cdot \sqrt{\frac{EI}{M \cdot L^3}} \tag{f}$$

This formula involves five variables, therefore to present it graphically we would need 216 curves on 36 charts (considering only six distinct values for each variable; Fig. 12-3 in Art. 12.1). However, if we use dimensionless variables defined in (a), then (f) can be written as

$$\pi_1 = \frac{\pi}{2} \cdot \frac{1}{(\pi_2)^{3/2}} \tag{g}$$

in which there are only two variables! Thus, the plot for this relation is a *single* curve—quite an improvement over the 216 curves on 36 charts obtained previously. Fig. 18-34 shows the plot of equation (g).

But suppose we do not know the analytical results of (f) and (g). Instead, we wish to determine the natural frequency of a given prototype by a modeling experiment. To this end, we first must establish the Model Law, which is done by the already derived dimensionless variables (a). Accordingly, the Model Law is

$$S_f = \sqrt{\frac{S_E}{S_M}} \cdot \sqrt[8]{S_I}; \qquad S_L = \sqrt[4]{S_I} \tag{h}$$

where S_f is the Frequency Scale Factor;
S_E is the Young's Modulus Scale Factor;
S_M is the Mass Scale Factor;
S_I is the Cross-section's Second Moment of Area Scale Factor.

The *prototype* is a tubular steel beam $L_1 = 3.2$ m long, outside and inside diameters of the cross-section are $D_0 = 0.25$ m, $D_i = 0.15$ m; Young's modulus $E_1 = 2.068 \times 10^{11}$ Pa, density $\rho_1 = 8300$ kg/m^3. It is now required to determine the transversal natural frequency of this *prototype* by way of an experiment on an aluminium *model* whose cross-section is a solid rectangle of dimensions 0.06 m by 0.08 m (Fig. 18-35).

586 APPLIED DIMENSIONAL ANALYSIS AND MODELING

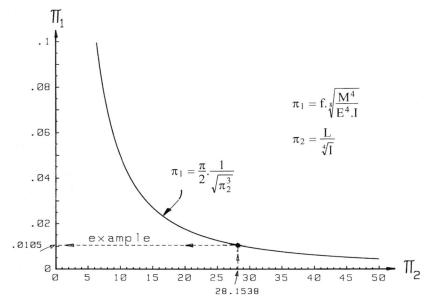

Figure 18-34
**Dimensionless plot to determine the natural frequency of a
laterally vibrating, simply supported beam of uniform cross-section**
See table in this example for definitions of physical variables;
π *without* subscript is 3.14159...

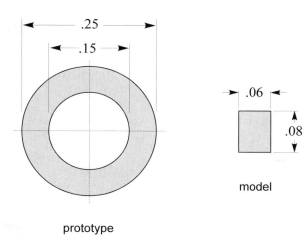

Figure 18-35
Cross-sections of the prototype and model
Note the geometric dissimilarity of cross-sections

For aluminium we have $E_2 = 6.9 \times 10^{10}$ Pa and $\rho_2 = 2770$ kg/m^3. With these data for the prototype $I_1 = 1.66897 \times 10^{-4}$ m^4, and for the model $I_2 = 2.56 \times 10^{-6}$ m^4. Therefore

$$S_I = \frac{I_2}{I_1} = \frac{2.56 \times 10^{-6}}{1.66897 \times 10^{-4}} = 1.53388 \times 10^{-2}$$

By the second relation of Model Law (h), the Length Scale Factor is

$$S_L = \sqrt[4]{S_I} = \sqrt[4]{1.53388 \times 10^{-2}} = 0.35192$$

But $S_L = L_2/L_1$, and hence the length of the model *must* be

$$L_2 = S_L \cdot L_1 = (0.35192) \cdot (3.2) = 1.12615 \text{ m}$$

Furthermore, from the given data the following determinations can be made:

prototype's mass: $M_1 = 834.399$ kg
model's mass: $M_2 = 14.973$ kg

Therefore the Mass Scale Factor is

$$S_M = \frac{M_2}{M_1} = \frac{14.973}{834.399} = 1.79447 \times 10^{-2}$$

and by the given Young's moduli

$$S_E = \frac{E_2}{E_1} = \frac{6.9 \times 10^{10}}{2.068 \times 10^{11}} = 0.33366$$

We now have all the ingredients to determine the Frequency Scale Factor as defined in Model Law (h). Accordingly,

$$S_f = \sqrt{\frac{S_E}{S_M}} \cdot \sqrt[8]{S_I} = \sqrt{\frac{0.33366}{1.79447 \times 10^{-2}}} \cdot \sqrt[8]{1.53388 \times 10^{-2}} = 2.55804 \quad \text{(i)}$$

Now we *build* the model and *measure* its natural frequency. Say we find $f_2 = 142.76$ 1/s. Next, we consider $S_f = f_2/f_1$, and (i), from which we find that the frequency of the prototype we are looking for is

$$f_1 = \frac{f_2}{S_f} = \frac{142.76}{2.55804} = 55.808 \text{ 1/s}$$

The Modeling Data Table in Fig. 18-36 summarizes all of the above inputs and results. Note the very close agreement (within 0.01%) of the dimensionless variables for the prototype and to the model. The dashed lines in Fig. 18-34 indicate the specific numerical values of π_1 and π_2 occurring in this application..

It is emphasized that in this instance there is no *geometric* similarity between the prototype and its model, yet there was no difficulty in constructing a *dimensionally* similar *model*, and then carrying out a successful experiment on it to determine the frequency of the *prototype*.

⇑

Example 18-30.   Buckling of a Vertical Rod Under its Own Weight

We have a vertical straight rod of uniform solid circular cross-section. The rod is clamped at its lower end, and free to move at the top. If we gradually increase the length from a

Variable					Scale factor S	Category	
name	symbol	dimension	prototype	model	model/prototype	prototype	model
beam's lateral frequency	f	1/s	55.808	142.76	2.55806	2	3
beam's length	L	m	3.2	1.12615	0.35192	1	2
Young's modulus	E	kg/(m·s²)	2.068E11	6.9E10	0.33366	1	1
beam's mass	M	kg	834.399	14.973	1.79447E-2	1	1
second moment of area	I	m⁴	1.66897E-4	2.56E-6	1.53388E-2	1	1
dimensionless	π_1	1	1.05148E-2	1.05149E-2	1.00001	—	—
dimensionless	π_2	1	28.15385	28.15375	1	—	—
categories of variables	1	freely chosen, *a priori* given, or determined independently					
	2	determined by application of Model Law					
	3	determined by measurement on the model					

Figure 18-36
Modeling Data Table for the experiment to determine the lateral frequency of a simply supported beam

conveniently small value, then at some point the rod will buckle under its own weight. This is the *critical length* L_c (Fig. 18-37).

It is now required to design and execute an experiment to determine the critical length of a prototype straight steel rod situated on Mars. The rod on Mars has a $D_1 = 0.045$ m diameter solid circle cross-section. The following variables are considered relevant:

Variable	Symbol	Dimension
critical length	L_c	m
Young's modulus of material	E	kg/(m·s²)
diameter of cross-section	D	m
material density	ρ	kg/m³
gravitational acceleration	g	m/s²

We have five variables and three dimensions, therefore there are $5 - 3 = 2$ dimensionless variables supplied by the Dimensional Set

	L_c	E	D	ρ	g
m	1	−1	1	−3	1
kg	0	1	0	1	0
s	0	−2	0	0	−2
π_1	1	0	−1	0	0
π_2	0	1	−1	−1	−1

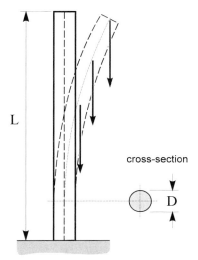

Figure 18-37
Vertical rod clamped at the bottom and of critical length buckles under its own weight

by which

$$\pi_1 = \frac{L_c}{D} \; ; \qquad \pi_2 = \frac{E}{D \cdot \rho \cdot g} \qquad (a)$$

Thus the Model Law is

$$S_{L_c} = S_D \; ; \qquad S_E = S_D \cdot S_\rho \cdot S_g \qquad (b)$$

where S_{L_c} is the Critical Length Scale Factor;
S_D is the Diameter of Cross-section Scale Factor;
S_E is the Young's Modulus Scale Factor;
S_ρ is the Density Scale Factor;
S_g is the Gravitational Acceleration Scale Factor.

We have $g_2 = 9.81$ m/s² for the model (on Earth) and $g_1 = 3.73$ m/s² for the prototype (on Mars). Moreover, since the material of the model and prototype is the same (steel), therefore $\rho_1 = \rho_2 = 8300$ kg/m³ and $E_1 = E_2 = 2 \times 10^{11}$ Pa. By these data, then

$$S_\rho = \frac{\rho_2}{\rho_1} = \frac{8300}{8300} = 1 \; ; \quad S_E = \frac{E_2}{E_1} = \frac{2 \times 10^{11}}{2 \times 10^{11}} = 1 \; ; \quad S_g = \frac{g_2}{g_1} = \frac{9.81}{3.73} = 2.63003 \quad (c)$$

and hence Model Law (b) simplifies to

$$S_{L_c} = S_D \; ; \qquad S_D \cdot S_g = 1 \qquad (d)$$

By the second part of (d) and third part of (c)

$$S_D = \frac{D_2}{D_1} = \frac{1}{S_g} = \frac{1}{2.63003} = 0.38022 \qquad (e)$$

590 APPLIED DIMENSIONAL ANALYSIS AND MODELING

and hence the diameter of the cross-section of the model (on Earth) *must* be

$$D_2 = D_1 \cdot S_D = (0.045) \cdot (0.38023) = 0.01711 \text{ m}$$

Now we build the model (made of steel) and by *experimentation* determine its critical length. Say we find that it buckles at $(L_c)_2 = 7.195$ m length (on Earth). By (e) and the first relation of Model Law (d)

$$S_{L_c} = \frac{(L_c)_2}{(L_c)_1} = S_D = 0.38022$$

from which

$$(L_c)_1 = \frac{(L_c)_2}{S_{L_c}} = \frac{7.195}{0.38023} = 18.923 \text{ m}$$

i.e., the critical length of our beam on Mars is 18.923 m. The Modeling Data Table (Fig. 18-38) summarizes all of the above input data and results.

name	symbol	dimension	Variable prototype	model	Scale factor S model/prototype	Category prototype	model
cross-section's diameter	D	m	0.045	0.01711	0.38023	1	2
gravitational acceleration	g	m/s²	3.73	9.81	2.63003	1	1
critical (buckling) length	L_c	m	18.923	7.195	0.38023	2	3
Young's modulus	E	kg/(m·s²)	2E11	2E11	1	1	1
density	ρ	kg/m³	8300	8300	1	1	1
dimensionless	π_1	1	420.51111	420.51432	1.00001	—	—
dimensionless	π_2	1	1.43559E8	1.43560E8	1.00001	—	—
categories of variables	1	freely chosen, *a priori* given, or determined independently					
	2	determined by application of Model Law					
	3	determined by measurement on the model					

Figure 18-38
Modeling Data Table for the experimental determination of the buckling length of a heavy vertical rod clamped at the bottom and free at the top

Example 18-31. Lateral Natural Frequency of a Cantilever (I)

By a modeling experiment we wish to determine the natural frequency of a *prototype* cantilever of following properties:

Length: $L_1 = 0.8$ m
Young's modulus of material: $E_1 = 2 \times 10^{11}$ Pa

Diameter of solid circle cross-section: $D_1 = 0.050$ m
Density of material: $\rho_1 = 7850$ kg/m^3

Accordingly, we first list the relevant variables.

Variable	Symbol	Dimension	Remark
natural frequency	f	1/s	lateral
length	L	m	
Young's modulus	E	kg/(m·s^2)	
second moment of area	I	m^4	uniform
density	ρ	kg/m^3	uniform

We have five variables and three dimensions, therefore there are $5 - 3 = 2$ dimensionless variables which are supplied by the Dimensional Set

	f	L	E	I	ρ
m	0	1	−1	4	−3
kg	0	0	1	0	1
s	−1	0	−2	0	0
π_1	1	0	$-\dfrac{1}{2}$	$\dfrac{1}{4}$	$\dfrac{1}{2}$
π_2	0	1	0	$-\dfrac{1}{4}$	0

from which

$$\pi_1 = f \cdot \sqrt[4]{\dfrac{I \cdot \rho^2}{E^2}} \quad ; \qquad \pi_2 = \dfrac{L}{\sqrt[4]{I}} \tag{a}$$

therefore the Model Law is

$$S_f = \dfrac{1}{\sqrt[4]{S_I}} \cdot \sqrt{\dfrac{S_E}{S_\rho}} \quad ; \qquad S_L = \sqrt[4]{S_I} \tag{b}$$

where S_f is the Frequency Scale Factor;
S_I is the Second Moment of Area Scale Factor;
S_E is the Young's Modulus Scale Factor;
S_ρ is the Density Scale Factor;
S_L is the Length Scale Factor.

For the model we *select* aluminium for material and solid circle cross-section with given diameter. Hence, for the model

Young's modulus of material: $E_2 = 6.7 \times 10^{10}$ Pa
Diameter of solid circle cross-section: $D_2 = 0.010$ m
Density of material: $\rho_2 = 2600$ kg/m^3

By the given data now

$$S_E = \dfrac{E_2}{E_1} = \dfrac{6.7 \times 10^{10}}{2 \times 10^{11}} = 0.335 \tag{c}$$

$$S_\rho = \frac{\rho_2}{\rho_1} = \frac{2600}{7850} = 0.33121 \tag{d}$$

$$S_I = \frac{I_2}{I_1} = \frac{(D_2^4 \cdot \pi)/64}{(D_1^4 \cdot \pi)/64} = \frac{4.90874 \times 10^{-10}}{3.06796 \times 10^{-7}} = 0.0016 \tag{e}$$

By the second relation of Model Law (b) and (e)

$$S_L = \frac{L_2}{L_1} = \sqrt[4]{S_I} = \sqrt[4]{0.0016} = 0.2 \tag{f}$$

Also, by given data, the Cross-section Diameter Scale Factor is

$$S_D = \frac{D_2}{D_1} = \frac{0.010}{0.050} = 0.2 \tag{g}$$

Therefore, by (f) and (g) we see that $S_D = S_L$, i.e., the cross-section and length must be reduced by the same rate. In other words, the model *must* be *geometrically* similar to the prototype.

Relation (f) yields the *mandatory* length of the model

$$L_2 = S_L \cdot L_1 = (0.2) \cdot (0.8) = 0.16 \text{ m} \tag{h}$$

Variable					Scale factor S	Category	
name	symbol	dimension	prototype	model	model/prototype	prototype	model
natural frequency	f	1/s	346.63	1743	5.02842	2	3
length	L	m	0.8	0.16	0.2	1	2
Young's modulus	E	kg/(m·s^2)	2E11	6.7E10	0.335	1	1
second moment of area	I	m^4	3.06796E-7	4.90874E-10	0.0016	1	1
density	ρ	kg/m^3	7850	2600	0.33121	1	1
dimensionless	π_1	1	1.61621E-3	1.61618E-3	0.99998	—	—
dimensionless	π_2	1	33.99207	33.99206	1	—	—
categories of variables		1	freely chosen, *a priori* given, or determined independently				
		2	determined by application of Model Law				
		3	determined by measurement on the model				

Figure 18-39
**Modeling Data Table for the experimental determination
of natural frequency of a cantilever**

By the first relation of (b) and the above obtained numerical values of scale factors, we have

$$S_f = \frac{1}{\sqrt[4]{S_I}} \cdot \sqrt{\frac{S_E}{S_\rho}} = \frac{1}{\sqrt[4]{0.0016}} \cdot \sqrt{\frac{0.335}{0.33121}} = 5.02853 \quad \text{(i)}$$

But $S_f = f_2/f_1$, hence the frequency of the *prototype* is

$$f_1 = \frac{f_2}{S_f} = \frac{f_2}{5.02853} = (0.19887) \cdot f_2 \quad \text{(j)}$$

Now we construct the model and measure its frequency. Say we find $f_2 = 1743$ Hz. By (j) then, the lateral frequency of the prototype is

$$f_1 = (0.19887) \cdot (1743) = 346.63 \text{ Hz}$$

The Modeling Data Table in Fig. 18-39 summarizes the above input and resultant values.

⇑

Example 18-32. Lateral Natural Frequency of a Cantilever (II)

The problem: The natural frequency of a steel constant cross-section cantilever (prototype) is to be determined on a model whose length is a third that of the prototype. The prototype's cross-section is an annulus of outside—inside diameter $j_1 = 0.15$ m and $k_1 = 0.14$ m, respectively. The model is aluminium and of constant solid circular cross-section. Relevant material properties are:

for the prototype: $E_1 = 2 \times 10^{11}$ Pa; $\rho_1 = 7950$ kg/m³
for the model: $E_2 = 7.32 \times 10^{10}$ Pa; $\rho_2 = 2900$ kg/m³

The solution: We first list the relevant variables.

Variable	Symbol	Dimension	Remark
frequency	f	1/s	lateral
length	L	m	
Young's modulus	E	kg/(m·s²)	
second moment of area	I	m⁴	uniform
density	ρ	kg/m³	uniform

We have five variables and three dimensions, therefore there are $5 - 3 = 2$ dimensionless variables supplied by the Dimensional Set

	f	L	E	I	ρ
m	0	1	−1	4	−3
kg	0	0	1	0	1
s	−1	0	−2	0	0
π_1	1	0	$-\frac{1}{2}$	$\frac{1}{4}$	$\frac{1}{2}$
π_2	0	1	0	$-\frac{1}{4}$	0

from which

$$\pi_1 = f \cdot \sqrt[4]{\frac{I \cdot \rho^2}{E^2}} \ ; \qquad \pi_2 = \frac{L}{\sqrt[4]{I}} \qquad (a)$$

Hence the Model Law is

$$S_f = \frac{1}{\sqrt[4]{S_I}} \cdot \sqrt{\frac{S_E}{S_\rho}} \ ; \qquad S_L = \sqrt[4]{S_I} \qquad (b)$$

where S_f is the Frequency Scales Factor;
S_I is the Second Moment of Area Scale Factor;
S_E is the Young's Modulus Scale Factor;
S_ρ is the Density Scale Factor;
S_L is the Length Scale Factor.

Based on the given data

$$S_E = \frac{E_2}{E_1} = \frac{7.3 \times 10^{10}}{2 \times 10^{11}} = 0.365 \ ; \quad S_\rho = \frac{\rho_2}{\rho_1} = \frac{2900}{7950} = 0.36478 \ ; \quad S_L = \frac{L_2}{L_1} = \frac{1}{3} \quad (c)$$

If the outside diameters of the prototype and model are denoted by j_1 and j_2, respectively, and, similarly, the inside diameters are k_1 and k_2, then we can write

$$S_I = \frac{I_2}{I_1} = \frac{(j_2^4 - k_2^4)/64}{(j_1^4 - k_1^4)/64} = \frac{j_2^4 - k_2^4}{j_1^4 - k_1^4}$$

from which the outside diameter of the model j_2 is

$$j_2 = \sqrt[4]{k_2^4 + S_I(j_1^4 - k_1^4)} \qquad (d)$$

But, by our condition $k_2 = 0$ (cross-section of the model is a solid circle), and by the second relation of Model Law (b)

$$S_I = S_L^4 \qquad (e)$$

Hence (d) can be written

$$j_2 = S_L \cdot \sqrt[4]{j_1^4 - k_1^4} \qquad (f)$$

and therefore from the given data

$$j_2 = \frac{1}{3} \cdot \sqrt[4]{0.15^4 - 0.14^4} = 0.035039 \text{ m}$$

Thus, if the model's cross-section is a solid circle, then it *must* have a diameter of 35.039 mm. By (e) now

$$S_I = S_L^4 = \left(\frac{1}{3}\right)^4 = \frac{1}{81}$$

and therefore, by the first relation of Model Law (b)

$$S_f = \frac{1}{\sqrt[4]{S_I}} \cdot \sqrt{\frac{S_E}{S_\rho}} = \frac{1}{\sqrt[4]{1/81}} \cdot \sqrt{\frac{0.365}{0.36478}} = 3.000905 \qquad (g)$$

Variable					Scale factor S	Category	
name	symbol	dimension	prototype	model	model/prototype	prototype	model
frequency	f	1/s	148.6	445.934	3.0009	2	3
length	L	m	1 (dummy)	1/3 (dummy)	1/3	1	1
Young's modulus	E	kg/(m·s²)	2E11	7.3E10	0.365	1	1
second moment of area	I	m⁴	5.99308E-6	7.39906E-8	1/81	1	2
density	ρ	kg/m³	7950	2900	0.36478	1	1
dimensionless	π_1	1	1.46588E-3	1.46589E-3	1.00001	—	—
dimensionless	π_2	1	20.21099	20.21085	0.99999	—	—
categories of variables	1	freely chosen, *a priori* given, or determined independently					
	2	determined by application of Model Law					
	3	determined by measurement on the model					

Figure 18-40
Modeling Data Table for the determination of the natural frequency of a prototype cantilever by measurement on a geometrically dissimilar model

Now we *build* the model and *measure* its frequency. Let us say we find f_2 = 445.934 Hz. Since $S_f = f_2/f_1$, therefore, by (g), the frequency of the *prototype* is

$$f_1 = \frac{f_2}{S_f} = \frac{445.934}{3.000905} = 148.6 \text{ Hz} \tag{h}$$

Note that although in this case the model is *dimensionally* completely *similar* to the prototype, it is *geometrically dissimilar* to it. For

$$\frac{L_2}{L_1} = \frac{1}{3} = 0.33333 = S_L > S_j = \frac{0.35039}{0.15} = 0.23359 \tag{i}$$

Thus the model is more slender than the prototype, and of course their cross-sections are very much different (the prototype's is and annulus, the model's is a solid circle).

All above input and results are conveniently compiled in the modeling data table (Fig. 18-40). Note that the lengths of the prototype and model are not given numerically, since only their *ratio* (i.e., Length Scale Factor) is necessary. Also note the close agreement of the two dimensionless variables in (a) between the prototype and model.

Example 18-33. Velocity of Sound in a Liquid

The velocity of sound is equivalent to the velocity of pressure propagation—or pressure wave. The following variables are relevant:

Variable	Symbol	Dimension
velocity of sound	v	m/s
bulk modulus of liquid	β	kg/(m·s²)
density of liquid	ρ	kg/m³

Here the bulk modulus of liquid β is defined as

$$\beta = \frac{\text{force per unit area (pressure)}}{\text{change of volume per volume}} = \frac{p}{\Delta V/V}$$

where p is the pressure, ΔV is the volume change this pressure causes, and V is the original volume. Therefore the dimension of bulk modulus is the same as that of pressure.

We have three variables and three dimensions. The dimensional matrix is

	v	β	ρ
m	1	−1	−3
kg	0	1	1
s	−1	−2	0

The rank of this matrix is 2, whereas the number of dimensions is 3. But the latter should not exceed the former. Therefore one dimension must go—say "s." The Dimensional Set is then

	v	β	ρ
m	1	−1	−3
kg	0	1	1
π_1	1	$-\frac{1}{2}$	$\frac{1}{2}$

from which

$$\pi_1 = v \cdot \sqrt{\frac{\beta}{\rho}} = \text{const} \qquad (a)$$

Thus,

$$v = \text{const} \cdot \sqrt{\frac{\beta}{\rho}} \qquad (b)$$

where the constant must be determined by either a single measurement or by analysis. In fact, the value of this constant is 1. For water at 15 °C we have $\beta = 2.09 \times 10^9$ kg/(m.s²) and $\rho = 999$ kg/m³, therefore (b) provides

$$v = \sqrt{\frac{2.09 \times 10^9}{999}} = 1446 \text{ m/s}$$

⇑

Example 18-34. Diameter of a Soap Bubble

The diameter of a soap bubble is obviously a function of the inside–outside pressure difference and the surface tension. The variables therefore are

Variable	Symbol	Dimension
diameter	D	m
pressure difference	Δp	kg/(m·s^2)
surface tension	σ	kg/s^2

As we see there are three variables and three dimensions, therefore the rank of the dimensional matrix must be less than 3, or there is no solution (Art. 7.6). The dimensional matrix is

	D	Δp	σ
m	1	−1	0
kg	0	1	1
s	0	−2	−2

whose rank is 2. Therefore the matrix is singular, thus we do have a solution. To find the solution we eliminate one dimension (see Fig. 7-4)—say "s." The resultant Dimensional Set is

	D	Δp	σ
m	1	−1	0
kg	0	1	1
π_1	1	1	−1

from which

$$\pi_1 = \frac{D \cdot \Delta p}{\sigma} = \text{const} \qquad (a)$$

or

$$D = \text{const} \cdot \frac{\sigma}{\Delta p} \qquad (b)$$

i.e., the diameter is proportional to the surface tension and inversely proportional to the inside–outside pressure difference. The latter conclusion is rather counterintuitive. For it implies that the lower the inside pressure in a soap bubble, the larger its diameter. When a boy blows a soap bubble, he actually decreases the pressure inside the bubble, since he increases its diameter. Or, to put it in even more astonishing way, the *more* air is blown into a soap bubble, the *smaller* its inside pressure will be!

From the condition of equilibrium of forces generated by surface tension and pressure difference, the constant in (b) can be determined to be 8. Also, for soap solution $\sigma = 0.026$ kg/s^2 (Ref. 4, p. 209). If the diameter is given, then by (b) and the above details we can determine the internal pressure. Accordingly,

$$\Delta p = 8 \cdot \frac{\sigma}{D} = 8 \cdot \frac{0.026}{D} = \frac{0.208}{D} \qquad (c)$$

To illustrate the extremely small pressure this formula represents, assume a $D = 0.05$ m (50 mm) bubble. Then, by (c),

$$\Delta p = \frac{0.208}{0.05} = 4.16 \text{ Pa } (= 0.000603 \text{ psi})$$

which is just 1/24357 of the atmospheric value.

Example 18-35. Velocity of Collapse of a Row of Dominoes
(adapted from Ref. 22, p. 208)

The problem: We wish to find the velocity of collapse (fall by leaning) of a row of uniformly placed dominoes. Fig. 18-41 shows the arrangement.

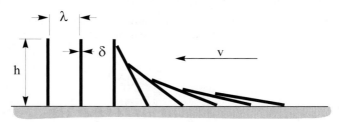

Figure 18-41
A collapsing row of uniformly placed dominoes

The solution: We first list the relevant variables

Variable	Symbol	Dimension
velocity of collapse	v	m/s
separation	λ	m
thickness	δ	m
height	h	m
gravitational acceleration	g	m/s^2

We have five variables and two dimensions, therefore there are $5 - 2 = 3$ dimensionless variables, supplied by the Dimensional Set

	v	λ	δ	h	g
m	1	1	1	1	1
s	−1	0	0	0	−2
π_1	1	0	0	$-\frac{1}{2}$	$-\frac{1}{2}$
π_2	0	1	0	−1	0
π_3	0	0	1	−1	0

from which

$$\pi_1 = \frac{v}{\sqrt{h \cdot g}} \ ; \qquad \pi_2 = \frac{\lambda}{h} \ ; \qquad \pi_3 = \frac{\delta}{h} \qquad \text{(a)}$$

Therefore we can write

$$\pi_1 = \Psi\{\pi_2, \pi_3\} \qquad \text{(b)}$$

where Ψ is a function. By (a) and (b) now

$$v = \sqrt{h \cdot g} \cdot \Psi\left\{\frac{\lambda}{h}, \frac{\delta}{h}\right\} \qquad \text{(c)}$$

If $\delta \ll h$, i.e., the dominoes are *thin*, then π_3 can be neglected, and (c) reduces to

$$v = \sqrt{h \cdot g} \cdot \Psi\left\{\frac{\lambda}{h}\right\} \qquad (d)$$

which states that *if λ/h is constant*, then the velocity of collapse varies as the square root of height h, or (which is the same thing since λ/h is constant) as the square root of separation λ. Of course, if δ is not negligible, then for the same conclusion it is necessary that the ratio δ/h also be constant.

It is remarkable that v varies as the square root of g, i.e., the same dominoes (set up identically) would collapse on the Moon at speed of only 40.3% of that on Earth. Also note that (d) does not allow us to draw any conclusion as to how speed varies with separation if the *same* dominoes are used. For, in this case, the ratio λ/h is no longer constant.

⇑

Example 18-36. Generated Pressure by an Underwater Explosion
(adapted from Ref. 22, p. 174)

We investigate the pressure caused by an underwater explosion at a given distance from the point of detonation. The following variables are considered:

Variable	Symbol	Dimension	Remark
pressure at distance u	p	kg/(m·s²)	
distance	u	m	from point of detonation
bulk modulus of water	β	kg/(m·s²)	Note 1
density of water	ρ	kg/m³	
mass of explosive	M	kg	Note 2

Note 1: Bulk modulus β of a liquid is defined by the relation

$$\Delta p = \beta \cdot \frac{\Delta V}{V} \qquad (a)$$

where ΔV is the volume change of original volume V due to pressure change Δp. Thus, the dimension of β is the same as that of pressure. For water $\beta \approx 2.07 \times 10^9$ Pa ($\approx 300{,}000$ psi), i.e., water is about 100 times more compressible than steel. (See also Example 18-33.)

Note 2: It is assumed that the energy liberated during explosion is proportional to the mass of the charge.

By the above table, we have five variables and three dimensions, therefore there are $5 - 3 = 2$ dimensionless variables supplied by the Dimensional Set

	Δp	u	β	ρ	M
m	−1	1	−1	−3	0
kg	1	0	1	1	1
s	−2	0	−2	0	0
π_1	1	0	−1	0	0
π_2	0	1	0	$\dfrac{1}{3}$	$-\dfrac{1}{3}$

600   APPLIED DIMENSIONAL ANALYSIS AND MODELING

from which

$$\pi_1 = \frac{\Delta p}{\beta}; \quad \pi_2 = u \cdot \sqrt[3]{\frac{\rho}{M}} \tag{b}$$

The Model Law therefore is

$$S_{\Delta p} = S_\beta; \quad S_u = \sqrt[3]{\frac{S_M}{S_\rho}} \tag{c}$$

where $S_{\Delta p}$ is the Pressure Increase Scale Factor;
S_β is the Bulk Modulus Scale Factor;
S_M is the Mass of Explosive Scale Factor;
S_ρ is the Density of Water Scale Factor;
S_u is the Distance Scale Factor.

To demonstrate the use of this Model Law, we conduct a model experiment to determine the pressure increase caused by the detonation of *prototype* charge $M_1 = 15$ kg at a point $u_1 = 320$ m from the explosion. The *model* consists of a $M_2 = 0.2$ kg explosive, and hence

$$S_M = \frac{M_2}{M_1} = \frac{0.2}{15} = \frac{1}{75} \tag{d}$$

Since both prototype and model explosions take place in the same medium (water), therefore

$$S_\rho = S_\beta = 1 \tag{e}$$

From these data the second relation of Model Law (c) provides

$$S_u = \sqrt[3]{\frac{S_M}{S_\rho}} = \sqrt[3]{\frac{1/75}{1}} = \frac{1}{\sqrt[3]{75}} = 0.23713 \tag{f}$$

Since $S_u = u_2/u_1$, therefore by (f) the model's *point of measurement* must be

$$u_2 = S_u \cdot u_1 = (0.23713) \cdot (320) = 75.882 \text{ m}$$

distance away from the detonation.

By (e) and the first relation of Model Law (c), we have $S_{\Delta p} = 1$. Hence $\Delta p_1 = \Delta p_2$. Say we measure on the *model* at $u_2 = 75.882$ m distance a $\Delta p_2 = 31030$ Pa (4.5 psi) pressure increase. By Model Law then, the same pressure increase Δp_1 will exist on the *prototype* at $u_1 = 320$ m distance away resulting from the explosion of a $M_1 = 15$ kg charge.

The Modeling Data Table (Fig. 18-42) summarizes the above details. Notice the close agreement between the values of the two dimensionless variables.

An important limitation of the expounded method in this example is now pointed out. Relations (b) and (c) do *not* provide us with any knowledge regarding the dependence of pressure increase Δp upon charge at distance u. For the deduced formulas merely make possible the *comparison* of the behaviors of two dimensionally similar systems. That is, *if* we know how one system functions, we may gain information about how the other system will function. The dimensional method offers nothing further!

Variable					Scale factor S	Category	
name	symbol	dimension	prototype	model	model/prototype	prototype	model
pressure increase	Δp	kg/(m·s²)	31030	31030	1	2	3
distance from explosion	u	m	320	75.882	0.23713	1	2
bulk modulus of water	β	kg/(m·s²)	2.07E9	2.07E9	1	1	1
density of water	ρ	kg/m³	1000	1000	1	1	1
mass of explosive	M	kg	15	0.2	1/75	1	1
dimensionless	π_1	1	1.49903E-5	1.49903E-5	1	—	—
dimensionless	π_2	1	1297.5364	1297.5297	0.99999	—	—
categories of variables	1	freely chosen, *a priori* given, or determined independently					
	2	determined by application of Model Law					
	3	determined by measurement on the model					

Figure 18-42
Modeling Data Table for the experimental determination of the pressure increase at a given distance from an underwater explosion

Example 18-37. Operational Characteristics of an Aircraft

A number of important operational characteristics of an aircraft can be easily derived by purely dimensional considerations. First, some basic assumptions must be made.

1. All aircraft are geometrically similar.
2. All aircraft have the same density ρ_p. Here ρ_p is defined by ρ_p = mass of aircraft/volume of aircraft.
3. Gravitational acceleration is constant.
4. The density of air ρ_a is constant.
5. All planes fly at subsonic speeds, thus both drag and lift are proportional to the square of airspeed u and the square of linear size L of the aircraft.
6. All aircraft have the same drag and lift coefficients.
7. All speed values are with respect to air.

Stall speed. Stall speed v_s is the minimum speed at which an aircraft must fly in order to stay aloft. To determine this value we consider the following variables:

602 APPLIED DIMENSIONAL ANALYSIS AND MODELING

Variable	Symbol	Dimension
stall speed	v_s	m/s
density of aircraft	ρ_p	kg/m³
characteristic length of aircraft	L	m
gravitational acceleration	g	m/s²
density of air	ρ_a	kg/m³

Note that neither the mass nor the weight of the aircraft is included. This is because mass is defined by L and ρ_p (both are included), and weight is defined by L, ρ_p, and g (all three are included).

We have five variables and three dimensions, therefore there are $5 - 3 = 2$ dimensionless variables supplied by the Dimensional Set

	v_s	ρ_p	L	g	ρ_a
m	1	−3	1	1	−3
kg	0	1	0	0	1
s	−1	0	0	−2	0
π_1	1	0	$-\frac{1}{2}$	$-\frac{1}{2}$	0
π_2	0	1	0	0	−1

yielding

$$\pi_1 = \frac{v_s}{\sqrt{L \cdot g}}; \qquad \pi_2 = \frac{\rho_p}{\rho_a} \tag{a}$$

The Model Law is then

$$S_{v_s} = \sqrt{S_L \cdot S_g}; \qquad S_{\rho_p} = S_{\rho_a} \tag{b}$$

where S_{v_s} is the Stall Speed Scale Factor;
S_L is the Characteristic Length Scale Factor;
S_g is the Gravitational Acceleration Scale Factor;
S_{ρ_p} is the Aircraft Density Scale Factor;
S_{ρ_a} is the Air Density Scale Factor.

Now if $S_{\rho_p} = S_{\rho_a} = S_g = 1$, then by assumptions (2), (4), and (3), Model Law (b) reduces to

$$S_{v_s} = \sqrt{S_L} \tag{c}$$

i.e., the larger the aircraft, the faster it has to fly to remain aloft. By (c), if one aeroplane is twice as large as another, i.e., $S_L = 2$, then the larger must fly $\sqrt{2} = 1.41$ times as fast as the smaller.

Since there are two dimensionless variables, we can write $\pi_1 = \Psi\{\pi_2\}$ or, by (a),

$$v_s = \sqrt{L \cdot g} \cdot \Psi\left\{\frac{\rho_p}{\rho_a}\right\} \tag{d}$$

where Ψ is some function.

Power. The power needed to fly (i.e., energy expended in unit time) is obviously a function of speed, air density, and size of aircraft. Accordingly, we have the list:

Variable	Symbol	Dimension
power needed to fly	P	$m^2 \cdot kg/s^3$
air density	ρ_a	kg/m^3
air speed	v	m/s
characteristic size of aircraft	L	m

The number of variables is four, the number if dimensions is three, therefore there is only one dimensionless variable, a constant (Theorem 7-4, Art. 7.10). The Dimensional Set is

	P	ρ_a	v	L
m	2	-3	1	1
kg	1	1	0	0
s	-3	0	-1	0
π_1	1	-1	-3	-2

from which

$$\pi_1 = \frac{P}{\rho_a \cdot v^3 \cdot L^2} = k_1 \tag{e}$$

where k_1 is a numeric constant. Hence

$$P = k_1 \cdot \rho_a \cdot v^3 \cdot L^2 \tag{f}$$

Now we define the ratio

$$q \equiv \frac{v}{v_s} \tag{g}$$

where v_s is the stall speed as in (d). By (e) and (g)

$$S_P = S_{\rho_a} \cdot S_v^3 \cdot S_L^2 \tag{h}$$

$$S_v = S_q \cdot S_{v_s} \tag{i}$$

where S_P is the Power Scale Factor
S_v is the Speed Scale Factor;
S_q is the Speed Ratio Scale Factor.

and the rest of the scale factors are as defined in (b). Substituting S_v from (i) into (h), we obtain

$$S_P = S_{\rho_a} \cdot S_q^3 \cdot S_{v_s}^3 \cdot S_L^2 \tag{j}$$

and by (b)

$$S_P = S_{\rho_a} \cdot S_q^3 \cdot (S_L \cdot S_g)^{1.5} \cdot S_L^2 = S_{\rho_a} \cdot S_q^3 \cdot S_L^{3.5} \cdot S_g^{1.5} \tag{k}$$

Since $S_{\rho_a} = S_g = 1$, (k) simplifies to

$$S_P = S_q^3 \cdot S_L^{3.5} \tag{ℓ}$$

604 APPLIED DIMENSIONAL ANALYSIS AND MODELING

This relation is most instructive: the power *needed* to fly an aeroplane is proportional to the cube of the speed ratio and to the 3.5 power of the size of the craft. For example, if a "big" plane is twice as large as a "small" one, and the big one flies 2.5 times its stall speed, while the small one flies at only 1.5 times faster than its stall speed, then we have $S_L = 2$, $S_q = 2.5/1.5 = \frac{5}{3}$, and by ($\ell$)

$$S_P = (\tfrac{5}{3})^3 \cdot 2^{3.5} = 52.378$$

i.e., the larger plane consumes 52.378 times more energy (i.e., fuel) in unit time than the smaller plane does. Relation (ℓ) is also interesting in that few would intuitively predict that, with Speed Ratio Scale Factor kept constant, the required power is proportional to the 3.5 power of the linear size of the craft.

Energy to fly unit distance. We consider the following variables:

Variable	Symbol	Dimension
energy to fly unit distance	E	m·kg/s²
air density	ρ_a	kg/m³
speed	v	m/s
characteristic length	L	m

We have four variables and three dimensions, therefore there is only one dimensionless variable supplied by the Dimensional Set

	E	ρ_a	v	L
m	1	−3	1	1
kg	1	1	0	0
s	−2	0	−1	0
π_1	1	−1	−2	−2

from which

$$\pi_1 = \frac{E}{\rho_a \cdot v^2 \cdot L^2} = k_2 \tag{m}$$

where k_2 is a numeric constant. The Model Law is then

$$S_E = S_{\rho_a} \cdot S_v^2 \cdot S_L^2 \tag{n}$$

where S_E, S_{ρ_a}, S_v, and S_L are the respective scale factors identified by the variables in the subscripts. By (i) and the first part of (b), relation (n) can be written

$$S_E = S_q^2 \cdot S_L^3 \tag{o}$$

since by assumptions (4) and (3), $S_{\rho_a} = S_g = 1$. This relation tells us that the energy (fuel) necessary to fly a *unit* distance varies with the cube of the plane's linear size, *provided* the speed relative to the individual stall speed remains the same, i.e., $S_q = 1$. If, however, planes fly with different relative speeds with respect to their individual stalling values, then $S_q \neq 1$. In either case, relation (o) supplies the correct *Consumed Energy Scale Factor*.

Cost of transporting a unit of goods over a unit distance. We assume that the cost proportional to the fuel consumed, which, in turn, is proportional to energy used. Moreover, we assume that the quantity of goods is proportional to the *volume* of the aeroplane, which is then proportional to the *cube* of its linear dimension. Therefore cost c of transporting 1 unit of goods over a unit distance is

$$c = k_3 \cdot \frac{E}{L^3} \tag{p}$$

where E is the energy necessary to fly unit distance, and k_3 is some numeric constant. The dimension of c can be obtained from (p)

$$[c] = \frac{[E]}{[L^3]} = \frac{(\text{m} \cdot \text{kg})/\text{s}^2}{\text{m}^3} = \frac{\text{kg}}{\text{s}^2 \cdot \text{m}^2}$$

We can now list the relevant variables.

Variable	Symbol	Dimension
transport cost of unit goods over unit distance	c	kg/(s²·m²)
density of air	ρ_a	kg/m³
speed of plane	v	m/s
characteristic length of plane	L	m

We have four variables and three dimensions, therefore there is only one dimensionless variable, a constant. The Dimensional Set is

	c	ρ_a	v	L
m	−2	−3	1	1
kg	1	1	0	0
s	−2	0	−1	0
π_1	1	−1	−2	1

from which

$$\pi_1 = \frac{c \cdot L}{\rho_a \cdot v^2} = k_4 \tag{q}$$

where k_4 is a numeric constant.
The Model Law is $S_c \cdot S_L = S_{\rho_a} \cdot S_v^2$, or, since $S_{\rho_a} = 1$ [assumption (4)]

$$S_c = \frac{S_v^2}{S_L} \tag{r}$$

where S_c is the Specific Cost of Energy Scale Factor;
S_v is the Velocity Scale Factor;
S_L is the Characteristic Linear Size Scale Factor.

By (i) and (c) we can write

$$S_c = \frac{(S_q \cdot S_{v_S})^2}{S_L} = \frac{(S_q \cdot \sqrt{S_L})^2}{S_L} = S_q^2 \tag{s}$$

Relations (r) and (s) offer some interesting inferences:

- For a given *size* of aircraft, the cost of transporting a unit of goods increases with the square of the speed.

- For a given speed of the aircraft, cost is inversely proportional to the size of the plane. This is the main reason why planes are getting larger and larger. For if flying speeds are kept constant, the cost of transporting one unit of goods can be reduced merely by increasing the size of the aircraft.
- The least expensive (and most dangerous!) form of transportation (minimum c) is achieved if the plane—irrespective of size—flies at its stall (i.e., minimum) speed.

Energy needed to fly for unit time. Since energy per unit time is power, therefore this quantity was already discussed above. The *Power Scale Factor* is given in relation (ℓ), which is repeated here for convenience

$$S_P = S_q^3 \cdot S_L^{3.5} \qquad \qquad \text{repeated } (\ell)$$

where S_q is the Speed Ratio Scale Factor;
S_L is the Characteristic Linear Size Scale Factor.

Of the assumptions listed at the beginning of this example, maybe (4)—the assumption of constancy of air density—is the most questionable. As planes fly at higher and higher altitudes, the air density decreases and so is the energy and power required for a given speed. Indeed, this is the principal reason for high altitude flying. In principle, there is no difficulty to include air density in the preceding discussions. This subject, however, shall not be pursued further in here.

⇑

Example 18-38. Volume of Fluid Flowing in a Horizontal Pipe

We examine the volume of an incompressible fluid flowing through a horizontal pipe, on a given pressure difference occurring in a unit length of the pipe. This problem was already dealt with in Example 16-2 and Example 16-14; but now we treat this subject differently.

The following variables are relevant:

Variable	Symbol	Dimension
fluid volume flow rate	Q	m³/s
fluid dynamic viscosity	μ	kg/(m·s)
fluid density	ρ	kg/m³
pipe inner diameter	D	m
pressure difference in unit length of pipe	Δp	kg/(m²·s²)

We have five variables and three dimensions, hence there are $5 - 3 = 2$ dimensionless variables supplied by the Dimensional Set

	μ	ρ	Q	D	Δp
m	-1	-3	3	1	-2
kg	1	1	0	0	1
s	-1	0	-1	0	-2
π_1	1	0	1	-4	-1
π_2	0	1	2	-5	-1

Note that contrary to our usual practice of putting the dependent variable (in this case Q) in the **B** matrix, we now place this variable in the **A** matrix. Why? Because this way we can put both μ and ρ in the **B** matrix, and thus the effects of these two variables on Q can be easily separated. This feature will be shown to be very beneficial.

By the Dimensional Set, the two dimensionless variables are

$$\pi_1 = \frac{\mu \cdot Q}{D^4 \cdot \Delta p}; \qquad \pi_2 = \frac{\rho \cdot Q^2}{D^5 \cdot \Delta p} \qquad \text{(a)}$$

Now we consider *laminar* flow, which is associated with *slow* speed, so that viscosity μ dominates and inertial effects, represented by density ρ, may be ignored. Accordingly, in (a) only the first relation is considered, and hence π_1 must be a constant. Thus,

$$\pi_1 = \frac{\mu \cdot Q}{D^4 \cdot \Delta p} = k_1 \qquad \text{(b)}$$

where k_1 is a constant. By (b)

$$Q = k_1 \cdot \frac{D^4 \cdot \Delta p}{\mu} \qquad \text{(c)}$$

This is the famous *Poiseuille* equation with $k_1 = \pi/128$.

Next, we consider *turbulent* flow. In this case inertial forces represented by the density of fluid ρ dominate. Thus viscosity may be ignored—i.e., the fluid is assumed *inviscid*—and in (a) we should only consider π_2. Accordingly,

$$\pi_2 = \frac{\rho \cdot Q^2}{D^5 \cdot \Delta p} = k_2 \qquad \text{(d)}$$

from which

$$Q = k_2 \cdot \sqrt{\frac{D^5 \cdot \Delta p}{\rho}} \qquad \text{(e)}$$

where k_2 is a "constant" depending on Reynolds number $\mathrm{Re} = (\rho \cdot D \cdot v)/\mu$, where v is the average velocity of the fluid. If $\mathrm{Re} < {\sim}2300$, then the flow is *laminar* and (c) applies; otherwise the flow is *turbulent* and (e) is applicable.

Although in this brief example we do not discuss how k_2 depends on the Reynolds number, from (c) and (e) we can draw the following general conclusions:

- If the flow is laminar, the volume flow rate increases with the fourth power of pipe diameter;
- In laminar flow the volume flow rate is *proportional* with the pressure difference between the two ends of the pipe; in a turbulent flow it varies only as the *square root* of this difference.

⇑

Example 18-39. Vertical Penetration of a Vehicle's Wheels into Soft Soil

The vertical penetration of a vehicle wheel treading on soft soil—like sand—is an important problem in military, agricultural, and recreational applications. The penetration is obviously a function of the vertical pressure on the soil imparted by the wheel, the weight, and characteristic length of the vehicle. Accordingly, we have the following variables to consider:

Variable	Symbol	Dimension
pressure on soil	p	kg/(m·s^2)
vertical penetration of wheel	z	m
mass of vehicle	M	kg
gravitational acceleration	g	m/s^2
vehicle's characteristic length	L	m

There are five variables and three dimensions, therefore we have 5 − 3 = 2 dimensionless variables supplied by the Dimensional Set

	p	z	M	g	L
m	−1	1	0	1	1
kg	1	0	1	0	0
s	−2	0	0	−2	0
π_1	1	0	−1	−1	2
π_2	0	1	0	0	−1

Thus

$$\pi_1 = \frac{p \cdot L^2}{M \cdot g} \; ; \qquad \pi_2 = \frac{z}{L} \tag{a}$$

and the Model Law is

$$S_p \cdot S_L^2 = S_M \cdot S_g \; ; \qquad S_z = S_L \tag{b}$$

where S_p is the Pressure on Soil Scale Factor;
S_L is the Vehicle Size Scale Factor;
S_M is the Vehicle Mass Scale Factor;
S_z is the Soil Penetration Scale Factor;
S_g is the Gravitational Acceleration Scale Factor.

At this point we cannot go further because we do not know the relationship between the pressure and the weight of the vehicle. However, an *experimentally* obtained relationship can now be considered. Schuring (Ref. 31, p. 245) quotes the following experimental formula:

$$p = k_0 \cdot z^n \tag{c}$$

where $n = 1.1$ and k_0 is a *dimensional* numeric constant whose dimension, by (c), is

$$[k_0] = \frac{[p]}{[z]^n} = \frac{\text{kg/(m·s}^2\text{)}}{\text{m}^{1.1}} = \frac{\text{kg}}{\text{m}^{2.1} \cdot \text{s}^2} \tag{d}$$

For the purpose of our inquiry, the *magnitude* of k_0 is unimportant. By (c) we now write

$$S_p = S_z^n \tag{e}$$

From this relation and (b)

$$S_p \cdot S_L^2 = S_p \cdot S_z^2 = S_z^n \cdot S_z^2 = S_z^{n+2} = S_M \cdot S_g \tag{f}$$

Thus

$$S_z = (S_M \cdot S_g)^{1/(n+2)} = (S_M \cdot S_g)^{1/3.1} = (S_M \cdot S_g)^{0.32258} \quad \text{(g)}$$

which is the answer to our question. This relation tells us that the vertical soil penetration caused by the *same* vehicle (i.e., $S_M = 1$) on the Moon ($g_1 = 1.6$ m/s^2) would be by

$$S_z = (S_M \cdot S_g)^{0.32258} = S_g^{0.32258} = \left(\frac{g_2}{g_1}\right)^{0.32258} = \left(\frac{9.81}{1.6}\right)^{0.32258} = 1.79493 = \frac{z_2}{z_1}$$

from which

$$z_1 = \left(\frac{1}{S_z}\right) \cdot z_2 = (0.55712) \cdot z_2$$

i.e., the penetration of the wheels on the Moon would be only 55.7% of that on Earth.

Note that on the same celestial body $S_g = 1$, therefore, by (g) and (b)

$$S_z = S_M^{0.32258} = S_L \quad \text{(h)}$$

and necessarily

$$S_M = S_L^3 \cdot S_{\rho V} \quad \text{(i)}$$

where $S_{\rho V}$ is the Effective Vehicle Density Scale Factor. Therefore, by (h) and (i),

$$S_M = S_L^{3.1} = S_L^3 \cdot S_{\rho V} \quad \text{(j)}$$

which yields

$$S_{\rho V} = S_L^{0.1} \quad \text{(k)}$$

i.e., ρ_V (i.e., the quotient of the mass and the volume of the vehicle) is not constant, but increases ever so slightly with the linear size. Thus, contrary to expectations, the larger the vehicle, the more "compact" it is.

⇑

Example 18-40. Deflection of a Simply Supported Beam upon a Centrally Dropped Mass

We design and execute a modeling experiment to determine the deflection of a simply supported $L_1 = 7.3$ m long prototype beam upon a mass $M_1 = 2950$ kg dropped on the beam's midpoint from a height of $h_1 = 3.15$ m on the Moon. The arrangement is shown in Fig. 18-43.

The beam is steel, Young's modulus $E_1 = 2 \times 10^{11}$ Pa and its cross-sectional second moment of area is $I_1 = 1.998 \times 10^{-5}$ m^4 (regular American Standard I beam, light series, size $8 \times 4 \times 0.245$ inches).

Of course, we build our model on Earth, where we do all the experimentation. For the model we select aluminium $E_2 = 6.7 \times 10^{10}$ Pa and a rectangular cross-section of $a = 0.2$ m width and $b = 0.040$ m height. What should be the model's length, the mass dropped on it and the height the mass dropped from? Further, how will the measured deflection of the *model* (on Earth) correlate with the deflection of the *prototype* (on the Moon)?

To proceed in an orderly manner, we first list the relevant variables.

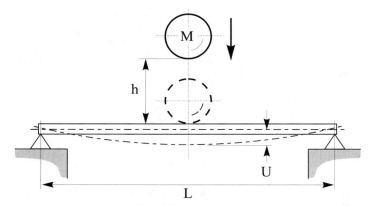

Figure 18-43
Deflection of a simply supported beam upon a mass dropped on its midpoint

Variable	Symbol	Dimension
deflection	U	m
second moment of area	I	m^4
height mass dropped from	h	m
gravitational acceleration	g	m/s^2
length of beam	L	m
mass dropped	M	kg
Young's modulus of beam	E	kg/(m·s^2)

We see that there are seven variables and three dimensions, therefore we have $7 - 3 = 4$ dimensionless variables furnished by the Dimensional Set

	U	I	h	g	L	M	E
m	1	4	1	1	1	0	−1
kg	0	0	0	0	0	1	1
s	0	0	0	−2	0	0	−2
π_1	1	0	0	0	−1	0	0
π_2	0	1	0	0	−4	0	0
π_3	0	0	1	0	−1	0	0
π_4	0	0	0	1	−2	1	−1

Note that U, h, and L have identical dimensions, therefore at the most only *one* of them may appear in the **A** matrix (Theorem 14-3 in Art. 14-2). From the above set

$$\pi_1 = \frac{U}{L} \; ; \qquad \pi_2 = \frac{I}{L^4} \; ; \qquad \pi_3 = \frac{h}{L} \; ; \qquad \pi_4 = \frac{g \cdot M}{L^2 \cdot E} \qquad \text{(a)}$$

and the Model Law therefore is

$$S_U = S_L \; ; \qquad S_I = S_L^4 \; ; \qquad S_h = S_L \; ; \qquad S_g \cdot S_M = S_L^3 \cdot S_E \qquad \text{(b)}$$

where S_U is the Deflection Scale Factor;
 S_L is the Length Scale Factor;
 S_I is the Second Moment of Area Scale Factor;
 S_h is the Height Mass Dropped from Scale Factor;
 S_g is the Gravitational Acceleration Scale Factor;
 S_E is the Young's Modulus Scale Factor;
 S_M is the Dropped Mass Scale Factor.

By the given data of the model's cross-section

$$I_2 = \frac{a \cdot b^3}{12} = \frac{(0.2) \cdot (0.04)^3}{12} = 1.06667 \times 10^{-6} \text{ m}^4 \quad \text{(c)}$$

hence

$$S_I = \frac{I_2}{I_1} = \frac{1.06667 \times 10^{-6}}{1.998 \times 10^{-5}} = 5.33867 \times 10^{-2} \quad \text{(d)}$$

Thus, by the second and third relations of Model Law (b)

$$S_L = \sqrt[4]{S_I} = \sqrt[4]{5.33867 \times 10^{-2}} = 0.48068 \quad \text{(e)}$$

But $S_L = L_2/L_1$, therefore the model's length *must* be

$$L_2 = L_1 \cdot S_L = (7.3) \cdot (0.48068) = 3.509 \text{ m}$$

Since gravitational acceleration on the Moon is $g_1 = 1.6$ m/s² and on Earth it is $g_2 = 9.81$ m/s²

$$S_g = \frac{g_2}{g_1} = \frac{9.81}{1.6} = 6.13125 \quad \text{(f)}$$

Moreover, by the given data

$$S_E = \frac{E_2}{E_1} = \frac{6.7 \times 10^{10}}{2 \times 10^{11}} = 0.335 \quad \text{(g)}$$

Next, the Mass Scale Factor is now supplied by (e), (f), (g), and the fourth relation of Model Law (b)

$$S_M = \frac{S_L^2 \cdot S_E}{S_g} = \frac{(0.48068^2) \cdot (0.335)}{6.13125} = 1.26243 \times 10^{-2} \quad \text{(h)}$$

But $S_M = M_2/M_1$, hence from the given data and (h), the mass to be dropped on the model *must* be

$$M_2 = S_M \cdot M_1 = (1.26243 \times 10^{-2}) \cdot (2950) = 37.2417 \text{ kg} \quad \text{(i)}$$

Finally, by (e) and the third relation of Model Law (b)

$$S_h = S_L = 0.48068 \quad \text{(j)}$$

Since $S_h = h_2/h_1$, by (j), the height from which the mass dropped onto the model *must* be

$$h_2 = h_1 \cdot S_h = (3.15) \cdot (0.48068) = 1.5141 \text{ m} \quad \text{(k)}$$

Now we have all the information necessary to build the model and perform the experiment on Earth. Accordingly, we drop an $M_2 = 37.2417$ kg mass from height $h_2 = 1.5141$ m onto an $L_2 = 3.509$ m long simply supported aluminium beam whose cross-section is a rectangle of 0.2 m width and 0.04 m height. By dropping the mass on this model, we measure on it a deflection of, say, $U_2 = 0.1180$ m. Hence, by (e) and the first relation of Model Law (b)

$$S_U = \frac{U_2}{U_1} = S_L = 0.48068$$

from which the sought-after deflection of the prototype (on the Moon) is

$$U_1 = \frac{U_2}{S_U} = \frac{0.1180}{0.48068} = 0.2455 \text{ m}$$

All the above input data and results are conveniently summarized in the Modeling Data Table (Fig. 18-44). Note the numerical equivalence (within rounding errors) of the four dimensionless variables defined in (a). This of course is expected and required for a modeling experiment in which the model is *dimensionally similar* to the prototype.

Variable					Scale factor S	Category	
name	symbol	dimension	prototype	model	model/prototype	prototype	model
length of beam	L	m	7.3	3.509	0.48068	1	2
Young's modulus	E	kg/(m·s^2)	2E11	6.7E10	0.335	1	1
second moment of area	I	m^4	1.998E-5	1.06667E-6	5.33869E-2	1	1
gravitational acceleration	g	m/s^2	1.6	9.81	6.13125	1	1
dropped mass	M	kg	2950	37.2417	1.26243E-2	1	2
height mass dropped from	h	m	3.15	1.5141	0.48067	1	2
deflection of beam	U	m	0.2455	0.118	0.48065	2	3
dimensionless	π_1	1	3.36411E-2	3.36278E-2	0.9996	—	—
dimensionless	π_2	1	7.03565E-9	7.03553E-9	0.99998	—	—
dimensionless	π_3	1	4.31507E-1	4.3149E-1	0.99996	—	—
dimensionless	π_4	1	4.4286E-10	4.4285E-10	0.99998	—	—
categories of variables	1	freely chosen, *a priori* given, or determined independently					
	2	determined by application of Model Law					
	3	determined by measurement on the model					

Figure 18-44
Modeling Data Table for the determination, by a modeling experiment, of the deflection of a simply supported beam by a centrally dropped mass (Version 1)

Also note that there are four quantities in the table that are determined by the application of Model Law (b). These are designated as *Category 2* variables. The Model Law comprises four relations, hence four unknowns can be determined. Actually, the relations constituting the Model Law should be viewed as *constraints* which are imposed upon the model. Thus, once the cross-section and the material of the model are fixed *a priori*, we have no control over its other characteristics. It follows, therefore, that the length of the beam, the dropped mass, and the height from which this mass is dropped are prescribed by the conditions of similarity.

The question now naturally presents itself: is there a way to increase our freedom in selecting characteristics for the model? Yes, there is: we have to *reduce* the number of relations in the Model Law. Since the number of such relations equals the number of dimensionless variables, we *must* reduce the number of dimensionless variables.

Some clear-head thinking is now in order (actually, clear-head thinking is always in order, but this diversionary subject shall not be further pursued here!). We note that both I and E appear in the set of relevant variables. Now in problems of statics and dynamics if both I and E are present in a relation, then they most likely do so as the *product I·E*. This suggests that a new variable—which we call "*rigidity*" (another name is "*flexural stiffness*")

$$z \equiv I \cdot E \qquad (\ell)$$

can be introduced. The dimension of z is by its definition (ℓ)

$$[z] = [I \cdot E] = [I] \cdot [E] = \text{m}^4 \cdot \frac{\text{kg}}{\text{m} \cdot \text{s}^2} = \frac{\text{m}^3 \cdot \text{kg}}{\text{s}^2} \qquad (\text{m})$$

With "rigidity," the number of variables is *reduced* to six, while the number of dimensions *remains* three. Hence we now have three dimensionless variables, instead of the former four. Accordingly, the new list of variables is

Variable	Symbol	Dimension
deflection	U	m
rigidity	z	m^3·kg/s^2
height mass dropped from	h	m
gravitational acceleration	g	m/s^2
length of beam	L	m
mass dropped	M	kg

which yields the Dimensional Set

	U	z	h	g	L	M
m	1	3	1	1	1	0
kg	0	1	0	0	0	1
s	0	−2	0	−2	0	0
π_1	1	0	0	0	−1	0
π_2	0	1	0	−1	−2	−1
π_3	0	0	1	0	−1	0

from which

$$\pi_1 = \frac{U}{L} \; ; \qquad \pi_2 = \frac{z}{g \cdot L^2 \cdot M} \; ; \qquad \pi_3 = \frac{h}{L} \qquad (\text{n})$$

614 APPLIED DIMENSIONAL ANALYSIS AND MODELING

and the Model Law is

$$S_U = S_L \ ; \qquad S_z = S_g \cdot S_L^2 \cdot S_M \ ; \qquad S_h = S_L \qquad \text{(o)}$$

where S_z is the Rigidity Scale Factor, and S_L, S_g, S_M, S_h, and S_U are as defined previously for relation (b). By the given data,

$$z_1 = E_1 \cdot I_1 = (2 \times 10^{11}) \cdot (1.998 \times 10^{-5}) = 3.996 \times 10^6 \ (\text{m}^3 \cdot \text{kg})/\text{s}^2 \qquad \text{(p)}$$

$$z_2 = E_2 \cdot I_2 = (6.7 \times 10^{10}) \cdot (1.06667 \times 10^{-6}) = 7.14673 \times 10^4 \ (\text{m}^3 \cdot \text{kg})/\text{s}^2 \qquad \text{(q)}$$

hence

$$S_z = \frac{z_2}{z_1} = \frac{7.14673 \times 10^4}{3.996 \times 10^6} = 1.78847 \times 10^{-2} \qquad \text{(r)}$$

Now we *select* mass M_2 to be dropped on the model. Note that this could *not* have been done in the previous arrangement, since the value of mass M_2 was defined for us by Model Law (b). Since we are now at liberty to choose the mass, we select a relatively smaller value to produce a smaller deflection. We select

$$M_2 = 17 \text{ kg} \qquad \text{(s)}$$

thus

$$S_M = \frac{M_2}{M_1} = \frac{17}{2950} = 5.76271 \times 10^{-3} \qquad \text{(t)}$$

By the second relation of Model Law (o)

$$S_L = \sqrt{\frac{S_z}{S_g \cdot S_M}} = \sqrt{\frac{1.78847 \times 10^{-2}}{(6.13125) \cdot (5.76271 \times 10^{-3})}} = 7.11464 \times 10^{-1} \qquad \text{(u)}$$

where scale factors S_z, S_g, and S_M are given in (r), (f), and (t), respectively.

Next, we write $S_L = L_2/L_1$, by which the length of the model beam *must* be

$$L_2 = S_L \cdot L_1 = (7.11464 \times 10^{-1}) \cdot (7.3) = 5.19369 \text{ m} \qquad \text{(v)}$$

By the third relation of the Model Law (o), $S_h = h_2/h_1 = S_L$, from which, by (u)

$$S_U = \frac{U_2}{U_1} = S_L = 7.11464 \times 10^{-1} \qquad \text{(w)}$$

from which the sought-after deflection of the prototype is

$$U_1 = \frac{U_2}{S_U} = \frac{0.1747}{7.11464 \times 10^{-1}} = 0.2456 \text{ m} \qquad \text{(x)}$$

All of the above data are summarized in the Modeling Data Table presented in Fig. 18-45. Note that in this version *Category 2* designation appears only three times—instead of four, as in Version 1. Also observe that three is the number of relations in Model Law (o), and the mass for the model is now "selectable," i.e., *Category 1*. This is our reward for knowing *a priori* that I and E appear only as a product in the expression of U.

Finally, observe again the very close numerical agreement of the dimensionless variables between the prototype and the model.

Variable					Scale factor S	Category	
name	symbol	dimension	prototype	model	model/prototype	prototype	model
length of beam	L	m	7.3	5.19369	7.11464E-1	1	2
rigidity of beam	z	m³·kg/s²	3.996E6	7.14673E4	1.78847E-2	1	1
gravitational acceleration	g	m/s²	1.6	9.81	6.13125	1	1
dropped mass	M	kg	2950	17	5.76271E-3	1	1
height mass dropped from	h	m	3.15	2.24111	7.11463E-1	1	2
deflection of beam	U	m	0.24555	0.1747	7.11464E-1	2	3
dimensionless	π_1	1	3.3637E-2	3.3637E-2	1	—	—
dimensionless	π_2	1	15.88685	15.88684	0.99999	—	—
dimensionless	π_3	1	4.31507E-1	4.31506E-1	0.99999	—	—
categories of variables	1	freely chosen, *a priori* given, or determined independently					
	2	determined by application of Model Law					
	3	determined by measurement on the model					

Figure 18-45
Modeling Data Table for the determination, by a modeling experiment, of the deflection of a simply supported beam by a centrally dropped mass (Version 2)

⇑

Example 18-41. Minimum Deflection Cantilevers*

In designing arms for manipulators and robots, the design engineer usually encounters the task of attaining the minimum lateral deflection of a constant cross-section horizontal tubular cantilever (Fig. 18-46). The lateral deflection of such a beam is determined by its geometric and flexural characteristics as well as its loading configuration. If we let the wall thickness of the annular cross-section to vary, while keeping all the other parameters (including the cross-section's outside diameter) constant, then—contrary to intuition—for a beam carrying *only* its own weight, the *thinner* the wall, the smaller the deflection; i.e., with decreasing wall thickness the beam becomes stiffer!

On the other hand, if the beam carries *only* a concentrated lateral load acting at the free end (as in Fig. 18-46)—i.e., the beam's own weight is negligible—then the *thicker* the wall, the smaller the deflection; with decreasing wall thickness the beam becomes less stiff.

Now if both own weight and payload are present—as is the case in most applications—then the question arises: for any given outside diameter and length, at what partic-

---

*A greatly condensed version of this material appeared in Ref. 142, p. 103.

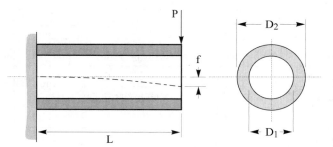

Figure 18-46
**Deflection of a heavy tubular cantilever loaded
at its free end by a concentrated lateral force**

ular wall thickness (or inner diameter) is the beam the stiffest i.e., is at its minimum deflection? To solve this problem, we first list the relevant variables:

Variable	Symbol	Dimension	Remark
lateral deflection	f	m	at free end
cross-section's inner diameter	D_1	m	constant
length	L	m	
specific weight	γ	N/m³	see Note
lateral force	P	N	at free end
cross-section's outer diameter	D_2	m	constant
Young's modulus	E	N/m²	

Note: specific weight is the product of density and gravitational acceleration.

By the above table, we have seven variables and two dimensions, hence there are 7 – 2 = 5 dimensionless variables. The Dimensional Set is

	f	D_1	L	γ	P	D_2	E
m	1	1	1	–3	0	1	–2
N	0	0	0	1	1	0	1
π_1	1	0	0	0	0	–1	0
π_2	0	1	0	0	0	–1	0
π_3	0	0	1	0	0	–1	0
π_4	0	0	0	1	0	1	–1
π_5	0	0	0	0	1	–2	–1

from which

$$\pi_1 = \frac{f}{D_2}; \quad \pi_2 = \frac{D_1}{D_2}; \quad \pi_3 = \frac{L}{D_2}; \quad \pi_4 = \frac{\gamma \cdot D_2}{E}; \quad \pi_5 = \frac{P}{D_2^2 \cdot E} \quad (a)$$

For a tubular cantilever loaded by its weight, the standard formula is (Ref. 133, vol. 1, p. 152) $f_q = (q \cdot L^4)/(8 \cdot E \cdot I)$, where q is the weight of *unit* length and I is the centroidal second moment of area of cross-section. For the present application this relation can be written

$$f_q = \frac{2 \cdot \gamma \cdot L^4}{E \cdot (D_1^2 + D_2^2)} \quad (b)$$

If the cantilever is loaded only by a lateral force P, then the deflection is (Ref. 133, vol. 1, p. 150) $f_P = (P \cdot L^3)/(3 \cdot I \cdot E)$, which in our case can be written

$$f_P = \frac{64}{3 \cdot \pi} \cdot \frac{P \cdot L^3}{(D_2^4 - D_1^4) \cdot E} \tag{c}$$

Since we consider small deformations only, the principle of superposition applies. Hence by (b) and (c)

$$f = f_q + f_P = \frac{2 \cdot L^3}{E \cdot (D_1^2 + D_2^2)} \cdot \left(\gamma \cdot L + \frac{32 \cdot P}{3 \cdot \pi \cdot (D_2^2 - D_1^2)} \right) \tag{d}$$

Using the dimensionless variables defined in (a), the above seven-variable expression can be reduced to a five-variable one. Accordingly, with the appropriate substitutions,

$$\pi_1 = \frac{2 \cdot \pi_3^4 \cdot \pi_4}{1 + \pi_2^2} + \frac{64}{3 \cdot \pi} \cdot \left(\frac{\pi_3^3 \cdot \pi_5}{1 - \pi_2^4} \right) \tag{e}$$

Note that π *without* subscript is *not* a dimensionless variable, but the numeric constant 3.14159.... To obtain the particular $\pi_2 = (\pi_2)_0$ at which π_1 (hence f) is *minimum*, we write the derivative of π_1 with respect to π_2 and equate the result to zero. Thus

$$\frac{d\pi_1}{d\pi_2} = \left(\frac{-2 \cdot \pi_3^4 \cdot \pi_4}{(1 + \pi_2^2)^2} \right) \cdot 2 \cdot \pi_2 + \left(\frac{64 \cdot \pi_3^3 \cdot \pi_5}{2 \cdot \pi \cdot (1 - \pi_2^4)^2} \right) \cdot 4 \cdot \pi_2^3 = 0$$

from which by simple, but careful, rearrangement we obtain

$$\pi_2 = (\pi_2)_0 = \sqrt{1 + \frac{32 \cdot \pi_5 \pm 8 \cdot \sqrt{16 \cdot \pi_5^2 + 3 \cdot \pi \cdot \pi_3 \cdot \pi_4 \cdot \pi_5}}{3 \cdot \pi \cdot \pi_3 \cdot \pi_4}} \tag{f}$$

Observe that π_2 must always be less than 1 (since the cross-section's inner diameter cannot reach the outer diameter), and that π_3, π_4, and π_5 must be all positive. Therefore in (f) it is the "−" sign that must be used in front of the number 8. We also notice that π_3 and π_4 appear only as the product $\pi_3 \cdot \pi_4$. Hence, merely for convenience, we *fuse* these two dimensionless variables into one

$$\pi_c \equiv \pi_3 \cdot \pi_4 = \left(\frac{L}{D_2} \right) \cdot \left(\frac{\gamma \cdot D_2}{E} \right) = \frac{L \cdot \gamma}{E} \tag{g}$$

where subscript "c" stands for "combined." By (g), relation (f) becomes

$$(\pi_2)_0 = \sqrt{1 + \frac{32 \cdot \pi_5 - 8 \cdot \sqrt{16 \cdot \pi_5^2 + 3 \cdot \pi \cdot \pi_c \cdot \pi_5}}{3 \cdot \pi \cdot \pi_c}} \tag{h}$$

This relation now only contains three variables, and hence it can be plotted in a single chart as presented in Fig. 18-47.

To demonstrate the use of the chart, a simple numerical example is presented. Consider a cantilever with the following characteristics:

Length:	$L = 1.016$ m
Concentrated lateral load:	$P = 822.921$ N
Specific weight of the material (aluminium):	$\gamma = 27144.71$ N/m^3
Young's modulus of material:	$E = 6.895 \times 10^{10}$ Pa
Outside diameter of cross-section:	$D_2 = 0.127$ m

618 APPLIED DIMENSIONAL ANALYSIS AND MODELING

$\pi_2 = \dfrac{D_1}{D_2}$

$\pi_3 = \dfrac{L}{D_2}$

$\pi_4 = \dfrac{\gamma \cdot D_2}{E}$

$\pi_5 = \dfrac{P}{D_2^2 \cdot E}$

$\pi_c = \pi_3 \cdot \pi_4$

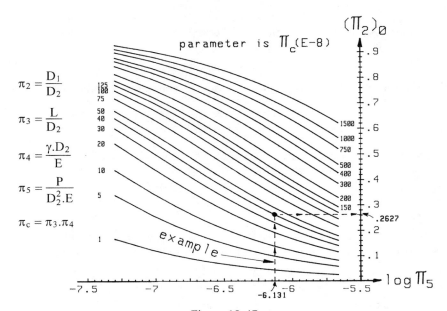

**Figure 18-47
Dimensionless plot to determine the inner diameter
yielding minimum deflection of a constant outer diameter
cross-section laterally loaded heavy cantilever**
Dotted lines represent the numeric example in text. The log is of base 10. For definition
of physical variables, see table in text and Fig. 18-46

By (a) and (g)

$$\pi_3 = \dfrac{L}{D_2} = 8;\ \pi_4 = \dfrac{\gamma \cdot D_2}{E} = 5 \times 10^{-8};\ \pi_5 = \dfrac{P}{D_2^2 \cdot E} = 7.4 \times 10^{-7};\ \pi_c = \pi_3 \cdot \pi_4 = 4 \times 10^{-7} \quad \text{(i)}$$

For these given π_5 and π_c the chart on Fig. 18-47 provides

$$(\pi_2)_0 = 0.263 \quad \text{(j)}$$

while the exact relation (h) yields 0.262671.... To calculate the minimum deflection, we determine $\pi_1 = (\pi_1)_{\min}$. We obtain this by substituting $(\pi_2)_0$ for π_2 in relation (e). Thus, by the values found in (i)

$$(\pi_1)_{\min} = \dfrac{2 \cdot (8^4) \cdot (5 \times 10^{-8})}{1 + 0.263^2} + \dfrac{(64) \cdot (8^3) \cdot (7.4 \times 10^{-7})}{3 \cdot \pi \cdot (1 - 0.263^4)} = 0.00297 \quad \text{(k)}$$

and the minimum deflection is by (a) and (k)

$$f_{\min} = D_2 \cdot (\pi_1)_{\min} = (0.127) \cdot (0.00297) = 0.000377 \text{ m} \quad (\ell)$$

The corresponding inner diameter D_1 of the cross-section is from $\pi_2 = D_1/D_2 = (\pi_2)_{\min}$, i.e.,

$$D_1 = (\pi_2)_{\min} \cdot D_2 = (0.263) \cdot (0.127) = 0.0334 \text{ m}$$

i.e., the wall thickness of the beam is $(D_2 - D_1)/2 = (0.127 - 0.0334)/2 = 0.0468$ m.

If we now plot the π_1 versus π_2 relation (e) for any trio of given π_3, π_4, and π_5 values, then we obtain a curve which has a minimum at $\pi_2 = (\pi_2)_0$. An instance of this curve is shown in Fig. 18-48 for $\pi_3 = 8$, $\pi_4 = 5 \times 10^{-8}$, and $\pi_5 = 7.4 \times 10^{-7}$, as given in relation (i).

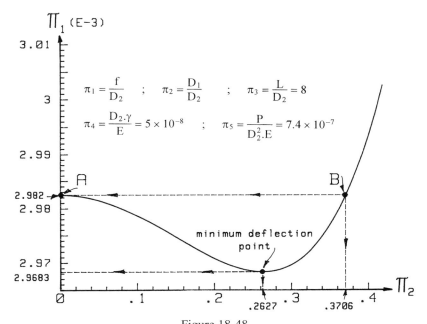

Figure 18-48
Dimensionless plot to determine the deflection of a laterally loaded heavy tubular cantilever of given $\pi_3 = 8$, $\pi_4 = 5 \times 10^{-8}$ and $\pi_5 = 7.4 \times 10^{-7}$ values
See table in text for definition of physical variables

For convenience we now define the following abbreviations:

$$a \equiv 2 \cdot \pi_3^4 \cdot \pi_4 \; ; \qquad b \equiv \left(\frac{64}{3 \cdot \pi}\right) \cdot \pi_3^3 \cdot \pi_5 \tag{m}$$

With these then, relation (e) can be written more compactly

$$\pi_1 = \frac{a}{1 + \pi_2^2} + \frac{b}{1 - \pi_2^4} \tag{n}$$

and therefore at point A on the curve of Fig. 18-48

$$\pi_2 = (\pi_2)_A = 0 \; ; \qquad \pi_1 = (\pi_1)_A = a + b \tag{o}$$

and at point B

$$\pi_2 = (\pi_2)_B = \sqrt{\frac{a}{a+b}} \; ; \qquad \pi_1 = (\pi_1)_B = (\pi_1)_A = a + b \tag{p}$$

In our concrete case depicted in Fig. 18-48, and using the definitions in (m)

$$a = 4.096 \times 10^{-4} \; ; \qquad b = 2.572827 \times 10^{-3} \tag{q}$$

According to these values and (p)

$$(\pi_1)_A = (\pi_1)_B = 2.98243 \times 10^{-3} ; \quad (\pi_2)_A = 0 ; \quad (\pi_2)_B = 0.370591 \quad (r)$$

with the corresponding values of

$$(D_1)_A = (\pi_2)_A \cdot D_2 = (0) \cdot (0.127) = 0$$

$$f_A = (\pi_1)_A \cdot D_2 = (2.98243 \times 10^{-3}) \cdot (0.127) = 0.0003788 \text{ m}$$

i.e., if the cantilever is a solid rod, its deflection under the specified external load *and* its own weight will be f_A. However, as we can see, the *same* deflection is obtained if the inside diameter is

$$(D_1)_B = (\pi_2)_B \cdot D_2 = (0.370591) \cdot (0.127) = 0.04707 \text{ m}$$

In fact, as the curve shows, if the inside diameter is *any* value between 0 and $(D_1)_B = 0.04707$ m, then it is *always* possible to select *two* D_1 inner diameter values which yield the *same* deflection.

By using the abbreviations defined in (m), for any π_1 between $(\pi_1)_{min}$ and $(\pi_1)_A$, relation (e) yields

$$\pi_2 = \sqrt{\frac{a \pm \sqrt{a^2 - 4 \cdot \pi_1 \cdot (a + b - \pi_1)}}{2 \cdot \pi_1}} \quad (s)$$

For example, if $\pi_1 = (\pi_1)_A = (\pi_1)_B = a + b$, then

$$(\pi_2)_A = \sqrt{\frac{a - a}{2 \cdot (a + b)}} = 0 \quad \text{and} \quad (\pi_2)_B = \sqrt{\frac{a + a}{2 \cdot (a + b)}} = \sqrt{\frac{a}{a + b}}$$

as indicated in (o) and (p). Substituting "*a*" and "*b*" from (m) into (f), we obtain—after some careful simplification—the $\pi_2 = (\pi_2)_0$ value at which *minimum* deflection occurs. Accordingly,

$$(\pi_2)_0 = \sqrt{\frac{(a + b) - \sqrt{b \cdot (b + 2 \cdot a)}}{a}} \quad (t)$$

which, if evaluated, provides $(\pi_2)_0 = 0.2627$—the same value as obtained previously in (j) from the chart in Fig. 18-47. If we now insert specific π_3, π_4, and π_5 data, as well as the obtained $(\pi_2)_0$ into (e), we get $(\pi_1)_{min}$ which confirms relation (k). These numeric examples are marked in Fig. 18-47 by dashed lines.

Finally, we mention that if the cantilever is not horizontal, but inclined to the horizontal by φ degrees, then the plots and formulas refer to lateral (i.e., perpendicular to the longitudinal axis of the beam) instead of the "vertical" direction. The procedure then is to substitute $P \cdot \cos \varphi$ for P, and $\gamma \cdot \cos \varphi$ for γ. Note that the payload still acts vertically; only the direction of the lateral deflection is changed. If the beam is horizontal, then of course the directions of the load and deflection are coincident.

Moreover, in the inclined case $(\pi_2)_0$ in (h) also remains unchanged since the factor $\cos \varphi$ appears in both the numerator and denominator under the square root. Therefore the optimized inner diameter of the tubular cantilever also remains unchanged.

Example 18-42. Choosing the Right Shock Absorber*

Often, a fast moving machinery of mass *M* must be stopped by an absorbing barrier—a *shock absorber*. The controlling parameter in selecting a shock absorber is usually the impact force which is to be *minimized*. More accurately, the *maximum* value of this force must be minimized (the impact force grows from zero to maximum during the deceleration of the impacting mass). The configuration in which we will study the problem is shown in Fig. 18-49.

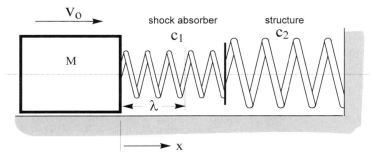

Figure 18-49
A conservative two-spring impact-absorbing system

We first dispel a popular—and false—belief that the "softer" a shock absorber, the smaller the maximum impacting force. Suppose we have a *very* soft absorber—like marmalade. Then, obviously, the impacting mass will go through it with little hindrance, and hence the (almost) full kinetic energy of the mass will be absorbed by the *hard* underlying structure; thus the impact force will be very large. From this argument, however, one must not conclude that the absorber should be very hard. For if the absorber is very hard, then the impacting (decelerating) mass would be resisted by a combination of hard absorber and hard structure—again resulting in a very large impact force.

It follows therefore that between these two extrema there must be a particular absorber stiffness which yields the *minimum* impact force. We will now determine this force and the associated system's characteristics.

For our study we make the following assumptions:

(1) The system is *conservative;* i.e., during the mass' deceleration from speed v_0 to zero, no energy is added or removed from the system. This means that the mass moves without frictional and air resistance, and that the internal friction of the springs (representing the shock absorber and the structure) is negligible. Furthermore, no force which is *external* to the system (e.g., a driving force) is acting on the mass during deceleration.

(2) The springs (representing the resistances by the shock absorber and structure) are *linear;* i.e., their deformations are *proportional* to the respective *x*-directional forces.

(3) There is a *given* maximum linear compression of the shock absorber, beyond which it becomes incompressible, i.e., infinitely stiff.

---

*A condensed version of this application of the dimensional method appeared in Ref. 145, p. 66.

The relevant variables are:

Variable	Symbol	Dimension
maximum impact force	F	m·kg/s^2
approach speed of mass	v_0	m/s
spring constant of shock absorber	c_1	kg/s^2
spring constant of structure	c_2	kg/s^2
maximum possible linear compression of shock absorber	λ	m
impacting mass	M	kg

We have six variables and three dimensions, therefore there are 6 − 3 = 3 dimensionless variables describing the behavior of the system. The Dimensional Set is

	F	v_0	c_1	c_2	λ	M
m	1	1	0	0	1	0
kg	1	0	1	1	0	1
s	−2	−1	−2	−2	0	0
π_1	1	0	0	−1	−1	0
π_2	0	1	0	−$\frac{1}{2}$	−1	$\frac{1}{2}$
π_3	0	0	1	−1	0	0

from which

$$\pi_1 = \frac{F}{c_2 \cdot \lambda}; \quad \pi_2 = \frac{v_0}{\lambda} \cdot \sqrt{\frac{M}{c_2}}; \quad \pi_3 = \frac{c_1}{c_2} \qquad (a)$$

Thus we have

$$\pi_1 = \Psi\{\pi_2, \pi_3\} \qquad (b)$$

where Ψ is an as-yet unknown function which we now shall determine.

If x, x_1, and x_2 are the movement of the mass (reckoned from contact point) and the respective compression of the shock absorber and the structure, then at contact $x = x_1 = x_2 = 0$, and the energy in the system is the kinetic energy of the moving mass

$$E_0 = \frac{1}{2} \cdot M \cdot v_0^2 \qquad (c)$$

At $x > 0$ we have $v < v_0$, and the sum of all energies (kinetic and potential) is

$$E_x = \frac{1}{2} \cdot M \cdot v^2 + \frac{1}{2} \cdot c_1 \cdot x_1^2 + \frac{1}{2} \cdot c_2 \cdot x_2^2 \qquad (d)$$

where v is the speed of the mass.

But at all times $E_x = E_0$, since the system is conservative [assumption (1)]. Hence by (c) and (d)

$$v = \sqrt{v_0^2 - \frac{1}{M} \cdot (x_1^2 + x_2^2)} \qquad (e)$$

The forces exerted by the two springs must be identical at all times, since the springs are connected in series. Therefore

$$c_1 \cdot x_1 = c_2 \cdot x_2 \tag{f}$$

thus

$$x_2 = \frac{c_1}{c_2} \cdot x_1 \tag{g}$$

By (e) and (g), then

$$v = \sqrt{v_0^2 - \frac{c_1 \cdot x_1^2}{M} \cdot \left(1 + \frac{c_1}{c_2}\right)} \tag{h}$$

At the instant when the mass comes to rest ($v = 0$), x_1 will be maximum. Hence, by (h)

$$(x_1)_{\max} = v_0 \cdot \sqrt{\frac{M}{c_1 \cdot \left(1 + \dfrac{c_1}{c_2}\right)}} \tag{i}$$

Two cases are now distinguished:

Case 1. Compression of the shock absorber is *less* than the maximum possible, i.e., $(x_1)_{\max} < \lambda$. In this case

$$F_{\max} = c_1 \cdot (x_1)_{\max} \tag{j}$$

Hence, by (j) and (i)

$$F_{\max} = v_0 \cdot \sqrt{M \cdot \frac{c_1 \cdot c_2}{c_1 + c_2}} \tag{k}$$

which, by (a), can be written

$$\pi_1 = \pi_2 \cdot \sqrt{\frac{\pi_3}{1 + \pi_3}} \tag{ℓ}$$

Case 2. Compression of the shock absorber is the maximum possible, i.e., $(x_1)_{\max} \geq \lambda$. Of course x_1 cannot physically exceed λ. If, by *calculation*, $(x_1)_{\max} > \lambda$, then $(x_1)_{\max}$ by definition is the compression that a *hypothetical* barrier (of identical stiffness c_1 and "sufficiently" long stroke λ) would experience. At the instant when the mass becomes stationary we have $v = 0$, $x_1 = \lambda$, and $x_2 = (x_2)_{\max}$. Hence, by (d),

$$E_x = E_0 = \frac{1}{2} \cdot M \cdot v_0^2 = \frac{1}{2} \cdot c_1 \cdot \lambda^2 + \frac{1}{2} \cdot c_2 \cdot (x_2)_{\max}^2 \tag{m}$$

from which

$$(x_2)_{\max} = \sqrt{\frac{1}{c_2} \cdot (M \cdot v_0^2 - c_1 \cdot \lambda^2)} \tag{n}$$

Thus, the maximum impact force is

$$F_{\max} = c_2 \cdot (x_2)_{\max} = \sqrt{c_2 \cdot (M \cdot v_0^2 - c_1 \cdot \lambda^2)} \tag{o}$$

which, by (a), can be written in the dimensionless form

$$\pi_1 = \sqrt{\pi_2^2 - \pi_3} \tag{p}$$

We see in this relation that an increase of π_3 will *decrease* π_1; the maximum π_1 occurs at $\pi_3 = 0$ and it is $(\pi_1)_{max} = \pi_2$. In contrast, in relation (ℓ) an increase of π_3 causes an *increase* of π_1 from zero to its maximum, which is again $(\pi_1)_{max} = \pi_2$ attained at infinite π_3.

If we plot relations (ℓ) and (p), the ascending and descending curves meet at a particular point (point A in Fig. 18-50) at which the generated impact force is *minimum*.

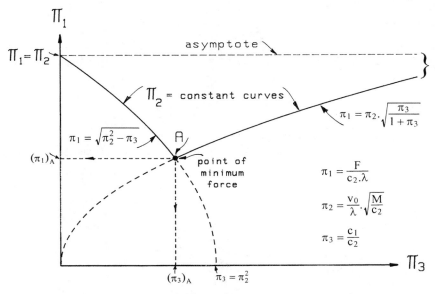

Figure 18-50
Explanatory dimensionless plot to determine the generated impact force in a two-element (absorber-structure) system shown in Fig. 18-49
At point A the impact force is minimum. Dashed parts of the curves are imaginary and cannot be realized. For definitions of physical variables see table in text.
The numerically informative plot is shown in Fig. 18-51. The present plot is not to scale.

To determine the location of this important point, we need merely equate the right sides of (ℓ) and (p). Accordingly, if the abscissa of point A is denoted as $(\pi_3)_A$, then

$$\pi_2 \cdot \sqrt{\frac{(\pi_3)_A}{1 + (\pi_3)_A}} = \sqrt{\pi_2^2 - (\pi_3)_A}$$

from which

$$(\pi_3)_A = \frac{1}{2} \cdot \left(\sqrt{1 + 4 \cdot \pi_2^2} - 1\right) \tag{q}$$

The corresponding $(\pi_1)_A$ will be furnished by either (ℓ) or (p). By (p), $(\pi_1)_A^2 = \pi_2^2 - (\pi_3)_A$, which, by (q) becomes—after some simple rearrangement

$$(\pi_1)_A^2 = \pi_2^2 + \frac{1}{2} - \frac{1}{2} \cdot \sqrt{1 + 4 \cdot \pi_2^2} \tag{r}$$

Again by (q)

$$(\pi_3)_A^2 = \frac{1}{4} \cdot \left(\sqrt{1 + 4 \cdot \pi_2^2} - 1\right)^2 = \pi_2^2 + \frac{1}{2} - \frac{1}{2} \cdot \sqrt{1 + 4 \cdot \pi_2^2} \qquad (s)$$

We now observe that the right sides of (r) and (s) are identical. Hence

$$(\pi_1)_A = (\pi_3)_A \qquad (t)$$

and therefore, by (q) and (a)

$$(\pi_1)_A = \frac{F_{\min}}{c_2 \cdot \lambda} = \frac{1}{2} \cdot \left(\sqrt{1 + \pi_2^2} - 1\right)$$

from which the minimum achievable force is

$$F_{\min} = \frac{c_2 \cdot \lambda}{2} \cdot \left(\sqrt{1 + \pi_2^2} - 1\right) \qquad (u)$$

As the plot in Fig. 18-50 shows, if $\pi_3 < (\pi_3)_A$, then the *solid* part of the curve—which is to the *left* of point A—should be used; if $\pi_3 > (\pi_3)_A$ then the *solid* part of the curve—which is to the *right* of point A—should be used.

In Fig. 18-51 a dimensionless plot is presented giving the detailed numerical solutions to relations (ℓ) and (p). To illustrate the facility of this plot, a numerical example is offered. Consider mass $M = 6.6$ kg approaching at speed $v_0 = 3.1$ m/s a shock-absorbing bar-

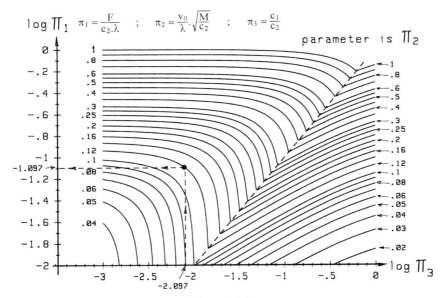

Figure 18-51
Dimensionless plot to determine the generated impact force in a two-element (absorber-structure) system shown in Fig. 18-49

The long inclined dashed line represents *minimum* impact force conditions.
The shorter dashed lines on the left represent the numerical example in text.
For definitions of variables see table in text. The base of log is 10.

rier of spring constant $c_1 = 13440$ N/m. The barrier is mounted on a much more rigid structure of spring constant $c_2 = 1.68 \times 10^6$ N/m. The stroke of the barrier is $\lambda = 0.0512$ m. What will the impact force be?

By (a)

$$\pi_2 = \frac{v_0}{\lambda} \cdot \sqrt{\frac{M}{c_2}} = \frac{3.1}{0.0512} \cdot \sqrt{\frac{6.6}{1.68 \times 10^6}} = 0.12 \quad \text{(v)}$$

and by (q)

$$(\pi_3)_A = \frac{1}{2} \cdot \left(\sqrt{1 + 4 \cdot \pi_2^2} - 1\right) = \frac{1}{2} \cdot \left(\sqrt{1 + 4 \cdot (0.12^2)} - 1\right) = 0.014198$$

By the given data

$$\pi_3 = \frac{c_1}{c_2} = \frac{13440}{1.68 \times 10^6} = 0.008$$

For this value (log $0.008 = -2.0969$) and with parameter $\pi_2 = 0.12$ [relation (v)] the chart gives log $\pi_1 = -1.0969$ (i.e., $\pi_1 = 0.08$). Therefore the impact force, by (a), is

$$F = \pi_1 \cdot c_2 \cdot \lambda = (0.08) \cdot (1.68 \times 10^6) \cdot (0.0512) = 6881.28 \text{ N}$$

The shock absorber's *theoretical* compression is, by (i)

$$(x_1)_{\max} = v_0 \cdot \sqrt{\frac{c_2 \cdot M}{c_1 \cdot (c_1 + c_2)}} = 0.068423 \text{ m}$$

which is *greater* than the maximum possible compression λ of the shock absorber. Therefore the shock absorber will be fully compressed. The above π_1, π_2, and π_3 values are marked by dashed lines in Fig. 18-51.

Now let us harden the shock absorber to a value which maintains the compressive force, but which results in an absorber's compression x_1 less than its full stroke λ. We solve this problem easily by extending the horizontal dashed line to the *right* until this extension (not shown in Fig. 18-51) intersects the $\pi_2 = 0.12$ curve. This happens at abscissa log $\pi_3 = -0.0969$, i.e., $\pi_3 = 0.8$. Thus, by the definition of π_3 in (a)

$$c_1 = \pi_3 \cdot c_2 = (0.8) \cdot (1.68 \times 10^6) = 1.344 \times 10^6 \text{ N/m}$$

Therefore, as shown here, the *same* impact force can be obtained by two *different* shock absorbers, although a *single* shock absorber (i.e., c_1 and λ) *uniquely* defines the generated impact force.

Finally, let us find the *minimum* achievable impact force. We recall that this force is represented by point A in Fig. 18-50. Therefore, with parameter $\pi_2 = 0.12$ for this point Fig. 18-51 provides log $(\pi_3)_A = -1.848$, from which $(\pi_3)_A = 0.0142$, and hence the spring constant of the shock absorber is

$$c_1 = (\pi_3)_A \cdot c_2 = (0.0142) \cdot (1.68 \times 10^6) = 23856 \text{ N/m}$$

and, by fundamental identity (t), we write $(\pi_1)_A = (\pi_3)_A = 0.0142$. Thus, by (a)

$$F_{\min} = (\pi_1)_A \cdot c_2 \cdot \lambda = (0.0142) \cdot (1.68 \times 10^6) \cdot (0.0512) = 1221.427 \text{ N}$$

and we see that compared with the two formerly considered variants, by using the *appropriate* barrier, we were able to achieve a 82.25 % (!) reduction in impact force.

⇑

Example 18-43.   Jamming a Circular Plug into a Nonsmooth Circular Hole*

A cylindrical plug is sliding in a nonsmooth circular hole under the influence of force P acting parallel with the axis of the hole, and torque M, whose axis is perpendicular to P. The configuration is shown in Fig. 18-52. Our task is to find relations among the geometric and force-related characteristics and the friction coefficient. Specifically, the conditions of jamming (binding, wedging) are to be formulated.

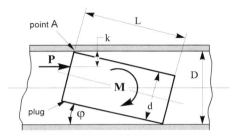

Figure 18-52
Cylindrical plug in hole is pushed by force and rotated by moment

First, we define jamming. Given torque M—as shown in Fig. 18-52—the magnitude of force P necessary to push the plug forward is determined by the above referred parameters. When, for any finite torque M, force P reaches infinite value, jamming has occurred; for the plug cannot be moved to the right, no matter how much P is applied.

It is important to realize that at jamming M can be very small (approaching zero), and P can be very large (approaching infinity). Therefore, strange as it may appear, as far as jamming is concerned, these two variables are irrelevant. Moreover, under any given set of *geometric* conditions, there is obviously a *particular* friction coefficient at which jamming occurs. If the *actual* friction coefficient is less than this particular value, then jamming does not occur, otherwise it does. This particular value the *critical friction coefficient* and we shall use the symbol μ_c for it.

Next, we observe in Fig. 18-52 that inclination φ of the plug in the hole is entirely defined by the given geometric data D, d, L—as shown. In particular,

$$\sin \varphi = \frac{L \cdot D - d \cdot \sqrt{d^2 + L^2 - D^2}}{d^2 + L^2} \tag{a}$$

as the inquisitive reader may wish to verify. Therefore φ is a dependent variable, and hence need not—in fact, must not—be included in the set of variables defining μ_c (which *is* the dependent variable in this enquiry). Accordingly, the following variables are relevant:

Variable	Symbol	Dimension	Remark
critical friction coefficient	μ_c	1	identical at both contact points
position of P force	k	m	defined in Fig. 18-52
diameter of plug	d	m	
length of plug	L	m	
diameter of hole	D	m	

---

*This material appeared in condensed form in Ref. 144. The problem itself was originally posed to the author by Robert Ferguson, then Senior Staff Engineer at SPAR Aerospace Ltd., Toronto, Ont., Canada, now President of Taiga Engineering Group, Inc., Bolton, Ont., Canada.

We have five variables and one dimension, therefore there are 5 − 1 = 4 dimensionless variables furnished by the Dimensional Set

	μ_c	k	d	L	D
m	0	1	1	1	1
π_1	1	0	0	0	0
π_2	0	1	0	0	−1
π_3	0	0	1	0	−1
π_4	0	0	0	1	−1

from which

$$\pi_1 = \mu_c; \quad \pi_2 = \frac{k}{D}; \quad \pi_3 = \frac{d}{D}; \quad \pi_4 = \frac{L}{D} \tag{b}$$

Hence we can write

$$\pi_1 = \Psi\{\pi_2, \pi_3, \pi_4\} \tag{c}$$

where Ψ is a function to be determined. This determination can be painlessly performed by considering the 3 equilibrium conditions for the plug:

- moments about point A (Fig. 18-52) is zero;
- algebraic sum of horizontal forces is zero;
- algebraic sum of vertical forces is zero.

and then calculating the *smallest* friction coefficient μ, which, for a finite M, makes P infinite. This particular μ is μ_c. The actual derivation yields (Ref. 144)

$$\mu_c = \frac{1}{1 - 2 \cdot \frac{k}{D}} \cdot \sqrt{\left(\frac{d}{D}\right)^2 + \left(\frac{L}{D}\right)^2 - 1} \tag{d}$$

which, by (b), can be written in dimensionless form

$$\pi_1 = \frac{1}{1 - 2 \cdot \pi_2} \cdot \sqrt{\pi_3^2 + \pi_4^2 - 1} \tag{e}$$

From this formula two important conclusions can be drawn: First, if $\pi_2 = \frac{1}{2}$ (i.e., $k = D/2$, meaning P is acting *exactly* along the centerline of the hole), then $\mu_c = \infty$. In this case jamming cannot occur, since *any* friction coefficient is smaller than infinite. Second, if $\pi_2 = 0$ (i.e., the load is at the perimeter of the plug), then $\pi_1 = \mu_c$ is *minimum*. Hence

$$(\pi_1)_{\min} = (\mu_c)_{\min} = \sqrt{\pi_3^2 + \pi_4^2 - 1} \tag{f}$$

defines the *smallest* critical friction coefficient, which—since it is the most conservative value—must be used for any prudent engineering design aiming to avoid jamming.

To illustrate, let $d = 0.080$ m, $D = 0.081$ m, $L = 0.025$ m. Then by (b)

$$\pi_3 = \frac{d}{D} = \frac{0.080}{0.081} = 0.98765 \quad \text{and} \quad \pi_4 = \frac{L}{D} = \frac{0.025}{0.081} = 0.30864$$

and by (f)

$$(\mu_c)_{min} = \sqrt{\pi_3^2 + \pi_4^2 - 1} = \sqrt{0.98765^2 + 0.30864^2 - 1} = 0.26593$$

i.e., if the friction coefficient is less than 0.26593, then jamming *cannot* occur, otherwise it *may*.

Relation (f) can be written (omitting subscript "min" for convenience)

$$\pi_1^2 = \pi_3^2 + \pi_4^2 - 1 \tag{g}$$

or

$$\frac{\pi_1^2}{\pi_4^2 - 1} - \frac{\pi_3^2}{\pi_4^2 - 1} = 1 \tag{h}$$

In the π_1 versus π_3 coordinate system the graph of this relation is a hyperbola with parameter π_4. The explanatory graph of (h) in Fig. 18-53 shows three typical curves: one curve for $\pi_4 > 1$, one for for $\pi_4 = 1$, and one for $\pi_4 < 1$. The detailed chart is presented in Fig. 18-54.

Relation (g) allows an easy and elegant *graphical* determination of critical friction co-

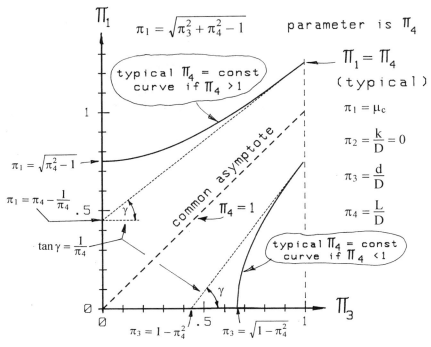

Figure 18-53
Explanatory dimensionless plot showing principal characteristics of three typical curves to determine critical friction coefficient μ_c for jamming a circular plug into a nonsmooth circular hole
If $\mu < \mu_c$ jamming cannot occur, otherwise it may. For definition of physical variables, see Fig. 18-52. For detailed plot, see Fig. 18-54

630 APPLIED DIMENSIONAL ANALYSIS AND MODELING

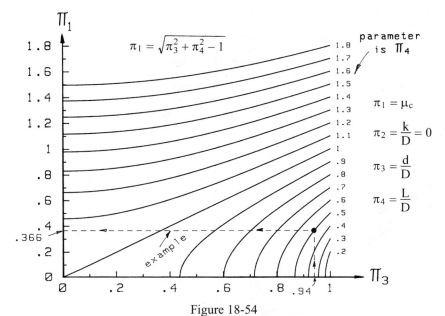

Figure 18-54
Dimensionless plot for the determination of critical friction coefficient μ_c for jamming a circular plug into a nonsmooth circular hole
Jamming cannot occur if $\mu < \mu_c$, otherwise it may. See Fig. 18-52 for definition of physical variables. To be conservative, $k = 0$ is assumed throughout (Fig. 18-52)

efficient $\mu_c \equiv \pi_1$. The *diagonal* distance between points A and B is denoted by U (Fig. 18-55).

We write

$$U^2 = d^2 + L^2 \tag{i}$$

and hence, by Fig. 18-55

$$\left(\frac{U}{D}\right)^2 = \frac{1}{\cos^2\gamma} = \left(\frac{d}{D}\right)^2 + \left(\frac{L}{D}\right)^2 = \pi_3^2 + \pi_4^2 \tag{j}$$

However, by (g)

$$\pi_3^2 + \pi_4^2 = \pi_1^2 + 1 \tag{k}$$

therefore by (j) and (k)

$$\pi_1^2 + 1 = \frac{1}{\cos^2\gamma} \tag{ℓ}$$

from which, by simple rearrangement,

$$\pi_1 = \sqrt{\frac{1}{\cos^2\gamma} - 1} = \sqrt{\frac{1-\cos^2\gamma}{\cos^2\gamma}} = \sqrt{\frac{\sin^2\gamma}{\cos^2\gamma}} = \tan\gamma \tag{m}$$

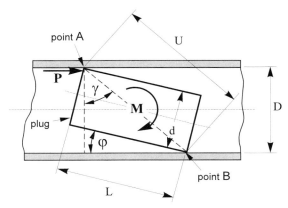

Figure 18-55
**Graphical determination of the critical friction coefficient
for jamming a circular plug into a nonsmooth circular hole**

and since $\pi_1 = \mu_c$, we can write

$$\mu_c = \tan \gamma \qquad (n)$$

i.e., the critical friction coefficient is just the tangent of the angle γ as defined in Fig. 18-55.

From these relations and charts the following important conclusions can be drawn:

(1) If $\pi_2 = \frac{1}{2}$ *exactly*, i.e., force P is acting *exactly* on the axis of the hole, then $\mu_c = \infty$ [relation (e)], i.e., irrespective of any other characteristic of the plug-hole system, no jamming can occur. However, the exactness of $\pi_2 = \frac{1}{2}$ in real life can never be assured, hence we should *not* say that jamming—for *this* reason—cannot occur.

(2) If $\pi_4 \cong 0$, i.e., $L \cong 0$ (we have a *very* thin disk acting as a plug), then by geometry (see Fig. 18-55) $\pi_3 \cong 1$ which is tantamount to $d \cong D$. This requirement is confirmed by relation (f) whereby for the expression to be real it is necessary that

$$\pi_3^2 + \pi_4^2 \geq 1 \qquad (o)$$

Hence, if $\pi_4 = 0$, then $\pi_3 = 1$ since $\pi_3 \leq 1$ at all times (the plug cannot be larger than the hole). If now $\pi_3 = 1$, then by (g) $\pi_1 = \mu_c = 0$, i.e., the critical friction coefficient is zero. In other words, no matter what, jamming *shall* occur. This is why it is *extremely* difficult to insert—or remove—a very thin disk to or from a closely fitting hole.

(3) If we know the upper limit of friction coefficient μ_{max} as well as the hole and plug diameters, then we can easily determine the *minimum* allowable length L_{min} of the plug that will assure the jam-free insertion or removal of the plug. For by (g)

$$\pi_4 = \sqrt{\pi_1^2 - \pi_3^2 + 1} \qquad (p)$$

and by (b) and (p)

$$\frac{L_{min}}{D} = \sqrt{\pi_1^2 - \pi_3^2 + 1} \qquad (q)$$

632   APPLIED DIMENSIONAL ANALYSIS AND MODELING

from which

$$L_{min} = D \cdot \sqrt{\pi_1^2 - \pi_3^2 + 1} \tag{r}$$

To illustrate, consider $\mu_{max} = 0.68$ (this means we are *sure* that μ does *not* exceed 0.68) and plug and hole diameters $d = 0.098$ m and $D = 0.100$ m. What is the minimum plug length to ensure jam-free movement of the plug in the hole? By (b)

$$\pi_1 = \mu_c = 0.68; \qquad \pi_3 = \frac{d}{D} = \frac{0.098}{0.100} = 0.98$$

and by (r)

$$L_{min} = (0.1) \cdot \sqrt{0.68^2 - 0.98^2 + 1} = 0.07085 \text{ m}$$

i.e., *if* the plug is longer than 0.07085 m, then jamming *cannot* occur.

Finally, we present a simple numerical example to demonstrate the use of the chart in Fig. 18-54. Suppose we have the following arrangement:

diameter of plug: $d = 0.0564$ m
diameter of hole: $D = 0.0600$ m
length of plug: $L = 0.0300$ m
friction coefficient: $\mu = 0.25$

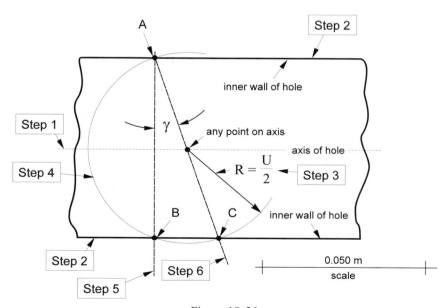

Figure 18-56
Step-by-step process to determine the critical friction coefficient for jamming a circular plug into a nonsmooth circular hole

We wish to find out if jamming may occur when pushing the plug into the hole. By (b) and the given data

$$\pi_3 = \frac{d}{D} = \frac{0.0564}{0.0600} = 0.94 \; ; \quad \pi_4 = \frac{L}{D} = \frac{0.0300}{0.0600} = 0.5$$

For these values the chart in Fig. 18-54 supplies $\pi_1 = 0.366$ which, by (b), is μ_c. That is, the critical friction coefficient is 0.366. But the *actual* friction coefficient is $\mu = 0.25$. Hence $\mu < \mu_c$ and jamming will *not* occur. This example is shown by dashed lines in Fig. 18-54.

The same answer can be obtained by a simple step-by-step graphical process now described with the aid of Fig. 18-56.

Step 1: Draw the axis (centerline) of the hole.

Step 2: Draw *to scale* two parallel lines representing the inner walls of the hole. The distance between these lines is the hole diameter; in our example it is $D = 0.060$ m.

Step 3: Calculate quantity $R = U/2$, where U is defined in Fig. 18-55. In our example it is $U = \sqrt{d^2 + L^2} = 0.063882$ m, hence $R = 0.031941$ m.

Step 4: Pick any point on the axis (drawn in Step 1) and draw an arc of radius R (Step 3) such that it intersects the lines (Step 2) representing the diameter of the hole. In this way, three points A, B, and C are generated.

Step 5: *Measure* the included angle γ of lines AB and AC. In the illustrated case $\gamma = 20$ deg.

By relation (n) now, the critical friction coefficient is

$$\mu_c = \tan \gamma = \tan 20° = 0.364$$

which is the same, within 0.6%, as provided by the chart of Fig. 18-54.

⇑

Example 18-44. Size of the Human Foot and Hand

Contrary to popular belief, the linear size (e.g. length) of a human foot is not proportional to the linear size (e.g. height) of the human body. So what is the relation? By applying dimensional analysis, one can easily determine this numerical link. The process is as follows:

As always, we first list the relevant variables, their symbols, and their dimensions:

Variable	Symbol	Dimension	Remark
size of foot	λ	m	linear measure (e.g. length)
body density	ρ	kg/m^3	constant
body size	L	m	linear measure (e.g. height)
gravitational acceleration	g	m/s^2	constant
pressure under foot	p	kg/(m·s^2)	constant

We have $N_v = 5$ variables and $N_d = 3$ dimensions. Thus, by Buckingham theorem, $N_p = N_v - N_d = 2$ dimensionless variables define the functioning of the system. These variables are obtained from the following Dimensional Set:

634 APPLIED DIMENSIONAL ANALYSIS AND MODELING

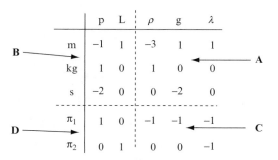

Dimensional Set

in which the **A, B, D** matrices are as shown, and the C matrix is by (8-19)

$$\mathbf{C} = -(\mathbf{A}^{-1} \cdot \mathbf{B})^T \qquad \text{repeated (8-19)}$$

The Dimensional Set now furnishes the dimensionless variables:

$$\pi_1 = \frac{p}{\rho \cdot g \cdot L}; \quad \pi_2 = \frac{L}{\lambda} \qquad (a)$$

By Buckingham's theorem

$$\pi_1 = \Psi\{\pi_2\} \qquad (b)$$

where Ψ is the sign of a function. If Ψ is monomial, which is plausible, then (b) can be written as

$$\pi_1 = c \cdot (\pi_2)^n \qquad (c)$$

where c and n are numeric constants. Substituting (a) into (c) yields $\frac{p}{\rho \cdot g \cdot L} = c \cdot \left(\frac{L}{\lambda}\right)^n$, and hence

$$p = c \cdot \rho \cdot g \cdot L^{(1+n)} \cdot \lambda^{-n} \qquad (d)$$

Now we undertake a modicum of *heuristic* reasoning: obviously p must be proportional to the body weight, which is proportional to the body volume, which is proportional to the *cube* of body height. Therefore p is proportional to the *cube* of the body height. Hence in (d) the exponent of L must be 3, therefore n = 2. Thus $p = c \cdot \rho \cdot g \cdot \frac{L^3}{\lambda^2}$, or

$$\lambda = \sqrt{\frac{c \cdot \rho \cdot g}{p}} \cdot L^{\frac{3}{2}} \qquad (e)$$

Because in this relation all of the quantities under the square root are constants, therefore the *size* (i.e. length) *of the human foot is proportional to the 3/2 power of the body height*; e.g., if X is 20% taller than Y, then X's foot is 31.5% larger than Y's foot.

But now we will show that by employing the elegant method of *directional* dimensions, by *dimensional analysis* the *same* conclusion can also be easily reached *without* any "heuristics." In employing this method, we observe that the linear dimension 'm' can be split into two directions: *axial* and *radial*, denoted by and m_A and m_R, respectively. Similarly for body density ρ and pressure under foot p. Thus the applicable—now directional—dimensions will be as given in the table below:

Variable	Symbol	Dimension	Remark
size of foot	λ	m_R	linear measure (e.g. length)
body density	ρ	$kg/[(m_R)^2 \cdot m_A]$	constant
body size	L	$(m_A)^{\frac{1}{3}} \cdot (m_R)^{\frac{2}{3}}$	linear measure (e.g. height)
gravitational acceleration	g	m_A/s^2	constant
pressure under foot	p	$m_A \cdot kg/[s^2 \cdot (m_R)^2]$	constant

Thus the Dimensional Set will be

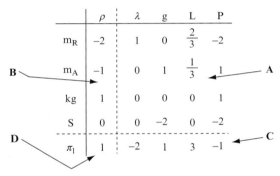

Dimensional Set

in which, again, the **A**, **B**, **D** matrices are as shown, and the **C** matrix is by (8-19). We still have $N_v = 5$ variables, but the number of dimensions has increased to $N_d = 4$. Therefore we only have $5 - 4 = 1$ dimensionless "variable"—actually now a constant. It is, by the above set:

$$\pi_1 = \frac{\rho \cdot g \cdot L^3}{\lambda^2 \cdot p} = \text{const.} = k \tag{f}$$

Therefore:

$$\lambda = \left(\sqrt{\frac{\rho \cdot g}{k \cdot p}}\right) \cdot L^{3/2} = c \cdot L^{3/2} \tag{g}$$

But in (g) all quantities under the square root are constants, hence c is constant. Thus we again may conclude that the *size (i.e. length) of the human foot is proportional to the 3/2 power of the height of the body*.

In the following table the relevant L and λ values of 13 male and 14 female subjects—selected randomly from among the author's acquaintances—are listed. It is seen that, as predicted by relation (g), for these 27 persons the c values calculated for the given L and

Subject no.	L	λ	c	Sex
1	167.6	26.7	0.012306	F
2	176.5	28.5	0.012154	M
3	178.0	27.0	0.011369	M
4	159.0	24.0	0.011971	F
5	161.3	26.0	0.012692	F
6	179.0	27.5	0.011483	F
7	157.5	25.0	0.012648	F
8	180.3	30.0	0.012392	M
9	175.2	28.0	0.012074	F
10	157.5	24.0	0.012142	F
11	177.8	30.0	0.012654	M
12	168.0	27.0	0.012399	F
13	160.0	25.5	0.012600	F
14	170.2	28.5	0.012835	F
15	173.0	25.0	0.010987	F
16	201.0	33.0	0.011580	M
17	167.6	27.0	0.012444	F
18	178.0	28.5	0.012001	M
19	188.0	31.0	0.012026	M
20	177.8	28.5	0.012021	M
21	167.6	26.5	0.012213	M
22	177.8	28.0	0.011810	M
23	172.2	28.5	0.012612	M
24	162.5	26.5	0.012793	F
25	170.2	26.5	0.011935	F
26	190.5	30.0	0.011410	M
27	172.7	27.0	0.011897	M

Symbols:
L = body height (cm); λ = foot length (cm); $c = \dfrac{\lambda}{L^{3/2}}$; M = male; F = female

λ data are very nearly constant, with a mean value of 0.01212767 cm$^{-1/2}$ and standard deviation 0.0004769 cm$^{-1/2}$.

FIFTY-TWO ADDITIONAL APPLICATIONS   **637**

Finally—and interestingly—because we were once quadrupeds, the size of human *hand* vs. human height also follows the same numerical relation!

⇑

Example 18-45.   Depression of an Inflated Balloon Floating on Oil

Given an inflated rubber balloon of diameter D floating on oil (Fig. 18-57). What is its radial depression h? We will determine this value by a modeling experiment.

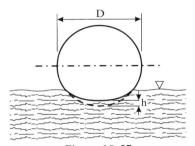

Figure 18-57
Depression of a rubber balloon floating on oil

But first, we list the names of the relevant variables, their symbols and dimensions:

Name	Symbol	Dimension	Remark
depression of balloon	h	m	radial, vertical
pressure in balloon	p	kg/(m.s²)	above atmospheric
density of balloon material	ρ_b	kg/m³	
density of oil	ρ_o	kg/m³	
diameter of balloon	D	m	non-deformed
gravitational acceleration	g	m/s²	

The given values of the *prototype* are:

Pressure: $p_1 = 68000$ kg/(m.s²)
Density of balloon material: $(\rho_b)_1 = 1393.24$ kg/m³
Density of oil: $(\rho_o)_1 = 869.7$ kg/m³
Diameter of balloon: $D_1 = 5$ m
Gravitational acceleration: $g_1 = 9.81$ m/s²

There are $N_V = 6$ variables and $N_d = 3$ dimensions. Accordingly, by Buckingham's theorem:

$$N_P = N_V - N_d = 6 - 3 = 3$$

dimensionless variables π_1, π_2, π_3 define the system. They are conveniently supplied by the Dimensional Set as follows:

	h	p	ρ_b	D	ρ_o	g
m	1	−1	−3	1	−3	1
kg	0	1	1	0	1	0
s	0	−2	0	0	0	−2
π_1	1	0	0	−1	0	0
π_2	0	1	0	−1	−1	−1
π_3	0	0	1	0	−1	0

Dimensional Set

where the **A**, **B**, and **D** matrices are as indicated, and the **C** matrix is by the *Fundamental Formula* (8-19):

$$\mathbf{C} = -\mathbf{D} \cdot (\mathbf{A}^{-1} \cdot \mathbf{B})^T \qquad \text{repeated (8-19)}$$

The 3 dimensionless variables can now be conveniently read off from the Dimensional Set as follows:

$$\pi_1 = \frac{h}{D}; \quad \pi_2 = \frac{p}{D \cdot \rho_o \cdot g}; \quad \pi_3 = \frac{\rho_b}{\rho_o} \qquad (a)$$

There are 3 dimensionless variables and 6 variables, therefore we have 6 Scale Factors and the Model Law consists of 3 relations as follows:

$$S_h = S_D; \qquad S_p = S_D \cdot S_{\rho o} \cdot S_g; \qquad S_{\rho b} = S_{\rho o} \qquad (b)$$

where:

S_h is the Depression Scale Factor,
S_D is the Balloon Diameter Scale Factor,
S_p is the Balloon Internal Pressure Scale Factor,
$S_{\rho o}$ is the Oil Density Scale Factor,
S_g is the Gravitational Acceleration Scale Factor,
$S_{\rho b}$ is the Balloon Material Scale Factor.

By (a) and (b) now, the following relations can be established:

$$\frac{h_2}{h_1} = \frac{D_2}{D_1}; \quad \frac{p_2}{p_1} = \left(\frac{D_2}{D_1}\right) \cdot \left(\frac{(\rho_o)_2}{(\rho_o)_1}\right) \cdot \left(\frac{g_2}{g_1}\right); \quad \frac{(\rho_b)_2}{(\rho_b)_1} = \frac{(\rho_o)_2}{(\rho_o)_1} \qquad (c)$$

where subscripts '1' and '2' represent, respectively, the *prototype* and the *model*.

Because there are 6 variables, we have a total of 12 variables for the prototype and the model. Of these, 1 can be *measured* on the model, and we can determine 3 by (c). Therefore, $12 - 4 = 8$ variables can be chosen. Since by the above specified *prototype* 5 are given, therefore 3 can be selected for the model, as follows:

Density of balloon material: $(\rho_b)_2 = 1393.24$ kg/m^3
Diameter of balloon: $D_2 = 0.8$ m
Gravitational acceleration: $g_2 = 9.81$ m/s^2

The required (imposed) p_2 air pressure in the model balloon is determined by the second formula in (c). Accordingly:

$$p_2 = p_1 \cdot \left(\frac{D_2}{D_1}\right) \cdot \left(\frac{(\rho_o)_2}{(\rho_o)_1}\right) \cdot \left(\frac{g_2}{g_1}\right) = 68000 \cdot \left(\frac{0.8}{5}\right) \cdot \left(\frac{869.7}{869.7}\right) \cdot \left(\frac{9.81}{9.81}\right) = 10880 \frac{\text{kg}}{\text{m.s}^2}$$

In other words, this is the pressure which *must* be applied in the model balloon.

Similarly, the required (imposed) $(\rho_b)_2$ density of the balloon material of the model is determined by the third relation of (c). Accordingly:

$$(\rho_b)_2 = (\rho_b)_1 \left(\frac{(\rho_o)_2}{(\rho_o)_1}\right) = 1393.24 \cdot \left(\frac{869.7}{869.7}\right) = 1393.24 \frac{\text{kg}}{\text{m}^3}$$

With these data, we can now construct the model, put it on oil, and *measure* on it the generated depression, which is, say, $h_2 = 0.036$ m. Use this value, and the first relation of (c) yields

$$h_1 = h_2 \cdot \left(\frac{D_1}{D_2}\right) = 0.036 \cdot \left(\frac{5}{0.8}\right) = 0.225 \text{ m}$$

which is the depression of the prototype; our desired result.

All the above can be conveniently summarized in the following *Modeling Data Table*:

Name	Variable Symbol	Variable Dimension	Variable Prototype	Scale Factor S Model	Scale Factor S Model/Prototype	Variable Category Prototype	Variable Category Model
balloon diameter	D_b	m	5	0.8	0.16	1	1
internal pressure	p	kg/(m/s²)	68,000	10,880	0.16	1	2
balloon density	ρ_b	kg/m³	1393.24	1393.24	1	1	1
oil density	ρ_o	kg/m³	869.7	869.7	1	1	2
gravitational accel.	g	m/s²	9.81	9.81	1	1	1
depression	h	m	0.225	0.036	0.16	2	3
dimensionless variable	π_1	1	0.045	0.045	1	not applicable	
dimensionless variable	π_2	1	1.594	1.594	1	not applicable	
dimensionless variable	π_3	1	1.602	1.602	1	not applicable	

Categories of Variables		
	1	Freely chosen, *a priori* given, or determined independently
	2	Determined by application of Model Law
	3	Determined by measurement on the model

Modeling Data Table

for the determination of depression of an inflated balloon floating on oil

640   APPLIED DIMENSIONAL ANALYSIS AND MODELING

We see that for both the model and prototype, all 3 dimensionless variables π_1, π_2, π_3 are pair-wise identical—as expected and required.

⇑

Example 18-46.   Resonant Frequency of an L,C Circuit

By dimensional analysis we now determine the resonant frequency of an L,C electric circuit seen in Fig. 18-58.

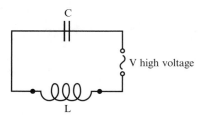

Figure 18-58
An L,C electric circuit

But first, we list the names of relevant variables, their symbols and dimensions:

Variable	Symbol	Dimension
frequency	v	1/s (Hz)
electric inductance	L	$m^2 \cdot kg/(s^2 \cdot A^2)$
electric capacitance	C	$A^2 \cdot s^4/(m^2 \cdot kg)$

The dimensional matrix is therefore:

	v	L	C
m	0	2	−2
kg	0	1	−1
s	−1	−2	4
A	0	−2	2

We now observe that the number of dimensions is 4 and the number of variables is 3. But by Theorem 7-6 (paragraph 7.11.2), the former must be at least 1 less than the latter—otherwise no relation among the variables can exist. Hence, we must eliminate 2 dimensions.

Let these 2 sacrificial dimensions be "m" and "kg." Therefore the Dimensional Set will now have $N_V = 3$ variables and $N_d = 2$ dimensions. By Buckingham's theorem this yields

$N_p = 3 - 2 = 1$ dimensionless "variable" which, by Theorem 7-11, must be a *constant*. The resultant Dimensional Set is

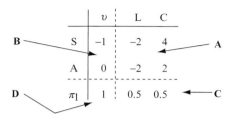

**Figure 18-59
Dimensional Set**

where the **A**, **B**, and **D** are matrices as given, and the **C** matrix is by the *Fundamental Formula* (8-19)

$$\mathbf{C} = -\mathbf{D} \cdot (\mathbf{A}^{-1} \cdot \mathbf{B})^T \qquad \text{repeated (8-19)}$$

From the Dimensional Set now, the single dimensionless "variable"—in this case, a constant—is

$$\pi_1 = v \cdot \sqrt{L \cdot C} = \text{const.}$$

from which

$$v = \text{const.} \frac{1}{\sqrt{L \cdot C}}$$

With constant = $\frac{1}{2 \cdot \pi}$ (Ref. 4, p. 498), this is our desired relation.

⇑

Example 18-47. Energy of Water Waves

We wish to determine the residing energy of water waves per unit of horizontal width—i.e. 1 m—perpendicular to the direction of propagation of waves.

The names, symbols, and dimensions of the physical variables of this energy are listed in the table below:

Variable	Symbol	Dimension	Remark
wave length	λ	m	
density of water	ρ	kg/m^3	
energy	Q	m·kg/s^2	for 1 m width of wave
amplitude of waves	a	m	
gravitational acceleration	g	m/s^2	

Note that the dimension of Q (energy) is not the usual $\dfrac{m^2 \cdot kg}{s^2}$, but $\dfrac{m \cdot kg}{s^2}$, because it is for *unit width* of the wave.

We have $N_V = 5$ variables and $N_d = 3$ dimensions, therefore by Buckingham's theorem $N_P = 5 - 3 = 2$ dimensionless variables—supplied by the Dimensional Set—define the system. Accordingly, the Dimensional Set is

		λ	ρ	Q	a	g
B	m	1	-3	1	1	1
	kg	0	1	1	0	0
	s	0	0	-2	0	-2
D	π_1	1	0	0	-1	0
	π_2	0	1	-1	3	1

Figure 18-60
Dimensional Set

In this set the **A**, **B**, and **D** matrices are as given, and the **C** matrix is by the *Fundamental Formula*

$$\mathbf{C} = -(\mathbf{A}^{-1} \cdot \mathbf{B})^T \qquad \text{repeated (8-19)}$$

The 2 dimensionless variables can now be easily read off from this set:

$$\pi_1 = \frac{\lambda}{a} \qquad (a)$$

$$\pi_2 = \frac{\rho \cdot a^3 \cdot g}{Q} \qquad (b)$$

Considering the likely *monomial* form of the relation between π_1 and π_2, we can now write

$$\pi_1 = c \cdot (\pi_2)^n \qquad (c)$$

where c and n are as-yet undetermined constants. Substituting the relevant values for π_1 and π_2 given in (a) and (b) into (c) will yield

$$\frac{\lambda}{a} = c \cdot \left(\frac{\rho \cdot a^3 \cdot g}{Q} \right)^n \qquad (d)$$

from which

$$Q = (\rho \cdot g \cdot a^3) \cdot \left(\frac{c \cdot a}{\lambda} \right)^{1/n}$$

Besides the knowledge that Q is proportional to the product $\rho \cdot g$, this equation does not tell us anything useful. Can we do better? Yes, by using *directional* dimensions for length m, density ρ, and energy Q. This way, the number of dimensions increases to 4, while the number of variables remains at 5. Therefore, by Buckingham's theorem, the number of dimensionless product will be 5 − 4 = 1, a constant. The result of this technique, as shown below, is very useful.

Accordingly, the dimensions of length "m," density ρ, and energy Q are split into *horizontal* and *vertical* parts designated by *subscripts* "h" and "v," respectively. This yields the following:

Variable	Symbol	Dimension	Remark
wave length	λ	m_h	
density of water	ρ	$kg/(m_h^2 \cdot m_v)$	
energy	Q	$kg \cdot m_v^2/(m_h \cdot s^2)$	for 1 m_h width of wave
amplitude of waves	a	m_v	
gravitational acceleration	g	m_v/s^2	

The corresponding Dimensional Set is:

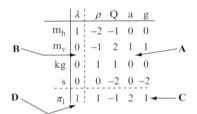

Figure 18-61
Dimensional Set

The resulting *single* dimensionless variable π_1—by Buckingham's theorem—must be a constant. Thus:

$$\pi_1 = \frac{\lambda \cdot \rho \cdot g \cdot a^2}{Q} = \text{const.} \tag{f}$$

from which

$$Q = \text{const.} \lambda \cdot \rho \cdot g \cdot a^2 \tag{g}$$

where—by Ref. 2, p. 95—the constant is 0.5 for "progressive" waves and 0.25 for "stationary waves" (in the former, water particles move horizontally as well; in the latter, they only move vertically).

This example shows the usefulness of employing *directional dimensions*. For—by dimensional considerations only—we now know what we formerly didn't: that wave-energy varies *linearly* with wavelength, water density, and gravitational acceleration, and with the *square* of wave amplitude.

⇑

Example 18-48. Vertical Oscillation of an Immersed Nicholson Hydrometer

Using only dimensional considerations, we will now find the expression for the vertical oscillation period of a Nicholson hydrometer immersed in a liquid. Oscillation can be easily generated by vertically slightly pushing then releasing the hydrometer neck from its equilibrium position. The arrangement is shown in Fig. 18-62.

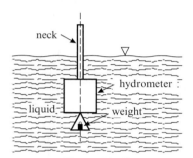

Figure 18-62
Nicholson hydrometer immersed in a liquid

The relevant variables and their dimensions are as follows:

Variable	Symbol	Dimension	Remark
oscillation period	t	s	
mass of hydrometer	M	kg	
cross-section area of "neck"	a	m²	uniform
density of liquid	ρ	kg/m³	uniform
gravitational acceleration	g	m/s²	

We have $N_V = 5$ variables and $N_d = 3$ dimensions. Therefore, by Buckingham's theorem, we have $N_P = 5–3 = 2$ dimensionless variables, supplied by the Dimensional Set.

FIFTY-TWO ADDITIONAL APPLICATIONS **645**

		t	M	a	ρ	g
	m	0	0	2	−3	1
	kg	0	1	0	1	0
	s	1	0	0	0	−2
	π_1	1	0	−1/4	0	1/2
	π_2	0	1	−3/2	−1	0

Figure 18-63
Dimensional Set

where the **A**, **B**, and **D** matrices are as given, and the **C** matrix is by (8-19)

$$\mathbf{C} = -\mathbf{D}(\mathbf{A}^{-1}.\mathbf{B})^T \qquad \text{repeated (8-19)}$$

Thus, by the Dimensional Set,

$$\pi_1 = t \cdot \sqrt[4]{\frac{g^2}{a}} \quad \text{and} \quad \pi_2 = \frac{M}{a^{1.5} \cdot \rho} \qquad (a)$$

In general, by Buckingham's theorem,

$$\pi_1 = \psi\{\pi_2\} \qquad (b)$$

where Ψ is an as-yet unknown function. By combining (a) and (b), we obtain

$$t \cdot \sqrt[4]{\frac{g^2}{a}} = \psi\left\{\frac{M}{a^{1.5} \cdot \rho}\right\} \qquad (c)$$

which, assuming a plausible monomial form, can be written

$$t = c \cdot \left(\frac{a}{g^2}\right)^{1/4} \cdot \left(\frac{M}{a^{1.5} \cdot \rho}\right)^n \qquad (d)$$

where c and n are as-yet unknown constants.

This relation tells us that the period t is inversely proportional to the square root of g. However, apart from this, (d) is not much help, because we do not know exponent n. Therefore we don't know how M, a, and ρ influence the period t. Can we do better?

Obviously, by Buckingham, in a relation, the fewer the *difference* between the numbers of variables and dimensions the better. Because this difference must be a positive number, it is now plain that the *least* number of an allowed difference is 1. In this case we have just 1 dimensionless variable, which must be a *constant*. The aim is to reduce the number of dimensionless variables as much as possible. But how do we this? By either *reducing* the number of variables N_V or *increasing* the number of dimensions N_d. We know to follow the second route.

One way to do this is to assign *directions* to dimensions. In this example, there are two principal directions: radial and axial. Obviously "a" is in the radial direction, whereas ρ is both radial and axial. Moreover, the *radial* direction for ρ must be counted twice, because it occurs both in the plane of the drawing and perpendicular to it. The *modified* Dimensional Set is in Fig. 18-64, in which the subscripts "r" and "z" denote *radial* and *axial* directions.

	t	M	a	ρ	g
m_r	0	0	2	−2	0
m_z	0	0	0	−1	1
kg	0	1	0	1	0
s	1	0	0	0	−2
π_1	1	$-\frac{1}{2}$	$\frac{1}{2}$	$\frac{1}{2}$	$\frac{1}{2}$

Figure 18-64
Modified Dimensional Set

where, again, the **A**, **B**, and **D** matrices are as given, and the **C** matrix is by (8-19).

We see that although the number of variables is still $N_V = 5$, the number of dimensions is increased from $N_d = 3$ to $N_d = 4$. Hence, $N_V - N_d = 1$; therefore, by Buckingham, we have but one dimensionless "variable": a constant!

Thus, by the Modified Dimensional Set

$$\pi_1 = t \cdot \sqrt{\frac{a \cdot \rho \cdot g}{M}} = \text{const}$$

from which

$$t = \text{const.} \sqrt{\frac{M}{a \cdot \rho \cdot g}} \qquad (e)$$

where the constant is 6.28, as can be determined by a *single* measurement (Ref. 2, p. 84).

We now see that relation (e) is vastly more informative than the previously obtained (d), since it now tells us how t varies not only by g, but also by M, a, and ρ.

Thus, by only 1 (!) measurement, and by the adroit use of the "dimensional method," we were able to establish an *explicit* relation among 5 variables and 1 constant!

(Adapted from Ref. 149, p. 222.)

Example 18-49. Relativistic Kinetic Energy

The kinetic energy in Newtonian (i.e. non-relativistic) mechanics is expressed as

$$E_N = \frac{1}{2} \cdot M \cdot v^2 \qquad (a)$$

where M is the mass of the body in motion, v is the linear speed, and E_N is the Newtonian kinetic energy. This formula of course assumes the constancy of mass. But, by the relativity theory, mass is not constant because it varies with speed. Hence correction is needed. We investigate this problem using the "dimensional method." First, we list the relevant variables:

Variable	Symbol	Dimension	Remark
relativistic kinetic energy	E_R	$m^2 \cdot kg/s^2$	
speed	v	m/s	linear
mass	M_0	kg	at rest
speed of light	c	m/s	in vacuum

The Dimensional Matrix is

	E_R	v	M_0	c
m	2	1	0	1
kg	1	0	1	0
s	−2	−1	0	−1

The rank of this matrix is $R_{DM} = 2$. The number of dimensions is $N_d = 3$, and by (7-22) $\Delta = N_d - N_{DM} = 3 - 2 = 1$, thus 1 dimension must be deleted. Let this dimension be "s." With this deletion, we will have 4 variables and 2 dimensions. It follows that, by (17-9), there will be $N_p = 2$ dimensionless variables, which can be determined by the Dimensional Set shown as:

	E_R	v	M_0	c
m	2	1	0	1
kg	1	0	1	0
π_1	1	0	−1	−2
π_2	0	1	0	1

Figure 18-65
Dimensional Set

in which the **A**, **B**, and **D** matrices are as given, and the **C** matrix is defined by (8-19)

$$\mathbf{C} = -\mathbf{D}(\mathbf{A}^{-1} \cdot \mathbf{B})^T \qquad \text{repeated (8-19)}$$

By the Dimensional Set, dimensionless variables are now

$$\pi_1 = \frac{E_R}{M_0 \cdot c^2}; \quad \pi_2 = \frac{v}{c} \tag{b}$$

Hence we write

$$\pi_1 = \psi\{\pi_2\} \tag{c}$$

where ψ is an as-yet undefined function.

Now, by Newton's law, with M being the mass, F the force, and v the speed

$$F = \frac{d}{dt}(M \cdot v) \tag{d}$$

Therefore, if dx is the infinitesimal distance, then the elementary energy dE will be

$$dE = F \cdot dx = \frac{d}{dt}(M \cdot v) \cdot dx \tag{e}$$

This can be written

$$dE = \frac{dM}{dt} \cdot v \cdot dx + \frac{dv}{dt} \cdot M \cdot dx = \frac{dM}{dv} \cdot v^2 \cdot dv + M \cdot v \cdot dv \tag{f}$$

By Einstein, the relativistic mass is

$$M = \frac{M_0}{\sqrt{1 - \left(\frac{v}{c}\right)^2}} \tag{g}$$

where M_0 is the mass at rest. Hence, by (g),

$$\frac{dM}{dv} = \frac{M_0 \cdot v}{c^2} \cdot \frac{1}{\left(1 - \frac{v^2}{c^2}\right)^{3/2}} \tag{h}$$

and therefore, by (f), (g), and (h),

$$E_R = \frac{M_0}{c^2} \cdot \int_0^v v^3 \left(1 - \frac{v^2}{c^2}\right)^{-\frac{2}{3}} dv + M_0 \int_0^v v \cdot \left(1 - \frac{v^2}{c^2}\right)^{-\frac{1}{2}} dv \tag{i}$$

Performing the required integration, we obtain, after some careful simplification

$$E_R = M_0 \cdot c^2 \cdot \left\{ \frac{1}{\sqrt{1 - \frac{v^2}{c^2}}} - 1 \right\} \tag{j}$$

By using the dimensionless variables derived in (b), relation (j) can be conveniently written as

$$\pi_1 = \frac{1}{\sqrt{1-\pi_2^2}} - 1 \qquad (k)$$

which is then the dimensionless expression for the relativistic kinetic energy. Note that (k) contains only 2 (!) variables, therefore it can be represented graphically by a *single* curve.

Relation (j) can also be written in the more user-friendly form

$$E_R = \tfrac{1}{2} \cdot M_0 \cdot v^2 \cdot Z = E_N \cdot Z \qquad (l)$$

where E_N is the Newtonian kinetic energy—as in (a)—and

$$Z = \frac{2}{\pi_2^2} \cdot \left(\frac{1}{\sqrt{1-\pi_2^2}} - 1 \right) \qquad (m)$$

is the *Relativistic Magnification Factor*. If $v \ll c$, then by (b) $\pi_2 \approx 0$. Thus by (m) $Z \cong 1$, and by (l) the relativistic and Newtonian kinetic energies nearly coincide. If, however, v approaches c, then Z increases without bound, and if $v = c$, then Z becomes infinite. This is the reason why no material object, however small, can travel at the speed of light. For to attain such a speed, an infinite amount of energy would be needed, which obviously is never available.

Fig. 18-66 below shows some representative values of Z as a function of v and π_2. For example, if an object moves at 99.99% of the speed of light, then its kinetic energy is 139.4528 times what the Newtonian relation (a) predicts.

$v \times 10^8$ m/s	$\pi_2 = \frac{v}{c}$	Z
0.003	0.001	1.00000075
0.03	0.01	1.000075
0.3	0.1	1.007563
1.5	0.5	1.237604
2.7	0.9	3.19545
2.97	0.99	12.42488
2.997	0.999	42.81814
2.9997	0.9999	139.4528
2.999997	0.999999	1412.217

Figure 18-66
$\pi_2 = \frac{v}{c}$ **ratios and Z values at various v velocities**
c is the speed of light; Z is defined in relations (l) and (m)

(Adapted from Ref. 149, p. 233.)

Example 18-50. Penetration of a Bullet

We now deal with the problem of the penetration of a bullet into an immovable aluminum block. The bullet has a pointed conical front and its material is considered hard enough so as to sustain only negligible deformation upon impact. Moreover, the frictional heat generated during penetration is also assumed to be negligible. Fig. 18-67 shows the arrangement.

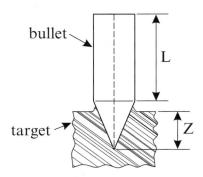

Figure 18-67
Hard bullet penetrating soft(er) target

Consequently, almost all of the kinetic energy of the bullet is transformed into the creation of *permanent* deformation of the target. It is assumed that all bullets are geometrically similar, so that the dimensions of a typical bullet are fully described by a single measure called *characteristic length*. The relevant variables are as follows:

Variable	Symbol	Dimension
penetration of bullet	Z	m
bullet speed at impact	v	m/s
ultimate stress of target	σ	kg/(m·s^2)
characteristic length of bullet	L	m
mass of bullet	M	kg

Note that *ultimate* stress, not yield stress, is used, because the target's deformation upon impact is permanent. We have $N_d = 5$ variables and $N_d = 3$ dimensions, thus, by Buckingham's theorem, $N_p = N_V - N_d = 5 - 3 = 2$ dimensionless variables—by the Dimensional Set—describe the system:

FIFTY-TWO ADDITIONAL APPLICATIONS   651

$$
\begin{array}{c|cc:ccc}
 & Z & v & \sigma & L & M \\
\hline
m & 1 & 1 & -1 & 1 & 0 \\
kg & 0 & 0 & 1 & 0 & 1 \\
s & 0 & -1 & 0 & -1 & 0 \\
\hdashline
\pi_1 & 1 & 0 & 0 & -1 & 0 \\
\pi_2 & 0 & 1 & -\frac{1}{2} & -\frac{3}{2} & \frac{1}{2}
\end{array}
$$

Dimensional Set

in which matrices **A**, **B**, and **D** are as shown, and the **C** matrix is by the *Fundamental Formula*

$$\mathbf{C} = -\mathbf{D}.(\mathbf{A}^{-1}.\mathbf{B})^{\mathrm{T}} \qquad \text{repeated (8-19)}$$

Now from the Dimensional Set

$$\pi_1 = \frac{Z}{L}; \quad \pi_2 = v.\sqrt{\frac{M}{\sigma.L^3}} \qquad (a)$$

Hence the Model Law is

$$S_Z = S_L; \quad S_v = \sqrt{S\sigma . \frac{S_L^3}{S_M}} \qquad (b)$$

where
- S_Z is the Bullet's Penetration Scale Factor;
- S_L is the Bullet's Size Scale Factor;
- S_v is the Bullet's Speed Scale Factor;
- S_M is the Bullet's Mass Scale Factor;
- S_σ is the Target's Ultimate Stress Scale Factor.

If the bullets are geometrically similar and are made of the same material, then

$$S_M = S_L^3 \qquad (c)$$

Therefore (b) is reduced to

$$S_Z = S_L; \quad S_v = \sqrt{S\sigma} \qquad (d)$$

which is now the resultant Model Law. This Model Law tells us that if we use the same target (i.e. $S_\sigma = 1$) and the projectile's geometry, material, and impact speed are not changed, then the penetration will vary *linearly* with the size of the bullet. Moreover, by (b), (c), and (d)

$$S_v^2 . S_M = S\sigma . S_L^3 = S\sigma . S_Z^3 \qquad (e)$$

Let us now introduce an additional scale factor, the Kinetic Energy Scale Factor S_Q defined as

$$S_Q = S_v^2 . S_M \qquad (f)$$

Thus by (e) and (f)

$$S_Q = S_\sigma \cdot S_Z^3 \qquad (g)$$

from which

$$\frac{S_Z}{\sqrt[3]{S_Q}} = \frac{1}{\sqrt[3]{S_\sigma}} \qquad (h)$$

Thus, if we aim at the same target ($S_\sigma = 1$), then the right side of relation (h) is constant, and so is its left side. This means that for any impact speed, the ratio of penetrations S_Z and the cubic root of impact energy S_Q is constant. Schuring confirms this fact on the basis of 15 firings into heavy blocks of aluminum-5083 with speeds ranging between 0.7 and 200 m/s (Ref. 31, p. 118).

(Adapted from Ref. 149, p. 236.)

Example 18-51. Sag of a Wire-rope Under General Load

Given a wire-rope assembly shown in Fig. 18-68.

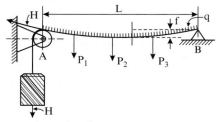

Figure 18-68
A wire-rope assembly under general load

A non-extensible, but flexible wire-rope is suspended between supports A and B. The rope is subjected to uniformly-distributed vertical load q, and a series of concentrated P_1, P_2, ...P_i forces acting vertically at fixed locations. If the relationships among the magnitudes of these forces are known, which is assumed to be the case here, then a single *hypothetical* force F—called the *representative* force—uniquely characterizes the P_1, P_2, ...P_i forces.

For the *prototype*, $L_1 = 52.4$ m, $q_1 = 85$ N/m, and $F_1 = 42,600$ N. The wire-rope serves as a high-frequency antenna, therefore its sag is *limited* to $f_1 = 1.85$ m. Question: what is the minimum tension force H_1 which assures that the sag will not exceed this permitted maximum value?

Because of the nature of the problem, the analytical approach is considered too complex; therefore we seek the solution by model experimentation. To this end, we first list the relevant variables:

FIFTY-TWO ADDITIONAL APPLICATIONS

Variable	Symbol	Dimension	Remark
sag	f	m	vertical measure
concentrated load	F	N	representative value
distributed load	q	N/m	vertical, uniform
wire tension	H	N	
span	L	m	horizontal measure

Accordingly, the dimensional matrix contains $N_V = 5$ variables and $N_d = 2$ dimensions.

$$\begin{array}{c|ccccc} & f & F & q & H & L \\ \hline m & 1 & 0 & -1 & 0 & 1 \\ N & 0 & 1 & 1 & 1 & 0 \end{array}$$

Therefore, by Buckingham's theorem, $N_p = N_v - N_d = 5 - 2 = 3$ dimensionless variables describe the system. These variables are supplied by the Dimensional Set shown in Fig. 18-69.

$$\begin{array}{c|ccc|cc} & f & F & q & H & L \\ \hline m & 1 & 0 & -1 & 0 & 1 \\ N & 0 & 1 & 1 & 1 & 0 \\ \hline \pi_1 & 1 & 0 & 0 & 0 & -1 \\ \pi_2 & 0 & 1 & 0 & -1 & 0 \\ \pi_3 & 0 & 0 & 1 & -1 & 1 \end{array}$$

Figure 18-69
Dimensional Set

In this Set, the **A**, **B**, and **D** matrices are as given, and the **C** matrix is by the *Fundamental Formula*

$$\mathbf{C} = -\mathbf{D}.(\mathbf{A}^{-1}.\mathbf{B})^T \qquad \text{repeated (8-19)}$$

Thus, by this Set, the dimensionless variables are:

$$\pi_1 = \frac{f}{L}; \quad \pi_2 = \frac{F}{H}; \quad \pi_3 = \frac{q.L}{H} \tag{a}$$

and therefore the Model Law is

$$S_f = S_L; \quad S_F = S_H; \quad S_q.S_L = S_H \tag{b}$$

where
  S_f is the Sag Scale Factor
  S_L is the Span Scale Factor

654   APPLIED DIMENSIONAL ANALYSIS AND MODELING

S_F is the Concentrated Load Scale Factor
S_H is the Tension Force Scale Factor
S_q is the Distributed Load Scale Factor

Now we *select* for the model $L_2 = 2.5$ m span and a $q_2 = 5.5$ N/m unit weight wire. Hence

$$S_L = \frac{L_2}{L_1} = \frac{2.5}{52.4} = 0.0477099; \qquad S_q = \frac{q_2}{q_1} = \frac{5.5}{85} = 0.0647059 \qquad (c)$$

where the subscripts "1" and "2" denote, respectively, the prototype and the model.

Therefore, by the third relation of the Model Law (b), the Tension Scale Factor must be

$$S_H = S_q \cdot S_L = (0.0647059)\cdot(0.0477099) = 0.00308711 \qquad (d)$$

By (c) and the first relation of the Model Law (b) now, the Deflection Scale Factor is

$$S_f = \frac{f_2}{f_1} = S_L = 0.0477099 \qquad (e)$$

Hence the sag on the *model* must not exceed

$$f_2 = f_1 \cdot S_f = f_1 \cdot S_L = (1.85)\cdot(0.0477099) = 0.0882633 \text{ m} \qquad (f)$$

Next, we consider the second relation of the Model Law (b). Thus $S_H = S_F = \frac{F_2}{F_1} = \frac{F_2}{42600}$ from which the representative concentrated load on the *model* is, by (d),

$$F_2 = F_1 \cdot S_H = (42600)\cdot(0.00308711) = 131.511 \text{ N} \qquad (g)$$

Next, we equip the model with a set of concentrated forces whose representative value is exactly $F_2 = 131.511$ N, and then *increase* the rope tension H_2 from a conveniently small value until the sag is *reduced* to the required $f_2 = 0.08826$ m.

Next, *this* tension on the *model* is *measured*. We find, say, $H_2 = 26.6$ N. By these datum and (d), the tension on the *prototype* will be

$$H_1 = \frac{H_2}{S_H} = \frac{26.6}{0.00308711} = 8616.47 \text{ N} \qquad (h)$$

Thus, for the prototype, *if* the tension force is *at least* $H_1 = 8616.47$ N, then its sag *will not be more* than the allowed $f_1 = 1.85$ m.

The required H_1 tension on the prototype can be easily provided by a 0.25 m diameter, 1.57 m long suspended lead cylinder, as illustrated in Fig. 18-68.

The Modeling Data Table (Fig. 18-70) conveniently summarizes all given and calculated data. Note that there are 3 category "2" variables in concordance with the fact that for Scale Factors the Model Law consists of 3 relations (conditions). Also note that, as required and expected, for the prototype and the model, the values of all 3 dimensionless variables are numerically identical pair-wise. This fact confirms that true *dimensional similarity* indeed existed in this experimentation.

(Adapted from Reference 149, p. 238.)

Variable					Scale Factor S	Category	
Name	Symb.	Dimen.	Prototype	Model	Model/Prototype	Prot.	Mod.
sag	f	m	1.85	0.0882633	0.0477099	1	2
span	L	m	52.4	2.5	0.0477099	1	1
concentrated load	F	N	42600	131.511	0.00308711	1	2
distributed load	q	N/m	85	5.5	0.0647059	1	1
rope tension force	H	N	8616.47	26.6	0.00308711	2	3
dimensionless	π_1	1	0.0353053	0.0353053	1	not applicable	
dimensionless	π_2	1	4.94402	4.94402	1		
dimensionless	π_3	1	0.516917	0.516917	1		
categories of variables		1	freely chosen, given a priori, or determined independently				
		2	determined by application of Model Law				
		3	determined by measurement on the model				

Figure 18-70
Modeling Data Table

Example 18-52. Proof that the Time Needed for a Mass-point to Descend on a Straight Line from the Top of a Vertical Circle to Any Point on its Perimeter is Independent of this Line's Slope

Given a vertical circle of diameter D (Fig. 18-71). With constant gravitation, a mass-point descends on a straight line chord A–B. Prove that the time it takes for this descent is independent of the slope φ of this chord.

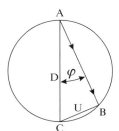

Figure 18-71
Point-mass descending without friction on line A–B

Although the analytical proof is simple, but we now present the proof by dimensional analysis—showing the great prowess of this method. In the figure, slope φ is defined by distance U. Thus if U = 0, then $\varphi = 0$; if U = D, then $\varphi = \dfrac{\pi}{2}$.

The variables, and their symbols and dimensions are given in the table below:

Name	Symbol	Dimension	Remark
time of descent	t	s	
characterizing distance	U	m	see Fig. 18-71
gravitational acceleration	g	m/s^2	
diameter of circle	D	m	see Fig. 18-71

There are $N_V = 4$ variables and $N_d = 2$ dimensions. Therefore by Buckingham's theorem, $N_P = N_V - N_d = 4 - 2 = 2$ dimensionless variables define the behaviour of the system. These 2 dimensionless variables are determined by the *Dimensional Set* (Fig. 18-72) as follows:

$$
\begin{array}{c|cc:cc}
 & t & U & g & D \\
\hline
m & 0 & 1 & 1 & 1 \\
s & 1 & 0 & -2 & 0 \\
\hdashline
\pi_1 & 1 & 0 & 0.5 & -0.5 \\
\pi_2 & 0 & 1 & 0 & -1
\end{array}
$$

with **B** and **D** on the left, **A** and **C** on the right.

Figure 18-72
Dimensional Set

in which the **A**, **B**, and **D** matrices are as shown, and the **C** matrix is by (8-19)

$$\mathbf{C} = -(\mathbf{A}^{-1}.\mathbf{B})^T \qquad \text{repeated (8-19)}$$

This Set now supplies the 2 dimensionless variables:

$$\pi_1 = t.\sqrt{\frac{g}{D}}\,; \quad \pi_2 = \frac{U}{D} \qquad (a)$$

If now we assume the plausible *monomial* form, then we have

$$t.\sqrt{\frac{g}{D}} = c.\left(\frac{U}{D}\right)^n \qquad (b)$$

where c and n are as yet undefined constants. From (b) now

$$t = c.\sqrt{\frac{D}{g}}.\left(\frac{U}{D}\right)^n \qquad (c)$$

which can be written as

$$t = c.U^n.D^{\frac{1}{2}-n}.g^{-\frac{1}{2}}. \qquad (d)$$

We now perform some elementary *heuristic* deliberation: obviously (d) must be true for all possible U values; therefore, it must be true when U = 0. In this case, our mass-point descends along the *vertical* diameter D. This mandates that the time of our mass-point journey is $t = \sqrt{2} \cdot \sqrt{\frac{D}{g}}$. Hence

$$t = \sqrt{2} \cdot D^{\frac{1}{2}} \cdot g^{-\frac{1}{2}} \qquad (e)$$

Therefore in (d), $c = \sqrt{2}$. Further, because U = 0, the expression U^n in (d) will only be *different* from 0 if n = 0. In this case, we have the indefinite form $\lim_{x \to 0} (x^x)$ which has the value of 1 (Ref. 150, p. 655 and 798, Exercise 20). Therefore $U^n = 1$. Putting these values into (d), we get

$$t = \sqrt{2} \cdot \sqrt{\frac{D}{g}} \qquad (f)$$

in which on the right-hand side U *does not appear*. Therefore U is a *dimensionally irrelevant* variable, and hence, by Theorem 11-3, it is also a *physically irrelevant* variable. Therefore, ***the time needed for a mass-point to descend along a chord A–B (Fig. 18-71) of a vertical circle is independent of the chord's slope*** φ (characterized by the U distance). This was to be proved.

REFERENCES IN NUMERICAL ORDER

1. Barenblatt, G. I., *Dimensional Analysis;* Gordon and Breach, New York, 1987.
2. Huntley, H. E., *Dimensional Analysis;* MacDonald & Co, London, 1952.
3. Wong, Young, *Theory of Ground Vehicles;* John Wiley and Sons, New York, 1978.
4. Semat, Henry, *Fundamentals of Physics;* Holt, Rinehart and Winston Inc., New York, 1966.
5. Fogiel, Max, *Essentials of Fluid Mechanics;* vol. 1. Research and Educational Association, Piscataway, NJ, 1987.
6. Fogiel, Max, *Essentials of Fluid Mechanics;* vol. 2. Research and Educational Association, Piscataway, NJ, 1987.
7. Muttnyánszky, Adám, *Statics and Strength of Materials* (in Hungarian); Technical Bookpublisher, Budapest, 1981.
8. Editors, *Canadian Metric Practice Guide;* National Standards of Canada (CAN/CSA-Z234.1-89), Canadian Standards Association, Toronto, 1989.
9. Feynman, Richard, *The Feynman Lectures on Physics;* in 3 volumes, Addison-Wesley, Reading, MA, 1964.
10. Jerrard, H. and McNeill, D., *Dictionary of Scientific Units;* 6th ed., Chapman & Hall, London, 1992.
11. Kern, Donald, *Process Heat Transfer;* McGraw-Hill Book Company, New York, 1950.
12. Editors, *CRC Handbook of Chemistry and Physics,* 69th ed., The Chemical Rubber Company Press, Cleveland, OH, 1989.
13. Szirtes, Thomas, *Some Human Communication Characteristics and the Principle of Least Effort;* Paper (No. I/11) presented at the Scientific Conference organized jointly by the Hungarian Academy of Sciences and Technical University of Budapest, Budapest, Hungary, 1989 August.
14. Khinchin, A. I., *Mathematical Foundation of Information Theory;* Dover Publications Inc., New York, 1957.
15. Timoshenko, Stephen P., *History of Strength of Materials;* Dover Publications Inc., New York, 1983.
16. Editors, *Van Nostrand's Scientific Encyclopedia;* 3rd ed. D. Van Nostrand Company Inc., Princeton, NJ, 1958.
17. Jensen, J. et al., *Design Guide to Orbital Flight;* McGraw-Hill Book Company, New York, 1962.

18. Teller, Edward, *Conversations on the Dark Secrets of Physics;* Plenum Press, New York, 1991.
19. Langhaar, Henry L., *Dimensional Analysis and Theory of Models;* Robert E. Krieger Publishing Co., Malabar, FL., 1980. This is the reissue of book published by John Wiley & Sons Inc., New York, in 1951.
20. Smart, W. M., *Textbook of Spherical Astronomy;* 4th ed., Cambridge University Press, Cambridge, 1960.
21. Douglas, John F., *An Introduction to Dimensional Analysis for Engineers;* Sir Isaac Pitman & Sons, London, 1969.
22. Isaacson de St Q, E. & M., *Dimensional Methods in Engineering and Physics;* Edward Arnold Ltd, London, 1975.
23. Porter, Alfred W., *The Method of Dimensions;* Methuen & Co., London, 1933.
24. Pedley, T. J. (Ed.), *Scale Effects in Animal Locomotion;* (a 3-volume collection of papers presented at a Cambridge University Symposium held in 1975 September), Academic Press, London, 1977.
25. Rashevsky, Nicolas, *Some Medical Aspects of Mathematical Biology;* Charles C Thomas Publisher, Springfield, IL, 1964.
26. Szücs, Ervin, *Similitude and Modelling;* Elsevier Scientific Publishing Company, Amsterdam, 1980.
27. David, F. and Nolle, H., *Experimental Modelling in Engineering;* Butterworth, London, 1982.
28. Pankhurst, R. C., *Dimensional Analysis and Scale Factors;* Chapman and Hall Limited, London, 1964.
29. Taylor, Edward S., *Dimensional Analysis for Engineers;* Clarendon Press, Oxford, 1974.
30. Zierep, Jürgen, *Similarity Laws and Modeling;* Marcel Dekker, New York, 1971 (text identical with Ref. 47).
31. Schuring, Dieterich J., *Scale Models in Engineering;* Pergamon Press, Oxford, 1977.
32. Venikov, V. A., *Theory of Similarity and Simulation;* Macdonald Technical & Scientific, London, 1969.
33. Staicu, Constantin I., *Restricted and General Dimensional Analysis;* Abacus Press, Tunbridge Wells, Kent, UK, 1982.
34. Baker, Wilfred et al., *Similarity Methods in Engineering Dynamics;* Elsevier, Amsterdam, 1991.
35. Johnstone, Robert et al., *Pilot Plants, Models and Scale-up Methods in Chemical Engineering;* McGraw-Hill Book Company, New York, 1957.
36. De Jong, Frits J., *Dimensional Analysis for Economists;* North Holland Publishing Co., Amsterdam, 1967.
37. Palacios, J., *Dimensional Analysis;* MacMillan and Co., London, 1964.
38. Sena, L. A., *Units of Physical Quantities and Their Dimensions;* Mir Publishers, Moscow, 1972.
39. Jupp, Edmund W., *An Introduction to Dimensional Method;* Cleaver-Hume Press Ltd., London, 1962.
40. Haynes, Robin, *A Introduction of Dimensional Analysis for Geographers;* Essay # 33 in the series *Concepts and Techniques in Modern Geography,* published in 1982 by Geo Abstracts Ltd., Norwich, UK.

41. Focken, C. M., *Dimensional Methods and their Applications;* Edward Arnold and Co., London, 1953.
42. Kurth, Rudolph, *Dimensional Analysis and Group Theory in Astrophysics;* Pergamon Press, Oxford, 1972 (see also Ref. 72).
43. Schepartz, Bernard, *Dimensional Analysis in the Biomedical Sciences;* Charles C Thomas Publishers, Springfield, IL, 1980.
44. Bridgeman, Percy W., *Dimensional Analysis;* Yale University Press, New Haven, CT, 1922 (reissued in paperbound in 1963).
45. Ipsen, D. C., *Units, Dimensions and Dimensionless Numbers;* McGraw-Hill Book Company, New York, 1960.
46. Skoglund, Victor J., *Similitude—Theory and Application;* International Textbook Co., Scranton, PA, 1967.
47. Zierep, Jürgen, *Similarity Laws and Modeling;* (text identical with Ref. 30).
48. Ramshaw, J. & Williams, T., *The Rolling Resistance of Commercial Vehicle Tyres;* Transport and Road Research Laboratory Report 701, ISSN 0305-1315, Crowthorne, Berkshire, UK, 1981.
49. Reinhold, Timothy A. (Ed.), *Wind Tunnel Modeling for Civil Engineering Applications;* Cambridge University Press, Cambridge, UK. Collection of papers presented at the International Workshop on Wind Tunnel Modeling in Civil Engineering, Gaithersburg, Maryland, 1982 April.
50. Schultz, Peter, *Atmospheric Effects on Ejecta Emplacement and Crater Formation on Venus from Magellan;* Journal of Geophysical Research, vol. 97, No. E10, p. 16.183, 1992 October 25.
51. Sedov, L. I., *Similarity and Dimensional Methods in Mechanics;* Academic Press, New York, 1959.
52. Blench, T., *Dimensional Analysis and Dynamical Similarity for Hydraulic Engineers;* University Bookstore, Edmonton, Alberta, 1969.
53. McAdams, William, *Heat Transmission;* McGraw-Hill Book Company, New York, 1954.
54. Reynolds, Osborne, *An Experimental Investigation of the Circumstances which determine whether the Motion of Water shall be Direct or Sinuous, and of the Law of Resistance in Parallel Channels;* Philosophical Transactions 1883, Part 3, p. 935.
55. White, R. & Guedelhoefer O., *Models of Concrete Structures;* two papers presented by White and Guedelhoefer, respectively, at the Annual Convention of the US Concrete Institute, Dallas, 1972.
56. Günther, B., *Dimensional Analysis and Theory of Biological Similarity;* Physiological Reviews, Vol. 55. No. 4, p. 659, 1975 October.
57. Naddor, Eliezer, *Dimensions in Operations Research;* (Journal of) Operations Research, IV, p. 508, 1966.
58. Costa, Fernando Vascos, *Directional Analysis in Model Design;* Journal of the Engineering Mechanics Division, Proceedings of the American Society of Civil Engineering, p. 519, 1971 April.
59. Stahl, Walter R., *Similarity and Dimensional Method in Biology;* Science, vol. 137, p. 205, 1962 July 20.
60. Stumpf, H. W., *Scale Models in Tyre Design;* Paper presented at the 23rd International Symposium on Automotive Technology and Automation, Vienna, Austria, 1990 December.

61. Macagno, Enzo O., *Historico-Critical Review of Dimensional Analysis;* Journal of The Franklin Institute, vol. 292, #6, p. 391, 1971 December.
62. Chen, Wai-Kai, *Algebraic Theory of Dimensional Analysis;* Journal of Franklin Institute, vol. 292, #6, p. 403, 1971 December.
63. Moran, M. J., *A Generalization of Dimensional Analysis;* Journal of Franklin Institute, vol. 292, #6, p. 423, 1971 December.
64. Pomerantz, Martin, *Foreword;* Journal of The Franklin Institute, vol. 292, #6, p. 389, 1971 December.
65. Barr, D. I. H., *The Proportionalities Method of Dimensional Analysis;* Journal of The Franklin Institute, vol. 292, #6, p. 441, 1971 December.
66. Staicu, Constantin I., *General Dimensional Analysis;* Journal of The Franklin Institute, vol. 292, #6, p. 433, 1971 December (see also Ref. 33).
67. Chakrabarti, Subrata K., *On the Constitutive Equations and Dimensional Analysis;* Journal of The Franklin Institute, vol. 296, #5, p. 329, 1973 November.
68. Buckingham, Edgar, *Letter to the Editor;* Nature, 1915 October 15.
69. Buckingham, Edgar, *Letter to the Editor;* Nature, 1915 December 9.
70. Buckingham, Edgar, *On Physically Similar Systems;* Physical Reviews, 1914, vol. 4, #4, 2nd series, p. 345.
71. Rayleigh, Lord, *The Principle of Similitude;* Nature, vol. 95, #2368, p. 66, 1915 March 18.
72. Kurth, Rudolph, *A Note on Dimensional Analysis;* American Mathematical Monthly, vol. 72, p. 965, 1965 November (see also Ref. 42).
73. Drobot, S., *On the Foundation of Dimensional Analysis;* Studia Mathematica, Polska Akademia, vol. 14, Fasc. 1, p. 84, Wroclaw, 1953.
74. Edwards, N. A., *Scaling of Renal Functions in Mammals;* Compendium of Biochemical Physiology, vol. 52A, p. 63, 1975.
75. Gilvarry, J. J., *Geometric and Physical Scaling of River Dimensions;* Journal of The Franklin Institute, vol. 292, #6, p. 499, 1971 December
76. Kleiber, Max, *Body Size and Metabolic Rate;* Physiological Reviews, vol. 27, #4, p. 511, 1947 October.
77. Alexander, R., *Estimates of Speeds of Dinosaurs;* Nature, vol. 261, p. 129, 1976 May 13.
78. Stahl, Walter R., *Dimensional Analysis in Mathematical Biology I;* (general discussion—continued in Ref. 79), Bulletin of Mathematical Biophysics, vol. 23, p. 355, 1961.
79. Stahl, Walter R., *Dimensional Analysis in Mathematical Biology II;* Bulletin of Mathematical Biophysics, vol. 24, p. 81, 1962.
80. Lehman, J. and Craig, E., *Dimensional Analysis in Applied Psychological Research;* Journal of Psychology, vol. 55, p. 223, 1963.
81. Kroon, R. P., *Dimensions;* Journal of The Franklin Institute, vol. 292, #1, p. 45, 1971 July.
82. Pankhurst, R. C., *Alternative Formulation of Pi-Theorem;* Journal of The Franklin Institute, vol. 292, #6, p. 451, 1971 December (see also Ref. 28).
83. Hainzl, J., *On Local Generalization of the Pi-Theorem of Dimensional Analysis;* Journal of The Franklin Institute, vol. 292, #6, p. 463, 1971 December.
84. Tonndorf, Juergen, *Dimensional Analysis of the Cochlear Models;* The Journal of the Acoustic Society of America, vol. 32, #4, p. 493, 1960 April.

85. Lodge, A., *The Multiplication and Division of Concrete Quantities;* Nature, p. 281, 1888 July 19.
86. Rücker, A. W., *On the Suppressed Dimensions of Physical Quantities;* Philosophical Magazine, Series 5, vol. 27, XIII, p. 104, 1889.
87. Van Driest, E. R., *On Dimensional Analysis and the Presentation of Data in Fluid-Flow Problems;* Journal of Applied Mechanics, p. A-34, 1946 March.
88. Fitzgerald, G. F., *On the Dimension of Electromagnetic Units,* Philosophical Magazine, Series 5, vol. 27, p. 323, 1889.
89. Sayao, Otavio J., *Physical Modeling of Pocket Beaches;* Proceedings of a Speciality Conference on Quantitative Approaches to Coastal Sediment Processes, p. 1625, Seattle, Washington, 1991 June.
90. Prasuhn, Alan L., *Fundamentals of Hydraulic Engineering;* Holt, Rinehart and Winston, New York, 1987.
91. Rao, Govinda N. S., *Hydraulics;* Chapter III, p. 35, ASIA Publishing House, Bombay.
92. Sayao, Otavio J. et al., *Dimensional Analysis of Littoral Transport;* Proceedings of the Canadian Coastal Conference, p. 241, St. John's, Newfoundland, 1985. Proceedings are published by the National Research Council of Canada, Ottawa, Ontario.
93. Bridgeman, Percy W., *Dimensional Analysis;* Encyclopedia Britannica, vol. 7, p. 439. Encyclopedia Britannica Inc., Chicago, IL.
94. Corsin, Stanley, *A Simple Geometrical Proof of Buckingham's π theorem;* American Journal of Physics, vol. 19, p. 180, 1951.
95. Ettinger, Morris B., *How to Plan an Inconsequential Research Project;* Journal of Sanitary Engineering Division (Proceedings of the American Society of Civil Engineers), p. 19, 1965 August.
96. Menkes, Aviva and Josh, *The Application of Dimensional Analysis to Learning Theory;* Psychological Review, vol. 64, No. 1, p. 8, 1957.
97. Gossage, R. J., *The Use of Shop Models Magnetically to Scale in Degaussing Research;* Journal of the Institute of Electrical Engineers, vol. 93, part 1, p. 447, 1946.
98. Ghista, Dhanjoo N. et al., *Measurement, Modeling, Control and Simulation—As Applied to the Human Left Ventricle for Purposeful Physiological Monitoring;* Journal of The Franklin Institute, vol. 292, # 6, p. 545, 1971 December.
99. Strahler, Arthur N., *Dimensional Analysis Applied to Fluvially Eroded Landforms;* Bulletin of the Geological Society of America, vol. 69 p. 279, 1958 March.
100. Murphy, Glenn, *Models with Incomplete Correspondence with the Prototype;* Journal of The Franklin Institute, vol. 292, #6, p. 513, 1971 December.
101. Blau, Garry E., *Optimization of Models Derived by Dimensional Analysis Using Generalized Polynomial Programming;* Journal of the Franklin Institute, vol. 292, #6, p. 519, 1971 December.
102. Happ, W. W., *Dimensional Analysis via Directed Graphs;* Journal of the Franklin Institute, vol. 292, #6, p. 527, 1971 December.
103. Riabouchinsky, D., *Letter to the Editor* (on the subject of The Principle of Similitude); Nature, vol. 95, # 2387, p. 591, 1915 July 29.
104. Rayleigh, Lord, *Letter to the Editor (on the subject of Fluid Friction on Even Surfaces);* Philosophical Magazine and Journal of Science, vol. 8, 6th series, p. 66, 1904.
105. Sinclair, George, *Theory of Models of Electromagnetic Systems;* Proceeding of the IRE, p. 1364, 1948 November.

106. Agassi, Joseph, *The Kirchoff–Planck Radiation Law;* Science, vol. 156, p. 30, 1967 April 7.
107. Bethe, Hans A., *Energy Production in Stars;* Science, vol. 161, p. 541, 1968 August.
108. Rayleigh, Lord, *The Size of Drops;* Philosophical Magazine and Journal of Science, Series 5, vol. 48, #293, p. 321, 1899 October.
109. Rayleigh, Lord, *Letter to the Editor;* Nature, #2389, vol. 95, p. 644, 1915, August 12 (reply to Dr. Riabouchinsky's letter, Ref. 103).
110. Thomson, D'Arcy W., *Letter to the Editor;* Nature, vol. 95, #2373, 1915 April.
111. Rayleigh, Lord, *Letter to the Editor;* Nature, vol. 95, #2373, 1915 April.
112. Hill, A. V., *The Dimension of Animals and their Muscular Dynamics;* Science Progress, vol. 38, #150, p. 209, 1950 April.
113. Dingle, Herbert, *On the Dimensions of Physical Magnitudes;* Philosophical Magazine, vol. 33, Ser. 7, #220 (1st article) p. 321, 1942 May.
114. Dingle, Herbert, *On the Dimensions of Physical Magnitudes;* Philosophical Magazine, vol. 34, Ser. 7, #236 (3rd article) p. 588, 1943 September.
115. Dingle, Herbert, *On the Dimensions of Physical Magnitudes;* Philosophical Magazine, vol. 35, Ser. 7, #248 (5th article) p. 616, 1944 May.
116. Dingle, Herbert, *On the Dimensions of Physical Magnitudes;* Philosophical Magazine, vol. 37, Ser. 7, #264 (6th article) p. 64, 1946 January.
117. Thomson, W. & Barton, V., *The Response of Mechanical Systems to Random Excitation;* ASME Journal of Applied Mechanics, 1957, June p. 248.
118. Rose, William K., *Astrophysics;* Holt, Rinehart and Winston Inc., New York, 1973.
119. Jeans, J. H., *On the Law of Radiation;* Proc. of The Royal Society, vol. 27, p. 545, 1905.
120. Farkas, K. & Lenard, J., *New Formulae for the Calculation of Roll Force and Torque in Hot Strip Rolling;* (private communication).
121. Johanson, J. R., *Modeling Flow of Bulk of Solids;* Powder Technology, Elsevier Sequoia S.A., Lusanne, vol. 5, p. 93, 1971/72.
122. Smith, Maynard J., *Mathematical Ideas in Biology;* Cambridge University Press, Cambridge, 1968.
123. Endrényi, John, *On Quantities, Dimensions and Choice of Units;* Internal Report # 61-27 of The Hydro-Electric Power Commission of Ontario, Research Division, 1961.
124. Hughes, W. and Brighton, J., *Fluid Dynamics;* Schaum outline series 2nd ed. McGraw-Hill Book Company, New York, 1991.
125. Editors, *Tire Rolling Losses and Fuel Economy;* Proceedings of R & D Planning Workshop at the Transportation Research Center, Cambridge, MA, 1977 October.
126. Mario, Thomas, *The Playboy Gourmet;* Castle Books, Secaucus, NJ, 1977.
127. O'Nan, Michael, *Linear Algebra;* Harcourt Brace Jovanovich, New York, 1971.
128. Kürschák, József (comp.), *Hungarian Problem Book,* vol. 1 (based on the Eötvös Competition, 1894–1905), New Mathematical Library Series #1, published by Random House and the L. W. Singer Co., Copyright Yale University, 1963.
129. Euclid, *Elements—Book 1;* Translated by Sir Thomas L. Heath. Encyclopedia Britannica Inc., Chicago, 1952.
130. Huntley, H. E., *The Divine Proportion—A Study in Mathematical Beauty;* Dover Publications Inc., New York, 1970.

131. Den Hartog, J. P., *Mechanics;* Dover Publication Inc., New York, 1961.
132. Thomson, William, *Theory of Vibration;* Prentice Hall Inc., Englewood Hills, NJ, 1972.
133. Timoshenko, Stephen P., *Strength of Materials* (in 2 volumes); 3rd ed., D. Van Nostrand Co., Princeton, NJ, 1955.
134. Timoshenko, S. & Gere, J., *Theory of Elastic Stability;* McGraw-Hill Book Company, New York, 1961.
135. Editors, *Hewlett Packard HP48SX Owner's Manual;* vol. 1, 4th ed. 1990 July.
136. Rózsa, Pál, *Linear Algebra and its Applications;* 3rd ed. (in Hungarian), Textbook Publishing Co., Budapest, 1991.
137. Wolverton, Raymond (ed.), *Flight Performance Handbook for Orbital Operations;* John Wiley and Sons, New York, 1963.
138. Taylor, Richard C., *The Energetics of Terrestrial Locomotion and Body Size in Vertebrates;* (included in Ref. 24, p. 127).
139. Bennet-Clark, H. C., *Scale Effect of Jumping Animals;* (included in Ref. 24, p. 185).
140. Haldane, Sanderson J. B., *On Being the Right Size;* in the *World of Mathematics,* a 4-volume collection of essays (James Newman, Editor), p. 952, Simon and Shuster, New York, 1956.
141. Timoshenko, S. & Young, D., *Engineering Mechanics;* 4th ed., McGraw-Hill Book Company, New York, 1956.
142. Szirtes, Thomas, *Minimizing the Deflection of Tubular Cantilevers;* Machine Design, vol. 57, #3, p. 103, 1985 February.
143. Kármán, von T. & Biot, M., *Mathematical Methods in Engineering;* McGraw-Hill Book Company, New York, 1940.
144. Szirtes, Thomas, *Conditions of Jamming a Circular Plug in a Non-Smooth Circular Hole;* SPAR Aerospace Ltd. Report, 1991 June.
145. Szirtes, Thomas, *Choosing the Right Shock Absorber;* Machine Design, vol. 61, #10, p. 66, 1989 May.
146. Ryabov, Y., *An Elementary Survey of Celestial Mechanics;* Dover Publications Inc., New York, 1961.
147. Waverman, Lucy, *The Ultimate Roast Turkey;* The Globe and Mail (Toronto, Ontario), 1995 December 13, p. A-24.
148. Szirtes, Thomas, *The Fine Art of Modeling;* SPAR Journal of Engineering and Technology, vol. 1, p. 37, 1992 May.
149. Szirtes, Thomas, *Some Interesting Applications of Dimensional Analysis;* Acta Technica, vol. 108, no. 1-2; Academic Publisher; Budapest, Hungary; 1999.
150. Thomas, George B. *Calculus and Analytic Geometry.* Addison-Wesley Publishing Co., Reading, California, USA; 1968.

REFERENCES IN ALPHABETICAL ORDER OF AUTHORS' SURNAMES

Sequential number appears in parentheses following the author's name.

Agassi, Joseph (106), *The Kirchoff-Planck Radiation Law;* Science, vol. 156, p. 30, 1967 April 7.

Alexander, R. (77), *Estimates of Speeds of Dinosaurs;* Nature, vol. 261, p. 129, 1976 May 13.

Baker, Wilfred et al. (34), *Similarity Methods in Engineering Dynamics;* Elsevier, Amsterdam, 1991.

Barenblatt, G. I. (1), *Dimensional Analysis;* Gordon and Breach, New York, 1987.

Barr, D. I. H. (65), *The Proportionalities Method of Dimensional Analysis;* Journal of The Franklin Institute, vol. 292, #6, p. 441, 1971 December.

Barton, V. (117), *The Response of Mechanical Systems to Random Excitation* (with Thompson, W.); ASME Journal of Applied Mechanics, 1957 June, p. 248.

Bennet-Clark, H. C. (139), *Scale Effect of Jumping Animals;* (included in Ref. 24, p. 185).

Bethe, Hans A. (107), *Energy Production in Stars;* Science, vol. 161, p. 541, 1968 August.

Biot, M. (143), *Mathematical Methods in Engineering* (with von Kármán, T.); McGraw-Hill Book Company, New York, 1940.

Blau, Garry E. (101), *Optimization of Models Derived by Dimensional Analysis Using Generalized Polynomial Programming;* Journal of the Franklin Institute, vol. 292, #6, p. 519, 1971 December.

Blench, T. (52), *Dimensional Analysis and Dynamical Similarity for Hydraulic Engineers;* University Bookstore, Edmonton, Alberta, 1969.

Bridgeman, Percy W. (44), *Dimensional Analysis;* Yale University Press, New Haven, CT, 1922 (reissued in paperbound in 1963).

Bridgeman, Percy W. (93), *Dimensional Analysis;* Encyclopedia Britannica, vol. 7, p. 439. Encyclopedia Britannica Inc., Chicago, IL.

Brighton, J. (124), *Fluid Dynamics* (with Hughes, W.); Schaum outline series 2nd ed. McGraw-Hill Book Company, New York, 1991.

Buckingham, Edgar (70), *On Physically Similar Systems;* Physical Reviews, 1914, vol. 4, #4, 2nd series, p. 345.

Buckingham, Edgar (69), *Letter to the Editor;* Nature, 1915 December 9.

Buckingham, Edgar (68), *Letter to the Editor;* Nature, 1915 October 15.

Chakrabarti, Subrata (67), *On the Constitutive Equations and Dimensional Analysis;* Journal of The Franklin Institute, vol. 296, #5, p. 329, 1973 November.

Chen, Wai-Kai (62), *Algebraic Theory of Dimensional Analysis;* Journal of The Franklin Institute, vol. 292, #6, p. 403, 1971 December.

Corsin, Stanley (94), *A Simple Geometrical Proof of Buckingham's π theorem;* American Journal of Physics, vol. 19, p. 180, 1951.

Costa, Fernando V. (58), *Directional Analysis in Model Design;* Journal of the Engineering Mechanics Division, Proceedings of the American Society of Civil Engineering, p. 519, 1971 April.

Craig, E. (80), *Dimensional Analysis in Applied Psychological Research* (with Lehman, J.); Journal of Psychology, vol. 55, p. 223, 1963.

David, F. (27), *Experimental Modelling in Engineering;* Butterworth, London, 1982 (with Nolle, H.)

De Jong, Frits J. (36), *Dimensional Analysis for Economists;* North Holland Publishing Co., Amsterdam, 1967.

Den Hartog, J. P. (131), *Mechanics;* Dover Publications Inc., New York, 1961.

Dingle, Herbert (113), *On the Dimensions of Physical Magnitudes;* Philosophical Magazine, vol. 33, Ser. 7, #220 (1st article) p. 321, 1942 May.

Dingle, Herbert (114), *On the Dimensions of Physical Magnitudes;* Philosophical Magazine, vol. 34, Ser. 7, #236 (3rd article) p. 588, 1943 September.

Dingle, Herbert (115), *On the Dimensions of Physical Magnitudes;* Philosophical Magazine, vol. 35, Ser. 7, #248 (5th article) p. 616, 1944 May.

Dingle, Herbert (116), *On the Dimensions of Physical Magnitudes;* Philosophical Magazine, vol. 37, Ser. 7, #264 (6th article) p. 64, 1946 January.

Douglas, John F. (21), *An Introduction to Dimensional Analysis for Engineers;* Sir Isaac Pitman & Sons, London, 1969.

Drobot, S. (73), *On the Foundation of Dimensional Analysis;* Studia Mathematica, Polska Akademia, vol. 14, Fasc. 1, p. 84, Wroclaw, 1953.

Editors (8), *Canadian Metric Practice Guide;* National Standards of Canada (CAN/CSA-Z234.1-89), Canadian Standards Association, Toronto, 1989.

Editors (12), *CRC Handbook of Chemistry and Physics,* 69th ed., The Chemical Rubber Company Press, Cleveland, OH, 1989.

Editors (125), *Tire Rolling Losses and Fuel Economy;* Proceedings of R & D Planning Workshop at the Transportation Research Center, Cambridge, MA, 1977 October.

Editors (135), *Hewlett Packard HP48SX Owner's Manual;* vol. 1, 4th ed. 1990 July.

Editors (16), *Van Nostrand's Scientific Encyclopedia;* 3rd ed. D. Van Nostrand Company Inc., Princeton, NJ, 1958.

Edwards, N. A. (74), *Scaling of Renal Functions in Mammals;* Compendium of Biochemical Physiology, vol. 52A, p. 63, 1975.

Endrényi, John (123), *On Quantities, Dimensions and Choice of Units;* Internal Report # 61-27 of The Hydro-Electric Power Commission of Ontario, Research Division, 1961.

Ettinger, Morris B. (95), *How to Plan an Inconsequential Research Project;* Journal of Sanitary Engineering Division (Proceedings of the American Society of Civil Engineers), p. 19, 1965 August.

Euclid (129), *Elements—Book 1;* Translated by Sir Thomas L. Heath. Encyclopedia Britannica Inc., Chicago, 1952.

Farkas, K. (120), *New Formulae for the Calculation of Roll Force and Torque in Hot Strip Rolling* (with Lenard, J); Private communication.

Feynman, Richard (9), *The Feynman Lectures on Physics;* in 3 volumes, Addison-Wesley, Reading, MA, 1964.

Fitzgerald, G. F. (88), *On the Dimension of Electromagnetic Units,* Philosophical Magazine, Series 5, vol. 27, p. 323, 1889.

Focken, C. M. (41), *Dimensional Methods and their Applications;* Edward Arnold and Co., London, 1953.

Fogiel, Max (5), *Essentials of Fluid Mechanics;* vol. 1. Research and Educational Association, Piscataway, NJ, 1987.

Fogiel, Max (6), *Essentials of Fluid Mechanics;* vol. 2. Research and Educational Association, Piscataway, NJ, 1987.

Gere, J. (134), *Theory of Elastic Stability* (with Timoshenko, S.); McGraw-Hill Book Company, New York, 1961.

Ghista, D. N. et al. (98), *Measurement, Modeling, Control and Simulation—As Applied to the Human Left Ventricle for Purposeful Physiological Monitoring;* Journal of The Franklin Institute, vol. 292, # 6, p. 545, 1971 December.

Gilvarry, J. J. (75), *Geometric and Physical Scaling of River Dimensions;* Journal of The Franklin Institute, vol. 292, #6, p. 499, 1971 December.

Gossage, R. J. (97), *The Use of Shop Models Magnetically to Scale in Degaussing Research;* Journal of the Institute of Electrical Engineers, vol. 93, part 1, p. 447, 1946.

Guedelhoefer, O. (55), *Models of Concrete Structures;* Paper presented at the Annual Convention of the US Concrete Institute, Dallas, 1972.

Günther, B. (56), *Dimensional Analysis and Theory of Biological Similarity;* Physiological Reviews, Vol. 55. No. 4, p. 659, 1975 October.

Hainzl, J. (83), *On Local Generalization of the Pi-Theorem of Dimensional Analysis;* Journal of The Franklin Institute, vol. 292, #6, p. 463, 1971 December.

Haldane, J. B. S. (140), *On Being the Right Size;* in the World of Mathematics, a 4-volume collection of essays (James Newman, Editor), p. 952, Simon and Shuster, New York, 1956.

Happ, W. W. (102), *Dimensional Analysis via Directed Graphs;* Journal of the Franklin Institute, vol. 292, #6, p. 527, 1971 December.

Haynes, Robin (40), *A Introduction of Dimensional Analysis for Geographers;* Essay # 33 in the series Concepts and Techniques in Modern Geography, published in 1982 by Geo Abstracts Ltd., Norwich, UK.

Hill, A. V. (112), *The Dimension of Animals and their Muscular Dynamics;* Science Progress, vol. 38, #150, p. 209, 1950 April.

Hughes, W. (124), *Fluid Dynamics* (with Brighton, J.); Schaum outline series 2nd ed. Mc-Graw-Hill Book Company, New York, 1991.

Huntley, H. E. (130), *The Divine Proportion—A Study in Mathematical Beauty;* Dover Publications Inc., New York, 1970.

Huntley, H. E. (2), *Dimensional Analysis;* MacDonald & Co, London, 1952.

Ipsen, D. C. (45), *Units, Dimensions and Dimensionless Numbers;* McGraw-Hill Book Company, New York, 1960.

Isaacson de St Q, E. (22), *Dimensional Methods in Engineering and Physics* (with Isaacson de St Q, M.); Edward Arnold Ltd, London, 1975.

Isaacson de St Q, M. (22), *Dimensional Methods in Engineering and Physics* (with Isaacson de St Q, E.); Edward Arnold Ltd, London, 1975.

Jeans, J. H. (119), *On the Law of Radiation;* Proc. of The Royal Society, vol. 27, p. 545, 1905.

Jensen, J. et al. (17), *Design Guide to Orbital Flight;* McGraw-Hill Book Company, New York, 1962.

Jerrard, H. (10), *Dictionary of Scientific Units* (with McNeill, D.); 6th ed., Chapman & Hall, London, 1992.

Johanson, J. R. (121), *Modeling Flow of Bulk of Solids;* Powder Technology, Elsevier Sequoia S.A., Lusanne, vol. 5, p. 93, 1971/72.

Johnstone, R. et al. (35), *Pilot Plants, Models and Scale-up Methods in Chemical Engineering;* McGraw-Hill Book Company, New York, 1957.

Jupp, Edmund W. (39), *An Introduction to Dimensional Method;* Cleaver-Hume Press Ltd., London, 1962.

Kármán, von T. (143), *Mathematical Methods in Engineering* (with Biot, M.); McGraw-Hill Book Company, New York, 1940.

Kern, Donald (11), *Process Heat Transfer;* McGraw-Hill Book Company, New York, 1950.

Khinchin, A. I. (14), *Mathematical Foundation of Information Theory;* Dover Publications Inc., New York, 1957.

Kleiber, Max (76), *Body Size and Metabolic Rate;* Physiological Reviews, vol. 27, #4, p. 511, 1947 October.

Kroon, R. P. (81), *Dimensions;* Journal of The Franklin Institute, vol. 292, #1, p. 45, 1971 July.

Kürschák, J. (comp.) (128), *Hungarian Problem Book,* vol. 1, (based on the Eötvös Competition, 1894–1905), New Mathematical Library Series #1, published by Random House and the L. W. Singer Co., Copyright Yale University, 1963.

Kurth, Rudolph (42), *Dimensional Analysis and Group Theory in Astrophysics;* Pergamon Press, Oxford, 1972 (see also Ref. 72).

Kurth, Rudolph (72), *A Note on Dimensional Analysis;* American Mathematical Monthly, vol. 72, p. 965, 1965 November (see also Ref. 42).

Langhaar, Henry L. (19), *Dimensional Analysis and Theory of Models;* Robert E. Krieger Publishing Co., Malabar, FL, 1980. This is the reissue of book published by John Wiley & Sons Inc., New York, in 1951.

Lehman, J. (80), *Dimensional Analysis in Applied Psychological Research* (with Craig, E.); Journal of Psychology, vol. 55, p. 223, 1963.

Lenard, J. (120), *New Formulae for the Calculation of Roll Force and Torque in Hot Strip Rolling* (with Farkas, K.); Private communication.

Lodge, A. (85), *The Multiplication and Division of Concrete Quantities;* Nature, p. 281, 1888 July 19.

Macagno, Enzo O. (61), *Historico-Critical Review of Dimensional Analysis;* Journal of The Franklin Institute, vol. 292, #6, p. 391, 1971 December.

Mario, Thomas (126), *The Playboy Gourmet;* Castle Books, Secaucus, NJ, 1977.

McAdams, William (53), *Heat Transmission;* McGraw-Hill Book Company, New York, 1954.

McNeill, D. (10), *Dictionary of Scientific Units* (with Jerrard, H); 6th ed., Chapman & Hall, London, 1992.

Menkes, A. and J. (96), *The Application of Dimensional Analysis to Learning Theory;* Psychological Review, vol. 64, No. 1, p. 8, 1957.

Moran, M. J. (63), *A Generalization of Dimensional Analysis;* Journal of Franklin Institute, vol. 292, #6, p. 423, 1971 December.

Murphy, Glenn (100), *Models with Incomplete Correspondence with the Prototype;* Journal of The Franklin Institute, vol. 292, #6, p. 513, 1971 December.

Muttnyánszky, Ádám, (7), *Statics and Strength of Materials* (in Hungarian); Technical Bookpublisher, Budapest, 1981.

Naddor, Eliezer (57), *Dimensions in Operations Research;* (Journal of) Operations Research, IV, p. 508, 1966.

Nolle, H. (27), *Experimental Modelling in Engineering;* Butterworth, London, 1982 (with David, F.)

O'Nan, Michael (127), *Linear Algebra;* Harcourt Brace Jovanovich, New York, 1971.

Palacios, J. (37), *Dimensional Analysis;* MacMillan and Co., London, 1964.

Pankhurst, R. C. (82), *Alternative Formulation of Pi-Theorem;* Journal of The Franklin Institute, vol. 292, #6, p. 451, 1971 December (see also Ref. 28).

Pankhurst, R. C. (28), *Dimensional Analysis and Scale Factors;* Chapman and Hall Limited, London, 1964.

Pedley, T. J. (Ed.) (24), *Scale Effects in Animal Locomotion;* (a 3-volume collection of papers presented at a Cambridge University Symposium held in 1975 September), Academic Press, London, 1977.

Pomerantz, Martin (64), *Foreword;* Journal of The Franklin Institute, vol. 292, #6, p. 389, 1971 December.

Porter, Alfred W. (23), *The Method of Dimensions;* Methuen & Co., London, 1933.

Prasuhn, Alan L. (90), *Fundamentals of Hydraulic Engineering;* Holt, Rinehart and Winston, New York, 1987.

Ramshaw, J. (48), *The Rolling Resistance of Commercial Vehicle Tyres* (with Williams, T.); Transport and Road Research Laboratory Report 701, ISSN 0305-1315, Crowthorne, Berkshire, UK, 1981.

Rao, Govinda N. S. (91), *Hydraulics;* Chapter III, p. 35, ASIA Publishing House, Bombay.

Rashevsky, Nicolas (25), *Some Medical Aspects of Mathematical Biology;* Charles C Thomas Publisher, Springfield, IL, 1964.

Rayleigh, Lord (111), *Letter to the Editor;* Nature, vol. 95, #2373, 1915 April.

Rayleigh, Lord (71), *The Principle of Similitude;* Nature, vol. 95, #2368, p. 66, 1915 March 18.

Rayleigh, Lord (108), *The Size of Drops;* Philosophical Magazine and Journal of Science, Series 5, vol. 48, #293, p. 321, 1899 October.

Rayleigh, Lord (104), *Letter to the Editor* (on the subject of Fluid Friction on Even Surfaces); Philosophical Magazine and Journal of Science, vol. 8, 6th series, p. 66, 1904.

Rayleigh, Lord (109), *Letter to the Editor;* Nature, #2389, vol. 95, p. 644, 1915 August 12 (reply to Dr. Riabouchinsky's letter, Ref. 103).

Reinhold, T. A. (Ed.) (49), *Wind Tunnel Modeling for Civil Engineering Applications;* Cambridge University Press, Cambridge, UK. Collection of papers presented at the International Workshop on Wind Tunnel Modeling in Civil Engineering, Gaithersburg, Maryland, 1982 April.

Reynolds, Osborne (54), *An Experimental Investigation of the Circumstances which determine whether the Motion of Water shall be Direct or Sinuous, and of the Law of Resistance in Parallel Channels;* Philosophical Transactions 1883, Part 3, p. 935.

Riabouchinsky, D. (103), *Letter to the Editor* (on the subject of The Principle of Similitude); Nature, vol. 95, # 2387, p. 591, 1915 July 29.

Rose, William K. (118), *Astrophysics;* Holt, Rinehart and Winston Inc., New York, 1973.

Rózsa, Pál (136), *Linear Algebra and its Applications;* 3rd ed. (in Hungarian), Textbook Publishing Co., Budapest, 1990.

Rücker, A. W. (86), *On the Suppressed Dimensions of Physical Quantities;* Philosophical Magazine, Series 5, vol. 27, XIII, p. 104.

Ryabov, Y. (146), *An Elementary Survey of Celestial Mechanics;* Dover Publications Inc., New York, 1961.

Sayao, Otavio J. (89), *Physical Modeling of Pocket Beaches;* Proceedings of a Speciality Conference on Quantitative Approaches to Coastal Sediment Processes, p. 1625, Seattle, Washington, 1991 June.

Sayao, Otavio J. et al. (92), *Dimensional Analysis of Littoral Transport;* Proceedings of the Canadian Coastal Conference, p. 241, St. John's, Newfoundland, 1985. Proceedings are published by the National Research Council of Canada, Ottawa, Ontario.

Schepartz, Bernard (43), *Dimensional Analysis in the Biomedical Sciences;* Charles C Thomas Publishers, Springfield, IL, 1980.

Schultz, Peter (50), *Atmospheric Effects on Ejecta Emplacement and Crater Formation on Venus From Magellan;* Journal of Geophysical Research, vol. 97, No. E10, p. 16.183, 1992 October 25.

Schuring, Dieterich J. (31), *Scale Models in Engineering;* Pergamon Press, Oxford, 1977.

Sedov, L. I. (51), *Similarity and Dimensional Methods in Mechanics;* Academic Press, New York, 1959.

Semat, Henry (4), *Fundamentals of Physics;* Holt, Rinehart and Winston Inc., New York, 1966.

Sena, L. A. (38), *Units of Physical Quantities and Their Dimensions;* Mir Publishers, Moscow, 1972.

Sinclair, George (105), *Theory of Models of Electromagnetic Systems;* Proceeding of the IRE, p. 1364, 1948 November.

Skoglund, Victor J. (46), *Similitude—Theory and Application;* International Textbook Co., Scranton, PA, 1967.

Smart, W. M. (20), *Textbook of Spherical Astronomy;* 4th ed., Cambridge University Press, Cambridge, 1960.

Smith, Maynard J. (122), *Mathematical Ideas in Biology;* Cambridge University Press, Cambridge, 1968.

Stahl, Walter R. (59), *Similarity and Dimensional Method in Biology;* Science, vol. 137, p. 205, 1962 July 20.

Stahl, Walter R. (78), *Dimensional Analysis in Mathematical Biology I;* (general discussion—continued in Ref. 79), Bulletin of Mathematical Biophysics, vol. 23, p. 355, 1961.

Stahl, Walter R. (79), *Dimensional Analysis in Mathematical Biology II;* Bulletin of Mathematical Biophysics, vol. 24, p. 81, 1962.

Staicu, Constantin I. (66), *General Dimensional Analysis;* Journal of The Franklin Institute, vol. 292, #6, p. 433, 1971 December (see also Ref. 33).

Staicu, Constantin I. (33), *Restricted and General Dimensional Analysis;* Abacus Press, Tunbridge Wells, Kent, UK, 1982.

Strahler, Arthur N. (99), *Dimensional Analysis Applied to Fluvially Eroded Landforms;* Bulletin of the Geological Society of America, vol. 69 p. 279, 1958 March.

Stumpf, H. W. (60), *Scale Models in Tyre Design;* Presented at the 23rd International Symposium on Automotive Technology and Automation, Vienna, Austria, 1990 December.

Szirtes, Thomas (13), *Some Human Communication Characteristics and the Principle of Least Effort;* Paper (No. I/11) presented at the Scientific Conference organized jointly by the Hungarian Academy of Sciences and Technical University of Budapest, Budapest, Hungary, 1989 August.

Szirtes, Thomas (148), *The Fine Art of Modeling;* SPAR Journal of Engineering and Technology, vol. 1, p. 37, 1992 May.

Szirtes, Thomas (145), *Choosing the Right Shock Absorber;* Machine Design, vol. 61, #10, p. 66, 1989 May.

Szirtes, Thomas (142), *Minimizing the Deflection of Tubular Cantilevers;* Machine Design, vol. 57, #3, p. 103, 1985 February.

Szirtes, Thomas (144), *Conditions of Jamming a Circular Plug in a Non-Smooth Circular Hole;* Spar Aerospace Ltd. Report, 1991 June.

Szirtes, Thomas (149), *Some Interesting applications of Dimensional Analysis.* Acta Technica; vol. 108, no. 1-2; Academic Publisher; Budapest, Hungary; 1999.

Szücs, Ervin (26), *Similitude and Modelling;* Elsevier Scientific Publishing Company, Amsterdam, 1980.

Taylor, Richard C. (138), *The Energetics of Terrestrial Locomotion and Body Size in Vertebrates;* (included in Ref. 24, p. 127).

Taylor, Edward S. (29), *Dimensional Analysis for Engineers;* Clarendon Press, Oxford, 1974.

Teller, Edward (18), *Conversations on the Dark Secrets of Physics;* Plenum Press, New York, 1991.

Thomas, George B. (150), *Calculus and Analytic Geometry.* Addison-Wesley Publishing Co., Reading, California, USA; 1968.

Thomson, D'Arcy W. (110), *Letter to the Editor;* Nature, vol. 95, #2373, 1915 April.

Thomson, William (132), *Theory of Vibration;* Prentice Hall Inc., Englewood Hills, NJ, 1972.

Thomson, W. (117), *The Response of Mechanical Systems to Random Excitation* (with Barton, V.); ASME Journal of Applied Mechanics, 1957 June, p. 248.

Timoshenko, Stephen (133), *Strength of Materials* (in 2 volumes); 3rd ed., D. Van Nostrand Co., Princeton, NJ, 1955.

Timoshenko, Stephen (134), *Theory of Elastic Stability* (with Gere, J.); McGraw-Hill Book Company, New York, 1961.

Timoshenko, Stephen (141), *Engineering Mechanics* (with Young, D.); 4th ed., McGraw-Hill Book Company, New York, 1956.

Timoshenko, Stephen (15), *History of Strength of Materials;* Dover Publications Inc., New York, 1983.

Tonndorf, Juergen (84), *Dimensional Analysis of the Cochlear Models;* The Journal of the Acoustic Society of America, vol. 32, #4, p. 493, 1960 April.

Van Driest, E. R. (87), *On Dimensional Analysis and the Presentation of Data in Fluid-Flow Problems;* Journal of Applied Mechanics, p. A-34, 1946 March.

Venikov, V. A. (32), *Theory of Similarity and Simulation;* Macdonald Technical & Scientific, London, 1969.

Waverman, Lucy (147), *The Ultimate Roast Turkey;* The Globe and Mail (Toronto, Ontario), 1995 December 13, p. A-24.

White, R. (55), *Models of Concrete Structures;* Paper presented at the Annual Convention of the US Concrete Institute, Dallas, 1972.

Williams, T. (48), *The Rolling Resistance of Commercial Vehicle Tyres* (with Ramshaw, J.); Transport and Road Research Laboratory Report 701, ISSN 0305-1315, Crowthorne, Berkshire, UK, 1981.

Wolverton, R. (ed.) (137), *Flight Performance Handbook for Orbital Operations;* John Wiley and Sons, New York, 1963.

Wong, Young (3), *Theory of Ground Vehicles;* John Wiley and Sons, New York, 1978.

Young, D. (141), *Engineering Mechanics* (with Timoshenko, S.); 4th ed., McGraw-Hill Book Company, New York, 1956.

Zierep, Jürgen (47), *Similarity Laws and Modeling;* (text identical with Ref. 30).

Zierep, Jürgen (30), *Similarity Laws and Modeling;* Marcel Dekker, New York, 1971 (text identical with Ref. 47).

APPENDICES

1. Recommended Names and Symbols for Some Physical Quantities　677

2. Some More-Important Physical Constants　681

3. Some More-Important *Named* Dimensionless Variables　683

4. Notes Attached to Figures　693

5. Acronyms　721

6. Solutions of Problems　723

7. Proofs of Selected Theorems and Equations　797

8. Blank Modeling Data Table　803

APPENDIX 1
RECOMMENDED NAMES AND SYMBOLS FOR SOME PHYSICAL QUANTITIES

Based on material published in the *CRC Handbook of Chemistry and Physics;* 69th Edition, 1988–89, pp. F-246.

Quantity	Symbol
length	L, l (small "el")
height	h
radius	r, R
diameter	d, D
path, length of arc	s
wavelength	λ
wavenumber	σ, ν
plane angle	$\theta, \phi, \alpha, \beta, \gamma, \varphi$
solid angle	ω, Ω
area	A, S
volume	V
time	t
frequency	ν, f
circular frequency	ω
period	T
time constant	τ
velocity	v, u, w, c
angular velocity	ω
acceleration	a
gravitational acceleration	g

Figure A1-1
Recommended names and symbols for space and time-related quantities

Quantity	Symbol
mass	M, m
density	ρ
moment of inertia	I
second moment of a plane area	I
momentum	p
angular momentum	L
force	F
weight	G, W
moment of force	M
energy	E, Q
power	P
pressure	p, P
mechanical normal stress	σ
mechanical shear stress	τ
linear strain	ϵ
volume strain	θ
modulus of elasticity, Young's modulus	E
shear modulus	G
bulk modulus	K
dynamic viscosity	μ
kinematic viscosity	ν
friction coefficient	μ
surface tension	γ, σ
diffusion coefficient	D
mass transfer coefficient	k_d

Figure A1-2
Recommended names and symbols for mechanics and related quantities

Quantity	Symbol
absolute temperature	T
Celsius temperature[a]	t, θ
gas constant	R
Boltzmann constant	k
heat[b]	q, Q
internal energy	U
enthalpy	H
Helmholtz energy	A
Planck function	Y
compression factor	Z
heat capacity	C
specific heat capacity	c
thermal conductivity	k
coefficient of heat transfer	h
linear expansion coefficient	α
volumetric expansion coefficient	γ
osmotic pressure	Π
osmotic coefficient	ϕ

Figure A1-3
Recommended names and symbols for thermodynamics and related quantities
Notes: (*a*) Where symbols are needed to represent both time and temperature, t is the preferred symbol for time and θ for temperature; (*b*) It is recommended that $q > 0$ and $Q > 0$ be used to indicate an *increase* of (heat) energy

Quantity	Symbol
elementary charge	e
quantity of electricity	Q
charge density	ρ
electric current	I
electric potential	V
electric potential difference	U, ΔV
electric field strength	E
capacitance	C
permittivity	ϵ
electric dipole moment	p
magnetic flux	ϕ
magnetic flux density	B
magnetic field strength	H
permeability	μ
permeability of vacuum	μ_0
magnetization	M
Bohr magneton	μ_B
resistance	R
resistivity	ρ
conductivity	κ
self-inductance	L
mutual inductance	M
reactance	X
impedance	Z
Faraday constant	F

Figure A1-4
Recommended names and symbols for electricity and magnetism-related quantities

APPENDIX 2
SOME MORE-IMPORTANT PHYSICAL CONSTANTS

Quantity	Symbol	Magnitude	Dimension	Source
gravitational acceleration	g	**9.80665**	m/s^2	by definition
Ampere constant	k	2 E7	m·kg/(s^2·A^2)	
Avogadro constant	N_A	6.0221367 E23	1/mol	CODATA
Bohr magneton	μ_B	9.2740154 E-24	m^2·A	CODATA
Boltzmann constant	k	1.380658 E-23	m^2·kg/(s^2·K)	CODATA
Coulomb constant	k	9 E9	kg·m^3/(A^2·s^4)	
electron radius	r_e	2.81794092 E-15	m	CODATA
elementary charge	e	1.60217733 E-19	A·s	CODATA
Faraday constant	F	9.6485309 E4	A·s/mol	CODATA
universal gravitational constant	k, G	6.67259 E-11	m^3/(kg·s^2)	CODATA
Kepler's constant	k	3.35 E18	m^3·s^2	
magnetic flux quantum	ϕ	2.06783461 E-15	m^2·kg/(A·s^2)	CODATA
mass of electron	m_e	9.1093897 E-31	kg	CODATA
mass of neutron	m_n	1.6749286 E-27	kg	CODATA
mass of proton	m_p	1.67262305 E-27	kg	CODATA
molar gas constant	R	8.31451	m^2·kg/(s^2·mol·K)	CODATA
molar volume of gas at 273.15 K, 101325 Pa	V_m	2.241410 E-2	m^3/mol	CODATA
nuclear magneton	μ_N	5.057866 E-27	m^2·A	CODATA
permeability of vacuum	μ_0	**4·π × 10^{-7}**	kg·m/(A^2·s^2)	by definition
Planck constant	h	6.6260755 E-34	m^2·kg/s	CODATA
Rydberg constant	R_∞	1.0973731534 E7	1/m	CODATA
speed of light in vacuum	c	**2.99792458 E8**	m/s	by definition
Stefan–Boltzmann constant	σ	5.67051 E-8	kg/(s^3·K^4)	CODATA
first radiation constant	c_1	3.7417749 E-16	kg·m^4/s^3	CODATA
second radiation constant	c_2	1.438769 E-2	m·K	CODATA

Figure A2-1
Some of the more-important physical constants
All data are from CODATA's (Appendix 5) *Adjustments of Fundamental Physical Constants,* 1986. Data in **bold** are exact. Excerpted from Ref. 8, p. 63

APPENDIX 3
SOME OF THE MORE-IMPORTANT NAMED DIMENSIONLESS VARIABLES

Unless specified otherwise, all dimensions are in SI.

API Gravity Degree

This number characterizes the density of fluidic petroleum products. It is defined as

$$\text{API degree} = \frac{1.415 \times 10^5}{\rho} - 131.5$$

where ρ is the density of fluid in kg/m^3, at 15.55 °C temperature. Hence the number 1.415×10^5 is a dimensional constant of dimension kg/m^3.

Beaford Number (*B*)

This number characterizes wind velocity. It was named after Admiral Sir Francis Beaford (1774–1857). The scale extends between $B = 0$ and $B = 15$. The speed of air movement (wind) is, by definition, $v = (1.87) \cdot B^{1.5}$ miles per hour, or $v = (0.836) \cdot B^{1.5}$ meter per second. From this it follows that the above coefficients are *dimensional* numbers, namely [1.87] = miles/hour; [0.836] = m/s.

Nusselt Number (*Nu*)

This is also called the Biot number. There are actually two Nusselt numbers: Nusselt number for heat transfer, and Nusselt number for mass transfer. The Nusselt number for *heat transfer* characterizes the heat energy transferred between a moving liquid and a solid surface (e.g., pipe wall). It is defined as

$$Nu = \frac{h \cdot D}{k}$$

where h = heat transfer coefficient between the liquid and the pipe; dimension: kg/(s³·°C)
  D = diameter of the pipe (or some other representative linear size); dimension: m
  k = thermal conductivity; dimension: (m·kg)/(s³·°C)

The Nusselt number for *mass transfer* is defined as

$$Nu^* = \frac{m \cdot L}{t \cdot A \cdot \rho \cdot D}$$

where m = mass transferred
  L = representative linear size
  t = time elapsed; dimension
  ρ = representative density
  D = diffusion coefficient
  A = area affected

H. Nusselt (1882–1957) was a German engineer who derived the Nusselt number; B. Biot (1774–1862) was a French physicist who was an early worker on the mathematical theories of convective heat transfer.

Fourier Number (*Fo*)

There are two Fourier numbers: one for *heat transfer* (*Fo*), and one for *mass transfer* (*Fo**). The former is defined as

$$Fo = \frac{k \cdot \Delta t}{\rho \cdot c \cdot r^2}$$

where k = thermal conductivity; dimension: (m·kg)/(s³·°C)
  Δt = time interval; dimension: s
  ρ = density; dimension: kg/m³
  c = the fluid's specific heat capacity; dimension: m²/(s²·°C)
  r = radius; dimension: m

The Fourier number for *mass transfer* is defined as

$$Fo^* = \frac{D \cdot \Delta t}{L^2}$$

where D = diffusion coefficient (diffusivity); dimension: m²/s
  Δt = time interval; dimension: s
  L = representative linear size; dimension: m

Joseph Fourier (1768–1830) was a French mathematician and scientist who did pioneering work on the theory of heat and numerical analysis.

Grätz Number (*Gz*)

This number is used for heat transfer expressions for laminar flow in pipes. The number is defined as

$$Gz = \frac{w \cdot c}{k \cdot L}$$

where w = mass flow rate of fluid; dimension: kg/s
  c = specific heat capacity of fluid; dimension: m²/(s²·°C)
  k = heat conductivity of fluid; dimension: (m·kg)/(s³·°C)
  L = representative linear size; dimension: m

L. Grätz (1856–1941) was a German investigator in the field of fluid mechanics.

Grashof Number (*Gr*)

This number is used in fluid mechanics to characterize free convection by gravitational and inertial effects; it is defined as

$$Gr = \frac{L^3 \cdot \rho^2 \cdot g \cdot \beta \cdot \Delta t}{\mu^2}$$

where L = characteristic linear size; dimension: m
  ρ = density of liquid (fluid or gas); dimension: kg/m³
  g = gravitational acceleration; dimension: m/s²
  β = volume expansion coefficient; dimension: °C⁻¹
  Δt = temperature difference; dimension: °C
  μ = dynamic viscosity of fluid; dimension: kg/(m·s)

The number is named after Franz Grashof (1826–1893), a German applied mathematician and engineer. Among his notable achievements was the development of the theory for drafts in chimneys.

Peclet Number (*Pe*)

The Peclet number has a role in characterizing forced heat convection and mass transfer. The former is defined as

$$Pe = \frac{\rho \cdot c \cdot v \cdot L}{k}$$

where ρ = density of fluid (or gas); dimension: kg/m³
  c = specific heat capacity of fluid; dimension: m²/(s²·°C)
  v = velocity of fluid; dimension: m/s
  L = characteristic linear size; dimension: m
  k = thermal conductivity of liquid; dimension: (m·kg)/(s³·°C)

Note that the Peclet number for heat transfer is the product of the Prandtl and Reynolds numbers, i.e., $Pe = Pr \cdot Re$.

The Peclet number for *mass transfer* is defined as

$$Pe^* = \frac{L \cdot v}{D}$$

where D is the diffusion coefficient (diffusivity) of dimension m²/s. Symbols L and v are as defined for Pe above. J. Peclet (1793–1857) was the first to apply Fourier's heat theories to solve engineering problems.

Prandtl Number (*Pr*)

The Prandtl number is used in fluid mechanics; it is defined as

$$Pr = \frac{c \cdot \mu}{k}$$

where c = specific heat capacity of the fluid; dimension: m²/(s²·°C)
 μ = dynamic viscosity of fluid; dimension: kg/m·s
 k = thermal conductivity of fluid; dimension: (m·kg)/(s³·°C)

This number was named after Ludwig Prandtl (1875–1953), a German physicist. However, it was H. Nusselt (1982–1957), a German engineer, who first derived this number in 1910. To his credit, Prandtl acknowledged Nusselt's priority in his book in 1952.

Rayleigh Number (*Ra*)

This number is used in problems of heat transfer by free convection. It is defined as

$$Ra = \frac{L^3 \cdot \beta \cdot g \cdot \rho^2 \cdot c_p \cdot \Delta t}{\mu \cdot k}$$

where L = representative linear size; dimension: m
 β = cubic expansion coefficient; dimension: 1/°C
 g = gravitational acceleration; dimension: m/s²
 ρ = density of fluid; dimension: kg/m³
 c_p = specific heat capacity of fluid (at constant pressure); dimension: m²/(s²·°C)
 Δt = temperature difference; dimension: °C
 μ = dynamic viscosity of fluid; dimension: kg/m·s
 k = thermal conductivity of fluid; dimension: (m·kg)/(s³·°C)

Note that the Rayleigh number is the product of the Grashof and Prandtl numbers. The Rayleigh number is named after Lord Rayleigh (John William Strutt), an English physicist, Third Baron (1842–1919).

Reynolds Number (*Re*)

This is one of the more important dimensionless *named* variables; it is used to characterize the laminar or turbulent nature of a fluid flow. If $Re < 2000$, then the flow is laminar, otherwise it is turbulent. The Reynolds number is defined as

$$Re = \frac{v \cdot L \cdot \rho}{\mu}$$

where v = velocity of fluid flow; dimension: m/s
L = representative linear size (e.g., inner diameter of a pipe); dimension: m
ρ = density of fluid; dimension: kg/m³
μ = dynamic viscosity of fluid; dimension: kg/(m·s)

The Reynolds number is named after Osborne Reynolds (1842–1912), an English physicist and engineer.

Schmidt Number (*Sc*)

The Schmidt number is used to characterize diffusive mass transports. It is defined as

$$Sc = \frac{\mu}{\rho \cdot D}$$

where μ = dynamic viscosity of fluid; dimension: kg/(m·s)
ρ = density of fluid; dimension: kg/m³
D = diffusion coefficient (diffusity); dimension: m²/s

The number was named after the German engineer E. Schmidt (1895–?).

Stanton Number (*St*)

There are two Stanton numbers: one for heat transfer, and one for mass transport. The heat transfer Stanton number (*St*) is used to characterize forced heat convection and is defined as

$$St = \frac{h}{\rho \cdot v \cdot c_p}$$

where h = heat transfer coefficient; dimension: kg/(s³·°C)
v = velocity; dimension: m/s
c_p = specific heat capacity of the fluidic medium at constant pressure; dimension: m²/(s²·°C)
ρ = density of the fluidic medium; dimension: kg/m³

The heat transfer Stanton number is equivalent to the Nusselt number (q.v.) divided by the product of the Prandtl (q.v.) and Reynolds (q.v.) numbers. Thus

$$St = \frac{Nu}{Pr \cdot Re}$$

The mass transfer Stanton number (*St**) is defined as

$$St^* = \frac{M}{\Delta t \cdot A \cdot \rho \cdot v}$$

where M = mass transferred; dimension: kg
Δt = time interval; dimension: s
A = area; dimension: m²
ρ and v are as defined above

The Stanton numbers were named after the British scientist Sir Thomas Stanton (1865–1931).

Cowling Number (*Co*)

The Cowling number is used in magneto-hydrodynamics; it is defined as

$$Co = \frac{B^2}{\mu \cdot \rho \cdot v^2}$$

where B = magnetic flux density; dimension: kg/(s²·A) (tesla)
μ = permeability; dimension: (m·kg)/(A²·s²)
ρ = density; dimension: kg/m³
v = fluid velocity; dimension: m/s

The number was named after Prof. T. Cowling of the University of Leeds (1906–?), who is (was) a pioneer in the study of magneto-hydrodynamics.

Euler Number (*Eu*)

This number is used in fluid dynamics, especially in the analysis of fluids flowing in a conduit. The Euler number is defined as

$$Eu = \frac{p}{\rho \cdot v^2}$$

where p = pressure; dimension: kg/(m·s²)
ρ = density; dimension: kg/m³
v = velocity; dimension: m/s

This number was named after Leonhard Euler (1707–1783), a Swiss mathematician, astronomer and engineering physicist nonpareil. He introduced natural logarithm (the base of which is "*e*"), *i* for $\sqrt{-1}$ and the notation *f*( ) for functions.

Froude Number (*Fr*)

This dimensionless variable is used to characterize fluid motion in open channels. It is defined as

$$Fr = \frac{v}{\sqrt{g \cdot L}}$$

where v = velocity of fluid; dimension: m/s
g = gravitational acceleration; dimension: m/s²
L = representative linear size; dimension: m

This number was named after William Froude (1810–1879), an English engineer and naval architect.

Hartmann Number (*Ha*)

This number is used in magneto-hydrodynamics to characterize the flow of a conducting fluid in transverse magnetic field. The Hartmann number is defined as

$$Ha = B \cdot L \cdot \sqrt{\frac{k_e}{\mu}}$$

where B = magnetic flux density; dimension: kg/(s²·A)
L = representative linear size; dimension m
k_e = electric conductivity; dimension: (A²·s³)/(m³·kg)
μ = dynamic viscosity; dimension: kg/(m·s)

Knudsen Number (*Kn*)

This number is used in the study of a very low pressure gas. In this circumstance the mean free path of a molecule is comparable with the size of the vessel in which the gas is contained. The Knudsen number is defined as

$$Kn = \frac{\lambda}{L}$$

where λ = free mean path of the molecule; dimension: m
L = linear size of the vessel; dimension: m

The number was named after M. Knudsen (1818–1943), the inventor of the pressure gauge for very low pressures.

Strouhal Number (*Sr*)

This parameter characterizes the frequency of vibrations in a taut wire produced by a liquid flowing perpendicular to the wire. The sounds produced are called *aeolian tones*. The Strouhal number is defined as

$$Sr = \frac{f \cdot d}{v}$$

where f = frequency of vibration; dimension: 1/s
d = diameter of the wire; dimension: m
v = velocity of the gas flowing across the wire; dimension: m/s

Weber Number (*We*)

This dimensionless variable is used in the study of surface tension and inertial forces in liquids (e.g., bubble formations). The number is defined as

$$We = \frac{\rho \cdot v^2 \cdot L}{\sigma}$$

where ρ = density of fluid; dimension: kg/m³
v = velocity of fluid; dimension: m/s
L = representative linear size; dimension: m
σ = surface tension; dimension: kg/s²

The number is named after the brothers Ernst (1795–1878) and Wilhelm (1804–1891) Weber. Ernst was a physiologist and Wilhelm a physicist, but together they studied the behavior of flowing liquids. Incidentally, the older Weber was the co-discoverer of the famous *Weber–Fechner* law, which states that the sensation is proportional to the logarithm of the stimulus.

Snellen Number (*Sn*) or Snellen

The Snellen number is a measure of visual acuity; it is defined as $Sn = a/b$,

where a = distance at which an object of size h can be recognized (Fig. A3-1); dimension: m
b = distance at which the *same* object can be *seen* at a subtended angle of 1 arcmin; dimension: m (see Fig. A3-1)

The normal eye can distinguish two points which form a subtended angle of 1 arcmin (= 1/60 deg) at the eye. Hence, by Fig. A3-1

$$\tan \frac{\varphi}{2} = 0.000145 = \frac{h}{2} \cdot b$$

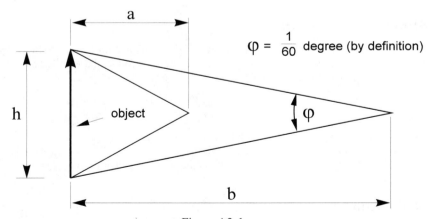

Figure A3-1
Explanatory sketch to define the Snellen number of visual acuity

from which

$$b = \frac{h}{2 \cdot \tan \frac{\varphi}{2}} = \frac{h}{0.000291}$$

Therefore

$$Sn = \frac{a}{b} = (0.000291) \cdot \frac{a}{h}$$

Perfect vision is $Sn = 1$ (this the often mentioned 20/20 vision). For example, if a man can recognize a $h = 0.00177$ m tall letter from $a = 4.22$ m distance, then he has a visual acuity of

$$Sn = (0.000291) \cdot \frac{4.22}{0.00177} = 0.694$$

i.e., his vision is 13.88/20.

The number was named after the Dutch ophtalmologist Herman Snellen (1834–1908).

Mach Number (*Ma*)

This dimensionless number is used in aerodynamics to characterize the speed of an object with respect to the velocity of sound in the same medium and under the same conditions (temperature, pressure). The Mach number is defined as

$$Ma = \frac{v_1}{v_2}$$

where v_1 = the velocity of the object (e.g., airplane); dimension: m/s
v_2 = the velocity of sound in the same medium; dimension: m/s

The number was named after Ernst Mach (1838–1916), Austrian physicist, who introduced it in 1887.

Cauchy Number (*Ca*)

The Cauchy number is used in compressible fluid flow characterization; it is the ratio of inertial to compression forces. The Cauchy number is definded as

$$Ca = \frac{\rho \cdot v^2}{\beta}$$

where ρ = density of fluid; dimenison kg/m^3
v = speed of fluid; dimension m/s
β = bulk modulus of fluid; dimension kg/(m·s^2)

The number was named after Baron Augustin Louis Cauchy (1789–1857), French mathematician and physicist.

APPENDIX 4
NOTES ATTACHED TO FIGURES

Brackets [ ] designate dimension, e.g., [time] = s.

Note 1 (Fig. 3-10)
The *becquerel* (Bq) is the SI dimension of radioactivity. 1 Bq represents 1 nuclear activity (i.e., disintegration) per second. Therefore the dimension of Bq is 1/s, which is the same as that of frequency Hz. The becquerel was approved by the 15th CGPM in 1975.

The unit was named after Antoine Henri Becquerel (French physicist, 1852–1908), who, in 1901 identified radioactive radiation by uranium. For this discovery he received the Nobel Prize in 1903.

Note 2 (Fig. 3-10)
The *coulomb* (C) is the quantity of electricity (charge) transported in 1 second by a current of 1 ampere intensity. From this it follows that 1 C = 1 A·s. The name coulomb was adopted for the electric charge by the International Electrotechnical Commission at its meeting in 1881.

The unit was named after Charles Augustin de Coulomb (French physicist, 1736–1806). In 1785 Coulomb showed that the force of electric attraction or repulsion between two bodies is proportional to the electric charge these bodies hold, and inversely proportional to the square of the distance separating them. This is called the Coulomb law.

Note 3 (Fig. 3-10)
Celsius is the commonly used temperature scale, except in scientific applications, where the Kelvin (i.e., absolute) scale is used. The relation between the *Celsius degree t* and *kelvin T* is $t = T - 273.15$. For a short biographic note on Celsius, see Art. 3.3.2 (a) under "Thermodynamic temperature."

Note 4 (Fig. 3-10)
The *farad* is the capacitance of an electrical capacitor having two surfaces holding 1 coulomb electric charge, and having 1 volt potential difference between them. Therefore $Q = C \cdot V$, where Q is the electric charge in coulombs, C is the capacitance in farads, and V is the potential difference in volts. Thus $C = Q/V$. Note that the *farad*, which is a derived dimension of capacitance in SI, is not the same as *Fara-*

day, which is a constant. The latter is defined as the charge required to deposit 1 mole of material (to a surface) from a conducting solution of that material. The value of *Faraday* is given in Appendix 2.

Both farad and Faraday were named after Michael Faraday (English physicist and chemist, 1791–1867). He was a pioneer in electrochemistry and electromagnetism. Among his many inventions are the electric generator, the transformer, and the electric motor. He was a supreme experimenter and yet a very modest man. He was buried—by his wishes—under a "gravestone of the most ordinary kind."

Note 5 (Fig. 3-10)
The 15th CGPM adopted the "gray" as the unit of *absorbed* ionizing radiation. This unit must be distinguished from the "sievert" (Note 15), which is the unit of *equivalent* ionizing radiation. The difference can be easily understood by considering the "gray" as the unit of the *input* radiation to a body disregarding body reaction to this input, whereas the "sievert" is the unit of the *effect* on the body caused by the entered radiation. Consequently, although the dimensions of both units are the same = m^2/s^2 (absorbed energy per kg body mass), numerically they can be very different. The following relation connects these two units:

$$H = q \cdot n \cdot D$$

where H = equivalent dose of ionizing radiation (sievert)
q = quality factor (dimensionless)
n = numeric factor (dimensionless)
D = absorbed dose of ionizing radiation (gray)

In the above, q is the *relative biological (effective) factor* (RBF) whose value can vary by as much as 20-fold; factor n reflects the distribution of energy of the inputted dose in the body. Both of these factors were set by the International Commission of Radiological Protection (ICRP) in its 33rd report, *Radiation Quantities and Units,* issued in 1980.

The unit "gray" was named after L. H. Gray (British cancer scientist, 1905–1965).

Note 6 (Fig. 3-10)
The "henry" was adopted by the International Electrotechnical Commission in its 1893 meeting held in Chicago; it is defined as the *inductance* of a closed circuit in which 1 volt potential is produced by changing the current (flowing in it) at the rate of one ampere per second.

The "henry" was named after Joseph Henry (American physicist and inventor, 1797–1878). He invented the electric relay, the electric telegraph (before Morse), and the electric motor. He also discovered self-induction and showed that sunspots are cooler than rest of the surface of the Sun. He was the first secretary of the Smithsonian Institution in Washington.

Note 7 (Fig. 3-10)
The "hertz" (Hz) was adopted as a unit of *frequency* in 1933 by the International Electric Commission. By definition, 1 hertz equals 1 oscillation per second, and hence the dimension of hertz is 1/s. The "hertz" is not to be confused with the *cir-*

cular frequency (ω) whose dimension is rad/s. Since a complete revolution is 2·π radians, therefore, if *f* is the frequency (Hz), then

$$\omega = 2\cdot\pi \cdot f \text{ rad/s}$$

The "hertz" was named after Heinrich Rudolf Hertz (German physicist, 1857–1894). He was the first to *observe* the photoelectric effect in 1888, for which Einstein, who developed its mathematical foundation and published it in 1905, received the Nobel Prize in 1921. Hertz's greatest scientific achievement was the discovery of electromagnetic waves. He was a protégé, close associate, and lifelong friend of the much older Helmholtz, who nevertheless outlived him by 8 months. Hertz died less than 2 months short of his 37th birthday.

Note 8 (Fig. 3-10)
The SI dimension of energy, work, and quantity of heat is "joule." By definition, 1 joule is the work done when 1 newton force acts over a distance of 1 meter in the direction of force. Therefore the SI dimension of energy (joule) expressed in base units is $m^2 \cdot kg/s^2$.

The unit was named after James Prescott Joule (English physicist, 1818–1889) who, independently of Julius Robert Mayer (German physician, 1814–1878), established the mechanical equivalent of heat in 1847. His paper describing this momentous discovery was rejected by all the learned journals, as well as by the Royal Society. As a result, Joule could present his findings only at a public lecture in Manchester, and then publish it in a local *newspaper* with the help of his brother who was the music critic there.

Joule, however, was not the first to determine the mechanical equivalent of heat. This feat had been accomplished 5 years earlier by Mayer who, unfortunately, did not publish his results in time, not even in a newspaper!

Note 9 (Fig. 3-10)
The "lumen" is the luminous energy *emitted* in unit time (flux) by a point-source of 1 candela intensity into a solid angle of 4·π steradian (i.e., radiating in all direction). It follows that the total flux emitted by a point-source of 1 candela intensity in all directions is 4·π lumens. Therefore the SI dimension of lumen is [candela]·[steradian], and since [steradian] = 1 (it is dimensionless), therefore [lumen] = cd·sr.

Note 10 (Fig. 3-10)
The "lux" is the unit of illuminance reaching its target. So that "lux" can be considered the unit of *illumination*. Accordingly, 1 lux is the luminous energy (lumen) falling onto a 1 m^2 area. Thus

$$[\text{lux}] = \frac{[\text{lumimous energy}]}{[\text{area}]} = \frac{\text{lumen}}{m^2} = \frac{\text{cd}\cdot\text{sr}}{m^2}$$

An associated quantity is the *exposure* which is the product of illumination and time. Thus

$$[\text{exposure}] = \text{lux}\cdot\text{s} = \frac{\text{cd}\cdot\text{sr}\cdot\text{s}}{m^2}$$

Note 11 (Fig. 3-10)
The unit "newton" (N) for force was adopted in the 9th CGPM in 1948. By definition, 1 newton is the force necessary to impart 1 m/s² acceleration to a 1 kg mass. Therefore 1 N = 1 (m·kg)/s², thus the dimension of newton (force) is [N] = (m·kg)/s².

The newton was named after Sir Isaac Newton (English scientist, astronomer and mathematician, 1642–1727).

Note 12 (Fig. 3-10)
The "ohm" (Ω) is the SI unit of electrical resistance. A conductor of a given length has 1 ohm resistance if 1 volt potential difference between its ends produces a 1 ampere electrical current.

The "ohm" was first accepted internationally in 1881 at the 1st meeting of the International Electrotechnical Commission. It has since undergone several revisions, the latest in 1990 January (Ref. 10, p. 117).

The "ohm" was named after Georg Simon Ohm (German physicist, 1787–1854). Ohm's significant contribution to science was the formulation of the law—now called Ohm's law—which states that the electric flow in a conductor is proportional to the potential difference and inversely proportional to the resistance. It is interesting to note that Henry Cavendish (English chemist and physicist, 1731–1810) discovered the same relationship maybe 50 years prior, but did not publish it.

A related unit—suggested by Lord Kelvin—is the unit of conductance, "*mho*." A mho is the reciprocal of ohm. The attentive reader will notice that mho is ohm spelled backward (see also Note 14).

Note 13 (Fig. 3-10)
The SI unit of pressure, "pascal" (Pa) was approved at the 14th meeting of CGPM in 1971. One pascal pressure occurs when 1 newton (force) acts over a surface area of 1 m². Numerically, 1 pascal is a very small unit; for example the standard atmospheric pressure equals 101,325 pascals.

This unit was named after Blaise Pascal (French mathematician, physicist and religious mystic, 1623–1662). Pascal lived for only 39 years, and he only occupied himself with mathematics and physics until age 31, when he became a devotee of religion. At age 16 he published a book on conic sections which was the best text on the subject since Apollonius 19 centuries earlier. Pascal was also instrumental together with Pierre de Fermat (French lawyer–mathematician, 1601–1665) in establishing the foundations of modern statistics and probability theory.

Note 14 (Fig. 3-10)
The unit "siemens" for electric conductance was adapted at the 14th CGPM in 1972. "Conductance" is by definition the reciprocal of "resistance," and hence its dimension is

$$[\text{siemens}] = \frac{1}{[\text{ohm}]} = \frac{s^3 \cdot A^2}{m^2 \cdot kg}$$

The unit was named after Sir William Siemens (German–British inventor, 1823–1883). Siemens was born in Germany, but emigrated to England in 1842 and in 1859 became a naturalized British citizen. He was instrumental in the development of the electric locomotive, the laying of transoceanic cables and the improvement of electric generators.

Note 15 (Fig. 3-10)
The "sievert" is the SI unit of dose equivalent of ionizing radiation H. It was proposed by the International Commission of Radiological Protection in 1977, and was adopted by the 16th CGPM in 1980. For relation between the dose equivalent and absorbed radiation see Note 5. The dimension "sievert" is easily derived from its definition, viz., absorbed energy per unit body mass. Accordingly:

$$[\text{sievert}] = \frac{[\text{energy}]}{[\text{mass}]} = \frac{\text{joule}}{\text{kg}} = \frac{(\text{m}^2 \cdot \text{kg})/\text{s}^2}{\text{kg}} = \frac{\text{m}^2}{\text{s}^2}$$

Note that this is also the dimension of absorbed ionizing radiation "gray" (see Note 5).

The unit "sievert" was named after Rolf Sievert (Swedish physicist, 1896–1966).

Note 16 (Fig. 3-10)
The unit "tesla" for *magnetic flux density* was approved by the International Electrotechnical Commission in 1954. One tesla represents the magnetic flux density of one weber per square meter.

The unit was named after Nicola Tesla (Croatian-American electrical engineer, 1856–1943). Tesla worked in the United States for the development of induction motors and, in particular, in the application of rotating magnetic fields. He was a pioneer in the design of high-voltage power transmission systems.

Note 17 (Fig. 3-10)
By definition 1 volt is the electrical potential between two points of a conductor which, when carrying a constant current of 1 ampere, dissipates 1 watt. In other words, if 1 watt is transported between two points of a conductor with a current of 1 ampere, then the potential difference between these points is 1 volt. Another definition of volt can be obtained from the fact that if between two points of a conductor the resistance is 1 ohm, and if the current flowing in this segment is 1 ampere, then the potential difference between the two points is 1 volt.

The unit "volt" was named after Count Alessandro Giuseppe Antonio Anastasio Volta (Italian physicist, 1745–1827). Volta was thought to be retarded as he did not utter a word until he was four. By seven, however, he caught up and by the time he was 14 he decided to become a physicist. He succeeded remarkably, for he invented the electric battery (voltaic pile), the electrophorus (a rotating device generating static electricity by induction) and the electroscope. He also contributed significantly to developing the condenser. He received the Copley Medal of the Royal Society in 1791 and was elected a member of that august body. But the greatest honor bestowed upon him by his fellow scientists was when they named the unit of potential difference and electromotive force after him.

Note 18 (Fig. 3-10)
"Watt" is the unit of mechanical, electrical, heat, etc. power. It is defined as the energy imparted, absorbed, transmitted, etc., in unit time. Since the SI unit of energy is

joule, therefore 1 watt = 1 joule/sec. The name for this unit was first proposed by William Siemens in his presidential address to the British Association in 1882.

The unit was named after James Watt (British mechanician, inventor, and engineer, 1736–1819).

He significantly improved Newcomen's steam engine by adding condensers to it and by letting the steam act on both sides of the piston. Watt also invented the centrifugal governor. These inventions laid the foundation of the Industrial Revolution.

Note 19 (Fig. 3-10)
The unit of magnetic flux "weber" was approved by the 8th CGPM in 1948. By definition, 1 weber is the magnetic flux produced by a coil of one turn generating 1 volt electromotive force when the flux is reduced uniformly to zero in 1 second. Hence we have the relation

$$\text{magnetic flux} = (\text{electromotive force}) \cdot (\text{time})$$

which provides the dimension

$$[\text{weber}] = [\text{volt}] \cdot [\text{time}] = \frac{m^2 \cdot kg}{s^3 \cdot A} \cdot s = \frac{m^2 \cdot kg}{s^2 \cdot A}$$

The unit "weber" was named after Wilhelm Eduard Weber (German physicist, 1804–1891). In 1846 Weber introduced a logical system of units for electricity involving length, mass, and time (Gauss did the same for magnetism). These units were officially accepted at the international congress in Paris in 1881. In cooperation with Gauss, Weber constructed the first practical telegraph in 1834. Together with his older brother Ernst, he also studied the behavior of flows of liquids through tubes (see also *Weber number* in Appendix 3).

Note 20 (Fig. 3-13a)
Density is the mass per unit volume (see also table attached to Note 83). *Specific volume* is the volume per unit mass (inverse of density). *Linear density* is the mass per unit length. *Area density* is the mass per unit area.

Note 21 (Fig. 3-13a)
Weight by definition is the force generated by gravitation. Hence the dimension of weight is the same as that of force $(m \cdot kg)/s^2$. The weight of 1 kg mass on Earth is g newton. Therefore the standard weight of an 1 kg mass on Earth is 9.80665 newton. The weight of the same mass on the Moon is only 1.6 newton, and on Mars it is 3.73 newton.

Note 22 (Fig. 3-13a)
Energy density is energy per unit volume. *Specific energy* is the energy per unit mass—formerly called *calorific value.*

Note 23 (Fig. 3-13a)
Angular velocity is the plane angle (in radians) travelled in unit time. *Angular acceleration* is the change of angular velocity in unit time.

Note 24 (Fig. 3-13a)
The *moment of inertia* of a material point P of mass M which is moving about an axis situated L distance away from P is defined as $M \cdot L^2$.

Note 25 (Fig. 3-13a)
Gravitational acceleration on Earth at latitude φ (with good approximation) is

$$g = (9.78073) \cdot (1 + (0.00524) \cdot \sin^2 \varphi) \cdot (1 - b \cdot h)$$

where h = altitude (positive) with respect to sea level; m
$b = 3.14961 \times 10^{-7}$ 1/m, a factor.

From the above it follows that at sea level

at the equator	$\varphi = 0$ deg,	$g = g_0 = 9.78073$ m/s²
at the poles	$\varphi = 90$ deg,	$g = g_p = 9.83198$ m/s²
at latitude of	$\varphi = 45.33$ deg,	$g = 9.80665$ m/s² (adopted value)

Hence, the numerical value of g at sea level can vary by 0.524%.

Note 26 (Fig. 3-13a)
For a viscous fluid, the *dynamic viscosity* is from Newton's equation $\tau = \mu \cdot (du/dy)$, in which du/dy is the velocity gradient in the direction normal to the flow of fluid, τ is the shear stress at the interface of the relatively moving lamina, and μ is the dynamic viscosity. From this relation

$$[\mu] = \frac{[\tau]}{[du/dy]} = \left(\frac{\text{kg}}{\text{s}^2\text{m}}\right) \cdot \frac{1}{(1/\text{s})} = \frac{\text{kg}}{\text{m} \cdot \text{s}}$$

For reference, the *dynamic viscosity* of water at 20 °C is about 0.001 kg/(m·s).

Kinematic viscosity ν is defined as $\nu = \mu/\rho$, where ρ is density. Hence, for dimension, $[\nu] = [\mu]/[\rho] = \text{m}^2/\text{s}$. The kinematic viscosity of water at 20 °C is about 10^{-6} m²/s.

Note 27 (Fig. 3-13a)
By Newton's second law $F = q \cdot (dv/dt)$, where q is mass, v is speed, and t is time. Thus we can write $F \cdot dt = q \cdot dv$. The left side of this relation is called *impulse*, the right side is called *momentum*. Their dimensions therefore are

$$[\text{impulse}] = [\text{momentum}] = \frac{\text{m} \cdot \text{kg}}{\text{s}}$$

Note 28 (Fig. 3-13a)
For rotary motion the second law of Newton is $M = I \cdot (d\omega/dt)$, where M is moment of force, I is moment of inertia, ω is angular speed and t is time. Hence $M \cdot dt = I \cdot d\omega$. But $I = q \cdot R^2$ and $dv = R \cdot d\omega$ where q is the (point) mass, R is the radius of rotation for q, and v is the tangential speed for q. Therefore, by these relations $M \cdot dt = (q \cdot dv) \cdot R$. The expression in parentheses is the elementary momentum, so the right side is the *mo-*

ment of momentum. Its dimension is therefore

$$[\text{moment of momentum}] = \frac{\text{m}^2 \cdot \text{kg}}{\text{s}}$$

Note 29 (Fig. 3-13a)
Specific heat capacity "*c*" of a substance is the amount of (heat) energy necessary to raise the temperature of the *unit* mass of that substance by 1 K (= 1 °C). Hence, for dimension

$$[c] = \frac{[\text{energy}]}{[\text{mass}] \cdot [\text{temperature}]} = \frac{\text{m}^2}{\text{s}^2 \cdot \text{K}}$$

Note that the term "specific heat" is no longer in use.

Heat capacity is the (heat) energy necessary to raise the temperature of a *given* mass by 1 K. Therefore

$$[\text{heat capacity}] = \frac{[\text{energy}]}{[\text{temperature}]} = \frac{\text{m}^2 \cdot \text{kg}}{\text{s}^2 \cdot \text{K}}$$

Note 30 (Fig. 3-13a)
Entropy (change) ΔS of a closed system between states 1 and 2 is by the Clausius equation

$$\Delta S = \int_1^2 \frac{dQ}{T}$$

where dQ is the (heat) energy and T is the absolute temperature. Hence the dimension of entropy is

$$[\text{entropy}] = \frac{[\text{heat energy}]}{[\text{temperature}]} = \frac{\text{m}^2 \cdot \text{kg}}{\text{s}^2 \cdot \text{K}}$$

The *specific entropy* is entropy per unit mass. Therefore its dimension is $\text{m}^2/(\text{s}^2 \cdot \text{K})$. The Clausius equation was named after Rudolf Julius Emanual Clausius (German physicist, 1822–1888).

By the second law of thermodynamics for a reversible process, $\Delta S = 0$, and for an irreversible process, $\Delta S > 0$. Since no process is totally reversible, the total entropy of the universe is continuously growing. The concept of entropy is used not only in thermodynamics, but in other branches of science as well. For example, in information theory and mathematical linguistics (Ref. 13, Transactions p. 342; Ref. 14. p. 3). In particular

$$H = \sum_{j=1}^{m} p_j \log p_j \; \frac{\text{bits of information}}{\text{message}}$$

where H is the entropy of a message (information), p_j is the probability the message is sent, m is the number of messages sent and the log is of base 2. Here, of course, the dimension of entropy differs from the SI unit. This relation is known as the Wiener–Shannon equation—after Norbert Wiener (American mathematician, 1894–1964) and Claude Elwood Shannon (American mathematician, 1916–2001).

Note 31 (Fig. 3-13a)
The *universal gravitational constant* is by Newton's equation governing the gravitational attraction of two massive bodies $F = k \cdot (M_1 \cdot M_2)/L^2$, in which F is the gravitational force, M_1 and M_2 are the two point-masses L distance apart, and k is the universal gravitational constant. By this formula the dimension is of k is

$$[k] = \frac{[F] \cdot [L^2]}{[M^2]} = \frac{\text{m}^3}{\text{s}^2 \cdot \text{kg}}$$

The accepted numerical value of this constant is $k = 6.6726 \times 10^{-11}$ m³/(s²·kg). It is interesting that Newton himself did not know value of this constant, for he died 71 years before Henry Cavendish (English chemist and physicist, 1731–1810) made the first determination of it in 1798. This was Cavendish's most spectacular achievement—now commonly referred to as the "Cavendish experiment."

Note 32 (Fig. 3-13a)
Thermal conductivity k of a conducting material is defined by the equation $Q = k \cdot (A \cdot \Delta t \cdot \Delta\theta)/L$, where A is the cross-section of the conductor, Δt is the temperature difference between two cross-sections of the conductor, $\Delta\theta$ is the time duration, L is the distance between the two cross-sections of the conductor, and Q is the transmitted heat. From this relation, the dimension of k is

$$[k] = \frac{[\text{energy}] \cdot [\text{length}]}{[\text{area}] \cdot [\text{temperature}] \cdot [\text{time}]} = \frac{\text{m} \cdot \text{kg}}{\text{s}^3 \cdot \text{K}}$$

If the heat goes through a film—i.e., a very thin barrier between two bodies—the thickness of which is unknown, then the ratio $h = k/L$ characterizes the transfer of energy. This ratio is called the *film heat transfer coefficient.* Consequently, its dimension is kg/(s³·K). *Thermal resistivity* is the inverse of *thermal conductivity.*

Note 33 (Fig. 3-13a)
Surface tension is defined as the force acting along the unit length of a demarcation line between two substances. Therefore the dimension is

$$[\text{surface tension}] = \frac{[\text{force}]}{[\text{length}]} = \frac{\text{kg}}{\text{s}^2}$$

The direction of surface tension is always perpendicular to the element of the separating line.

Note 34 (Fig. 3-13a)
The *wavelength* of a periodic phenomenon is the linear distance between any two of its characteristically identical points. The dimension is therefore [wavelength] = m.
 The *wave number* is the reciprocal of the wave length. Hence its dimension is 1/m. See also Note 61.

Note 35 (Fig. 3-13a)
The *second moment of area* of a plane figure with respect to an axis, perpendicular to or parallel with the plane, is defined by $I = \int r^2 \cdot dA$, where r is the distance of the area element dA from the axis. Therefore the dimension of I is m⁴.

Note 36 (Fig. 3-13a)
The stress *normal* to or *parallel* with an elementary area is the ratio of the force generating the stress on the area and the area itself. These stresses are respectively called *normal* or *shear* stress. In both cases the dimension is [stress] = [force]/[area] = kg/(s²·m).

Young's modulus (modulus of elasticity) is the ratio of normal stress and the linear deformation per unit length (i.e., linear *strain*) this stress generates. Consequently, the dimension is

$$[\text{Young's modulus}] = \frac{[\text{stress}]}{[\text{linear elongation}]/[\text{length}]} = \frac{\text{kg}}{\text{s}^2 \cdot \text{m}}$$

Young's modulus was named after Thomas Young (British physician and physicist, 1773–1829).

The *modulus of shear* is entirely analogous with Young's modulus to the extent that shear stress replaces normal stress, and shear strain replaces linear strain. Here shear strain can be visualized as the *angle* (in radians) through which an imaginary cube—under the effect of shear stress acting on its opposing sides—is deformed into a rhomboid. Consequently, the dimensions of the modulus of shear and of Young's modulus are the same kg/(s²·m).

Note 37 (Fig. 3-13a)
Compressibility is the inverse of Young's modulus. Hence its dimension is (m·s²)/kg.

Note 38 (Fig. 3-13a)
Linear thermal expansion coefficient α is defined by the relation $\Delta L = \alpha \cdot L \cdot \Delta t$, where ΔL is the expansion of the original length L by temperature difference Δt. Consequently, the dimension is

$$[\alpha] = \frac{[\Delta L]}{[L] \cdot [\Delta t]} = \frac{1}{\text{K}}$$

Note that the "expansion" coefficient may also indicate contraction.

Volumetric thermal expansion coefficient γ is defined analogously to linear expansion coefficient. Hence $\Delta V = \gamma \cdot V \cdot \Delta t$, where ΔV is the change of the original volume V by temperature difference Δt. Thus the dimension of γ is

$$[\gamma] = \frac{[\Delta V]}{[V] \cdot [\Delta t]} = \frac{1}{\text{K}}$$

Note 39 (Fig. 3-13a)
Enthalpy H is a thermodynamic property. It is defined by $H = U + p \cdot V$, where U is the internal energy of a system with respect to a reference (i.e., zero) level, p is the pressure, and V is the volume. Hence the dimension is

$$[\text{enthalpy}] = [\text{energy}] = \frac{\text{m}^2 \cdot \text{kg}}{\text{s}^2}$$

Here U may include heat of fusion, heat of evaporation, etc., depending on the temperature.

Specific enthalpy is the enthalpy content of a substance per unit mass. Hence the dimension is

$$[\text{specific enthalpy}] = \frac{[\text{enthalpy}]}{[\text{mass}]} = \frac{m^2}{s^2}$$

Note 40 (Fig. 3-13a)
Heat flux is the heat energy transmitted per unit time. Therefore the dimension is

$$[\text{heat flux}] = \frac{[\text{heat energy}]}{[\text{time}]} = \frac{m^2 \cdot kg}{s^3}$$

Heat flux density is the heat flux per unit area. Therefore the dimension is

$$[\text{heat flux density}] = \frac{[\text{heat flux}]}{[\text{area}]} = \frac{kg}{s^3}$$

Note 41 (Fig. 3-13b)
Electric charge density with respect to surface area is the amount of electric charge per unit surface area. Hence the dimension is

$$[\text{electric charge density (area)}] = \frac{[\text{electric charge}]}{[\text{area}]} = \frac{s \cdot A}{m^2}$$

Electric charge density with respect to volume is the amount of electric charge per unit volume. Hence the dimension is

$$[\text{electric charge density (volume)}] = \frac{[\text{electric charge}]}{[\text{volume}]} = \frac{s \cdot A}{m^3}$$

Note 42 (Fig. 3-13b)
Surface current density in a conductor is the current carried by the conductor through unit cross-sectional area. Hence the dimension is

$$[\text{current density}] = \frac{[\text{current}]}{[\text{area}]} = \frac{A}{m^2}$$

Note 43 (Fig. 3-13b)
Electric field strength (intensity) is the force generated at a point by a unit charge. Hence the dimension is

$$[\text{electric field strength}] = \frac{[\text{force}]}{[\text{electric charge}]} = \frac{m \cdot kg}{s^3 \cdot A}$$

Note 44 (Fig. 3-13b)
Permittivity ϵ of a medium is defined by the Coulomb equation

$$F = \frac{1}{4\cdot\pi\cdot\epsilon} \cdot \frac{q_1 \cdot q_2}{r^2}$$

where F is the force by charges q_1 and q_2 in a medium r distance apart. Therefore the dimension of permittivity is

$$[\text{permittivity}] = \frac{[\text{electric charge}]^2}{[\text{force}] \cdot [\text{distance}]^2} = \frac{A^2 \cdot s^4}{m^3 \cdot kg}$$

Note 45 (Fig. 3-13b)
Electrical resistivity ρ of a conductor is defined by $R = \rho \cdot L/a$, where R is its electrical resistance, L its length and "a" its cross-sectional area. Accordingly, the dimension of ρ is

$$[\text{electrical resistivity}] = \frac{[\text{electrical resistance}] \cdot [\text{area of cross-section}]}{[\text{length}]} = \frac{m^3 \cdot kg}{s^3 \cdot A^2}$$

where we used the dimension of electrical resistance (ohm) derived previously in Note 12. *Conductivity* κ is the inverse of *resistivity* ρ.

Note 46 (Fig. 3-13b)
Magnetic field strength (intensity) H at a point is force F (produced by a magnet) divided by its magnetic flux p (see Note 19). Thus $H = F/p$, and hence the dimension is

$$[\text{magnetic field strength}] = \frac{[\text{force}]}{[\text{pole strength}]} = \frac{A}{m}$$

Note 47 (Fig. 3-13b)
Magnetic dipole moment M_m is defined for a magnet of length L and pole strength p (Note 19) by $M_m = p \cdot L$. Hence its dimension is

$$[\text{magnetic dipole moment}] = [\text{magnetic flux}] \cdot [\text{length}] = \frac{m^3 \cdot kg}{A \cdot s^2}$$

Note 48 (Fig. 3-13b)
Reactance can be inductive X_L or capacitive X_C. *Inductive reactance* is defined by $X_L = 2 \cdot \pi \cdot f \cdot L$, where f is the frequency and L is the inductance (Note 6). Hence the dimension is

$$[\text{inductive reactance}] = [\text{frequency}] \cdot [\text{inductance}] = \frac{m^2 \cdot kg}{A^2 \cdot s^3}$$

Capacitive reactance is defined by $X_C = 1/(2 \cdot \pi \cdot f \cdot C)$, where f is the frequency and C is the capacitance (Note 4). Consequently, the dimension is

$$[\text{capacitive reactance}] = \frac{1}{[\text{frequency}] \cdot [\text{capacitance}]} = \frac{m^2 \cdot kg}{A^2 \cdot s^3}$$

which is, again, the same dimension as that of electrical resistance (Note 12).

Note 49 (Fig. 3-13b)
Electrical impedance Z in a circuit is defined by $Z = \sqrt{R^2 + (X_L - X_C)^2}$ (Ref. 4, p. 488), where R is the ohmic resistance (Note 12), and X_L and X_C the inductive and capacitive reactances (Note 48). Therefore the dimension is

$$[\text{impedance}] = [\text{resistance}] = \frac{m^2 \cdot kg}{A^2 \cdot s^3}$$

Note 50 (Fig. 3-13b)
Magnetomotive force $\mathscr{F}$ is measured in ampere-turns. Thus a coil of 50 turns carrying a current of 5 amperes has a magnetomotive force of $\mathscr{F} = 250$ ampere-turns. The dimension is

$$[\text{magnetomotive force}] = [\text{electric current}] \cdot [\text{number of turns}] = A \cdot 1 = A$$

since the number of turns is a dimensionless number (i.e., it has a dimension of 1).

Note 51 (Fig. 3-13b)
Reluctance in a magnetic circuit corresponds to *resistance* in an electric circuit. The latter equals the ratio of electromotive force to the current. Accordingly, in a magnetic circuit, reluctance equals the ratio of the magnetomotive force (Note 50) to the magnetic flux (Note 19). The dimension therefore is

$$[\text{reluctance}] = \frac{[\text{magnetomotive force}]}{[\text{magnetic flux}]} = \frac{A^2 \cdot s^2}{m^2 \cdot kg}$$

This dimension is the inverse of *inductance* (Note 6).

Note 52 (Fig. 3-13b)
Magnetic permeance is the inverse of *reluctance* (Note 51). Hence its dimension is

$$[\text{magnetic permeance}] = \frac{1}{[\text{reluctance}]} = \frac{m^2 \cdot kg}{A^2 \cdot s^2}$$

This dimension is the same as that of *inductance* (Note 6).

Note 53 (Fig. 3-13a)
Material permeance k_m is used to characterize the rate of mass transport; it is defined by the relation $k_m = Q/(p \cdot \Delta t \cdot a)$, where Q is the mass of the transported material (e.g., water vapor) through a barrier (surface), p is the pressure, Δt is the time duration, and "a" is the surface area of the barrier. Thus the dimension is

$$[\text{material permeance}] = \frac{[\text{mass of transported material}]}{[\text{pressure}] \cdot [\text{time}] \cdot [\text{area}]} = \frac{kg}{\left(\frac{kg}{m \cdot s^2}\right) \cdot s \cdot m^2} = \frac{s}{m}$$

Note 54 (Fig. 3-13b)
Susceptance in an electrical circuit can be inductive B_L or capacitive B_C. Inductive susceptance is defined by $B_L = 1/(\omega \cdot L)$, where ω is the circular frequency (Note 7) and L is the inductance (Note 6). Hence the dimension is

$$[\text{inductive susceptance}] = \frac{1}{[\text{circular frequency}]\cdot[\text{inductance}]} = \frac{s^3 \cdot A^2}{m^2 \cdot kg}$$

Capacitive susceptance is defined by $B_C = \omega \cdot C$, where C is capacitance (Note 4). Hence

$$[\text{capacitive susceptance}] = [\text{circular frequency}]\cdot[\text{capacitance}] = \frac{s^3 \cdot A^2}{m^2 \cdot kg}$$

As seen, the dimensions of both types of susceptances are identical and they are the inverse of that of *resistance* (Note 12).

Note 55 (Fig. 3-13b)
Admittance of an electrical circuit is the reciprocal of its *impedance* (Note 49). Hence the dimension is

$$[\text{admittance}] = \frac{1}{[\text{impedance}]} = \frac{A^2 \cdot s^3}{m^2 \cdot kg}$$

Note 56 (Fig. 3-13b)
Electric flux is the quantity of electricity (charge) flowing through a medium. Therefore, by Note 2, the dimension is

$$[\text{electric flux}] = [\text{electric charge}] = A \cdot s$$

Electric flux density is the quantity of electricity (charge) flowing through a medium per unit cross-sectional area of that medium. Thus, the dimension is

$$[\text{electric flux density}] = \frac{[\text{electric flux}]}{[\text{area}]} = \frac{A \cdot s}{m^2}$$

Note 57 (Fig. 3-13b)
Electric dipole moment is the product of the individual electric charge (Note 2) of the dipole and the distance between these charges (i.e., the length of the dipole). Hence the dimension is

$$[\text{electric dipole moment}] = [\text{electric charge}]\cdot[\text{length}] = A \cdot s \cdot m$$

Note 58 (Fig. 3-13b)
The *electric dipole potential* at a point P is defined by the relation

$$\Phi = \frac{1}{4 \pi \epsilon_0} \cdot \frac{p \cos \varphi}{r^2}$$

where p is the dipole moment (Note 57), ϵ_0 is the permittivity (Note 44), r is the distance between the dipole and point P (assuming $r \gg$ length of dipole), and φ is the angle between the axis of the dipole and the radius vector pointing to point P. Consequently, the dimension is

$$[\text{electric dipole potential}] = \frac{[\text{dipole moment}]}{[\text{length}]^2} = \frac{m^2 \cdot kg}{A \cdot s^3}$$

APPENDIX 4

Note 59 (Fig. 3-13b)
Permeability μ of a substance is defined by

$$F = \frac{1}{4\cdot\pi\cdot\mu} \cdot \frac{p_1 \cdot p_2}{r^2}$$

where p_1 and p_2 are magnetic pole strengths (Note 19), F is the force generated by p_1 and p_2, and r is the distance between the magnetic poles. Accordingly, the dimension is

$$[\text{permeability}] = \frac{[\text{pole strength}]^2}{[\text{force}]\cdot[\text{distance}]^2} = \frac{\text{m}\cdot\text{kg}}{\text{A}^2\cdot\text{s}^2}$$

Note 60 (Fig. 3-13b)
Electric power—like any power—is the energy imparted or absorbed per unit time (Note 18). Hence the dimension is $(\text{m}^2\cdot\text{kg})/\text{s}^3$.

Note 61 (Fig. 3-13c)
Wave number k for a sinusoidal wave is defined by $k = \omega/c$, where ω is the circular frequency and c is the velocity of propagation of the wave. Hence the dimension is

$$[\text{wave number}] = \frac{[\text{circular frequency}]}{[\text{speed}]} = \frac{\text{rad}}{\text{m}} = \frac{1}{\text{m}}$$

Now we know that if λ is the wavelength, then $c = (\lambda\cdot\omega)/(2\cdot\pi)$, from which $\omega/c = (2\cdot\pi)/\lambda = k$. Hence the wave number is proportional to the reciprocal of the wavelength. See also Note 34.

Note 62 (Fig. 3-13c)
Radiant energy, since it is energy, has the dimension of energy $(\text{m}^2\cdot\text{kg})/\text{s}^2$.

Note 63 (Fig. 3-13c)
Radiant power (flux), since it is power, has the dimension of power $(\text{m}^2\cdot\text{kg})/\text{s}^3$.

Note 64 (Fig. 3-13c)
Radiant intensity is the power radiated (Note 63) in one steradian solid angle. Hence the dimension is

$$[\text{radiant intensity}] = \frac{[\text{radiant power}]}{[\text{solid angle}]} = \frac{\text{m}^2\cdot\text{kg}}{\text{s}^3\cdot\text{sr}}$$

Note 65 (Fig. 3-13c)
The *radiance* of a source is the radiating intensity (Note 64) in unit radiating area of the source. Therefore the dimension is

$$[\text{radiance}] = \frac{[\text{radiant intensity}]}{[\text{area}]} = \frac{\text{kg}}{\text{s}^3\cdot\text{sr}}$$

Note 66 (Fig. 3-13c)
The *irradiance* of a receiver (i.e., irradiated surface) is the radiant power reaching a unit area of that surface. Therefore the dimension is

$$[\text{irradiance}] = \frac{[\text{radiated power reaching the surface}]}{[\text{area of surface}]} = \frac{\text{kg}}{\text{s}^3}$$

Note 67 (Fig. 3-13c)
Quantity of light is the light energy emitted by the source over a time. Hence, the dimension is

$$[\text{quantity of light emitted}] = [\text{luminous intensity}] \cdot [\text{time}] = \text{cd} \cdot \text{s}$$

The *quantity of light per unit solid angle* is measured in lumens, since 1 lumen = $1/(4 \cdot \pi)$ cd (Note 9). Hence

$$[\text{quantity of light emitted per unit solid angle}] = \frac{[\text{quantity of light emitted}]}{[\text{solid angle}]} = \text{lm} \cdot \text{s}$$

Note 68 (Fig. 3-13c)
Light exposure of a surface is the quantity of light (Note 67) reaching a unit area of that surface for a given time duration. Since the unit of illumination is the lux (Note 10), therefore

$$[\text{light exposure}] = \frac{[\text{quantity of light reaching the surface}]}{[\text{area of surface}]} = [\text{lux}] \cdot \text{s} = \frac{\text{cd} \cdot \text{s} \cdot \text{sr}}{\text{m}^2}$$

Note 69 (Fig. 3-13c)
Luminous efficacy is the ratio of the visible power (in form of light) emitted (Note 9) to the power input necessary for this emission. Hence

$$[\text{luminous efficacy}] = \frac{[\text{power output}]}{[\text{power input}]} = \frac{[\text{luminous flux}]}{[\text{power input}]} = \frac{[\text{lumen}]}{[\text{watt}]} = \frac{\text{cd} \cdot \text{s}^3}{\text{m}^2 \cdot \text{kg}}$$

Note 70 (Fig. 3-13c)
Luminance is the luminous intensity per unit area normal to the impinging radiation. Therefore

$$[\text{luminance}] = \frac{[\text{luminous intensity}]}{[\text{perpendicular illuminated area}]} = \frac{\text{cd}}{\text{m}^2}$$

Note 71 (Fig. 3-13c)
Sound pressure, since it is a pressure, is defined as a normal force acting upon a unit area (Note 13). Thus, the dimension is

$$[\text{sound pressure}] = \frac{[\text{normal force}]}{[\text{area}]} = \frac{\text{kg}}{\text{m} \cdot \text{s}^2}$$

Note 72 (Fig. 3-13c)
Sound pressure level L_p is a function of the ratio of the *actual* sound pressure p (Note 71) to a *reference* sound pressure p_r; it is defined by $L_p = 20 \cdot \log(p/p_r)$, where the log is of base 10. The value of L_p is given in *decibels* (abbreviated "db") and the reference pressure for air usually is $p_r = 2 \times 10^{-5}$ kg/(m·s^2). Thus, if the sound pressure is $p = 0.002$ kg/(m·s^2), then the *sound pressure level* is

$$L_p = 20 \cdot \log \frac{0.002}{0.00002} = 20 \cdot \log 100 = 40 \text{ db}.$$

Since the argument of the logarithmic function is dimensionless, and since the factor 20 is also dimensionless, therefore the sound pressure level is a *dimensionless quantity;* i.e., its dimension is 1. For this reason it does not qualify to be in the table of Fig. 3-13c, which contains (by its title) *dimensional* SI units. However, this entry is included in this table for reasons of importance and completeness.

Note 73 (Fig. 3-13c)
Sound energy flux is sound energy represented in unit time, i.e., *sound power.* Accordingly,

$$[\text{sound energy flux}] = [\text{sound power}] = \frac{\text{m}^2 \cdot \text{kg}}{\text{s}^3}$$

Note 74 (Fig. 3-13c)
Sound intensity is sound energy flux per area exposed. Therefore,

$$[\text{sound intensity}] = \frac{[\text{sound energy flux}]}{[\text{area}]} = \frac{\text{kg}}{\text{s}^3}$$

Note 75 (Fig. 3-13c)
Acoustic impedance is the ratio of sound pressure to sound velocity. Therefore its dimension is

$$[\text{acoustic impedance}] = \frac{[\text{sound pressure}]}{[\text{sound velocity}]} = \frac{\text{kg}}{\text{m}^2 \cdot \text{s}}$$

Note 76 (Fig. 3-13c)
Mechanical impedance is the ratio of the force acting on a body to the speed of that body in the direction of that force. Hence the dimension is

$$[\text{mechanical impedance}] = \frac{[\text{force}]}{[\text{speed}]} = \frac{\text{kg}}{\text{s}}$$

Note 77 (Fig. 3-13d)
Molar mass is the mass of one mole of a substance (see Art. 3.3.2a). The following formula gives the molar mass of substance x:

$$\text{molar mass of } x = \left(\frac{\text{mass of } x}{\text{mass of carbon 12}} \right) \cdot (0.012)$$

where the expression in the first pair of parentheses on the right is the ratio of the molecular weight of substance *x* to carbon 12. The dimension therefore is

$$[\text{molar mass}] = \frac{[\text{mass}]}{[\text{amount of substance}]} = \frac{\text{kg}}{\text{mol}}$$

See also table attached to Note 83.

Note 78 (Fig. 3-13d)
Molar volume is the volume of one unit of a substance—which is the mole. Therefore

$$[\text{molar volume}] = \frac{[\text{volume}]}{[\text{amount of substance}]} = \frac{m^3}{mol}$$

See also table attached to Note 83.

Note 79 (Fig. 3-13d)
Molar energy—precisely: *molar internal energy*—is the energy content of one mole of a substance. Therefore

$$[\text{molar internal energy}] = \frac{[\text{energy}]}{[\text{amount of substance}]} = \frac{m^2 \cdot kg}{s^2 \cdot mol}$$

Note 80 (Fig. 3-13d)
Molar heat capacity is the amount of (heat) energy necessary to raise the temperature of 1 mole of substance by 1 K. Therefore

$$[\text{molar heat capacity}] = \frac{[\text{energy}]}{[\text{amount of substance}] \cdot [\text{temperature}]} = \frac{m^2 \cdot kg}{s^2 \cdot mol \cdot K}$$

Note 81 (Fig. 3-13d)
Molar entropy is the entropy (change) of a closed heat system whose mass is 1 mole. Hence

$$[\text{molar entropy (change)}] = \frac{[\text{entropy (change)}]}{[\text{amount of substance}]} = \frac{m^2 \cdot kg}{s^2 \cdot mol \cdot K}$$

For the definition of entropy, see Note 30.

Note 82 (Fig. 3-13d)
Concentration is the amount of substance of solute (solute = material which is dissolved) in unit volume of solution. Therefore the dimension is

$$[\text{concentration}] = \frac{[\text{amount of substance of solute}]}{[\text{volume of solution}]} = \frac{mol}{m^3}$$

In older texts "molality" is sometimes used for "concentration." See table attached to Note 83.

Note 83 (Fig. 3-13d)
Molality is the amount of solute (material which is dissolved) in unit mass of solution. Therefore

$$[\text{molality}] = \frac{[\text{amount of substance of solute}]}{[\text{mass of solution}]} = \frac{mol}{kg}$$

The table below summarizes the subtle differences in dimensions of molality, concentration, etc.

		numerator		
		mol	m³	kg
denominator	mol	mole fraction	molar volume	molar mass
	m³	concentration	volume fraction	density
	kg	molality	specific volume	mass fraction

Dimensions of density and concentration-related quantities
example: the dimension of *specific volume* is $\frac{m^3}{kg}$

In the table, those quantities where the numerator and denominator are identical are, of course, dimensionless. These quantities are in the *main diagonal* (↘) of the table.

Note 84 (Fig. 3-13d)
Diffusion coefficient D, relating to two different substances, is a numerical measure of molecular "intermingling" as a result of their different random thermal motions. Although these motions are random, their "aimless" progress nevertheless favors one particular direction across the boundary between the two substances. This movement of material is expressed by Fick's law

$$\frac{dM}{dt} = -D \cdot a \cdot \frac{dc}{dx} \cdot \Phi$$

where dM/dt = the transfer of material x across the boundary in unit time (dimension: (kg of x)/s)
a = area of boundary surface (dimension: m²)
dc/dx = concentration gradient across the boundary (dimension: (mol of x)/m⁴)
Φ = molar mass of material x (dimension: (kg of x)/(mol of x))

By the formula and by the indicated dimensions, therefore, the dimension of diffusion coefficient D is

$$[\text{diffusion coefficient}] = \frac{[dM/dt]}{[dc/dx] \cdot [a] \cdot [\Phi]} = \frac{m^2}{s}$$

Note 85 (Fig. 3-13d)
Transport diffusion coefficient k is defined by $M = k \cdot \Delta p \cdot a \cdot \Delta t$ (Ref. 11, p. 341),

where M = amount of substance transported (mol)
Δp = pressure difference (kg/(m·s²))
a = surface area of the barrier (m²)
Δt = time duration of transfer (s)

therefore the dimension of k is

$$[\text{transport diffusion coefficient}] = \frac{[\text{amount of substance}]}{[\text{pressure}] \cdot [\text{area}] \cdot [\text{time}]} = \frac{mol \cdot s}{m \cdot kg}$$

Note 86 (Fig. 3-13d)
Elementary charge e is the amount of electricity residing in an electron. It is a fundamental constant of nature, determined first by Robert Andrews Millikan (American physicist, 1868–1953) who got the Nobel Prize for this feat in 1923. The accepted value of *e* is in Appendix 2. The dimension—as defined in Note 2—is $[e] = A \cdot s$.

Note 87 (Fig. 3-13d)
Electronvolt $e \cdot V$ is the work done on an electron (Note 86) moving through an electric field from one point to another under the influence of 1 volt potential difference (Note 17). Hence this energy is $E = e \cdot V = e \cdot 1$ joules; its dimension is therefore

$$[\text{electronvolt}] = [\text{electric charge}] \cdot [\text{potential difference}] = (A \cdot s) \cdot \left(\frac{m^2 \cdot kg}{s^3 \cdot A} \right) = \frac{m^2 \cdot kg}{s^2}$$

which is, of course, the dimension of energy (Note 8). The value of 1 electronvolt is given in Fig. 3-14.

Note 88 (Fig. 3-13d)
Energy flux density or *fluence rate* is the energy migrating through a unit area in unit time. Therefore

$$[\text{energy flux density}] = \frac{[\text{energy}]}{[\text{time}] \cdot [\text{area}]} = \frac{kg}{s^3}$$

Note 89 (Fig. 3-13d)
Linear attenuation coefficient μ characterizes the ability of a material to *diffuse* or *absorb* radiation. It is defined by

$$\Phi = \Phi_0 \cdot e^{-\mu \cdot x}$$

where Φ = luminous or radiant flux (power) at penetration x
Φ_0 = luminous or radiant flux (power) at penetration $x = 0$, i.e., the incident power
x = depth of penetration

Since the argument of any transcendental function must be dimensionless—i.e., its dimension must be 1—therefore $\mu \cdot x$ must be dimensionless. Hence the dimension of μ is $1/m$. At this point it should be remarked that, astonishingly, the dimension of radiant energy, the intensity of which is characterized by the subject attenuation coefficient, does not influence the latter's dimension! This is of course due to the presence of the transcendental function, which alone establishes the dimension of the coefficient.

Note 90 (Fig. 3-13d)
Ion number density is the number of ions in unit volume of a substance. Therefore

$$[\text{ion number density}] = \frac{[\text{number of ions}]}{[\text{volume}]} = \frac{1}{m^3}$$

The numerator of the dimension is 1 since the "number of ions" is a cardinal number.

Note 91 (Fig. 3-13d)
The *mean free path* ℓ of a molecule in Brownian (i.e., thermal) motion is defined by $v = \ell/\tau$ where v is the mean velocity of a molecule and τ is the average time between successive collisions. Therefore the dimension is

$$[\text{mean free path of a molecule}] =$$
$$[\text{mean velocity}] \cdot [\text{average time between collisions}] = \frac{m}{s} \cdot s = m$$

Note 92 (Fig. 3-13d)
Activity in this context means the average number of disintegrating atoms in unit time (Note 1). Therefore the dimension is

$$[\text{activity}] = \frac{[\text{number of atoms disintegrating}]}{[\text{time}]} = \frac{1}{s}$$

since the numerator is a cardinal number.

Note 93 (Fig. 3-13d)
The *absorbed dose rate* is the absorbed dose of ionizing radiation (Note 5) in unit time. Hence

$$[\text{absorbed dose rate}] = \frac{[\text{absorbed dose}]}{[\text{time}]} = \frac{(m^2/s^2)}{s} = \frac{m^2}{s^3}$$

Note that the *absorbed dose* is the absorbed energy by unit body mass; consequently, the dose rate is the *absorbed power* by unit body mass.

Note 94 (Fig. 3-13d)
The *dose equivalent rate* is the dose equivalent of ionizing radiation (Note 15) in unit time. Hence the dimension is

$$[\text{dose equivalent rate}] = \frac{[\text{dose equivalent}]}{[\text{time}]} = \frac{(m^2/s^2)}{s} = \frac{m^2}{s^3}$$

Note that the *dose equivalent* is the *energy* absorbed by unit body mass, modified by two numerical factors (Note 5). Therefore the dose equivalent rate is the absorbed *power* (modified by the above-cited two numerical factors) in unit body mass.

Note 95 (Fig. 3-13d)
Radiant exposure is the measure of radiant energy incident upon a unit area surface. Hence

$$[\text{radiant exposure}] = \frac{[\text{incident radiant energy}]}{[\text{area}]} = \frac{(m^2 \cdot kg/s^2)}{m^2} = \frac{kg}{s^2}$$

Note 96 (Fig. 3-13d)
Particle fluence is the number of (elementary) particles incident upon a unit area surface. Hence the dimension is

$$[\text{particle fluence}] = \frac{[\text{number of particles}]}{[\text{area}]} = \frac{1}{m^2}$$

since the number of particles is a cardinal number whose dimension is 1.

Note 97 (Fig. 3-13d)
Particle fluence rate is the particle fluence (Note 96) in unit time. Hence the dimension is

$$[\text{particle fluence rate}] = \frac{[\text{particle fluence}]}{[\text{time}]} = \frac{(1/m^2)}{s} = \frac{1}{m^2 \cdot s}$$

Note 98 (Fig. 3-13d)
Particle flux density is the same as *particle fluence rate* (Note 97) and thus its dimension is $1/(m^2 \cdot s)$.

Note 99 (Fig. 3-14)
Mean solar times.

Note 100 (Fig. 3-14)
The value is based on "calendar" year, which is exactly 365 days. There are other types of years whose lengths are (within 1 s):

1 sidereal year   = 31,558,150 s = 365 day 6 hr 9 min 9 s
1 tropical year   = 31,556,926 s = 365 day 5 hr 48 min 46 s
1 anomalistic year = 31,558,433 s = 365 day 6 hr 13 min 53 s

The *sidereal year* is the duration of time between two successive passages of the Earth from one particular position of its orbit to the same position *with respect to distant stars*.

The *tropical year* is the duration of time between two successive passages of the Earth through the *vernal equinox* (the point in the orbit where the Earth passes the equatorial plane northbound).

The *anomalistic* year is the duration of time between two successive passages of Earth through *perihelion* (point in the orbit where the Earth is nearest to the Sun).

Note 101 (Fig. 3-14)
There should be no space between the symbol and the last digit of the number. For example, 26°, not 26 °.

Note 102 (Fig. 3-14)
The designation "revolution per minute" (acronym: RPM) is widely used in mechanical engineering.

Note 103 (Fig. 3-14)
Used in surveying and agriculture.

Note 104 (Fig. 3-14)
Care must be taken in the interpretation of some Canadian texts where "tonne" implies 2000 pounds (= 907.185 kg).

Note 105 (Fig. 3-14)
"Tex" is used only in the textile industry; it expresses the mass (kg) of 10^6 m long yarn.

Note 106 (Fig. 3-14)
The definition of *electronvolt* is given in Note 87.

Note 107 (Fig. 3-14)
The unified atomic mass unit is the 1/12 part of the mass of the ^{12}C atom.

Note 108 (Fig. 3-14)
Astronomical unit does not have an international symbol. Some abbreviations (acronyms) are AU (or au) in English, UA in French, AE in German, etc. The astronomical unit is the length of the radius of the circular orbit of a negligible mass circling the Sun such that its orbital period is 1 tropical year (Note 100). In short, it is almost the same as the Earth's mean distance from the Sun.

Note 109 (Fig. 3-14)
A *light-year* is the *distance* that light travels in vacuum in a tropical year (Note 100). Hence the dimension is m. 1 light-year is 9.46053×10^{15} m.

Note 110 (Fig. 3-14)
Parsec is the *distance* at which 1 astronomical unit (Note 108) is seen at an angle of 1 second of arc. 1 parsec = 3.261634 light-years = 206,264.806247 astronomical units = 3.08568×10^{16} m.

Note 111 (Fig. 3-15)
By definition, 1 *nautical mile* is 1 minute of arc on a meridian of Earth at 48 deg latitude. There is no accepted international symbol for nautical mile, although the International Hydrographic Organization (IHO) recommended M in its first conference in Monaco in 1929. The above definition mandates a radius of Earth (at 48 deg latitude) of 6,366,707 m.

Note 112 (Fig. 3-15)
By definition, the *knot* is a unit of *speed;* 1 knot = 1 nautical mile per hour, exactly. There is no universally recognized symbol for knot, although "kn" was suggested by the IHO (Note 111).

Note 113 (Fig. 3-15)
The *millibar* is 1/1000 of a bar. 1 bar = 100,000 pascal, exactly. The millibar may continue to be used temporarily, but only for international meteorological work.

Note 114 (Fig. 3-15)
The barn is an exceedingly small unit of area; 1 barn equals (approximately) the cross-sectional area of an atomic nucleus. It is interesting to note that an atom itself is about 10,000 times larger than its nucleus, so the volume of the atom is larger than that of its nucleus by a factor of 10^{12} (trillion). According to Ref. 19, p. 18, the barn was proposed by H. Holloway and C. Baker in Chicago in 1942. However, the Council of Ministers of the European Economic Community in 1976 July proposed its abolition.

Note 115 (Fig. 3-16a)
The given SI equivalent is not exact; it can vary $\pm\ 3 \times 10^{-15}$ m with respect to the indicated SI value.

Note 116 (Fig. 3-16a)
The *stere* was introduced in France in 1798 to measure bundles of cordwood. The unit is now obsolete.

Note 117 (Fig. 3-16a)
This *carat* means "metric carat" which is mass; 5000 carats = 1 kg. The metric carat is used exclusively for weighing precious stones (e.g., diamonds). The metric carat should not be confused with "goldsmith carat," which expresses the purity of a gold alloy. The latter is the ratio of the mass of gold to the mass of alloy. Material which is 100% gold is "24 carat gold." An alloy that is 50% gold and 50% impurity (e.g., silver) is "12 carat gold." Neither the metric carat nor the goldsmith carat has a universally accepted abbreviation.

Note 118 (Fig. 3-16a)
1 *kilogram-force* is the weight of a mass of 1 kg under standard gravitational acceleration of $g = 9.80665$ m/s². Therefore we can write: (weight) = (mass)·(gravitational acceleration), or 1 kgf = (1 kg)·(9.80665 m/s²) = 9.80665 (m·kg)/s² = 9.80665 newton (see also Art. 3.4.2).

Note 119 (Fig. 3-16a)
Kilopond (kp) is not to be confused with *kilopound* (kip). The former is 1 kgf = 9.80665 N, while the latter is 1000 lbf (pound-force) = 4448.22161526 N.

Note 120 (Fig. 3-16a)
The *millibar* is also permitted temporarily (Note 113).

Note 121 (Fig. 3-16a)
This is the so-called *IT calorie* (International Table). In addition, there are the following calories:

(a) $cal_{15} = 4.1855$ joules; this is the energy necessary to increase the temperature of 1 kg water from 15 °C to 16 °C. This value was determined experimentally.

(b) cal_{US} = 4.1858 joules; this is the accepted value in North America; it was determined by the United States National Bureau of Standards in 1939.

(c) cal_t = 4.184 joules; this is the thermodynamic calorie.

Note 122 (Fig. 3-16a)
Erg is the energy necessary to impart a 0.01 m/s² acceleration to a 0.001 kg mass over a distance of 0.01 m.

Note 123 (Fig. 3-16a)
Dyn is a force necessary to impart a 0.01 m/s² acceleration to a 0.001 kg mass.

Note 124 (Fig. 3-16a)
The relation between dynamic and kinematic viscosity (Note 26) is

$$\text{kinematic viscosity} = \frac{\text{dynamic viscosity}}{\text{density}}$$

Note 125 (Fig. 3-16b)
The unit *mho* is the reciprocal of *ohm*. See Note 12.

Note 126 (Fig. 3-16b)
Oersted is the magnetic strength which in vacuum exerts a 1 dyne force (Note 123) on a unit magnetic pole. The unit was named after Hans Christian Oersted (Danish physicist, 1777–1851).

Note 127 (Fig. 3-16b)
The unit *maxwell* is the magnetic flux which is generated by an *n*-turn coil carrying a current of $n \cdot 10^{-8}$ volts (Note 17). The SI equivalent of the *maxwell* is the *weber* (Note 19); 1 maxwell = 10^{-8} weber. The unit was named after James Clerk Maxwell (Scottish mathematician and physicist, 1831–1879).

Note 128 (Fig. 3-16b)
The *gauss* is the magnetic flux density, i.e., 1 gauss = 1 maxwell per 1 cm². The corresponding SI unit is the *tesla* (Note 16). 1 gauss = (0.0001) tesla. The unit was named after Johann Karl Friedrich Gauss (German mathematician, 1777–1855)

Note 129 (Fig. 3-16b)
1 phot = 10^4 lux (see Note 10).

Note 130 (Fig. 3-16b)
1 stilb = 10^4 cd/m².

Note 131 (Fig. 3-16b)
The *curie* "Ci" is the former unit of radioactive disintegration, 1 Ci = 3.7×10^{10} disintegrations. In SI, 1 disintegration = 1 becquerel "Bq," therefore 1 Ci = 3.7×10^{10} Bq (Note 1). 1 Ci is the activity of approximately 0.001 kg radium.

The unit curie was named after Pierre Curie (French chemist, 1859–1906) and Mme. Curie (Polish–French chemist, 1867–1934). They were awarded the Nobel Prize for physics in 1903, together with Henri Becquerel (Note 1).

Note 132 (Fig. 3-16b)
The *röntgen* (not "roentgen") is the formerly used unit of x-ray and gamma ray radiation-induced ionization of air. 1 *röntgen* is the quantity (dose) of radiation which produces a 2.58×10^{-4} coulomb charge in 1 kg air (Note 2).

The unit *röntgen* was named after Wilhelm Konrad Röntgen (German physicist, 1845–1923). He discovered x-ray radiation in 1895 (he named it as such because he did not know what it was). This discovery created such an effect that within a year more than one thousand papers were published on the subject. Röntgen was awarded the first-ever Nobel Prize in physics in 1901.

Note 133 (Fig. 3-16b)
Rad is the former unit of dose of *absorbed* radiation. It corresponds to the SI unit "gray" (Note 5); 1 rad = 0.01 gray.

Note 134 (Fig. 3-16b)
Rem is the former dose equivalent of radiation exposure; it was replaced by the SI unit "sievert." 1 rem = 0.01 sievert (Note 15).

Note 135 (Fig. 4-1b)
1 *Imperial gallon* (UK gallon) is 4.546092 liter, exactly, whereas 1 *Canadian gallon* is 4.54609 liter, exactly. Therefore, 1 Canadian gallon is only 0.9999996006 Imperial gallon.

Note 136 (Fig. 4-1b)
The *US gallon* is defined as 231 in^3, exactly, 1 inch being 0.0254 m, exactly (Fig. 4-1a). Hence

$$1 \text{ m}^3 = \frac{1}{(0.0254)^3} \text{ in}^3 = 61023.7440947 \text{ in}^3, \text{ exactly}$$

Or

$$1 \text{ in}^3 = (0.0254)^3 \text{ m}^3 = 1.6387064 \times 10^{-5} \text{ m}^3, \text{ exactly}$$

Therefore,

$$1 \text{ USgallon} = (231) \cdot (1.6387064 \times 10^{-5}) \text{ m}^3$$
$$= 0.003785411784 \text{ m}^3 = 3.785411784 \text{ liters, exactly.}$$

Note 137 (Fig. 4-1c)
In the UK 1 lb (mass) by definition equals 0.45359237 kg (mass), exactly.

Note 138 (Fig. 4-1e)
1 *British Thermal Unit* (BTU) is the energy required to raise the temperature of 1 lb (mass) water by 1 °F (Note 144). However, since the specific heat of water varies

with temperature, it is necessary to identify *which* BTU we use. The most accepted value is the BTU IT (International Table); 1 BTU IT = 1055.05585262 J.

Note 139 (Fig. 4-1e)
Kilowatt-hour (kW·h) is energy, since watt is power (Note 18). Hence 1 W·s = 1 J and 1 W·h = 1 W·(3600 s), thus

$$1 \text{ kW·h} = 1000 \text{ W·h} = (1000)\cdot(3600) \text{ W·s} = 3.6 \times 10^6 \text{ W·s} = 3.6 \times 10^6 \text{ J}$$

Note 140 (Fig. 4-1e)
This is the energy represented by Einstein's equation $E = M \cdot c^2$, where M is the mass, and c is the speed of light in vacuum (Appendix 2). To illustrate the enormous magnitude of energy which can be obtained—if we know how—by converting mass to energy, consider that *less* than half a gram of mass represents the total *chemical* energy contained in 1,000,000 kg gasoline.

Note 141 (Fig. 4-1f)
1 *metric horse-power* "HP_m" is 75 m·kgf energy expended in 1 s time. Since 1 kgf = 9.80665 N, therefore

$$1 \text{ HP}_m = (75)\cdot(9.80665) = 735.49875 \; \frac{\text{m}^2 \cdot \text{kg}}{\text{s}^3} = 735.49875 \text{ W}$$

exactly (Note 18).

Note 142 (Fig. 4-1f)
1 (US) *horse-power* "HP" is 550 lbf · ft energy expended in 1 s time. Since 1 lbf = 4.44822 N, and 1 ft = 0.3048 m, therefore

$$1 \text{ HP} = (550)\cdot(4.44822)\cdot(0.3048) = 745.69987 \; \frac{\text{m}^2 \cdot \text{kg}}{\text{s}^3} = 745.69987 \text{ W}$$

Therefore, considering Note 141,

$$1 \text{ HP} = \frac{745.69987}{735.49875} = 1.01387 \text{ HP}_m.$$

Note 143 (Fig. 4-1g and Fig. 3-16a)
Torr is the length of column in millimeters in a mercury barometer if the density of mercury is 13595.1 kg/m^3 at 0 °C and $g = 9.80665$ m/s^2.

The *torr* was named after Evangeliste Torricelli (Italian physicist, 1608–1647). He invented the mercury barometer as a solution to a problem posed to him by Galilei (why is it that water can not be sucked up beyond 10 m height?). Torricelli worked with Galilei during the last three months of the latter's life.

Note 144 (Fig. 4-2a)
The *Fahrenheit* temperature scale was invented by Gabriel Daniel Fahrenheit (German–Dutch physicist, 1686–1736). He invented the modern thermometer by substi-

tuting alcohol for mercury. Fahrenheit mixed water, salt, and ice to the lowest freezing point he could get and called the temperature of this mixture zero degree. He then designated the boiling temperature of water at sea level as 212 degrees. Thus the first reliable thermometer was created. For this feat Fahrenheit was elected a member of the Royal Society in 1724.

Note 145 (Fig. 4-2a)
The *Rankine* temperature scale was invented by and named for William John Macquorn Rankine (Scottish engineer, 1820–1872). The Rankine is an absolute scale, i.e., its zero is the absolute zero, but unlike the Kelvin scale (Art. 3.3.2a), the Rankine is graduated in Fahrenheit degrees (Note 144).

Note 146 (Fig. 4-1d)
The *poundal* (pdl) is the unit of force in the American British Mass System (Art. 3.4.4). By definition 1 pdl is the force necessary to impart 1 ft/s² acceleration to a mass of 1 pound. Since (force) = (mass)·(acceleration), hence 1 pdl = (1 lb)·(1 ft/s²) = 1(lb·ft/s²). But 1 lb = 0.45359237 kg (Note 137) and 1 ft/s² = 0.3048 m/s² (Fig. 4-1a), hence 1 pdl = (0.45359237)·(0.3048) = 0.13825 N.

Note 147 (Fig. 4-1d)
A *pound-force* (lbf) is the force—appearing as a weight—acting on a 1 lb mass attracted by Earth at a place where the gravitational acceleration is $g = 9.80665$ m/s². We have 1 lb = 0.45359237 kg (Note 137) and (weight) = (mass)·(acceleration), hence 1 lbf = (0.45359)·(9.80665) = 4.44822 N.

Note 148 (Fig. 3-13a)
The linear jerk is the third derivative of the linear (tangential) distance traveled with respect to time. Hence its dimension is

$$[\text{linear jerk}] = \frac{[\text{linear distance}]}{[\text{time}]^3} = \frac{m}{s^3} \quad \text{(see Example 5.1 in Chapter 5)}.$$

Note 149 (Fig. 3-13a)
The angular jerk is the third derivative of the angular distance travelled (angle rotated) with respect to time. Hence its dimension is

$$[\text{angular jerk}] = \frac{[\text{angular distance}]}{[\text{time}]^3} = \frac{[\text{rad}]}{[\text{time}]^3} = \frac{1}{s^3}$$

APPENDIX 5
ACRONYMS

For foreign names English translations only are given.

API	=	American Petroleum Institute
ASME	=	American Society of Mechanical Engineers
BIPM	=	International Bureau of Weights and Measures
BTU	=	British Thermal Unit
CGPM	=	General Conference of Weights and Measures
CIPM	=	International Committee of Weights and Measures
CODATA	=	Committee on Data for Science and Technology
DM	=	dimensional matrix
EDS	=	economics dimensional system
EEC	=	European Economic Community
ICRP	=	International Commission of Radiological Protection
IEEE	=	Institute of Electrical and Electronics Engineers
IHO	=	International Hydrographic Organization
IRE	=	Institute of Radio Engineers (superseded by IEEE)
IT	=	International Table
OR	=	operations research
psi	=	poundforce per square inch (pressure)
RBF	=	relative biological (effective) factor
RDM	=	reduced dimensional matrix
RPM	=	revolution per minute
SI	=	International System of Units

APPENDIX 6
SOLUTIONS TO PROBLEMS

Art. 3.3.4.1

3/1 Here newton is a derived named unit, and therefore it should not be capitalized (Rule c).

3/2 There is no space between "1.93" and "m" (Rule f). Also, there is no period after "m" although there should be because it is at the end of a sentence (Rule e).

3/3 "Celsius" should have a capital C (Rule h).

3/4 The adjective "gauge" (g) should modify "pressure" not the dimension "kPa" (Rule j). The sentence should be written "The gauge pressure in the tire is at least 206 kPa."

3/5 Names and symbols for dimensions should not be mixed. The sentence should read ". . . 29 m/s," or ". . . 29 meters per second" (Rule i).

3/6 Multiple prefixes must not be used. The sentence should be written ". . . 2.3 nm." (Rule ℓ)

3/7 The product of two dimensions should be indicated by a dot between them. The dimension in question should be written "km·h^{-1}" or "km/h" (Rule o).

3/8 A sentence should not start with a symbol (dimension). The sentence should be recast to read, e.g., "The dimension of mass in SI is kg" (Rule g).

3/9 A solidus "/" should not be used in a text where the names of units (not symbols) are used. The offending phrase should be changed to ". . . kilometers per hour," or ". . . km/h" (Rule r).

3/10 No more than one unit (dimension) should be used to define a quantity. The sentence therefore should read "My body's mass is 85.3 kg" (Rule m).

3/11 No more than one prefix should be used, and this prefix should be in the numerator—if at all possible. In the present case this is possible, therefore "85 cm/ks" should be written as "8.5 × 10^{-4} m/s," or "0.85 mm/s," or "0.085 cm/s" (Rule n).

3/12 Symbols for dimensions should only be printed (typed) upright (roman), not in *italics*. The sentence should be written ". . . g = 9.80665 m/s^2" (Rule b).

3/13 No abbreviations for symbols (dimensions) are permitted. Hence the sentence should read ". . . was 260 cm^3" (Rule a).

3/14 Symbols for dimensions must be written without full stops, unless they appear at the end of sentence. Therefore "kg" in the sentence should not have a period (Rule e).

3/15 Temperature in the Kelvin scale is not a temperature degree. There is no such a thing as °K. The correct text therefore is ". . . was 4.5 K" (see Art. 3.3.2.a).

3/16 There must be no space between the prefix and the symbol for a dimension. Therefore ". . . 20 M s" is wrong. The correct form is ". . . 20 Ms" (Rule k).

3/17 There must be no space between the prefix and the symbol for a dimension. Therefore the correct form is "2×10^{12} Ekg" (exakilogram). Better still, "2×10^{30} kg," since an exponent must be present anyway (Rule k).

3/18 Multiple prefixes are not allowed. So, the double prefix "TE" (teraexa) is illegal (Rule ℓ).

3/19 Multiple units are not allowed to define a quantity. Therefore the phrase ". . . 1 m 77 cm" should be written as ". . . 1.77 m" or ". . . 177 cm" or ". . . 1770 mm" (Rule m).

3/20 As it stands, it reads "millikilogram" and therefore it not only violates Rule l (multiple prefixes are not allowed), but is also totally purposeless since "milli" and "kilo" mean factors of 10^{-3} and 10^3, and hence the two cancel out. So what we have is simply "kg/s²." If we now consider "m" not as a prefix, but as a dimension, then the dot indicating the multiplication of "m" and "kg" is missing (Rule o). Therefore this dimension should have been written "m·kg/s²."

3/21 A solidus "/" is repeated without the use of parentheses thereby creating ambiguous expression (Rule p). Is it "5.5 (m/s)/s" or "5.5 m/(s/s)"? The first results in "m/s²," the second in "m."

3/22 A symbol should not begin a sentence. Therefore this text should be recast, for example "In SI, three of the seven fundamental dimensions are m, kg, and s" (Rule g).

3/23 The dimension is ambiguous because of double use of solidus "/"; it could be read as either (A²·s⁴)/(m²·kg) or as (A²·s⁴·kg)/m². The sentence violates Rule p.

3/24 Solidus "/" should not be used in text where names (as opposed of symbols) of dimensions are used (Rule r). Therefore the sentence should read ". . . pressure is force per area."

3/25 The prefix should appear in the *numerator,* not in the denominator (Rule n). Therefore, the expression should be modified to read "p = 30000 N/m²" or "p = 30 kN/m²" or "30 kPa."

3/26 A space must be placed between the last digit of a number and the symbol for the dimension. Therefore the expression should be modified to read "0.03 kPa" (Rule f).

3/27 Symbols of dimensions must remain unaltered in the plural. Therefore the expression should be changed to ". . . of 93.5 kg" (Rule d).

3/28 The dimension of a quantity may not replace its name. The correct sentence is "The distance between the Earth and the Sun is about 4.86×10^{12} megaparsec" (Rule q).

3/29 Symbols for *named* units are capitalized (first letter only). Hence it is ". . . 49 Wb" (Rule c).

3/30 Symbols for dimensions should not be mixed with the dimensions themselves. In the present case it should be ". . . 43 kg·m^2" (Rule i).

3/31 Abbreviations of symbols for dimensions are not allowed. Hence this expression should be written ". . . 144.3 m^2" (Rule a).

3/32 Symbols of dimensions must be printed upright, not in *italics* (Rule b).

3/33 Symbols of dimensions should remain unchanged in plural. Therefore here we should have ". . . his dainty 140 kg frame . . ." (Rule d).

3/34 The prefix should be in the numerator, not in the denominator. In this case 4.5 m/ks^2 is wrong (Rule n). The expression should be altered (to eliminate the prefix in the denominator) to

$$4.5 \text{ m/(ks)}^2 = 4.5 \text{ m/(1000·s)}^2 = 4.5 \text{ m/(10}^6\text{·s}^2) = 4.5 \times 10^{-6} \text{ m/s}^2$$

3/35 There must not be full stop after a symbol for a dimension, except at the end of a sentence (Rule e). Hence ". . . was 6 m.—not less!" is erroneous, ". . . was 6 m—not less!" is correct.

3/36 Named SI units should be written in lower-case letters (Rule *h*). Therefore ". . . 3600 Watt" is incorrect; it should be ". . . 3600 watt."

Art. 4.4

4/1 Planck's constant in SI is $h = 6.6260755 \times 10^{-34}$ J·s or $6.6260755 \times 10^{-34}$ (m^2·kg)/s. Hence

(a) Using the *American/British Mass System* we can write

$$6.6260775 \times 10^{-34} \text{ m}^2\text{·kg·s}^{-1} = x \text{·ft}^2\text{·lb·s}^{-1}$$

so by (4-1)

$Q \quad = 6.6260775 \times 10^{-34}$
1 m $= 3.28084$ ft; therefore $k_1 = 3.28084$, $e_1 = 2$
1 kg $= 2.20462$ lb; therefore $k_2 = 2.20462$, $e_2 = 1$
1 s $\ = 1$ s; therefore $k_3 = 1$, $e_3 = -1$

and therefore, by (4-4),

$$x = Q \cdot (k_1^{e_1} \cdot k_2^{e_2} \cdot k_3^{e_3}) = 6.6260775 \times 10^{-34} \cdot (3.28084)^2 \cdot (2.20462)^1 \cdot (1)^{-1} = 1.57239 \times 10^{-32}$$

therefore, by (4-4), $h = 1.57239 \times 10^{-32}$ (ft^2·lb)/s.

(b) Using the *American/British Force System* we can write

$$6.6260775 \times 10^{-34} \text{ J·s} = 6.6260755 \times 10^{-34} \text{ N·m·s}$$

so, by (4-1),

$$6.6260775 \times 10^{-34} \text{ N·m·s} = x \cdot \text{lbf·ft·s}$$

and therefore

$Q \quad = 6.6260755 \times 10^{-34}$
$1 \text{ N} = 0.22481 \text{ lbf (Fig. 4-1d)}; \quad$ therefore $k_1 = 0.22481, e_1 = 1$
$1 \text{ m} = 3.28084 \text{ ft (Fig. 4-1a)}; \quad$ therefore $k_2 = 3.28084, e_2 = 1$
$1 \text{ s} \ = 1 \text{ s} \quad$ therefore $k_3 = 1, e_3 = 1$

By (4-4) now

$$x = Q \cdot (k_1^{e_1} \cdot k_2^{e_2} \cdot k_3^{e_3}) = (6.6260755 \times 10^{-34}) \cdot (0.22481)^1 \cdot (3.28084)^1 \cdot (1)^1 = 4.88717 \times 10^{-34}$$

so that $h = 4.88717 \times 10^{-34}$ lbf·ft·s

4/2 The mechanical equivalent of heat is $q = 4.18680$ J/cal. Using now the *American/British Force System,* we write

$$4.18680 \text{ J·cal}^{-1} = x \cdot \text{lbf·ft·(BTU)}^{-1}, \text{ or } 4.18680 \text{ N·m·cal}^{-1} = x \cdot \text{lbf·ft·(BTU)}^{-1}$$

so that, by (4-1),

$Q \quad = 4.1868$
$1 \text{ N} = 0.22481 \text{ lbf (Fig. 4-1d)}; \quad$ therefore $k_1 = 0.22481, e_1 = 1$
$1 \text{ m} = 3.28084 \text{ ft (Fig. 4-1a)}; \quad$ therefore $k_2 = 3.28084, e_2 = 1$
$1 \text{ cal} = 0.00396832 \text{ BTU (Fig. 4-1e)}; \quad$ therefore $k_3 = 0.00396832, e_3 = -1$

and hence

$$x = Q \cdot (k_1^{e_1} \cdot k_2^{e_2} \cdot k_3^{e_3}) = (4.1868) \cdot (0.22481)^1 \cdot (3.28084)^1 \cdot (0.00396832)^{-1} = 778.16926$$

so that 4.1868 J/cal = 778.16926 (lbf·ft)/BTU.

4/3 The efficiency of technical writers is defined by $\eta = n_w/Q_m$, where n_w is the number of words written and Q_m is the energy expended in doing so in units of m·kgf. On the other hand, the *z-factor* is defined by $z = Q_{cal}/n_?$, where $n_?$ is the number of question marks written and Q_{cal} is the energy expended in calories in doing so. We also know that in an average written essay there are 3.8 question marks per page of 662.5 words. Hence we can write $n_? = (3.8) \cdot n_p$ and $n_w = (662.5) \cdot n_p$, where n_p is the number of pages and n_w is the number of words. By these relations $n_? = 3.8/662.5 \cdot n_w$. We now substitute this value into the relation for the *z*-factor. This yields

$$z = \frac{Q_{cal}}{n_w} \cdot \frac{662.5}{3.8}$$

and therefore

$$\eta \cdot z = \frac{n_w}{Q_m} \cdot \frac{Q_{cal}}{n_w} \cdot \frac{662.5}{3.8} = \frac{Q_{cal}}{Q_m} \cdot (174.342)$$

By Fig. 4-1e

$$\frac{Q_{cal}}{Q_m} = 2.34228$$

and therefore $\eta \cdot z = (2.34228) \cdot (174.342) = 408.358$, so our sought-after relation is $z = 408.358/\eta$, or equivalently $\eta = 408.358/z$.

4/4 The dimension of dynamic viscosity in SI is $[\mu] = m^{-1} \cdot kg \cdot s^{-1}$ (Note 26, Appendix 4). Here, of course, kg is the dimension of mass. In MKS force-based system, however, kg is the dimension of force (designated kgf to avoid ambiguity). By Newton's law, the dimension of mass in MKS is

$$[mass] = \frac{[force]}{[acceleration]} = \frac{kgf}{m/s^2} = m^{-1} \cdot kgf \cdot s^2$$

so that

$$[\mu] = m^{-1} \cdot kg \cdot s^{-1} = (m)^{-1} \cdot (m^{-1} \cdot kgf \cdot s^2)^1 \cdot (s)^{-1} = m^{-2} \cdot kgf \cdot s$$

is the dimension of dynamic viscosity in the MKS Force System.

4/5 We can write the problem—by (4-1)—as

$$40 \text{ ft}^{3.3} \cdot s^{2.2} \cdot lbf^{-1.1} = x \cdot m^{3.3} \cdot s^{2.2} \cdot N^{-1.1}$$

thus we have

Q = 40
1 ft = 0.3048 m (Fig. 4-1a); therefore $k_1 = 0.3048$, $e_1 = 3.3$
1 s = 1 s; therefore $k_2 = 1$, $e_2 = 2.2$
1 lbf = 4.44822 N (Fig. 4-1d); therefore $k_3 = 4.44822$, $e_3 = -1.1$

so that by (4-4) we have

$$x = Q \cdot (k_1^{e_1} \cdot k_2^{e_2} \cdot k_3^{e_3}) = (40) \cdot (0.3048)^{3.3} \cdot (1)^{2.2} \cdot (4.44822)^{-1.1} = 0.15357$$

and hence

$$40 \, \frac{\text{ft}^{3.3} \cdot s^{2.2}}{\text{lbf}^{1.1}} = 0.15357 \, \frac{m^{3.3} \cdot s^{2.2}}{N^{1.1}}$$

4/6 We write the problem in the form of (4-1)

$$M = (0.004) \cdot \text{ton} \cdot \text{year}^2 \cdot \text{parsec}^{-1} = x \cdot N \cdot s^2 \cdot m^{-1}$$

728 APPENDICES

thus we have

$Q = 0.004$

1 ton $= 1000$ kgf $= 9806.65$ N (Fig. 4-1d), thus $k_1 = 9806.65$; $e_1 = 1$

1 year $= 31,556,926$ s (tropical year, Note 100, App. 4), thus $k_2 = 31,556,926$; $e_2 = 2$

1 parsec $= 3.08568 \times 10^{16}$ m (Fig.4-1.a), thus $k_3 = 3.08568 \times 10^{16}$; $e_3 = -1$

and hence by (4-4)

$$x = Q \cdot (k_1^{e_1} \cdot k_2^{e_2} \cdot k_3^{e_3}) = (0.004) \cdot (9806.65)^1 \cdot (31,556,926)^2 \cdot (3.08568 \times 10^{16})^{-1} = 1.26596$$

which yields the result $M = 1.26596$ (N·s²)/m $= 1.26596$ kg.

4/7 The Broggg (correctly spelled) formula will yield identical results if the members on the right side remain individually unchanged by the transformation of dimensions. We have three members in the formula, and we handle them separately as follows:

First member: To simplify writing we use the temporary designations: z_1 if A is in ft² and λ is in degrees, and z_2 if A is in m² and λ is in radians. So we have

$$z_1 \text{ ft}^{4.4} \cdot \text{deg}^{-2.2} = z_2 \text{ m}^{4.4} \cdot \text{rad}^{-2.2}$$

But

1 ft $= 0.3048$ m (Fig. 4-1a), therefore $k_1 = 0.3048$; $e_1 = 4.4$

1 deg $= 1.74533 \times 10^{-2}$ rad, therefore $k_2 = 1.74533 \times 10^{-2}$; $e_2 = -2.2$

and hence by (4-4)

$$z_2 = z_1 \cdot (k_1^{e_1} \cdot k_2^{e_2}) = z_1 \cdot (0.3048)^{4.4} \cdot (1.74533 \times 10^{-2})^{-2.2} = z_1 \cdot (39.58542)$$

so the first member U_1 can be written

$$U_1 = 39 \cdot z_1 = 39 \cdot \frac{z_2}{39.58542} = (0.98521) \cdot z_2 = (0.98521) \cdot \left(\frac{A}{\lambda}\right)^{2.2}$$

in which A and λ are now expressed in SI.

Second member: Again, for the sake of expediency we write z_1 if T is in days and M is in pounds, and z_2 if T is in seconds and M is in kilograms. So we have

$$z_1 \cdot \text{day}^{0.7} \cdot \text{lb}^{1.9} = z_2 \cdot \text{s}^{0.7} \cdot \text{kg}^{1.9}$$

We now write

1 day $= 86,400$ s, therefore $k_1 = 86,400$; $e_1 = 0.7$

1 lb $= 0.45359$ kg (Fig. 4-1c), therefore $k_2 = 0.45359$; $e_2 = 1.9$

and by (4-4)

$$z_2 = z_1 \cdot (k_1^{e_1} \cdot k_2^{e_2}) = z_1 \cdot (86400)^{0.7} \cdot (0.45359)^{1.9} = z_1 \cdot (635.65317)$$

from which $z_1 = z_2/635.65317$. Thus the second member U_2 on the right side of the Broggg (correctly spelled) formula can be written

$$U_2 = (0.74) \cdot \frac{z_2}{635.65317} = (1.16415 \times 10^{-3}) \cdot z_2 = (1.16415 \times 10^{-3}) \cdot (T^{0.7} \cdot M^{1.9})$$

in which T is now in seconds and M is in kilograms, as required.

Third member: Here we first notice that n (number of lockers) is a dimensionless number, therefore it can be temporarily ignored. We (temporarily) set z_1 if w is in lbf, and z_2 if w is in N. Therefore we can write $z_1 \cdot \text{lbf}^{-1.1} = z_2 \cdot \text{N}^{-1.1}$ and

$$1 \text{ lbf} = 4.44822 \text{ N (Fig. 4-1d), therefore } k_1 = 4.44822;\ e_1 = -1.1$$

from which, by (4-4), we have $z_2 = z_1 \cdot k_1^{e_1} = z_1 \cdot (4.44822)^{-1.1} = (0.19364) \cdot z_1$. Thus $z_1 = (5.16422) \cdot z_2$. Therefore the third member U_3 on the right side of the Broggg formula is

$$U_3 = \frac{10^{16}}{n^2} \cdot z_1 = \frac{10^{16}}{n^2} (5.16422) \cdot z_2 = (5.16422 \times 10^{16}) \cdot \frac{z_2}{n^2} = (5.16422 \times 10^{16}) \cdot \frac{1}{n^2 \cdot w^{1.1}}$$

To sum up, the Broggg formula in SI is

$$C = (U_1 + U_2 - U_3)^{0.52}$$

or

$$C = \left[(0.98521) \cdot \left(\frac{A}{\lambda}\right)^{2.2} + (1.16415 \times 10^{-3}) \cdot (T^{0.7} \cdot M^{1.9}) - (5.16422 \times 10^{16}) \cdot \frac{1}{n^2 \cdot w^{1.1}} \right]^{0.52}$$

To verify this formula, assume the set of input values in the second column resulting in the data in the third column:

Quantity	Original Value and Dimension	Equivalent in SI
A	24,220 ft²	2250.112 m²
λ	79 deg	1.37881 rad
T	33 days	285,1200 s
M	536,100 lb	243,170.87 kg
n	172	172
w	166.2 lbf	739.29443 N
C	1,395,786.33 $	1,395,789.26 $

The end results (last row) coincide within 0.00021%!

4/8 We write the problem, by (4-1), as

$$900 \text{ BTU} \cdot \text{h}^{-1} \cdot \text{ft}^{-2} \cdot {}^\circ\text{F}^{-1} = x \cdot \text{J} \cdot \text{s}^{-1} \cdot \text{m}^{-2} \text{K}^{-1}$$

and consider that

Q	= 900		
1 BTU	= 1055.05585 J (Fig. 4-1e),	therefore $k_1 = 1055.05585$;	$e_1 = 1$
1 h	= 3600 s,	therefore $k_2 = 3600$;	$e_2 = -1$

1 ft   = 0.3048 m (Fig. 4-1a),   therefore $k_3 = 0.3048$; $e_3 = -2$
1 °F interval = $\frac{5}{9}$ K interval (Fig. 4-2b),   therefore $k_4 = \frac{5}{9}$; $e_4 = -1$

Therefore, by (4-4),

$$x = Q \cdot k_1^{e_1} \cdot k_2^{e_2} \cdot k_3^{e_3} \cdot k_4^{e_4} = 900 \cdot (1055.05585)^1 \cdot (3600)^{-1} \cdot (0.3048)^{-2} \cdot \left(\frac{5}{9}\right)^{-1} = 5110.43701$$

so that

$$900 \, \frac{\text{BTU}}{\text{h} \cdot \text{ft}^2 \cdot °\text{F}} = 5110.43701 \, \frac{\text{J}}{\text{s} \cdot \text{m}^2 \cdot \text{K}} = 5110.43701 \, \frac{\text{kg}}{\text{s}^3 \cdot \text{K}}$$

since 1 J = 1(m²kg)/s² (Fig. 3-10).

4/9 We write the problem thus c_1 lbf$^{0.7}$·cm$^{-4.3}$·°F^{-2}·h$^{-2.8}$ = c_2 N$^{0.7}$·m$^{-4.3}$·K^{-2}·s$^{-2.8}$, where c_1 and c_2 are the magnitudes of q in the respective dimensional systems. Following (4-1) we have

1 lbf   = 4.44822 N (Fig. 4-1d),   therefore $k_1 = 4.44822$; $e_1 = 0.7$
1 cm   = 0.01 m,   therefore $k_2 = 0.01$; $e_2 = -4.3$
1 °F interval = $\frac{5}{9}$ K (Fig. 4-2b),   therefore $k_3 = \frac{5}{9}$; $e_3 = -2$
1 h   = 3600 s,   therefore $k_4 = 3600$; $e_4 = -2.8$

so by (4-4)

$$c_2 = c_1 \cdot k_1^{e_1} \cdot k_2^{e_2} \cdot k_3^{e_3} \cdot k_4^{e_4} = c_1 \cdot (4.44822)^{0.7} \cdot (0.01)^{-4.3} \cdot \left(\frac{5}{9}\right)^{-2} \cdot (3600)^{-2.8} = c_1 \cdot (0.40423)$$

Therefore if $c_1 = 99$, then $c_2 = (99) \cdot (0.40423) = 40.01877$.

4/10 We write the formula:

$$C = 150 \cdot A + (0.04) \cdot v^2 \cdot A \qquad (*)$$

For the *first member* on the right, we let $z_1 = A$ if A is in square feet, and $z_2 = A$ if A is in square meters. We then can write z_1 ft² = z_2 m², and following (4-1) and (4-4)

1 ft = 0.3048 m (Fig. 4-1a), therefore $k_1 = 0.3048$; $e_1 = 2$

thus $z_2 = z_1 \cdot k_1^{e_1} = z_1 \cdot (0.3048)^2 = z_1 \cdot (0.092903)$, which yields $z_1 = (10.76391) \cdot z_2$. Therefore the first member of the formula (*) is

$$U_1 = 150 \cdot z_1 = (150) \cdot (10.76391) \cdot z_2 = (1614.58656) \cdot A$$

where A on the right is in square meters.

For the *second member* on the right side of (*) now we write $z_1 = v^2 \cdot A$, where v is the speed in miles per hour and A is the area of the throat in square feet. Similarly, we let $z_2 = v^2 \cdot A$, where v is in meter per second and A is in square meters. Consequently, we can write

$$z_1 \text{ miles}^2 \cdot \text{h}^{-2} \cdot \text{ft}^2 = z_2 \text{ m}^2 \cdot \text{s}^{-2} \cdot \text{m}^2$$

and, by (4-1) and (4-4)

1 mile = 1609.344 m (Fig. 4-1a),   therefore $k_1 = 1609.344$; $e_1 = 2$
1 h = 3600 s,   therefore $k_2 = 3600$; $e_2 = -2$
1 ft = 0.3048 m (Fig. 4-1a),   therefore $k_3 = 0.3048$; $e_3 = 2$

and further

$$z_2 = z_1 \cdot k_1^{e_1} \cdot k_2^{e_2} \cdot k_3^{e_3} = z_1 \cdot (1609.344)^2 \cdot (3600)^{-2} \cdot (0.3048)^2 = z_1 \cdot (0.0185662)$$

and therefore $z_1 = (53.86136) \cdot z_2$, which then yields the second member of (*)

$$U_2 = (0.04) \cdot z_1 = (0.04) \cdot (53.86136) \cdot z_2 = (2.15445) \cdot v^2 \cdot A$$

where of course v and A are now expressed in SI. The Wood formula (*) in SI therefore is

$$C = U_1 + U_2 = (1614.58656) \cdot A + (2.15445) \cdot v^2 A$$

or

$$C = A \cdot (1614.58656 + 2.15445 \times v^2) \qquad (**)$$

To check, let us consider $A = 1033$ ft^2 = 95.96884 m^2, and $v = 77.6$ miles/h = 34.6903 m/s. According to formulas (*) and (**), $C = 403769.12$ \$ and $C = 403768.56$ \$, respectively. These two values are identical within 0.00014%.

4/11 The number appearing in the formula is *not* dimensionless; its dimension is

$$[46300] = \frac{[\text{force}]}{[\text{second moment of area}]} = \frac{\text{lbf}}{\text{in}^4}$$

Therefore, by using the method described in Chapter 4, our task is to find x in

$$46300 \cdot \text{lbf} \cdot \text{in}^{-4} = x \cdot \text{N} \cdot \text{m}^{-4}$$

We have

Q = 46300
1 lbf = 4.44822 N (Fig. 4-1d),   therefore $k_1 = 4.44822$; $e_1 = 1$
1 in = 0.0254 m (Fig. 4-1a),   therefore $k_2 = 0.0254$; $e_2 = -4$

and by (4-4)

$$x = Q \cdot k_1^{e_1} \cdot k_2^{e_2} = (46300) \cdot (4.44822)^1 \cdot (0.0254)^{-4} = 4.94803 \times 10^{11}$$

which yields the critical force

$$P_{cr} = (4.94803 \times 10^{11}) \cdot I$$

where P_{cr} is in newtons and I is in m^4, as required.

4/12 To combine *importance factor* i with the *unimportance factor* u, we let $z_1 = t_{TV}/t_D$, where t_{TV} is the TV "coverage" time in seconds, and t_D is the demonstration time in minutes. Similarly, we let $z_2 = t_{TV}/t_D$, where everything is in seconds. Therefore we can write $z_1 \cdot s \cdot min^{-1} = z_2 \cdot s \cdot s^{-1}$ and therefore, by (4-1) and (4-4),

1 s = 1 s,   therefore $k_1 = 1$; $e_1 = 1$
1 min = 60 s, therefore $k_2 = 60$; $e_2 = -1$

so with $Q = z_1$

$$z_2 = Q \cdot k_1^{e_1} \cdot k_2^{e_2} = z_1 \cdot (1)^1 \cdot (60)^{-1} = \frac{1}{60} \cdot z_1$$

from which $z_1 = 60 \cdot z_2$. Therefore the importance factor i in SI can be written

$$i = \left(\frac{2 \times 10^4}{n} \cdot 60 \cdot \frac{t_{TV}}{t_D}\right)^{1.5} = \left(\frac{1.2 \times 10^6}{n} \cdot \frac{t_{TV}}{t_D}\right)^{1.5}$$

where both t_{TV} and t_D are now expressed in seconds.

Now for the *unimportance factor* we let $z_1 = t_D/t_{TV}$, where t_D is in hours and t_{TV} is in seconds. Analogously, we let $z_2 = t_D/t_{TV}$, where both times are in seconds. Hence $z_1 \cdot h \cdot s^{-1} = z_2 \cdot s \cdot s^{-1}$ so that

Q = z_1
1 h = 3600 s,   therefore $k_1 = 3600$; $e_1 = 1$
1 s = 1 s,   therefore $k_2 = 1$; $e_2 = -1$

and hence, by (4-4)

$$z_2 = Q \cdot k_1^{e_1} \cdot k_2^{e_2} = z_1 \cdot (3600)^1 \cdot (1)^{-1} = 3600 \cdot z_1$$

from which

$$z_1 = \frac{1}{3600} \cdot z_2$$

Thus the *unimportance factor* in SI will be

$$u = \left(\frac{n}{1000} \cdot \frac{t_D}{3600 \cdot t_{TV}}\right)^{1.5} = \left(\frac{n}{3.6 \times 10^6} \cdot \frac{t_D}{t_{TV}}\right)^{1.5}$$

where both times are expressed in seconds. To get the *relation* now between the factors i and u, we generate their *product*. Accordingly,

$$i \cdot u = \left(\frac{1.2 \times 10^6}{n} \cdot \frac{t_{TV}}{t_D}\right)^{1.5} \cdot \left(\frac{n}{3.6 \times 10^6} \cdot \frac{t_D}{t_{TV}}\right)^{1.5} = \left(\frac{1}{3}\right)^{1.5} = \frac{1}{\sqrt{27}} = 0.19245$$

Therefore we finally have

$$i = \frac{0.19245}{u} \quad \text{or} \quad u = \frac{0.19245}{i}$$

For example, let $t_{TV} = 170$ s, $n = 2600$, $t_D = 300$ min $= 5$ h. This yields the *importance factor* $i = 9.10074$ and the *unimportance factor* $u = 0.021147$. Their product is indeed $i \cdot u = (9.10074) \cdot (0.021147) = 0.19245$.

4/13 From the formula presented, the dimension of the constant on the right side is

$$[3.132] = [\omega_e] \cdot [r]^{0.5} = \frac{\text{rad}}{\text{s}} \cdot \text{m}^{0.5} = \text{rad} \cdot \text{s}^{-1} \cdot \text{m}^{0.5}$$

since tan φ is dimensionless.

(a) We write, by (4-1), $(3.132) \cdot \text{rad} \cdot \text{s}^{-1} \cdot \text{m}^{0.5} = x \cdot \text{rev} \cdot \text{min}^{-1} \cdot \text{in}^{0.5}$, and by (4-2),

$Q = 3.132$
1 rad = $1/(2 \cdot \pi)$ revolution, therefore $k_1 = 1/(2 \cdot \pi)$; $e_1 = 1$
1 s = 1/60 min, therefore $k_2 = 1/60$; $e_2 = -1$
1 m = 39.37008 in (Fig. 4-1a), therefore $k_3 = 39.37008$; $e_3 = 0.5$

So, by (4-4)

$$x = Q \cdot k_1^{e_1} \cdot k_2^{e_2} \cdot k_3^{e_3} = 3.132 \times \left(\frac{1}{2 \cdot \pi}\right)^1 \cdot \left(\frac{1}{60}\right)^{-1} \cdot (39.37008)^{0.5} = 187.66198$$

Therefore

$$3.132 \text{ rad} \cdot \text{s}^{-1} \cdot \text{m}^{0.5} = 187.66198 \text{ rev} \cdot \text{min}^{-1} \cdot \text{in}^{0.5}$$

Thus, our formula is $\omega_e = 187.66198/\sqrt{r \cdot \tan \varphi}$ rev/min, in which r is in inches.

(b) Similarly to (a), we write

$$3.132 \text{ rad} \cdot \text{s}^{-1} \cdot \text{m}^{0.5} = x \cdot \text{deg} \cdot \text{millennium}^{-1} \cdot \text{parsec}^{0.5} \qquad (*)$$

Again, by (4-1)

$Q = 3.132$
1 rad = $180/\pi$ deg, therefore $k_1 = 180/\pi$; $e_1 = 1$
1 s = 3.16888×10^{-11} millennium (Fig. 3-14), thus $k_2 = 3.16888 \times 10^{-11}$; $e_2 = -1$
1 m = 3.24078×10^{-17} parsec (Fig. 3-14), thus $k_3 = 3.24078 \times 10^{-17}$; $e_3 = 0.5$

so by (*), the above relations and (4-4)

$$x = Q \cdot k_1^{e_1} \cdot k_2^{e_2} \cdot k_3^{e_3} = 3.132 \times \left(\frac{180}{\pi}\right)^1 \cdot (3.16888 \times 10^{-11})^{-1} \cdot (3.24078 \times 10^{-17})^{0.5}$$

$$= 32237.64936$$

and the formula then

$$\omega_e = \frac{32237.64936}{\sqrt{r \cdot \tan \varphi}} \frac{\text{deg}}{\text{millennium}}$$

in which r is now in *parsecs* (1 parsec = 3.08568×10^{13} km).

4/14 The SI numerical values of the given constants are as follows:

Stefan–Boltzmann constant $\qquad \sigma = 5.67051 \times 10^{-8}$ kg/(s$^3 \cdot$K^4)
speed of light in vacuum $\qquad c = 2.99792 \times 10^8$ m/s

universal gravitational constant $\quad k = 6.67259 \times 10^{-11}$ m^3/(kg·s^2)
charge of electron $\quad e = 1.60218 \times 10^{-19}$ A·s
gravitational acceleration (sea level) $\quad g = 9.80665$ m/s^2

We employ the process detailed in Example 4-16. The following symbols are used: μ_σ, μ_c, μ_k, μ_e, and μ_g are the *magnitudes* of σ, c, k, e, and g expressed in SI; x, y, z, u, and w are the *dimensions* of length, mass, time, temperature, and electric current in the new system.

We now write

$$\mu_\sigma \text{ kg·s}^{-3}\text{·K}^{-4} = 1\cdot y\cdot z^{-3}\cdot u^{-4}$$

$$\mu_c \text{ m·s}^{-1} = 1\cdot x\cdot z^{-1}$$

$$\mu_k \text{ m}^3\text{·kg}^{-1}\text{·s}^{-2} = 1\cdot x^3\cdot y^{-1}\cdot z^{-2}$$

$$\mu_e \text{ s·A} = 1\cdot z\cdot w$$

$$\mu_g \text{ m·s}^{-2} = 1\cdot x\cdot z^{-2}$$

By these then

$$\mu_\sigma = \left(\frac{y}{\text{kg}}\right)\cdot\left(\frac{z}{s}\right)^{-3}\cdot\left(\frac{u}{K}\right)^{-4} \quad \text{or} \quad \ln\mu_\sigma = \ln\frac{y}{\text{kg}} - 3\cdot\ln\frac{z}{s} - 4\cdot\ln\frac{u}{K}$$

$$\mu_c = \left(\frac{x}{m}\right)\cdot\left(\frac{z}{s}\right)^{-1} \quad \text{or} \quad \ln\mu_c = \ln\frac{x}{m} - \ln\frac{z}{s}$$

$$\mu_k = \left(\frac{x}{m}\right)^3\cdot\left(\frac{y}{\text{kg}}\right)^{-1}\cdot\left(\frac{z}{s}\right)^{-2} \quad \text{or} \quad \ln\mu_k = 3\cdot\ln\frac{x}{m} - \ln\frac{y}{\text{kg}} - 2\cdot\ln\frac{z}{s}$$

$$\mu_e = \left(\frac{z}{s}\right)\cdot\left(\frac{w}{A}\right) \quad \text{or} \quad \ln\mu_e = \ln\frac{z}{s} + \ln\frac{w}{A}$$

$$\mu_g = \left(\frac{x}{m}\right)\cdot\left(\frac{z}{s}\right)^{-2} \quad \text{or} \quad \ln\mu_g = \ln\frac{x}{m} - 2\cdot\ln\frac{z}{s}$$

and therefore

$$\begin{bmatrix} \ln\mu_\sigma \\ \ln\mu_c \\ \ln\mu_k \\ \ln\mu_e \\ \ln\mu_g \end{bmatrix} = \mathbf{T}\cdot\begin{bmatrix} \ln\frac{x}{m} \\ \ln\frac{y}{\text{kg}} \\ \ln\frac{z}{s} \\ \ln\frac{u}{K} \\ \ln\frac{w}{A} \end{bmatrix} \quad \text{where } \mathbf{T} = \begin{bmatrix} 0 & 1 & -3 & -4 & 0 \\ 1 & 0 & -1 & 0 & 0 \\ 3 & -1 & -2 & 0 & 0 \\ 0 & 0 & 1 & 0 & 1 \\ 1 & 0 & -2 & 0 & 0 \end{bmatrix}$$

is the *transformation matrix* whose determinant is –4 (not zero), so we have a unique solution. We write

$$\begin{bmatrix} \ln\dfrac{x}{m} \\ \ln\dfrac{y}{kg} \\ \ln\dfrac{z}{s} \\ \ln\dfrac{u}{K} \\ \ln\dfrac{w}{A} \end{bmatrix} = \boldsymbol{\tau}^{-1} \cdot \begin{bmatrix} \ln\mu_\sigma \\ \ln\mu_c \\ \ln\mu_k \\ \ln\mu_e \\ \ln\mu_g \end{bmatrix} = \dfrac{1}{4} \cdot \begin{bmatrix} 0 & 8 & 0 & 0 & -4 \\ 0 & 16 & -4 & 0 & -4 \\ 0 & 4 & 0 & 0 & -4 \\ -1 & 1 & -1 & 0 & 2 \\ 0 & -4 & 0 & 4 & 4 \end{bmatrix} \cdot \begin{bmatrix} -16.68540 \\ 19.51860 \\ -23.43043 \\ -43.27775 \\ 2.28306 \end{bmatrix}$$

which yields, by the indicated multiplication,

$$\begin{bmatrix} \ln\dfrac{x}{m} \\ \ln\dfrac{y}{kg} \\ \ln\dfrac{z}{s} \\ \ln\dfrac{u}{K} \\ \ln\dfrac{w}{A} \end{bmatrix} = \begin{bmatrix} 36.75414 \\ 99.22177 \\ 17.23554 \\ 16.05014 \\ -60.513294 \end{bmatrix}$$

and hence we obtain

$$x = (9.16472 \times 10^{15}) \cdot m$$
$$y = (1.23443 \times 10^{43}) \cdot kg$$
$$z = (3.05703 \times 10^{7}) \cdot s$$
$$u = (9.34300 \times 10^{6}) \cdot K$$
$$w = (5.24097 \times 10^{-27}) \cdot A$$

For example, in the new system the dimension of length is 9.16472×10^{15} m, and the temperature scale has a unit 9.343×10^{6} times as large as kelvin, etc. To check, let us find the value of g in the new system. We have $g = 9.80665$ m·s^{-2}. But

$$1\ m = \dfrac{x}{9.16472 \times 10^{15}} \quad \text{and} \quad 1\ s = \dfrac{z}{3.05703 \times 10^{7}}$$

Therefore

$$9.80665 \cdot \text{m} \cdot \text{s}^{-2} = 9.80665 \cdot \left(\frac{1}{9.16472 \times 10^{15}}\right) \cdot \left(\frac{1}{3.05703 \times 10^{7}}\right)^{-2} \cdot \frac{x}{z^2} = 1 \cdot \frac{x}{z^2}$$

as expected and required. The other constants can be verified similarly.

4/15 The constants of nature Kapitza uses as units for his dimensional systems are (in SI):

radius of electron	$r_e = 2.81794 \times 10^{-15}$ m
mass of electron	$m_e = 9.10939 \times 10^{-31}$ kg
electron's rest-mass relativistic energy	$E = 8.18711 \times 10^{-14}$ (m^2·kg)/s^2

The last unit was calculated by Einstein's formula $E = m_e \cdot c^2$, where $c = 2.99792 \times 10^8$ m/s is the speed of light in vacuum. We shall use the symbols μ_{r_e}, μ_E, μ_{m_e} for the *magnitudes* of physical constants r_e, E, m_e in SI, and x, y, z for the *dimensions* of length, mass and time in the Kapitza system. By following the process outlined in the previous problem, we have

$$\mu_{r_e} \cdot \text{m} = 1 \cdot x; \qquad \mu_E \cdot \text{m}^2 \cdot \text{kg} \cdot \text{s}^{-2} = 1 \cdot x^2 \cdot y \cdot z^{-2}; \qquad \mu_{m_e} \cdot \text{kg} = 1 \cdot y$$

which can be written in matrix form as

$$\begin{bmatrix} \ln \mu_{r_e} \\ \ln \mu_E \\ \ln \mu_{m_e} \end{bmatrix} = \mathbf{T} \cdot \begin{bmatrix} \ln \dfrac{x}{\text{m}} \\ \ln \dfrac{y}{\text{kg}} \\ \ln \dfrac{z}{\text{s}} \end{bmatrix} \qquad \text{where } \mathbf{T} = \begin{bmatrix} 1 & 0 & 0 \\ 2 & 1 & -2 \\ 0 & 1 & 0 \end{bmatrix}$$

is the *transformation matrix* between the Kapitza system and SI. From the above relation

$$\begin{bmatrix} \ln \dfrac{x}{\text{m}} \\ \ln \dfrac{y}{\text{kg}} \\ \ln \dfrac{z}{\text{s}} \end{bmatrix} = \mathbf{T}^{-1} \cdot \begin{bmatrix} \ln \mu_{r_e} \\ \ln \mu_E \\ \ln \mu_{m_e} \end{bmatrix} = \begin{bmatrix} 1 & 0 & 0 \\ 0 & 0 & 1 \\ 1 & -\dfrac{1}{2} & \dfrac{1}{2} \end{bmatrix} \cdot \begin{bmatrix} -33.50277 \\ -30.13363 \\ -69.17083 \end{bmatrix}$$

Performing the indicated matrix multiplication, we get

$$\begin{bmatrix} \ln \dfrac{x}{\text{m}} \\ \ln \dfrac{y}{\text{kg}} \\ \ln \dfrac{z}{\text{s}} \end{bmatrix} = \begin{bmatrix} -33.50277 \\ -69.17083 \\ -53.02137 \end{bmatrix}$$

yielding

$$x = 2.81794 \times 10^{15} \cdot m; \qquad y = 9.10939 \times 10^{-31} \cdot kg; \qquad z = 9.39964 \times 10^{-24} \cdot s$$

which means, for instance, that the mass unit in the Kapitza system is 9.10939×10^{-31} kg. If the above results are correct, then the magnitudes of r_e, E, and m_e in the Kapitza system will all be unity. Let us verify this. From the above, then

$$r_e = 2.81794 \times 10^{-15} \cdot m = 2.81794 \times 10^{-15} \cdot \frac{x}{2.81794 \times 10^{-15}} = 1 \cdot x$$

$$m_e = 9.10939 \times 10^{-31} \cdot kg = 9.10939 \times 10^{-31} \cdot \frac{y}{9.10939 \times 10^{-31}} = 1 \cdot y$$

$$E = 8.18711 \times 10^{-14} \cdot \frac{m^2 \cdot kg}{s^2} = 8.18711 \times 10^{-14} \cdot$$

$$\cdot \left(\frac{x}{2.81794 \times 10^{-15}}\right)^2 \left(\frac{y}{9.10939 \times 10^{-31}}\right) \left(\frac{z}{9.39964 \times 10^{-24}}\right)^{-2} = 1 \cdot \frac{x^2 y}{z^2}$$

So, in the Kapitza system the radius, mass and the relativistic rest-mass energy equivalent of electron are indeed unity.

4/16 From the Clark formula, the dimension of the constant is

$$[2273] = [f] \cdot [L]^{0.81} = \frac{\text{beats}}{\text{min}} \cdot \text{in}^{0.81}$$

where f is the heartbeat frequency per minute and L is the linear size of the animal in inches. By (4-1) now we write

$$2273 \cdot \text{beat} \cdot \text{min}^{-1} \cdot \text{in}^{0.81} = x \cdot \text{beat} \cdot s^{-1} \cdot m^{0.81} \qquad (*)$$

and by (4-2)

$Q \quad = 2273$
1 beat = 1 beat,   therefore $k_1 = 1$; $e_1 = 1$
1 min = 60 s,   therefore $k_2 = 60$; $e_2 = -1$
1 in = 0.0254 m,   therefore $k_3 = 0.0254$; $e_3 = 0.81$

By (4-4), x in (*) will be

$$x = Q \cdot k_1^{e_1} \cdot k_2^{e_2} \cdot k_3^{e_3} = 2273 \cdot (1)^1 \cdot (60)^{-1} \cdot (0.0254)^{0.81} = 1.933$$

Therefore the Clark formula in SI is $f = (1.933) \cdot L^{-0.81}$, where f is the heartbeat per second, and L is the linear size of the animal in meters. For example, an $L = 1.8$ m tall man's resting heart rate is about

$$f = (1.933) \cdot (1.8)^{-0.81} = 1.2 \, \frac{\text{beat}}{s} \left(= 72 \, \frac{\text{beat}}{\text{min}} \right)$$

Art. 6.3

6/1 First we establish the dimensions of the variables in the equation given.

gas concentration $\quad [C] = \dfrac{\text{mol}}{\text{m}^3}$

heat content per mole $\quad [H] = \dfrac{[\text{heat}]}{[\text{quantity of substance}]} = \dfrac{\text{m}^2 \cdot \text{kg}}{\text{s}^2 \cdot \text{mol}}$

universal gas constant $\quad [R] = \dfrac{\text{m}^2 \cdot \text{kg}}{\text{s}^2 \cdot \text{K} \cdot \text{mol}} \quad$ (see Appendix 2)

The dimensions of constants k_2 and k_3 will be determined shortly, but we first let $q' = \ln q$ and then place all the logarithmic members on the left

$$\ln C - \dfrac{k_1}{R} \cdot \ln T - \ln q = \dfrac{k_2}{R} \cdot T + \dfrac{k_3}{2 \cdot R} \cdot T^2 - \dfrac{H}{R \cdot T}$$

or

$$\ln \dfrac{C}{T^{k_1/R} \cdot q} = \dfrac{k_2}{R} \cdot T + \dfrac{k_3}{2 \cdot R} \cdot T^2 - \dfrac{H}{R \cdot T} \qquad (*)$$

This equation is now dimensionally homogeneous, provided that

$$[q] = \dfrac{[C]}{[T]^{k_1/R}} = \dfrac{[C]}{[T]^n} = \dfrac{\text{mol}}{\text{m}^3 \cdot \text{K}^n}$$

where n is a dimensionless number numerically equal to k_1/R. By the rules of dimensional homogeneity (Art. 6.1), all members on the right side of (*) must be dimensionless. This fact now makes it possible to determine the dimensions of k_2 and k_3. Accordingly,

$$[k_2] = \dfrac{[R]}{[T]} = \dfrac{\text{m}^2 \cdot \text{kg}}{\text{s}^2 \cdot \text{K}^2 \cdot \text{mol}} \quad \text{and} \quad [k_3] = \dfrac{[R]}{[T]^2} = \dfrac{\text{m}^2 \cdot \text{kg}}{\text{s}^2 \cdot \text{K}^3 \cdot \text{mol}}$$

and with this, the solution of the problem is complete.

6/2 For a right circular cylinder of length L and diameter D, surface area A is

$$A = \pi \cdot D \cdot L + \dfrac{D^2 \cdot \pi}{2} \qquad (*)$$

(a) Designating L as independent variable, and A as dependent variable, we see that D can be considered a parameter, which is not dimensionless. Hence, the chart in Fig. A6-1 depicting relation (*) is dimensionally nonhomogeneous. For example, if D (parameter) is 0.75 m, and $L = 3.4$ m, then $A = 8.89463$ m^2—as illustrated.

(b) We divide both sides of (*) by D^2. This will yield

$$\dfrac{A}{D^2} = \pi \cdot \dfrac{L}{D} + \dfrac{\pi}{2} \qquad (**)$$

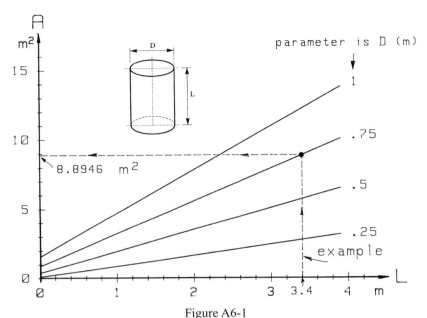

Figure A6-1
Dimensionally nonhomogeneous chart giving surface area A of a right circular cylinder
Parameter is D (diameter)

Let us introduce *dimensionless variables* $\pi_1 = A/D^2$ and $\pi_2 = L/D$. By using these

$$\pi_1 = \pi \cdot \pi_2 + \frac{\pi}{2} \qquad (\text{***})$$

Note that π *without* a subscript designates the dimensionless number $3.14159\ldots$, while π *with* a subscript—viz., π_1, π_2, etc.—designates dimensionless variables, such as shown above.

Since (***) contains only dimensionless variables, its plot is *homogeneous* (Fig. A6-2). Note that in part (a) of this problem we had a dimensionally *nonhomogeneous* plot involving *dimensional* variables and a *family* of curves, here in part (b), we have a dimensionally *homogeneous* plot involving *dimensionless* variables and a *single* curve. The reason for this very advantageous reduction of the number of variables is that the number of *dimensionless* variables describing a relation is always less than the number of generating *dimensional* variables. This subject is thoroughly treated in Chapter 7.

Consider again a cylinder of $L = 3.4$ m and $D = 0.75$ m. Thus $\pi_2 = L/D = 4.53333$, and from the figure $\pi_1 = A/D^2 = 15.81$. Hence $A = \pi_1 \cdot D^2 = (15.81) \cdot (0.75)^2 = 8.89$ m^2, the same value as obtained previously.

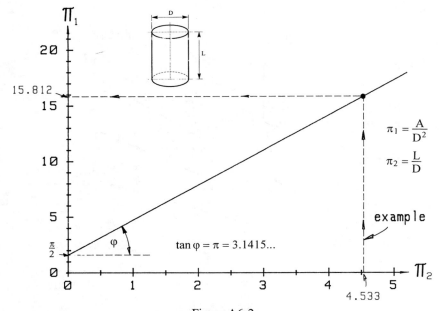

Figure A6-2
Dimensionally homogeneous plot giving surface area A of a right circular cylinder

6/3 For convenience we repeat the subject *empirical* equation here.

$$D = \frac{46800}{e^{a/5}} + 6000 \cdot e \qquad (*)$$

where D is the yearly income; $[D] = \$/\text{year}$

a is the age of worker in years; $[a] = \text{year}$

e is the time worker spent for postsecondary education in years; $[e] = \text{year}$

Since $a/5$ is in the exponent, it must be dimensionless (Rule 4), thus "a" itself must be dimensionless, but it is not. To make the exponent dimensionless the number "5" must have an identical dimension to that of "a." Thus $[5] = \text{year}$. Now by Rule 1 it is necessary that $[D] = [6000] \cdot [e]$. Hence $[6000] = [D]/[e] = \$/\text{year}^2$. Finally, by Rule 2, the two members of the right side of (*) must have identical dimensions. From this condition $[46800/e] = [6000 \cdot e]$ since the exponent of "e" is dimensionless. Thus, it follows that

$$[46800] = [6000] \cdot [e]^2 = \frac{\$}{\text{year}^2} \cdot \text{year}^2 = \$$$

6/4 By Theorem 5-7 (Chapter 5), the dimension of the left side is $[x]^3$. By Rule 4, $[a] = [x]^{-1}$. Moreover, by Rule 1 and Rule 2, *all* members on the right *must* have the dimension $[x]^3$. Since the arguments of all trigonometric functions must be dimensionless (Rule 4), we only have to scrutinize the coefficients on the right side. Accordingly,

The *first* member's dimension is $\left[\dfrac{2 \cdot x}{a^2}\right] = \dfrac{[x]}{[a]^2} = \dfrac{[x]}{[x]^{-2}} = [x]^3$. This is correct.

The *second* member's dimension is $\left[\dfrac{2 \cdot x}{a^3}\right] = \dfrac{[x]}{[a]^3} = \dfrac{[x]}{[x]^{-3}} = [x]^4$. This is *not* correct.

The *third* member's dimension is $\left[\dfrac{x^2}{a}\right] = \dfrac{[x]^2}{[a]} = \dfrac{[x]^2}{[x]^{-1}} = [x]^3$. This is correct.

Therefore only the second member is erroneous. Let us write this member thus $(2 \cdot x^p)/a^q$, where p and q are as-yet undefined. However, we must have

$$\left[\dfrac{2 \cdot x^p}{a^q}\right] = \dfrac{[x]^p}{[x]^{-q}} = [x]^{p+q} = [x]^3$$

Thus, $p + q = 3$. Therefore $p = 3 - q$. But q is correct by the condition stated, consequently $q = 3$. Hence $p = 3 - 3 = 0$, thus the *correct second* member on the right is $(2/a^3) \cdot \cos(a \cdot x)$.

6/5 We consider the fact that a dimensionally homogeneous equation must have identical dimensions on both sides (Rule 1). Therefore we can write $[v] = [k \cdot \sqrt{R^{n_1} \cdot \rho^{n_2} \cdot g^{n_3}}]$, or in SI

$$\text{m} \cdot \text{s}^{-1} = \text{m}^{1.5} \cdot \text{kg}^{-1.5} \cdot \text{m}^{n_1/2} \cdot \left(\dfrac{\text{kg}}{\text{m}^3}\right)^{n_2/2} \cdot \left(\dfrac{\text{m}}{\text{s}^2}\right)^{n_3/2}$$

which yields, by equating the exponents of dimensions of both sides

for "m" $1 = 1.5 + (0.5) \cdot n_1 - (1.5) \cdot n_2 + (0.5) \cdot n_3$
for "kg" $0 = -0.5 + (0.5) \cdot n_2$
for "s" $-1 = -n_3$

Solving the above for n_1, n_2, and n_3, we obtain $n_1 = n_2 = n_2 = 1$. Thus our equation becomes $v = k \cdot \sqrt{R \cdot \rho \cdot g}$. Dimensional constant k, of course, incorporates not only the shape of the falling object (a sphere in this case), but also the air density.

6/6 By Rule 2 the dimensions of the *first* and *third* members on the left must be identical. Hence

$$\left[\dfrac{\partial^2 \Psi}{\partial x^2}\right] = \left[\dfrac{M_0}{h^n}\right] \cdot [(E + m_0 \cdot c^2) \cdot \Psi]$$

in which

$$\left[\frac{\partial^2 \Psi}{\partial x^2}\right] = \frac{[\Psi]}{m^2} \qquad \text{by Theorem 5-6, Chapter 5}$$

$$[E + M_0 \cdot c^2] = \frac{m^2 \cdot kg}{s^2} \qquad \text{energy}$$

$$[h] = \frac{m^2 \cdot kg}{s} \qquad \text{Planck's constant, Appendix 2}$$

Therefore, by Rule 2, $[\Psi] \cdot m^{-2} = kg \cdot (m^2 \cdot kg \cdot s^{-1})^{-n} \cdot (m^2 \cdot kg \cdot s^{-2}) \cdot [\Psi]$. We now observe that the power of dimension "kg" on the left side is zero, and hence it must also be zero on the right side. Thus, $0 = 1 - n + 1$, from which $n = 2$. Note that in this case, to determine exponent n, it was not necessary to know the dimensions of all quantities in the equation. In particular, the dimension of variable Ψ is unknown.

6/7 (a) Since the left side of the given formula has the dimension m^3 (in SI), therefore *every* member on the right must also have this dimension (Rules 1 and 2). Hence the expression on the right, $h \cdot (8 \cdot R^2 - 4 \cdot R \cdot D + D^{n_1})$, must have the dimension m^3. Since $[h] = m$, therefore $[8 \cdot R^2 - 4 \cdot R \cdot D + D^{n_1}] = m^2$. Now by Rule 2 $[D]^{n_1} = m^2$, hence $n_1 = 2$. Moreover, since $[4 \cdot h^{n_2}] = m^3$, therefore $n_2 = 3$. Finally, on the right, for $h \cdot \sqrt{R^2 - h^2} + R^{n_3} \cdot \arcsin(h/R)$ to be dimensionally homogeneous, it is necessary that $[h \cdot \sqrt{R^2 - h^2}] = [R]^{n_3}$ (the arcsin function is dimensionless). Since the dimension on the left side of this relation is m^2, hence $n_3 = 2$.

With these results, our equation for volume of the barrel has the following concrete form

$$V = \frac{\pi}{6} \cdot \left(3 \cdot h \cdot (8 \cdot R^2 - 4 \cdot R \cdot D + D^2) - 4 \cdot h^3 - 6 \cdot (2 \cdot R - D) \cdot \left(h \cdot \sqrt{R^2 - h^2} + R^2 \cdot \arcsin \frac{h}{R}\right)\right) \quad (*)$$

As an example, consider a barrel with the following geometry:

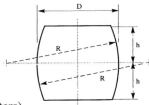

largest diameter (at "waist")	$D = 0.8$ m
half height	$h = 0.8$ m
side-generating radius	$R = 4.96$ m

Thus, equation (*) yields $V = 0.72165$ m^3 (= 721.65 liters).

(b) There are many ways formula (*) can be expressed by dimensionless variables. Here we select the one which is obtained by dividing both sides by D^3. Thus we will have

$$\frac{V}{D^3} = \frac{\pi}{6} \cdot \left(\left(3 \cdot \frac{h}{D}\right) \cdot \left(8 \cdot \left(\frac{R}{D}\right)^2 - 4 \cdot \frac{R}{D} + 1\right) - 4 \cdot \left(\frac{h}{D}\right)^3 \right.$$
$$\left. - 6 \cdot \left(2 \cdot \frac{R}{D} - 1\right) \cdot \left(\frac{h}{D} \cdot \sqrt{\left(\frac{R}{D}\right)^2 - \left(\frac{h}{D}\right)^2} + \left(\frac{R}{D}\right)^2 \cdot \arcsin \frac{(h/D)}{(R/D)}\right)\right)$$

Let us now use the following symbols for dimensionless variables:

$$\pi_1 = \frac{V}{D^3} \quad ; \quad \pi_2 = \frac{R}{D} \quad ; \quad \pi_3 = \frac{h}{D}$$

with this notation our dimensionless formula becomes—after some rearrangement—

$$\pi_1 = \frac{\pi}{6} \cdot \left((3 \cdot \pi_3) \cdot (8 \cdot \pi_2^2 - 4 \cdot \pi_2 + 1) - 4\pi_3^3 \right.$$

$$\left. - 6 \cdot (2 \cdot \pi_2 - 1) \cdot \left(\pi_3 \cdot \sqrt{\pi_2^2 - \pi_3^2} + \pi_2^2 \cdot \arcsin \frac{\pi_3}{\pi_2} \right) \right) \qquad (**)$$

Note that while in (*) the *number* of variables is four (viz., V, D, R, h), in (**) we only have three variables (viz., π_1, π_2, π_3) and yet this latter formula is just as informative as the former one. This reduction in the number of variables is one notable advantage of using dimensionless variables.

To continue our numeric example, we have (based on the geometric data specified above) $\pi_2 = R/D = 4.96/0.8 = 6.2$ and $\pi_3 = h/D = 0.8/0.8 = 1$. By this, (**) provides $\pi_1 = V/D^3 = 1.40948$, thus $V = \pi_1 \cdot D^3 = (1.40948) \cdot (0.8)^3 = 0.72165$ m³, which reassuringly is the same as before.

(c) The plot of (**) is easily constructed; we designate π_1 the *dependent* variable (since it contains the volume), π_2 the *independent* variable, and π_3 the *parameter* (Fig. A6-3). The plot is dimensionally homogeneous since it uses only dimensionless variables. Therefore it is valid without any changes regardless of the dimensional system used. The numerical example appearing in parts (a) and (b) is illustrated by the dashed lines in the figure.

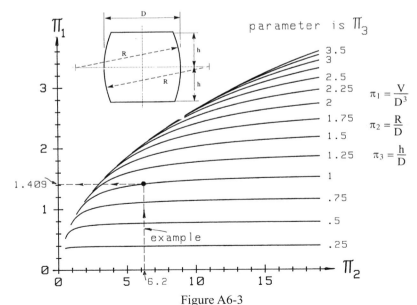

Figure A6-3
Dimensionally homogeneous plot to determine the volume V of a barrel

(d) If formulas (*) and (**) give correct results for a number of specific cases, then this greatly increases our confidence in those formulas. Nevertheless, it is emphasized that "confirmations"—however numerous—do *not* prove the formulas; they only make them *less likely* to be wrong.

(i) *Barrel degenerates into a cylinder.* In this case D and h remain constant, but $R \to \infty$. Hence $\pi_2 \to \infty$, i.e., $\pi_2 \gg \pi_3$. Therefore with good approximation (using the first two members of the Taylor series in each case)

$$\sqrt{\pi_2^2 - \pi_3^2} = \left(1 - \frac{1}{2}\cdot\left(\frac{\pi_3}{\pi_2}\right)^2\right)\cdot\pi_2$$

and

$$\arcsin\frac{\pi_3}{\pi_2} = \frac{\pi_3}{\pi_2} + \frac{(\pi_3/\pi_2)^3}{6}$$

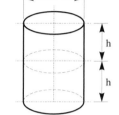

With these, (**) can now be written

$$\pi_1 = \frac{\pi}{6}\cdot\left(24\cdot\pi_2^2\cdot\pi_3 - 12\cdot\pi_2\cdot\pi_3 + 3\cdot\pi_3 - 4\cdot\pi_3^3 - 24\cdot\pi_2^2\cdot\pi_3 + 2\cdot\pi_3^3\cdot\left(2 - \frac{1}{\pi_2}\right) + 12\cdot\pi_2\cdot\pi_3\right)$$

Since $1/\pi_2$ is negligible in comparison with 2, the above formula can be simplified to

$$\pi_1 = \frac{\pi}{6}\cdot(3\cdot\pi_3) = \frac{\pi}{2}\cdot\pi_3 \qquad (***)$$

which, by the definitions of π_1 and π_2, can be written $V/D^3 = (\pi/2)\cdot(h/D)$, and thus $V = (\pi/2)\cdot(h\cdot D^2)$, which is the volume of a right circular cylinder of diameter D and half-height h. Therefore (**) yields the correct result in the *special* case where the barrel degenerates to a cylinder.

Incidentally, formula (***) can be readily verified by Fig. A6-3. If π_2 is very large, then every π_3 = const curve has a horizontal asymptote of ordinate $(\pi/2)\cdot\pi_3$. Thus, if $\pi_3 = 0.5$, then the asymptote is at $\pi_1 = 0.785$.

(ii) *Barrel degenerates into a sphere.* In this case since $R = D/2$ and $h = R$, we have $\pi_2 = R/D = 0.5$ and $\pi_3 = h/D = 0.5$. If these values now are substituted into (**), we obtain

$$\pi_1 = \frac{\pi}{6}\cdot\left(3\cdot\frac{1}{2}\cdot\left(8\cdot\frac{1}{4} - 4\cdot\frac{1}{2} + 1\right) - 4\cdot\frac{1}{8}\right) = \frac{\pi}{6} \qquad (****)$$

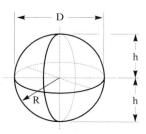

since the factor $2 \cdot \pi_2 - 1$ of the third member on the right side of (**) vanishes. From the definition of π_1, (****) can now be written $V = (\pi/6) \cdot D^3$, which, of course, is the volume of a sphere of diameter D. Therefore (**) yields the correct result in the *special* case where the barrel degenerates into a sphere. In the Fig. A6-3 chart, the point representing a sphere is at $\pi_2 = \pi_3 = 0.5$. At this point $\pi_1 = \pi/6 = 0.523$, as one can see.

6/8 Both sides of this equation must have the same dimension. Therefore the dimension of the first member on the left must be the same as that of the first member on the right. But in the former, the dimension "kg" does not appear, therefore it must also be absent from the latter. The dimension of the latter is

$$\frac{1}{(\text{kg/m}^3)^{n_1}} \cdot \frac{\text{kg/(m·s}^2)}{\text{m}} = \text{m}^{3 \cdot n_1 - 2} \cdot \text{kg}^{1-n_1} \cdot \text{s}^{-2}$$

Hence $1 - n_1 = 0$, from which $n_1 = 1$. Next, we determine n_2. The dimension on the left side of the Navier–Stokes equation is

$$[\text{left side}] = [U] \cdot \frac{[U]}{[x]} = \frac{\text{m}}{\text{s}} \cdot \frac{\text{m/s}}{\text{m}} = \text{m} \cdot \text{s}^{-2}$$

For the right side we have

for kinematic viscosity $[\nu] = \text{m}^2 \cdot \text{s}^{-1}$ (Fig.3-13.a, Art. 3.3.2.d)
gravitational acceleration $[g] = \text{m} \cdot \text{s}^{-2}$

$$\left[\frac{\partial^2 U}{\partial x^2} \right] = \frac{[U]}{[x]^2} = \frac{\text{m/s}}{\text{m}^2} = \text{m}^{-1} \cdot \text{s}^{-1} \quad \text{(Chapter 5)}$$

Therefore

$$[\text{right side}] = [\nu] \cdot [g]^{n_2} \cdot \left[\frac{\partial^2 U}{\partial x^2} \right] = (\text{m}^2 \cdot \text{s}^{-1}) \cdot (\text{m} \cdot \text{s}^{-2})^{n_2} \cdot (\text{m}^{-1} \cdot \text{s}^{-1}) = \text{m}^{1+n_2} \cdot \text{s}^{-2 - 2 \cdot n_2}$$

But, obviously, [left side] = [right side], therefore every dimension on both sides must have the same exponents. Accordingly, $1 + n_2 = 1$ and $-2 - 2 \cdot n_2 = -2$. Hence, uniformly, $n_2 = 0$. Since n_2 is the exponent of the *gravitational acceleration g*, this result means that g does *not* appear in the Navier–Stokes equation; it is a *physically irrelevant* variable.

6/9 (a) It is obvious that we must have a dimension for quantities of *money*. Let us use the symbol $ for this dimension (the fact that $ is also used for dollar is irrelevant here). Another dimension we need is to express the quantity (mass) of *goods*, whether they are cars, books on mathematics, cosmetics, barley, etc. Let us use the symbol γ for this dimension. Finally, we need to express *time*. We choose the dimension "day," represented by the symbol d.

Therefore our dimensional system—called *Economics Dimensional System*, EDS (a thing to be respectable, it must have an acronym!)—shall have the following set of fundamental dimensions: $ = dimension of *money*, γ = dimension of *goods*, d = dimension of *time*.

(b) With the set of dimensions established in (a), we can now define the dimensions of variables constituting the Fisher equation.

Variable	Symbol	Dimension
quantity of money	M	\$
velocity of circulation of money	V_m	\$/d
price of goods	P	\$/γ
velocity of circulation of goods	$V_γ$	γ/d

(c) If the Fisher equation is dimensionally homogeneous—as it should be—then its sides must have the same dimensions. From this condition $[M]\cdot[V_m] = [c]\cdot[P]\cdot[V_γ]$, and hence

$$[c] = \frac{[M]\cdot[V_m]}{[P]\cdot[V_γ]} = \frac{\$\cdot(\$/d)}{(\$/γ)\cdot(γ/d)} = \$$$

If this dimension is now substituted into the Fisher equation for c, then both sides will have the *same* dimension (viz., \$), fulfilling the required dimensional homogeneity.

6/10 (a) Using EDS (see Problem 6.9) the variables and dimensions of the *Cobb–Douglas production function* are as follows:

Variable	Symbol	Dimension	Remark
rate of production of goods	U	γ/d	
number of persons working	n	person	
amount of capital in use	K	\$	
goods product by one worker	P	γ/(person·d)	in unit time

(b) Based on the condition of dimensional homogeneity of the given equation and the dimensions listed above, we can write

$$[c] = \frac{[U]\cdot[K]^{\alpha-1}}{[n]^\alpha\cdot[P]^\alpha} = \frac{(γ/d)\cdot\$^{\alpha-1}}{(\text{person})^\alpha\cdot\left(\dfrac{γ}{\text{person}\cdot d}\right)^\alpha} = \left(\frac{γ}{\$\cdot d}\right)^{1-\alpha}$$

Thus, we have the dimension of the *technology parameter* as a function of α, as required.

6/11 Generally we can write

$$\varphi = q\cdot M\cdot k^{n_1}\cdot c^{n_2}\cdot R^{n_3} \qquad (*)$$

where the symbols are as defined in the following table:

Variable	Symbol	Dimension	Remark
angle of deflection	φ	1 (rad)	
mass of celestial body	M	kg	
speed of light	c	m/s	in vacuum
radius of celestial body	R	m	
universal gravitational constant	k	m³·kg⁻¹·s⁻²	
a constant	q	1	dimensionless
exponents	n_1, n_2, n_3	? ? ?	to be determined

From the condition of dimensional homogeneity of (*), we can write

$$1 = \text{kg} \cdot (\text{m}^3 \cdot \text{kg}^{-1} \cdot \text{s}^{-2})^{n_1} \cdot (\text{m} \cdot \text{s}^{-1})^{n_2} \cdot (\text{m})^{n_3}$$

or with rearrangement

$$\text{kg}^{-1} = \text{m}^{3n_1+n_2+n_3} \cdot \text{kg}^{-n_1} \cdot \text{s}^{-2 \cdot n_1 - n_2}$$

From this it is at once evident that $n_1 = 1$, and $n_2 = -2$, since $0 = -2 \cdot n_1 - n_2 = -2 - n_2$. Also, the exponent of "m" on the left of the above formula is zero, hence on the right it must also be zero. Thus, $0 = 3 \cdot n_1 + n_2 + n_3 = 3 - 2 + n_3 = 1 + n_3$, and therefore $n_3 = -1$. Thus (*) will have the form

$$\varphi = q \cdot \frac{M \cdot k}{c^2 \cdot R} \quad (**)$$

where q is a constant to be determined.

This type of dimensional consideration—performed here in a rather *ad hoc* fashion—is systematically developed in Art. 13.5.

(b) First, we determine q. From relation (**) we have $q = (\varphi \cdot c^2 \cdot R)/(M \cdot k)$. All quantities on the right side of this formula are known for the Sun (given in the problem). Substitution of these into the formula yields

$$q = \frac{(8.48424 \times 10^{-6}) \cdot (2.998 \times 10^8)^2 \cdot (6.955 \times 10^8)}{(1.987 \times 10^{30}) \cdot (6.673 \times 10^{-11})} = 4$$

Hence we now have (**) in the concrete form

$$\varphi = 4 \cdot \frac{M \cdot k}{c^2 \cdot R} \quad (***)$$

which gives the relativistic deflection of light caused by the gravitational field of *any* celestial body.

Now we consider Jupiter (data are in the text of the problem). Input of the values into (***) will yield

$$\varphi = \frac{4 \cdot (1.897 \times 10^{27}) \cdot (6.673 \times 10^{-11})}{(2.998 \times 10^8)^2 \cdot (6.898 \times 10^7)} = 8.167 \times 10^{-8} \text{ rad} = 0.017 \text{ arcsec}$$

which shows that by prevalent techniques (capable of measuring to about 1 arcsec) the deviation caused by Jupiter cannot be detected. The deflection of 1.75 arcsec caused by the Sun was, however, measured by Eddington in 1919 (Sir Arthur Stanley Eddington, English astronomer and physicist, 1882–1944). This measurement was one of the three cardinal confirmations of Einstein's relativity theory (the other two were the gravitational red-shift and the migration of the perihelion of Mercury).

6/12 (a) Since the given equation (repeated here for convenience)

$$\ln R = k_1 \cdot T^{-0.8} + (0.8) \cdot \ln ((43.8) \cdot \rho + 1) \quad (*)$$

is dimensionally non-homogeneous, the dimension of the constant k_1 is not defined. However, since for any number U we have $U = U \cdot 1 = U \cdot \ln e = \ln e^U$, therefore (*) can be written as $\ln R = \ln e^{k_1 \cdot T^{-0.8}} + \ln ((43.8) \cdot \rho + 1)^{0.8}$. Hence

$$R = e^{k_1 \cdot T^{-0.8}} \cdot ((43.8) \cdot \rho + 1)^{0.8} \quad (**)$$

By Rule 4, the argument of e (i.e., exponent) must be dimensionless, that is $[k_1] \cdot [T]^{-0.8} = 1$, and from this condition $[k_1] = [T]^{0.8} = K^{0.8}$, where K (kelvin) is the absolute temperature.

(b) By Rule 2, $[(43.8) \cdot \rho] = [1]$, which cannot be, since $[\rho] \neq 1$ and if both 43.8 and 1 are dimensionless numbers. Therefore at least one of these must be a *dimensional* constant, say k_2. Let us select the number "43.8" to be this constant whose magnitude and dimension are therefore 43.8 and $[\rho]^{-1}$, respectively. Hence $[k_2] = [\rho]^{-1} = 1/(m/day) = day/m$. The right side of (**) has the dimension 1 (i.e., it is dimensionless), so we must have $[R] = 1$, which cannot be, since $[R] = \$/day \neq 1$. Therefore we must introduce yet another dimensional constant, say k_3, whose magnitude is 1 and dimension that of R. Hence $[k_3] = [R] = \$/day$. Finally, with this, our battered (**) becomes dimensionally homogeneous and assumes its legitimate form

$$R = k_3 \cdot e^{k_1 \cdot T^{-0.8}} \cdot (k_2 \cdot \rho + 1)^{0.8} \qquad (***)$$

in which—in summary—the dimensional constants are:

$$k_1 = 1\ K^{0.8}; \qquad k_2 = 43.8\ day/m; \qquad k_3 = 1\ \$/day$$

(c) Despite the fact that our equation (***) is now dimensionally homogeneous, we do not have much confidence in its legitimacy. The reason for this is that the formula does not seem to have a supporting *theory*. For although one could argue that decreasing temperature increases the demand for taxis (negative exponent of T), since people do not like to walk in cold weather, but the same argument applies to increasing temperatures as well. Yet the formula does not reflect this fact. Moreover, the dimensional constants do not have any *meaning*. This fact alone should be a serious warning against acceptance of the formula.

Finally, we mention the important difference between *causation* and *correlation*. The former gives raise to theoretically correct relationships, while the latter (usually) does not. If event A causes the occurrence of *both* events B and C, then these are *causative* relations. But from this fact we should not infer that event B causes event C, although both B and C do occur. Here we say that events B and C have *correlative* relations. To illustrate, it is observed that in children the level of mathematical proficiency increases with their running speed. Does it mean that to enhance Johnny's standing in mathematics, we should train him to be a good runner? The truth is that as children grow older, *both* their mathematical skill and running speeds increase. Therefore—in this example—increased age is the cause of *both* increased mathematical skills *and* running ability. So that age versus mathematical skill, and age versus running speed are *causally* connected and hence *derivation* of their relations are (theoretically) possible. In contrast, the characteristics of mathematical skill and running speed are only *correlatively* connected, and hence *derivation* of their relation is not possible.

Usually, a correlative relation can be recognized by the rather grotesque dimensions of the embedded dimensional constants (introduced only to make the equation dimensionally homogeneous). For example, in our formula (***) k_1 has the dimension $[temperature]^{0.8}$, which makes no sense at all.

To sum up, no matter how well an empirical relationship fits the "real world data," if it is dimensionally awkward, then it probably describes *correlation, not causation.*

APPENDIX 6   **749**

6/13 (a) In the given formula (repeated here for convenience)

$$v = \sqrt{3 \cdot g \cdot h} \qquad (*)$$

the dimension on the left side is m/s, and on the right side is $[g \cdot h]^{1/2} = [m/s^2]^{1/2} \cdot [m]^{1/2} = $ m/s. Therefore the dimensions of both sides are identical, and hence formula (*) is dimensionally homogeneous.

(b) Although the formula is dimensionally homogeneous—i.e., dimensionally "correct"—it is *numerically incorrect*. To be numerically correct, the constant "3" under the square root should be replaced by "2."

This underlines the important fact that dimensional homogeneity is a *necessary, but not sufficient* condition for the overall correctness of a formula. In other words, if a formula is dimensionally nonhomogeneous, then it is *certainly* false; if it is dimensionally homogeneous, then it *may* be correct (but may not be). In this regards, it is emphasized that dimensional considerations do not provide information concerning the veracity of dimensionless quantities, such as absolute numbers.

6/14 To express normal stress σ in terms of bending moment M and cross-section diameter, we write

$$\sigma = c \cdot M^p \cdot D^q \qquad (*)$$

where c is a numerical constant and p and q, are as-yet unknown exponents. Now, if (*) is to be dimensionally homogeneous, then dimensions of its left and right sides must be identical. Hence $[\sigma] = [M]^p \cdot [D]^q$. It follows that $kg/(s^2 \cdot m) = ((m^2 \cdot kg)/s^2)^p \cdot m^q$. By inspection $p = 1$, otherwise the dimension "kg" could not have the same exponent on both sides. Moreover, for "m" we must have $m^{-1} = m^{2 \cdot p + q} = m^{2+q}$, from which $q = -3$. Hence (*) becomes

$$\sigma = c \cdot \frac{M}{D^3} \qquad (**)$$

which is then the solution of the problem. The constant c can be determined by either a single experiment or by analysis; its value is 0.0982 ($= \pi/32$).

6/15 We have the equation (repeated here for convenience)

$$F \cdot x^2 \cdot \frac{\partial^2 F}{\partial x \partial y} = K \cdot \left(\left(x \cdot \frac{\partial F}{\partial x} \right)^2 - \left(y \cdot \frac{\partial F}{\partial y} \right)^2 \right) \qquad (*)$$

By the Rule of Differentials (Chapter 5)

$$\left[\frac{\partial^2 F}{\partial x \partial y} \right] = \frac{[F]}{[x][y]} \; ; \quad \left[\frac{\partial F}{\partial x} \right] = \frac{[F]}{[x]} \; ; \quad \left[\frac{\partial F}{\partial y} \right] = \frac{[F]}{[y]}$$

Hence, the dimension on the left side of (*) is

$$[F] \cdot [x]^2 \cdot \frac{[F]}{[x] \cdot [y]} = \frac{[F]^2 \cdot [x]}{[y]} \qquad (**)$$

By Rule 2, the two terms on the right side of (*) must have identical dimensions. Thus the dimension on the right side of (*) is

$$\left[K\cdot\left(x\cdot\frac{\partial F}{\partial x}\right)^2\right] = [K]\cdot[x]^2\cdot\frac{[F]^2}{[x]^2} = [K]\cdot[F]^2$$

which—by Rule 1—must be identical with the dimension given in (**). Therefore

$$\frac{[F]^2\cdot[x]}{[y]} = [K]\cdot[F]^2$$

from which, after simplification, we get

$$\frac{[K]\cdot[y]}{[x]} = 1$$

Consequently,

$$[U] = \frac{[K]\cdot[y]}{[x]} = 1$$

so that U is a *dimensionless* quantity.

6/16 (a) By the formula, the dimension of k is

$$[k] = \frac{[A]}{[W]^{0.425}\cdot[L]^{0.725}} = \frac{\text{ft}^2}{\text{lbf}^{0.425}\cdot\text{in}^{0.725}}$$

(b) By (4-1), we can write

$$0.10864 \text{ ft}^2\cdot\text{lbf}^{-0.425}\cdot\text{in}^{-0.725} = x\cdot\text{m}^2\cdot\text{kgf}^{-0.425}\cdot\text{m}^{-0.725}$$

and by (4-2)

Q = 0.10864
1 ft = 0.3048 m (Fig. 4-1a),   therefore k_1 = 0.3048; e_1 = 2
1 lbf = 0.45359 kgf (Fig.4-1d),   therefore k_2 = 0.45359; e_2 = −0.425
1 in = 0.0254 m (Fig. 4-1a),   therefore k_3 = 0.0254; e_3 = −0.725

Now, by (4-4) and by the above data

$$x = Q\cdot k_1^{e_1}\cdot k_2^{e_2}\cdot k_3^{e_3} = (0.10864)\cdot(0.3048)^2\cdot(0.45359)^{-0.425}\cdot(0.0254)^{-0.725} = 0.2025$$

so that $A = (0.2025)\cdot W^{0.425}\cdot L^{0.725}$ in which A is now in square meters, W is in units of "kgf," and L is in meters.

It was proven in Example 4-6, that the weight (force) of a body expressed in "kgf" is numerically identical to the mass of that body expressed in "kg." Therefore the DuBois formula in SI is $A = k\cdot M^{0.425}\cdot L^{0.725}$, where M is now the mass of the body (of male ectomorphs) in kilograms, L is height in meters, and A is the body's surface area in square meters. Therefore k is a dimensional constant whose magnitude and dimension are 0.2025 and $[k] = \text{m}^2/(\text{kg}^{0.425}\cdot\text{m}^{0.725}) = \text{m}^{1.275}\cdot\text{kg}^{-0.425}$, respectively.

6/17 The *first* condition entails that $Q = k_1\cdot M^{0.75}$, where M is the mass of the animal and Q is the amount of food per day the animal consumes. This relation is true, otherwise metabolic equilibrium could not be maintained. Hence k_1 is a dimensional constant whose dimension is

$$[k_1] = \frac{[Q]}{[M]^{0.75}} = \frac{\text{kg}_{\text{meal}}}{(\text{kg}_{\text{animal}})^{0.75} \cdot \text{day}}$$

The *second* condition implies that $q = k_2 \cdot M$, where M is again the mass of the animal and q is the quantity of food intake per one meal. Hence k_2 is a dimensional constant whose dimension is $[k_2] = \text{kg}_{\text{meal}}/\text{kg}_{\text{animal}}$.

Now the frequency of meals f (the number of meals per day) must be a function of k_1, k_2, and M. Hence we can write

$$f = k_1^{n_1} \cdot k_2^{n_2} \cdot M^{n_3} \tag{*}$$

where n_1, n_2, and n_3 are as-yet underdetermined exponents. Since the dimensions on both sides of an equation must be identical (Rule 1), therefore $[f] = [k_1]^{n_1} \cdot [k_2]^{n_2} \cdot [M]^{n_3}$, from which—considering the specific dimensions

$$\text{day}^{-1} = (\text{kg}_m \cdot \text{day}^{-1} \cdot \text{kg}_a^{-0.75})^{n_1} \cdot (\text{kg}_m \cdot \text{kg}_a^{-1})^{n_2} \cdot (\text{kg}_a)^{n_3}$$

Equating the exponents of like dimensions, we obtain three linear equations

$$-1 = -n_1 \quad ; \quad 0 = n_1 + n_2 \quad ; \quad 0 = (-0.75) \cdot n_1 - n_2 + n_3$$

the unique solution of which is $n_1 = 1$; $n_2 = -1$; $n_3 = -0.25$. Therefore (*) becomes

$$f = \frac{k_1}{k_2} \cdot \frac{1}{\sqrt[4]{M}} = k_3 \cdot \frac{1}{\sqrt[4]{M}} \tag{**}$$

where k_3 is a *dimensional* constant whose dimension is $[k_3] = [k_1]/[k_2] = (\text{kg}_a)^{0.25}/\text{day}$.

Equation (**) implies that the number of meals per day should be inversely proportional to the fourth root of the body mass of the animal. If a 80 kg man eats three times a day, then a 0.02 kg mouse should eat 24 times in the same period—i.e., it should eat continuously—which it usually does during its waking hours.

Incidentally, relation (**) could also be arrived at *without* dimensional considerations. Since, obviously, $f \cdot q = Q$, therefore $f \cdot k_2 \cdot M = k_1 \cdot M^{0.75}$. From this it follows that $f = (k_1/k_2) \cdot M^{-0.25}$, the same as (**). This proves that, contrary to the general tone of this book, the "dimensional method" is not *always* the most efficient!

Art. 7.12

7/1 The *dimensional matrix* (for constants k, c, and m_e)

	k	c	m_e
m	3	1	0
kg	−1	0	1
s	−2	−1	0

is nonsingular, hence its columns are independent, hence k, c, and m_e are independent, hence *none* of them can be expressed as a function of the other two.

7/2 The Dimensional Set (for constants h, k, c, and m_e) is

	h	k	c	m_e
m	2	3	1	0
kg	1	−1	0	1
s	−1	−2	−1	0
π_1	1	−1	1	−2

from which $\pi_1 = (h{\cdot}c)/(k{\cdot}m_e^2)$ = constant. By substituting the relevant values from Fig. 7-6, we obtain constant = 3.5876×10^{45}. Hence $h = (3.5876 \times 10^{45}){\cdot}(k{\cdot}m_e^2)/c$.

7/3 To solve this problem somewhat elegantly we recast the question: Can, say, e expressed by ϵ_0, c, and h (notice the absence of m_e) uniquely? If "yes," then the answer to the original question is "no," since m_e is obviously not needed. Contrarily, if the answer to the "recast" question is "no," then m_e is relevant and hence the answer to the original question is affirmative. Thus, we write the Dimensional Set

	e	ϵ_0	c	m_e	h
m	0	−3	1	0	2
kg	0	−1	0	1	1
s	1	4	−1	0	−1
A	1	2	0	0	0
π_1	1	$-\dfrac{1}{2}$	$-\dfrac{1}{2}$	0	$-\dfrac{1}{2}$

from which $\pi_1 = e/\sqrt{\epsilon_0{\cdot}c{\cdot}h}$ = constant. Note the conspicuous absence of m_e in this relation. Therefore m_e cannot be expressed in terms of ϵ_0, e, c, and h.

7/4 Let us try to express ϵ_0 in terms of e, c, and h. The dimensional matrix is

	ϵ_0	e	c	h
m	−3	0	1	2
kg	−1	0	0	1
s	4	1	−1	−1
A	2	1	0	0

The rank of this matrix is three, therefore in order to compose the Dimensional Set we must delete one dimension, say "A." Thus the Dimensional Set will be

	ϵ_0	e	c	h
m	−3	0	1	2
kg	−1	0	0	1
s	4	1	−1	−1
π_1	1	−2	1	1

from which $\pi_1 = (\epsilon_0{\cdot}c{\cdot}h)/e^2$ = constant, or $\epsilon_0 = \text{constant} \cdot e^2/(c{\cdot}h)$. Therefore ϵ_0, e, c, and h are not independent, since any one of them can be expressed by the other three. Incidentally, the constant in the above relations is 68.517995.

Another way to solve this problem—maybe more elegantly, but less instructively—is to see whether the dimensional matrix is singular. If it is singular, the columns (the universal constants in this case) are *not* independent, otherwise they *are* independent. In the present case the matrix is singular, therefore the universal constants constituting its columns are *dependent*.

7/5 We construct the dimensional matrix

	N_A	k_B	V_m	R	h	F
m	0	2	3	2	2	0
kg	0	1	0	1	1	0
s	0	−2	0	−2	−1	1
K	0	−1	0	−1	0	0
A	0	0	0	0	0	1
mol	−1	0	−1	−1	0	−1

whose rank is 5. Thus we have to eliminate one row (dimension), say "kg." Then the Dimensional Set will be

	N_A	k_B	V_m	R	h	F
m	0	2	3	2	2	0
s	0	−2	0	−2	−1	1
K	0	−1	0	−1	0	0
A	0	0	0	0	0	1
mol	−1	0	−1	−1	0	−1
π_1	1	1	0	−1	0	0

from which $\pi_1 = (N_A \cdot k_B)/R$ = constant. Therefore N_A, k_B, and R are connected, and hence any two determine the third. For example we can write $k_B = $ constant$\cdot R/N_A$, where the constant is 1.00000014576.

The remaining V_m, h, and F are clearly independent. This fact is evident from their "separate" dimensional matrix

	V_m	h	F
m	3	2	0
kg	0	1	0
s	0	−1	1
K	0	0	0
A	0	0	1
mol	−1	0	−1

We see that F is the only constant containing A (ampere), hence F cannot be connected to either V_m or h. Moreover, the latter two constants cannot be connected either, since h has the dimension "s," but V_m does not. Hence all three constants are independent.

Therefore the answer to the original question is: of the 6 universal constants given only 5 are independent. They are V_m, h, F, and any two of the trio k_B, R, N_A.

7/6 The Dimensional Set is

	c	g	k	h	m_e
m	1	1	3	2	0
kg	0	0	−1	1	1
s	−1	−2	−2	−1	0
π_1	1	0	−1	1	−2
π_2	0	1	−3	4	−7

from which

$$\pi_1 = \frac{c \cdot h}{k \cdot m_e^2}; \qquad \pi_2 = \frac{g \cdot h^4}{k^3 \cdot m_e^7} \qquad (*)$$

It is emphatically pointed out that here π_1 and π_2 are *not* dimensionless variables, since they are composed entirely of *constants*. Therefore they are constants themselves. In particular, by (*) and the table given in Fig. 7-6

$$\pi_1 = 3.5876 \times 10^{45}; \qquad \pi_2 = 1.222469 \times 10^{109} \qquad (**)$$

By the Dimensional Set above we can express c and g in terms of h, k, and m_e. For example, by (*) and (**), we can write

$$c = (3.5876 \times 10^{45}) \cdot \frac{k \cdot m_e^2}{h} \qquad \text{and} \qquad g = (1.222469 \times 10^{109}) \cdot \frac{k^3 \cdot m_e^7}{h^4}$$

It therefore follows that any two of the given universal constants can be expressed by the other three. To illustrate, suppose we wish to express m_e and k in terms of h, g, and c. For this all we have to do is to construct a Dimensional Set in which m_e and k are in the **B** matrix (by construction, variables in matrix **B** appear *only* in *one* dimensionless product). Accordingly,

	m_e	k	h	c	g
m	0	3	2	1	1
kg	1	−1	1	0	0
s	0	−2	−1	−1	−2
π_1	1	0	−1	3	−1
π_2	0	1	1	−7	2

from which, considering $g = 9.80665$ m/s²,

$$\pi_1 = \frac{m_e \cdot c^3}{h \cdot g} = 3.77724 \times 10^{27} = \text{constant}; \quad \pi_2 = \frac{k \cdot h \cdot g^2}{c^7} = 1.95365 \times 10^{-101} = \text{constant}$$

Therefore $m_e = (3.77724 \times 10^{27}) \cdot h \cdot g / c^3$ and $k = (1.95365 \times 10^{-101}) \cdot c^7 / (h \cdot g^2)$.

Art. 10.6

10/1 By Theorem 10-1, if two variables in matrix **B** are interchanged, then the generated new dimensionless variables remain equivalent to the original ones. By Theorem 10-2, the same is true for matrix **A**. But if **A** = **B**, as is the case here, then interchanging of two variables between **A** and **B** is clearly the same as interchanging the corresponding two variables *within* **A**, or *within* **B**. This latter activity, however, according to the two quoted, yields an equivalent set of dimensionless variables. This completes the *general* proof.

In the *particular* case where the number of variables is $N_V = 6$, then both **A** and **B** are 3 × 3 matrices, thus obviously the number of dimensions is $N_d = 3$, and the number of dimensionless variables is $N_P = 3$. Hence, by Theorem 9-14, the **C** matrix is −**I**, namely

$$\mathbf{C} = \begin{bmatrix} -1 & 0 & 0 \\ 0 & -1 & 0 \\ 0 & 0 & -1 \end{bmatrix}$$

The k value is by (10-19) or by Fig. 10-4

$$k = \binom{N_V}{N_d} = \binom{6}{3} = 20$$

while the number of *impossible* combinations of variables U_R (by Fig. 10-2) for rows 1, 2, and 3 of **C** is $U_R = 4 + 4 + 4 = 12$, since all three rows of **C** contain $N_z = 2$ zeros. The number of *duplications* of pairs of zeros of identical columns in **C** is $\vartheta = 0$. The number of *equivalent* sets is calculated by considering that $n_1 = n_2 = n_3 = 1$. Hence

$$U_c = n_1 + n_2 + n_3 + n_1 \cdot n_2 + n_1 \cdot n_3 + n_2 \cdot n_3 + n_1 \cdot n_2 \cdot n_3 = 7$$

Thus, by (10-20) and the above determined data, the number of *distinct* sets is

$$N_S = k - U_R - U_c + \vartheta = 20 - 12 - 7 + 0 = 1$$

That is, there is only *one* distinct set. This completes the proof.

10/2 (a) distinct, (b) equivalent, (c) equivalent, (d) distinct, (e) equivalent.

To illustrate, let us show how results (a) were obtained. We know that by the "base" set

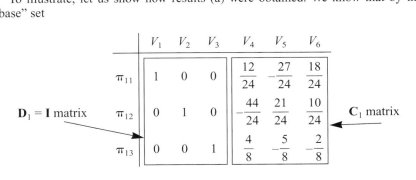

and by set (a)

		V_1	V_2	V_3	V_4	V_5	V_6
	π_{21}	1	0	3	2	-3	0
$\mathbf{D}_2$ matrix	π_{22}	2	1	0	$\dfrac{20}{24}$	$-\dfrac{33}{24}$	$\dfrac{46}{24}$	$\mathbf{C}_2$ matrix
	π_{23}	2	1	3	$\dfrac{8}{12}$	$\dfrac{39}{12}$	$\dfrac{14}{12}$

Next, submatrices $\mathbf{S}_1$ and $\mathbf{S}_3$ of the *Shift Matrix* $\mathbf{S}$ is constructed

	V_1	V_2	V_3	
V_1	1	0	0	
V_2	0	1	0	$\mathbf{S}_1$
V_3	0	0	1	
V_4	0	0	0	
V_5	0	0	0	$\mathbf{S}_3$
V_6	0	0	0	

Thus $\mathbf{S}_1 = \mathbf{I}$ and $\mathbf{S}_3 = \mathbf{0}$. By (9-53) the transformation matrix $\boldsymbol{\tau}$ is

$$\boldsymbol{\tau} = \mathbf{D}_2 \cdot [\mathbf{D}_1 \cdot \mathbf{S}_1 + \mathbf{C}_1 \cdot \mathbf{S}_3]^{-1} = \mathbf{D}_2 \cdot [\mathbf{D}_1 \cdot \mathbf{I} + \mathbf{C}_1 \cdot \mathbf{0}]^{-1} = \mathbf{D}_2$$

Therefore, for set (a)

$$\boldsymbol{\tau} = \mathbf{D}_2 = \begin{bmatrix} 1 & 0 & 3 \\ 2 & 1 & 0 \\ 2 & 1 & 3 \end{bmatrix}$$

and we see that $\boldsymbol{\tau}$ does not fulfil Definition 19-1 for equivalent sets, since it is not a matrix which has exactly one nonzero element in each of its rows and columns. Therefore set (a) is *not equivalent* to the base set and hence it is *distinct*.

10/3 For (a) $\mathbf{C}_2 = \begin{bmatrix} \dfrac{1}{2} & 0 & 0 \\ \dfrac{1}{4} & 2 & -\dfrac{3}{2} \end{bmatrix}$ and for (b) $\mathbf{C}_3 = \begin{bmatrix} \dfrac{1}{2} & \dfrac{1}{4} & -\dfrac{3}{4} \\ 0 & -2 & 0 \end{bmatrix}$.

Therefore $x = \frac{1}{2}$, $y = -2$.

10/4 (a) There are $U_R = 2$ *impossible* combinations of variables.
   (b) There is $U_c = 1$ *equivalent* set—with respect to the "base" set given.
   (c) There are $N_S = 17$ *distinct* sets of dimensionless variables.

10/5 (a) Number of variables is $N_V = 8$, number of dimensions is $N_d = 5$.
   (b) There are $N_P = 3$ dimensionless variables.

APPENDIX 6   757

(c) Not counting duplications, there are $U_R = 44$ impossible combinations of variables.
(d) There are $\delta = 4$ duplications in the above U_R combinations.
(e) There are $U_c = 15$ possible equivalent sets of dimensionless variables.
(f) There is only $N_S = 1$ distinct set of dimensionless variables.
(g) There are $k = 56$ possible combinations of $N_V = 8$ variables whose number of dimensions is $N_d = 5$.

10/6 The relation between the two sets is

$$\pi_{21} = \frac{\pi_{11}^2}{\pi_{12}} \; ; \quad \pi_{22} = \frac{\pi_{12}}{\sqrt{\pi_{11}^3}} \; ; \quad \pi_{23} = \frac{\pi_{12}^2 \cdot \pi_{13}}{\pi_{11}^4}$$

and the relevant *transformation matrix* is $\mathbf{T} = \begin{bmatrix} 2 & -1 & 0 \\ -\dfrac{3}{2} & 1 & 0 \\ -4 & 2 & 1 \end{bmatrix}$.

10/7 Variables V_5 and V_7 are dimensionally irrelevant.

10/8 There are $N_S = 7$ *distinct* sets of dimensionless variables.

10/9 There are $N_S = 17$ *distinct* sets of dimensionless variables.

10/10 There are $N_S = 10$ *distinct* sets of dimensionless variables.

Art. 11.3

11/1 Period of oscillation of a fluid in a U tube. Suppose the involved variables are

Variable	Symbol	Dimension
period of oscillation	T	1/s
density of fluid	ρ	kg/m^3
gravitational acceleration	g	m/s^2
total length of fluid column	L	m

We see that the density of fluid ρ is the *only* variable having "kg" dimension. Therefore, by Theorem 11-1, ρ is a *dimensionally irrelevant* variable, hence it is, by Theorem 11-3, a *physically irrelevant* variable.

11/2 Velocity of disturbance along a stretched wire. Suppose the variables are

Variable	Symbol	Dimension	Remark
velocity of disturbance	v	m/s	longitudinal
tension	F	m·kg/s^2	
mass of wire	q	kg/m	of unit length
amplitude of disturbance	a	m	lateral

This yields the Dimensional Set

	v	F	q	a
m	1	1	−1	1
kg	0	1	1	0
s	−1	−2	0	0
π_1	1	$-\frac{1}{2}$	$\frac{1}{2}$	0

↑ all-zero column in matrix **C**

The column in matrix **C** under "a" is all zero, therefore "a" is a *dimensionally irrelevant* variable (Theorem 10-4), and hence it is a *physically irrelevant* variable (Theorem 11-3). Thus the amplitude does not influence the velocity of propagation of the disturbance.

11/3 Geometry of a catenary. The named relevant variables with their dimensions are as follows:

Variable	Symbol	Dimension
span	λ	m
sag	U	m
length	L	m
density	ρ	kg/m³
gravitational acceleration	g	m/s²

Therefore the dimensional matrix is

	λ	U	L	ρ	g
m	1	1	1	−3	1
kg	0	0	0	1	0
s	0	0	0	0	−2

It is seen that ρ is the only variable with dimension "kg," and g is the only variable with dimension "s." Therefore, by Theorem 11-1, both ρ and g are *physically irrelevant* variables, i.e., they have no effect on the shape of the catenary (provided of course that $g > 0$). Indeed, by removing ρ and g from the dimensional matrix, together with dimensions "kg" and "s," we can construct the Dimensional Set composed only by the relevant variables

	λ	U	L
m	1	1	1
π_1	1	0	−1
π_2	0	1	−1

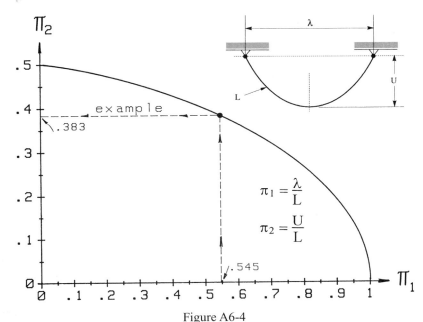

Figure A6-4
Dimensionless plot to determine the sag of a catenary from its length and span

Hence we have two dimensionless variables

$$\pi_1 = \frac{\lambda}{L}; \qquad \pi_2 = \frac{U}{L} \qquad (*)$$

and therefore the $\pi_2 = \Psi\{\pi_1\}$ function entirely defines the shape of the catenary. Fig. A6-4 presents the plot of this function by which we can determine the sag from known length and span. To illustrate, assume $L = 94.5$ m and $\lambda = 51.5$ m. Hence

$$\pi_1 = \frac{\lambda}{L} = \frac{51.5}{94.5} = 0.545$$

For this, the plot provides $\pi_2 = 0.383$. Now, from the second relation of (*), the sag is

$$U = \pi_2 \cdot L = (0.383) \cdot (94.5) = 36.19 \text{ m}$$

This example is indicated on the plot by dashed lines.

11/4 Stress in glass windows by wind. As the problem states, we are only dealing with *geometrically similar* windows. Therefore all the linear dimensions of the window pane (length, height, thickness) are proportional to a representative size we call *characteristic length* L_c. Therefore the following variables may be relevant in this problem:

Variable	Symbol	Dimension
normal stress	σ	N/m^2
characteristic length	L_c	m
wind pressure	p	N/m^2

Hence the dimensional set is

	σ	L_c	p
m	−2	1	−2
N	1	0	1
π_1	1	0	−1

↑ all-zero column in matrix **C**

in which the **C** matrix has an all-zero column under variable L_c. Therefore L_c is a *dimensionally irrelevant* variable (Theorem 10-4), thus it is also a *physically irrelevant* variable (Theorem 11-3). By the Dimensional Set

$$\pi_1 = \frac{\sigma}{p} = \text{constant}$$

in which of course L_c does not appear. This relation tells us that wind-induced stresses in geometrically similar windows are independent of the size of the windows, a quite astonishing result if one considers the popular—and false—belief that large windows are more prone to wind-caused failure. Maybe the discrepancy between fact and belief is due to the condition "geometrically similar," which of course includes thickness as well, an attribute often ignored.

11/5 Energy of a laterally vibrating stretched wire. The given variables and their dimensions are as follows:

Variable	Symbol	Dimension	Remark
energy	Q	m^2·kg/s^2	
length of wire	L	m	
linear density	$\bar{\rho}$	kg/m	
amplitude of vibration	a	m	lateral
tension force	F	m·kg/s^2	longitudinal

from which the Dimensional Set is

	Q	L	$\bar{\rho}$	a	F
m	2	1	−1	1	1
kg	1	0	1	0	1
s	−2	0	0	0	−2
π_1	1	0	0	−1	−1
π_2	0	1	0	−1	0

↑ column in **C** matrix is all zeros

We see that the column in matrix **C** under variable $\bar{\rho}$ is all zeros, therefore by Theorem 10-4 this variable *dimensionally irrelevant* and hence, by Theorem 11-3, it is also *physically irrelevant*. The fact that density does not influence the energy of a vibrating wire may be astonishing first, but a little reasoning convinces us that indeed must be so. The wire stretched against the tensile force F, thus the deforming energy is independent of the mass (density) of the string.

11/6 Sloshing frequency of fuel in missile tanks during take-off.

(a) From the listed variables we have the following table:

Variable	Symbol	Dimension	Remark
sloshing frequency of fuel in tank	ν	1/s	
tank's linear size	L	m	characteristic value
fuel density	ρ	kg/m³	
resultant acceleration	a	m/s²	

by which the Dimensional Set is

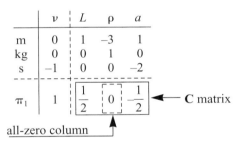

We see that the column of **C** matrix under ρ is all zeros, hence the density of fuel is a *dimensionally irrelevant* variable (Theorem 10-4), therefore it is also a *physically irrelevant* variable (Theorem 11-3).

(b) By the Dimensional Set $\pi_1 = \nu \cdot \sqrt{L/a}$ = constant. Thus ν = constant $\cdot \sqrt{a/L}$. Therefore the frequency varies as the square root of the resultant acceleration of the missile, and as the inverse square root of the characteristic length of the fuel tank.

11/7 Time scale of the universe.

(a) By the suggested variables we set up the following list:

Variable	Symbol	Dimension
period	T	s
density	ρ	kg/m³
length	L	m
universal gravitational constant	k	m³/(kg·s²)

Therefore the Dimensional Set is

	T	ρ	L	k
m	0	−3	1	3
kg	0	1	0	−1
s	1	0	0	−2
π_1	1	$\frac{1}{2}$	0	$\frac{1}{2}$

C matrix ← the boxed portion

all-zero column ← (under L)

from which the sole dimensionless variable—a constant in this case—is

$$\pi_1 = T \cdot \sqrt{\rho \cdot k} = \text{const} \quad \text{or} \quad T = \frac{\text{const}}{\sqrt{\rho \cdot k}} \tag{*}$$

We now notice that the column in matrix **C** under variable L is all zeros. Hence, by Theorem 10-4, L is a *dimensionally irrelevant* variable and, as a consequence, it is also a *physically irrelevant* variable (Theorem 11-3). Consequently, if density ρ remains constant, then the time scale of the universe remains unchanged, regardless of any *uniform* change to lengths—a rather astonishing result indeed.

(b) We introduce the notations $T_2/T_1 = S_T$ and $\rho_2/\rho_1 = S_\rho$, where S_T is the *Time Scale Factor* and S_ρ is the *Density Scale Factor*. By (*) we write $T_2 = \text{const}/\sqrt{\rho_2 \cdot k}$; $T_1 = \text{const}/\sqrt{\rho_1 \cdot k}$, and hence $T_2/T_1 = \sqrt{\rho_1/\rho_2}$. This latter can be expressed with the introduced scale factors as

$$S_T = \frac{1}{\sqrt{S_\rho}} \tag{**}$$

i.e., the time scale would vary as the inverse square root of density, e.g., if *all* densities doubled, all time duration would *decrease* by a factor of $\sqrt{2} = 1.414$, i.e., every activity would take place 41.4% *faster* than in the unchanged universe.

11/8 Linear momentum of a quantum.

(i) By Theorem 10-4, a dimensionally irrelevant variable cannot appear in any of the dimensionless variables. But in our case every one of the five listed physical variables does appear in at least one dimensionless variable as defined in relation (a) of the problem. Therefore *none* of the 5 variables listed is dimensionally irrelevant.

(ii) By relations (a) and (b) of the problem

$$q \cdot \sqrt{\frac{k}{h \cdot c^3}} = u \cdot v \cdot \sqrt{\frac{k \cdot h}{c^5}} \quad (u = \text{constant})$$

which can be simplified to

$$q = \left(\frac{u \cdot h}{c}\right) \cdot v = \text{const} \cdot v$$

since both h and c are constants. In this relation k does not appear, therefore the universal gravitational constant does not affect the momentum of a quantum.

(iii) Also, by the last relation, momentum q of the quantum is *proportional* to frequency v.

11/9 A general physical system. (i) Since the plot of the π_1 versus π_2 function is a rectangular hyperbola whose asymptotes are the coordinate axes, the *product* $\pi_1 \cdot \pi_2$ must be a constant. Hence, by relation (a) of the problem

$$\pi_1 \cdot \pi_2 = \left(\frac{\omega}{g} \cdot \sqrt{\frac{\sigma}{\rho}}\right) \cdot \left(\frac{L \cdot g \cdot \rho}{\sigma}\right) = \text{constant}$$

This expression can be simplified to

$$\omega \cdot L \cdot \sqrt{\frac{\rho}{\sigma}} = \text{constant} \qquad (*)$$

in which g does not appear. Thus, gravitational acceleration is *physically irrelevant*.

(ii) By relation (*) above, $\sigma = \text{const} \cdot L^2 \cdot \omega^2 \cdot \rho$. Therefore σ varies as the square of L.

(iii) By relation (*) above, $\omega = \text{const} \cdot (1/L) \cdot \sqrt{\sigma/\rho}$. Therefore ω varies as the square root of σ.

(iv) By relation (*) above, $\rho = \text{const} \cdot \sigma/(L^2 \cdot \omega^2)$. Therefore ρ varies as the inverse square of ω.

11/10 Stresses generated by the collision of two steel balls.
(i) The dimensional matrix is

	σ	ρ	D	v	E
m	−1	−3	1	1	−1
kg	1	1	0	0	1
s	−2	0	0	−1	−2

The rank of this matrix is $R_{DM} = 3$ and there are $N_V = 5$ variables. Therefore, by the first part of (7-26), the number of independent dimensionless variables is $N_P = 5 - 3 = 2$.

(ii) We construct the two dimensionless variables by the Dimensional Set

	σ	ρ	D	v	E
m	−1	−3	1	1	−1
kg	1	1	0	0	1
s	−2	0	0	−1	−2
π_1	1	0	0	0	−1
π_2	0	1	0	2	−1

← **C** matrix

↑ column is all zeros

Hence the two dimensionless variables are $\pi_1 = \sigma/E$ and $\pi_2 = (\rho \cdot v^2)/E$.

(iii) In the Dimensional Set we see that in matrix **C** the column under variable D is all zeros. Therefore, by Theorem 10-4, the diameter of the balls is a *dimension-*

ally irrelevant variable, and further, by Theorem 11-3, it is a *physically irrelevant* variable.

(iv) Since mass is a function of diameter and density only and, by (iii) above, diameter does not affect the generated stresses, therefore for any chosen mass we can always select a diameter which satisfies the given density. Consequently, mass does not influence the generated stresses either. This fact is unexpected, for one could easily argue that larger balls represent more energy and hence the create correspondingly larger impact stresses. But this is not so, as just demonstrated.

Art. 12.2

12/1 The equation for the volume of a right circular cone is

$$V = \frac{\pi}{12} \cdot D^2 \cdot h \tag{*}$$

where D is the diameter of the base and h is the altitude. Therefore the variables and dimensions are

Variable	Symbol	Dimension
volume	V	m³
base diameter	D	m
altitude	h	m

Thus we have three variables and one dimension, and hence there are $3 - 1 = 2$ dimensionless variables.

(a) By Fig. 12-3, we need one curve and hence one chart to plot the relation with dimensionless variables.

(b) The dimensionless variables are supplied by the Dimensional Set

	V	D	h
m	3	1	1
π_1	1	0	-3
π_2	0	1	-1

from which

$$\pi_1 = \frac{V}{h^3} \; ; \quad \pi_2 = \frac{D}{h} \tag{**}$$

(c) Formula (*) can be converted to the dimensionless form

$$\pi_1 = \frac{\pi}{12} \cdot \pi_2^2 \tag{***}$$

by dividing both of its sides by h^3, and then using dimensionless variables π_1 and π_2 as defined in (b) above. The plot for relation (***) is presented in Fig. A6-5.

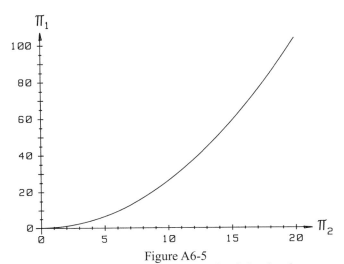

Figure A6-5
Dimensionless plot for the volume of a right circular cone
The curve is the graph of relation (***); symbols are defined in relation (**)
and in the table in the text

12/2 (a) Since we have $N_V = 6$ variables and $k = 6$, therefore, by Fig. 12-3, the number of curves is $N_{\text{curv}} = 1296$ and the number of charts is $N_{\text{chart}} = 216$.

(b) The rank of the Dimensional Matrix cannot be 3 since the third row is three times the first row minus the second row. Hence the rank is $R_{\text{DM}} = 2$. Therefore one dimension must go. Thus we have $N_d = 2$ dimensions and, by (7-19), $N_P = N_V - N_d = 6 - 2 = 4$ dimensionless variables. By (12-5) using dimensionless variables we have to plot $N_{\text{curv}} = k^{N_V - N_d - 2} = 6^{6-2-2} = 6^2 = 36$ curves on $N_{\text{chart}} = k^{N_V - N_d - 3} = 6^{6-2-3} = 6^1 = 6$ charts.

12/3 Power of a dynamo.

(a) We have $k = 8$, and the table gives us five variables. Therefore, by Fig. 12-3, we need 512 curves on 64 charts to plot the quoted *Carvallo* formula.

(b) By the given physical variables, the dimensional matrix is

	W	I	E	L	T
m	2	0	2	2	0
kg	1	0	1	1	0
s	−3	0	−3	−2	1
A	0	1	−1	−2	0

The rank of this matrix is 3, whereas the number of dimensions is 4. Therefore 1 dimension must be deleted. Let this sacrificial dimension be "m." Thus $N_d = 3$ and $N_V = 5$. Hence $N_P = N_V - N_d = 5 - 3 = 2$, i.e., 2 independent dimensionless variables can be formed.

(c) To find the two dimensionless variables we construct the Dimensional Set

	W	I	E	L	T
kg	1	0	1	1	0
s	-3	0	-3	-2	1
A	0	1	-1	-2	0
π_1	1	0	-2	1	-1
π_2	0	1	-1	1	-1

from which $\pi_1 = (W \cdot L)/(E^2 \cdot T)$ and $\pi_2 = (I \cdot L)/(E \cdot T)$.

(d) We have the original *Carvallo* formula

$$W^2 = E^2 \cdot I^2 - 4 \cdot \pi^2 \cdot \frac{L^2 \cdot I^4}{T^2} \qquad (*)$$

From the result of (c) we have $W = (\pi_1 \cdot E^2 \cdot T)/L$ and $I = (\pi_2 \cdot E \cdot T)/L$, which are now substituted into (*). After some simplification, this will yield

$$\pi_1 = \pi_2 \cdot \sqrt{1 - 4 \cdot \pi^2 \cdot \pi_2^2} \qquad (**)$$

(e) Using the table in Fig. 12-3, relation (**) can now be graphically represented on one chart by one curve—a *dramatic* improvement over (a)!

(f) The single-curve plot of formula (**) is as presented in Fig. A6-6.

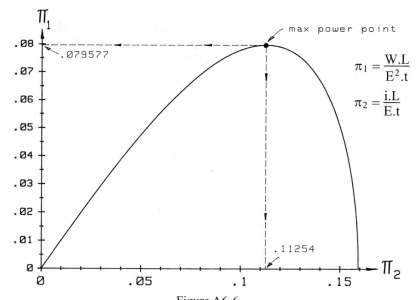

Figure A6-6
Dimensionless plot to determine the power of a dynamo
For definition of variables, see table in the text of the problem

Note that the curve has a maximum-power point. The coordinates of this point can be obtained easily by differentiating (**) with respect to π_1, and equating the result to zero. Accordingly, $(\pi_1)_{max} = 1/(4\cdot\pi) = 0.079577$ which occurs at $(\pi_2)_0 = 1/(\pi\cdot\sqrt{8}) = 0.11254$.

12/4 Critical axial load on columns.

(a) The variables relevant are:

Variable	Symbol	Dimension	Remark
critical load	F_c	N	axial
length of column	L	m	
cross-section's second moment of area	I	m^4	uniform
Young's modulus	E	N/m^2	

(b) The dimensional matrix is

	F_c	L	I	E
m	0	1	4	−2
N	1	0	0	1

(c) The number of curves and charts necessary to plot the relation with $k = 8$ distinct values for variables (parameters) are 64 and 8, respectively (Fig. 12-3).

(d) The complete set of dimensionless variables is supplied by the Dimensional Set

	F_c	L	I	E
m	0	1	4	−2
N	1	0	0	1
π_1	1	0	$-\frac{1}{2}$	−1
π_2	0	1	$-\frac{1}{4}$	0

from which

$$\pi_1 = \frac{F_c}{E\cdot\sqrt{I}}; \qquad \pi_2 = \frac{L}{\sqrt[4]{I}} \qquad (*)$$

(e) Since by (*) we have two dimensionless variables, we need but *one* curve on *one* chart to plot the relation

$$\pi_1 = \Psi\{\pi_2\} \qquad (**)$$

(f) Assume a monomial form for (**), which is plausible. Then $\pi_1 = c\cdot\pi_2^n$, where c and n are constants. Using (*) this formula can be written

$$F_c = c\cdot E\cdot\sqrt{I}\cdot\left(\frac{L}{\sqrt[4]{I}}\right)^n \qquad (***)$$

Now assume the strong likelihood that F_c is *proportional* to I. Thus, by (***), $1/2 - n/4 = 1$ from which $n = -2$. Therefore $F_c = c\cdot E\cdot\sqrt{I}\cdot(L/\sqrt[4]{I})^{-2} = c\cdot(E\cdot I)/L^2$, where $c = \pi^2/4 = 2.4674$, as can be determined by analysis or a *single* experiment. By (*), this last relation can be written $\pi_1 = \pi^2/(4 \cdot \pi_2^2)$, the plot of which is presented in Fig. A6-7.

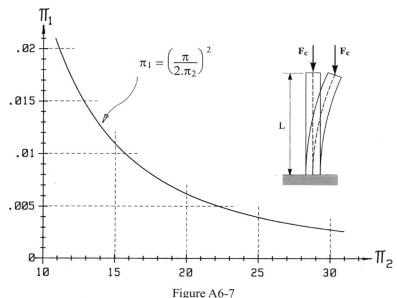

Figure A6-7
Dimensionless plot for the critical buckling load of a column
Relation (*) defines π_1, π_2; other symbols are defined in the table in the text

12/5 Relativistic mass.

(a) In the Einstein equation there are three variables and one constant. Thus, with $k = 8$ distinct values for each variable, eight curves on one chart are necessary to plot the relation (Fig.12-3).

(b) The dimensional matrix for the given variables and dimensional constant is

	M	v	M_0	c
m	0	1	0	1
kg	1	0	1	0
s	0	−1	0	−1

The rank of this matrix cannot be 3 since the third row is −1 times the first row. Thus, the rank is 2 and therefore either "m" or "s" must be jettisoned. Say we delete "s." Then the Dimensional Set is

	M	v	M_0	c
m	0	1	0	1
kg	1	0	1	0
π_1	1	0	−1	0
π_2	0	1	0	−1

from which the dimensionless variables are

$$\pi_1 = \frac{M}{M_0}; \quad \pi_2 = \frac{v}{c} \tag{*}$$

(c) Since there are two dimensionless variables, we need one curve to plot their relation.

(d) Assuming monomial form for the relation among the dimensionless variables (*), we write

$$\pi_1 = \text{const} \cdot \pi_2^n \tag{**}$$

where n is a constant. By (*),

$$M = \text{const} \cdot M_0 \cdot \left(\frac{v}{c}\right)^n$$

Obviously, n must be positive since M must grow with v. But if n is positive, then $v = 0$ entails $M = 0$, which is impossible, since at zero speed $M = M_0$ (the rest-mass). Therefore the monomial form (**) cannot exist.

(e) The given Einstein equation is (repeated here for convenience) $M = (M_0 \cdot c)/\sqrt{c^2 - v^2}$ which, by (*), can be expressed as $\pi_1 = 1/\sqrt{1 - \pi_2^2}$. Fig. A6-8 presents the graph of this latter formula. Note the curious fact that the tangent at $\pi_2 = 0.8$ intersects the abscissa at precisely 0.35.

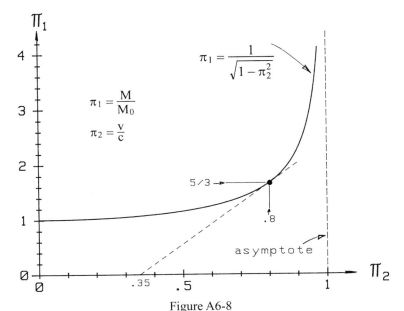

Figure A6-8
Dimensionless plot to determine the relativistic mass of a moving object
M = relativistic mass, M_0 = rest-mass, c = speed of light, v = speed of object

Art. 13.6

13/1 Deflection of a simply supported beam loaded by a lateral force. We first list the relevant variables (see Fig. 13-15).

Variable	Symbol	Dimension	Remark
deflection	U	m	maximum value
position of load	a	m	from left support
rigidity	Z	m$^2\cdot$N	defined $Z = I\cdot E$
length of beam	L	m	
load	F	N	lateral, concentrated

There are five variables and two dimensions, hence we have $5 - 2 = 3$ dimensionless variables supplied by the Dimensional Set

	U	a	Z	L	F
m	1	1	2	1	0
N	0	0	1	0	1
π_1	1	0	0	−1	0
π_2	0	1	0	−1	0
π_3	0	0	1	−2	−1

from which

$$\pi_1 = \frac{U}{L}; \quad \pi_2 = \frac{a}{L}; \quad \pi_3 = \frac{Z}{L^2\cdot F} \qquad (*)$$

From test results and the constants given in the problem, the values of these dimensionless variables can be determined in all four test-points.

Test #	a	U	π_1	π_2	π_3
1	0.2	0.0040157	0.0011811	0.0588235	0.8650519
2	0.4	0.0141176	0.0041522	0.1176471	0.8650519
3	0.7	0.0350206	0.0103002	0.2058824	0.8650519
4	1	0.0564706	0.016609	0.2941176	0.8650519

Let us assume that π_1 can be expressed by a polynomial of π_2 of a degree not higher than 4, i.e., there exists a relation of the form

$$\pi_1 = c_0 + c_1\cdot\pi_2 + c_2\cdot\pi_2^2 + c_3\cdot\pi_2^3 + c_4\cdot\pi_2^4 \qquad (**)$$

where $c_0, \ldots, c_4$ are constant coefficients to be determined.

It is now evident that $c_0 = 0$, since if $\pi_2 = 0$ (i.e., $a = 0$), deflection U must be zero, and thus $\pi_1 = 0$. With this condition, and with the data given in the above table, we can determine the c coefficients by solving the matrix equation

$$\begin{bmatrix} c_1 \\ c_2 \\ c_3 \\ c_4 \end{bmatrix} = \mathbf{P}^{-1}\cdot\mathbf{Q} = \begin{bmatrix} 0.00000244 \\ 0.38528100 \\ -0.77035600 \\ 0.38479000 \end{bmatrix} \qquad (***)$$

where

$$\mathbf{P} = \begin{bmatrix} 0.0588235 & 0.0588235^2 & 0.0588235^3 & 0.0588235^4 \\ 0.1176471 & 0.1176471^2 & 0.1176471^3 & 0.1176471^4 \\ 0.2058824 & 0.2058824^2 & 0.2058824^3 & 0.2058824^4 \\ 0.2941176 & 0.2941176^2 & 0.2941176^3 & 0.2941176^4 \end{bmatrix} ; \mathbf{Q} = \begin{bmatrix} 0.0011811 \\ 0.0041522 \\ 0.0103002 \\ 0.0166090 \end{bmatrix}$$

The elements of **P** are the appropriate powers of π_2 [by relation (**)], and the elements of **Q** are the corresponding π_1 values also supplied by (**).

We see in (***) that c_1 is five orders of magnitude smaller than the other coefficients, so it is safe to assume that $c_1 = 0$. Next, we consider that if π_2 = constant, then for small deflections, U must be proportional to F; in other words, π_1 is proportional to the inverse of π_3 (since F is in the denominator of π_3). Therefore we can write

$$c_2 = \frac{k_2}{\pi_3}, \text{ or } k_2 = c_2 \cdot \pi_3 ; \quad c_3 = \frac{k_3}{\pi_3}, \text{ or } k_3 = c_3 \cdot \pi_3 ; \quad c_4 = \frac{k_4}{\pi_3}, \text{ or } k_4 = c_4 \cdot \pi_3$$

where the c coefficients are given in (***) and $\pi_3 = 0.8650519$ = constant, as seen in the table above. Therefore the k values are determined as follows:

$$k_2 = (0.385281) \cdot (0.86505) = 0.333288 \cong \frac{1}{3}$$

$$k_3 = (-0.770356) \cdot (0.86505) = -0.666398 \cong -\frac{2}{3}$$

$$k_4 = (0.38479) \cdot (0.86505) = 0.332863 \cong \frac{1}{3}$$

Consequently, relation (**) can be written

$$\pi_1 = \frac{1}{3 \cdot \pi_3} \cdot \pi_2^2 - \frac{2}{3 \cdot \pi_3} \cdot \pi_2^3 + \frac{1}{3 \cdot \pi_3} \cdot \pi_2^4$$

or

$$\pi_1 = \frac{\pi_2^2}{3 \cdot \pi_3} \cdot (1 - 2 \cdot \pi_2 + \pi_2^2) = \frac{\pi_2^2}{3 \cdot \pi_3} \cdot (1 - \pi_2)^2$$

into which the physical variables of (*) are now substituted. This will yield, after simplification and considering $Z = I \cdot E$

$$U = \frac{F \cdot a^2}{3 \cdot I \cdot E \cdot L} \cdot (L - a)^2$$

which is the formula we sought. We succeeded in deriving the explicit exact relation among six variables on the basis of only four measurements!

13/2 Fundamental lateral frequencies of geometrically similar cantilevers.

First list the relevant variables

Variable	Symbol	Dimension	Remark
frequency	f	1/s	lateral mode
length	L	m	characteristic value
density of material	ρ	kg/m³	
Young's modulus	E	kg/(s²·m)	

Since we are dealing with geometrically similar beams, *one* linear variable defines any particular beam's size. This variables is then the *characteristic length*. We have four variables and three dimensions, therefore there is 4 − 3 = 1 dimensionless "variable"—a constant, because its singleness. The Dimensional Set is

	f	L	ρ	E
m	0	1	−3	−1
kg	0	0	1	1
s	−1	0	0	−2
π_1	1	1	$\dfrac{1}{2}$	$-\dfrac{1}{2}$

from which $\pi_1 = f \cdot L \cdot \sqrt{\rho/E}$ = constant, or, by this relation, $f = \text{constant} \cdot (1/L) \cdot \sqrt{E/\rho}$. Therefore the frequencies of geometrically similar cantilevers vary *inversely* with their linear sizes, provided the material remains the same.

13/3 Area of an elliptic segment. (a) The relevant variables and their dimensions are

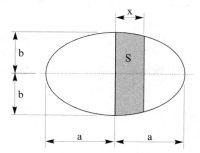

Variable	Symbol	Dimension
area of segment	S	m²
segment's thickness	x	m
length of semimajor axis	a	m
length of semiminor axis	b	m

(b) We have four variables and one dimension, therefore there are three dimensionless variables obtained by the Dimensional Set

	S	x	b	a
m	2	1	1	1
π_1	1	0	0	−2
π_2	0	1	0	−1
π_3	0	0	1	−1

from which

$$\pi_1 = \dfrac{S}{a^2}; \qquad \pi_2 = \dfrac{x}{a}; \qquad \pi_3 = \dfrac{b}{a} \qquad (*)$$

APPENDIX 6   **773**

(c) If a *monomial* form connecting the dimensionless variables (*) is assumed, then

$$\pi_1 = k \cdot \pi_2^p \cdot \pi_3^q$$

where k, p, and q are constants. This formula, by (*), can be written

$$S = k \cdot a^2 \cdot \left(\frac{x}{a}\right)^p \cdot \left(\frac{b}{a}\right)^q \qquad (**)$$

Obviously, if $x = a$ (segment is half the ellipse), then $S = (\pi/2) \cdot a \cdot b$. This, when considering (**), mandates $q = 1$ and $k = \pi/2$. Thus (**) becomes

$$S = \frac{\pi}{2} \cdot a^2 \cdot \left(\frac{x}{a}\right)^p \cdot \left(\frac{b}{a}\right) = \frac{\pi}{2} \cdot a^{1-p} \cdot b \cdot x^p$$

which we now differentiate with respect to x. This yields $dS/dx = \frac{1}{2} \cdot \pi \cdot a^{1-p} \cdot b \cdot p \cdot x^{p-1}$. Since the area is *increasing* with x for all values of x within the range $0 \leq x < a$, therefore we can state that $(dS/dx) > 0$ is mandatory if $x = 0$. But, by the same condition, the exponent of x must be positive. If it is positive, then $(dS/dx) = 0$ at $x = 0$. Thus we see that the above *two* derivatives at $x = 0$ are contradictory. Therefore the *monomial* form (**) is untenable.

(d) By the dimensionless variables of (*), the relation given in the problem can be written in dimensionless form

$$\pi_1 = \pi_3 \cdot (\pi_2 \cdot \sqrt{1 - \pi_2^2} + \arcsin \pi_2) \qquad (***)$$

which is definitely not a monomial. Now if $x = 7$ m, $a = 12$ m, and $b = 6$ m, then the direct relation given in the problem yields $S = 78.95729$ m². Therefore, by (*), $\pi_1 = 0.54831$, $\pi_2 = 0.58333$, and $\pi_3 = 0.5$ and, as the reader can verify, this trio of numbers satisfies relation (***).

(e) The graph of the dimensionless formula (***) is shown in Fig. A6-9 (parameter is π_3). The above numerical example is indicated on the plot by dashed lines.

13/4 Torus volume—normal, degenerating, degenerated. (a) The relevant variables and their dimensions are

Variable	Symbol	Dimension
volume	V	m³
distance h	h	m
cross-section radius	R	m

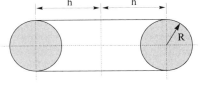

(b) We have three variables and one dimension. Therefore there are $3 - 1 = 2$ dimensionless variables obtained from the Dimensional Set

	V	h	R
m	3	1	1
π_1	1	0	-3
π_2	0	1	-1

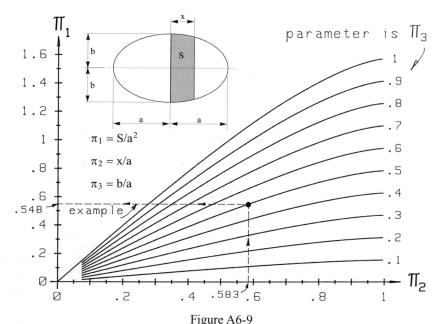

Figure A6-9
Dimensionless plot to determine the area of an elliptic segment
The top curve, $\pi_3 = 1$, represents a circle. Dashed lines indicate numerical example.

from which

$$\pi_1 = \frac{V}{R^3}; \qquad \pi_2 = \frac{h}{R} \qquad (*)$$

(c) The equation $V = 2 \cdot \pi^2 \cdot h \cdot R^2$ can be converted to the dimensionless form

$$\pi_1 = 2 \cdot \pi^2 \cdot \pi_2 \qquad (**)$$

Note that the first π on the right is *not* a dimensionless variable; it is the constant 3.1415.... Obviously, form (**) is a monomial.

(d) Suppose we write relation (**) as $V = k \cdot R^3 \cdot (h/R)^n$, where k and n are constants. If n is positive and $R > 0$, then $h = 0$ implies zero volume. But $h = 0$ also implies a sphere (see Fig. 13-18d) and a sphere does not have zero volume if $R > 0$. Hence n cannot be positive. If n is negative, then $h = 0$ implies (if $R > 0$) infinite volume, which is absurd. Finally, if $n = 0$, then V is independent of h, which is equally nonsense. Therefore n cannot be any number (including zero) and hence the *monomial* form (**) in the case of $h < R$ is untenable.

(e) The formula given in the problem can be expressed by the dimensionless variables (*) as

$$\pi_1 = 2 \cdot \pi^2 \cdot \pi_2 + 2 \cdot \pi \cdot \left[\pi_2^2 \cdot \sqrt{1 - \pi_2^2} - \pi_2 \cdot \arccos \pi_2 + \frac{2}{3} \cdot (1 - \pi_2^2)^{3/2} \right] \qquad (***)$$

If the torus is degenerated into a *sphere*, then $h = 0$, $\pi_2 = 0$, and, by (***), $\pi_1 = (4 \cdot \pi)/3$. Hence $\pi_1 = V/R^3 = (4 \cdot \pi)/3$, or $V = (4 \cdot \pi \cdot R^3)/3$, which is the volume of a sphere of radius R.

(f) The graph of (**) and (***) are presented in Fig. A6-10. The dashed line represents the separation between the *normal* and the *degenerating* cases (see Fig. 13-18). Specifically, at case (b) $\pi_2 = 1$ and $\pi_1 = 2 \cdot \pi^2 = 19.73921$.

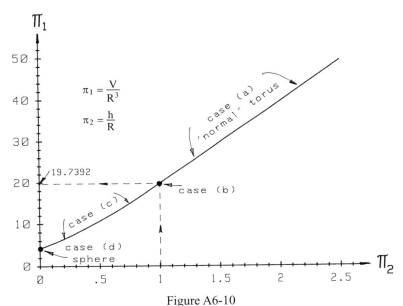

Figure A6-10
**Dimensionless plot to determine the volume of a torus
in four characteristic configurations**
See Fig. 13-18 for these configurations

13/5 Bolometric luminosity of a star.

(a) By the given variables and constants, the Dimensional Set is

	L	R	k	T	c	h
m	2	1	2	0	1	2
kg	1	0	1	0	0	1
s	-3	0	-2	0	-1	-1
K	0	0	-1	1	0	0
π_1	1	0	-2	-2	0	1
π_2	0	1	1	1	-1	-1

We have six variables and constants, and four dimensions, therefore there are $6 - 4 = 2$ dimensionless variables defined by the above Dimensional Set

$$\pi_1 = \frac{L \cdot h}{k^2 \cdot T^2} \; ; \qquad \pi_2 = \frac{R \cdot k \cdot T}{c \cdot h} \tag{*}$$

(b) If we have a monomial form for Ψ in $\pi_1 = \Psi\{\pi_2\}$, then $\pi_1 = b \cdot \pi_2^n$, where b and n are numerical constants. By (*) this relation can be written

$$L = b \cdot \frac{k^2 \cdot T^2}{h} \cdot \left(\frac{R \cdot k \cdot T}{c \cdot h}\right)^n \qquad (**)$$

The surface area of a star is $A = 4 \cdot \pi \cdot R^2$, therefore A is proportional to R^2. If now L is proportional to the surface area (as it must be), then L is proportional to R^2. Hence in (**) exponent n must be 2. Thus, (**) can be written $L = b \cdot (R^2 \cdot k^4 \cdot T^4)/(c^2 \cdot h^3)$. Consequently:
 (i) L varies as the *fourth power* of temperature T (Stefan–Boltzmann law).
 (ii) L varies as the *inverse square* of speed of light c.

13/6 Volume of a conical wedge.

(a) The following variables are relevant:

Variable	Symbol	Dimension
wedge angle	γ	1 (rad)
half apex angle	φ	1 (rad)
volume of wedge	V	m³
base diameter	D	m

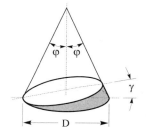

(b) Accordingly, the Dimensional Set is

	γ	φ	V	D
m	0	0	3	1
π_1	1	0	0	0
π_2	0	1	0	0
π_3	0	0	1	−3

We have four variables and one dimension, therefore there are $4 - 1 = 3$ dimensionless variables obtained from the above set

$$\pi_1 = \gamma; \qquad \pi_2 = \varphi; \qquad \pi_3 = \frac{V}{D^3} \qquad (*)$$

(c) The *monomial* form is $\pi_3 = c \cdot \pi_1^p \cdot \pi_2^q$, where $c, p,$ and q are as-yet undefined fixed numbers. Thus, by (*)

$$V = c \cdot D^3 \cdot \gamma^p \cdot \varphi^q \qquad (**)$$

Now, obviously, if apex angle φ increases, then the volume decreases, hence q must be negative. But if it is negative, then $\varphi = 0$ yields an infinite volume, which is absurd. Reason: $\varphi = 0$ means a *cylinder*, and the volume of a cylindrical wedge of given diameter D and wedge angle γ is $V_{cyl} = (\pi/8) \cdot D^3 \cdot \tan \gamma$, which is certainly finite for any $\gamma < \pi/2$ angle. Therefore the *monomial* form (**) to express the volume of a conical wedge is untenable.

(d) As the astute reader may want to derive (it is a rewarding exercise!), the relation for the volume of the conical wedge is

$$V = D^3 \cdot \frac{\pi}{24 \cdot \tan \varphi} \cdot \left[1 - \left(\frac{1 - \tan \varphi \cdot \tan \gamma}{1 + \tan \varphi \cdot \tan \gamma}\right)^{3/2}\right]$$

By using the dimensionless variables in (*), this formula can be written

$$\pi_3 = \frac{\pi}{24} \cdot \frac{1}{\tan \pi_2} \cdot \left[1 - \left(\frac{1 - \tan \pi_2 \cdot \tan \pi_1}{1 + \tan \pi_2 \cdot \tan \pi_1}\right)^{3/2}\right] \quad (***)$$

Consider the following example:

diameter of base $\quad D = 0.15$ m
half apex angle $\quad \varphi = 0.2$ rad ($\cong 11.46$ deg)
wedge angle $\quad \gamma = 1.2$ rad ($\cong 68.76$ deg)

Hence, by (*) $\pi_1 = 1.2$ and $\pi_2 = 0.2$, and by (***) $\pi_3 = 0.53181$. Direct evaluation provides now $V = 0.0017949$ m³, thus

$$\pi_3 = \frac{V}{D^3} = \frac{0.0017949}{0.15^3} = 0.53182$$

which checks to within 0.002 %.

(d) The plot with parameter π_2 is shown in Fig. A6-11. Note that if $\gamma = \pi/2 - \varphi$, the "wedge" becomes the cone itself, i.e., $\pi_1 = \pi/2 - \pi_2$, or $\pi_2 = \pi/2 - \pi_1$. In this

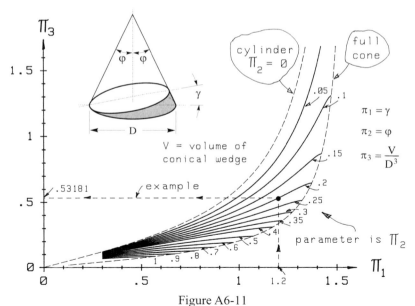

Figure A6-11
Dimensionless plot to determine the volume of a conical wedge
All angles are in radians

case (***) simplifies to $\pi_3 = (\pi/24) \cdot (1/\tan \pi_2) = (\pi/24) \cdot \tan \pi_1$. Also, if $\pi_2 = 0$, then the cone becomes a *cylinder* and (***) transforms into $\pi_3 = (\pi/8) \cdot \tan \pi_1$.

13/7 Volume of the frustum of a right circular cone.

(a) The relevant variables and their dimensions are as follows:

Variable	Symbol	Dimension
volume	V	m³
height	h	m
top diameter	d_1	m
base diameter	d_2	m

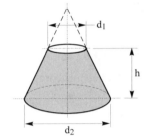

(b) We have four variables and one dimension, therefore there are $4 - 1 = 3$ dimensionless variables obtained from the Dimensional Set

	V	h	d_1	d_2
m	3	1	1	1
π_1	1	0	0	-3
π_2	0	1	0	-1
π_3	0	0	1	-1

from which

$$\pi_1 = \frac{V}{d_2^3}; \qquad \pi_2 = \frac{h}{d_2}; \qquad \pi_3 = \frac{d_1}{d_2} \qquad (*)$$

(c) The monomial form, by (*), is

$$V = c \cdot d_2^3 \cdot \left(\frac{h}{d_2}\right)^p \cdot \left(\frac{d_1}{d_2}\right)^q \qquad (**)$$

where c, p, and q are numeric constants. If $q > 0$, then $d_1 = 0$ implies zero volume, which cannot be since if the top diameter is zero, then the frustum becomes a cone whose volume is not zero. If $q = 0$, then the volume is independent of the top diameter, which obviously cannot be, either. If $q < 0$, then the volume decreases with increasing top diameter, which is impossible. Therefore q cannot be any number, including zero, and hence the *monomial* form (**) is untenable.

(d) The given formula (repeated here for convenience) is $V = (\pi/12) \cdot (d_2^2 + d_1 \cdot d_2 + d_1^2) \cdot h$, which in dimensionless form becomes, using the dimensionless variables (*),

$$\pi_1 = \frac{\pi}{12} \cdot \pi_2 \cdot (1 + \pi_3 + \pi_3^2) \qquad (***)$$

APPENDIX 6   779

(e) In case of a *cone*, $\pi_3 = 0$ and hence (***) becomes $\pi_1 = (\pi/12) \cdot \pi_2$. This, by (*), can be written

$$\frac{V}{d_2^3} = \frac{\pi}{12} \cdot \frac{h}{d_2}$$

or simplified $V = (\pi/12) \cdot h \cdot d_2^2$, which is the volume of a cone of base diameter d_2 and altitude h.

In the case of a *cylinder*, $\pi_3 = 1$ and hence (***) becomes $\pi_1 = (\pi/4) \cdot \pi_2$. This, by (*), can be written

$$\frac{V}{d_2^3} = \frac{\pi}{4} \cdot \frac{h}{d_2}$$

or simplified $V = (\pi/4) \cdot h \cdot d_2^2$, which is the volume of a cylinder of diameter d_2 and length h.

13/8 Discharge of a capacitor. (a) The physical variables and their dimensions are as follows:

Variable	Symbol	Dimension
current	i	A
time	t	s
potential	V_0	$m^2 \cdot kg \cdot s^{-3} \cdot A^{-1}$
ohmic resistance	R	$m^2 \cdot kg \cdot s^{-3} \cdot A^{-2}$
capacitance	C	$m^{-2} \cdot kg^{-1} \cdot s^4 \cdot A^2$

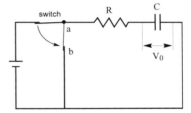

(b) The dimensional matrix is

	i	t	V_0	R	C
m	0	0	2	2	−2
kg	0	0	1	1	−1
s	0	1	−3	−3	4
A	1	0	−1	−2	2

Obviously the rank of this matrix cannot be 4 since the first row is twice the second row. Therefore one of these rows must be deleted. Let us delete the first row (dimension "m"). Thus, the Dimensional Set is

	i	t	V_0	R	C
kg	0	0	1	1	−1
s	0	1	−3	−3	4
A	1	0	−1	−2	2
π_1	1	0	−1	1	0
π_2	0	1	0	−1	−1

From the above set

$$\pi_1 = \frac{i \cdot R}{V_0} \quad ; \quad \pi_2 = \frac{t}{R \cdot C} \tag{*}$$

(c) The monomial power form is $\pi_1 = k \cdot \pi_2{}^n$ which, by (*), can be written

$$i = k \cdot \frac{V_0}{R} \cdot \left(\frac{t}{R \cdot C} \right)^n \tag{**}$$

where k and n are numeric constants. If n is positive, then with increasing time t, the current grows without bounds—which is impossible. If n is negative, then at zero time, the current is infinite—which again is absurd. If n is zero, then the current is independent of both time and capacitance as well—an obvious nonsense. Therefore n cannot be any number, therefore the *monomial* (**) is untenable.

(d) By (*), the given relation $i = (V_0/R) \cdot e^{-t/(R \cdot C)}$ can be converted into $\pi_1 = e^{-\pi_2}$.

(e) Fig A6-12 shows the plot of the above dimensionless relation. Note that since we have five variables, to graphically represent the relation considering only six distinct values for each variable, we would need 216 curves plotted on 36 charts (see Fig. 12-3). But by using dimensionless variables, the relation can be presented by a *single* curve!

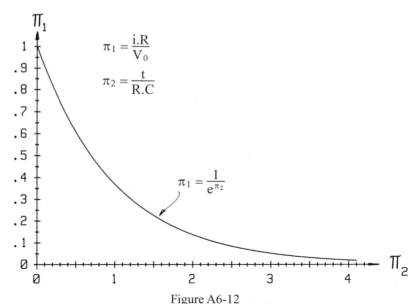

Figure A6-12
**Dimensionless plot to determine the current
by a discharging capacitor in an RC circuit**
See table in text for definition of physical variables

13/9 Area of a triangle whose side lengths form an arithmetic progression.
(a) The relevant variables and their dimensions are as follows:

Variable	Symbol	Dimension
area of triangle	T	m^2
common difference of sides	d	m
length of shortest side	a	m

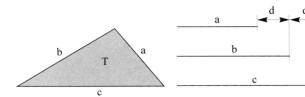

(b) We have three variables and one dimension, therefore there are $3 - 1 = 2$ dimensionless variables determined by the Dimensional Set

	T	d	a
m	2	1	1
π_1	1	0	-2
π_2	0	1	-1

from which

$$\pi_1 = \frac{T}{a^2} \; ; \quad \pi_2 = \frac{d}{a} \qquad (*)$$

(c) The monomial has the form $\pi_1 = k \cdot \pi_2^n$, where k and n are constants. This relation, by (*), can be written

$$T = k \cdot a^2 \cdot \left(\frac{d}{a}\right)^n \qquad (**)$$

Assume that n is positive. Then $d = 0$ implies $T = 0$, which is impossible since $d = 0$ implies an *equilateral* triangle whose area with $a > 0$ is not zero. Next, assume that n is negative. Then $d = 0$ implies infinite T, which is again absurd. Finally, if $n = 0$, then T is independent of d, which is equally nonsense. Therefore n cannot be any number—including zero—and hence the *monomial* form (**) cannot exist.

Indeed—as the inquisitive reader might want to confirm (hint: use *Heron's* formula)—

$$T = \frac{\sqrt{3}}{4}(a + d) \cdot \sqrt{a^2 + 2 \cdot a \cdot d - 3 \cdot d^2}$$

or in dimensionless form using (*)

$$\pi_1 = \frac{\sqrt{3}}{4} \cdot (1 + \pi_2) \cdot \sqrt{1 + 2 \cdot \pi_2 - 3 \cdot \pi_2^2}.$$

which is obviously a *nonmonomial*.

Art. 14.4

14/1 Terminal speed of a mass sliding down a nonfrictionless inclined surface.

(a) Since both φ and μ are *dimensionless physical* variables, they must be in the **B** matrix, thus they cannot be in the **A** matrix (Theorem 14-1, Art. 14.1).

(b) Of all the named physical variables: φ, μ, M, and h are independent; V and E_f are dependent.

(c) A possible Dimensional Set is

	V	E_f	φ	μ	g	h	M
m	1	2	0	0	1	1	0
kg	0	1	0	0	0	0	1
s	-1	-2	0	0	-2	0	0
π_1	1	0	0	0	$-\frac{1}{2}$	$-\frac{1}{2}$	0
π_2	0	1	0	0	-1	-1	-1
π_3	0	0	1	0	0	0	0
π_4	0	0	0	1	0	0	0

Note that both dependent variables (V, E_f) and both dimensionless physical variables (φ, μ) are in the **B** matrix (Theorem 14-1, Art. 14.1).

(d) The dimensionless variables (by the above Dimensional Set) are

$$\pi_1 = \frac{V}{\sqrt{g \cdot h}}; \qquad \pi_2 = \frac{E_f}{g \cdot h \cdot M}; \qquad \pi_3 = \varphi; \qquad \pi_4 = \mu \qquad (*)$$

(e) Since there are two dependent physical variables (V, E_f), each appearing in only one dimensionless variable (V in π_1 and E_f in π_2), therefore we have two relations among the dimensionless variables defined in (*). Namely $\pi_1 = \Psi_1\{\pi_3, \pi_4\}$ and $\pi_2 = \Psi_2\{\pi_3, \pi_4\}$, where Ψ_1 and Ψ_2 designate functions to be determined by either analyses or tests.

14/2 The matrix given in the problem is repeated here for convenience

$$\mathbf{C} = \begin{bmatrix} 2 & 1 & 3 & 2 \\ 4 & -3 & 2 & 0 \\ 0 & 0 & 0 & 0 \end{bmatrix}$$

(a) The **C** matrix has N_P rows and N_d columns (Art. 8.1), where N_P is the number of dimensionless variables and N_d is the number of dimensions. Here we have $N_P = 3$ and $N_d = 4$. The **D** matrix in a Dimensional Set is always a square (Fig. 8-2, Art. 8.1) and hence it has as many columns as rows. As it has N_P rows, it has N_P columns. By the construction of any Dimensional Set, the number of variables N_V is the *sum* of columns of **D** and **C** matrices (Fig. 8-2 and Fig. 8-3). From this rationale then, our present sum is $N_V = N_P + N_d = 3 + 4 = 7$, thus we have $N_V = 7$ physical variables.

(b) Also by the above reasoning, we have $N_d = 4$ dimensions.

(c) Also by the above reasoning, we have $N_P = 3$ dimensionless variables.

(d) The third *row* of matrix **C** is all zeros. Therefore the third *column* of matrix **B** is all zeros (Theorem 14-2, Art. 14.1). Hence there is *one* dimensionless physical variable and it occupies the third column of matrix **B**.

APPENDIX 6   **783**

(e) If any two variables in the **A** matrix were identical, then two of its columns would be identical, in which case **A** would be singular and matrix **C** would not exist. But matrix **C** does exist. Hence no two variables in **A** are identical.

(f) In view of the result of (e), if there were two identical variables—say V_i, and V_j—in the Dimensional Set, then there could be only two possibilities:

Possibility 1. Both V_i and V_j are in the **B** matrix. In this case two rows of **C** would be identical (Theorem 14-4, Art. 14.2). But no two rows of **C** are identical. Hence Possibility 1 cannot exist.

Possibility 2. One of V_i or V_j is in matrix **B**, the other is in matrix **A**. In this case one row of matrix **C** would consist of all zeros except one "–1" element (Theorem 9-11, Art. 9.1). But there is no such a row in the **C** matrix. Hence Possibility 2 cannot exist either.

In summation, no two or more physical variables involved are identical.

14/3 Surface area, volume and weight of a right circular cylinder.
(a) First, we list the physical variables and their dimensions

Variable	Symbol	Dimension	Remark
volume	V	m³	
surface area	A	m²	
weight	G	m·kg/s²	
diameter of cross-section	D	m	cross-section is solid circle
gravitational acceleration	g	m/s²	
height	h	m	
density of material	ρ	kg/m³	

The relations are

$$V = \frac{\pi \cdot D^2 \cdot h}{4} \; ; \qquad A = \frac{\pi \cdot D^2}{2} + \pi \cdot D \cdot h \; ; \qquad G = \frac{\pi \cdot D^2 \cdot h \cdot \rho \cdot g}{4} \qquad (*)$$

(b) The Dimensional Set is constructed by the table given in (a).

	dependent			independent			
	V	A	G	D	g	h	ρ
m	3	2	1	1	1	1	–3
kg	0	0	1	0	0	0	1
s	0	0	–2	0	–2	0	0
π_1	1	0	0	0	0	–3	0
π_2	0	1	0	0	0	–2	0
π_3	0	0	1	0	–1	–3	–1
π_4	0	0	0	1	0	–1	0

Variables V, A, and G are *dependent*; D, g, h, and ρ are *independent*.

784 APPENDICES

(c) From above Dimensional Set, the dimensionless variables are

$$\pi_1 = \frac{V}{h^3} \; ; \qquad \pi_2 = \frac{A}{h^2} \; ; \qquad \pi_3 = \frac{G}{g \cdot h^3 \cdot \rho} \; ; \qquad \pi_4 = \frac{D}{h} \qquad (**)$$

(d) From (**), relations (*) can be written

$$\pi_1 = \frac{\pi}{4} \cdot \pi_4^2 \; ; \qquad \pi_2 = \pi \cdot \pi_4 \cdot \left(\frac{\pi_4}{2} + 1 \right); \qquad \pi_3 = \frac{\pi}{4} \cdot \pi_4^2 \qquad (***)$$

(e) By (***) we can write

$$\frac{\pi_1 \cdot \pi_2}{\pi_3} = \frac{\pi \cdot \pi_4 \cdot (\pi_4 + 2)}{2}$$

from which

$$\pi_1 = \frac{\pi \cdot \pi_3 \cdot \pi_4 \cdot (\pi_4 + 2)}{2 \cdot \pi_2} \qquad (****)$$

in which all four dimensionless variables appear.

(f) For the right side of (****) let us *select* $\pi_2 = 2$, $\pi_3 = 3$, $\pi_4 = 4$. Then relation (****) provides $\pi_1 = 56.54867$. However, the first, second, and third relations of (***) render, in order, $\pi_1 = 12.56637$, $\pi_2 = 37.69911$, and $\pi_3 = 12.56637$, all *contradicting* the above selected and obtained, from (****), values. Therefore relation (****) is *wrong*—the reason being the presence of *more than one dependent* dimensionless variable in it.

Art. 16.4

16/1 Terminal velocity of a sphere slowly descending in a viscous liquid (III).
For easy reference we list the variables and their dimensions in SI.

Variable	Symbol	Dimension	Remark
terminal speed	v	m/s	
sphere diameter	D	m	
differential density	$\Delta\rho$	kg/m³	between sphere and liquid
liquid viscosity	μ	kg/(m·s)	dynamic
gravitational acceleration	g	m/s	

(a) In this case, relations (16-7), (16-9), and (16-10) are applicable. Thus

$[\mu] = \text{m}^{-1} \cdot \text{kg}^1 \cdot \text{s}^{-1}$, therefore $a = -1, b = 1, c = -1$
$[g] = \text{m}^1 \cdot \text{kg}^0 \cdot \text{s}^{-2}$, therefore $a = 1, b = 0, c = -2$

and hence for μ

$$\begin{bmatrix} x \\ y \\ z \end{bmatrix} = \mathbf{T}_m \cdot \begin{bmatrix} a \\ b \\ c \end{bmatrix} = \begin{bmatrix} 1 & 0 & 0 \\ -1 & 1 & 0 \\ 2 & 0 & 1 \end{bmatrix} \cdot \begin{bmatrix} -1 \\ 1 \\ -1 \end{bmatrix} = \begin{bmatrix} -1 \\ 2 \\ -3 \end{bmatrix}$$

where x, y, and z are the exponents of N, kg, and s, respectively. Thus $[\mu] = N^{-1} \cdot kg^2 \cdot s^{-3}$. A similar process for g yields $[g] = N^1 \cdot kg^{-1} \cdot s^0$. Thus the Dimensional Set will be

	v	D	$\Delta\rho$	μ	g
m	1	1	−3	0	0
kg	0	0	1	2	−1
N	0	0	0	−1	1
s	−1	0	0	−3	0
π_1	1	0	$\dfrac{1}{3}$	$-\dfrac{1}{3}$	$\dfrac{1}{3}$

We observe that the column in matrix **C** under D is all zeros. Therefore, by Theorem 10-4, the diameter of the ball is *dimensionally irrelevant,* and hence, by Theorem 11-3, it is also *physically irrelevant.* But obviously this cannot be since the terminal speed *must* be a function of the diameter. Thus the result is *wrong,* which was to be proven.

(b) In this case relations (16-12), (16-14), and (16-15) are applicable. Thus

$[D] = m^1 \cdot kg^0 \cdot s^0$, therefore $a = 1$, $b = 0$, $c = 0$
$[\mu] = m^{-1} \cdot kg^1 \cdot s^{-1}$, therefore $a = -1$, $b = 1$, $c = -1$

and hence for D

$$\begin{bmatrix} x \\ y \\ z \end{bmatrix} = \mathbf{T}_{kg} \cdot \begin{bmatrix} a \\ b \\ c \end{bmatrix} = \begin{bmatrix} 1 & -1 & 0 \\ 0 & 1 & 0 \\ 0 & 2 & 1 \end{bmatrix} \cdot \begin{bmatrix} 1 \\ 0 \\ 0 \end{bmatrix} = \begin{bmatrix} 1 \\ 0 \\ 0 \end{bmatrix}$$

where x, y, and z are the exponents of m, N, and s, respectively. Thus $[D] = m^1 \cdot N^0 \cdot s^0$. A similar process for μ yields $[\mu] = m^{-2} \cdot N^1 \cdot s^1$. From the above, the dimensional matrix is

	v	D	$\Delta\rho$	μ	g
m	1	1	−3	−2	1
kg	0	0	1	0	0
N	0	0	0	1	0
s	−1	0	0	1	−2
π_1	1	$-\dfrac{1}{2}$	0	0	$-\dfrac{1}{2}$

We see that there are zeros in the **C** matrix under variables $\Delta\rho$ and μ. Therefore these variables are *dimensionally,* and hence also *physically,* irrelevant. But this is absurd since the terminal speed of the ball *must* obviously depend on these characteristics. Thus, the dimensionless variable $\pi_1 = v/\sqrt{D \cdot g} = $ const (obtained from the Dimensional Set) is *wrong.* This was to be proven.

(c) In this case, relations (16-17), (16-19), and (16-20) are applicable. Thus

$[v] = \text{m}^1 \cdot \text{kg}^0 \cdot \text{s}^{-1}$, therefore $a = 1, b = 0, c = -1$

$[\mu] = \text{m}^{-1} \cdot \text{kg}^1 \cdot \text{s}^{-1}$, therefore $a = -1, b = 1, c = -1$

and hence for V

$$\begin{bmatrix} x \\ y \\ z \end{bmatrix} = \mathbf{T}_s \cdot \begin{bmatrix} a \\ b \\ c \end{bmatrix} = \begin{bmatrix} 1 & 0 & \frac{1}{2} \\ 0 & 1 & \frac{1}{2} \\ 0 & 0 & -\frac{1}{2} \end{bmatrix} \cdot \begin{bmatrix} 1 \\ 0 \\ -1 \end{bmatrix} = \begin{bmatrix} \frac{1}{2} \\ -\frac{1}{2} \\ \frac{1}{2} \end{bmatrix}$$

where x, y, and z are the exponents of m, kg, and N, respectively. Hence $[V] = \text{m}^{1/2} \cdot \text{kg}^{-1/2} \cdot \text{N}^{1/2}$. A similar process for μ yields $[\mu] = \text{m}^{-3/2} \cdot \text{kg}^{1/2} \cdot \text{N}^{1/2}$. Therefore the Dimensional Set is

	v	D	$\Delta\rho$	μ	g
m	$\frac{1}{2}$	1	-3	$-\frac{3}{2}$	1
kg	$-\frac{1}{2}$	0	1	$\frac{1}{2}$	0
N	$\frac{1}{2}$	0	0	$\frac{1}{2}$	0
s	0	0	0	0	-2
π_1	1	1	1	-1	0

We see that there is a zero under variable g in the **C** matrix. Therefore the gravitational acceleration is a *dimensionally* and hence *physically* irrelevant variable. But this cannot be, since terminal speed is affected by gravity. Thus, the dimensionless variable (derived from the dimensional set) $\pi_1 = (v \cdot D \cdot \Delta\rho)/\mu = $ const is *wrong*. This was to be proven.

(d) In this case relations (16-17), (16-19), and (16-20) apply. Thus

$[\Delta\rho] = \text{m}^{-3} \cdot \text{kg}^1 \cdot \text{s}^0$, therefore $a = -3, b = 1, c = 0$

$[\mu] = \text{m}^{-1} \cdot \text{kg}^1 \cdot \text{s}^{-1}$, therefore $a = -1, b = 1, c = -1$

and so for $\Delta\rho$

$$\begin{bmatrix} x \\ y \\ z \end{bmatrix} = \mathbf{T}_s \cdot \begin{bmatrix} a \\ b \\ c \end{bmatrix} = \begin{bmatrix} 1 & 0 & \frac{1}{2} \\ 0 & 1 & \frac{1}{2} \\ 0 & 0 & -\frac{1}{2} \end{bmatrix} \cdot \begin{bmatrix} -3 \\ 1 \\ 0 \end{bmatrix} = \begin{bmatrix} -3 \\ 1 \\ 0 \end{bmatrix}$$

where x, y, and z are the exponents of m, kg, and N, respectively. Therefore, since $[\mu] = m^{-3/2} \cdot kg^{1/2} \cdot s^{1/2}$ from (c), the Dimensional Set is

	v	D	$\Delta\rho$	μ	g
m	1	1	-3	$-\dfrac{3}{2}$	1
kg	0	0	1	$\dfrac{1}{2}$	0
N	0	0	0	$\dfrac{1}{2}$	0
s	-1	0	0	0	-2
π_1	1	$-\dfrac{1}{2}$	0	0	$-\dfrac{1}{2}$

There are two all-zero columns in the **C** matrix under variables $\Delta\rho$ and μ—meaning that these two variables are *dimensionally* and hence *physically* irrelevant. But this is absurd, since terminal speed is obviously affected by differential density and fluid viscosity. Hence the dimensionless variable $\pi_1 = v/\sqrt{D \cdot g}$ = const (from the set) is *false*. This was to be proven.

(e) Again, relations (16-17), (16-19), and (16-20) are applicable. Thus

$[V] = m^1 \cdot kg^0 \cdot s^{-1}$, therefore $a = 1$, $b = 0$, $c = -1$
$[g] = m^1 \cdot kg^0 \cdot s^{-2}$, therefore $a = 1$, $b = 0$, $c = -2$

and therefore, as in (c) above, $[V] = m^{1/2} \cdot kg^{-1/2} \cdot N^{1/2}$. For g we write

$$\begin{bmatrix} x \\ y \\ c \end{bmatrix} = \mathbf{T}_s \cdot \begin{bmatrix} a \\ b \\ c \end{bmatrix} = \begin{bmatrix} 1 & 0 & \dfrac{1}{2} \\ 0 & 1 & \dfrac{1}{2} \\ 0 & 0 & -\dfrac{1}{2} \end{bmatrix} \cdot \begin{bmatrix} 1 \\ 0 \\ -2 \end{bmatrix} = \begin{bmatrix} 0 \\ -1 \\ 1 \end{bmatrix}$$

where x, y, and z are the exponents of m, kg, and N, respectively. Thus $[g] = m^0 \cdot kg^{-1} \cdot N^1$ and hence the Dimensional Set is

	v	D	$\Delta\rho$	μ	g
m	$\dfrac{1}{2}$	1	-3	-1	0
kg	$-\dfrac{1}{2}$	0	1	1	-1
N	$\dfrac{1}{2}$	0	0	0	1
s	0	0	0	-1	0
π_1	1	$-\dfrac{1}{2}$	0	0	$-\dfrac{1}{2}$

We see that there are two all-zero columns in the **C** matrix under variables $\Delta\rho$ and μ. Therefore these two variables are *dimensionally* and hence *physically irrelevant*—which is nonsense. It follows that the dimensionless variable $\pi_1 = v/\sqrt{D \cdot g}$ = const (obtained from the above set) is *in error*. This proves what was required.

16/2 (a) yes; (b) no; (c) yes; (d) yes; (e) no; (f) no; (g) yes; (h) yes; (i) no; (j) yes; (k) yes; (ℓ) yes.

16/3 None of the indicated fusions results in the reduction of variables.

16/4 By Rule 4 of dimensional homogeneity (Art. 6.1), the argument $V_2 \cdot V_3$ of the cosine function in $V_1 = V_2 \cdot V_3 \cdot V_5^2/(V_4 \cdot \cos(V_2 \cdot V_3))$ must be dimensionless. Hence $(V_1 \cdot V_4)/V_5^2$ must also be dimensionless. Therefore we must have at least *two* dimensionless variables. It follows that there cannot be any *fusion* of variables resulting in a *single* dimensionless variable.

16/5 Zero.

16/6 Curvature of a bimetallic thermometer.
(a) The following variables are relevant:

Variable	Symbol	Dimension
subtended angle	α	$m_t \cdot m_r^{-1}$
temperature change	Δt	K
thickness of each strip	h	m_r
differential expansion coefficient	$\Delta\beta$	1/K
length	L	m_t

Note that we have *directed* linear dimensions, namely *tangential* m_t and *radial* m_r. Thus the plane angle, which is ordinarily dimensionless, now assumes the dimension m_t/m_r. Similarly, thickness h of the strip is radial, while the length is tangential, as obviously from the depicted geometry.

We have five variables and three dimensions, therefore there are $5 - 3 = 2$ dimensionless variables determined by the Dimensional Set

	α	Δt	h	$\Delta\beta$	L
m_t	1	0	0	0	1
m_r	-1	0	1	0	0
K	0	1	0	-1	0
π_1	1	0	1	0	-1
π_2	0	1	0	1	0

from which

$$\pi_1 = \frac{\alpha \cdot h}{L}; \qquad \pi_2 = \Delta t \cdot \Delta\beta \qquad (*)$$

Therefore we can write

$$\alpha = c \cdot \frac{L}{h} \cdot \Psi\{\Delta t \cdot \Delta\beta\} \qquad (**)$$

where c is a constant, and Ψ is an as-yet unknown function.

(b) By the monomial form of Ψ in (**), we can write

$$\alpha = c \cdot \frac{L}{h} \cdot (\Delta t \cdot \Delta \beta)^n \qquad (***)$$

where n is a constant. We now consider that α is proportional to temperature change Δt (as stipulated in the problem). From this condition $n = 1$. Hence (***) becomes

$$\alpha = c \cdot \frac{L}{h} \cdot \Delta t \cdot \Delta \beta$$

We have thus derived the important results that subtended angle a is proportional to the product $L \cdot \Delta t \cdot \Delta \beta$ and inversely proportional to strip thickness h. The constant c can be easily determined by calibration, i.e., by performing a *single* measurement (of course in *practice* a series of measurements would be made from which a linear approximation of c would be obtained).

Art. 17.6

17/1 To crack a window.

(a) From the assumed relevant physical variables, the following list can be prepared:

Variable	Symbol	Dimension	Remark
mass of stone	M	kg	
stress in window	σ	kg/(m·s²)	at breaking level
speed of stone	v	m/s	normal to window
linear size of window	L	m	characteristic value
Young's modulus of window	E	kg/(m·s²)	

We have five variables and three dimensions, therefore there are $5 - 3 = 2$ dimensionless variables obtained from the Dimensional Set

	M	σ	v	L	E
m	0	−1	1	1	−1
kg	1	1	0	0	1
s	0	−2	−1	0	−2
π_1	1	0	2	−3	−1
π_2	0	1	0	0	−1

by which $\pi_1 = (M \cdot v^2)/(L^3 \cdot E)$; $\pi_2 = \sigma/E$, and therefore the *Model Law* is

$$S_M S_v^2 = S_L^3 \cdot S_E; \qquad S_\sigma = S_E \qquad (*)$$

where S_M is the Mass of Stone Scale Factor
S_v is the Speed of Stone Scale Factor
S_L is the Size of Window Scale Factor
S_σ is the Breaking Strength of Window Scale Factor
S_E is the Young's Modulus Scale Factor

Since the window materials are the same, $S_\sigma = S_E = 1$. Thus Model Law (*) can be simplified to

$$S_M S_v^2 = S_L^3 \qquad (**)$$

From the given conditions $S_L = L_2/L_1 = 1.4/0.7 = 2$ and $S_M = M_2/M_1 = 0.1/0.2 = 0.5$. Thus by (**) $S_v = v_2/v_1 = \sqrt{S_L^3/S_M} = \sqrt{2^3/0.5} = 4$, from which $v_1 = v_2/S_v = 12.6/4 = 3.15$ m/s.

(b) To construct the *Modeling Data Table* for this case we must to *decide* which is the "prototype" and which is the "model." Since it is the *lighter* stone that is *measured* for speed, we shall call this the "model," and hence its variables will bear the subscript "2." Correspondingly, the heavier stone shall be the "prototype" and will bear the subscript "1" throughout. Accordingly, the Modeling Data Table is presented in Fig. A6-13.

Variable					Scale factor S	Category	
name	symbol	dimension	prototype	model	model/prototype	prototype	model
mass of stone	M	kg	0.2	0.1	0.5	1	1
stress in window	σ	kg/(m·s²)	1 (dummy)	1 (dummy)	1	1	2
speed of stone	v	m/s	3.15	12.6	4	2	3
size of window	L	m	0.7	1.4	2	1	1
Young's modulus	E	kg/(m·s²)	1 (dummy)	1 (dummy)	1	1	1
dimensionless	π_1	1	5.78571	5.78571	1		
dimensionless	π_2	1	1	1	1		
categories of variables	1	freely chosen, *a priori* given, or determined independently					
	2	determined by application of Model Law					
	3	determined by measurement on the model					

Figure A6-13
Modeling Data Table for the cracking window experiment

By (17-13), the number of *freely chosen* variables (Category 1) is

$$(N_V)_1 = N_V + N_d - 1 = 5 + 3 - 1 = 7$$

By (17-14), the number of variables *imposed* by Model Law (Category 2) is

$$(N_V)_2 = N_V - N_d = 5 - 3 = 2$$

By (17-12), the number of *measured* (Category 3) variables is $(N_V)_3 = 1$.

As seen, these numbers confirm the content of the Modeling Data Table (Fig. A6-13). Note that the equality of stresses of the "prototype" and "model" is not by free choice; it was *imposed* by the Model Law, which stipulates that the ratio of Young's moduli must be identical to the ratio of stresses [see second relation of Model Law (*)].

(c) The two dimensionless variables are indeed identical (see table), as expected and required for two *dimensionally similar* systems.

17/2 The Cartesian equation of the hyperbola is

$$\frac{x^2}{a^2} - \frac{y^2}{b^2} = 1 \qquad (*)$$

where a and b are positive numbers. Fig A6-14 shows the graph of a hyperbola defined by (*).

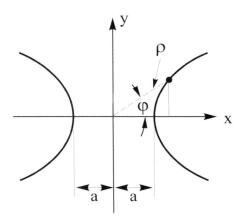

Figure A6-14
A hyperbola in a Cartesian coordinate system

By the notation of this figure, $y = \rho \cdot \sin \varphi$ and $x = \rho \cdot \cos \varphi$. Thus, (*) can be written

$$\rho = \frac{a \cdot b}{\sqrt{b^2 \cdot \cos^2 \varphi - a^2 \cdot \sin^2 \varphi}} \qquad (**)$$

and if $\varphi = 0$, then $\rho = \rho_0$. Therefore, by (**), $\rho_0 = a$ and also

$$\frac{\rho}{\rho_0} = \frac{\rho}{a} = \frac{b}{\sqrt{b^2 \cdot \cos^2 \varphi - a^2 \cdot \sin^2 \varphi}} = \frac{1}{\sqrt{\cos^2 \varphi - (a/b)^2 \cdot \sin^2 \varphi}}$$

Thus, by (17-2), the condition of *similarity* for two hyperbolas is the *equality* of their a/b ratios.

17/3 (a) The four physical variables and their dimensions are given in the table below.

Variable	Symbol	Dimension	Remark
slope of incline	φ	1 (rad)	to horizontal
time of travel	t	s	
travel distance	h	m	vertical
gravitational acceleration	g	m/s²	

(b) The dependent variable is t, the independent variables are φ, h, and g. Therefore, there are one dependent and three independent variables.

(c) There are $N_V = 4$ variables and $N_d = 2$ dimensions. Thus, there are

by relation (17-13), $(N_V)_1 = N_V + N_d - 1 = 5$ Category 1 (freely selectable) variables;
by relation (17-13), $(N_V)_2 = N_V - N_d = 2$ Category 2 (imposed by Model Law) variables;
by relation (17-12), $(N_V)_3 = 1$ Category 3 (measured) variable.

(d) The number of independent dimensionless variables is $N_P = N_V - N_d = 4 - 2 = 2$; they are determined by the Dimensional Set

	φ	t	h	g
m	0	0	1	1
s	0	1	0	−2
π_1	1	0	0	0
π_2	0	1	$-\frac{1}{2}$	$\frac{1}{2}$

yielding

$$\pi_1 = \varphi; \qquad \pi_2 = t \cdot \sqrt{\frac{g}{h}} \qquad (*)$$

(e) The Model Law, by (*), is $S_\varphi = 1$; $S_t = \sqrt{S_h/S_g}$,

where S_φ is the Slope Scale Factor
S_t is the Time Scale Factor
S_h is the Vertical Travel Scale Factor
S_g is the Gravitational Acceleration Scale Factor

17/4 Contact time of impacting balls.
(a) The relevant variables and their dimensions are as follows:

Variable	Symbol	Dimension	Remark
contact time	t	s	
speed of balls	v	m/s	absolute
radii of balls	R	m	identical
density of balls	ρ	kg/m³	identical
Young's modulus	E	kg/(m·s²)	identical

(b) By the above table, we have $N_V = 5$ variables and $N_d = 3$ dimensions. Therefore we have by relations (17-13), (17-14), and (17-12), in order

$(N_V)_1 = N_V + N_d - 1 = 5 + 3 - 1 = 7$ Category 1 (freely selectable) variables;
$(N_V)_2 = N_V - N_d = 5 - 3 = 2$ Category 2 (imposed by Model Law) variables;
$(N_V)_3 = 1$ Category 3 (measured on the model) variable.

(c) For the prototype, four variables are given (Category 1) and one variable (t_1) is determined by Model Law (Category 2). For the model, one variable is measured (Category 3), three are given (Category 1) and one (v_2) is determined by Model Law (Category 2).

(d) The number of dimensionless variables is $N_P = 5 - 3 = 2$; and they are by the Dimensional Set

	t	v	R	ρ	E
m	0	1	1	-3	-1
kg	0	0	0	1	1
s	1	-1	0	0	-2
π_1	1	0	-1	$-\frac{1}{2}$	$\frac{1}{2}$
π_2	0	1	0	$\frac{1}{2}$	$-\frac{1}{2}$

yielding

$$\pi_1 = \frac{t}{R}\cdot\sqrt{\frac{E}{\rho}}; \qquad \pi_2 = v\cdot\sqrt{\frac{\rho}{E}} \qquad (*)$$

(e) The Model Law is by (*).

$$S_t = S_R\cdot\sqrt{\frac{S_\rho}{S_E}}; \qquad S_v = \sqrt{\frac{S_E}{S_\rho}} \qquad (**)$$

where S_t is the Contact Time Scale Factor
S_R is the Radius of Balls Scale Factor
S_ρ is the Density Scale Factor
S_E is the Young's Modulus Scale Factor
S_v is the Speed Scale Factor

(f) By data in the problem

$$S_E = \frac{E_2}{E_1} = \frac{6.7 \times 10^{10}}{2 \times 10^{11}} = 0.335$$

and

$$S_\rho = \frac{\rho_2}{\rho_1} = \frac{2600}{7850} = 0.33121.$$

Hence, by the second relation of Model Law (**)

$$S_v = \frac{v_2}{v_1} = \sqrt{\frac{S_E}{S_\rho}} = \sqrt{\frac{0.335}{0.33121}} = 1.00571$$

Thus, $v_2 = S_v \cdot v_1 = (1.00571) \cdot (2.8) = 2.816$ m/s, where $v_1 = 2.8$ m/s is the given speed of the prototype.

(g) We have $S_R = R_2/R_1 = 0.05/0.6 = 1/12$, and hence, by (**), and S_E and S_ρ given in (f),

$$S_t = \frac{t_2}{t_1} = S_R \cdot \sqrt{\frac{S_\rho}{S_E}} = \frac{1}{12} \cdot \sqrt{\frac{0.33121}{0.335}} = 8.28606 \times 10^{-2}$$

from which the sought-after contact time, t_1, for the prototype is

$$t_1 = \frac{t_2}{S_t} = \frac{4.412 \times 10^{-5}}{8.28606 \times 10^{-2}} = 5.3246 \times 10^{-4} \text{ s}$$

where contact time t_2 of the *model* is as given in the problem.

(h) Dimensionless variables π_1 and π_2 are defined in (*). Numerically they are, for the prototype:

$$\pi_1 = \frac{5.3246 \times 10^{-4}}{0.6} \cdot \sqrt{\frac{2 \times 10^{11}}{7850}} = 4.47936 \; ; \quad \pi_2 = (2.8) \cdot \sqrt{\frac{7850}{2 \times 10^{11}}}$$

$$= 5.54725 \times 10^{-4}$$

and for the model:

$$\pi_1 = \frac{4.412 \times 10^{-5}}{0.05} \cdot \sqrt{\frac{6.7 \times 10^{10}}{2600}} = 4.47936 \; ; \quad \pi_2 = (2.816) \cdot \sqrt{\frac{2600}{6.7 \times 10^{10}}}$$

$$= 5.54730 \times 10^{-4}$$

We see that they are identical (within rounding error)—as they should be.

(i) The modeling data table is as presented in Fig. A6-15.

17/5 Gravitational collapse of a star. To solve this problem, we must establish the Model Law. To establish the Model Law, we must determine a complete set of dimensionless variables. To determine this set, we must construct the Dimensional Set. To construct this set, we must list the relevant variables and their dimensions. This list is as follows:

Variable	Symbol	Dimension	Remark
diameter of star	D	m	
density of star	ρ	kg/m^3	
internal stress	σ	kg/(m·s^2)	critical value
universal gravitational constant	k	m^3/(s^2·kg)	

APPENDIX 6

Variable					Scale factor S	Category	
name	symbol	dimension	prototype	model	model/prototype	prototype	model
contact time	t	s	5.3246E-4	4.412E-5	0.082861	2	3
impact speed	v	m/s	2.8	2.816	1.00571	1	2
ball radius	R	m	0.6	0.05	0.08333	1	1
ball density	ρ	kg/m³	7850	2600	0.33121	1	1
Young's modulus	E	kg/(m·s²)	2E11	6.7E10	0.335	1	1
dimensionless	π_1	1	4.47936	4.47936	1		
dimensionless	π_2	1	5.54725E-4	5.5473E-4	1.00001		
categories of variables	1	freely chosen, *a priori* given, or determined independently					
	2	determined by application of Model Law					
	3	determined by measurement on the model					

Figure A6-15
Modeling Data Table for determination of contact time of impacting balls

We have four variables and three dimensions, therefore there is only one dimensionless variable—a constant. We obtain this sole "variable" by the Dimensional Set

	D	ρ	σ	k
m	1	−3	−1	3
kg	0	1	1	−1
s	0	0	−2	−2
π_1	1	1	$-\dfrac{1}{2}$	$\dfrac{1}{2}$

yielding $\pi_1 = D \cdot \rho \cdot \sqrt{k/\sigma}$. The Model Law is therefore $S_D \cdot S_\rho = \sqrt{S_\sigma/S_k}$,

where S_D is the Diameter Scale Factor
S_ρ is the Density Scale Factor
S_σ is the Critical Stress Scale Factor
S_k is the Universal Gravitational Constant Scale Factor

Since for all cases $S_k = 1$, the Model Law simplifies to

$$S_\sigma = S_D^2 \cdot S_\rho^2 \qquad (*).$$

By the given data, $S_D = D_2/D_1 = 20$ and $S_\rho = \rho_2/\rho_1 = 0.08$, therefore (*) becomes

$$S_\sigma = \frac{\sigma_2}{\sigma_1} = (20^2) \cdot (0.08^2) = 2.56$$

from which $\sigma_2 = S_\sigma \cdot \sigma_1 = (2.56) \cdot \sigma_1$, i.e., the second star's critical stress, under which it collapsed, was 2.56 times that of the first star.

APPENDIX 7
PROOFS FOR SELECTED THEOREMS AND EQUATIONS

For each proof the numbering of equations starts with "1."

Example 11-24 in Art. 11.2.2

Proof that in an elliptical orbit the average distance of the orbiting body (planet) from either of the foci (Sun) is half the major-axis of the ellipse.

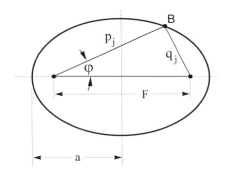

Figure A7-1
An ellipse with semimajor axis and general radius vectors defined
"a" is semimajor axis; p_j and q_j are radius vectors to an arbitrary point B

Consider the ellipse shown in Fig. A7-1. By the definition of an ellipse, sum S of lengths of radius vectors (pointing to B from the foci) is constant. Hence S is independent of angle φ. If $\varphi = 0$, then according to Fig. A7-1,

$$S = \left(\frac{F}{2} + a\right) + \left(a - \frac{F}{2}\right) = 2 \cdot a \tag{1}$$

By symmetry, we can write for the *average* distance $\bar{\rho}$

$$\bar{\rho} = \frac{1}{n} \cdot \sum_{j=1}^{n} p_j = \frac{1}{n} \cdot \sum_{j=1}^{n} q_j \tag{2}$$

797

where n is a large positive integer. In view of (1) and (2) now

$$\sum_{j=1}^{n} S_j = \sum_{j=1}^{n} (p_j + q_j) = \sum_{j=1}^{n} p_j + \sum_{j=1}^{n} q_j = \sum_{j=1}^{n} p_j + \sum_{j=1}^{n} p_j = 2 \cdot \sum_{j=1}^{n} p_j = 2 \cdot a \cdot n$$

Therefore $\sum_{j=1}^{n} p_j = a \cdot n = \bar{\rho} \cdot n$, hence $\bar{\rho} = a$. This was to be proven.

Relation (f) of Example 13-8 in Art. 13.3

Let τ_b and τ_g designate the arrival (point of) times of the boy and girl, respectively, in the time interval 0 to T. Then obviously $0 \leq \tau_b \leq T$ and $0 \leq \tau_g \leq T$ since neither the boy nor the girl will arrive outside the specified time interval. If the waiting times for the boy and girl are Δt_b and Δt_g, respectively, then they will meet if—and only if—*both* of the following relations are satisfied

$$\tau_g - \tau_b \leq \Delta t_b \, ; \qquad \tau_b - \tau_g \leq \Delta t_g \tag{1}$$

Consider Fig. A7-2. In this plot every *point* within the square of side length T represents a particular *pair* of τ_b, τ_g values. By (1)

$$\tau_g \leq \tau_b + \Delta t_b \, ; \qquad \tau_g \geq \tau_b - \Delta t_g \tag{2}$$

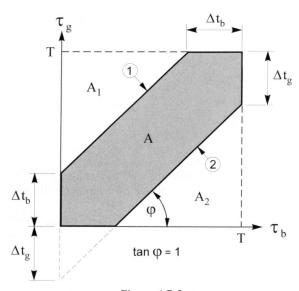

Figure A7-2
Geometric interpretation of probability of boy meets girl at an agreed location and within an agreed time interval T
Symbols: τ_b and τ_g are boy's and girl's *arrival* times;
Δt_b and Δt_g are boy's and girl's *waiting* times

i.e., to satisfy these two inequalities, the *point* representing a pair of τ_b and τ_g (arrival) values must lie *below* line 1 and *above* line 2; i.e., it must be in the shaded area marked A in the figure.

Since every point in the square is equally likely, and since the favorable outcome (i.e., meeting) occurs only if the point is in area A, therefore the probability that a randomly placed point falls in A is the same as the probability that our boy meets our girl; i.e., p. Accordingly

$$p = \frac{A}{A_1 + A_2 + A} = \frac{T^2 - A_1 - A_2}{T^2} \tag{3}$$

But by the figure

$$A_1 = \frac{(T - \Delta t_b)^2}{2} \; ; \quad A_2 = \frac{(T - \Delta t_g)^2}{2} \tag{4}$$

which, if substituted into (3), yields (after some simplification)

$$p = \frac{\Delta t_b}{T} - \frac{1}{2} \cdot \left(\frac{\Delta t_b}{T}\right)^2 + \frac{\Delta t_g}{T} - \frac{1}{2} \cdot \left(\frac{\Delta t_g}{T}\right)^2 \tag{5}$$

Considering the definitions of π_1, π_2, and π_3 given in relation (a) of Example 13-8, we can write this as

$$\pi_1 = \pi_2 + \pi_3 - \frac{1}{2} \cdot (\pi_2^2 + \pi_3^2)$$

which is relation (f) of Example 13-8. Thus the proof is complete.

Relations (9-53) and (9-64) in Art. 9.3 (The proofs of these relations are provided by Prof. Pál Rózsa)

To be a little less abstract, for these proofs we shall consider a case of $N_V = 5$ variables and $N_d = 2$ dimensions (the process can be extended to any acceptable number of these "ingredients"). The number of dimensionless variables is then $N_P = N_V - N_d = 5 - 2 = 3$. Thus we consider two systems, each with three dimensionless variables; in Set 1 they are π_{11}, π_{12}, and π_{13}; in Set 2 they are π_{21}, π_{22}, and π_{23}. These two sets are connected by relation (9-50), which is (repeated here for convenience)

$$\begin{bmatrix} \ln \pi_{21} \\ \ln \pi_{22} \\ \ln \pi_{23} \end{bmatrix} = \mathbf{\tau} \cdot \begin{bmatrix} \ln \pi_{11} \\ \ln \pi_{12} \\ \ln \pi_{13} \end{bmatrix} \tag{1}$$

where $\mathbf{\tau}$ is the *transformation matrix*. We now consider two *sequences* of variables: V_1, V_2, V_3, V_4, V_5 in Dimensional Set 1 and V_4, V_5, V_1, V_3, V_2 in Dimensional Set 2 (these sequences are the same as in the illustrative example in Art 9.3). For convenience, we also introduce the notation

$$\mathbf{P}_1 \equiv \begin{bmatrix} \ln \pi_{11} \\ \ln \pi_{12} \\ \ln \pi_{13} \end{bmatrix} \;;\quad \mathbf{P}_2 \equiv \begin{bmatrix} \ln \pi_{21} \\ \ln \pi_{22} \\ \ln \pi_{23} \end{bmatrix} \;;\quad \mathbf{W}_1 \equiv \begin{bmatrix} \ln V_1 \\ \ln V_2 \\ \ln V_3 \\ \ln V_4 \\ \ln V_5 \end{bmatrix} \;;\quad \mathbf{W}_2 \equiv \begin{bmatrix} \ln V_4 \\ \ln V_5 \\ \ln V_1 \\ \ln V_3 \\ \ln V_2 \end{bmatrix}$$

Therefore, by the definition of *Shift Matrix* **S**, we can write

$$\mathbf{W}_1 = \mathbf{S}\cdot\mathbf{W}_2 = \begin{bmatrix} 0 & 0 & 1 & 0 & 0 \\ 0 & 0 & 0 & 0 & 1 \\ 0 & 0 & 0 & 1 & 0 \\ 1 & 0 & 0 & 0 & 0 \\ 0 & 1 & 0 & 0 & 0 \end{bmatrix}\cdot\mathbf{W}_2 \tag{2}$$

If $\mathbf{D}_1$, $\mathbf{C}_1$, and $\mathbf{D}_2$, $\mathbf{C}_2$ are the respective submatrices of Dimensional Sets 1 and 2 (see Fig. 8-3), then

$$\mathbf{P}_1 = [\mathbf{D}_1 \quad \mathbf{C}_1]\cdot\mathbf{W}_1 \tag{3}$$

and

$$\mathbf{P}_2 = [\mathbf{D}_2 \quad \mathbf{C}_2]\cdot\mathbf{W}_2 \tag{4}$$

By (2) and (3)

$$\mathbf{P}_1 = [\mathbf{D}_1 \quad \mathbf{C}_1]\cdot\mathbf{S}\cdot\mathbf{W}_2 \tag{5}$$

and by (1) and (4)

$$\mathbf{T}\cdot\mathbf{P}_1 = [\mathbf{D}_2 \quad \mathbf{C}_2]\cdot\mathbf{W}_2 \tag{6}$$

We now *premultiply* both sides of (5) by **T**. This yields

$$\mathbf{T}\cdot\mathbf{P}_1 = \mathbf{T}\cdot[\mathbf{D}_1 \quad \mathbf{C}_1]\cdot\mathbf{S}\cdot\mathbf{W}_2 \tag{7}$$

and we see that the left sides of (6) and (7) are identical. Therefore their right sides must be also identical. Hence

$$\mathbf{T}\cdot[\mathbf{D}_1 \quad \mathbf{C}_1]\cdot\mathbf{S}\cdot\mathbf{W}_2 = [\mathbf{D}_2 \quad \mathbf{C}_2]\cdot\mathbf{W}_2 \tag{8}$$

Since the rightmost column-matrices of both sides of this equation are identical, and since the variables in them can be in any *arbitrary* order, therefore

$$\mathbf{T}\cdot[\mathbf{D}_1 \quad \mathbf{C}_1]\cdot\mathbf{S} = [\mathbf{D}_2 \quad \mathbf{C}_2] \tag{9}$$

or, using the partitioned Shift Matrix **S**, as defined in Fig. 9-3 (Art. 9.3),

$$\mathbf{T}\cdot[\mathbf{D}_1 \quad \mathbf{C}_1]\cdot\begin{bmatrix} \mathbf{S}_1 & \mathbf{S}_2 \\ \mathbf{S}_3 & \mathbf{S}_4 \end{bmatrix} = [\mathbf{D}_2 \quad \mathbf{C}_2] \tag{10}$$

from which we obtain two matrix equations

$$\mathbf{T} \cdot (\mathbf{D}_1 \cdot \mathbf{S}_1 + \mathbf{C}_1 \cdot \mathbf{S}_3) = \mathbf{D}_2 \tag{11}$$

and

$$\mathbf{T} \cdot (\mathbf{D}_1 \cdot \mathbf{S}_2 + \mathbf{C}_1 \cdot \mathbf{S}_4) = \mathbf{C}_2 \tag{12}$$

By (11) now

$$\mathbf{T} = \mathbf{D}_2 \cdot (\mathbf{D}_1 \cdot \mathbf{S}_1 + \mathbf{C}_1 \cdot \mathbf{S}_3)^{-1} \tag{13}$$

Next, we write (2) thus

$$\mathbf{W}_2 = \mathbf{S}^T \cdot \mathbf{W}_1 \tag{14}$$

since, by construction, $\mathbf{S}$ is an *orthogonal* matrix, i.e., its *inverse* equals its *transpose*. Substituting (14) into (4), we obtain

$$\mathbf{P}_2 = [\mathbf{D}_2 \quad \mathbf{C}_2] \cdot \mathbf{S}^T \cdot \mathbf{W}_1 \tag{15}$$

Comparison this with (1) and (3) yields

$$[\mathbf{D}_2 \quad \mathbf{C}_2] \cdot \mathbf{S}^T \cdot \mathbf{W}_1 = \mathbf{T} \cdot \mathbf{P}_1 = \mathbf{T} \cdot [\mathbf{D}_1 \quad \mathbf{C}_1] \cdot \mathbf{W}_1 \tag{16}$$

which is true for an *arbitrary* sequence of V variables and hence for *any* $\mathbf{W}_1$. Therefore we can write

$$[\mathbf{D}_2 \quad \mathbf{C}_2] \cdot \mathbf{S}^T = \mathbf{T} \cdot [\mathbf{D}_1 \quad \mathbf{C}_1] \tag{17}$$

or

$$[\mathbf{D}_2 \quad \mathbf{C}_2] \cdot \begin{bmatrix} \mathbf{S}_1^T & \mathbf{S}_3^T \\ \mathbf{S}_2^T & \mathbf{S}_4^T \end{bmatrix} = \mathbf{T} \cdot [\mathbf{D}_1 \quad \mathbf{C}_1] \tag{18}$$

which can be resolved into two matrix equations

$$\mathbf{D}_2 \cdot \mathbf{S}_1^T + \mathbf{C}_2 \cdot \mathbf{S}_2^T = \mathbf{T} \cdot \mathbf{D}_1 \tag{19}$$

and

$$\mathbf{D}_2 \cdot \mathbf{S}_3^T + \mathbf{C}_2 \cdot \mathbf{S}_4^T = \mathbf{T} \cdot \mathbf{C}_1 \tag{20}$$

By postmultiplying (19) by the inverse of $\mathbf{D}_1$, we get

$$\mathbf{T} = (\mathbf{D}_2 \cdot \mathbf{S}_1^T + \mathbf{C}_2 \cdot \mathbf{S}_2^T) \cdot \mathbf{D}_1^{-1} \tag{21}$$

We see that (13) and (21) are the two formulas (9-53) in Art. 9.3 of the text. Furthermore, (12) is formula (9-64), also in Art. 9.3. Thus the proofs for all three formulas are now complete.

APPENDIX 8
BLANK MODELING DATA TABLE

Variable					Scale factor S	Category	
name	symbol	dimension	prototype	model	model/prototype	prototype	model
categories of variables	1	freely chosen, *a priori* given, or determined independently					
	2	determined by application of Model Law					
	3	determined by measurement on the model					

Figure 8.1
Modeling Data Table (blank)

SUBJECT INDEX

"ff." = "*and on the following page(s)*"

A matrix (see Dimensional Set)
Acceleration
  angular 57, 698
  gravitational (see Gravitational . . .)
  linear 57, 677
Acoustic impedance 58, 709
Acre 60, 715
Acronyms 721
Activity (nuclear) 59, 713
Admittance 58, 706
Aircraft operational characteristics 601
Alternate dimensions 401 ff.
Altitude dependent gravity 321
Ampere 48
Ampere constant 681
Angle, plane 677
Angle, solid 677
Angström 61
Animals falling to the ground 364 ff.
API gravity degree 683
Appendices xxix, 675 ff.
Applications (43 additional) 527
Are 61
Area 57, 677
  elliptic segment 374
  triangle 341, 379, 791
Arithmetic of dimensions 95 ff.
  association 96
  differentials 97
  exponents 96
  integrals 98
  products 95
  quotients 96

Articles xxiii
Astronomical unit 60, 715
Atmosphere (standard) 61
Atmospheric pressure vs. altitude 107
Attenuation coefficient
  atomic 59
  linear 59, 712
  mass 59
atto (SI prefix) 62
Avogadro constant (number) 450, 681
Axial thrust of a propeller (see
    Propeller's . . .)
Axially loaded bar 391

B matrix (see Dimensional Set)
Ballet on the Moon 515 ff.
Balloon depression 637
Bar 61, 715, 716
Barn 60, 716
Barrel volume 127
Beach characteristics 208
Beaford number 683
Beam deflection upon a dropped mass
    609
Becquerel 54, 693
Bending of light 129
Bi-metallic thermometer 461, 788
Big Bang 420
Biot number 683
Black-body radiation law 324
Black hole's existence criteria 548
Body surface area of slim individuals 131

Bohr
    atom xv, 547
    magneton 161, 680, 681
    radius 161
Bolometric luminosity of a star 376
Boltzmann constant 324, 376, 679, 681
Boussinesq's problem xvii, 551
Boy meets girl 343
Brick falling 350
British Association 698
Brownian (thermal) motion 713
Buckingham's theorem 148, 473, 493
Buckling of columns 330, 543, 587
Bulk modulus 302, 596, 599, 678
Bullet
    ejected vertically 346, 580
    landing on a horizontal plane 436
    landing on an inclined plane 533
    penetration 650
Bursting speed of a flywheel 389

C matrix (see Dimensional Set)
Calorie (IT) 61, 716
Calorifer (see Heat transfer in . . .)
Cantilever loaded laterally 78, 116, 131, 141, 143, 146, 149, 150, 422, 474, 490, 495
Capacitance 54, 680, 693
Capillary tube (see Meniscus . . .)
Capstan drive 348
Carat 61, 716
Carvallo formula 765, 766
Cascading effect (see Relevance of variables)
Catenary force characteristics 566 ff.
Catenary geometry 312, 758 ff.
Cauchy number 472, 558, 691
Cavendish experiment 701
Cavendish Laboratory 94
Celsius scale (see Temperature . . .)
centi (SI prefix) 62
Centre for Research in Earth and Space Technology xxiii
Centrifugal force acting on a point 544
Chapters xxix
Charge density 680
Charts, number of (see Number of . . .)

Circuit frequency 640
Circular frequency 677
Circulation of money 128, 698, 745
Clark formula 94, 737
Clausius equation 700
Cobb-Douglas production function 129, 746
Coefficient of heat transfer (see Film heat . . .)
Coefficients 34
Coffee warmer 356, 386
Compound interest 403
Compressibility 57, 702
Compressing ideal gas 546
Compression factor 679
Concentration 59, 710
Conductivity, electric 58, 656, 705
Cone volume 318
Conical pendulum (see Pendulum . . .)
Conical wedge volume 376, 776
Constants 32, 35
Copley Medal 697
Coulomb
    constant 681
    definition 693
    equation 704
    symbol and dimension 54
Cowling number 688
Critical sliding friction 270
Curie 62, 717
Current, electric 48, 680
Curved beam reaction forces 476
Curves, number of (see Number of . . .)
Cylinder (see Right circular . . .)

D matrix (see Dimensional Set)
Day 60
deca (SI prefix) 62
deci (SI prefix) 62
Deflection
    beam 370, 373, 770 ff.
    curved bar 483
    elastic foundation 326 ff.
    radome 284 ff.
    rope 652
    semi-circular ring 322 ff.
Degree (angle) 60, 714
Density 678, 711

Density and size of universe (see
 Universe ...)
Determinants xxx, 1 ff.
 cofactors 3
 expansion 4
 of a matrix 3
 properties 5
Diameter 677
Diffusion coefficient 59, 678, 711
Dimensional analysis xxvi
Dimensional
 homogeneity (see Homogeneity ...)
 irrelevancy (see Relevance of variables)
 matrix 134
  reduced 147
 method xxv
 modeling (see Modeling ...)
 Set xxvi, 163 ff.
  definition 163
  distinct and equivalent Sets 229 ff., 231 ff., 238
   independent/dependent 389
  structure of 167
  submatrices of
   A matrix 136
    singular **A** 144
     elimination of singularity 145 ff.
   B matrix 136
   C matrix 165, 166
   D matrix 165
  transformation between Sets 181 ff., 211 ff.
   by Shift Matrix **S** (see **S** matrix)
   of different **D** matrices 202
  variables in
   dimensionless 381
   identical 385
 systems 37 ff.
  a note on classification 68
  coherence 43
  monodimensional 38
  multidimensional 38, 42
   American/British Force (Imperial) 42, 43, 67
   American/British Mass 43, 67
   CGS 66

Economics Dimensional (EDS) 745
   impossible 89
  Kapitza 94, 736
  Kiang 66
  MKS 43, 67
  SI 35, 43, 45 ff.
   fundamental dimensions 45 ff.
   structure 44
  unusual 85, 87, 94, 733
  omnidimensional 38, 41
 variables (see Variables)
Dimensionless relations
 monomials 333 ff.
  impossible 338, 348
Dimensionless variables (see Variables)
Dimensions (general)
 derived 43
  with specific names 53, 54
  without specific names 53, 55 ff.
 fundamental 37, 44, 45 ff.
 magnitude of 37
 number of 37
Dipole moment, electric 680
Dirac constant 161
Discharge of a capacitor 378, 779
Distinct and Equivalent Dimensional Sets 229 ff.
Dogs' tails 106
Dolphins' length vs. weight 84
Dominoes' velocity of collapse 598
Dose rate, absorbed 59, 713
Dose rate, equivalent 59, 713
Drag on a body moving in a fluid 336
Drag on a flat plate in a fluid 582
Drop of liquid 178
DuBois formula 131
Dyne 61, 691

E matrix 137
Eddington's formula 422
Edsel (automobile) 464
Einstein's relativity theory (see Relativity)
Einstein equation 158, 719, 736, 768
Electric
 dipole moment 58, 706
 dipole potential 58, 706
 field by a dipole 537

field strength 58, 703
flux 58, 706
  density 58, 706
  power 58, 707
Electrochemically deposited material 381
Electromagnetic moment 58
Electron
  velocity in a vacuum tube 538
  charge (see Elementary charge)
  mass (see Mass of electron)
  radius 681
Electronvolt 59, 60, 712, 715
Elementary charge (electron) 59, 680, 681, 712
Elongation of a suspended bar 271
Energy
  density 57, 698
  fluence rate 59, 712
  flux density 59, 712
  internal 679
  levels in Bohr atom (see Bohr)
  moving ball 227
  symbol 678
  vibrating wire (see Vibrating . . .)
  water waves 641
Enthalpy 57, 679, 702
Entropy 57, 700
Eötvös experiment 68
Epsilon "ϵ" matrix 165
Equations xxix
Equivalence principle 68
Equivalent Dimensional Sets (see Distinct and . . .)
Erg 61, 717
Ergodic phenomena 557
Ether-wind 68
Euler number 472, 557, 688
Exa (SI prefix) 62
Examples xxix
Expansion coefficient, linear and volumetric 57, 679, 702
Exponent matrix (see **E** matrix)

Fahrenheit scale (see Temperature . . .)
Farad 54, 693
Faraday constant 381, 680, 681
Fee electrons in a conductor 449

Feeding frequency of animals 132, 750
femto (SI prefix) 62
Ferguson's problem xi, 93
Fermi 61
Fick's law 711
Field strength, electric 58, 680, 703
Field strength, magnetic 58, 680, 704
Figures xxix
Film heat transfer coefficient 57, 679, 701
First radiation constant (see Radiation . . .)
Fisher formula 128, 745, 746
Flat surface moving on water 175
Flexural stiffness (see Rigidity)
Flow of fluid over a spillway 335
Flow rate (mass) 57
Flow rate (volume) 57
Fluid flow characteristics 557
Fluid flowing through a tube 415, 443, 606
Flux, magnetic 680
Flux density, magnetic 680
Foot size (human) 633
Force
  on wires carrying current 541, 542
  SI unit (see Newton)
  symbol 678
Formats and elements of relations 27 ff.
  elements
    dimension 29
    equality sign 28
    magnitude 29
    name 27
  formats
    mixed 30 ff.
    numeric 27 ff.
    symbolic 29
Fourier
  equation 555
  number 684
  theory of heat 686
Free electrons in a wire 402
Free fall 395
Frequency 54, 667, 694
Froude number 472, 557, 688
Frustum of a cone, volume and surface 377, 393, 778
Fuel consumption of cars 80
Fundamental Formula xxix, 176, 177

Gal 61
Gamma, magnetic flux density 62
Gamma, mass 61
Gas constant 679
Gauss 62, 717
Giga (SI prefix) 62
Government of Canada xxiii
Grashof number 685, 686
Grätz number 685
Gravitational
   acceleration 57, 321, 677, 681, 699
   collapse of a star 525, 794
   pull by a plate 280
   pull by a solid sphere 275, 282 ff.
Gray 54, 694, 718

Hand size 633
Hartman number 689
Heart rate of mammals 94, 737
Heat
   capacity 57, 700
   flux 57, 703
   loss through a pipe wall 105
   symbol 679
   transfer coefficient 57, 701
   transfer in a calorifer 535
   transfer to a fluid in a pipe 558
Hectare 60, 715
hecto (SI prefix) 62
Height 677
Helmholtz energy 679
Henry 54, 694
Heron's formula 781
Hertz 50, 54, 694
Hill's relation 84
Homogeneity, dimensional 99 ff.
   for equations 99 ff.
      advices and rules 102, 103, 104, 108, 109
   for graphs 110 ff.
      rules 113, 120, 121, 124
      scales 111
      variations (3) 115
Homology 467
Hooke's law 114, 397
Hour 60
Hubble time 66

Hydrometer oscillation 668
Hydroplaning of tires 277
Identical variables 385
Impact velocity of a meteorite (see Meteorite . . .)
Impacting balls 524, 792
Impedance 58, 680, 705
Impossible dimensional system 89
Impulse 57, 699
Inductance 680
Industrial Revolution 698
Internal energy (see Energy . . .)
Invariance of dimensionless products 225 ff.
Ion number density 59, 712
Irradiance 58, 707
Irrelevancy of variables (see Relevance)

Jamming a plug into a hole 627
Jerk, angular 57, 720
Jerk, linear 57, 97, 720
Joule 54, 695
Jumping heights of animals 124, 290, 428

Kapitza dimensional system xiii, 94, 736, 737
Kelvin temperature scale (see Temperature . . .)
Kepler's constant 681
Kepler's laws 308, 309, 360, 362
Kettledrum (see Pitch of . . .)
Kiang system (see Dimensional systems)
kilo (SI prefix) 62
Kilogram-force 61, 716
Kilogram (mass) 45
Kilopond 61, 716
Kilopound 716
Kinetic energy—Newtonian, relativistic 646
Kinetic friction coefficient 440
Knot 60, 715
Knudsen number 689

Labocetta's problem 313
Land price 410
L,C circuit frequency 640
Length 677
Length of arc 677

Light-year 37, 60, 715
Light quantity 58, 708
Lincoln Center (New York) 517
Linear equation systems 19 ff.
   augmented matrix of 24
   coefficient matrix of 19
   homogeneous 19, 23
   nonhomogeneous 23, 24
   solutions
      general 24
      linearly dependent 20, 22, 23
      linearly independent 20, 22
      nontrivial 20
      particular 24
      trivial 20
   unknowns, selectable and nonselectable 20, 22
Linear thermal expansion coefficient (see Expansion coeff . . .)
Liter 60
Litre (see Liter)
Load-carrying capacity of animals 363
Lumen 54, 695
Luminance 58, 708
Lux 54, 695

Mach number 34, 35, 691
Magnetic
   dipole moment 58, 704
   field strength 58, 704
   flux quantum 681
   permeance 58, 705
   polarization 58
   potential difference 58
   vector potential 58
Magnetization 680
Magnetomotive force 58, 705
Mass
   atom (unified) 60, 715
   electron 681
   on a weightless spring 264, 452
   proton 681
   symbol 678
   transfer coefficient 678
Material permeance 57, 705
Matrices (general) xxx, 1 ff.
   addition 6
   columns 1
   dyadic decomposition (minimal) 15
   elements 1
   main diagonal 2
   minors 14
   order of square matrix 2
   partitioning 2, 8
   product 7
   properties
      compatibility 6
      distributivity 6, 8
   rank 14 ff.
      determination 16
      properties 18
   rows 1
Matrices (special)
   adjoint 10, 11
   diagonal 2
   identity 2
   inverse 10 ff.
   null (zero) 2
   orthogonal 13
   permutation 12
   singular and nonsingular 10
   skew-symmetric 6
   square xxx, 2
   sub 2
   symmetric 6
   transpose 1
Maxwell 62, 717
McGill University, Montreal xxv
McGraw-Hill xxix
Mean free path 59, 713
Mechanical impedance 58, 709
Medicinal bathhouse 91, 728
Mega (SI prefix) 62
Megaparsec (see Parsec)
Meniscus of fluid in a capillary tube 434
Meteorite size and velocity 505
Meteorite velocity 339, 505
Mho 62, 696, 717
Michelson-Morley experiment 68
micro (SI prefix) 62
Micron 61
Mile (nautical) 60, 715
Mile (statute) 69
milli (SI prefix) 62

Millibar 60, 715
Minimum deflection cantilevers 615
Minute (time) 60, 714
Minute (angle) 60, 714
Model Law 479
Modeling
   categories and relations of variables 489
   characteristics, interdependence 492 ff., 500, 501
   dimensional XII, 463 ff.
      applications of 465
      benefits of 464
      cost 466
      not recommended in 465
      sequential tasks of 466
   garment and cosmetic 463, 464
   mathematical 463, 464
   mock-up 463, 464
Modeling Data Table 495 ff.
   blank 803
   construction 497
Modulus of elasticity (see Young's modulus)
Modulus of shear 57, 702
Molality 59, 710
Molar
   energy 59, 710
   entropy 59, 710
   gas constant 681
   gas volume 681
   heat capacity 59, 710
   mass 59, 683, 710 ff.
   volume 59, 710 ff.
Mole 50
Moment
   of force 57, 678
   of inertia 57, 678, 699
   of inertia of a lamina 417
   of momentum 57, 699
Momentum 57, 654, 699
Momentum of a quantum 313, 762
Monomial power form 133 ff.
Monomials (see Dimensionless relations and Reconstructions)
Monte Carlo simulation 345, 410
Most comfortable walking speed 518
Mutual inductance (see Inductance)

Named dimensionless variables (numbers) 683 ff.
nano (SI prefix) 62
Nautical mile (see Mile)
Navier-Stoke equations 128, 745
Nernst equation 126
Newcomen's engine 698
Newton's laws 38, 44, 46, 47, 66, 67, 75, 76 275, 282, 284, 308, 699, 701, 727
Newton (force) 46, 54, 696
Nicholson hydrometer oscillation 644
Nobel prize 693, 695, 712, 718
Nonmonomials (see Reconstructions)
Nonselectable dimensions 151
Non-SI units
   permitted (long term) 56, 60
   permitted (short term) 59, 60
   prohibited 59, 61, 62
Nuclear magneton (see Bohr magneton)
Number 34
Number of curves and charts 317 ff.
Numerical equivalencies of dimensions 69 ff.
   energy 71
   force 71
   length 69, 70
   magnetism 72
   mass 70
   power 71
   pressure 72
   temperature (see also Temperuture . . .)
      fixed points 73
      intervals 72
      scales 72
   volume 70
Nusselt number 559, 683

Ohm 54, 696
Ohmic resistance 705
Oscillation of a fluid in a U tube 312, 757
Oscillation of a torsional dipole 539
Oscillation of Nicholson hydrometer 644
Osmotic coefficient 679
Osmotic pressure 679

P matrix 140
Paint on a sphere 109

Paint on an inclined surface 383
Parameter 32, 35
Partheon 343
Parsec 60, 715
Particle
  fluence 59, 714
  fluence rate 59, 714
  flux density 59, 714
Pascal 54, 696
Path length 677
Path of an electron 418
Peclet number 685
Pendulum
  conical 305 ff., 438
  simple xvii, 40, 250
  torsional 487
Penetration of a bullet 650
Permeability 58, 707
Permeability of vacuum 680, 681
Permittivity 58, 680, 704
Peta (SI prefix) 62
Phot 62, 717
Physical constants (numerical values) 161, 681
Physical irrelevancy (see Relevance of variables)
Physical quantities 32 ff.
  classification 32
  dimensionality 32, 33
  names 677 ff.
  symbols 677 ff.
  variability 32
Physical system characteristics 314, 763
pico (SI prefix) 62
Pitch of a kettledrum 563
Planck
  constant 127, 161, 324, 376, 681, 725, 742
  function 679
  law 324
Poise (see also Viscosity) 61, 717
Poiseuille equation 417, 444, 607
Poisson's ratio 35, 381
Potential, electric 680
Potential difference, electric 680
Pouring hot liquid into a tank 450
Power
  needed to lift a mass 367
  needed to tow a barge 480
  of a dynamo 330, 765
Prandtl number 82, 559, 685, 686, 687
Prefixes (SI) 59, 62
Pressure 678
Pressure
  in a radome 297, 304
  symbol 678
  underwater explosion 599
Principle of similitude xxv
Problems xxix
  solutions 723 ff.
  statements 65, 91, 126, 160, 259, 312, 329, 373, 399, 460, 523
Products of variables of given dimensions 135
  completeness 143
  construction 135 ff.
  number 139
Proofs of selected theorems and equations 797 ff.
Propeller's axial thrust 550
Proton mass (see Mass of proton)
Public demonstrations 92, 731 ff.
Pythagorean theorem xvii, 528, 530

q matrix 165
Quantity of electricity 680

Rad (radiation) 62, 718
Radiance 58, 707
Radiation
  constants, first and second 681
  energy 58, 707
  exposure 59, 713
  intensity 50, 58, 707
  law (see Black body . . .)
  power (flux) 58, 707
  pressure on satellites 177
Radius 677
Raindrop velocity 414
Range of an ejected bullet (see Bullet . . .)
Rankine temperature scale (see Temperature scale)
Ratio 34
Rayleigh number 686
RCA Victor, Montreal xxv

Reactance 680
Reconstructions 353 ff.
  monomials 353 ff.
  nonmonomials 366 ff.
Reduced dimensional matrix (see Dimensional matrix)
Reduction of number of dimensionless variables 413 ff.
  by decreasing the number of physical variables 414
  by dimension importation 449
  by dimension splitting 433
  by fusion of dimensionless variables 427
  by hybrid dimensioning 454
Relativistic energy-mass equivalence 71, 157
Relativistic kinetic energy 646
Relativistic mass 331
Relativistic red shift 358
Relativity, Theory of 46, 68, 129, 157, 331, 358, 747
Relevance of variables 263 ff.
  dimensional irrelevancy 263 ff.
    cascading effect 268 ff.
    condition 263
  physical irrelevancy 274 ff.
    analytic forms 304
    condition 274
    identification 263
    scale effects 291
    tests 296
Reluctance 58, 705
Rem (radiation) 62, 718
Resistance, electric 680
Resistivity, electric 58, 680, 704
Respiratory frequency of animals 485
Retail shop density and locations 406
Reverberation in a room 532
Revolution (rotation) 60, 714
Reynolds number 34, 286, 472, 686
Right circular cylinder 30, 400, 783
Rigidity 370, 371, 613, 770
Roasting time for turkey 508
Rolling resistance of tires 555
Röntgen 62, 718
Rope deflection 676

Rotating-blade mixer 172
Royal Society 695, 697, 720
Rules of writing dimensions 63 ff.
Running speeds of animals 289
Rydberg constant 161, 681

S matrix (Shift Matrix) 213 ff.
  submatrices S_1, S_2, S_3, S_4 213, 214
Scale effects (see also Relevance of variables) 511 ff.
Scale Factors 479
Schmideg Engineering Inc. 297
Schmidt number 687
Schwarzschild radius 549
Second (angle) 60, 714
Second moment of area 57, 678, 701
Second radiation constant (see Radiation . . .)
Selectable dimensions (see Dimensions . . .)
Self-inductance 680
Seneca College of Toronto xxiii
Sequence of variables (see Dimensional Set)
Shear modulus 678
Shift Matrix (see **S** matrix)
Shock absorber selection 621
Siemens 54, 696
Sievert 54, 694, 697, 718
Similarities 468
  dimensional 472
  dynamic 471
  geometric 468
  kinematic 471
  thermal 472
Simple pendulum (see Pendulum)
Simply supported beam 85, 354
Singular **A** matrix (see **A** matrix)
Sinking ball (see Velocity of . . .)
Size of a meteorite (see Meteorite . . .)
Skeleton mass of mammals 513
Sleeping Beauty, The 515, 518
Sliding on a circle's chord 655
Sloshing of fuel in missile tanks 313, 761
Smithsonian Institution (Washington) 694
Snellen number 690
Soap bubble diameter 596

Sound
  energy flux 58, 709
  impedance (see Acoustic impedance)
  intensity 58, 709
  pressure 58, 708
  velocity 58
  velocity in a liquid 595
SPAR Aerospace xxiii, 627
Specific
  energy 57, 698
  enthalpy 57, 702
  entropy 57, 700
  heat capacity 57, 679, 700
  volume 57, 698, 711
Speed of a mass sliding down 399, 782
Speed of a sphere rolling down 299
Speed of light (in vacuum) 681
Spherical segment 564
Stanton number 687
Statute mile (see Mile)
Stefan-Boltzmann constant and law 94, 161, 324, 681, 733, 776
Stere 61, 716
Stilb 62, 717
Stokes 61, 717
Strain
  angular 698
  linear 678, 702
  volumetric 678
Stress
  by collision of 2 balls 314, 763
  in a window by wind 313, 759
  in structures 397
  mechanical, normal 57, 678
  mechanical, shear 678
Strouhal number 689
Surface
  charge density 58, 703
  current density 58, 703
  tension 57, 678, 701
Susceptance 58, 705
Symbols of physical quantities (see Physical quantities)

Taiga Engineering xxiii, 627
Taylor expansion (functions) 104, 744
Technical University of Budapest xxiii

Temperature scale
  absolute 48, 679, 693, 720
  Celsius 49, 54, 679, 693
  Fahrenheit 719
  Kelvin 48, 693
  Rankine 720
Tension in a rotating wire ring 530
Tera (SI prefix) 62
Tesla 54, 697, 717
Thermal conductivity 57, 679, 701
Thermal resistivity 57, 701
Thermally generated forces 287
Time
  constant, symbol 677
  scale of the universe 313, 761
  symbol 677
  to fall a distance 530
Tires' rolling resistance (see Rolling . . .)
Ton (or Tonne, metric) 60, 715
Torr 61, 719
Torsion of a prismatic bar 444
Torsional pendulum (see Pendulum)
Torus of arbitrary cross-section 561
Torus volume 375, 773
Towing a barge (see Power needed . . .)
Transformation between
  Dimensional Sets 181 ff.
    by Shift Matrix $\mathbf{S}$ 211, 213 ff., 390, 391
    having different $\mathbf{D}$ matrices 202.
    of arbitrary constructions 211, 213ff.
  dimensional systems 73 ff.
  hybrid dimensional systems 454 ff.
Transformation matrix between
  Dimensional Sets (see also $\mathbf{S}$ matrix) 183, 205, 207, 211, 214 ff., 217, 219, 224, 757
  dimensional systems 87 ff., 90, 230, 455, 456, 457, 458, 734, 735, 784, 785, 786, 787, 799
Transport diffusion coefficient 59, 711
Triple point 48, 49
Turkey roasting (see Roasting . . .)

Underground explosion 279
Underwater explosion (see Pressure by . . .)

SUBJECT INDEX

Universal gravitational constant 57, 681, 701
Universe, density and size 420
Unusual dimensional systems 85, 87, 94, 733
Urine secretion rate of animals 125

Variable 33, 35
  dependent 33, 389 ff.
  dimensional 33
  dimensionless xii, 34
  independent 33, 389 ff.
  magnitude 29
  name 27
  relevance of (see Relevance . . .)
  sequence in Dimensional Set 381 ff.
Vectors
  column 2
  linear combination 17
    dependence 16, 17
    independence 16, 17
  products, inner and outer 9
  row 2
Velocity
  angular 57, 677, 698
  disturbance in a liquid 531
  disturbance in wire 312, 757
  fluid through an orifice 302
  linear 57, 677
  meteorite (see Meteorite . . .)
  raindrop (see Raindrop . . .)
  sinking ball 425, 430, 458, 460, 784 ff.
  sound (see Sound)
  surface waves 291
Venn diagram 276
Vertically ejected bullet (see Bullet . . .)
Vibrating
  cantilevers 374, 590 ff., 593 ff., 771
  liquid sphere 294
  mass on a spring 264, 452

  simply supported beams 584
  wire 310 ff., 313, 760
Viscosity, dynamic 57, 678, 699
Viscosity, kinematic 57, 678, 699
Visual acuity 690
Volt 54, 697
Volume 57, 677
Volume charge density 58, 703
Volumetric thermal expansion coefficient (see Expansion . . .)

Watt 44, 54, 697
Wave length 57, 677, 701
Wave number 57, 58, 701, 707
Wavefront in an atomic explosion 154 ff.
Waves energy 641
Weber-Fechner law 690
Weber 54, 698, 717
Weber number 472, 689
Weight 57, 678, 698
Wheels' penetration into soft soil 607
Wien's displacement constant 161
Wiener-Shannon equation 700
Window cracking 523, 789
Wire ring (see Tension in . . .)
Wood's formula 92, 730
Writers' efficiency 91, 726

x unit 61, 716

Year
  anomalistic 694
  mean solar 60, 694
  sidereal 694
  tropical 694
Young's modulus 28, 33, 35, 42, 57, 81, 114, 315, 678, 702, 790

Z matrix 140

SURNAME INDEX

Bold italics indicate included biographical details

Adams, Walter *360*
Agassi, Joseph 664, 667
Alexander, R. 662, 667
Ampère, André *48*, 681
Anton, H. 26
Appolonius [of Perga] 696
Avogadro, Amadeo, Count 450, 681

Baekeland 464
Baker, C. 716
Baker, Wilfred 148, 464, 660, 667
Barenblatt, G. 94, 154, 155, 659, 667
Barnett, S. 26
Barr, D. 662, 667
Barton, V. 664, 667
Beaford, Sir Francis *683*
Becquerel, Antoine 54, 693, 718
Bennet-Clark, H. 665, 667
Bethe, Hans 664, 667
Biot, B. 683, *689*
Biot, M. 665, 667
Birkhoff, J. 148
Blau, Garry 663, 667
Blench, T. 661, 667
Bohr, Niels xvii, 161, 547, 681
Boltzmann, Ludwig 94, 161, *324*, 376, 679, 733, 776
Boussinesq xvii, 551
Brach, Eugene xxiii
Brahe, Tycho *308*
Brand, L.148
Brenkert, K. 148

Bridgeman, Percy 126, 148, 460, 661, 663, 667
Brighton, J. 664, 667
Broggg, James (spelled right) 91, 728
Brown, Robert 713
Buckingham, Edgar VI, 148, 473, 493, 551, 662, 667, 668

Carvallo, E. 330, 765, 766
Cauchy, Augustin, Baron 472, 558, 691
Cavendish, Henry 94, 696, 701
Celsius, Anders 49, 50, 679, 693
Chakrabarty, Subrata 662, 668
Chen, Wai-Kai 662, 668
Clark, A. 94, 737
Clausius, Rudolf *700*
Cobb, W. 129, 746
Connor. John xxiii
Copley, Sir Godfrey 697
Corsin, Stanley 663, 668
Costa, Fernando 661, 668
Coulomb, Charles 54, 681, *693*, 704
Cowling, T. *688*
Craig, E. 662, 669
Curie, Pierre and Marie 62, 717, *718*

David, F. 660, 668
De Jong, Frits 128, 129, 660, 668
Den Hartog, J. 665, 668
Dingle, Herbert 38, 664, 668
Dirac, Paul 161
Douglas, John 660, 668

Douglas, P. 129, 746
Drobot, S. 148, 662, 668
DuBois, E. 131, 750
Dunham, William 36
Durand, W. 148

Eddington, Sir Arthur *421*, 422, 747
Edwards, N. 125, 662, 668
Einstein, Albert 129, 157, 158, *331*, 358, 695, 719, 736, 747, 768
Endrényi, John 664, 668
Eötvös, Loránd *68*
Ettinger, Morris 663, 668
Euclid [Euclides] 341, 664, 669
Euler, Leonhard 36, *102*, 472, 557, *688*

Fahrenheit, Gabriel 719, 720
Faraday, Michael 54, 381, 680, 681, *694*
Farkas, K. 664, 669
Fechner, Gustav 690
Ferguson, Robert xi, xix, *93*, 627
Fermat, Pierre *696*
Feynman, Richard 659, 669
Fick, Adolf 711
Fisher, Irving 128, 745, 746
Fitzgerald, G. 663, 669
Focken, C. 661, 669
Fogiel, Max 659, 669
Foldes, Peter xxv
Ford, Edsel 464
Ford, Henry 464
Fourier, Joseph *102*, 555, *684*
Froude, William 472, 557, 688, *689*

Galilei, Galileo *396*, 530, 719
Gauss, Karl 62, 698, *717*
Gere, J. 26, 665, 669
Ghista, Dhanjoo 663, 669
Gilvarry, J. 662, 669
Gossage, R. 663, 669
Grashof, Franz *685*, 686
Grätz, L. 685
Gray, L. 54, *694*, 697, 718
Guedelhoefer, O. 661, 669
Günther, B. 661, 669

Hainzl, J. 662, 669
Haldane, Sanderson *366*, 665, 669
Happ, W. 663, 669
Hartman 689
Haynes, Robin 660, 669
Helmholtz, Hermann 679, 695
Henry, Joseph 54, *694*
Heron [the Younger] 781
Hertz, Heinrich 50, 54, 694, *695*
Hill, A. 84, 94, 664, 669
Holloway, H. 716
Hooke, Robert 114, 397
Houserman, Robert xxiii
Hubble, Edwin 66, *420*
Hughes, W. 664, 669
Huntley, H. 659, 664, 669

Ipsen, D. 661, 669
Isaacson, E. 660, 670
Isaacson, Michael vii, xix, 551, 660, 670

Jeans, J. 664, 670
Jensen, J. 659, 670
Jerrard, H. 44, 659, 670
Johanson, J. 664, 670
Johnstone, Robert 660, 670
Joule, James 54, *695*
Jupp, Edmund 660, 670

Kapitza, Peter xi, *94*, 736, 737
Karman, Theodore, von 443, 665, 670
Kelvin, 1st Baron 48, *49*, 693, 696, 720
Kepler, Johann *308*, 309, 360, 362, 681
Kern, Donald 659, 670
Khinchin, A. 659, 670
Kiang, T. 66
Kleiber, Max 131, 661, 670
Knudsen, M. *689*
Kroon, R. 662, 670
Kürschák, József 664, 670
Kurth, Rudoph 661, 662, 670

Labocetta 313
Langhaar, Henry xxv, 148, 660, 670
Lavosier, Antoine *43*
Lehman, J. 660, 670
Lenard, John 662, 670

Lincoln, Abraham 517
Lodge, A. 661, 670

Macagno, Enzio 102, 148, 330, 662, 670
MacDuffee, C. 26
Mach, Ernst 34, 35, **691**
Margaria, R. 520
Mario, Thomas 662, 670
Maxwell, James 62, **717**
Mayer, Julius **695**
McAdams, William 661, 670
McGill, James xxv
McNeill, D. 659, 671
Menkes, Aviva 663, 671
Menkes, Josh 663, 671
Michelson, Albert 68
Millikan, Robert **712**
Milne 38
Moran, M. 662, 671
Morley, Edward 68
Morse, Samuel 694
Murphy, Glenn 663, 671
Muttnyánszky, Ádám v, 659, 671

Naddor, Eliezer 661, 671
Navier, Claude 128, 745
Nelson, Horatio, Viscount 343
Nernst, Hermann 126
Newcomen, Thomas 698
Newton, Sir Isaac 38, 44, 46, 47, 54, 66, 67, 75, 76, 275, 282, 284, 308, **696**, 699, 701, 727
Nicholas, Keith 26
Nobel, Alfred 693, 695, 712, 718
Noble, Ben 26
Nolle, H. 660, 671
Nusselt, W. 559, 683, **684**, 686, 687

O'Nan, Michael 664, 671
Oersted, Hans 62, **717**
Ohm, Georg 54, **696**, 704, 705

Palacios, J. 313, 660, 671
Pankhurst, R. 127, 660, 662, 671
Pascal, Blaise 44, 54, **696**
Peclet, J. 685, **686**
Pedley, T. 660, 671

Planck, Max 127, 161, 324, 376, 681, 725, 742
Poiseuille, Jean *417*, 444, 607
Poisson, Simeon 35, 381
Pomerantz, Martin 662, 671
Porter, Alfred 660, 671
Prandtl, Ludwig 559, 685, **686**, 687
Prasuhn, Alan 663, 671
Pythagoras [of Samos] xvii, **528**, 530

Ramshaw, J. 557, 661, 671
Rankine, William **720**
Rao, Govinda 663, 671
Rashevsky, Nicolas 660, 671
Rayleigh, 3rd Baron xxv, 179, 295, 551, **553**, 554, 555, 662, 663, 664, 671, 686
Reinhold, Timothy 661, 671
Reynolds, Osborne 34, 286, 472, 558, 559, 607, 661, 672, 685, 686, **687**
Riabouchinsky 554, 555, 663, 672
Röntgen, Wilhelm 62, **718**
Rose, William 664, 672
Rózsa, Pál vii, xxiii, xxvi, 1, 665, 672, 799
Rücker, A. 663, 672
Russel, Bertrand, 3rd Earl xxiii
Ryabov, Y. 665, 672
Rydberg, J. 161, 681

Sayao, Otavio 208, 209, 210, 663, 672
Sayer, Bruce xxiii
Schepartz, Bernard 289, 523, 661, 672
Schmideg, George 29, 297
Schmidt, E. **687**
Schultz, Peter 340, 661, 672
Schuring, Dieterich 295, 608, 660, 672, 677
Schwarzschild, Karl **549**
Sedov, L. 148, 673, 678
Semat, Henry 671, 678
Sena, L. 660, 672
Shannon, Claude **700**
Siemens, Sir William 54, **696**, 698
Sievert, Rolf 54, 694, **697**, 718
Sinclair, George 675, 678
Skoglund, Victor 673, 678
Smart, W. 672, 678
Smith, Maynard 289, 664, 672

Smith, Mrs. 87
Smithson, James 694
Snellen, Herman *691*
Stahl, Walter 661, 662, 672
Staicu, Constantin 313, 660, 662, 672, 673
Stanton, Sir Thomas 687, *688*
Stefan, Josef 94, 161, 324, 681, 733, 776
Stokes, Sir George 128, 745
Strahler, Arthur 663, 673
Strouhal, V. 689
Strutt, John (see Rayleigh)
Stumpf, H. 661, 673
Szirtes, Mrs. Penny xxiii
Szirtes, Thomas v, xix, 659, 665, 673
Szűcs, Ervin 660, 673

Taylor, Brook 104, 744
Taylor, Edward 660, 673
Taylor, Richard 520, 665, 673
Tchaikovsky, Peter 515, 518
Teller, Edward 660, 673
Tesla, Nicola 54, *697*, 717
Thompson, William 664, 665, 673
Thomson, D'Arcy W. 664, 673
Thomson, William (see Kelvin)
Timoshenko, Stephen 440, 659, 665, 673

Tonndorf, Juergen 662, 673
Torricelli, Evangelista *719*

Van Driest, E. 663, 673
Vaschy, A. 148
Venikov, V. 660, 673
Venn, John 276
Volta, Count Allessandro 54, *697*

Watt, James 44, 54, 697, *698*, 719
Waverman, Lucy 509, 511, 665, 673
Weaver, William Jr. 26
Weber, Ernst 689, *690, 698*
Weber, Wilhelm 54, 472, 689, *690*, 697, *698*, 717
White, R. 661, 674
Wien, Wilhelm 161
Wiener, Norbert *700*
Williams, T. 557, 661, 674
Wolverton, Raymond 665, 674
Wong, Young 278, 659, 674
Wood, K. 92, 731

Yiu, Simon xxiii
Young, Thomas 28, 33, 35, 42, 57, 81, *114*, 315, 613, 678, *702*, 790, 791

Zierep, Jürgen 660, 661, 674